Lecture Notes in Computer Science   1685

Edited by G. Goos, J. Hartmanis and J. van Leeuwen

W0107472

Springer-Verlag Berlin Heidelberg GmbH

Patrick Amestoy   Philippe Berger
Michel Daydé   Iain Duff   Valérie Frayssé
Luc Giraud   Daniel Ruiz (Eds.)

Euro-Par'99 Parallel Processing

5th International Euro-Par Conference
Toulouse, France, August 31 – September 3, 1999
Proceedings

Springer

Series Editors

Gerhard Goos, Karlsruhe University, Germany
Juris Hartmanis, Cornell University, NY, USA
Jan van Leeuwen, Utrecht University, The Netherlands

Volume Editors

Patrick Amestoy
Philippe Berger
Michel Daydé
Daniel Ruiz
ENSEEIHT, 2, Rue Camichel, F-31071 Toulouse Cedex 7, France
E-mail: {amestoy,berger,dayde,ruiz}@enseeiht.fr

Iain Duff
Valérie Fraysse
Luc Giraud
CERFACS, 42, Av. Gaspard Coriolis, F-31057 Toulouse Cedex 1, France
E-mail: {duff,fraysse,giraud}@cerfacs.fr

Cataloging-in-Publication data applied for

Die Deutsche Bibliothek - CIP-Einheitsaufnahme

Parallel processing : proceedings / Euro-Par '99, 5th International
Euro-Par Conference, Toulouse, France, August 31 - September 3,
1999. Patrick Amestoy ... (ed.). [ACM ; IFIR]. - Berlin ; Heidelberg ;
New York ; Barcelona ; Hong Kong ; London ; Milan ; Paris ;
Singapore ; Tokyo : Springer, 1999
    (Lecture notes in computer science ; Vol. 1685)
    ISBN 978-3-540-66443-7

CR Subject Classification (1998): C.1-4, D.1-4, F.1-2, G.1-2, E.1, H.2

ISSN 0302-9743
ISBN 978-3-540-66443-7   ISBN 978-3-540-48311-3 (eBook)
DOI 10.1007/978-3-540-48311-3

© Springer-Verlag Berlin Heidelberg 1999
Originally published by Springer-Verlag Berlin Heidelberg New York in 1999

Typesetting: Camera-ready by author
SPIN: 10704397   06/3142 – 5 4 3 2 1 0   Printed on acid-free paper

Topic 09: Parallel Computer Architecture

Global Chair
Chris Jesshope   Massey University, New Zealand
Local Chair
Daniel Litaize   Université Paul Sabatier, France
Vice Chairs
Karl Dieter Reinartz   University of Erlangen, Germany
Per Stenstrom   Chalmers University, Sweden

Topic 10: Distributed Systems and Algorithms

Global Chair
André Schiper   EPFL, Lausanne, Switzerland
Local Chair
Gérard Padiou   ENSEEIHT-IRIT, Toulouse, France
Vice Chairs
Roberto Baldoni   University of Rome, Italy
Jerzy Brzezinski   Fac. of Electrical Eng., Poznan, Poland
Friedemann Mattern   TU-Darmstadt, Germany
Luis Rodrigues   Faculty of Sciences, Univ. of Lisbon, Portugal

Topic 11: Parallel Programming: Models, Methods and Languages

Global Chair
Luc Bougé   ENSL, Lyon, France
Local Chair
Mamoun Filali   IRIT, Toulouse, France
Vice Chairs
Bill McColl   Oxford Univ. Computing Lab., UK
Henk J. Sips   TU Delft, The Netherlands

Topic 12: Architectures and Algorithms for Vision and other Senses

Global Chair
Virginio Cantoni   University of Pavia, Italy
Local Chair
Alain Ayache   ENSEEIHT-IRIT, Toulouse, France
Vice Chairs
Concettina Guerra   Purdue Univ., USA, and Padoa Univ., Italy
Pieter Jonke   TU Delft, The Netherlands

Topic 05: Parallel and Distributed Databases

Global Chair
Burkhard Freitag   University of Passau, Germany
Local Chair
Kader Hameurlain   IRIT, Toulouse, France
Vice Chairs
László Böszörmenyi   University of Klagenfurt, Austria
Waqar Hasan   DB WIZARD, CA, USA

Topics 06 + 20: Fault Avoidance and Fault Removal in Real-Time Systems

Global Chairs
Ravi Iyer   Univ. of Illinois, USA
Tomasz Szmuc   St. Staszic TU, Krakow, Poland
Local Chairs
Gilles Motet   INSA, Toulouse, France
David Powell   LAAS-CNRS, Toulouse, France
Vice Chairs
Wolfgang Halang   FernUniversität, Hagen, Germany
Andras Pataricza   TU Budapest, Hungary
Joao Gabriel Silva   University of Coimbra, Portugal
Janusz Zalewski   Univ. of Central Florida, Orlando, USA

Topic 07: Theory and Models for Parallel Computation

Global Chair
Michel Cosnard   LORIA-INRIA, Nancy, France
Local Chair
Afonso Ferreira   CNRS-I3S-INRIA, Sophia Antipolis, France
Vice Chairs
Sajal Das   University of North Texas, USA
Frank Dehne   Carleton University, Canada

Topic 08: High-Performance Computing and Applications

Global Chair
Horst Simon   NERSC, L. Berkeley Lab., CA, USA
Local Chair
Michael Rudgyard   Oxford Univ. and CERFACS, England
Vice Chairs
Wolfgang Gentzsch   GENIAS Software, Germany
Jesus Labarta   CEPBA, Spain

Euro-Par'99 Programme Committee

Topic 01: Support Tools and Environments

Global Chair
Ian Foster   Argonne Nat. Lab., USA
Local Chair
Frédéric Desprez   LIP-INRIA, Lyon, France
Vice Chairs
Jean-Marie Garcia   LAAS, Toulouse, France
Thomas Ludwig   TU Munich, Germany

Topic 02: Performance Evaluation and Prediction

Global Chair
Aad van der Steen   Utrecht Univ., The Netherlands
Local Chair
Jean-Marc Vincent   IMAG, Grenoble, France
Vice Chairs
Erich Strohmaier   Univ. of Tennessee at Knoxville, USA
Jerzy Wasniewski   Uni-C, Denmark

Topic 03: Scheduling and Load Balancing

Global Chair
Jean-Marc Geib   LIFL, Lille, France
Local Chair
Jean Roman   LaBRI, Bordeaux, France
Vice Chairs
Bruce Hendrickson   SNL, Albuquerque, USA
Pierre Manneback   PIP, Mons, Belgium

Topic 04: Compilers for High Performance Systems

Global Chair
Barbara Chapman   University of Southampton, UK
Local Chair
François Bodin   INRIA-IRISA, Rennes, France
Vice Chairs
Louis Féraud   IRIT, Toulouse, France
Chris Lengauer   FMI, University of Passau, Germany

Euro-Par Steering Committee

Chair
Ron Perrott   Queen's University Belfast, UK
Vice Chair
Christian Lengauer   University of Passau, Germany
Committee
Makoto Amamiya   Kyushu University, Japan
Luc Bougé   ENS Lyon, France
Helmar Burkhart   University of Basel, Switzerland
Pierrick Fillon-Ashida   European Commission, Belgium
Paul Feautrier   University of Versailles, France
Ian Foster   Argonne National Laboratory, USA
Jeff Reeve   University of Southampton, UK
Henk Sips   Technical University Delft, The Netherlands
Paul Spirakis   CTI, Greece
Marian Vajtersic   Slovak Academy, Slovakia
Marco Vanneschi   University of Pisa, Italy
Jens Volkert   Johannes Kepler University Linz, Austria
Emilio Zapata   University of Malaga, Spain

Euro-Par'99 Local Organisation

The conference has been jointly organised by CERFACS (Centre Européen de Recherche et de Formation Avancée en Calcul Scientifique) and ENSEEIHT-IRIT (Ecole Nationale Supérieure d'Electronique, d'Electrotechnique, d'Informatique et d'Hydraulique de Toulouse - Institut de Recherche en Informatique de Toulouse).

General Chair
Iain Duff   CERFACS and RAL
Vice Chair
Michel Daydé   ENSEEIHT-IRIT
Committee
Patrick Amestoy   ENSEEIHT-IRIT
Philippe Berger   ENSEEIHT-IRIT
Valérie Frayssé   CERFACS
Luc Giraud   CERFACS
Informatics Chair
Daniel Ruiz   ENSEEIHT-IRIT
Secretary
Brigitte Yzel   CERFACS

Acknowledgments

Knowing the quality of the past Euro-Par conferences makes the task of organising one daunting indeed and we have many people to thank. Ron Perrott, Christian Lengauer, Luc Bougé, Jeff Reeve and David Pritchard have given us the benefit of their experience and helped us generously throughout the past 18 months. The topic structure of the conference means that we must depend on the goodwill and enthusiasm of all the 90 programme committee members listed below. Their professionalism makes this the most academically rigorous conference in the field worldwide. The programme committee meeting at Toulouse in April was well attended and, thanks to sound preparation by everyone and Ron Perrott's guidance, resulted in a coherent, well-structured conference. The smooth running of the organisation of the conference can be attributed to a few individuals. Firstly the software for the submission and refereeing of papers that we inherited from Lyon via Passau was significantly enhanced by Daniel Ruiz. This attracted many compliments from those who benefited. Valérie Frayssé and Luc Giraud spent copious hours checking, printing and correcting papers. Finally, Brigitte Yzel, secretary to the conference and the CERFACS Parallel Algorithms Project, has been invaluable in monitoring the conference organisation and seeing to the myriad of tasks that invariably arise.

June 1999   Patrick Amestoy
                                                  Philippe Berger
                                                  Michel Daydé
                                                  Iain Duff
                                                  Valérie Frayssé
                                                  Luc Giraud
                                                  Daniel Ruiz

Preface

Euro-Par is an international conference dedicated to the promotion and advancement of all aspects of parallel computing. The major themes can be divided into the broad categories of hardware, software, algorithms and applications for parallel computing. The objective of Euro-Par is to provide a forum within which to promote the development of parallel computing both as an industrial technique and an academic discipline, extending the frontier of both the state of the art and the state of the practice. This is particularly important at a time when parallel computing is undergoing strong and sustained development and experiencing real industrial take-up. The main audience for and participants in Euro-Par are seen as researchers in academic departments, government laboratories and industrial organisations. Euro-Par's objective is to become the primary choice of such professionals for the presentation of new results in their specific areas. Euro-Par is also interested in applications which demonstrate the effectiveness of the main Euro-Par themes.

There is now a permanent Web site for the series `http://brahms.fmi.uni-passau.de/cl/europar` where the history of the conference is described. Euro-Par is now sponsored by the Association of Computer Machinery and the International Federation of Information Processing.

Euro-Par'99

The format of Euro-Par'99 follows that of the past four conferences and consists of a number of topics each individually monitored by a committee of four. There were originally 23 topics for this year's conference. The call for papers attracted 343 submissions of which 188 were accepted. Of the papers accepted, 4 were judged as distinguished, 111 as regular and 73 as short papers. Distinguished papers are allowed 12 pages in the proceedings and 30 minutes for presentation, regular papers are allowed 8 pages and 30 minutes for presentation, short papers are allowed 4 pages and 15 minutes for presentation. Distinguished papers are indicated in the table of contents with a superscribed asterisk after the title. Four extra pages could be purchased for distinguished papers, two for regular papers and one for short papers. There were on average 3.5 reviews per paper. Submissions were received from 44 countries, 34 of which are represented at the conference. The principal contributors by country are France with 39 papers, the U.S.A. with 22, Germany with 19, and the U.K., Italy and Spain with 13 papers each. This year's conference, Euro-Par'99, features new topics such as fault avoidance and fault removal in real-time systems, instruction-level parallelism and uniprocessor architecture, and global environment modelling.

The Web site for the conference is `http://www.enseeiht.fr/europar99/`.

Topics 13+19: Numerical Algorithms for Linear and Nonlinear Algebra

Global Chairs
Jack Dongarra — Univ. of Tennessee and ORNL, USA
Bernard Philippe — INRIA-IRISA, Rennes, France
Local Chairs
Patrick Amestoy — ENSEEIHT-IRIT, Toulouse, France
Valérie Frayssé — CERFACS, Toulouse, France
Vice Chairs
Françoise Chaitin-Chatelin — Univ. Toulouse I and CERFACS, France
Donato Trigiante — University of Florence, Italy
Marian Vajtersic — Slovak Academy, Bratislava
Henk van der Vorst — Utrecht University, The Netherlands

Topic 14: Emerging Topics in Advanced Computing in Europe

Global Chair
Renato Campo — European Commission, DG XIII, Belgium
Local Chair
Luc Giraud — CERFACS, Toulouse, France
Vice Chair
Pierrick Fillon-Ashida — European Commission, DG XIII, Belgium
Rizos Sakellariou — CRPC, Rice University, Houston, USA

Topic 15: Routing and Communication in Interconnection Networks

Global Chair
Ernst W. Mayr — TU Munich, Germany
Local Chair
Pierre Fraigniaud — LRI, Orsay, France
Vice Chairs
Bruce Maggs — Carnegie Mellon Univ., Pittsburg, USA
Jop Sibeyn — MPI, Saarbrücken, Germany

Topic 16: Instruction-Level Parallelism and Uniprocessor Architecture

Global Chair
Mateo Valero — CEPBA, Polytech. Univ. of Catalunya, Spain
Local Chair
Pascal Sainrat — IRIT, Toulouse, France
Vice Chairs
D. K. Arvind — Edinburgh Univ., Computer Science Dept., UK
Stamatis Vassiliadis — TU Delft, The Netherlands

Topic 17: Concurrent and Distributed Programming with Objects

Global Chair
Akinori Yonezawa   University of Tokyo, Japan
Local Chair
Patrick Sallé   ENSEEIHT-IRIT, Toulouse, France
Vice Chairs
Joe Armstrong   Ericsson, Sweden
Vasco T. Vasconcelos   University of Lisbon, Portugal
Marc Pantel   ENSEEIHT-IRIT, Toulouse, France

Topic 18: Global Environment Modelling

Global Chair
David Burridge   ECMWF, Reading, UK
Local Chair
Michel Déqué   Météo-France, Toulouse, France
Vice Chairs
Daniel Cariolle   Météo-France, Toulouse, France
Jean-Pascal van Ypersele   Louvain la Neuve Inst. of Astrophysics,
                                Belgium

Topic 22: High-Performance Data Mining and Knowledge Discovery

Global Chair
David Skillicorn   Queens University, Kingston, Canada
Local Chair
Domenico Talia   ISI-CNR, Rende, Italy
Vice Chairs
Vipin Kumar   Univ. of Minnesota, USA
Hannu Toivonen   RNI, University of Helsinki, Finland

Topic 23: Symbolic Computation

Global Chair
John Fitch   University of Bath, UK
Local Chair
Mike Dewar   NAG Ltd, Oxford, UK
Vice Chairs
Erich Kaltofen   North Carolina State Univ., Raleigh, USA
Anthony D. Kennedy   Edinburgh University, Scotland

Euro-Par'99 Referees

(excluding members of the programme and organisation committees)

Adelantado, Martin
Akl, Selim
Albanesi, Maria Grazia
Alvarez-Hamelin, I.
Amir, Yair
Amodio, Pierluigi
Ancourt, Corinne
Arbab, Farhad
Arbenz, Peter
Arnold, Marnix
Aronson, Leon
Asenjo, Rafael
Avresky, D.
Aydt, Ruth
Ayguade, Eduard
Aykanat, Cevdet
Baden, Scott
Bagchi, Saurabh
Bai, Z.
Baker, Mark
Bal, Henri
Baquero, Carlos
Barreiro, Nuno
Bartoli, Alberto
Baudon, Olivier
Becchetti, Luca
Béchennec, Jean-Luc
Becka, Martin
Beemster, Marcel
Bellegarde, Françoise
Bellosa, Frank
Benzi, Michele
Beraldi, Roberto
Bernon, Carole
Berthomé, Pascal
Bétourné, Claude
Biancardi, Alberto
Birkett, Nick
Bodeveix, Jean-Paul
Boichat, Romain
Bolychevsky, Alex
Bos, André

Boukerche, Azzedine
Boulet, Pierre
Bourzoufi, Hafid
Braconnier, Thierry
Bradford, Russell
Brandes, Thomas
Braunl, Thomas
Broggi, Alberto
Broom, Bradley
Browne, Shirley
Brugnano, Luigi
Brun, Olivier
Bungartz, Hans-Joachim
Burkhart, Helmar
Buvry, Max
Caires, Luis
Calamoneri, Tiziana
Cameron, Stephen
Caniaux, Guy
Cappello, Franck
Carissimi, Alexandre
Caromel, Denis
Carter, Larry
Cassé, Hugues
Castro, Miguel
Catthoor, Francky
Cavalheiro, Gerson
Chartier, Philippe
Charvillat, Vincent
Chassin de
Kergommeaux, Jacques
Chauhan, Arun
Chaumette, Serge
Cheng, Ben-Chung
Chesneaux, Jean-Marie
Chich, Thierry
Chmielewski, Rafal
Cilio, Andréa
Cinque, Luigi
Ciuffoletti, Augusto
Clocksin, William
Coelho, Fabien

Cole, Murray
Conter, Jean
Corbal, Jesus
Correia, Miguel
Costa, Antonio
Costa, Ernesto
Cotofana, Sorin
Coudert, David
Coulette, Bernard
Counilh, Marie-Christine
Courtier, Philippe
Cristal, Adrian
Crouzet, Yves
Crégut, Xavier
Cubero-Castan, Michel
Cucchiara, Rita
Cung, Van-Dat
Dackland, Krister
Daoudi, El Mostafa
Dechering, Paul
Defago, Xavier
Dekeyser, Jean-Luc
Delmas, Olivier
Delord, Xavier
Denissen, Will
Denneulin, Yves
Di, Vito
Di Martino, Beniamino
Di Stefano, Luigi
Doblas, Francisco Javier
Dominique, Orban
Doreille, Mathias
Dörfel, Mathias
Douville, Hervé
Drira, Khalil
Duato, José
Duchien, Laurence
Durand, Bruno
Eigenmann, Rudolf
Eijkhout, Victor
Eisenbeis, Christine
Eldershaw, Craig

Ellmenreich, Nils
Endo, Toshio
Erhel, Jocelyne
Etiemble, Daniel
Evans., D.J.
Even, Shimon
Fabbri, Alessandro
Faber, Peter
Farcy, Alexandre
Feautrier, Paul
Fent, Alfred
Ferragina, Paolo
Ferreira, Maria
Fidge, Colin
Fimmel, Dirk
Fischer, Claude
Fischer, Markus
Fleury, Eric
Fonlupt, Cyril
Fraboul, Christian
Fradet, Pascal
Fragopoulou, Paraskevi
Frietman, Edward E.E.
Fünfrocken, Stefan
Galilee, François
Gallopoulos, Efstratios
Gamba, Paolo
Gandriau, Marcel
Gaspar, Graca
Gatlin, Kang Su
Gauchard, David
Gautama, Hasyim
Gavaghan, David
Gavoille, Cyril
Geist, Al
Geleyn, Jean-François
van Gemund, Arjan
Gengler, Marc
Genius, Daniela
Gerndt, Michael
Getov, Vladimir
Girau, Bernard
Goldman, Alfredo
Gonzalez, Antonio
Gonzalez, José

Gonzalez-Escribano, A.
Goossens, Bernard
Grammatikakis, Miltos
Gratton, Serge
Greiner, Alain
Griebl, Martin
Grislin-Le Strugeon, E.
Gropp, William
Guedes, Paulo
Guivarch, Ronan
Guo, Katherine
Gupta, Anshul
Gustedt, Jens
Guyennet, Hervé
Hains, Gaetan
Hains, Gaétan
Halleux, Jean-Pierre
Han, Eui-Hong
Haquin, Thierry
Harald, Kosch
Hart, William
Hatcher, Phil
Hegland, Markus
Helary, Jean-Michel
Henty, David
Herley, Kieran
Herrmann, Christoph
Higham, Nicholas J.
Hitz, Markus
Hoisie, Adolfy
Houzet, Dominique
Hruz, Tomas
Hurfin, Michel
Ierotheou, Cos
Izu, Cruz
Jabouille, Patrick
Jeannot, Emmanuel
Jégou, Yvon
Jin, Guohua
Jonker, Pieter
Joshi, Mahesh
Jourdan, Stephan
Juanole, Guy
Juurlink, Ben
Kågström, Bo

Kakugawa, Hirotsugu
Kalbarczyk, Zbigniew
Kale, Laxmikant
Katayama, Takuya
Kaufman, Linda
Kehr, Roger
Keleher, Peter
Kelly, Paul
Kesselman, Carl
Killijian, Marc-Olivier
Kindermann, Stephan
Kiper, Ayse
Klasing, Ralf
Knijnenburg, Peter
Knoop, Jens
Koelbel, Charles
Kolda, Tamara
Kosch, Harald
Kossmann, Donald
Koster, Jacko
Kredel, Heinz
Kruse, Hans-Guenther
Kuchen, Herbert
Kucherov, Gregory
Kurmann, Christian
Kvasnicka, Dieter
de Laat, Cees
Ladagnous, Philippe
Lahlou, Chams
Langlois, Philippe
Larriba-Pey, Josep L.
Laure, Erwin
Lavenier, Dominique
LeCun, Bertrand
Lecomber, David
Lecomber, david
Lecussan, Bernard
Leese, Robert
Legall, Françoise
Leuschel, Michael
L'Excellent, Jean-Yves
Li, Xiaoye S.
Lindermeier, Markus
van Lohuizen, Marcel P.
Lombardi, Luca

Santos, Eunice
Saouter, Yannick
Savoia, Giancarlo
Schaerf, Andréa
Schreiber, Robert S.
Schreiner, Wolfgang
Schulz, Martin
Schuster, Assaf
Schwertner-Charao, A.
Seznec, André
Siegle, Markus
Sieh, Volkmar
Silva, Luis Moura
Silva, Mario
Slomka, Frank
Smith, Jim
Smith, Warren
Sobaniec, Cezary
Spezzano, Giandomenico
Stanca, Marian

Stanton, Jonathan
Stathis, Pyrrhos
Stein, Benhur
Steinbach, Michael
Stricker, Thomas M.
Strout, Michelle Mills
Stunkel, Craig
Subhlok, Jaspal
Sumii, Eijiro
Surlaker, Kapil
Swarbrick, Ian
Swierstra, Doaitse
Sybein, Jop
Szychowiak, Michal
Talbi, El ghazali
Talcott, Carolyn
Taylor, Valerie
Temam, Olivier
Thakur, Rajeev
Thiesse, Bernard

Thuné, Michael
Tiskin, Alexandre
Toetenel, Hans
Torng, Eric
Tortorici, Adèle
Trehel, Michel
Trichina, Eléna
Trystram, Denis
Tsikas, Themos
Tuecke, Steve
Tůma, Miroslav
Tvrdik, Pavel
Ubeda, Stephane
Ueberhuber, C. W.
Ugarte, Asier
Vakalis, Ignatis
Zomaya, Albert
Zory, Julien
Zwiers, Job

Table of Contents

Invited Talks

Topic 01
Support Tools and Environments

Topic 02
Performance Evaluation and Prediction **163**

Topic 03

Topic 04
Compilers for High Performance Systems 373

Topic 11
Parallel Programming: Models, Methods and Languages 831

Topic 13+19
Numerical Algorithms for Linear and Nonlinear Algebra 1023

Topic 14

Topic 16

Topic 17

Topic 18
Global Environment Modelling 1393

Some Parallel Algorithms for Integer Factorisation

Richard P. Brent

Oxford University Computing Laboratory,
Wolfson Building, Parks Road,
Oxford OX1 3QD, UK
rpb@comlab.ox.ac.uk
http://www.comlab.ox.ac.uk/
oucl/people/richard.brent.html

Abstract. Algorithms for finding the prime factors of large composite numbers are of practical importance because of the widespread use of public key cryptosystems whose security depends on the presumed difficulty of the factorisation problem. In recent years the limits of the best integer factorisation algorithms have been extended greatly, due in part to Moore's law and in part to algorithmic improvements. It is now routine to factor 100-decimal digit numbers, and feasible to factor numbers of 155 decimal digits (512 bits). We describe several integer factorisation algorithms, consider their suitability for implementation on parallel machines, and give examples of their current capabilities.

1 Introduction

Any positive integer N has a unique *prime power decomposition*

$$N = p_1^{\alpha_1} p_2^{\alpha_2} \cdots p_k^{\alpha_k}$$

($p_1 < p_2 < \cdots < p_k$ primes, $\alpha_j > 0$). This result is easy to prove, but the standard proof gives no hint of an efficient algorithm for computing the prime power decomposition. In order to compute it, we need –

1. An algorithm to test if an integer N is prime.
2. An algorithm to find a nontrivial factor f of a composite integer N.

Given these components there is a simple recursive algorithm to compute the prime power decomposition.

Fortunately or unfortunately, depending on one's point of view, problem 2 is generally believed to be hard. There is no known deterministic or randomised polynomial-time[1] algorithm for finding a factor of a given composite integer N. This empirical fact is of great interest because the most popular algorithm for

---

[1] For a polynomial-time algorithm the expected running time should be a polynomial in the length of the input, i.e. $O((\log N)^c)$ for some constant c.

P. Amestoy et al. (Eds.): Euro-Par'99, LNCS 1685, pp. 1–22, 1999.
© Springer-Verlag Berlin Heidelberg 1999

public-key cryptography, the RSA algorithm [75], would be insecure if a fast integer factorisation algorithm could be implemented [62].

In this paper we survey some of the most successful integer factorisation algorithms. Since there are already several excellent surveys emphasising the number-theoretic basis of the algorithms, we concentrate on the computational aspects, and particularly on parallel/distributed implementations of the algorithms.

1.1   Primality Testing

There are deterministic primality testing algorithms whose worst-case running time on a sequential computer is $O((\log N)^{c \log \log \log N})$, where c is a moderate constant. These algorithms are practical for numbers N of several hundred decimal digits [2, 20, 55]. If we are willing to accept a very small probability of error, then faster (polynomial-time) probabilistic algorithms are available [36, 57, 72]. Thus, in this paper we assume that primality testing is easy and concentrate on the more difficult problem of factoring composite integers.

1.2   Public Key Cryptography

As we already observed, large primes have a significant practical application – they can be used to construct *public key* cryptosystems[2]. The best-known is the *RSA system*, named after its inventors Rivest, Shamir and Adleman [75]. The security of RSA depends on the (assumed) difficulty of factoring the product of two large primes. This is the main practical motivation for the current interest in integer factorisation algorithms. Of course, mathematicians have been interested in factorisation algorithms for hundreds of years, but until recently it was not known that such algorithms were of "practical" importance.

In the RSA system we usually take $N = p_1 p_2$, where p_1, p_2 are large primes, each approximately equal, but not too close, to $n^{1/2}$. The product N is made public but the factors p_1, p_2 are kept secret. There is an implementation advantage in using a product of three large primes, $N = p_1 p_2 p_3$, where each p_i is approximately $N^{1/3}$. Some of the computations can be done mod p_i and the results (mod N) deduced via the Chinese remainder theorem. This is faster if we use three primes instead of two. On the other hand, the security of the system may be compromised because N, having smaller prime factors, may be easier to factor than in the two-prime case.

1.3   The Discrete Logarithm Problem

The difficulty of the discrete logarithm problem [61, 46] was used by Diffie and Hellman [27] to construct the *Diffie-Hellman key agreement protocol*. This well-known protocol allows two parties to establish a secret key through an exchange

---

[2] A concept introduced by Diffie and Hellman [27]. Also known as *asymmetric* or *open encryption key* cryptosystems [77, 85].

of public messages. Related public-key algorithms, such as the El Gamal algorithm [30, 31, 77], also depend on the difficulty of the discrete logarithm problem. These public-key algorithms provide practical alternatives to the RSA algorithm. Although originally considered in the setting of the multiplicative group of $\mathbf{GF}(p)$ (the finite field with a prime number p of elements), they generalise to any finite group G. There may be advantages (increased speed or security for a fixed size) in choosing other groups. Neal Koblitz [37] and Victor Miller independently proposed using the group of points on an elliptic curve, and this is a subject of much current research.

We do not consider algorithms for discrete logarithms in this paper. However, it is interesting to note that in some cases integer factorisation algorithms have analogues which apply to the discrete logarithm problem [61, 88, 89]. This does not seem to be the case for discrete logarithms over elliptic curves, which is one reason for the popularity of elliptic curves in cryptographic applications [47, 48].

1.4   Parallel Algorithms

When designing parallel algorithms we hope that an algorithm which requires time T_1 on a computer with one processor can be implemented to run in time $T_P \sim T_1/P$ on a computer with P independent processors. This is not always the case, since it may be impossible to use all P processors effectively. However, it is true for many integer factorisation algorithms, provided P is not too large.

The *speedup* of a parallel algorithm is $S = T_1/T_P$. We aim for a linear speedup, i.e. $S = \Theta(P)$.

2   Multiple-Precision Arithmetic

Before describing some integer factorisation algorithms, we comment on the implementation of multiple-precision integer arithmetic on vector processors and parallel machines. Multiple-precision arithmetic is necessary because the number N which we want to factor may be much larger than can be represented in a single computer word (otherwise the problem is trivial).

2.1   Carry Propagation and Redundant Number Representations

To represent a large positive integer N, it is customary to choose a convenient *base* or *radix* β and express N as

$$N = \sum_0^{t-1} d_j \beta^j,$$

where $d_0, \ldots, d_{t-1}$ are "base β digits" in the range $0 \leq d_j < \beta$. We choose β large, but small enough that $\beta - 1$ is representable in a single word [4, 36]. Consider multiple-precision addition (subtraction and multiplication may be handled in a similar way). On a parallel machine there is a problem with *carry propagation* because a carry can propagate all the way from the least to the most significant digit. Thus an addition takes worst case time $\Theta(t)$, and average time $\Theta(\log t)$, independent of the number of processors.

The carry propagation problem can be reduced if we permit digits d_j outside the normal range. Suppose that we allow $-2 \le d_j \le \beta + 1$, where $\beta > 4$. Then possible carries are in $\{-1, 0, 1, 2\}$ and we need only add corresponding pairs of digits (in parallel), compute the carries and perform one step of carry propagation. It is only when comparisons of multiple-precision numbers need to be performed that the digits have to be reduced to the normal range by fully propagating carries. Thus, redundant number representation is useful for speeding up multiple-precision addition and multiplication. On a parallel machine with sufficiently many processors, such a representation allows addition to be performed in constant time.

2.2 High Level Parallelism

Rather than trying to perform individual multiple-precision operations rapidly, it is often more convenient to implement the multiple-precision operations in bit or word-serial fashion, but perform many independent operations in parallel. For example, a *trial* of the elliptic curve algorithm (§7) involves a predetermined sequence of additions and multiplications on integers of bounded size. Our implementation on a Fujitsu VPP 300 performs many trials concurrently (on one or more processors) in order to take advantage of each processor's vector pipelines.

2.3 Use of Real Arithmetic

Most supercomputers were not designed to optimise the performance of exact (multiple-precision) integer arithmetic. On machines with fast floating-point hardware, e.g. pipelined 64-bit floating point units, it may be best to represent base β digits in *floating-point* words. The upper bound on β is imposed by the multiplication algorithm – we must ensure that β^2 is exactly representable in a (single or double-precision) floating-point word. In practice it is convenient to allow some slack – for example, we might require $8\beta^2$ to be exactly representable. On machines with IEEE standard arithmetic, we could use $\beta = 2^{24}$.

2.4 Redundant Representations Mod N

Many integer factorisation algorithms require operations to be performed modulo N, where N is the number to be factored. A straightforward implementation would perform a multiple-precision operation and then perform a division by N to find the remainder. Since N is fixed, some precomputation involving N (e.g. reciprocal approximation) may be worthwhile. However, it may be faster to avoid explicit divisions, by taking advantage of the fact that it is not usually necessary to represent the result uniquely.

For example, consider the computation of $x * y \bmod N$. The result is $r = x * y - q * N$ and it may be sufficient to choose q so that $0 \le r < 2N$

(a weaker constraint than the usual $0 \leq r < N$). To compute r we multiply x by the digits of y, most significant digit first, but modify the standard "shift and add" algorithm to subtract single-precision multiples of N in order to keep the accumulated sum bounded by $2N$. Formally, a partial sum s is updated by $s \leftarrow \beta * s + y_j * x - q_j * N$, where q_j is obtained by a division involving only a few leading digits of $\beta * s + y_j * x$ and N.

Alternatively, a technique of Montgomery [49] can be used to speed up modular arithmetic.

2.5   Computing Inverses Mod N

In some factorisation algorithms we need to compute inverses mod N. Suppose that x is given, $0 < x < N$, and we want to compute z such that $xz = 1 \bmod N$. The extended Euclidean algorithm [36] applied to x and N gives u and v such that

$$ux + vN = GCD(x, N).$$

If $GCD(x, N) = 1$ then $ux = 1 \bmod N$, so $z = u$. If $GCD(x, N) > 1$ then $GCD(x, N)$ is a nontrivial factor of N. This is a case where failure (in finding an inverse) implies success (in finding a factor) !

3   Integer Factorisation Algorithms

There are many algorithms for finding a nontrivial factor f of a composite integer N. The most useful algorithms fall into one of two classes –

A. The run time depends mainly on the size of N, and is not strongly dependent on the size of f. Examples are –

 - Lehman's algorithm [39], which has worst-case run time $O(N^{1/3})$.
 - The Continued Fraction algorithm [56] and the Multiple Polynomial Quadratic Sieve (MPQS) algorithm [67], which under plausible assumptions have expected run time $O(\exp(c(\ln N \ln \ln N)^{1/2}))$, where c is a constant (depending on details of the algorithm). For MPQS, $c \approx 1$.
 - The Number Field Sieve (NFS) algorithm [41, 40], which under plausible assumptions has expected run time $O(\exp(c(\ln N)^{1/3}(\ln \ln N)^{2/3}))$, where c is a constant (depending on details of the algorithm and on the form of N).

B. The run time depends mainly on the size of f, the factor found. (We can assume that $f \leq N^{1/2}$.) Examples are –

 - The trial division algorithm, which has run time $O(f \cdot (\log N)^2)$.
 - Pollard's "rho" algorithm [66], which under plausible assumptions has expected run time $O(f^{1/2} \cdot (\log N)^2)$.
 - Lenstra's Elliptic Curve (ECM) algorithm [45], which under plausible assumptions has expected run time $O(\exp(c(\ln f \ln \ln f)^{1/2}) \cdot (\log N)^2)$, where $c \approx 2$ is a constant.

In these examples, the time bounds are for a sequential machine, and the term $(\log N)^2$ is a generous allowance for the cost of performing arithmetic operations on numbers which are $O(N^2)$. If N is very large, then fast integer multiplication algorithms [24, 36] can be used to reduce the $(\log N)^2$ term.

Our survey of integer factorisation algorithms in §§4–10 below is necessarily cursory. For more information the reader is referred to the literature [7, 13, 52, 69, 74].

3.1 Quantum Factorisation Algorithms

In 1994 Shor [79, 80] showed that it is possible to factor in polynomial expected time on a quantum computer [25, 26]. However, despite the best efforts of several research groups, such a computer has not yet been built, and it remains unclear whether it will ever be feasible to build one. Thus, in this paper we restrict our attention to algorithms which run on classical (serial or parallel) computers [86]. The reader interested in quantum computers could start by reading [71, 87].

4 Pollard's "rho" Algorithm

Pollard's "rho" algorithm [5, 66] uses an iteration of the form

$$x_{i+1} = f(x_i) \bmod N, \ \ i \geq 0,$$

where N is the number to be factored, x_0 is a random starting value, and f is a nonlinear polynomial with integer coefficients, for example

$$f(x) = x^2 + a \ \ \ (a \neq 0 \bmod N) \,.$$

Let p be the smallest prime factor of N, and j the smallest positive index such that $x_{2j} = x_j \pmod{p}$. Making some plausible assumptions, it is easy to show that the expected value of j is $E(j) = O(p^{1/2})$. The argument is related to the well-known "birthday" paradox – the probability that $x_0, x_1, \ldots, x_k$ are all distinct mod p is approximately

$$(1 - 1/p) \cdot (1 - 2/p) \cdots (1 - k/p) \sim \exp\left(\frac{-k^2}{2p}\right) ,$$

and if $x_0, x_1, \ldots, x_k$ are not all distinct mod p then $j \leq k$.

In practice we do not know p in advance, but we can detect x_j by taking greatest common divisors. We simply compute GCD $(x_{2i} - x_i, N)$ for $i = 1, 2, \ldots$ and stop when a GCD greater than 1 is found.

4.1 Pollard Rho Examples

An early example of the success of a variation of the Pollard "rho" algorithm is the complete factorisation of the Fermat number $F_8 = 2^{2^8} + 1$ by Brent and Pollard [11]. In fact

$$F_8 = 1238926361552897 \cdot p_{62},$$

where p_{62} is a 62-digit prime.

The *Cunningham project* [12] is a collaborative effort to factor numbers of the form $a^n \pm 1$, where $a \le 12$. The largest factor found by the Pollard "rho" algorithm during the Cunningham project is a 19-digit factor of $2^{2386} + 1$ (found by Harvey Dubner on a Dubner Cruncher [14]). Larger factors could certainly be found, but the discovery of ECM (§7) has made the Pollard "rho" algorithm uncompetitive for factors greater than about 10 decimal digits [6, Table 1].

4.2   Parallel Rho

Parallel implementation of the "rho" algorithm does not give linear speedup[3]. A plausible use of parallelism is to try several different pseudo-random sequences (generated by different polynomials f). If we have P processors and use P different sequences in parallel, the probability that the first k values in each sequence are distinct mod p is approximately $\exp(-k^2 P/(2p))$, so the speedup is $\Theta(P^{1/2})$. Recently Crandall [23] has suggested that a speedup $\Theta(P/(\log P)^2)$ is possible, but his proposal has not yet been tested.

5   The Advantages of a Group Operation

The Pollard rho algorithm takes $x_{i+1} = f(x_i) \bmod N$ where f is a polynomial. Computing x_n requires n steps. Suppose instead that $x_{i+1} = x_0 \circ x_i$ where "$\circ$" is an associative operator, which for the moment we can think of as multiplication. We can compute x_n in $O(\log n)$ steps by the *binary powering* method [36].

Let m be some bound assigned in advance, and let E be the product of all maximal prime powers q^e, $q^e \le m$. Choose some starting value x_0, and consider the cyclic group $<x_0>$ consisting of all powers of x_0 (under the associative operator "$\circ$"). If this group has order g whose prime power components are bounded by m, then $g|E$ and $x_0^E = I$, where I is the group identity.

We may consider a group defined mod p but work mod N, where p is an unknown divisor of N. This amounts to using a redundant representation for the group elements. When we compute the identity I, its representation mod N may allow us to compute p via a GCD computation (compare Pollard's rho algorithm). We give two examples below: Pollard's $p-1$ algorithm and Lenstra's elliptic curve algorithm.

6   Pollard's $p-1$ Algorithm

Pollard's "$p-1$" algorithm [65] may be regarded as an attempt to generate the identity in the multiplicative group of $\mathbf{GF}(p)$. Here the group operation "$\circ$" is just multiplication mod p, so (by Fermat's theorem) $g|p-1$ and

$$x_0^E = I \Rightarrow x_0^E = 1 \pmod{p} \Rightarrow p | \text{GCD}\ (x_0^E - 1, N)$$

---

[3] Variants of the "rho" algorithm can be used to solve the discrete logarithm problem. Recently, van Oorschot and Wiener [63, 64] have shown that a linear speedup is possible in this application.

6.1   Rho Example

The largest factor found by the Pollard "$p-1$" algorithm during the Cunningham project is a 32-digit factor

$$p_{32} = 49858990580788843054012690078841$$

of $2^{977} - 1$. In this case

$$p_{32} - 1 = 2^3 \cdot 5 \cdot 13 \cdot 19 \cdot 977 \cdot 1231 \cdot 4643 \cdot 74941 \cdot 1045397 \cdot 11535449$$

6.2   Parallel $p - 1$

Parallel implementation of the "$p - 1$" algorithm is difficult, because the inner loop seems inherently serial. At best, parallelism can speed up the multiple precision operations by a small factor (depending on $\log N$ but not on p).

6.3   The Worst Case for $p - 1$

In the worst case, when $(p - 1)/2$ is prime, the "$p - 1$" algorithm is no better than trial division. Since the group has fixed order $p - 1$ there is nothing to be done except try a different algorithm. In the next section we show that it is possible to overcome the main handicaps of the "$p - 1$" algorithm, and obtain an algorithm which is easy to implement in parallel and does not depend on the factorisation of $p - 1$.

7   Lenstra's Elliptic Curve Algorithm

If we can choose a "random" group G with order g close to p, we may be able to perform a computation similar to that involved in Pollard's "$p - 1$" algorithm, working in G rather than in $\mathbf{GF}(p)$. If all prime factors of g are less than the bound m then we find a factor of N. Otherwise, repeat with a different G (and hence, usually, a different g) until a factor is found. This is the motivation for H. W. Lenstra's *elliptic curve algorithm (or method)* (ECM).

A curve of the form

$$y^2 = x^3 + ax + b \tag{1}$$

over some field F is known as an *elliptic curve*. A more general cubic in x and y can be reduced to the form (1), which is known as the Weierstrass normal form, by rational transformations, provided $\operatorname{char}(F) \neq 2$.

There is a well-known way of defining an Abelian group $(G, \circ)$ on an elliptic curve over a field. Formally, if $P_1 = (x_1, y_1)$ and $P_2 = (x_2, y_2)$ are points on the curve, then the point $P_3 = (x_3, y_3) = P_1 \circ P_2$ is defined by –

$$(x_3, y_3) = (\lambda^2 - x_1 - x_2, \ \lambda(x_1 - x_3) - y_1), \tag{2}$$

where
$$\lambda = \begin{cases} (3x_1^2 + a)/(2y_1) & \text{if } P_1 = P_2 \\ (y_1 - y_2)/(x_1 - x_2) & \text{otherwise.} \end{cases}$$

The identity element in G is the "point at infinity", (∞, ∞).

From now on we write "$\circ$" as "$+$", since this is standard in the elliptic curve literature. Thus (∞, ∞) is the "zero" element of G, and is written as 0.

The geometric interpretation of $P_1 + P_2$ is straightforward: the straight line $P_1 P_2$ intersects the elliptic curve at a third point $P_3' = (x_3, -y_3)$, and P_3 is the reflection of P_3' in the x-axis. We refer the reader to a suitable text [19, 35, 38, 81] for an introduction to the theory of elliptic curves.

In Lenstra's algorithm [45] the field F is the finite field $\mathbf{GF}(p)$ of p elements, where p is a prime factor of N. The multiplicative group of $\mathbf{GF}(p)$, used in Pollard's "$p-1$" algorithm, is replaced by the group G defined by (1–2). Since p is not known in advance, computation is performed in the ring Z/NZ of integers modulo N rather than in $\mathbf{GF}(p)$. We can regard this as using a redundant group representation.

A *trial* is the computation involving one random group G. The steps involved are –

1. Choose x_0, y_0 and a randomly in $[0, N)$. This defines $b = y_0^2 - (x_0^3 + ax_0) \bmod N$. Set $P \leftarrow P_0 = (x_0, y_0)$.
2. For prime $q \leq m$ set $P \leftarrow q^e P$ in the group G defined by a and b, where e is an exponent chosen as in §5. If $P = 0$ then the trial succeeds as a factor of N will have been found during an attempt to compute an inverse mod N. Otherwise the trial fails.

The work involved in a trial is $O(m)$ group operations. There is a tradeoff involved in the choice of m, as a trial with large m is expensive, but a trial with small m is unlikely to succeed.

Given $x \in \mathbf{GF}(p)$, there are at most two values of $y \in \mathbf{GF}(p)$ satisfying (1). Thus, allowing for the identity element, we have $g = |G| \leq 2p + 1$. A much stronger result, the *Riemann hypothesis for finite fields*, is known –
$$|g - p - 1| < 2p^{1/2} .$$

Making a plausible assumption about the distribution of prime divisors of g, one may show that the optimal choice of m is $m = p^{1/\alpha}$, where
$$\alpha \sim (2 \ln p / \ln \ln p)^{1/2} .$$

It follows that the expected run time is
$$T = p^{2/\alpha + o(1/\alpha)} . \tag{3}$$

For details, see Lenstra [45]. The exponent $2/\alpha$ in (3) should be compared with 1 (for trial division) or $1/2$ (for Pollard's "rho" method). Because of the overheads involved with ECM, a simpler algorithm such as Pollard's "rho" is preferable for finding factors of size up to about 10^{10}, but for larger factors the asymptotic advantage of ECM becomes apparent. The following examples illustrate the power of ECM.

7.1   ECM Examples

1. In 1995 we completed the factorisation of the 309-decimal digit (1025-bit) Fermat number $F_{10} = 2^{2^{10}} + 1$. In fact

$$F_{10} = 45592577 \cdot 6487031809 \cdot$$
$$4659775785220018543264560743076778192897 \cdot p_{252}$$

where $46597 \cdots 92897$ is a 40-digit prime and $p_{252} = 13043 \cdots 24577$ is a 252-digit prime. The computation, which is described in detail in [10], took about 240 Mips-years.

2. The largest factor known to have been found by ECM is the 53-digit factor

$$53625112691923843508117942311516428173021903300344567$$

of $2^{677} - 1$, found by Conrad Curry in September 1998 using a program written by George Woltman and running on 16 Pentiums (for more details see [9]). Note that if the RSA system were used with 512-bit keys and the three-prime variation, as described in §1.2, the smallest prime would be less than 53 decimal digits, so ECM could be used to break the system.

7.2   The Second Phase

Both the Pollard "$p-1$" and Lenstra elliptic curve algorithms can be speeded up by the addition of a second phase. The idea of the second phase is to find a factor in the case that the first phase terminates with a group element $P \neq 0$, such that $|\langle P \rangle|$ is reasonably small (say $O(m^2)$). (Here $\langle P \rangle$ is the cyclic group generated by P.) There are several possible implementations of the second phase. One of the simplest uses a pseudorandom walk in $\langle P \rangle$. By the birthday paradox argument, there is a good chance that two points in the random walk will coincide after $O(|\langle P \rangle|^{1/2})$ steps, and when this occurs a nontrivial factor of N can usually be found. Details of this and other implementations of the second phase may be found in [6, 10, 28, 50, 51, 83].

The use of a second phase provides a significant speedup in practice, but does not change the asymptotic time bound (3). Similar comments apply to other implementation details, such as ways of avoiding most divisions and speeding up group operations, ways of choosing good initial points, and ways of using preconditioned polynomial evaluation [6, 50, 51].

7.3   Parallel/Distributed Implementation of ECM

Unlike the Pollard "rho" and "$p-1$" methods, ECM is "embarrassingly parallel", because each trial is independent. So long as the expected number of trials is much larger than the number P of processors available, linear speedup is possible by performing P trials in parallel. In fact, if T_1 is the expected run time on one processor, then the expected run time on a MIMD parallel machine with P processors is

$$T_P = T_1/P + O(T_1^{1/2+\epsilon}) \tag{4}$$

The bound (4) applies on a SIMD machine if we use the Montgomery-Chudnovsky form [18, 50]

$$by^2 = x^3 + ax^2 + x$$

instead of the Weierstrass normal form (1) in order to avoid divisions.

In practice, it may be difficult to perform P trials in parallel because of storage limitations. The second phase requires more storage than the first phase. Fortunately, there are several possibilities for making use of parallelism during the second phase of each trial. One parallel implementation performs the first phase of P trials in parallel, but the second phase of each trial sequentially, using P processors to speed up the evaluation of the high-degree polynomials which constitute most of the work during the second phase.

8   Quadratic Sieve Algorithms

Quadratic sieve algorithms belong to a wide class of algorithms which try to find two integers x and y such that $x \neq \pm y \pmod{N}$ but

$$x^2 = y^2 \pmod{N}. \tag{5}$$

Once such x and y are found, then GCD $(x - y, N)$ is a nontrivial factor of N.

One way to find x and y satisfying (5) is to find a set of *relations* of the form

$$u_i^2 = v_i^2 w_i \pmod{N}, \tag{6}$$

where the w_i have all their prime factors in a moderately small set of primes (called the *factor base*). Each relation (6) gives a row in a matrix M whose columns correspond to the primes in the factor base. Once enough rows have been generated, we can use sparse Gaussian elimination in $\mathbf{GF}(2)$ (Weidemann [90]) to find a linear dependency (mod 2) between a set of rows of M. Multiplying the corresponding relations now gives an expression of the form (5). With probability at least $1/2$, we have $x \neq \pm y \bmod N$ so a nontrivial factor of N will be found. If not, we need to obtain a different linear dependency and try again.

In quadratic sieve algorithms the numbers w_i are the values of one (or more) polynomials with integer coefficients. This makes it easy to factor the w_i by *sieving*. For details of the process, we refer to [15, 43, 67, 70, 73, 82]. The inner loop of the sieving process has the form

```
while j < bound do
    begin
        s[j] ← s[j] + c;
        j ← j + q;
    end
```

Here *bound* depends on the size of the (single-precision real) sieve array s, q is a small prime or prime power, and c is a single-precision real constant depending on q ($c = \Lambda(q) = \log p$ if $q = p^e$, p prime). The loop can be implemented efficiently on a pipelined vector processor. It is possible to use scaling to avoid floating point additions, which is desirable on a small processor without floating-point hardware.

In order to minimise cache misses on a machine whose memory cache is too small to store the whole array s, it may be desirable to split the inner loop to perform sieving over cache-sized blocks.

The best quadratic sieve algorithm, the *multiple polynomial quadratic sieve algorithm* MPQS [67, 82] can, under plausible assumptions, factor a number N in time $\Theta(\exp(c(\ln N \ln \ln N)^{1/2}))$, where $c \sim 1$. The constants involved are such that MPQS is usually faster than ECM if N is the product of two primes which both exceed $N^{1/3}$. This is because the inner loop of MPQS involves only single-precision operations.

Use of "partial relations", i.e. incompletely factored w_i, in MPQS is analogous to the second phase of ECM and gives a similar performance improvement [3]. In the "one large prime" (P-MPQS) variation w_i is allowed to have one prime factor exceeding m (but not too much larger than m). In the "two large prime" (PP-MPQS) variation w_i can have two prime factors exceeding m – this gives a further performance improvement at the expense of higher storage requirements [44], and does not seem to have an analogue applicable to ECM.

8.1   Parallel/Distributed Implementation of MPQS

Like ECM, the sieving stage of MPQS is ideally suited to parallel implementation. Different processors may use different polynomials, or sieve over different intervals with the same polynomial. Thus, there is a linear speedup so long as the number of processors is not much larger than the size of the factor base. The computation requires very little communication between processors. Each processor can generate relations and forward them to some central collection point. This was demonstrated by A. K. Lenstra and M. S. Manasse [43], who distributed their program and collected relations via electronic mail. The processors were scattered around the world – anyone with access to electronic mail and a C compiler could volunteer to contribute[4]. The final stage of MPQS – Gaussian elimination to combine the relations – was not so easily distributed. In practice it is only a small fraction of the overall computation, but it may become

---

[4] This idea of using machines on the Internet as a "free" supercomputer has recently been adopted by several other computation-intensive projects

a limitation if very large numbers are attempted by MPQS (a similar problem is discussed below in connection with NFS).

8.2 MPQS Examples

MPQS has been used to obtain many impressive factorisations [12, 73, 82]. Arjen Lenstra and Mark Manasse [43] (with many assistants scattered around the world) have factored several numbers larger than 10^{100}. For example, a recent factorisation was the 116-decimal digit number $(3^{329} + 1)/$(known small factors) into a product of 50-digit and 67-digit primes. The final factorisation is

$$3^{329} + 1 = 2^2 \cdot 547 \cdot 16921 \cdot 256057 \cdot 36913801 \cdot 177140839 \cdot 1534179947851 \cdot$$
$$2467707882284001426665277903676806291837269743524 1 \cdot p_{67}$$

Such factorisations require many years of CPU time, but a real time of only a month or so because of the number of different processors which are working in parallel.

At the time of writing, the largest number factored by MPQS is the 129-digit "RSA Challenge" [75] number RSA129. It was factored in 1994 by Atkins *et al* [1]. It is certainly feasible to factor larger numbers by MPQS, but for numbers of more than about 110 decimal digits GNFS is faster [32, 33, 34]. For example, it is estimated in [21] that to factor RSA129 by MPQS required 5000 Mips-years, but to factor the slightly larger number RSA130 by GNFS required only 1000 Mips-years [22].

9   The Special Number Field Sieve (SNFS)

The *number field sieve* (NFS) algorithm was developed from the *special number field sieve* (SNFS), which we describe in this section. The *general number field sieve* (GNFS or simply NFS) is described in §10.

Most of our numerical examples have involved numbers of the form

$$a^e \pm b \, , \tag{7}$$

for small a and b, although the ECM and MPQS factorisation algorithms do not take advantage of this special form.

The *special number field sieve* (SNFS) is a relatively new algorithm which does take advantage of the special form (7). In concept it is similar to the quadratic sieve algorithm, but it works over an algebraic number field defined by a, e and b. We refer the interested reader to Lenstra *et al* [40, 41] for details, and merely give an example to show the power of the algorithm. For an introduction to the relevant concepts of algebraic number theory, see Stewart and Tall [84].

9.1 SNFS Examples

Consider the 155-decimal digit number

$$F_9 = N = 2^{2^9} + 1$$

as a candidate for factoring by SNFS. Note that $8N = m^5 + 8$, where $m = 2^{103}$. We may work in the number field $Q(\alpha)$, where α satisfies

$$\alpha^5 + 8 = 0,$$

and in the ring of integers of $Q(\alpha)$. Because

$$m^5 + 8 = 0 \pmod{N},$$

the mapping $\phi : \alpha \mapsto m \bmod N$ is a ring homomorphism from $Z[\alpha]$ to Z/NZ.

The idea is to search for pairs of small coprime integers u and v such that both the algebraic integer $u + \alpha v$ and the (rational) integer $u + mv$ can be factored. (The factor base now includes prime ideals and units as well as rational primes.) Because

$$\phi(u + \alpha v) = (u + mv) \pmod{N},$$

each such pair gives a relation analogous to (6).

The prime ideal factorisation of $u + \alpha v$ can be obtained from the factorisation of the *norm* $u^5 - 8v^5$ of $u + \alpha v$. Thus, we have to factor simultaneously two integers $u + mv$ and $|u^5 - 8v^5|$. Note that, for moderate u and v, both these integers are much smaller than N, in fact they are $O(N^{1/d})$, where $d = 5$ is the degree of the algebraic number field. (The optimal choice of d is discussed below.)

Using these and related ideas, Lenstra *et al* [42] factored F_9 in June 1990, obtaining

$$F_9 = 2424833 \cdot 7455602825647884208337395736200454918783366342657 \cdot p_{99},$$

where p_{99} is an 99-digit prime, and the 7-digit factor was already known (although SNFS was unable to take advantage of this). The collection of relations took less than two months on a network of several hundred workstations. A sparse system of about 200,000 relations was reduced to a dense matrix with about 72,000 rows. Using Gaussian elimination, dependencies (mod 2) between the rows were found in three hours on a Connection Machine. These dependencies implied equations of the form $x^2 = y^2 \bmod F_9$. The second such equation was nontrivial and gave the desired factorisation of F_9.

More recently, considerably larger numbers have been factored by SNFS. The current record is the 211-digit number $10^{211} - 1$, factored early in 1999 by a collaboration called "The Cabal" [17]. In fact, $10^{211} - 1 = p_{93} \cdot p_{118}$, where

$$p_{93} = 6926245573243896206627823226773367111381084825$$
$$8828173973437557050649239193184952463673186879$$

and p_{118} may be found by division. The factorisation of $N = 10^{211} - 1$ used two polynomials

$$f(x) = x - 10^{35}$$

and

$$g(x) = 10x^6 - 1$$

with common root $m = 10^{35}$ mod N. Details of the computation can be found in [17]. To summarise: after sieving and reduction a sparse matrix over $\mathbf{GF}(2)$ was obtained with about 4.8×10^6 rows and weight (number of nonzero entries) about 2.3×10^8, an average of about 49 nonzeros per row. Montgomery's block Lanczos program (see §10) took 121 hours on a Cray C90 to find 64 dependencies. Finally, the square root program needed 15.5 hours on one CPU of an SGI Origin 2000, and three dependencies to find the two prime factors.

10   The General Number Field Sieve (GNFS)

The *general number field sieve* (GNFS or just NFS) is a logical extension of the special number field sieve (SNFS). When using SNFS to factor an integer N, we require two polynomials $f(x)$ and $g(x)$ with a common root m mod N but no common root over the field of complex numbers. If N has the special form (7) then it is usually easy to write down suitable polynomials with small coefficients, as illustrated by the two examples given in §9. If N has no special form, but is just some given composite number, we can also find $f(x)$ and $g(x)$, but they no longer have small coefficients.

Suppose that $g(x)$ has degree $d > 1$ and $f(x)$ is linear[5]. d is chosen empirically, but it is known from theoretical considerations that the optimum value is

$$d \sim \left(\frac{3 \ln N}{\ln \ln N} \right)^{1/3}$$

We choose $m = \lfloor N^{1/d} \rfloor$ and write

$$N = \sum_{j=0}^{d} a_j m^j$$

where the a_j are "base m digits" and $a_d = 1$. Then, defining

$$f(x) = x - m, \quad g(x) = \sum_{j=0}^{d} a_j x^j \; ,$$

it is clear that $f(x)$ and $g(x)$ have a common root m mod N. This method of polynomial selection is called the "base m" method.

---

[5] This is not necessary. For example, Montgomery found a clever way (described in [32]) of choosing two quadratic polynomials.

In principle, we can proceed as in SNFS, but many difficulties arise because of the large coefficients of $g(x)$. For details, we refer the reader to [32, 33, 53, 54, 59, 68, 69, 91]. Suffice it to say that the difficulties can be overcome and the method works! Due to the constant factors involved it is slower than MPQS for numbers of less than about 110 decimal digits, but faster than MPQS for sufficiently large numbers, as anticipated from the theoretical run times given in §3.

Some of the difficulties which had to be overcome to turn GNFS into a practical algorithm are:

1. Polynomial selection. The "base m" method is not very good. Brian Murphy [58, 59, 60] has shown how a very considerable improvement (by a factor of more than ten for number of 140 digits) can be obtained.
2. Linear algebra. After sieving a very large, sparse linear system over $\mathbf{GF}(2)$ is obtained, and we want to find dependencies amongst the rows. It is not practical to do this by Gaussian elimination [90] because the "fill in" is too large. Montgomery [54] showed that the Lanczos method could be adapted for this purpose. (This is nontrivial because a nonzero vector x over $\mathbf{GF}(2)$ can be orthogonal to itself, i.e. $x^T x = 0$.) To take advantage of bit-parallel operations, Montgomery's program works with blocks of size dependent on the wordlength (e.g. 64).
3. Square roots. The final stage of GNFS involves finding the square root of a (very large) product of algebraic numbers[6]. Once again, Montgomery [53] found a way to do this.

At present, the main obstacle to a fully parallel implementation of GNFS is the linear algebra. Montgomery's block Lanczos program runs on a single processor and requires enough memory to store the sparse matrix. In principle it should be possible to distribute the block Lanczos solution over several processors of a parallel machine, but the communication/computation ratio will be high.

It should be noted that if special hardware is built for sieving, as pioneered by Lehmer and recently proposed (in more modern form) by Shamir [78], the linear algebra will become relatively more important[7].

10.1   RSA140

At the time of writing, the largest number factored by GNFS is the 140-digit RSA Challenge number RSA140. It was split into the product of two 70-digit primes in February, 1999, by a team coordinated from CWI, Amsterdam. For details see [16]. To summarise: the amount of computer time required to find the factors was about 2000 Mips-years. The two polynomials used were

---

[6] An idea of Adleman, using quadratic characters, is essential to ensure that the square root exists.

[7] The argument is similar to Amdahl's law: no matter how fast sieving is done, we can not avoid the linear algebra.

$$f(x) = x - 34435657809242536951779007$$

and

$$
\begin{aligned}
g(x) = &+439682082840x^5 + 390315678538960x^4 \\
&-7387325293892994572x^3 - 190271532437429887148242x^2 \\
&-63441025694464617913930613x + 31855391707147435039222350749 4\,.
\end{aligned}
$$

The polynomial $g(x)$ was chosen to have a good combination of two properties: being unusually small over the sieving region and having unusually many roots modulo small primes (and prime powers). The effect of the second property alone makes $g(x)$ as effective at generating relations as a polynomial chosen at random for an integer of 121 decimal digits (so in effect we have removed at least 19 digits from RSA140 by judicious polynomial selection). The polynomial selection took 2000 CPU-hours on four 250 MHz SGI Origin 2000 processors. Sieving was done on about 125 SGI and Sun workstations running at 175 MHz on average, and on about 60 PCs running at 300 MHz on average. The total amount of CPU time spent on sieving was 8.9 CPU-years.

The resulting matrix had about 4.7×10^6 rows and weight about 1.5×10^8 (about 32 nonzeros per row). Using Montgomery's block Lanczos program, it took almost 100 CPU-hours and 810 MB of memory on a Cray C916 to find 64 dependencies among the rows of this matrix. Calendar time for this was five days.

10.2   RSA155

At the time of writing, an attempt to factor the 512-bit number RSA155 is well underway. We confidently predict that it will be factored before the year 2000!

11   Conclusion

We have sketched some algorithms for integer factorisation. The most important are ECM, MPQS and NFS. The algorithms draw on results in elementary number theory, algebraic number theory and probability theory. As well as their inherent interest and applicability to other areas of mathematics, advances in public key cryptography have lent them practical importance.

Despite much progress in the development of efficient algorithms, our knowledge of the complexity of factorisation is inadequate. We would like to find a polynomial time factorisation algorithm or else prove that one does not exist. Until a polynomial time algorithm is found or a quantum computer capable of running Shor's algorithm [79, 80] is built, large factorisations will remain an interesting challenge.

A survey similar to this one was written in 1990 (see [8]). Comparing the examples there we see that significant progress has been made. This is partly due to Moore's law, partly due to the use of many machines on the Internet, and partly due to improvements in algorithms (especially GNFS). The largest number factored by MPQS at the time of writing [8] had 111 decimal digits. According to [21], the 110-digit number RSA110 was factored in 1992, and the 120-digit number RSA120 was factored in 1993 (both by MPQS). In 1996 GNFS was used to factor RSA130, and in February 1999 GNFS also cracked RSA140. We have predicted that the 512-bit number RSA155 will be factored before the year 2000. Progress seems to be accelerating, but this is due in large part to algorithmic improvements which are unlikely to be repeated.

512-bit RSA keys are now definitely insecure. 1024-bit keys should remain secure for many years, barring the unexpected (but unpredictable) discovery of a completely new algorithm which is better than GNFS, or the development of a practical quantum computer.

Acknowledgements

Thanks are due to Peter Montgomery, Brian Murphy, Herman te Riele, Sam Wagstaff, Jr. and Paul Zimmermann for their assistance.

References

[1] D. Atkins, M. Graff, A. K. Lenstra and P. C. Leyland, The magic words are squeamish ossifrage, *Advances in Cryptology: Proc. Asiacrypt'94*, *LNCS* **917**, Springer-Verlag, Berlin, 1995, 263–277.

[2] A. O. L. Atkin and F. Morain, Elliptic curves and primality proving, *Math. Comp.* **61** (1993), 29–68. Programs available from `ftp://ftp.inria.fr/INRIA/ecpp.V3.4.1.tar.Z` .

[3] H. Boender and H. J. J. te Riele, *Factoring integers with large prime variations of the quadratic sieve*, Experimental Mathematics, **5** (1996), 257–273.

[4] R. P. Brent, A Fortran multiple-precision arithmetic package, *ACM Transactions on Mathematical Software* **4** (1978), 57–70.

[5] R. P. Brent, An improved Monte Carlo factorisation algorithm, *BIT* **20** (1980), 176–184.

[6] R. P. Brent, Some integer factorisation algorithms using elliptic curves, *Australian Computer Science Communications* **8** (1986), 149–163. `ftp://ftp.comlab.ox.ac.uk/pub/Documents/techpapers/Richard.Brent/rpb102.dvi.gz` .

[7] R. P. Brent, Parallel algorithms for integer factorisation, in *Number Theory and Cryptography* (edited by J. H. Loxton), London Mathematical Society Lecture Note Series **154**, Cambridge University Press, 1990, 26–37.

[8] R. P. Brent, Vector and parallel algorithms for integer factorisation, *Proceedings Third Australian Supercomputer Conference* University of Melbourne, December 1990, 12 *pp.* `ftp://ftp.comlab.ox.ac.uk/pub/Documents/techpapers/Richard.Brent/rpb122.dvi.gz` .

[9] R. P. Brent, *Large factors found by ECM*, Oxford University Computing Laboratory, May 1999. `ftp://ftp.comlab.ox.ac.uk/pub/Documents/techpapers/Richard.Brent/champs.txt` .

[10] R. P. Brent, Factorization of the tenth Fermat number, *Math. Comp.* **68** (1999), 429-451. Preliminary version available as *Factorization of the tenth and eleventh Fermat numbers*, Technical Report TR-CS-96-02, CSL, ANU, Feb. 1996, 25*pp.* ftp://ftp.comlab.ox.ac.uk:/pub/Documents/techpapers/Richard.Brent/rpb161tr.dvi.gz .

[11] R. P. Brent and J. M. Pollard, Factorisation of the eighth Fermat number, *Math. Comp.* **36** (1981), 627–630.

[12] J. Brillhart, D. H. Lehmer, J. L. Selfridge, B. Tuckerman and S. S. Wagstaff, Jr., *Factorisations of $b^n \pm 1, b = 2, 3, 5, 6, 7, 10, 11, 12$ up to high powers*, American Mathematical Society, Providence, Rhode Island, second edition, 1988. Updates available from http://www/cs/purdue.edu/homes/ssw/cun/index.html .

[13] D. A. Buell, Factoring: algorithms, computations, and computers, *J. Supercomputing* **1** (1987), 191–216.

[14] C. Caldwell, The Dubner PC Cruncher – a microcomputer coprocessor card for doing integer arithmetic, review in *J. Rec. Math.* **25** (1), 1993.

[15] T. R. Caron and R. D. Silverman, Parallel implementation of the quadratic sieve, *J. Supercomputing* **1** (1988), 273–290.

[16] S. Cavallar, B. Dodson, A. K. Lenstra, P. Leyland, W. Lioen, P. L. Montgomery, B. Murphy, H. te Riele and P. Zimmermann, *Factorization of RSA-140 using the number field sieve*, announced 4 February 1999. Available from ftp://ftp.cwi.nl/pub/herman/NFSrecords/RSA-140 .

[17] S. Cavallar, B. Dodson, A. K. Lenstra, P. Leyland, W. Lioen, P. L. Montgomery, H. te Riele and P. Zimmermann, *211-digit SNFS factorization*, announced 25 April 1999. Available from ftp://ftp.cwi.nl/pub/herman/NFSrecords/SNFS-211 .

[18] D. V. and G. V. Chudnovsky, Sequences of numbers generated by addition in formal groups and new primality and factorization tests, *Adv. in Appl. Math.* **7** (1986), 385–434.

[19] H. Cohen, *A Course in Computational Algebraic Number Theory*, Springer-Verlag, Berlin, 1993.

[20] H. Cohen and H. W. Lenstra, Jr., Primality testing and Jacobi sums, *Math. Comp.* **42** (1984), 297–330.

[21] S. Contini, The factorization of RSA-140, RSA Laboratories Bulletin **10**, 8 (March 1999). Available from http://www.rsa/com/rsalabs/html/bulletins.html .

[22] J. Cowie, B. Dodson, R. M. Elkenbracht-Huizing, A. K. Lenstra, P. L. Montgomery and J. Zayer, A world wide number field sieve factoring record: on to 512 bits, *Advances in Cryptology: Proc. Asiacrypt'96*, *LNCS* **1163**, Springer-Verlag, Berlin, 1996, 382–394.

[23] R. E. Crandall, *Parallelization of Pollard-rho factorization*, preprint, 23 April 1999.

[24] R. Crandall and B. Fagin, Discrete weighted transforms and large-integer arithmetic, *Math. Comp.* **62** (1994), 305–324.

[25] D. Deutsch, Quantum theory, the Church-Turing principle and the universal quantum computer, *Proc. Roy. Soc. London, Ser. A* **400** (1985), 97–117.

[26] D. Deutsch, Quantum computational networks, *Proc. Roy. Soc. London, Ser. A* **425** (1989), 73–90.

[27] W. Diffie and M. Hellman, New directions in cryptography, *IEEE Trans. Inform. Theory* **22** (1976), 472–492.

[28] B. Dixon and A. K. Lenstra, Massively parallel elliptic curve factoring, *Proc. Eurocrypt '92*, *LNCS* **658**, Springer-Verlag, Berlin, 1993, 183–193.

[29] B. Dodson and A. K. Lenstra, NFS with four large primes: an explosive experiment, *Proc. Crypto'95*, *LNCS* **963**, Springer-Verlag, Berlin, 1995, 372–385.

[30] T. El Gamal, A public-key cryptosystem and a signature scheme based on discrete logarithms, *Advances in Cryptology: Proc. CRYPTO'84*, Springer-Verlag, Berlin, 1985, 10–18.

[31] T. El Gamal, A public-key cryptosystem and a signature scheme based on discrete logarithms, *IEEE Trans. on Information Theory* **31** (1985), 469–472.

[32] M. Elkenbracht-Huizing, An implementation of the number field sieve, *Experimental Mathematics*, **5** (1996), 231–253.

[33] M. Elkenbracht-Huizing, *Factoring integers with the number field sieve*, Doctor's thesis, Leiden University, 1997.

[34] M. Elkenbracht-Huizing, A multiple polynomial general number field sieve *Algorithmic Number Theory – ANTS III, LNCS* **1443**, Springer-Verlag, Berlin, 1998, 99–114.

[35] K. F. Ireland and M. Rosen, *A Classical Introduction to Modern Number Theory*, Springer-Verlag, Berlin, 1982.

[36] D. E. Knuth, *The Art of Computer Programming*, Vol. 2, Addison Wesley, third edition, 1997.

[37] N. Koblitz, *A Course in Number Theory and Cryptography*, Springer-Verlag, New York, 1994.

[38] S. Lang, *Elliptic Curves – Diophantine Analysis*, Springer-Verlag, Berlin, 1978.

[39] R. S. Lehman, Factoring large integers, *Math. Comp.* **28** (1974), 637–646.

[40] A. K. Lenstra and H. W. Lenstra, Jr. (editors), The development of the number field sieve, *Lecture Notes in Mathematics* **1554**, Springer-Verlag, Berlin, 1993.

[41] A. K. Lenstra, H. W. Lenstra, Jr., M. S. Manasse and J. M. Pollard, *The number field sieve*, Proc. 22nd Annual ACM Conference on Theory of Computing, Baltimore, Maryland, May 1990, 564–572.

[42] A. K. Lenstra, H. W. Lenstra, Jr., M. S. Manasse, and J. M. Pollard, The factorization of the ninth Fermat number, *Math. Comp.* **61** (1993), 319–349.

[43] A. K. Lenstra and M. S. Manasse, Factoring by electronic mail, *Proc. Eurocrypt '89, LNCS* **434**, Springer-Verlag, Berlin, 1990, 355–371.

[44] A. K. Lenstra and M. S. Manasse, Factoring with two large primes, *Math. Comp.* **63** (1994), 785–798.

[45] H. W. Lenstra, Jr., Factoring integers with elliptic curves, *Annals of Mathematics (2)* **126** (1987), 649–673.

[46] K. S. McCurley, The discrete logarithm problem, in *Cryptography and Computational Number Theory*, C. Pomerance, ed., *Proc. Symp. Appl. Math.*, Amer. Math. Soc., 1990.

[47] A. Menezes, *Elliptic Curve Public Key Cryptosystems*, Kluwer Academic Publishers, Boston, 1993.

[48] A. Menezes, Elliptic curve cryptosystems, *CryptoBytes* **1**, 2 (1995), 1–4. Available from http://www.rsa.com/rsalabs/pubs/cryptobytes .

[49] P. L. Montgomery, Modular multiplication without trial division, *Math. Comp.* **44** (1985), 519–521.

[50] P. L. Montgomery, Speeding the Pollard and elliptic curve methods of factorisation, *Math. Comp.* **48** (1987), 243–264.

[51] P. L. Montgomery, *An FFT extension of the elliptic curve method of factorization*, Ph. D. dissertation, Mathematics, University of California at Los Angeles, 1992. ftp://ftp.cwi.nl/pub/pmontgom/ucladissertation.psl.Z .

[52] P. L. Montgomery, A survey of modern integer factorization algorithms, *CWI Quarterly* **7** (1994), 337–366. ftp://ftp.cwi.nl/pub/pmontgom/cwisurvey.psl.Z .

[53] P. L. Montgomery, Square roots of products of algebraic numbers, *Mathematics of Computation 1943 - 1993, Proc. Symp. Appl. Math.* **48** (1994), 567–571.

[54] P. L. Montgomery, A block Lanczos algorithm for finding dependencies over $GF(2)$, *Advances in Cryptology: Proc. Eurocrypt'95, LNCS* **921**, Springer-Verlag, Berlin, 1995, 106–120.

[55] F. Morain, *Courbes elliptiques et tests de primalité*, Ph. D. thesis, Univ. Claude Bernard – Lyon I, France, 1990. ftp://ftp.inria.fr/INRIA/publication/Theses/TU-0144.tar.Z .

[56] M. A. Morrison and J. Brillhart, A method of factorisation and the factorisation of F_7, *Math. Comp.* **29** (1975), 183–205.

[57] R. Motwani and P. Raghavan, *Randomized Algorithms*, Cambridge University Press, 1995.

[58] B. A. Murphy, Modelling the yield of number field sieve polynomials, *Algorithmic Number Theory – ANTS III, LNCS* **1443**, Springer-Verlag, Berlin, 1998, 137–150.

[59] B. A. Murphy, *Polynomial selection for the number field sieve integer factorisation algorithm*, Ph. D. thesis, Australian National University, 1999.

[60] B. A. Murphy and R. P. Brent, On quadratic polynomials for the number field sieve, *Australian Computer Science Communications* **20** (1998), 199–213.

[61] A. M. Odlyzko, Discrete logarithms in finite fields and their cryptographic significance, *Advances in Cryptology: Proc. Eurocrypt '84, LNCS* **209**, Springer-Verlag, Berlin, 1985, 224–314.

[62] A. M. Odlyzko, The future of integer factorization, *CryptoBytes* **1**, 2 (1995), 5–12. Available from http://www.rsa.com/rsalabs/pubs/cryptobytes .

[63] P. C. van Oorschot and M. J. Wiener, Parallel collision search with application to hash functions and discrete logarithms, *Proc 2nd ACM Conference on Computer and Communications Security*, ACM, New York, 1994, 210–218.

[64] P. C. van Oorschot and M. J. Wiener, Parallel collision search with cryptanalytic applications, *J. Cryptology* **12** (1999), 1–28.

[65] J. M. Pollard, Theorems in factorisation and primality testing, *Proc. Cambridge Philos. Soc.* **76** (1974), 521–528.

[66] J. M. Pollard, A Monte Carlo method for factorisation, *BIT* **15** (1975), 331–334.

[67] C. Pomerance, The quadratic sieve factoring algorithm, *Advances in Cryptology, Proc. Eurocrypt '84, LNCS* **209**, Springer-Verlag, Berlin, 1985, 169–182.

[68] C. Pomerance, The number field sieve, *Proceedings of Symposia in Applied Mathematics* **48**, Amer. Math. Soc., Providence, Rhode Island, 1994, 465–480.

[69] C. Pomerance, A tale of two sieves, *Notices Amer. Math. Soc.* **43** (1996), 1473–1485.

[70] C. Pomerance, J. W. Smith and R. Tuler, A pipeline architecture for factoring large integers with the quadratic sieve algorithm, *SIAM J. on Computing* **17** (1988), 387–403.

[71] J. Preskill, *Lecture Notes for Physics 229: Quantum Information and Computation*, California Institute of Technology, Los Angeles, Sept. 1998. http://www.theory.caltech.edu/people/preskill/ph229/ .

[72] M. O. Rabin, Probabilistic algorithms for testing primality, *J. Number Theory* **12** (1980), 128–138.

[73] H. J. J. te Riele, W. Lioen and D. Winter, Factoring with the quadratic sieve on large vector computers, *Belgian J. Comp. Appl. Math.* **27** (1989), 267–278.

[74] H. Riesel, *Prime numbers and computer methods for factorization*, 2nd edition, Birkhäuser, Boston, 1994.

[75] R. L. Rivest, A. Shamir and L. Adleman, A method for obtaining digital signatures and public-key cryptosystems, *Communications of the ACM* **21** (1978), 120–126.

[76] RSA Laboratories, Information on the RSA challenge, `http://www.rsa/com/rsalabs/html/challenges.html` .

[77] B. Schneier, *Applied Cryptography*, second edition, John Wiley and Sons, 1996.

[78] A. Shamir, *Factoring large numbers with the TWINKLE device* (extended abstract), preprint, 1999. Announced at Eurocrypt'99.

[79] P. W. Shor, Algorithms for quantum computation: discrete logarithms and factoring, *Proc. 35th Annual Symposium on Foundations of Computer Science*, IEEE Computer Society Press, Los Alamitos, California, 1994, 124–134. CMP 98:06

[80] P. W. Shor, Polynomial time algorithms for prime factorization and discrete logarithms on a quantum computer, *SIAM J. Computing* **26** (1997), 1484–1509.

[81] J. H. Silverman, *The arithmetic of elliptic curves*, Graduate Texts in Mathematics **106**, Springer-Verlag, New York, 1986.

[82] R. D. Silverman, The multiple polynomial quadratic sieve, *Math. Comp.* **48** (1987), 329–339.

[83] R. D. Silverman and S. S. Wagstaff, Jr., A practical analysis of the elliptic curve factoring algorithm, *Math. Comp.* **61** (1993), 445–462.

[84] I. N. Stewart and D. O. Tall, *Algebraic Number Theory*, second edition, Chapman and Hall, 1987.

[85] D. Stinson, *Cryptography – Theory and Practice*, CRC Press, Boca Raton, 1995.

[86] A. M. Turing, On computable numbers, with an application to the Entscheidungsproblem, *Proc. London Math. Soc.* (2) **42** (1936), 230–265. Errata *ibid* **43** (1937), 544–546.

[87] U. Vazirani, Introduction to special section on quantum computation, *SIAM J. Computing* **26** (1997), 1409–1410.

[88] D. Weber, Computing discrete logarithms with the number field sieve, *Algorithmic Number Theory – ANTS II, LNCS* **1122**, Springer-Verlag, Berlin, 1996, 99–114.

[89] D. Weber, *On the computation of discrete logarithms in finite prime fields*, Ph. D. thesis, Universität des Saarlandes, 1997.

[90] D. H. Wiedemann, Solving sparse linear equations over finite fields, *IEEE Trans. Inform. Theory* **32** (1986), 54–62.

[91] J. Zayer, *Faktorisieren mit dem Number Field Sieve*, Ph. D. thesis, Universität des Saarlandes, 1995.

MERCATOR, the Mission

Philippe Courtier

Météo-France

Abstract. Mercator, a project initiated by the oceanographic community in France, is described. The objective is to develop an eddy resolving data assimilation system for the ocean.

1 Introduction

The Mercator project whose aim is the development of an eddy resolving global data assimilation system is a proposal elaborated by the scientific oceanographic community and endorsed by the French institutes interested in operational oceanography, Centre National d'Etudes Spatiales (CNES), Centre National de la Recherche Scientifique / Institut National des Sciences de l'Univers (CNRS/INSU), Institut Français de Recherche pour l'Exploitation de la MER (IFREMER), Météo-France, Office de la Recherche Scientifique et Technique Outre Mer (ORSTOM) and Service Hydrographique et Océanographique de la Marine (SHOM).

Two workshops took place in 1995 in order to draw the general lines of the project while CLS-Argos (Collecte, Localisation par Satellite) was in charge of the synthesis and as such of the phase 0. 1996 was the year where the project was set-up by Jean-François Minster (CNRS / INSU) and the phase A of feasibility was decided (and funded) by the above institutes February 3rd, 1997. CERFACS (Centre Européen de Recherche et Formation Avancés en Calcul Scientifique) is the focal point for Mercator and Jean-Claude André has been appointed as executive secretary in order, among other things, to ensure an efficient feedback between the project and the institutes.

Mercator is a contribution to the Global Ocean Data Assimilation Experiment (GODAE) planned for the years 2003--2005.

The following section describes the rationale of the project and the mission objectives. Some perspectives relevant for operational oceanography are then outlined in conclusion.

2 Mercator Project Rationale and Mission Objectives

The goal of Mercator is the operational implementation within five to seven years of a system which simulates the global ocean with a primitive equation high-resolution model and which assimilates satellite and in situ data. In the longer term the system has to contribute to the development of a climatic prediction system relying on a

P. Amestoy et al. (Eds.): Euro-Par'99, LNCS 1685, pp. 23-29, 1999.

coupled ocean atmosphere model. In addition the system has to be useful for the military and commercial applications of oceanography.

Three users of Mercator have been clearly identified during the set-up of the project:

Seasonal Prediction. The meteorological community is considering the operational implementation of seasonal prediction systems. The predictions rely on the integration of a coupled ocean atmosphere model. Initial conditions are required both for the atmosphere and the ocean over the whole globe. Mercator will provide the initial conditions for the ocean whereas the atmospheric data assimilation systems used for numerical weather forecasts will provide the initial conditions for the atmosphere.

At this stage, it is not envisaged to operate coupled ocean atmosphere data assimilation systems. A fundamental difficulty would be the difference of time scale of the two fluids, which for the ocean goes beyond the deterministic predictability limit of the atmosphere. This issue will be reconsidered in the future if unstable coupled modes are identified which would then need to be properly controlled during the assimilation process.

The European Centre for Medium-Range Weather Forecasts (ECMWF) is presently operating a seasonal prediction system in experimental mode. It is foreseen the Mercator will be developed in collaboration with ECMWF. The main short-term priority is the availability of a global system with the emphasis being put on the tropical oceans.

Marine Applications. The French Navy is presently operating an experimental data assimilation system for the Eastern Atlantic. High spatial resolution, better than a tenth of a degree, is deemed necessary. The primary area of interest is the Eastern Atlantic with also a high priority for the Mediterranean Sea. The prediction range should be of 4 to 6 weeks ahead. Error bars will have to be provided which imply that ensemble prediction may have to be considered at some stage.

Coastal applications are not part stricto sensu of the Mercator project. Nevertheless Mercator has to be able to provide lateral boundary conditions for coastal models.

The UK Meteorological Office (UKMO) expressed interest in developing collaboration with the Mercator project.

Scientific Applications. Besides the operational implementation, which produces oceanic fields made available to the scientific community, Mercator will be used as a research tool for producing level 3 (gridded) fields at the occasion of measurement campaigns. Consequently flexibility has to be part of the overall design in order to be able to accommodate specific requirements like being able to concentrate on a given area.

2.1   Mercator Main Components

The Mercator system consists of three main components: firstly, an ocean general circulation model, secondly a data assimilation algorithm and lastly a data stream (observations, forcing fields, initialisation fields, validation data ...).

The Ocean General Circulation Model. The ocean general circulation model of Mercator is a primitive equation model based on OPA, which has been developed at Laboratoire d'Océanographie Dynamique et de Climatologie (LODYC). It is recognised that the OGCM will be in constant evolution. At the time of GODAE the model will have been validated at the global scale and with a typical horizontal resolution of $1/12°$.

In order to achieve this goal, two lines of development will be followed in parallel:

Global, low resolution. The main priority here is to concentrate on the global aspect of the problem, and in particular on the validation in coupled ocean-atmosphere mode. Mercator will benefit from and rely on the effort made by the French research community in the field of climate simulations and co-ordinated under the " groupe GASTON ". Two atmospheric models are being used, the code developed by Laboratoire de M\'et\'eorologie Dynamique (LMD) and the code whose kernel has been jointly developed by M\'et\'eo-France and ECMWF under the framework of the Arpege-IFS collaboration. The coupling module is OASIS developed by CERFACS.

Furthermore, OPA has been implemented at ECMWF and the sensitivity of seasonal predictions to the OGCM used will be assessed by comparisons with the present experimental seasonal prediction system of ECMWF which makes use of the OGCM HOPE developed by the Max Planck Institute, Hamburg.

Basin, high resolution.

The CLIPPER project which involves scientists from LPO (Brest), LEGI (Grenoble), LEGOS (Toulouse) and LODYC (Paris) aims at simulating the Atlantic ocean circulation at a resolution of a sixth of a degree. Several scientific issues particularly relevant for Mercator are being addressed.

The following step is to build on the experience acquired during CLIPPER and to operate in 1999 a prototype system based on OPA, which will simulate the North Atlantic and Mediterranean circulations at a resolution of a twelfth of a degree, and with TBD vertical levels. Temperature, salinity and horizontal velocity are prognostic variables. The rigid lid hypothesis is made. A mixing layer based on a TKE closure is used

Data Assimilation. Data assimilation methods are currently in quick evolution, particularly under the impulse of the oceanographic community. It is therefore necessary to allow for flexibility in the algorithms, which can be accommodated as well as to offer various possibilities for future evolution. On the other hand, (pre-) operational requirements imply to be rather conservative in the choice for initial implementation. ECMWF has implemented for the seasonal prediction project an existing bidimensional Optimal Interpolation scheme that is not in the position, at the present stage of development, to use altimeter data. The basin, high resolution, initial implementation of Mercator will rely on the already validated code SOFA that uses empirical orthogonal functions of the vertical covariance matrix to decouple the vertical and horizontal directions. SOFA does not assimilate yet in-situ observations, this is a short-term development scheduled for 1998.

Data assimilation attempts at combining various information sources (observations, dynamic's of evolution, ...) and as such is a statistical estimation problem. The methods, which may be used within Mercator, will rely on a specification of the statistics of error (either static or adaptive) in order to be able to improve the system by a careful study of the innovation vector. Besides the theoretical motivation, this choice is motivated by the lesson drawn from the meteorological operational experience that the only Centres still at the cutting edge of weather forecasts have followed this path.

Mercator has to offer the possibility to accommodate the main "modern" data assimilation methods, Optimal Interpolation since it is the initial implementation, variational approach (3D and 4D) in the direct and dual formulation (representers), simplified Kalman filters and adaptive filtering. In order to save computational time, space reduction techniques have to be accounted for in the design. A plausible implementation for the GODAE era would be an adaptive filtering which evaluates key aspects of the statistics of estimation error while the analysis is solved on a limited time window with a variational approach formulated in a reduced space.

It can be shown that all the above methods use more or less the same operators (observation operator, model, projector on the reduced space, covariance matrices of the various errors, gain, ...) or solver (minimisation, eigen values), which are in limited number, but with a very different calling sequence. In order to accommodate the above flexibility requirement, the algebraic component of the data assimilation methods and the scientific content of the operators will be made independent. A generalised coupler, PALM, will solve the algebraic part; it builds on the technical experience acquired with OASIS. The scientific content of the operators will depend on the experience acquired during the experimental phases of the basin, high resolution and global, low-resolution implementations, and on the results obtained by the research community.

While the initial experimental implementation may rely on a univariate formulation, a multivariate formulation in order to be able to accommodate several different data types will be implemented quickly afterward.

In the initial experimental phase, the computer resources used for a week of assimilation should be less than twice the cost of a week of the assimilating model integration. At the time of GODAE, the cost of a week of assimilation should not exceed five time the cost of the assimilating model integration.

Data Stream

Ocean observations, space segment. The requirement of Mercator for oceanic space based data consists of sea surface temperature and altimetry. In the longer term, ocean colour information may be used in coupled physical biological assimilation system, but this is not foreseen as an operational requirement for Mercator at the time of GODAE. Nevertheless this will most likely become soon a research requirement, and as such be thought of within Mercator. There is a requirement of Mercator for global salinity observations, therefore the space agencies are invited to develop, if feasible, the required technologies.

Sea surface temperature fields may be produced using observations from various instruments. The main operational source is the visible/infrared imager AVHRR (Advanced Very High Resolution Radiometer) on board the National Oceanic and Atmospheric Administration (NOAA) operational satellites. AVHRR is also part of the METOP/EPS payload as approved by the Eumetsat Council in June 1996.

Altimetric data are necessary for Mercator. Beyond Topex-Poseïdon, the Jason collaborative NASA/CNES programme will provide high precision observations. In parallel, observations of lower precision like onboard the ERS-1 and ERS-2 satellites (and ENVISAT for the near future) of the European Space Agency provide a significant complement, which with its improved sampling is essential for mesoscale applications. It has to be pointed out that no altimetric operational programmes have been decided yet, it is one of the goal of Mercator to demonstrate the operational need. A fast delivery product (a very few days) is necessary for some of the Mercator operational applications, but high quality products remain necessary for scientific applications.

Oceanic in-situ observations. Electromagnetic waves do not penetrate beyond a few meters in the ocean it is therefore necessary for Mercator to assimilate in-situ observations, which provide access to the vertical structure of the ocean. Temperature, currents and salinity are the parameters to be assimilated. Eulerian observations are useful in that under some sort of stationarity hypothesis, they allow to identify model biases. Nevertheless there is no difficulty for assimilating Lagrangian observations.

Tomography may provide access to the mean thermal structure of the ocean with access to a few degrees of freedom in the vertical. In a project like Mercator, the meteorological experience shows that once the physical interpretation of remote-sensed data is mature enough at the research level, it is still necessary to have one or two persons dedicated to the implementation in the assimilation suite. Therefore assimilation of tomographic data is not considered as a first short-term priority for Mercator. The issue will have to be reconsidered a couple of years before GODAE, resources permitting.

Forcings. Reanalysis as performed at the ECMWF are essential for providing consistent thermodynamical and momentum forcings for Mercator. It only covers the period 1979 to 1993 and has not used as surface wind speed observations the passive microwave radiometer SSM/I from the DMSP satellite, neither the scatterrometer wind observations (with their 180° directional ambiguity) from the satellites ERS-1 and ERS-2. A reanalysis of the "Scatt years" at higher resolution (T213) would be extremely useful, a bonus would be to cover the whole of the "Topex years", such an effort is taking place at ECMWF within the ERA-40 reanalysis project, nominally at a T106 resolution, possibly at T213, resources permitting. At T213, the smallest half wavelength resolved is 95 km and the shortest scales are heavily dissipated, a significant variability of the wind field is then not resolved. Mercator will benefit from the effort to produce as representative as possible wind forcings, conducted in the research community.

The latent heat fluxes at the surface of the ocean can be estimated from satellite data with some accuracy, at least under the Semaphore conditions. This is not the case for the sensible heat flux, which is significantly smaller. Precipitations are largely unknown. Radiation fluxes are heavily dependent upon clouds whose representation is not the strongest feature of the atmospheric models. It is not a priori obvious that an improvement of the latent heat fluxes of the reanalysis by using satellite data but without the possibility of improving the other fluxes will result in an improved oceanic circulation since the overall consistency will have been lost. Mercator will have to assess its requirements in view of the results obtained during the experimental phases.

Real time. Real time is a key aspect of the various data streams for the operational applications of oceanography. The requirement is of the order of 3 days, and in that respect Jason will be a significant improvement upon Topex-Poseïdon.

2.2   Mercator Technical Requirements

Considering the computer resources involved in running in a high resolution global oceanic model, the assimilation suite will be have to run on the most powerful computers available on the market. In the coming few years, it can be expected that such platforms will rely on moderately high (O100) to high (O1000) parallelism and on distributed memory.

The observational data flow as well as the oceanic fields data flow require a significant amount of Input/Output and of archiving capabilities, while the bandwidth remain limited. Mercator will build on the experience acquired by the ECMWF reanalysis team for the development of the assimilation suite.

A secured data flow has to be designed for sensitive data.

2.3   Some Steps toward Operational Oceanography, Context of Mercator

France has significantly contributed to the emergence of operational oceanography through an investment in the international observing research programmes. As an illustration, France has been the main contributor to the ERS-1/2 programmes, and beyond the ERS altimetric mission is the partner of NASA in Topex/Poseïdon and in the forthcoming Jason. France will also be the main contributor of the space segment of Metop. Concerning in situ data, the involvement in the observing phase of WOCE has also been significant.

In order to valorise this already significant investment, France has to be present in the three aspects of operational oceanography as foreseen in GODAE.

Space Segment. Assuming the Metop programme decided, a long-term commitment of Europe for the measurement of sea surface temperature and the surface momentum fluxes will have been achieved. Operational oceanography further requires a long-term commitment for high precision altimetry and therefore a follow-on of Jason. As of today, altimetry has been funded by development agencies like CNES, NASA or ESA. Following the example of scatterometry and the transfer of the mission responsibility from the development agency ESA to the user agency Eumetsat, the emergence of operational oceanography will probably imply a similar transfer.

In Situ Observations. At the international level, the ARGO proposal of the deployment of an in situ network of Lagrangian profiler will allow Mercator to be fed with the necessary observations. The contribution of France to the in situ part of GODAE is currently being discussed in the Coriolis working group chaired by Ifremer. It could be along 4 actions, which build on already existing or planned systems. First an effort similar as during WOCE, typically 5000 per year, should be performed for maintaining the XBT measurements. Second, Lagrangian profilers of the class of the US Palace or the French Provor has to be deployed on a large-scale basis, research and development is necessary for the measurement of salinity. Third, Eulerian profilers need to be developed and deployed, with a typical number similar to the Lagrangian profilers. The cost of a profile has to be less than 1000FF for providing a viable long term observing system. Lastly moored and drifting buoys have to be maintained and/or systematically instrumented for oceanographic measurements. The Pomme experiment will be a significant milestone.

Modelling and Assimilation. The Mercator project is currently conducting the research and development effort in this area. In order to implement operationally the data assimilation suite to be developed by Mercator, two additional elements are required besides the Mercator team. First, high performance computing facilities and the corresponding archiving system are mandatory. This corresponds to resources comparable to a significant fraction of operational numerical weather prediction requirements. Second a team needs to be in place whose responsibility is to manage and scientifically monitor the incoming data flow (observations, forcings...) and the outcoming data flow (products). In other words, an Operational Oceanographic Centre (CO^2) is required.

Adaptive Scheduling for Task Farming with Grid Middleware

Henri Casanova[1], MyungHo Kim[2], James S. Plank[3], and Jack J. Dongarra[3,4]

[1] Department of Computer Science and Engineering, University of California
at San Diego, La Jolla, CA 92093-0114, USA
[2] School of Computing, SoongSil University, Seoul, 156-743 Korea
[3] Department of Computer Science, University of Tennessee,
Knoxville, TN 37996-1301, USA
[4] Mathematical Science Section, Oak Ridge National Laboratory,
Oak Ridge, TN 37831, USA

Abstract. Scheduling in metacomputing environments is an active field
of research as the vision of a Computational Grid becomes more concrete.
An important class of Grid applications are long-running parallel computations with large numbers of somewhat independent tasks (Monte-Carlo
simulations, parameter-space searches, etc.). A number of Grid middleware projects are available to implement such applications but scheduling strategies are still open research issues. This is mainly due to the
diversity of both Grid resource types and of their availability patterns.
The purpose of this work is to develop and validate a general adaptive
scheduling algorithm for task farming applications along with a user interface that makes the algorithm accessible to domain scientists. Our
algorithm is general in that it is not tailored to a particular Grid middleware and that it requires very few assumptions concerning the nature
of the resources. Our first testbed is NetSolve as it allows quick and easy
development of the algorithm by isolating the developer from issues such
as process control, I/O, remote software access, or fault-tolerance.

Keywords: Farming, Master-Slave Parallelism, Scheduling, Metacomputing, Grid Computing.

1 Introduction

The concept of a *Computational Grid* envisioned in [1] has emerged to capture
the vision of a network computing system that provides broad access not only
to massive information resources, but to massive computational resources as
well. Such computational grids will use high-performance network technology to
connect hardware, software, instruments, databases, and people into a seamless
web that supports a new generation of computation-rich problem solving environments for scientists and engineers. Grid resources will be ubiquitous thereby
justifying the analogy to the Power Grid.

Those features have generated interest among many domain scientists and
new classes of applications arise as being potentially *griddable*. Grid resources

P. Amestoy et al. (Eds.): Euro-Par'99, LNCS 1685, pp. 30–43, 1999.
© Springer-Verlag Berlin Heidelberg 1999

and their access policies are inherently very diverse, ranging from directly accessible single workstations to clusters of workstations managed by Condor [2], or MPP systems with batch queuing management. Furthermore, the availability of these resources changes dynamically in a way that is close to unpredictable. Lastly, predicting networking behavior on the grid is an active but still open research area. Scheduling applications in such a chaotic environment according the end-users' need for fast response-time is not an easy task. The concept of a universal scheduling paradigm for any application at the current time is intractable and the current trend in the scheduling research community is to focus on schedulers for broad *classes* of applications. Given the characteristics of the Grid, it is not surprising that even applications with extremely simple structures raise many challenges in terms of scheduling.

In this paper we address applications that have simple task-parallel structures (master-slave) but require large number of computational resources. We call such applications *task farming applications* according to the terminology introduced in [3]. Examples of such applications include Monte-Carlo simulations and parameter-space searches. Our goal is not only to design a scheduling algorithm but also to provide a convenient user interface that can be used by domain scientists that have no knowledge about the Grid structure.

Section 2 shows how some of the challenges can be addressed by using a class of grid middleware projects as underlying operating environments, while others need to be addressed specifically with adaptive scheduling algorithms. Section 3 gives an overview of related research work and highlights the original elements of this work. Section 4 contains a brief overview of NetSolve, the grid middleware that we used as a testbed. Sections 5 and 6 describe the implementation of the task farming interface and the implementation of the adaptive scheduling algorithm underneath that interface. Section 7 presents experimental results to validate the scheduling strategy. Section 8 concludes with future research and software design directions.

2   Motivation and Challenges for Farming

Our intent is to design and build an easily accessible computational framework for task farming applications. An obvious difficulty then is to isolate the users from details such as I/O, process control, connections to remote hosts, fault-tolerance, etc. Fortunately an emerging class of Grid middleware projects provides the necessary tools and features to transparently handle most of the low-level issues on behalf of the user. We call these middleware projects *functional metacomputing environments*, the user's interface to the Grid is a functional remote procedure call (i.e a call without side effects). The middleware intercepts the procedure call and treats it as a request for service. The procedure call arguments are wrapped up and sent to the Grid resources that are currently best able to service the request, and when the request has been serviced, the results are shipped back to the user, and his or her procedure call returns. The middleware is responsible for the details of managing the service on the Grid – resource selection and allocation, data movement, I/O, and fault-tolerance.

There are two main functional metacomputing environments available today. These are NetSolve (see Section 4) and Ninf [4]. Building our framework on top of such architectures allows us to focus on meaningful issues like the scheduling algorithm rather than building a whole system from the ground up. Our choice of NetSolve as a testbed is motivated by the authors' experience with that system. Section 5 describes our first attempts at an API.

Of course, the main challenge is scheduling. Indeed, for long running farming applications it is to be expected that the availability and workload of resources within the server pool will change dynamically. We must therefore design and validate an adaptive scheduling algorithm (see Section 6). Furthermore, that algorithm should be general and applicable not only for a large class of applications but also for any operating environment. The algorithm is therefore designed to be portable to other metacomputing environments [4, 5, 6, 2, 7].

3   Related Work

Nimrod [7] is targeted to computational applications based on the "exploration of a range of parameterized scenarios" which is similar to our definition of task farming. The user interfaces in Nimrod are at the moment more evolved than the API described in Section 5. However, we believe that our API will be a building block for high-level interfaces (see Section 8). The current version of Nimrod (or Clustor, the commercial version available from [8]) does not use any metacomputing infrastructure project whereas our task farming framework is built on top of grid middleware. However, a recent effort, Nimrod/G [9], plans to build Nimrod directly on top of Globus [5]. We believe that a project like NetSolve (or Ninf) is a better choice for this research work. First, NetSolve is freely available. Second, NetSolve provides a very simple interface, letting us focus on scheduling algorithms rather than Grid infrastructure details. Third, NetSolve can and probably will be implemented on top of most Globus services and will then leverage the Grid infrastructure without modifications of our scheduling algorithms. Another distinction between this work and Nimrod is that the latter does not contain adaptive algorithms for scheduling like the one described in Section 6. In fact, it is not inconceivable that the algorithms eventually produced by this work could be incorporated seamlessly into Nimrod.

Calypso [10] is a programming environment for a loose collection of distributed resources. It is based on C++ and shared memory, but exploits task-based parallelism of relatively independent jobs. It has an eager scheduling algorithm and like the functional metacomputing environments described in this paper, uses the idempotence of the tasks to enable a replication-based fault-tolerance.

A system implemented on top of the Helios OS that allows users to program master-slave programs using a "Farming" API is described in[3, 11]. Like in

Calypso, the idempotence of tasks is used to achieve fault-tolerance. They do not focus on scheduling.

The AppLeS project [12, 13] develops metacomputing scheduling agents for broad classes of computational applications. Part of the effort targets scheduling master-slave applications [14] (task farming applications with our terminology). A collaboration between the NetSolve and the AppLeS team has been initiated and and integration of AppLeS technology, the Network Weather Service (NWS) [15] NetSolve-like systems, and the results in this document is underway.

As mentioned earlier, numerous ongoing projects are trying to establish the foundations of the *Computational Grid* envisioned in [1]. Ninf [4] is similar to NetSolve in that it is targeted to domain scientists. Like NetSolve, Ninf provides simple computational services and and the development teams are collaborating to make the two system interoperate and standardize the basic protocols. At a lower-level are Globus [5] and Legion [6] which aim at providing basic infrastructure for the grid. Condor [2, 16] defines and implements a powerful model for grid components by allowing the idle cycles of networks of workstation to be harvested for the benefit of grid users without penalizing local users.

4 Brief Overview of NetSolve

The NetSolve project is under development at the University of Tennessee and the Oak Ridge National Laboratory. Its original goal is to alleviate the difficulties that domain scientists usually encounter when trying to locate/install/use numerical software, especially on multiple platforms. With NetSolve, the user does not need to be concerned with the location/type of the hardware resources

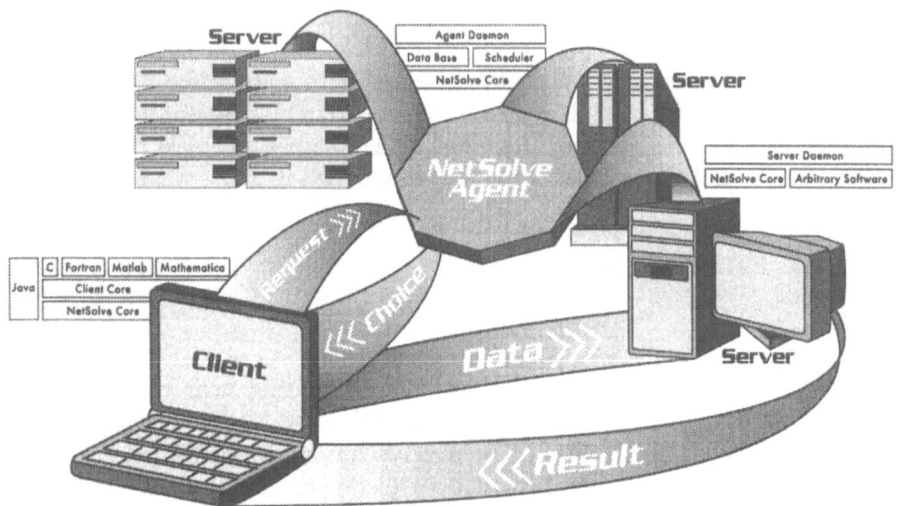

Fig. 1. The NetSolve System

being used or with the software installation. Furthermore, NetSolve provides transparent fault-tolerance mechanisms and implements scheduling algorithms to minimize overall response time. As seen on Figure 1 NetSolve has a three-tiered design in that a *client* consults an *agent* prior to sending requests to a *server*. Let us give basic concepts about those three components as well as information about the current status of the project.

The NetSolve Server: A NetSolve server can be started on any hardware resource (single workstation, cluster of workstations, MPP). It can then provide access to arbitrary software installed on that resource (NetSolve provides mechanisms to integrate any software component into a server so that it may become available to NetSolve clients [17]).

The NetSolve Agent: The NetSolve agent is the key to the computation-resource mapping decisions as it maintains a database about the statuses and capabilities of servers. It uses that database to make scheduling decisions for incoming user requests. The agent is also the primary participant in the fault-tolerance mechanisms. Note that there can be multiple instances of the NetSolve agent to manage a confederation of servers.

The NetSolve Client: The user can submit (possibly simultaneous) requests to the system and retrieve results with one of the provided interfaces (C, Fortran, Matlab [18], Mathematica [19], Java APIs or Java GUI).

Current Status of NetSolve: At this time, a pre-version of NetSolve 1.2, containing full-fledge software for all UNIX flavors, Win32 C, and Matlab APIs, can be downloaded from the homepage at:

`http://www.cs.utk.edu/netsolve`

The NetSolve Users' Guide [20] contains general purpose information and examples. Details about the NetSolve agent can be found in [21]. Recent developments and applications of NetSolve are described in [22]. Lastly, technical details about the current NetSolve implementation are to be found in [23].

5   Task Farming API

5.1   Basics

In this work, we assume that a functional metacomputing environment is available (see Section 2). That environment provides an API that contains two functions: (i)submit() to send a request asynchronously for computation and (ii)poll() to poll asynchronously for the completion of a request. Polling returns immediately with the status of the request. If the computation is complete, the result is returned as well. The NetSolve and Ninf APIs satisfy these requirements. In addition, the environment provides access to pre-installed software and hardware resources. The user just provides input data and a way to identify which software should be used to process that data. Again, both NetSolve and Ninf comply.

A farming job is one composed of a large number of independent requests that may be serviced simultaneously. This is sometimes referred to as the "bag-of-tasks" model [24, 25]. Farming jobs fall into the class of "embarrassingly parallel" programs, for which it is very clear how to partition the jobs for parallel programming environments. Many important classes of problems, such as Monte-Carlo simulations (e.g. [26]) and parameter-space searches (e.g. [7]) fall into this category.

Without a farming API, the user is responsible for managing the requests himself. One possibility would be to submit all the desired requests at once and let the system schedule them. However, we have seen that scheduling on the grid is a challenging issue and as a result, the available grid middleware projects implement only minimal scheduling capabilities that do not optimize even this simple class of parallel programs. A second possibility is for the user to manually manage a ready queue by having at most n requests submitted to the system at any point in time. This solution seems more reasonable; however the optimal value of n depends on Grid resource availability, which is beyond the user's control and is dynamic.

5.2   API

It is difficult to design an API that is both convenient for the end-user and sophisticated enough to handle many real applications. Our farming API contains one function, farm(), with which the user specifies all data for all the computation tasks. The main idea is to replace multiple calls to submit() by one call to farm() whose arguments are *lists* of arguments to submit(). In this first implementation, we assume that arguments to submit() are either integers or pointers (which is consistent with the NetSolve specification). Extending the call to support other argument types would be trivial. The first argument to farm() specifies the number of request by declaring an induction variable and defining its range. The syntax if "i=%d,%d" (see example below). The second argument is the identifier for the computational functionality in the metacomputing environment (a string with Ninf and NetSolve). Then follow a (variable) number of argument lists. Our implementation provides three functions that need to be called to generate such lists:(i)expr() allows an argument to computation i to be an integer computed as an arithmetic expression containing i; (ii)int_array() allows an integer argument to computation i to be an element of an integer array indexed by the value of an arithmetic expression containing i; (iii)ptr_array() is similar to int_array() but handles pointer arguments. Arithmetic expressions are specified with Bourne Shell syntax (accessing the value of i with '$i').

Let us show an example assuming that the underlying metacomputing environment provides a computational function called "foo". The code:

```
double x[10],y[30],z[10];

submit("foo",2,x,10);
submit("foo",4,y,30);
submit("foo",6,z,10);
```

makes three requests to run the "foo" software with the sets of arguments: $(2,x,10)$, $(4,y,30)$ and $(6,z,10)$. Note that x, y and z will hold the results of the NetSolve call. With farming these calls are replaced by

needs to be replaced by

```
void *ptrs[3];
int  *ints[3];

ptrs[0] = x;  ptrs[1] = y;  ptrs[2] = z;
ints[0] = 10; ints[1] = 30; ints[2] = 10;

farm("i=0,2","foo",expr("2*($i+1)"),ptr_array(ptrs,"$i"),
                    int_array(ints,"$i"));
```

We expect to use this API as a basis for more evolved interfaces (e.g. graphical or Shell-based). So far, we have used the API directly to implement basic example computations (2D block-cyclic matrix-multiply, Mandelbrot set computation) and to build a Shell interface to MCell (see Section 7). Section 8 describes how we plan to generalize this work to automatically generate high-level interfaces.

6 Scheduling Strategy

6.1 The Scheduling Algorithm

The main idea behind the scheduling algorithm has already been presented in Section 5.1: managing a ready queue. We mentioned that the user had no elements on which to base the choice for n, the size of the ready queue. Our farming algorithm manages a ready queue and adapts to the underlying metacomputing environment by modifying the value of n dynamically according to constant computation throughput measurement. The algorithm really sees the environment as an opaque entity that gives varying responses (request response times) to repeated occurrence of the same event (the sending of a request).

Let us go through the algorithm shown in Figure 2. First, the algorithm chooses the initial value of n. That choice can be arbitrary but it may benefit from additional information provided by the underlying metacomputing environment. NetSolve provides a way to query the agent about the number of available servers for a given computation and that number is the initial guess for n in this first implementation. Second, the algorithm sets the *scheduling factor* α which takes values in $[0,1)$ and determines the behavior of the algorithm. Indeed the value of n may be changed only when more than n tasks completed during one iteration of the outermost while loop. A value of $\alpha = 1$ causes the algorithm to be extremely

```
n = initial guess on the queue size;
α = scheduling factor;
δ = 1;
while (tasks remaining) {
        while (number of pending tasks < n) {
                submit();
        }
        foreach (pending task) {
                poll();
        }
        if (n− number of pending tasks ≥ n × α) {
                if (average task response time has improved) {
                        n = n + δ;
                        δ = δ + 1;
                }
                else {
                        n = n − δ;
                        δ = 1;
                }
        }
}
```

Fig. 2. Adaptive scheduling algorithm

conservative (only when all n requests are completed instantly may the value of n be changed). The smaller α the more often will the algorithm try to modify n. The algorithm keeps a running history of the average request response times for all requests in the queue. That history is used to detect improvements or deterioration in performance and modify the value of n accordingly.

This algorithm is rather straightforward at the moment but it will undoubtedly be improved after more experiments have been conducted. However, early experimental results shown in Section 7 are encouraging.

6.2   Current Implementation

In our testbed implementation of farming for NetSolve, we implement farm() as an additional layer on top of the traditional NetSolve API, exactly as detailed in Section 5.2. A similar implementation would be valid for a system like Ninf. In other metacomputing environments, placing the scheduling algorithm within the client library might not be feasible, in which case the algorithm needs to be implemented in other parts of the system (central scheduler, client proxy, etc..). However, the algorithm is designed to rest on top of the metacomputing system, rather than merged with the internals of system.

6.3   Possible Extensions

The NetSolve farming interface is very general and we believe that it can serve as a low-level building-block for deploying various classes of applications. However, this generality leads to shortcomings. The embedded scheduler cannot take advantage of application-specific features, such as exploitable data patterns. Real applications are likely to manipulate very large amounts of data and it may be possible for the scheduler to make decisions based on I/O requirements. For instance, one can imagine that a subset of the tasks to farm make uses of one or more constant input data. This is a frequent situation in MCell (see Section 7.1) for example. Such input data could then be *shared* (via files for instance) by multiple resources as opposed to being replicated across all the resources. Another possibility would be for the farming application to contain simple data dependences between tasks. In that case, our framework could detect those dependences and schedule the computations accordingly. Another shortcomings of the farming interface that is a direct cause of its generality is that the call to farm() is completely atomic. This is an advantage from the point of view of ease-of-use, but it prevents such things as visualization of results as they become available for instance. Once again, such a feature would be desirable for MCell. Section 8 lays ground for research in these directions and work is under way in the context of MCell.

7   Preliminary Experimental Results

7.1   MCell

MCell [26, 27] is a general Monte Carlo simulator of cellular microphysiology. MCell uses Monte Carlo diffusion and chemical reaction algorithms in 3D to simulate the complex biochemical interactions of molecules inside and outside of living cells. MCell is a collaborative effort between the Terry Sejnowski lab at the Salk Institute, and the Miriam Salpeter lab at Cornell University. Like any Monte Carlo simulation, MCell must run large numbers of identical, independent simulations for different values of its random number generator seed. It therefore qualifies as a task farming application and was our first motivation to develop a farming API along with a scheduling algorithm.

As mentioned earlier, we developed for MCell a Shell-based interface on top of the C farming API. This interface takes as input a user-written *script* and automatically generates the call to farm(). The script is very intuitive as it follows the MCell command-line syntax by just adding the possibility for *ranges* of values as opposed to fixed values. For instance, instead of calling MCell as:

```
mcell foo1 1
mcell foo1 2
....
mcell foo1 100
```

it is possible to call MCell as

```
mcell foo1 [1-100]
```

which is simpler, uses Grid computational resources from NetSolve and ensures good scheduling with the use of the algorithm described in Section 6.

7.2   Results

The results presented in this section were obtained by using a NetSolve system spanning 5 to 25 servers on a network of Sun workstations (Sparc ULTRA 1) interconnected via 100Mb Ethernet. The farming application uses MCell to compute the shape of the parameter space which describes the possible modes of operation for the process of synaptic transmission at the vertebrate neuromuscular junction. Since MCell's results include the true stochastic noise in the system the signal must be averaged at each parameter space point. This is done by running each point 10 times with 10 different values of the random number generator seed. In this example, three separate 3-D parameter spaces are sampled, each parameter-space is of dimension $3 \times 3 \times 3$. The number of tasks to farm is therefore $3 \times 3 \times 3 \times 3 \times 10 = 810$ and each task generates 10 output files.

These preliminary experiments were run on a dedicated network. However, we simulated a dynamically changing resource pool by linearly increasing and decreasing the number of available NetSolve computational servers. Results are shown in Table 1 for our adaptive scheduling, a fixed queue size of $n = 25$ and a fixed queue size of $n = 5$.

Scheduling	Time	Resource Availability	Relative performance
adaptive	3982 s	64 %	100 %
$n = 25$	4518 s	62 %	85 %
$n = 5$	10214 s	63 %	38 %

Table 1. Preliminary experimental results

The resource availability measures the fraction of servers available during one run of the experiment. As this number changes throughout time, the availability is defined as the sum of the number of servers available servers over all time steps (10 seconds). We compare scheduling strategies by measuring *relative performance* which we define as a ratio of *adjusted elapsed times*, taking the adaptive scheduling as a reference. Adjusted elapsed times are computed by assuming a 100% availability and scaling the real elapsed times accordingly. One can see that the adaptive strategy performs 15% better than the strategy with $n = 25$. Of course, the strategy with $n = 5$ performs very poorly since it does not take advantage of all the available resources.

These first results are encouraging but not as satisfactory as expected. This is due to the implementation of NetSolve and the way the experiment was set-up. Indeed, NetSolve computational tasks are not interrupted when a NetSolve server is terminated. Terminating a server only means that no further requests will be answered but that pending requests are allowed to terminate. Thus, this experiment does not reflect the worst case scenario of machines being shutdown causing all processes to terminate. We expect our adaptive strategy to perform even better in an environment where tasks are terminated prematurely and need to be restarted from scratch on remaining available resources. Due to time constraints, this article does not contain results to corroborate this assumption, but experiments are underway.

8   Conclusion and Future Work

In this paper we have motivated the need for schedulers tailored to broad classes of applications running on the Computational Grid. The extreme diversity of Grid resource types, availabilities and access policies makes the design of schedulers a difficult task. Our approach is to build on existing and available metacomputing environments to access the Grid as easily as possible and to implement a interface and scheduling algorithm for task farming applications. An adaptive scheduling algorithm was described in Section 6. That algorithm is independent from the internal details of the Grid and of the metacomputing environment of choice. We chose NetSolve as a testbed for early experiments with the MCell application. The effectiveness of our scheduler is validated by preliminary experimental results in Section 7. Thanks to our framework for farming, a domain scientist can easily submit large computations to the Grid in a convenient manner and have an efficient adaptive scheduler manage execution on their behalf.

There are many ways in which this work can be further extended. We already mentioned in Section 6.3 that it is possible to use the farming API to detect data dependencies or shared input data between requests. The adaptive scheduling algorithm could be augmented to take into account such patterns. A possibility is for the farming interface to take additional arguments that describe domain-specific features and that may activate more sophisticated scheduling strategies if any. A first approach would be to consider only input or output coming from files (which is applicable to MCell and other applications) and partition the request space such as to minimize the number of file transfers and copies. This will require that the underlying metacomputing environment provide feature to describe such dependences. Work is being done in synergy with the NetSolve project to take into account data locality and the farming interface will undoubtedly take advantage of these developments [28]. This will be fertile ground for scheduling and data logistic research. The scheduling algorithm can also be modified to incorporate more sophisticated techniques. For instance, if the metacomputing environment provides and API to access more details about the status of available resources, it might be the case that n, the size of the ready queue, can be tuned effectively. The danger however is to lose portability as the requirements for the metacomputing environment (see Section 5.1) would be more stringent. Experiments will be conducted in order to investigate whether such requirements can be used to significantly improve scheduling.

The farming API can be enhanced so that certain tasks may be performed upon submitting each request and receiving each result. For instance, the user may want to visualize the data as it is coming back as opposed to have to wait for completion of all the requests. This is not possible at the moment as the call to `farm()` is atomic and does not provide control over each individual request. A possibility would be to pass pointers to user defined functions for `farm()` and execute them for events of interest (e.g. visualization for each reception of a result). Such functions could take arbitrary arguments for the sake of versatility. Some of the available metacomputing environments provide attractive interactive interface to which a farming call could be contributed. Examples include Matlab (NetSolve) and Mathematica (NetSolve,Ninf). In order to make our task farming framework easily accessible to a growing number of domain scientists, we need to develop ways to use the C farming API as a basis for more usable high-level interfaces. Steps in that direction have already been taken with the Shell-interface for MCell (see Section 7.1). It would be rather straightforward to design or use an existing specification language to describe specific farming applications and automatically generate custom Shell-based of graphical interfaces like the ones in [7].

9 Acknowledgments

This material is based upon work supported by the National Science Foundation under grant CCR-9703390, by NASA/NCSA Project Number 790NAS-1085A, under Subaward Agreement #790 and by the National Science Foundation Science and Technology Center Cooperative Agreement No. CCR-8809615.

References

[1] Ian Foster and Carl Kesselman, editors. *The Grid, Blueprint for a New computing Infrastructure*. Morgan Kaufmann Publishers, Inc., 1998.

[2] M. Litzkow, M. Livny, and M.W. Mutka. Condor - A Hunter of Idle Workstations. In *Proc. of the 8th International Conference of Distributed Computing Systems*, pages 104–111. Department of Computer Science, University of Winsconsin, Madison, June 1988.

[3] L. Silva, B. Veer, and J. Silva. How to Get a Fault-Tolerant Farm. In *World Transputer Congress*, pages 923–938, Sep. 2993.

[4] S. Sekiguchi, M. Sato, H. Nakada, S. Matsuoka, and U. Nagashima. Ninf : Network based Information Library for Globally High Performance Computing. In *Proc. of Parallel Object-Oriented Methods and Applications (POOMA), Santa Fe*, 1996.

[5] I. Foster and K Kesselman. Globus: A Metacomputing Infrastructure Toolkit. In *Proc. Workshop on Environments and Tools*. SIAM, to appear.

[6] A. Grimshaw, W. Wulf, J. French, A. Weaver, and P. Jr. Reynolds. A Synopsis of the Legion Project. Technical Report CS-94-20, Department of Computer Science, University of Virginia, 1994.

[7] D. Abramson, I. Foster, J. Giddy, A. Lewis, R. Sosic, and R. Sutherst. The Nimrod Computational Workbench: A Case Study in Desktop Metacomputing. In *Proceedings of the 20th Autralasian Computer Science Conference*, Feb. 1997.

[8] http://www.activetools.com.

[9] D. Abramson and J. Giddy. Scheduling Large Parametric Modelling Experiments on a Distributed Meta-computer. In *PCW'97*, Sep. 1997.

[10] A. Baratloo, P. Dasgupta, and Z. Kedem. Calypso: A Novel Software System for Fault-Tolerant Parallel Processing on Distributed Platforms. In *4th IEEE International Symposium on High Performance Distributed Computing*, Aug. 1995.

[11] L. M. Silva, J. G. Silva, S. Chapple, and L. Clarke. Portable checkpointing and recovery. In *Proceedings of the HPDC-4, High-Performance Distributed Computing*, pages 188–195, Washington, DC, August 1995.

[12] F. Berman, R. Wolski, S. Figueira, J. Schopf, and G. Shao. Application-Level Scheduling on Distributed Heterogeneous Networks. In *Proc. of Supercomputing'96, Pittsburgh*, 1996.

[13] F. Berman and R. Wolski. The AppLeS Project: A Status Report. In *Proc. of the 8th NEC Research Symposium, Berlin, Germany*, 1997.

[14] F. Berman, R. Wolski, and G. Shao. Performance Effects of Scheduling Strategies for Master/Slave Distributed Applications. Technical Report TR-CS98-598, U. C., San Diego, 1998.

[15] R. Wolski. Dynamically forecasting network performance using the network weather service. Technical Report TR-CS96-494, U.C. San Diego, October 1996.

[16] M. Litzkow and M. Livny. Experience with the Condor Distributed Batch System. In *Proc. of IEEE Workshop on Experimental Distributed Systems*. Department of Computer Science, University of Winsconsin, Madison, 1990.

[17] H. Casanova and J. Dongarra. Providing Uniform Dynamic Access to Numerical Software. In M. Heath, A. Ranade, and R. Schrieber, editors, *IMA Volumes in Mathematics and its Applications, Algorithms for Parallel Processing*, volume 105, pages 345–355. Springer-Verlag, 1998.

[18] The Math Works Inc. *MATLAB Reference Guide*. The Math Works Inc., 1992.

[19] S. Wolfram. *The Mathematica Book, Third Edition*. Wolfram Median, Inc. and Cambridge University Press, 1996.

[20] H. Casanova, J. Dongarra, and K. Seymour. Client User's Guide to Netsolve. Technical Report CS-96-343, Department of Computer Science, University of Tennessee, 1996.

[21] H Casanova and J. Dongarra. NetSolve: A Network Server for Solving Computational Science Problems. *The International Journal of Supercomputer Applications and High Performance Computing*, 1997.

[22] H. Casanova and J. Dongarra. NetSolve's Network Enabled Server: Examples and Applications. *IEEE Computational Science & Engineering*, 5(3):57–67, September 1998.

[23] H. Casanova and J. Dongarra. NetSolve version 1.2: Design and Implementation. Technical Report to appear, Department of Computer Science, University of Tennessee, 1998.

[24] D. E. Bakken and R. D. Schilchting. Supporting fault-tolerant parallel programming in linda. *IEEE Transactions on Parallel and Distributed Systems*, 6(3):287–302, March 1995.

[25] D. Gelernter and D. Kaminsky. Supercomputing out of recycled garbage: Preliminary experience with piranha. In *International Conference on Supercomputing*, pages 417–427, Washington, D.C., June 1992. ACM.

[26] J.R. Stiles, T.M. Bartol, E.E. Salpeter, , and M.M. Salpeter. Monte Carlo simulation of neuromuscular transmitter release using MCell, a general simulator of cellular physiological processes. *Computational Neuroscience*, pages 279–284, 1998.

[27] J.R. Stiles, D. Van Helden, T.M. Bartol, E.E. Salpeter, , and M.M. Salpeter. Miniature end-plate current rise times <100 microseconds from improved dual recordings can be modeled with passive acetylcholine diffusion form a synaptic vesicle. In *Proc. Natl. Acad. Sci. U.S.A.*, volume 93, pages 5745–5752, 1996.

[28] M. Beck, J. Plank, T. Moore, and W. Elwasif. Why IBP Now. *The International Journal of Supercomputer Applications and High Performance Computing*, to appear.

Applying Human Factors to the Design of Performance Tools

Cherri M. Pancake

Oregon State University, Department of Computer Science, Corvallis, OR 97331, USA
pancake@cs.orst.edu
http://www.cs.orst.edu/~pancake

Abstract. After two decades of research in parallel tools for performance tuning, why are users still dissatisfied? This paper outlines the human factors issues that determine how performance tools are perceived by users. The information provides insight into why current performance visualizations are not as well received as they should be — and what must be done in order to develop tools that are more closely aligned to user needs and preferences. Specific mechanisms are suggested for improving three aspects of performance visualizations: how the user explores the performance space, how the user compares different aspects of program behavior, and how the user navigates through complex source code.

1 The Performance Tuning Problem

For over two decades, a great deal of research effort has been directed at tools for improving the performance of parallel applications. Significant progress has been made, as can be seen by comparing some of the surveys of parallel tools from that time period [22, 23, 2]. Why, then, are parallel performance tools still so hard to "sell" to the user community?

Part of the problem is that parallel tools can be extremely difficult to implement. The tool developer must copy with an inherently unstable execution environment, where it may be impossible to reproduce program events or timing relationships. Monitoring and other tool activities, intended to observe program behavior, in fact perturb that behavior, sometimes causing errors or performance problems to appear or disappear in unpredictable ways. Further, the notoriously short lifetime of most parallel computers means that there is an extremely small "window of opportunity." Tool development must often begin before the hardware or operating system is stable, but must become available very soon after the computers are first deployed in order to acquire any significant user base [20].

Technological challenges are just one aspect of the problem, however. The most common complaints about parallel performance tools concern their lack of usability [17, 21]. They are criticized for being too hard to learn, too complex to use in most programming tasks, and unsuitable for the size and structure of real-world parallel applications. Users are skeptical about how much value tools

P. Amestoy et al. (Eds.): Euro-Par'99, LNCS 1685, pp. 44–60, 1999.

can really provide in typical program development situations. They are reluctant to invest time in learning to apply tools effectively because they are uncertain that the investment will pay off.

The situation is paradoxical. As we have established elsewhere, users do not turn to parallel processing unless they need to increase the size, complexity, resolution or speed of their applications.[19] This means that all parallel users are concerned to some degree with application performance. They should be predisposed in favor of any tools that could help automate the processes involved in measuring, analyzing, and improving performance. Instead, they lament the lack of suitable tools.

The fact is that a tool is viewed as just that, a *tool*, something that should facilitate a human process and enable the human to achieve his/her intended goal. Tools are not used in isolation, nor are they appreciated as standalone elements. A tool is only perceived as valuable when it clearly assists the user to accomplish the tasks to which it is applied. Thus, a tool's propensity for being misapplied or its failure to support the kinds of tasks users wish to perform constitute serious impediments to tool acceptance. This is where human factors come into play.

Human factors refers to the characteristics, capabilities, and limitations of human beings and how these affect our use of technology. *Human factors engineering*, then, seeks to improve system effectiveness and safety by explicitly addressing human factors in the design of new systems. (For a general overview of human factors applied to software development, see [5].)

This paper outlines some of the basic human factors that influence the usability of parallel tools. In particular, we examine the human factors environment within which a user attempts to tune the performance of a parallel application. This is followed by a discussion of how current tools support, or fail to support, the human factors involved. The information provides insight into why current performance visualizations are not as well received as they should be — and what must be done in order to develop tools that are more closely aligned to user needs and preferences. Specific mechanisms are suggested for improving three aspects of performance visualizations: how the user explores the performance space, how the user compares different aspects of program behavior, and how the user navigates through complex source code.

2   How Users Approach Performance Tuning

The best way to understand what kinds of tool support are important is to consider how application developers actually go about the process of performance tuning. Over the past decade, we have carried out *in situ* studies of hundreds of scientists and engineers involved in developing parallel applications.[17] Based on this work, we propose that the tuning process has two dimensions, distinguished here as the conceptual framework and the task structure.

2.1   Conceptual Framework — What the User Seeks

As noted earlier, all application developers are concerned, at least on a superficial level, with program performance. Typically, performance measurement and analysis activities are interspersed with periods of code development and debugging.[18] During a performance tuning interval, the user consciously or unconsciously poses a series of questions that address different aspects of the application's behavior.

Those questions establish a *conceptual framework*, or a series of subgoals that must be met in order to accomplish the more general goal of improving performance. Any dependencies among the subgoals establish the order in which they will have to be accomplished. The following is an example of a typical conceptual framework:

1. *Identification:* Is there a performance problem? What are the symptoms, and how serious are they?
2. *Localization:* At what point in execution is performance going wrong? What is causing the problem to occur?
3. *Repair:* What about the application must be changed to fix the problem? [Perform the repair.]
4. *Verification:* Did the "fix" improve performance? [If not, optionally undo the repair, then go back to (2).]
5. *Validation:* Is there still a performance problem? [If so, return to (1).]

Subgoals are indicated in italics, and the ordering and repetition of particular subgoals has also been shown explicitly. In real life, we have observed that users rarely stop to think about the steps they are following or the rationale behind them. That is, the process is largely intuitive in nature.

2.2   Task Structure — How the User Seeks It

While the user may not recognize the underlying conceptual framework, he/she carries it out through a series of deliberate actions. This is the *task structure*, a sequence of individual activities that the user plans and executes to achieve a particular subgoal.

Problem stabilization. The subgoals of identification, verification, and validation are generally accomplished using a single task structure, which we call *problem stabilization*. The performance of the application is benchmarked to obtain some sort of timing information, either overall execution time for the application or a series of timings reflecting the duration of individual phases of the application. Execution will be timed repeatedly in order to obtain a series of "representative" data — that is, reflecting inputs, parameter settings, system load, etc., that are considered typical of the application's intended use.

The timings must then be compared to determine how consistent they are. The performance data are also examined in terms of what the user anticipated (e.g., the computational kernel may be expected to occupy roughly 80% of the total execution time, to complete in about 70 minutes, or to achieve roughly 10% of aggregate peak speed). From the consistency of these timings, the user infers whether the performance is acceptable. Note that this really is a process of inference, since there is no simple way of determining whether an arbitrary application's performance is "good" or "bad," and even less way of determining *a priori* how much effort will be needed to improve upon that performance.

Search space reduction. The subgoal of localization involves a second set of tasks, which we will call *search space reduction*. In a typical scenario, the user first makes an educated guess about which aspect of the application's execution (e.g., memory use, communications, load balance) is responsible for the perceived behavior. That hypothesis is generally based on intuition and recollections of previous experiences with performance tuning, and is therefore highly individual. The user then tests the hypothesis by adding hand-coded instrumentation (or using a performance monitoring tool) in an attempt to isolate where the problem occurs during program execution. The objective is to narrow down which region(s) of the source code is responsible for the problem (e.g., in the part of the program where cells exchange values with their neighbors). It should be noted that this may be a very rough approximation, and may later turn out to be erroneous.

Selective modification. The subgoal of repair is achieved with a third set of tasks, applied in iterative cycles, that we'll call *selective modification*. Focusing attention on one potential problem location at a time, the application developer uses manual inspection to study the code and attempt to determine what is causing the problem. The difficulty — and the effectiveness — of this step is highly dependent on the user's experience level, since it is largely a matter of intuition and judgement. The code is modified in an attempt to alleviate the performance problem, and the user proceeds to the verification stage. If the code change is deemed to be successful, attention is then focused on the next problem location. If, on the other hand, performance does not appear to have improved, the user is likely to un-do the changes and try again.

The task set is called selective because we have found few instances where experienced users attempt simultaneous changes in different regions of the program. When asked why, they say they have learned that it is difficult to accurately assess and balance the effects of multiple, possibly counteracting "improvements." It is perceived as more productive to narrow their focus to small problems that can be solved incrementally, rather than attempting to develop and apply a more global strategy.

2.3   Implications for Performance Tools

An understanding of typical approaches to performance tuning explains much of the frustration that users feel about current tools support. Most tools begin by popping up a series of windows, each revealind detailed information about a particular performance metric. Yet users prefer to begin by gaining an in-the-large perspective on application behavior (where it is spending its time, major hotspots for memory use, etc). A fine-grained summary really does nothing to help the user assess whether performance is generally "good" or "bad," or if tuning efforts are likely to pay off. As more than one user has commented, "Don't tool developers know how awful it is to bring up a tool and have it try to show me all 100,000 communications that took place?" [17] For problem stabilization, the user needs to be able to *explore* the total program space at a higher level, looking for evidence of anomalies.

Current tools are somewhat better when applied to search space reduction. Many allow the user to view different aspects of program behavior, such as memory use, communications traffic, CPU utilization, etc. However, they fail to support the core activity — deciding which aspect is responsible for poor performance. What is needed here is support for *comparing* different metrics. Such comparisons (e.g., the fact that CPU utilization drops in the same places that memory use soars, but appears unrelated to message traffic)would yield meaningful suggestions for directing programmer effort. It is also essential to provide source-code clickback or some other mechanism for identifying which source statements correspond to a particular pattern of behavior.

In terms of selective modification, current tools are singularly lacking in support. What users need are mechanisms for *navigating* through complex source code hierarchies and pinpointing the regions of code that have similar performance behavior. Existing tools do not even provide mechanisms for flagging portions of code as interesting so that they can be revisited or compared with others. Users typically resort to printed source listings and highlighting pens to carry out the activities associated with this set of tasks. Further, current tools do not provide any guidance about what type of code modifications are called for. As users are apt to describe it, tools "don't tell me anything I can actually *use* to go back and fix my code." [17]

Finally, users can be alienated by what a tool does, as well as by what it fails to do. Application developers are trying to explore the behavior of their code, not the behavior of the tool. When a tool shows particular information without indicating why that information is shown or how it might be used to tune application performance, for example, users quickly lose patience. The single most persistent complaint we have recorded is that tools "keep popping up windows I don't need in places I don't like."

3   Applying Human Factors to Tool Design

The human factors issues just discussed can be applied directly to the design of parallel performance tools, specifically in designing graphical representations of performance data. This section discusses how that can be accomplished.

3.1   How Graphical Representations Relate to Task Structure

Most parallel tools have adopted graphical representations in order to portray performance data or other tool information. In many other types of software, graphics have been shown to be useful for a variety of reasons (cf. [10, 28, 1, 4]). In addition to making large quantities of quantitative data coherent, effective graphics encourage the eye to compare and contrast elements, revealing patterns or exposing anomalies in the data that would not be discernible if the representations were numeric or textual. Graphical techniques are also capable of revealing information at varying levels or detail, capitalizing on human familiarity with how the appearance of physical objects changes when they are seen from different distances. Graphics can facilitate the problem-solving process by providing memorable, easily manipulated symbols representing a wide range of concepts, even highly encapsulated information.

The taxonomy of visualization goals proposed by [13] is useful in clarifying how graphics can be exploited by software tools. The authors define the types of activities that can be facilitated through appropriate visualizations:

- **Identify**: establish the collective characteristics by which an object is distinctly recognizable.
- **Locate**: determine specific position, either absolute or relative.
- **Distinguish**: recognize as different or distinct (independent of being able to identify the object).
- **Categorize**: place in specifically defined divisions within a classification scheme.
- **Cluster**: join into (conceptual or physical) groups of the same, similar, or related type.
- **Rank**: give an object order or position with respect to other objects of like type.
- **Compare**: examine so as to notice similarities and differences (independent of ordering).
- **Associate**: link or join in a relationship.
- **Correlate**: establish a direct causal, complementary, parallel, or reciprocal connection.

All of these activities are important in performance tuning, where the user must make sense of large quantities of interrelated performance measurements, draw conclusions about their relative importance, and infer the effects that will occur when behavior is modified.

From the standpoint of parallel performance tools, visualization offers three particular advantages. First, it provides a way to manage the voluminous and complex data associated with parallel program execution. Second, it can capitalize on the user's pattern recognition capabilities. Third, it can assist the user to explore and "interpret'" program behavior by providing alternate views, reflecting different aspects or levels of behavior.

It is not easy to implement graphical displays that exploit these capabilities, however. Consider the problem of visualizing a large and complex set of performance data. The most straightforward techniques result in a series of windows, each showing a different portion of the data,[8] but as previously described, this is overwhelming to most users. The problem is that this type of representation simply uses graphics to expose the dimensions of performance behavior. In applying human factors to tools design, the key is to concentrate on what users need to do. We must structure the presentation of information so that it specifically supports user tasks.

Recall that our discussion of the user task structure indicated three primary requirements:

1. Support for exploring the total performance space. This is particularly important for problem stabilization activities.
2. Support for comparing different aspects of program behavior. This is needed primarily during search space reduction.
3. Support for navigating through complex source code. While navigation is useful during all phases of performance tuning, it is particularly germane to selective modification.

In the sections which follow, heuristics are suggested for integrating each of these features into the kinds of graphical representations used by parallel performance tools. Recall that the goal is to make the operations, and the information revealed through those operations, fit more "naturally" into the user's task structure.

3.2   Facilitating Exploration

Exploration requires mechanisms that allow the user to view program performance from a high level, while still being able to identify anomalous behavior. Three heuristic techniques particularly lend themselves to the exploration of performance visualizations.

1. Add filtering capabilities to zoom operations. Zooming operations are commonly employed to accommodate scalability in performance visualizations. When the program is lengthy or complex, a performance visualization can grow arbitrarily large in size; it cannot be expected to fit in its entirety on the screen. Zooming operations allow the user to increase the apparent magnification of an image, so that the window is filled by what was formerly a small area of the image. The converse operation, panning, reduces the size of the image, making it possible to fit a larger area within the window boundaries.

Human factors considerations indicate that zooming operations should be much more powerful than simple magnification. Instead, they should mimic the human's change in perspective as he/she moves closer to an object by showing more information with each successive zoom (rather than just a magnified version of the same information). Consider, for example, the System Performance Visualization Tool (SPV), developed by Intel for its Paragon series of distributed-memory multiprocessors.[12] The results of its zooming operations are shown in Figure 1. At the upper right is the "zoomed out" view, which portrays the entire system at the level of nodes and interconnets. The nodes are colored to indicate the general level of activity on each CPU. Zooming in on the display (middle window) focuses on a smaller number of nodes and shows specific figures for CPU utilization as well as the amount of traffic with neighboring nodes. Zooming in again (lower right) focuses attention on a particular CPU, revealing details such as memory bus and communication processor usage. This was found to be a very intuitive mechanism for supporting both high-level, overall views of the system and the ability to see a detailed perspective when behavior seemed to be anomalous.

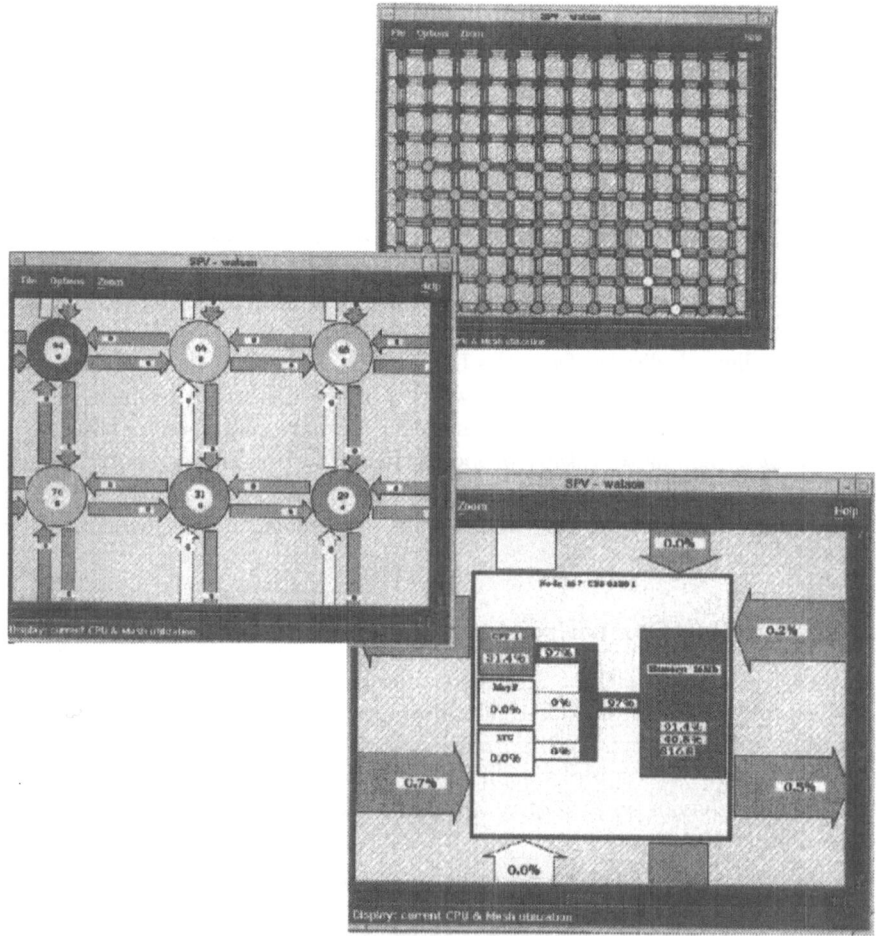

Fig. 1. Effects of zoom operation in **SPV** [12]; each successive zoom reveals more detail about a smaller portion of the system.

2. Make data labels more adaptive. Another problem with exploration is that of providing the user with adequate textual labels. Color, shape, and other graphical characteristics can be very useful for encoding many dimensions of information into a single visualization. With no textual labels, however, the user must being with a mental picture of what he/she is looking for. For example, in the middle display of Figure 1, the user must know ahead of time what the numbers in the circles refer to and recognize the difference between numbers in two different positions (the upper figure refers to compute processor utilization and the lower to communication processor utilization). A label would clarify this.

The problem with textual labels is that they occupy considerable screen real estate and are difficult to position so as not to overlap one another. However, recent human factors studies have shown that by displaying only a small subset of labels within the immediate vicinity of the cursor, it is possible to facilitate exploration of complex displays without overwhelming the user with unwanted detail.[7] In our example, when the user moves the cursor over a node, labels would appear to indicate the meaning of each number. The labels would disappear as the cursor moved to another location.

3. Make it possible to focus attention selectively. Another technique that human factors studies have demonstrated to be effective for exploration is the use of deformation-based focusing, commonly called "fisheye" focus.[24] With this mechanism, the user can magnify just a selective area of a large display — most typically a circular area around the cursor position. For example, Figure 2 shows a callgraph display from the Lightweight Corefile Browser (LCB),[16] a Parallel Tools Consortium project, to which a fisheye distortion has been applied. Only the node labels closest to the cursor position are completely visible.

A recent human factors study proposes that distortion focusing could be even more effective if multiple focal points are used in the same display.[27] In the SPV displays, for example, this would make it possible, to see close-ups of two different nodes simultaneously.

3.3   Facilitating Comparison

As noted previously, much of the user's time is typically spent comparing performance information from different executions, different portions of the same execution, or different types of information about the same area of execution. To facilitate these activities, tool visualizations should underscore the differences in performance behavior so that they are more readily identified and understood by the user.

4. Support display cloning. Current tools do not provide minimal support for comparing across program executions or different portions during a single execution. It is up to the user to bring up two instantiations of the tool and experiment with the settings of both of them simultaneously. In fact, existing tools do little to facilitate comparison of different metrics or types of performance data. Typically, the user may only view one particular region of the program and one metric at a time. **CXperf**, a performance analysis tool developed by Hewlett Packard for their Exemplar series, does allow the user to combine arbitrary pairs of metrics and view them with respect to different granularities of source code (routines, basic blocks).[9] Figure 3 illustrates the display when the user chooses to view CPU time x thread x source routine. Note, however, that to compare

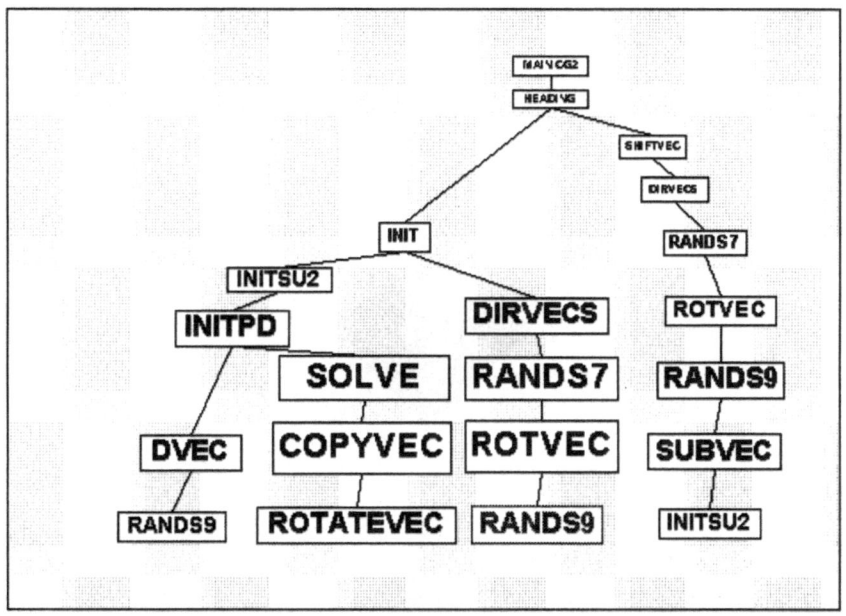

Fig. 2. Effects of deformation focus (fisheye) on a callgraph display from **LCB** [16]; objects near the cursor are magnified, while others remain obscure.

the memory usage or instruction counts for those same threads and routines, the user must switch to a different display. It becomes very difficult to determine cause-and-effect relationships, such as the possible impact of communications congestion on CPU utilization. If the user could clone the window, then change from CPU utilization to communications and the metric, it would be very easy to determine if the two behaviors were interrelated.

5. Use color to enhance discrimination. All too often, parallel tools appear to choose colors based on the ease with which they can be specified. Yet human factors specialists have known for decades that poorly chosen colors actually hinder the user's ability to find targets or recognize patterns (see [3] for a survey of studies on the impact of color on human performance). Given that the visualizations used in performance tools are likely to containvery dense information, coloring is particularly important.

Unfortunately, it is not possible for this paper to show clear examples of how color can be modified to improve usability. We have prepared Web pages listing colors that can be discriminated well on most types of screens and using even fairly simple color models (like that supported by Web browsers). Since color perception is strongly affected by the background and surrounding colors, lists are shown for display against both white and black backgrounds. (See **http://www.cs.orst.edu/p̄ancake/colors.html** and **colors2.html**.) There

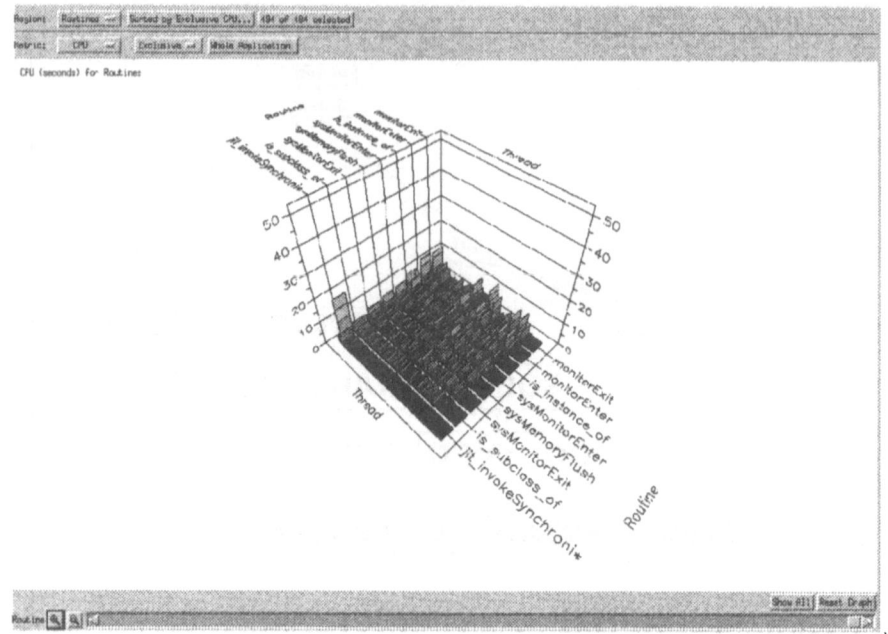

Fig. 3. CXperf [9] display plotting CPU time x thread x code region. Different combinations of metric/location information can be shown, but just one at a time.

are also a number of references that provide guidance on how to use color more effectively; two particularly clear ones are [14] and [26].

 6. Make it easier to bring up related information. Current tools are also lacking in their capability for recognizing when different types of information might be related. Consider a typical set of metrics: CPU utilization, communications traffic, memory traffic, etc. Rather than simply presenting data as it is requested by the user, the tool could pre-analyze the range of values along each performance dimension. Then, when a user is examining an area where CPU utilization is high, it would be possible to for him/her to request to see those areas where some other metric is "high."

 The **Paradyn**[15] performance monitoring tools from University of Illinois operate in something of this vein, searching through different regions of execution for evidence of bottlenecks. Only a single type of information is presented to the user, however; the tool suppresses what it has determined about other areas that might have performance problems. Human factors studies of other settings indicate that the ability to gain quick access to possibly related information can play a major factor in improving human performance and increasing the user's sense of command over the software.[6]

3.4 Facilitating Navigation

7. Provide a context for navigation. Given the volume of information available through parallel performance tools, it's all too easy for the user to become lost in large displays. Other types of electronic displays have found it useful to maintain some sort of overview representation on the screen at all times, so that the user can keep some sense of where he/she is in terms of the overall information space.

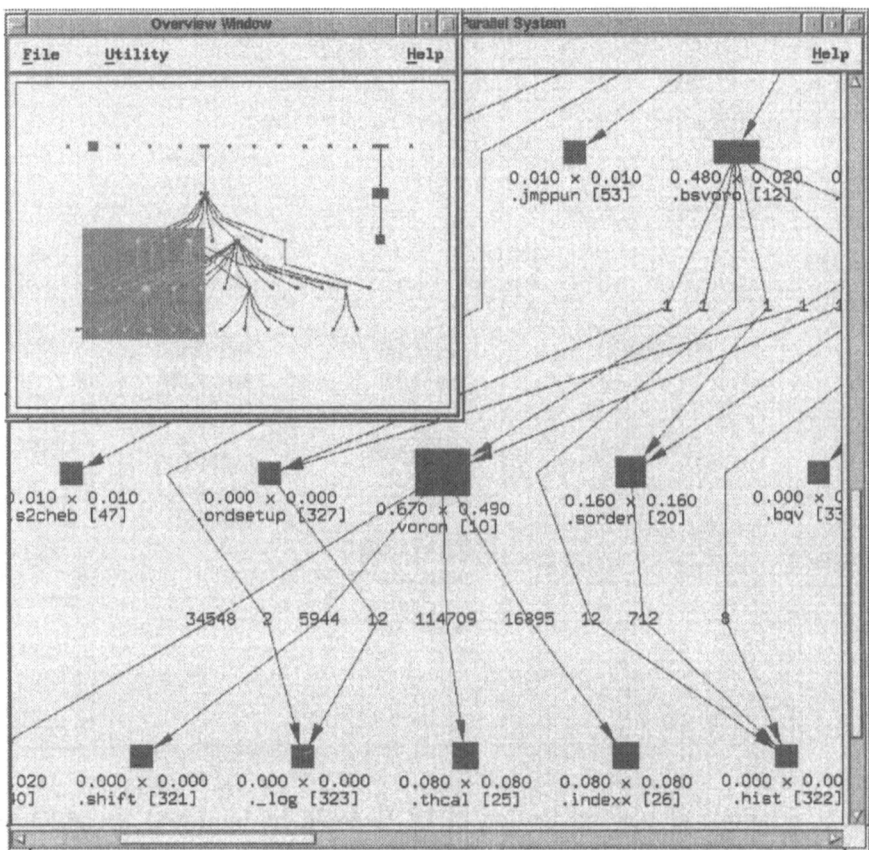

Fig. 4. Xprofiler [11] uses thumb-nail images to help the user navigate through large amounts of performance data.

One performance tool does this through the use of a thumb-nail sketch. **Xprofiler** [11], developed for IBM's SP/2 computers, maintains a very small-scale image that is highlighted to show which portion of the overall program graph is being viewed. As may be seen in Figure 4, the user can quickly grasp a sense of where the detailed information fits into the overall view. An alternate way of providing a sense of context is portrayed in Figure 5. Here, the call graph representation from **LCB** has been augmented with a miniaturized representation of the source code. The user can control which area is zoomed in by sliding the rectangular box that is superimposed on the miniaturized code, displayed along the left side of the screen. The display at the right shows a zoomed-in area of execution, revealing the actual calling structure.

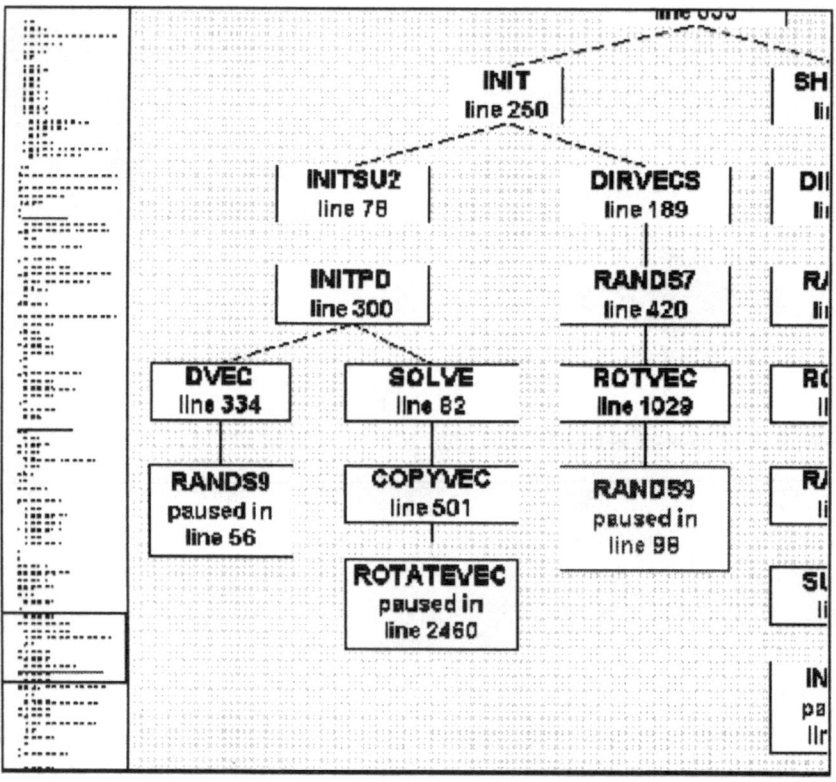

Fig. 5. Example of how navigation in **LCB** could be improved by adding a minitiarized program listing.

8. Show hierarchical relationships explicitly. Most performance tool developers have come to recognize that they must provide links from execution data back to the source code routines or lines where the behavior occurred. However, the typical source-code-clickback mechanism simply pops up a window that has scrolled automatically to the statement or the first line of the corresponding routine. This is not sufficient for most real-world applications, which tend to be very large, organized in complex ways, and developed by whole teams of people over long periods of time. Scrolling to a message-sending line, for example, may provide little clue as to where in the program logic a problem has arisen.

Here, too, other human factors work indicate a strategy for facilitating user interaction with performance tools. Shneiderman's work with "treemaps" (e.g., [25]) indicate that rectangular maps can be very effective in representing hierarchical structures. A containment relationship applies; that is, information that is subordinate to other information is embedded within its superordinate's area. Figure 5 applies this concept to the representation of the source code structure from Figure 4. Note that very little space is needed to represent the call tree, but no information is sacrificed. Like trees, treemaps require that the informa-

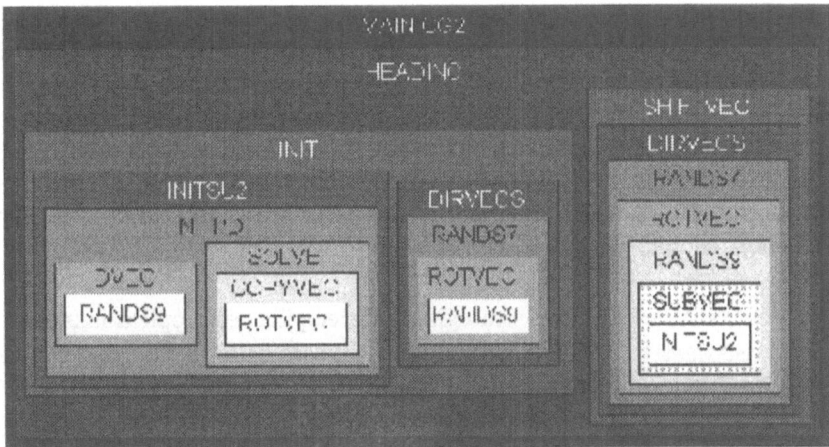

Fig. 6. Example of how navigation in **LCB** could be improved by making the hierarchical relationships of the source code explicit.

tion to be represented contain no cycles or recursion. The approach could also be combined with zooming or fisheye techniques described earlier, in order to provide control over the amount of detail portrayed.

9. Support user-inserted "landmarks". A final navigational aid is the addition of mechanisms whereby a user can "mark," or visually flag, regions of the display. This does not affect the data in any way; it simply makes it easier for the user to return to an area for later comparison or study. Consider the displays shown in Figures 3 and 4, or any of the traditional time-line diagrams so common in performance tools. The user ultimately navigates through a considerable information space, so returning to a point typically requires a great deal of searching. If the tool allowed the user to superimpose visual landmarks (such as small colored flags) directly on the display, it would be very simple to re-locate those positions.

4   Conclusions

Parallel tools suffer in comparison to the usability of software on personal computers because the resources available for development are much greater in the desktop world and the problems to be solved are much less complex. Many usability issues remain unresolved for parallel tools, which are often the implementation of an untried solution. As a result, the ways in which performance tools can be applied effectively during application development remain obscure to the user community.

The developers of parallel tools appear unaware of many basic human factors studies which could help them improve the usability of their visualizations and interaction mechanisms. Nine examples were presented in this paper, with descriptions of how each would improve a typical performance tool scenario.

User acceptance of parallel tools will not increase appreciably until such tools are usable within the framework of typical application development strategies. This will require that tool developers come to an understanding of how experienced users go about developing and tuning large-scale applications. The goals of this effort should be:

- to identify successful user strategies in developing real applications;
- to devise ways to apply knowledge of those strategies so that tool functionality can be presented in an intuitive, usable, and familiar manner; and
- to use this functionality in the development of new tools.

Tools that do not mesh well with user goals and task structures are not considered by users to be worth the time that must be invested to learn them. We will not see tools appreciated properly until they can address the human factors issues associated with parallel performance tuning.

References

[1] Card, S., J. Mackinlay and B. Shneiderman, *Readings in Information Visualization: Using Vision to Think*, Morgan Kaufman, 1999.
[2] Casavant, T. L., "Tools and Methods for Visualization of Parallel Systems and Computations: Gues Editor's Introduction," *Journal of Parallel and Distributed Computing*, 1993. 18 (2): 103–104.
[3] Christ, R. E., "Review and Analysis of Color Coding Research for Visual Displays," *Human Factors*, 1975, 17 (6): 542-570.
[4] Cleveland, W., *Visualizing Data*, Hobart Press, 1993.
[5] Curtis, B., *Human Factors in Software Development: A Tutorial*, 2nd Edition, IEEE Computer Society Press, Washington, DC, 1985.
[6] Czerwinski, M., et al., "Visualizing Implicit Queries for Information Management and Retrieval," *Proceedings ACM Conference on Human Factors in Computing Systems (CHI'99)*, 1999, pp. 560-567.

[7] Fekete, J-D. and C. Palisant, "Excentric Labeling: Dynamic Neighborhood Labeling for Data Visualization," *Proceedings ACM Conference on Human Factors in Computing Systems (CHI'99)*, 1999, pp. 512–519.

[8] Heath, M. T., A. D. Malony and D. T. Rover, "The Visual Display of Parallel Performance Data," *IEEE Computer*, 1995, 28 (11): 21–28.

[9] Hewlett-Packard Corporation, *CXperf User's Guide*, Publication B6323-96001, available online at **http://docs.hp.com:80/dynaweb/hpux11/dtdcen1a/0449/@Generic__BookView**, 1998.

[10] Horton, W., *Illustrating Computer Documentation: The Art of Presenting Information Graphically on Paper and Online*, John Wiley & Sons, 1991.

[11] IBM Corporation, *IBM AIX Parallel Environment: Operation and Use*, IBM Corporation publication SH26-7231, 1996.

[12] Intel Corporation, *System Performance Visualization Tool User's Guide*, Intel Corporation, publication 312889-001, 1993.

[13] Keller, P. R. and M. M. Keller, *Visual Cues: Practical Data Visualization*, IEEE Computer Society Press, 1993.

[14] Marcus, A., *Graphic Design for Electronic Documents and User Interfaces*, ACM Press, Tutorial Series, 1992.

[15] Miller, B. P. *et al.*, "The Paradyn Parallel Performance Measurement Tools," *IEEE Computer*, 1995, 28 (11): 37–46.

[16] Muddarangegowda, M. and C. M. Pancake, "Basing Tool Design on User Feedback: The Lightweight Corefile Browser," Technical Report, Oregon State University, available online at **http://www.CS.ORST.EDU/ pancake/papers/lcb/lcb.html**, 1995.

[17] Pancake, C. M., unpublished interview notes from field studies and group interviews at user sites in the U.S., 1987–1999.

[18] Pancake, C. M. and D. Bergmark, "Do Parallel Languages Respond to the Needs of Scientific Researchers?" *IEEE Computer*, 1990, 23 (12): 13–23.

[19] Pancake, C. M. and C. Cook, "What Users Need in Parallel Tool Support: Survey Results and Analysis," *Proc. Scalable High Performance Computing Conference*, 1994, pp. 40–47.

[20] Pancake, C. M., "Establishing Standards for HPC Systems Software and Tools," *NHSE Review*, 1997, 2 (1). Available online at **nhse.cs.rice.edu/NHSEreview**.

[21] Pancake, C. M. "Exploiting Visualization and Direct Manipulation to Make Parallel Tools More Communicative," in *Applied Parallel Computing*, ed. B. Katstrom et al., Springer Verlag, Berlin, 1998, pp. 400–417.

[22] Pancake, C. M., M. L. Simmons and J. C. Yan, "Guest Editor's Introduction: Performance Evaluation Tools for Parallel and Distributed Systems," *IEEE Computer*, 1995, 28 (11): 16–19.

[23] Pancake, C. M., M. L. Simmons and J. C. Yan, "Guest Editor's Introduction: Performance Evaluation Tools for Parallel and Distributed Systems," *IEEE Parallel and Distributed Technology*, 1995, 3 (4): 14–19.

[24] Sarkar, M. and M. H. Brown, "Graphical Fisheye Views of Graphs," *Proceedings 1997 IEEE Symposium on Visual Languages*, 1997, pp. 76–83.

[25] Shneiderman, B., "Tree Visualization with Tree-maps: A 2-D Space-Filling Approach," *ACM Transactions on Graphics*, 1992, 11 (1): 92–99.

[26] Thorell, L. G. and W. J. Smith, *Using Computer Color Effectively, An Illustrated Reference*, Prentice Hall, 1990.

[27] Toyoda, M. and E. Shibayama, "Hyper Mochi Sheet: A Predictive Focusing Interface for Navigating and Editing Nested Networks through a Multi-focus Distortion-Oriented View," *Proceedings ACM Conference on Human Factors in Computing Systems (CHI'99)*, 1999, pp. 504–510.

[28] Tufte, E. R., *Visual Explanations: Images and Quantities, Evidence and Narrative*, Graphics Press, 1997.

Building the Teraflops/Petabytes
Production Supercomputing Center

Horst D. Simon, William T.C. Kramer, and Robert F. Lucas

National Energy Research Scientific Computing Center (NERSC), Mail Stop 50B-4230,
Lawrence Berkeley National Laboratory, Berkeley, CA 94720
{hdsimon, wtkramer, rflucas}@lbl.gov

Abstract. In just one decade, the 1990s, supercomputer centers have undergone
two fundamental transitions which require rethinking their operation and their
role in high performance computing. The first transition in the early to mid-
1990s resulted from a technology change in high performance computing
architecture. Highly parallel distributed memory machines built from
commodity parts increased the operational complexity of the supercomputer
center, and required the introduction of intellectual services as equally
important components of the center. The second transition is happening in the
late 1990s as centers are introducing loosely coupled clusters of SMPs as their
premier high performance computing platforms, while dealing with an ever-
increasing volume of data. In addition, increasing network bandwidth enables
new modes of use of a supercomputer center, in particular, computational grid
applications. In this paper we describe what steps NERSC is taking to address
these issues and stay at the leading edge of supercomputing centers.

1   Introduction

In just one decade, the 1990s, supercomputer centers have undergone two
fundamental transitions which require a rethinking of the basic tenets of their
operation and their role in the high performance computing (HPC) world. The first
transition in the early to mid 1990s was a result of a technology change in high
performance computing architecture. The introduction of highly parallel distributed
memory machines built from commodity parts increased the operational complexity
of the supercomputer center, and required the introduction of intellectual services as
equally important components of the center.

We have only recently completed this transition and developed the tools necessary
to bring the revolution of the mid-1990s to a successful conclusion. Now three new
developments are appearing which will again force us to step up to new challenges in
supercomputer center management: (1) yet another change in the architecture of
supercomputing platforms, (2) the increasing importance of managing large volumes
of scientific data, and (3) the deployment of a new generation of high-speed wide-area
networks. After reviewing the two transitions in Section 2, in the remainder of this
paper we will discuss these three technology developments and their likely impact on
high performance computing. We will describe how the U.S. Department of Energy's

P. Amestoy et al. (Eds.): Euro-Par'99, LNCS 1685, pp. 61-77, 1999.
© Springer-Verlag Berlin Heidelberg 1999

National Energy Research Scientific Computing Center (NERSC) [1] is preparing itself to meet these challenges.

The first issue we are facing is another technology change in high performance computing systems. The next generation of supercomputers will be built from commodity components, in all likelihood from shared memory multiprocessor (SMP) systems. These cluster-of-SMP systems not only add an additional level of complexity for user applications development, but also lack most of the robust systems software that a high-quality, production-oriented center relies on. Furthermore, these systems also require an order of magnitude increase in available floor space and power consumption. NERSC is planning to install such a system during 1999, and we will report in Section 3 on our initial experiences.

The second issue we are facing is increasing rates of data generation both from computer simulations and from experiments. For example, the high-energy physics community is bringing new experiments on line that will generate data at rates exceeding 250 terabytes per year in 1999 and 1 petabyte per year in 2005. The bioinformatics community points out that the number of base pairs output by the various genome projects is growing faster than Moore's law. These projected data rates force us not only to reevaluate tertiary storage and bandwidth requirements, but also to develop new tools in scientific data management. NERSC has developed a data-intensive computing strategy to deal with these issues in a comprehensive way. We will describe this strategy and some of its technology elements in Section 4.

Lastly, we are facing incredible increases in the available bandwidth for wide-area networks. This enables new modes of using both compute and data resources at the centers. Computational steering and remote visualization and exploration of data will become commonplace. While many of these technologies are already at the prototype or advanced development stage, substantial work needs to be invested to provide these tools on a routine basis. These developments are often summarized under the term «grids.» We will discuss NERSC's potential role in the data grid and how it may benefit our user community in Section 5.

2   The Two Technology Transitions of the 1990s

NERSC recently announced the successful completion of the NERSC-3 procurement, resulting in the acquisition of an IBM SP-3 with more than 3 Tflop/s peak performance. The machine will arrive at NERSC in two phases.

Phase I installation, scheduled to begin in June 1999, will consist of an RS/6000 SP with 304 of the two-CPU POWER3 SMP nodes that were recently announced by IBM. This system will be the first implementation of the POWER3 microprocessor, with two processors per node. The 64-bit POWER3 can perform up to two billion operations per second and is more than twice as powerful as its predecessor. In all, Phase I will have 512 processors for computing, 256 gigabytes of memory, and 10 terabytes of disk storage for scientific computing. The other 48 nodes will be used for interactive, network, parallel file system, and other services. The system will have a peak performance of 410 Gflop/s, or 410 billion calculations per second.

Phase II, slated for installation no later than December 2000 (but likely much sooner), will consist of 152 16-CPU POWER3+ SMP nodes, utilizing an enhanced POWER3 microprocessor. The entire system will have 2,048 processors dedicated to large-scale scientific computing. The system will have a peak performance capability of more than 3 Tflop/s.

While this configuration may appear as a logical continuation of the trend toward highly parallel systems, established during the mid 1990s at many U.S. supercomputer installations, there are significant differences between NERSC-3, the new IBM SP system, and NERSC-2, a 640-processor SGI-Cray T3E/900. These differences are summarized in Table 1.

Table 1. Comparison of typical high-end platforms in the 1990s

	NERSC-1 Cray C90	NERSC-2 Cray T3E	NERSC-3 IBM SP-3
Year of Installation	1991	1996	2000
Number of Processors	16	640	2048
Processor Technology	Custom ECL	Commodity CMOS	Commodity CMOS
Peak System Performance	16 Gflop/s	580 Gflop/s	3000 Gflop/s
Measured System Performance	3.75 Gflop/s	29.6 Gflop/s	365 Gflop/s
Architecture	Shared memory, parallel vector	Distributed memory (shared address space)	128 nodes with 16-processor SMPs
System	Fully integrated custom system	Fully integrated custom system with commodity CPU and memory	Loosely integrated system with commodity system components
System Software	Vendor supplied, ready on delivery	Vendor supplied, completed after nearly three years development	Vendor supplied, contract for delivery in about three years
Footprint	588 ft	360 ft	1440 ft
Power Consumption	500 kW	288 kW	<1 MW

The three major NERSC systems of the 1990s are representative of the changes in high-end technology which were experienced in this decade. Between each successive generation, a major technology shift occurred that had an immediate impact on the production supercomputer centers. The first transition, which happened from about 1994 to 1996, is commonly characterized as the transition to massively parallel computing based on commodity microprocessors. This transition was widely anticipated as the «attack of the killer micros» and has been well documented, e.g., in various analyses of the TOP500 list [2]. The two important consequences of this transition for the supercomputer center were a reinvention of the center as a balance of production capabilities and intellectual services, together with an increase in the effort to develop system tools and software for highly parallel machines. The reinvention of NERSC has been discussed previously [3], and we will review some of the system software efforts at NERSC in Section 3.

The second transition is the result of an attempt to exploit not just commodity processors and memory, but whole commodity systems as the building blocks of the supercomputer. These cluster-of-SMP supercomputers have higher peak performance, greater memory capacity, and better cost performance than their predecessors. They also have higher inter-node communication latency, a larger footprint as they are air-cooled, and greater power requirements. They must be integrated with external disk caches and tape archives to manipulate the increasing volume of scientific data. They must also be closely coupled to wide-area networks to provide seamless access to a nationwide user base as well as other large computational assets. What is not yet fully appreciated is that this second technology transition (represented by NERSC-3) will be as fundamental as the previous one, and potentially even more dramatic (Table 2).

Table 2. Impact of the two technology transitions of the 1990s

	1994–1996 transition	**1998–2000 transition**
Economic Driver	Price performance of commodity processors and memory	16–64 CPU «sweet spot» for SMP technology in the commercial market place
Advantages of Transition	Higher performance and better price performance	Higher performance
Challenges of Transition	1. Applications transition to distributed memory, message passing model (MPI) 2. More complex system software (scheduling, checkpoint restarting)	1. Applications transition to hierarchical, distributed memory model (threads + MPI) 2. New development efforts for even more complex systems software 3. Increased cost of facilities

The changes are threefold

- applications need to be written that can tolerate an increase in communication latency and parallelism as well as a distributed, hierarchical memory model;
- system software will have to be developed anew for increasingly complex, more difficult to manage, one-of-a-kind systems; and
- center management will be forced to take creative new approaches to solve the space and power requirements for the new systems.

These changes in the high-end platform technology will be accompanied by the previously mentioned increases in data storage requirements and the integration of the supercomputer center into a computational grid utilizing high bandwidth, wide-area networks.

3   Challenges of Clusters of SMPs as a Teraflop Production Platform

This section will discuss three challenges: the increase in parallelism and a distributed hierarchical memory model, the need for new system software for high-end platforms, and increased space and power requirements.

3.1   Increase in Parallelism and a Distributed Hierarchical Memory Model

In order to facilitate the transition to the new programming paradigm of massively parallel computing, NERSC in 1996 changed its service model to include both excellent facilities and excellent intellectual services. In this change, several new groups were added to NERSC, so that the computer center occupies the tension field between computational science and computer science (Fig. 1). For both areas, a strategy was developed for bringing the latest research results to bear on the effectiveness of the center.

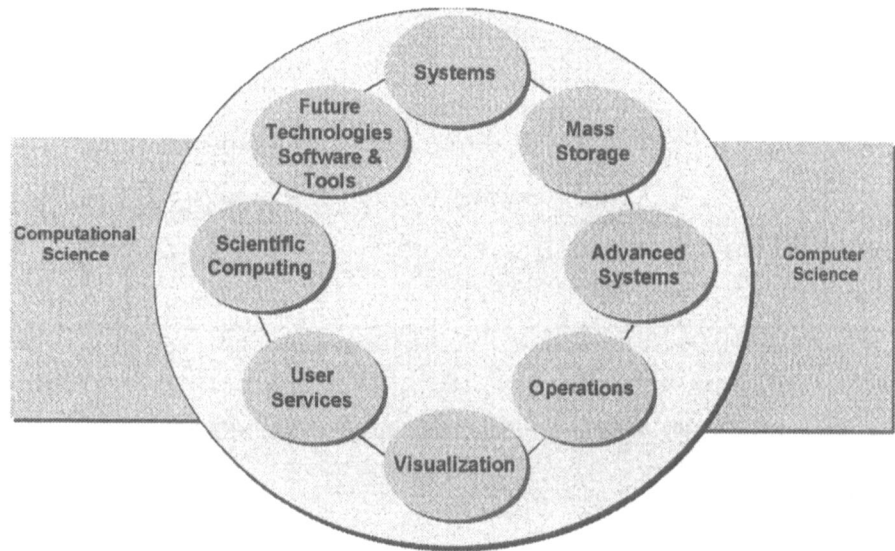

Fig. 1. The intellectual home of NERSC

One new group for which this change in strategy was particularly important was the Scientific Computing Group [4]. In the early 1990s, focus was on an applications group with experts in parallel computing representing the different applications areas, whereas the Scientific Computing Group consists of experts in computational techniques, which are relevant to a wider set of applications. This model is shown in Fig. 2.

The Scientific Computing Group at NERSC has been operating under this principle since 1996, and staff have been involved as research partners in many of the Grand Challenge applications at NERSC. One example of this successful work is the Gordon Bell Prize-winning collaboration between NERSC and Oak Ridge National Laboratory. In this collaboration, NERSC staff developed a new implementation of fast Fourier transforms (FFTs) on spherical data for distributed memory machines. The group was recognized for their simulation of metallic magnet atoms (Fig. 3), which was run on increasingly powerful Cray T3E supercomputers. They started with NERSC's 512-processor machine, and once the code was optimized, moved to ever larger T3Es at other centers. They won the prize with a sustained performance of 657

Gflop/s using a 1024-processor T3E-1200. The group later topped that performance by achieving 1.02 Tflop/s on a 1480-processor T3E-1200. This was the first complete scientific application to have exceeded a sustained 1.0 Tflop/s performance [5].

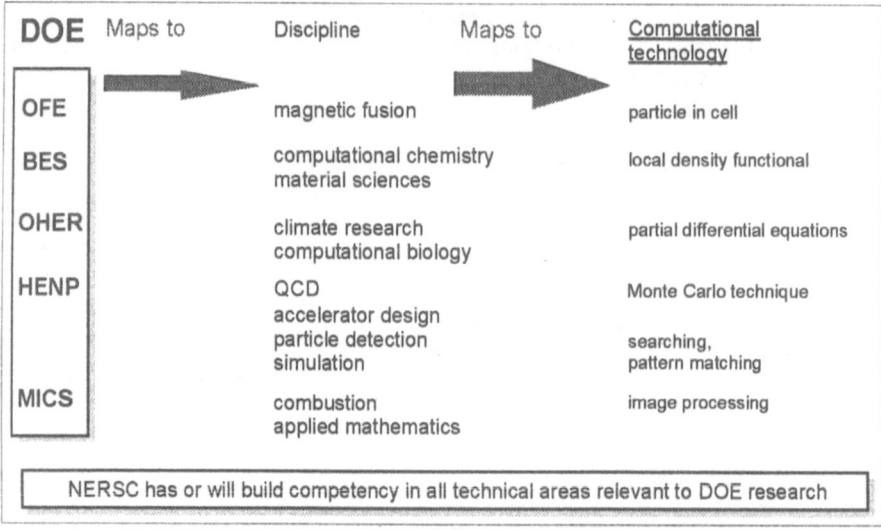

Fig. 2. Building computational competency at NERSC

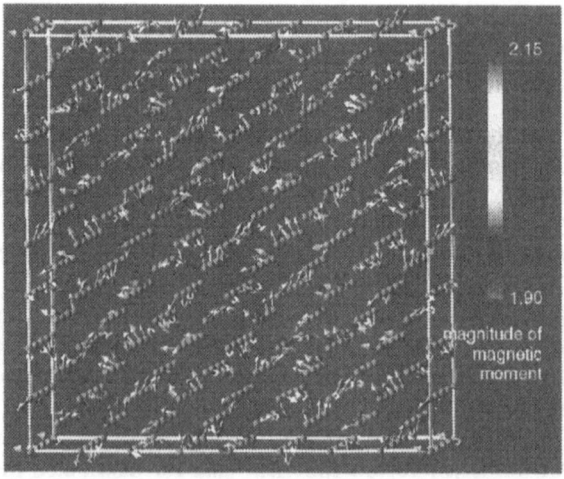

Fig. 3. Modeling of metallic magnetism – 1998 Gordon Bell Prize winner

This and many other success stories indicate that the Scientific Computing Group at NERSC is on the right track. For the next couple of years, the responsibilities and organization of the group will stay the same. The technology focus will be on the continued development of generic algorithmic techniques, which will support large-scale applications on platforms with larger numbers of processors and with a hierarchical memory structure. The group is well positioned to exploit the next generation architecture and produce algorithmic techniques that will benefit the NERSC client community at large.

3.2   System Software for High-End Platforms

The second challenge brought about by the next generation of high performance computing platforms is the decreasing maturity and lack of production quality of the system software. In the past NERSC has made pioneering contributions to bring first vector and later massively parallel machines into a full production environment. Based on work carried out in the Computational Systems [6] and Advanced Systems [7] groups, NERSC was the first site to demonstrate checkpoint/restart on a highly parallel system in the fall of 1997 [8]. Continued work in collaboration with SGI/Cray led to the first installation of the complete Psched software system in the spring of 1999. Psched combines a number of software components that allow NERSC to manage the T3E very effectively. In early April 1999, NERSC demonstrated a utilization of more than 93% of the T3E for a sustained period of time (Fig. 4) [9].

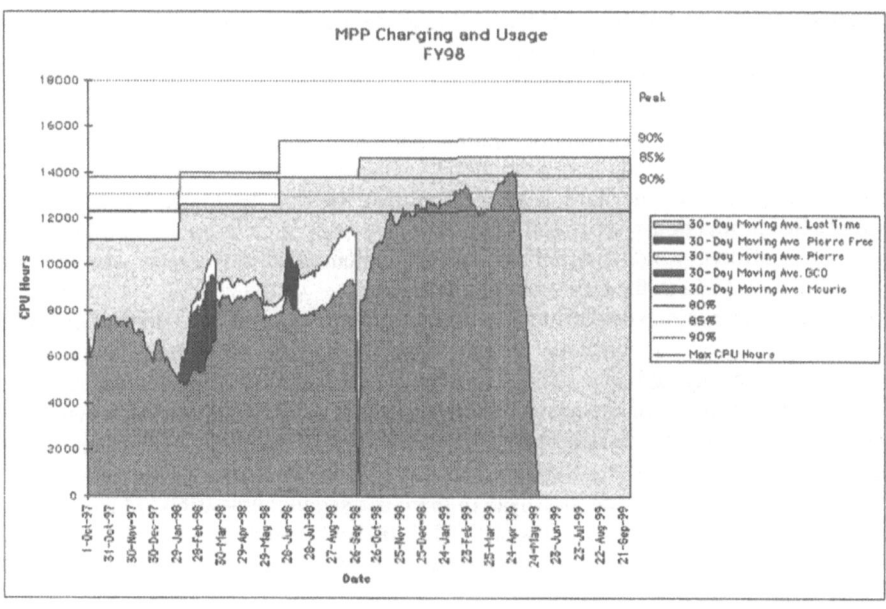

Fig. 4. Utilization of the Cray T3E at NERSC

Again this was a first for highly parallel machines. This is even more remarkable considering that the NERSC operating environment includes a wide range of jobs, ranging from interactive and debugging jobs to 512-processor Grand Challenge runs of up to 12 hours. The combination of these development efforts has made a cumulative impact on the utilization of the T3E, which has been consistently increasing over the last three years.

After these successes with the T3E, the question needs to be raised, can NERSC continue with a similar strategy on the new IBM platform? A first evaluation of the IBM system software reveals that it is far less mature than the Cray T3E. Many features which have been developed with great effort on the T3E over the last couple of years do not exist yet, and may become available only halfway through the expected lifetime of the IBM SP3. For example, checkpoint/restarting, which NERSC believes to be critical for the success for a production parallel machine, will only arrive in late 2001. This is symptomatic for the high performance computing industry in the U.S., which has focused most development efforts on 16 to 64 processor platforms. These constitute a «sweet spot» in the market and can be sold with larger margins to industrial and commercial users. Consequently software development efforts that are of interest only to a handful of high-end users, such as the government labs, have taken a backseat.

Looking a few years into the future, Paul Messina [10] has characterized the situation for high-end software by the «above the line–below the line» argument. He believes that vendors will only deliver system software up to the line of profitability, that is up to the «single box» SMP. Any software beyond the single system SMP, e.g., schedulers, file systems, communication protocols, accounting tools, etc. will have to be developed by the HPC community. The Department of Energy's Accelerated Strategic Computing Initiative (ASCI) will take the lead here, since they will have the first and largest of these configurations.

While NERSC is certainly willing to participate in any joint development effort, we believe that the «line» may be drawn too low. By lowering our expectations too much, we risk letting our vendor deliver hardware without the system software necessary to integrate it into a true production computing system. For the current IBM SP-3 system, NERSC and IBM have jointly developed an approach for developing high-quality production system software. Here we only want to mention one of our many joint efforts, which we believe has the potential to set a new standard for measuring utilization in the HPC community.

Currently there is no standard test to measure improvements in system utilization. While we have in general terms argued that developments such as checkpoint/restarting and scheduling algorithms will increase utilization, there is no quantitative assessment of the achieved improvements available today. As a matter of fact, most performance measurements today focus simply on improving speed of applications, i.e., on how much scientific work can be done for a given quantum of CPU time. In terms of a production system there is a second dimension, which we call effectiveness, i.e., how many quanta of CPU time can be made available to scientific programs. In order to maximize utilization, we must improve performance *and* effectiveness (Fig. 5).

The tool NERSC is planning to use to measure the impact of system software improvements on utilization is the SUPER benchmark [11]. This test uses a mix of NERSC test codes that run in a random order, testing standard system scheduling. There are also full configuration codes, I/O tests, and typical system administration activities. The setup of the SUPER benchmark is explained in Fig. 6. We expect at least 5% improvement per year on the IBM system.

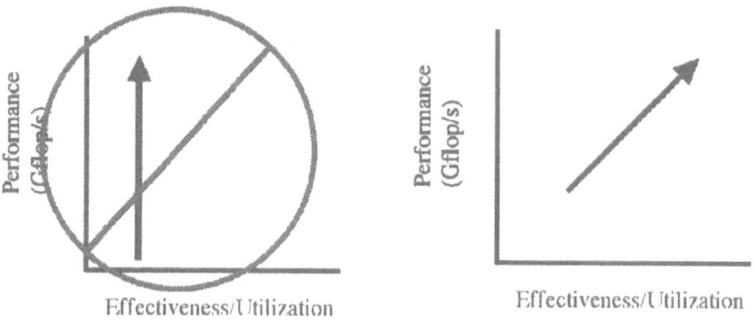

Fig. 5. The goal of high quality system software: improving both performance and utilization

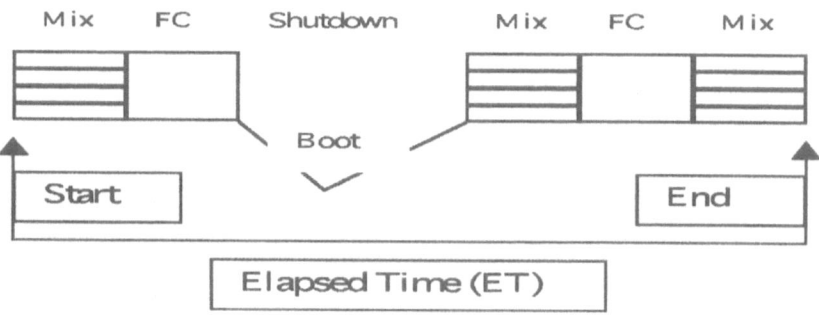

Fig. 6. Schematic organization of the SUPER benchmark

NERSC continues on an aggressive acquisition strategy, which will lead to the installation of NERSC-4 in 2002 and NERSC-5 in 2005. One recent development that we are tracking closely is the rapid development of cluster computing with commodity processors and open software based on Linux. The Future Technologies Group at NERSC [12] has entered a CRADA (cooperative research and development agreement) with Intel for the development of M-VIA. M-VIA is an implementation of the Virtual Interface Architecture (VIA) for Linux [13]. VIA is a new standard being promoted by Intel, Compaq, and Microsoft that enables high performance communication on clusters. M-VIA is being developed as part of the NERSC PC Cluster Project [14]. The goal of this project is to enable NERSC and NERSC clients to build PC clusters for scientific computing. Small clusters are useful for parallel code development, special-purpose applications, and small- to medium-sized problems. Large clusters show promise of replacing MPPs such as the NERSC T3E for certain applications. While we are not yet ready to claim that NERSC-4 or NERSC-5 will be a cluster-based platform, the rapid changes of the last decade teach us to anticipate yet another technology transition. Work done in the Future

Technologies group will help us to position ourselves for change in the three to five years.

An example of NERSC's M-VIA work, shown in Fig. 7, is a benchmark comparison of the NERSC PC cluster versus the T3E. For small numbers of processors (up to 32), the cluster is clearly competitive; but for larger numbers of processors, the well-designed interconnection network of the T3E shows its strength. Our data show that PC processors are even today not far from MPP processors in performance, and PC clusters with weak interconnects (fast Ethernet) are already a good alternative for some of our applications.

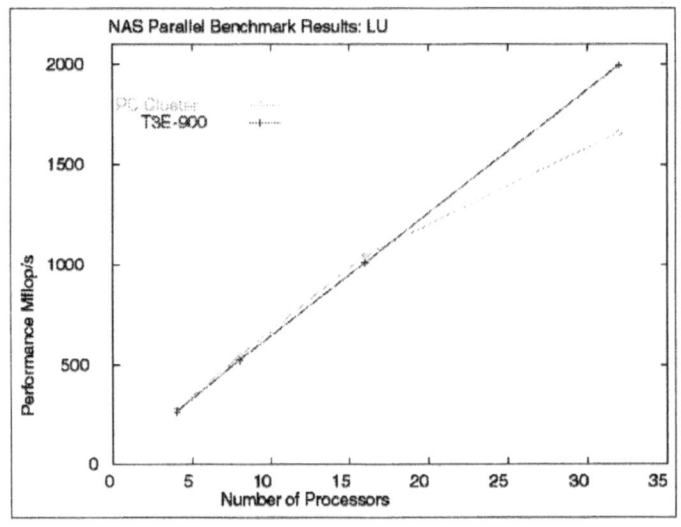

Fig. 7. PC cluster performance with M-VIA versus Cray T3E

3.3   Space and Power Requirements

The transition to commodity systems in the next generation of HPC platforms comes at a significant cost, which is usually hidden and is rarely mentioned when discussing the merits of different architectural approaches. In the transition from NERSC-2 to NERSC-3, we face a threefold increase in the machine's footprint and a twofold increase in its power requirements. These translate into substantial increases in ongoing facility costs for leasing space and purchasing electric power. NERSC is approaching this challenge by expanding to a new building. In an urban environment with a well-developed commercial real-estate market, this is a costly but manageable task. Our colleagues in the ASCI program have been forced to resort to new buildings at a scale previously unheard of for supercomputer centers. New buildings at LANL and LLNL are planned with a price tag in the $100M range. While NERSC will be able to meet the challenge at only a fraction of this cost, these high costs raise an interesting perspective on the transition to systems composed of commodity SMPs. Would not the HPC community in the U.S. be better off paying a higher price to our computer vendors in exchange for more tightly integrated systems that can be maintained within our existing facilities? As things stand today, we are spending tens of millions of dollars for non-technology-related expenses such as new buildings.

4   The Petabyte Data Challenge

While the challenges on the computing side are already quite formidable, supercomputer centers must also cope with an ever-increasing amount of data. In the past it has been correct to say that the amount of data generated *by computer simulations* was usually limited by the available computational technology. Thus the increase in archival storage was comparable to the increase in computational capability. In 1999 this view is no longer correct. What has changed is the fact that we will have to deal increasingly with *experimental data* which are generated from new technologies. One such example is the recent success in automating gene-sequencing technology. The rate at which data from the various genome projects will become available for further analysis, and the corresponding time it takes to search the genome database, are increasing faster than Moore's Law (Fig. 8).

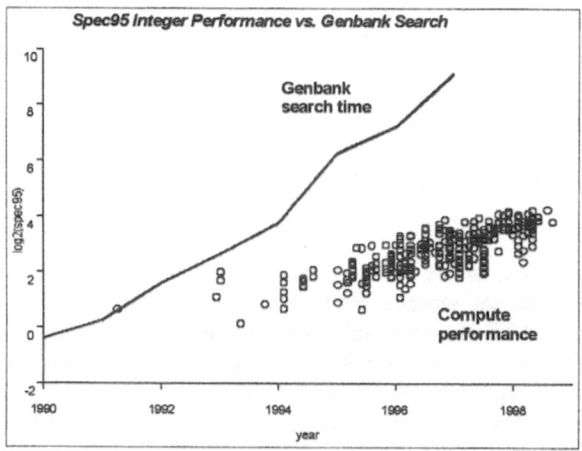

Fig. 8. Genome database search time is increasing at a faster rate than Moore's Law [15]

NERSC expects the increasing data it will be handling to come from several sources:

- genome data, e.g., the Human Genome Project at the DOE Joint Genome Institute;
- climate data, e.g., from the Program for Climate Model Diagnosis and Intercomparison (PCMDI) at Lawrence Livermore National Laboratory;
- high-energy physics data from new experiments such as the STAR experiment at Brookhaven's Relativistic Heavy Ion Collider or the ATLAS experiment at CERN's Large Hadron Collider.

Estimates of the high energy physics data produced by the Large Hadron Collider at CERN are on the order of a petabyte per year in 2005.

There are two efforts at NERSC to respond to this challenge. The File Storage Group [16] will continue to provide the storage media and baseline technology for large amounts of data. This group has increased the tertiary storage capacity at NERSC at an exponential rate, and so far has done an outstanding job of keeping our available storage capacity ahead of the demand (Fig. 9).

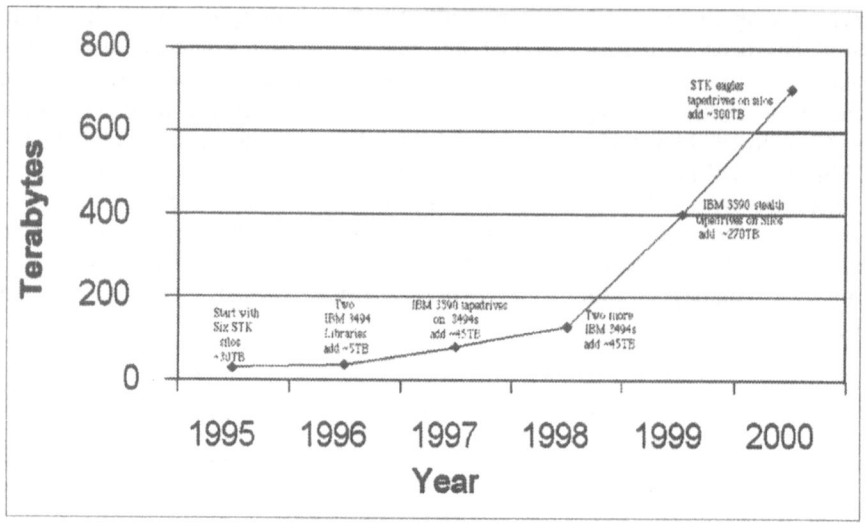

Fig. 9. NERSC storage capacity (raw)

At the same time, while raw capacity was increasing at an exponential rate, NERSC transitioned its storage management system completely to HPSS. As a developer site, NERSC is able to influence the HPSS consortium to provide tools to meet the requirements of our data intensive applications. Given the flood of future data, this will be a significant advantage for NERSC clients.

The second thrust in meeting the petabyte data challenge at NERSC is to provide tools for scientists to manage their data more effectively. There are two groups that work in this area. The Center for Bioinformatics and Computational Genomics (CBCG) [17] provides tools for the analysis of biological sequences, protein structure and function prediction, and large-scale genome annotation, as well as tools for access to biological information (database integration, data mining). The Scientific Data Management Group [18] is involved in various projects including tertiary storage management for high energy physics applications, data management tools, and efficient access to mass storage data.

Here we can highlight just one of the many projects of the Scientific Data Management Group (SDM). Typically, climate simulations and assimilated observational climate data are large multidimensional datasets in space (represented as meshes) and time. Accurate models require that these meshes are as dense as possible. When scientists return to analyze these datasets, they often need to access only one of the fifty or more parameters associated with each node in the mesh. If the entire dataset must be accessed from tape in order to extract the fields of interest, the time required can be many minutes or hours. This slows down the effectiveness of data analysis to the point that much of the data is never analyzed. Increasing access time of subsets from hours to minutes is the key to effective analysis (Fig. 10).

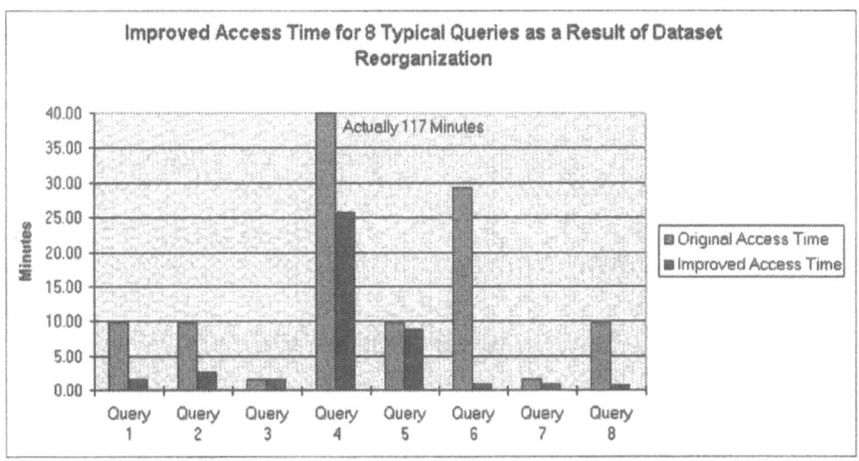

Fig. 10. Improved access time for climate data from mass storage

The NERSC SDM group has developed optimization algorithms for reorganizing the original datasets to match their intended usage. Further, the SDM group has designed enhancements to current storage server protocols to permit control over physical placement of data on storage devices. At the analysis level, this involves the application of clustering algorithms and organization of data into bins.

The work of the SDM group is unique among supercomputing centers, and we are not aware of a comparable research effort elsewhere. Projects in the SDM group and in CBCG will result in efficient new tools that NERSC clients can use to deal with their petabyte datasets.

5   Preparing for the Data Grid

In the last two years, the vision of a computational grid has gained broad acceptance. The grid is envisioned as a unified collection of geographically dispersed supercomputers, storage devices, scientific instruments, workstations, and advanced user interfaces (Fig. 11). The recent book *The Grid* [19] is an excellent summary of the current status of efforts to build such a grid.

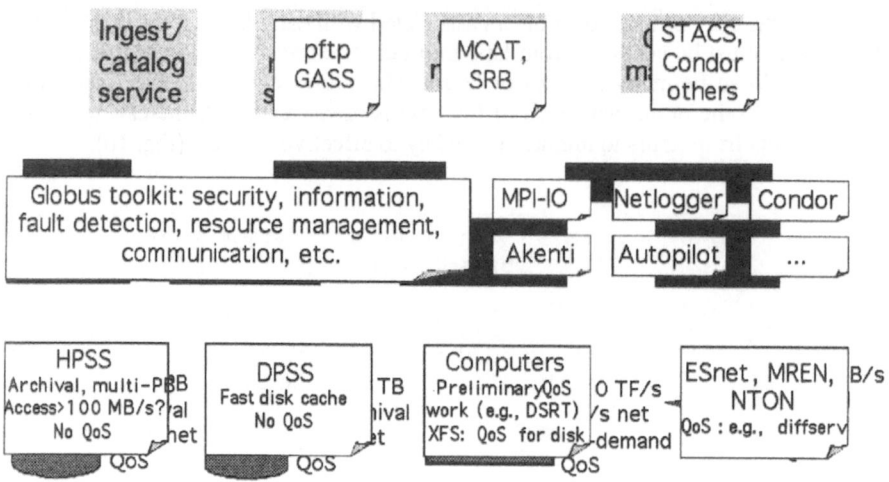

Fig. 11. Components of the data grid [20]

The most significant aspect of the grid for a supercomputer center like NERSC is the new concept of a data grid, enabling transparent access to data by scientists widely distributed across the United States. The petabyte datasets discussed in the previous section are community resources, which will be shared by researchers who are geographically distributed yet participating in collaborative projects. We do not expect these data to reside exclusively at one site, nor do we expect access to be restricted to a local set of users. Therefore, NERSC is investigating research issues related to large datasets distributed over a wide-area network. An example is the Distributed Parallel Storage System (DPSS) [21]. DPSS is a distributed disk cache which provides high-performance data handling as well as an architecture for building high-performance storage systems from low-cost commodity hardware components. This technology has been quite successful in providing an economical, high-performance, widely distributed, and highly scalable architecture for caching large amounts of data that can potentially be used by many different users.

One recent project that builds on DPSS, and which can be considered a prototype instantiation of data-grid technology, is the China Clipper Project (Fig. 12) [22]. In this project, high energy physics data which are generated at the Stanford Linear Accelerator Center (SLAC) are shared among storage systems at SLAC, NERSC, and Argonne National Laboratory. One of the early successes was a sustained data transfer rate of 58 Mbyte/sec from SLAC to the data archive in Berkeley [23].

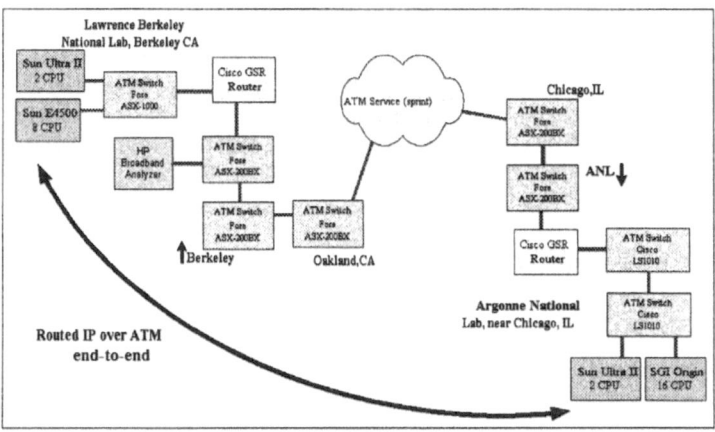

Fig. 12. The China Clipper testbed

6   The High Performance Organization

One of the driving factors for the continuous change in high performance computing is Moore's Law. Moore's Law postulates exponential growth in technology—a performance doubling of microprocessors every 18 months. Normally Moore's Law is plotted on a semi log scale and appears to us as a straight line, as in Fig. 13.

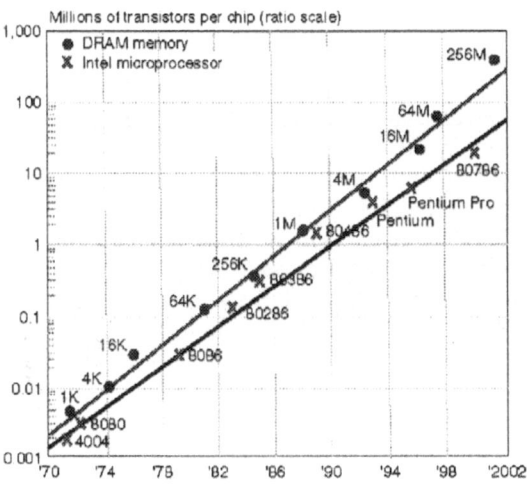

Fig. 13. Moore's Law – the customary straight-line view

However, human experience cannot deal well with a logarithmic scale, or intuitively grasp well the true effects of exponential growth. In the real world, we are sitting at the «bend» of an exponential curve. From our perspective, Moore's Law appears more like a straight wall, climbing steeply up into the infinite (Fig. 14). The significance of the «wall» is that in a few years, technology will be again completely different, and we have no clue what the future will be. For anyone not accustomed to change, Moore's Wall will be impenetrable and bewildering.

Fig. 14. Moore's Wall – the true exponential point of view (16th century version)

Thus the last and grandest challenge for a high performance computing center is to be a high performance organization. By this we mean an organization where staff can thrive and perform well under the stress of constant technology change and an unpredictable future.

Acknowledgements. We would like to thank all NERSC group leaders and NERSC staff, who helped in building an excellent organization. In particular, we acknowledge contributions by David Bailey, Andrew Canning, Jim Craw, Brent Draney, Keith Fitzgerald, Bill Saphir, Arie Shoshani, Brian Tierney, Mike Welcome, Tammy Welcome, and Manfred Zorn (all at NERSC) as well as contributions by Ian Foster (ANL) and Bill Johnston (LBNL and NASA). Their ideas, research results, and data were used in writing this paper. This work was supported by the Director, Office of Advanced Scientific Computing Research, Division of Mathematical, Information, and Computational Sciences of the U.S. Department of Energy under contract number DE-AC03-76SF00098.

References

1. http://www.nersc.gov
2. Horst D. Simon: High Performance Computing in the U.S. in 1994. Supercomputer **11** (1995) 21–31
3. Horst D. Simon: Site Report: Reinventing the Supercomputer Center at NERSC. IEEE Computational Science and Engineering Vol. 4 No. 3 (1997)

4. Esmond Ng et al.: http://www.nersc.gov/research/SCG/
5. B. Ujfalussy, X. Wang, X. Zhang, D. M. C. Nicholson, W. A. Shelton, G. M. Stocks, A. Canning, Y. Wang, and B. L. Gyorffy: High Performance First Principles Method for Complex Magnetic Properties. IEEE Proceedings of SC98, Orlando, Florida (1998)
6. James Craw et al.: http://www.nersc.gov/aboutnersc/sys.html
7. Tammy Welcome et al.: http://www.nersc.gov/research/annrep98/29advanced.html
8. Jon Bashor: NERSC First to Reach Goal of Seamless Shutdown, Restart of Supercomputer.   http://www.lbl.gov/Science-Articles/Archive/Cray-checkpointing.html (1997)
9. Jon Bashor: NERSC Achieves Breakthrough 93% Utilization on Cray T3E. http://www.nersc.gov/news/t3e-utilization4-19-99.html (1999)
10. Paul Messina: The 30Tflops Procurement at Los Alamos. Presentation (January 1999)
11. David H. Bailey et al.: System Utilization Performance Effectiveness Rating (SUPER) (in preparation, 1999)
12. Bill Saphir et al.: http://www.nersc.gov/research/FTG/
13. Bill Saphir et al.: http://www.nersc.gov/research/FTG/via/
14. Bill Saphir et al.: http://www.nersc.gov/research/FTG/pcp/
15. David J. States: http://www.ibc.wustl.edu/~states/mooreslaw.html
16. Keith Fitzgerald et al.: http://www.nersc.gov/groups/FSG/
17. Sylvia Spengler and Manfred Zorn: http://www.nersc.gov/cbcg/
18. Arie Shoshani et al.: http://gizmo.lbl.gov/DM.html
19. Ian Foster and Carl Kesselman (eds.): The Grid: Blueprint for a New Computing Infrastructure. Morgan-Kauffman (1998)
20. Ian Foster: Large-Scale Data Grids. DOE Conference on High Speed Computing, Salishan Lodge, OR (1999)
21. Brian Tierney et al.: http://www-itg.lbl.gov/DPSS/
22. William Johnston et al.: http://www-itg.lbl.gov/Clipper/
23. Jon Bashor: New Technology Demonstrates Priority Service for Internet Traffic Between Lawrence Berkeley and Argonne National Laboratories. http://www.lbl.gov/CS/Archive/headlines4-14-98.html (1998)

A Coming of Age for Beowulf-Class Computing

Thomas Sterling and Daniel Savarese

Center for Advanced Computing Research
California Institute of Technology
1200 E. California Blvd. MC 158-79
Pasadena, CA 91125 USA
tron@cacr.caltech.edu, dfs@cacr.caltech.edu

1 Introduction

Beowulf-class systems along with other forms of PC clustered systems have matured to the point that they are becoming the strategy of choice for some areas of high performance applications. A Beowulf system is a cluster of mass market COTS personal computers interconnected by means of widely available local area network (LAN) technology. Beowulf software is based on open source code Unix-like operating systems that, in a majority of cases, is Linux. The API for Beowulf is based on message passing semantics and mechanisms including explicit models such as PVM and MPI or implicit models such as BSP of HPF. Since its introduction in 1994, Beowulf-class computing has gone through five generations of PCs from multiple microprocessor vendors including the Intel x86 family, DEC's Alpha, and the PowerPC from IBM and Motorola. Originally, Beowulfs were implemented as small clusters in the range of 4 to 32 nodes. Larger clusters of 48 to 96 processors were deployed two and a half years ago. Today there are many systems of 100 to 300 processors with systems of over a thousand processors in the planning stage for implementation over the next year.

The earliest Beowulf systems had a peak floating point performance of approximately 320 Mflops sustaining approximately 70 Mflops on floating point intensive problems. By the third generation, peak performance systems of 3.2 Gflops were delivering sustained performance of 1.2 Gflops on non-trivial real world applications. A year later, larger systems were reported sustaining performance in excess of 10 Gflops. Shortly thereafter, the first Beowulfs were credited with being among the 500 largest computers in the world. In the last two years, Beowulf-class systems were recognized as providing the best price-performance of general-purpose high-end systems by winning the Gordon Bell Prize for supercomputer price-performance two years in a row. From a few experimental systems four years ago, many hundreds of systems have been deployed across the nation and around the world with as many as a thousand such systems operational today.

From this experience and level of acceptance, it is clear that Beowulf-class systems are having a significant impact on the field of high-end computing. But

P. Amestoy et al. (Eds.): Euro-Par'99, LNCS 1685, pp. 78–88, 1999.
© Springer-Verlag Berlin Heidelberg 1999

the value of this technology extends beyond simply the numbers of systems deployed. These systems are enabling applications work that might not have been conducted otherwise or would have been done at a substantially lower level. Also they are providing an exceptional vehicle for educational programs in parallel computing hardware, software, and parallel programming techniques, even down to the high school level. Many people are entering the field of parallel computing through the opportunity door opened by the availability of Beowulf systems. The roles and range of this class of systems are proving highly diverse; more than was originally imagined by those conducting the earliest path-finding work in this area.

Beowulf-class systems can no longer be ignored as inconsequential and relegated to the niche of novelty and hobbyist by their detractors. But nor can they be represented as the replacement for all other high-end system types for all computing environments and purposes by their advocates. The controversy surrounding Beowulf-class systems is important because it will determine future directions for advanced development and applied research in a number of related fields including system software, computational techniques, and system area network technology. The results of such work will determine the applicability and utility of this alternative approach to scalable computer systems implementation. This paper addresses the important issues that are at the center of the controversy and that will determine future directions and acceptance of Beowulf-class systems to satisfy near term computing requirements in the medium to high end of the performance spectrum. An attempt is made to clarify three critical questions: applicability, advantages/disadvantages, and economics. Most of these issues are subject to varied interpretation based on the underlying assumptions. The biggest problem with the continued contentions is that the arguments are driven by different expectations and tolerances. In essence, they are an example of attempting to compare apple and oranges. It is the intent of this paper to resolve, at least in part, the existing confusion.

2   Advantages: Impact Beyond Dollars

The selection of a Beowulf over a conventional vendor-offered parallel system is rarely driven simply by a desire to pay a lower price for a specified performance level. Should this be the case, the level of computing capability acquired would remain constant while freed dollars from the system procurement budget would be reallocated to some other part of an organization's needs. But an important aspect of the qualitative advantage of Beowulfs stems from their dramatically lower cost, not to save money but rather to acquire much more capability for a limited budget. While at first this may appear to be a distinction without a difference, its impact is substantial. In the first case, nothing changed other than the saving of some money. In this case, an approximate price-performance advantage of an order of magnitude will deliver a Beowulf-class system of more than ten times the peak performance of a vendor supplied system. In science and technology, an order of magnitude is an important differentiator. In the transportation industry, society supports two entirely separate infrastructures (actually three) because of the times ten difference in speed between the automobile with its interstate highway system and parking garages and the commercial jet aircraft with its FAA traffic control and airports. (railroads are the third independent infrastructure)

On a per node peak floating point performance basis, comparisons between comparable generation systems of equal number of processors and memory capacity show that tightly coupled vendor provided distributed shared memory systems cost between 8 and 12 times that of a Beowulf-class system. This does not include vendor supplied clusters of workstations where the price differential may be only about a factor of 4 (even less in specific cases), still in favor of the Beowulfs. For certain classes of applications, this can translate into a performance advantage of an order of magnitude. Problems that might have taken a full workday can be performed in less than an hour. Or a problem can be performed in a single user's day that would have taken two user work weeks (here a user day is defined as 8 hours, a user week is 40 hours) Such a difference qualitatively alters the way science and engineering can be performed and the impact that computation can have on it.

Perhaps more importantly are those problems that can be addressed due to the possibility of a Beowulf implementation that would not have been attacked by parallel processing at all. The entry level cost of vendor supplied parallel systems is often too high for some environments, principally those in the pure sciences, the social sciences, and many educational institutions. Under these circumstances before Beowulf, these problems were relegated to desktop workstations because the next higher cost system, a small SMP was much too high. Beowulfs gave low cost scalability and a graceful way for low budgeted computations to benefit from higher performance systems. But its not just shorter execution time that yields the qualitative advantage. Also, problem size, resolution, or fidelity from the modeling of additional phenomenology can enhance the quality of the science being performed, yielding new insights, otherwise unattainable under restricted budgets.

User's have greater control over their Beowulfs systems than they do with those provided by vendors while being less vulnerable to changes in vendor marketing plans. High performance computing through the late 80's and early 90's suffered from rapid changes in available system architectures. Many vendors went out of business with customers stuck with their rapidly obsolescent and unsupported orphan systems. Dramatic examples include the Intel Paragon, the TMC CM-5, the Maspar MP-2, and the KSR-1. Even when the company does not meet its demise, radical change in architecture can leave long term customers with no viable migration path. TMC switched from its SIMD class CM-2 to the MIMD class CM-5, Convex went from its vector C-4 to the DSM SP-1000. CRI is abandoning its vector family and its T3 family as well after multiple versions of both. In marked contrast, Beowulfs are forever. As long as there are low cost desk-top and server PC based systems, and third party LAN or SAN network technology, users will always be able to acquire and assemble the necessary subsystems to create future generation Beowulfs.

Ironically, not only can the specific component types change over time, Beowulf customers benefit from this technology dynamic. Again ironically, the most rapid advances in processor, memory, and networking technologies first impact the lowest end systems; those used within Beowulf clusters. The reasons are first that the biggest market opportunity is in the mass market arena (i.e. PCs) so the new devices are first incorporated in them, and second that large MPP class systems require considerable development time to integrate the new devices prior to delivering the completed parallel systems to their customers. In contrast, as soon as a new processor is packaged on a PC motherboard, it is ready to be installed in a Beowulf 24 hours later. Frequently, new Beowulfs have the hottest microprocessors available. This trend is likely to continue with the near term release of the Compaq DEC EV-6 and the AMD K-7, both of which will deliver more than 1 Gflops peak performance. However, more than processor performance, Beowulfs require advances in communication interconnect technology. For a majority of Beowulf-class systems, Fast-Ethernet at 100 Mbps has been tolerable and cost less than 25% of the total system price. For those applications sensitive to network characteristics, Myrianet has provided order of magnitude advantage in bandwidth and latency, but at an effective cost of about 50% of the total system price. At the same time, large system switches at acceptable per port price are enabling multi-hundred node configurations to be assembled. New techniques based on zero-copy methods such as VIA will greatly reduce software overhead in the next year. The new PCI bus will reduce the effect of this interface as a bottleneck. These trends in communications will match the advances in processor performance to retain if not improve the relative balance of overall system properties.

The flexibility permits custom system configurations on site by users. This just-in-place configuration capability allows structures of PCs to be organized such that the resulting system topology conforms to the data flow paths of the anticipated workload. Thus, Beowulf-class technology provides the means to establish semi-custom systems to meet particular requirements. Furthermore, such configurations can be altered at the site by local personnel to adapt to changing requirements or to incorporate technology upgrades. This evolvability provides a dynamic capability that extends lifetime and utility of Beowulfs and optimizes their applicability to user-specific workloads.

Beowulf-class systems leverage the community investment in open source software including Linux and the Gnu tools. Many experts around the world have contributed to this large and robust sophisticated software environment. Equally important, many users have provided a continued stream of feedback that has driven the maturing and debugging of the package. The result is that a very low cost highly capable environment is being employed by Beowulf. Although a small detail, the cost of Linux for a cluster system does not increase with the number of nodes comprising the system as would be the case with at least one widely distributed commercial operating system. The use of PVM and MPI have provided a means of creating highly portable applications software that will also run on many vendor parallel platforms. Indeed, a role of Beowulfs has been as development platforms in preparation to run on larger commercial systems. Amusingly, in some cases at least, the computational scientists have never bothered to make the transition to the larger, often contended, commercial systems.

3   Disadvantages: Limitations in Application and Usability

Beowulf-class systems are the third wave to performing high speed computations. The first wave is very high speed processors or a few such processors such as would be found in a CRI vector computer or an IBM mainframe. The second wave is the massively parallel systems incorporating tightly coupled microprocessors with custom high bi-section bandwidth internal networks and low overhead hardware mechanisms for at least some aspects of resource management (such as cache coherence). Beowulf-class systems do not fall into either of these categories and where applications demand the particular strengths of these first two genre of high-end computers, Beowulfs often may not perform well. Beowulfs, along with clusters of workstations, and other forms of loosely coupled systems provide a third path to high end computing that delivers very high price-performance for those problems that do not require (and should not pay for) the tightly coupled support of the first two system families. A disadvantage of Beowulf-class systems is that they are not as general as some vendor offerings.

The software environments currently available on most Beowulfs are primitive compared to the environments provided by vendors. While the basic node operating system, usually Linux, is very powerful and complete, and the equal of any commercial node operating system (perhaps better in some cases), the set of tools provided to manage the ensemble of nodes is usually superior for the vendor supplied systems. For some user environments, this is very important. While advances are being made and experimental tools are being tested at a number of organizations, a common framework has as yet not been adopted by the community as a whole. New users of Beowulf-class systems will not find readily available the wide array of useful services that exist, for example, on an SGI SMP.

The customer-engineer approach to maintenance of vendor supplied hardware and software at the customer site is not available to Beowulf users. The turnkey hands-off support that is provided with, for example, the IBM SP-2 is not the normal service that Beowulf sites can anticipate. When something goes wrong, there isn't a single one-stop-shopping 1-800 number that the user can call to solve the problem. There are many environments where this is not acceptable. As one colleague, somewhat exasperated by what he called the Beowulf Hype Factory put it: there is a price on low-cost. Beowulf is not suitable for all computing environments. Only those organizations with some in-house expertise, like systems administrators, can expect to employ Beowulfs effectively.

Floor space and packaging is an unfortunate problem with Beowulf-class systems. The two approaches to packaging Beowulfs is to develop some custom approach which can package motherboards tightly or to use the mass market per node tower boxes with power supplies that are incredibly inexpensive and robust. Unfortunately, the low cost attractiveness of the mass market packaging takes up a premium in floor space. Even with strong/tall utility shelves, the space required can be as much as twice that of custom packaging. For some machine rooms, such waste is unacceptable, especially as Beowulfs expand in scale through a number of hundreds of nodes. While custom packaging experiences have demonstrated significant improvements for space utilization, the cost of such packaging could be as much as half the cost of the system.

4   Comparative Economics

Controversy surrounds the economics of Beowulf-class systems. Users and general proponents simply quantify the costs of the component ensembles, sometimes with necessary physical packaging. Detractors identify a number of support services expected of conventional vendor provided systems that are not directly available to Beowulf users. The contention is that the cost of providing such support would eliminate or at least severely narrow the putative cost gap. There is merit to both arguments but it is based on a contradiction of assumptions. Here, the principle issues are examined that determine the real economics of Beowulf clusters deployment.

4.1   Hardware Costs

In fact, it is hard to say exactly what the difference in hardware costs is between Beowulfs and vendor systems. For Beowulfs to a first order, it can be considered the combined purchase costs of the subsystems comprising the total system ensemble. But this will vary among systems, even of equivalent form, dependent on negotiated purchase price between procuring institute and supplying distributor. It will also vary with time, even with short spans, as the market moves aggressively forward in the presence of constant change in product offerings and in the context of mass market competitive pricing.

It is harder to determine the actual cost of the vendor offerings. The vendors correctly assert that their hardware prices include hidden support services and delivered base level software. However, there are also additional costs for annual maintenance contracts and unbundled software. Finally, the list prices are often substantially higher than the final negotiated price based on various discounts that are applied to win the sale.

The third issue is that the comparison of hardware costs relates systems that are not actually equivalent. Often the processors are not the same. And almost always, the Beowulf networks are lower in bandwidth and higher in latency than the more tightly coupled vendor systems. Beowulf comparisons based on peak performance to purchase price are therefore a distortion. Routinely factors of 10 to 40 are reported on a per node basis of peak performance to procurement cost. It is not to be assumed that all Beowulf nodes are necessarily lower in performance than vendor provided systems. Some recent Beowulfs are employing Compaq DEC Alpha microprocessors packaged for the lower-end market. These have the highest peak performance of any microprocessors. The EV-6 and AMD K7 are likely to continue the availability of high performance micros to Beowulf ensembles. But the networks are routinely of lower capability. It would be interesting to build a comparative table based on bi-section bandwidth versus system cost and see if the order of magnitude difference between vendor systems and Beowulf systems for comparable nodes would be retained; probably not. While as not severe a gap, memory capacities are also not always equivalent. Finally, I/O bandwidth is another area in which some vendor systems significantly exceed comparable Beowulf systems.

The first conclusion must be that just from a hardware perspective, comparisons are difficult to make. The systems are rarely comparable against various metrics even if the number of nodes between the two is the same. The second conclusion is that as always, the real cost advantage is on a per application basis. This varies dramatically in a number of dimensions. Memory capacity requirements can vary by three orders of magnitude across applications using the same level of performance. Inter-node communication bandwidth and latency requirements can also vary by three orders of magnitude or more across applications for the same scale of system. Some applications are embarrassingly parallel with essentially no data sharing and intra-program synchronization points. Some have restricted synchronization and partitioned data and task sets that for sufficient granularity can permit efficient execution in spite of limited communications framework. These two classes of applications demonstrate excellent price-performance, even when the communication costs are much higher for Beowulfs than some commercial machines. The communication demands if small enough allow Beowulfs to operate with excellent price-performance even with relatively weak system area network hardware and sometimes superior raw performance. The third conclusion is that there are some highly sequential applications that will not yield significantly better price-performance for Beowulfs than vendor systems, and sometimes worse, due to the high communications demands.

4.2   System Management Costs

The cost of Beowulf systems does not reflect the cost of system support and maintenance. This is true. But the impact of this varies dramatically depending on the environment in which a new Beowulf is deployed. At a new site having no support for PCs or other computing facilities, additional full time support personnel have to be hired to manage the software and perform routine hardware maintenance. Under these circumstances, the cost should be attributed to that of the Beowulf. But in many other instances, Beowulfs are deployed in environments that are MPP/Unix/PC capable. For example, any site that already manages a farm of PCs already has in place the necessary talent to manage the hardware resources of a Beowulf including the standard maintenance. It is a point of confusion that there is no vendor maintenance support. All of the hardware components come with warrantees. Beowulf users do the same thing with broken Beowulf nodes that any owner of a PC does; ship it back to the supplier. The same is true with the hardware maintenance of the system area networks. The software management costs must also be considered. At many Beowulf sites, there is already a strong presence of Unix operating system workstations and even high-end computers with Unix-like software environments. With the proliferation of Linux, many more sites already have staff who have direct familiarity with Linux installation procedures. Beowulf-system operation under these circumstances leverages these existing talent resources. Often, the user and the systems administration are the same person. The additional overhead in doing it this way is relatively small after an initial installation period of a few days.

The primary discrepancy in viewpoint about the economics of Beowulf clusters is the expectation of system services. Beowulfs do not require the machine room mentality typical of vendor provided large systems because they are often used by one or a few people. The complex management infrastructure required of an expensive heavily shared MPP is simply unnecessary for many Beowulf installations. Because of their very low cost, they can be owned by a small organization and treated as a local resource with informal management; no accounting, no job control, not even automated partitioning. In some cases, Beowulfs are organized as single application systems. These run the same program continuously on newly supplied data sets; the application is the system. Some applications run for days or weeks without interruption. Again, due to their low cost, they can be dedicated to single applications of this sort.

Then it is clear that both sides of the debate about the economics of clusters versus vendor offerings are valid depending on the expectations and underlying assumptions. What has to be understood by the conventional computing community is that Beowulfs do open up an alternative approach for accomplishing some of the important computation that had been performed previously solely on commercial systems. And that furthermore, there is additional work being done that would not have been done at all. Cost advantage depends on the nature of the problem and the environment in which it is being performed.

5   Trading Peak Performance for Total Work

Beowulf-class systems can be dedicated to the execution of a single application for a long period of time because of their low cost. This opens a new important approach to accomplishing large scale computation at exceptional cost. But it also puts some additional responsibilities on the support software.

Historically, large applications required large systems with high peak performance. Such systems are routinely expensive and shared among many users. At any one time, a user is allocated some part of the machine (assuming a multihead system that can be simultaneously divided among multiple users) for a limited amount of time. At some later time, the same user may be allocated the next installment of time and resources for the same problem. The average amount of work accomplished by this shared system on behalf of the user's application is the product of the allotted resources and time divided by the duty cycle (how frequently the user gets on the machine). This can often be surprisingly low while the cost of these premium resources can be very high.

It is a myth that large problems demand large systems. It is true that some classes of algorithms require tightly coupled systems with high bi-section bandwidth as discussed earlier. But for problems that are latency tolerant, a time space trade-off is possible. Large problems demand large work where work is the time-space product. A few processors working for a long time can accomplish the same amount of work as a large number of processors working for a short time. Beowulf-class systems provide an acceptable medium for achieving this trade-off between time and resources. A moderate scale Beowulf of 32 Compaq DEC Alpha EV-6 processors in one week would perform the same work as a thousand processor T3E in a day (actually a little less than a day) with a cost difference of approximately a factor of a hundred.

But the additional problem of memory capacity of the two systems implies that the T3E would store far more data during the execution. And for some problems, this is particularly important. However, studies have shown a wide variance in memory requirements for applications of equivalent performance. While some problems require memory in bytes comparable to performance in flops (a 100 Gflops computation requires 100 Gbyte memory), the majority of science and engineering codes require less than one percent of this capacity. For these problems, the 32 EV-6 machine would have sufficient main memory to support the storage of the required data set. But for those problems that demand the larger memory capacities, a second opportunity exists. Because these applications are constructed using latency tolerant methods, the use of out-of-core computing techniques can address them memory discrepancy challenge. Out-of-core techniques allow secondary storage to replace part of main memory, using memory as a kind of cache and windowing across the larger data set held on disk. For some problems and algorithm types, swapping data between disk and memory can be done effectively in spite of the long delays to disk. As a result, we find that large scale science and engineering computations can be enabled by moderate scale low cost Beowulfs.

One problem does occur as the duration of computation is extended. Although reliability for Beowulf class systems is very high, nonetheless, longer computations incur increased probability that they will fail to complete correctly. The common solution for this is checkpoint-restarting. Other reasons for checkpointing are important beyond the possibility of a hardware failure. Sometimes even a Beowulf needs to be shared among a couple of users. It is nice to be able to migrate a running program to a modified configuration. The means is the same as that for checkpointing; save the data set and have the means for reallocating the intermediate data and control-state to a different configuration. Currently, all checkpointing is done by the applications programmers with little help from the system software support tools. Over time this will need to change.

6   An Agenda for Beowulf

Future Beowulf implementations will require a richer set of software tools to improve usability, portability, reliability, and reconfigurability. Some of the key requirements are discussed below. In each case, there are example and experimental tools that have been implemented at various institutions. Where appropriate, these need to be generalized, packaged, and distributed in a way that will make them generally useful to the Beowulf community.

6.1   Resource Discovery

Beowulf systems currently lack a standard means of determining the identities of their constituent elements and possible sub-elements for the purposes of executing tasks, reconfiguring the system, and supporting higher level functions such as system monitoring and scheduling. A telling example of why such a mechanism is needed is the way current system software is forced to provide this functionality for itself. Parallel programming libraries such as PVM and MPI require the definition of host files that are located in different parts of the file system. Schedulers such as PBS require similar custom support. Beyond not being able to readily identifying the nodes in a system, it is not possible for arbitrary processes to determine if a user may gain access to a given subset of the system or what system resources are available. Basic functionality, such as determining what nodes are part of the system, what nodes a user can use, and what sub-partitions of the system exist can provide a common foundation for general system software for Beowulf systems.

6.2   Resource Allocation/Assignment

Organizations that rely on commodity clustered computing as the backbone of their computational science research programs often share Beowulf clusters between multiple research groups. They require the ability to dynamically reconfigure access to portions of a cluster based on user requirements. Batch scheduling and job queues do not adequately meet the resource management needs of these organizations. They want to be able to arbitrarily allocate sets of resources to sets of users for periods of time, not only to run applications, but to perform development work and perform general experimentation. Current generation Beowulf system software does not meet these requirements. The ability to create arbitrary named groups and subgroups of resources, perhaps managed through an LDAP server, with associated property lists will add a new level of flexibility in the management of Beowulf clusters and provide underlying infrastructure for system software such as schedulers.

6.3   System Status Monitoring

Closely tied to both resource allocation and discovery is system status monitoring. A function as simple as tracking what nodes are usable and unusable in a cluster is integral to the aforementioned tasks. Although many monitoring tools have been independently developed, it is not currently possible to export the information provided by these tools in a generally usable fashion to other programs that may wish to make use of such data. Either a common set of system information gathering APIs or a common protocol for speaking to monitoring daemons will eventually have to emerge. The purpose of such an approach is twofold. The first is to allow future system software to leverage the valuable information provided by system monitors. And the second is to allow the interchangeability of system monitoring software in the same way it is possible to replace a mail or ftp server from one vendor with that of another.

6.4   Remote Dynamic Process Instantiation

Beowulf clusters can only perform useful work by executing processes on their component nodes. This essential function is currently served by one of the most inefficient possible mechanisms: rshell. A fundamental problem for large Beowulf clusters is that rshell does not scale well due to its use of reserved ports. The number of processes that can be started using rshell from a single node is limited to the number of reserved ports available to the command, which in the best case is 512. Not only does rshell not scale to larger systems, but it requires the persistent establishment of one or two TCP connections depending on whether or not standard error is required. Most PVM and MPI programs do not require this and would prefer to instantiate processes on a remote node in a fire and forget manner. In addition, some system software is more logically designed using something akin to an rfork() mechanism. We are already seeing signs of improvement in this area as the Beowulf community begins to experiment with alternative remote execution approaches, including an rfork(), currently under development at NASA Goddard.

Topic 01
Support Tools and Environments

Frédéric Desprez

Local Chair

Developing large applications on parallel machines is a difficult task, especially for newcomers or people without a deep knowledge of parallelism. Currently, we notice that parallelism not only enters industrial areas but also other fields of science. The aim of European projects like EUROPORT 1 and 2 is clear: to prove the validity of a parallel solution for large industrial codes. Moreover, in order to solve larger problems, scientists need more and more powerful computers both in terms of Mflops, memory size and I/O bandwidth. During the last five years, big efforts have been put in the definition of portable libraries and languages like MPI, HPF and OpenMP. Some parallel libraries like ScaLAPACK or PETSc start to be widely used. Finally, many tools for the development of applications have been developed.

Parallel architectures have also known many changes. Clusters of PCs connected with high speed networks are now well used both in academia and industry. Distributed shared memory machines are also main actors on the market.

Research in the domain of "Support Tools and Environments" is of course a core area of high-performance computing. Even if environments and tools are now able to help a scientist in getting good performances on a parallel machine or a cluster of workstations, a lot of work remains to be done around debugging, portability, heterogeneity, and many other fields.

We tried in this workshop to provide a meeting place for specialists in these important domains but also for users that need to know the very last developments that may change their way to program parallel machines or clusters. Eleven interesting papers will be presented that will start, as we hope, discussions around the future of parallel programming and debugging.

P. Amestoy et al. (Eds.): Euro-Par'99, LNCS 1685, pp. 89–89, 1999.
© Springer-Verlag Berlin Heidelberg 1999

Systematic Debugging of Parallel Programs in DIWIDE Based on Collective Breakpoints and Macrosteps[1]

P. Kacsuk, R. Lovas, and J. Kovács

MTA SZTAKI
Hungarian Academy of Sciences
H1518 Budapest, P.O. Box 63, Hungary
E-mail: kacsuk@sztaki.hu
URL: http://www.lpds.sztaki.hu

Abstract. The paper introduces the concept of collective breakpoints and macrosteps. Based on the collective breakpoints the macrostep-by-macrostep execution mode has been defined. After introducing the concept of the execution tree and meta-breakpoints the systematic debugging of message passing parallel programs is explained. The main features and distributed structure of DIWIDE, a macrostep debugger is described. The integration of DIWIDE into the GRADE and WINPAR parallel programming environments is outlined..

1 Introduction

In case of parallel programming novel techniques and methods are needed as well as tools that can support debugging in the inherently nondeterministic environment of parallel systems. The control&replay debugging method introduced by [2] facilitates the systematic testing and debugging of parallel programs. In the control&replay debugging every execution is under control just like in the replay phases of the monitor&replay debugging. The control&replay mechanism can systematically test all the possible timing conditions and after the exhaustive test of the program it would be sure that no hidden hazardous condition remained untested [6].

To make sure that the testing is really exhaustive we introduced several new breakpoint concepts like collective breakpoints, meta-breakpoints as well as a new debugging methodology called as macrostep-by-macrostep debugging. Macrosteps are determined by global communication breakpoint sets called collective breakpoints. Parallel debugging techniques analogous to the traditional cyclic

---

[1] The work described in the paper is partially supported by the ESPRIT project WINPAR No. 23516, the Hungarian-German Intergovernmental TéT project GRADE-OMIS No. D-62/96, the Hungarian-Portuguese Intergovernmental TéT project Graphical Quality No. P-12/97 and by the OMFB project WINPAR-DEBUGGER No. E151/98.04.03.

P. Amestoy et al. (Eds.): Euro-Par'99, LNCS 1685, pp. 90-97, 1999.

debugging techniques can be applied in the framework of macrostep debugging. First, the macrostep-by-macrostep execution mode is a generalisation of the step-by-step debugging method of sequential programs. Second, debugging of parallel programs based on meta-breakpoints is equivalent with the debugging of sequential programs based on traditional breakpoints. Finally, on-line construction and exhaustive search of the Execution Tree [4] ensures the exhaustive testing of message passing parallel programs. The new macrostep-by-macrostep debugging method is supported by the DIWIDE (DIstributed WIndows DEbugger) and has been introduced into two parallel programming environments: GRADE [3] and WINPAR [1].

2 Elements of the Macrostep Debugging Methodology

The main ideas of the macrostep debugging methodology can be summarised by the following new concepts: (i) collective breakpoints, (ii) macrosteps, (iii) macrostep-by-macrostep execution mode, (iv) execution tree, (v) meta-breakpoints.

Collective breakpoints. In order to support systematic debugging of message passing parallel programs and to eliminate the lousy effect of single global breakpoints STEPS introduced the concept of global breakpoint sets that was applied in DDBG [6] . They also introduced the notion of step that means to execute the parallel program until the next global breakpoint set that is generated by the tool STEPS. However, they did not give a precise definition of the global breakpoint sets and even more they realised them as a set of local breakpoints. In [4], we have put a restriction on the global breakpoint sets and introduced a special version of them called **collective breakpoints**. When all the breakpoints of the global breakpoint set are placed on communication instructions, the global breakpoint set is called as collective breakpoint. In order to handle the beginning and finishing events of a process, special breakpoints belonging to the collective breakpoint set are placed on the "begin" and "end" instructions of each process. A formal definition of the collective breakpoints can be found in [4].

If there is at least one breakpoint for each process of the parallel program within the collective breakpoint, we say that the collective breakpoint is **complete**. If there is at least one breakpoint for each alternative execution path of every process, the collective breakpoint is called **strongly complete**. The collective breakpoint B_i is **partial** if at least one process exists for which no global breakpoint has been defined in B_i.

Macrosteps. The set of executed code regions between two consecutive collective breakpoints is called a **macrostep**. Precise definition of macrosteps are given in [4]. A macrostep is called **pure** if it contains communication instructions only as the last instructions of its regions. A macrostep is called **compound** if it contains at least one communication instruction inside some 'of its regions. Provided that sequential program parts between communication instructions are already tested, a systematic debugging of a parallel program requires to debug the parallel program by pure macrosteps, i.e. instrumenting all the communication instructions by global breakpoints.

A breakpoint of the collective breakpoint is called **active** if it was hit in a macrostep and its associated instruction has been completed. A breakpoint is called **sleeping** if it was hit in a macrostep and its associated instruction has not been completed (for example, receive instruction waiting for a message). Those breakpoints that were either active or sleeping in a macrostep are jointly called **effective** breakpoints.

Macrostep-by-macrostep execution mode. After the definitions given above we can define the macrostep-by-macrostep execution mode of message passing parallel programs. In each step either the user or the debugger generates the next strongly complete collective breakpoint and then runs the program until the collective breakpoint is hit. Under these conditions the parallel program will be executed macrostep-by-macrostep. The boundaries of the macrosteps are defined by a series of effective global breakpoint sets. In such cases the user is interested only in checking the program state at the well defined boundary conditions.

There is a clear analogy between the step-by-step execution mode of sequential programs realised by (hidden) local breakpoints and the macrostep-by-macrostep execution mode of parallel programs. The macrostep-by-macrostep execution mode enables to check the progress of the parallel program at the points that are relevant from the point of view of parallel execution, i.e. at the message passing points. What we should ensure is that the macrostep-by-macrostep execution mode should work deterministically just like the step-by-step execution mode works in case of sequential programs. In order to ensure it, the debugger should store the history of collective breakpoints, the acceptance order of messages at receive instructions and the result of input operations.

Sequential program → (Local breakpoints) → step-by-step execution
Parallel program → Collective breakpoints → macrostep-by-macrostep execution

---

Fig. 1. Equivalence of step-by-step and macrostep-by-macrostep executions

At replay, the progress of processes are controlled by the stored collective breakpoints and the program is automatically executed again macrostep-by-macrostep as in the execution phase. Ensuring the deterministic macrostep-by-macrostep execution in parallel programs, we achieve a generalisation of the step-by-step execution of sequential programs. The equivalence of the two systems is demonstrated by Figure 1.

Execution tree. The **execution path** is a graph whose nodes represent macrosteps and the directed arcs connect the consecutive macrosteps. The **execution tree** is a generalisation of the execution path, it contains all the possible execution paths of a parallel program. Nodes of the execution tree can be of four types: (i) Root node, (ii) Alternative nodes, (iii) Fork nodes, (iv) Deterministic nodes.

The Root node represents the starting conditions of the parallel program. Alternative nodes indicate wildcard receive instructions which can choose a message non-deterministically from several processes. Fork nodes indicate the presence of if-then-else branches where the choice of the branch data dependent. Both alternative and fork nodes can create new execution paths in the execution tree and hence they are called choice nodes. Deterministic nodes cannot create any new execution path in the execution tree.

The execution tree is incrementally built (and in the case of a graphical system is incrementally drawn) as the various possible program paths are debugged. An execution path inside the execution tree represents either a previously debugged execution path of the parallel program or the currently debugged execution path. Notice that different execution paths represent different timing conditions among the component processes of the program or different data range conditions inside the processes.

Meta-breakpoints. Breakpoints can be placed at the nodes of the execution tree. Such breakpoints are called **meta-breakpoints**. The role of meta-breakpoints is analogous with the role of the breakpoints of sequential programs. A breakpoint in a sequential program means to run the program until the breakpoint is hit. Similarly, a meta-breakpoint at a node of the execution tree means to place the collective breakpoint belonging to that node and run the parallel program until the collective breakpoint is hit. Replay guarantees that the collective breakpoint will be hit and the parallel program will be stopped at the requested node.

3 Systematic Macrostep Debugging

The task of a systematic debugging or testing is to exhaustively traverse the complete execution tree with all the possible execution paths in it. Therefore, the execution tree represents a search space that should be completely explored by the debugging method. Accordingly, systematic testing and debugging of a parallel program require (i) generation of its execution tree (ii) exhaustive traverse of its execution tree.

Generation of the execution tree is equivalent with the generation of possible macrosteps from every node of the execution tree. The determination of the next node is straightforward, it simple means to place an element of the new strongly complete collective breakpoint on each branch of every process that can be accessed in the next step from the current collective breakpoint.

If the new collective breakpoint contains at least one breakpoint on an alternative receive instruction, i.e. the new node is an alternative node, the user or the debugger should decide which path to follow in the execution tree. It means that before executing the macrostep defined by the new alternative node, the alternative receive instruction should be notified from which process it should accept the message. Similarly, in the case of a fork node the data range determining the selection of the if-then-else branch should be defined before the macrostep is to be executed.

Exhaustive traverse of the execution tree. In order to exhaustively explore the execution tree the debugger should maintain a choice point administration for each choice node. After finishing the current execution path in the execution tree the debugger should backtrack to the last choice node and select an alternative branch that has not been tested yet. The method is very similar to the choice point handling and backtracking of Prolog interpreters. However, instead of real backtracking the debugger uses the replay mechanism to replay the program until the last choice node where the debugger should choose another path of the execution tree. The next questions therefore are:

1. What to store in the logfile to replay the program to any node of the execution tree?
2. How to use the information in the logfile to replay the program until a certain node of the execution tree?

Information in the execution tree logfile. In a deterministic node, it is sufficient to store the active breakpoints of the collective breakpoint that defines the corresponding macrostep. Non-effective breakpoints are obviously superfluous to store in the logfile since they cannot contribute to the definition of the given macrostep. Even sleeping breakpoints can be omitted from the logfile since they were not executed in the current macrostep and will be repeated in the collective breakpoint of the next macrostep.

Similarly, non effective breakpoints and sleeping breakpoints are not needed in choice nodes either. This fact means that the size of the logfile can be manageable even for realistic size parallel programs. In alternative nodes, besides the active breakpoints the possible sender processes and the already selected sender processes should be stored in order to select another one after backtracking. In fork nodes, possible value ranges and the already checked value ranges should be stored.

Replay in the execution tree. Replay can happen in three basic ways: (i) Macrostep-by-macrostep replay, (ii) Application of meta-breakpoints, (iii) Automatic backtracking.

Macrostep-by-macrostep replay is applied when the user wants to see in a detailed way how the program was running under a certain condition. As it was mentioned it is analogous with the step-by-step debugging of sequential programs. In a deterministic node it simply means to take the next macrostep (node) from the execution tree logfile and to place its collective breakpoint in the program. If the node is a choice node, then the additional information defining the selected path is taken as well. In both cases the program should run until the newly placed collective breakpoint is hit.

Meta-breakpoints are applied if the replay is necessary to a certain macrostep of the program execution. A meta-breakpoint at a node of the execution tree means to place the collective breakpoint belonging to that node and run the parallel program until the collective breakpoint is hit. Replay guarantees that the collective breakpoint will be hit and the parallel program will be stopped at the requested node. Hidden from the user, reaching the meta-breakpoint is realised by several steps. The number of these steps depends on the number of choice nodes between the current node of the execution tree and the selected node (where the meta-breakpoint is placed).

Automatic backtracking is applied when the user wants to build up systematically the complete execution tree. In such a case after finishing an execution path, the debugger automatically backtracks to the last choice node where unexplored branches are available. The realisation is equivalent with placing a meta-breakpoint at the last choice node and the current node is the beginning of the execution tree. After reaching the last choice node the debugger asks about the parameters that determine the next yet unexplored path of the execution tree.

4 The Concept of DIWIDE

DIWIDE (DIstributed WIndows DEbugger) is a macrostep debugger which is under construction in the framework of the WINPAR (WINdows for PARallel computing) ESPRIT project. DIWIDE is aimed at implementing all the principles explained in Section 2 and 3. In order to do that DIWIDE is implemented in a distributed manner. On each processor a local Debug Controller (DC) process is responsible to control the debugging, replaying and monitoring of the local application processes. A standard sequential debugger is attached to each application process on every processor. The sequential debugger is responsible to realise the pure debugging functions (for example set breakpoint, step, etc.). The application processes are compiled with a monitor that can collect data both for replaying and performance debugging. The monitored data are collected in local memory during a macrostep. At the end of a macrostep the monitored data are passed by the DC process to the central Master Debugger (MD) process running on the console host machine. The MD provides the graphical user interface and is responsible for all interactions with the user. It is also used for controlling all the DC processes based on either user commands or monitored replay information. The MD maintains a logfile for controlling the replay phases of program execution. The structure of DIWIDE is shown is Figure 2.

The MD process communicates with each DC process via a pair of channels. The REQUEST channels are used to transfer control commands and variable values from MD to DCs. The ANSWER channels deliver the monitored data as well as the values of inspected variables [5] .

The complete description of the work of DIWIDE is out of the scope of this paper. As an example to illustrate the work distribution among the components of DIWIDE we describe some of the most important issues.

The initialisation of the debugger starts with MD that brings DCs to live on the processors where at least one user process can be launched. Depending on the communication layout mapping information can be different type, but each must contain at least the set of clusters the can be used by the user processes. During startup phase the DCs store information for process identification and places the breakpoints in their process description tables that are built up by messages coming from MD. At launching the user processes SDs start, built the INPUT and OUTPUT channels to their DC, load user process and download the breakpoint table from DC. At this point the system is ready for debugging and the structure can be seen on Figure 2.

The execution of a macrostep is based on the decision of MD and is carried out by the underlying sequential debuggers. The user can also define the next collective breakpoint by the support of MD through the Graphical User Interface. Notice that the GUI and MD are logically two components, but in the NT version of DIWIDE they are implemented together. After the next collective breakpoint is defined the breakpoints are placed by the sequential debuggers into the code of their associated process. One macrostep instruction coming from the user executes the processes until they hit the next breakpoint that can belong to the collective breakpoint or can be an additional local breakpoint placed by the user before. When hitting a local breakpoint the variables of the process can be examined and then the process can be continued. After hitting all the breakpoints belonging to the collective breakpoint set MD collects replay data from the DCs and stores them in the replay logfile for the purpose of being able to execute the application later on the same sequence concerning the communication.

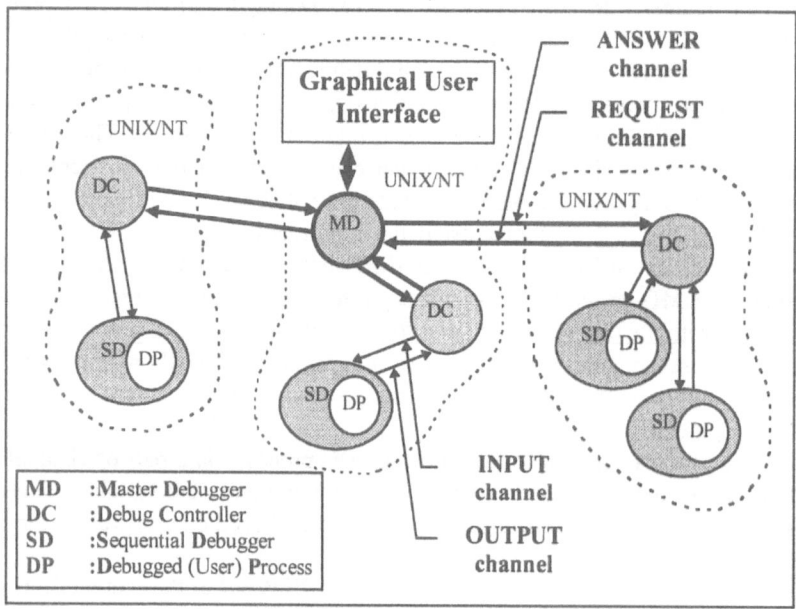

Fig. 2. Structure of DIWIDE

DIWIDE provides a rich set of window types to the user: Main Window, Application Window, Process Window, Host Connection Dialog Window, Stack Window, Watch Window and Structure Browser Window [5].

5 Integration of DIWIDE into Parallel Programming Environments

DIWIDE has been integrated into two parallel programming environments: (i) GRADE (GRAphical Development Environment) [3], (ii) WINPAR (WINdows for PARallel computing) [1]

Both have the objective to support the whole life-cycle of parallel program development by a graphical programming environment. GRADE (GRaphical Application Development Environment) has been developed in the framework of the SEPP (Software Engineering for Parallel Processing) and HPCTI (High Performance Computing Tools for Industry) COPERNICUS projects and its prototype is already available for workstation clusters and distributed memory supercomputers. The WINPAR (WINdows for PARallel computing) project is a recently finished ESPRIT project in which the main target parallel computing platform was the local cluster of Windows NT computers. Both environments can support the development of both PVM and MPI programs.

There are some essential differences between the implementations of the NT and UNIX versions of the DIWIDE debugger. The GUI and MD are implemented in one executable on NT but they are built in separate processes on UNIX for being able to be integrated easily into the GRADE environment. The UNIX version provides a library module which is a set of functions for accessing different services of DIWIDE and this library can be used in any developing environment. The UNIX version applies "rsh" or "ssh" for starting remote components but, for this purpose, we implemented a Service on NT. The UNIX version of DC contains a recognition module for the underlying sequential debugger however, on NT we implemented our own SD because there is no command-line source-level Microsoft compliant base debugger.

6 Conclusions

Based on the results of the current paper point-to-point message passing parallel programs can be systematically tested and debugged by the new macrostep-by-macrostep debugging method which is under implementation in the DIWIDE debugger. A version of DIWIDE has been built both into the GRADE and into the WINPAR parallel programming environments.

References

[1] Bäcker, A. et al.: WINPAR, Windows-Based Parallel Computing, in Proc. of the ParCo'97 Conference, Bonn, 1997
[2] Carver, R.H. and Tai, K.-C.: Replay and Testing for Concurrent Programs, IEEE Software, March, 1991, 66-74
[3] Kacsuk, P. et al.: A Graphical Development and Debugging Environment for Parallel Programs, *Parallel Computing*, Elsevier, 22(13), 1997, 1747-1770.
[4] Kacsuk, P.: Systematic Debugging of Parallel Programs based on Collective Breakpoints, Int. Symp. on Software Eng. for Parallel and Distributed Systems, Los Angeles, 1999,83-96
[5] Kovács, J. et al.: User's Guide of DIWIDE, ESPRIT Project 23516 WINPAR Deliverable, 1999
[6] Lourenco, J. et al.: An Integrated Testing and Debugging Environment for Parallel and Distributed Programs, Proc. of the 23rd Euromicro Conf., Budapest, 1997, 291-298

Project Workspaces for Parallel Computing - The TRAPPER Approach

Dino Ahr and Andreas Bäcker

SCAI, Institute for Algorithms and Scientific Computing
GMD, German National Research Center for Information Technology
Sankt Augustin, Germany
{Dino.Ahr|Andreas.Baecker}@gmd.de
http://www.gmd.de/SCAI/

Abstract. PC clusters running Windows NT have been investigated as a low-cost alternative to parallel computers. One important reason why NT clusters are not broadly accepted yet is the lack of powerful development environments, which are a crucial success factor for software development projects. This paper presents the new Windows NT version of TRAPPER, an integrated development and visualization environment for parallel systems. Emphasis is put on a new component called the *project workspace tool*. Project workspaces enable user-friendly interaction with all components and settings required for the development of a parallel application. They automate many repetitive tasks that are typical to parallel system development.

1 Introduction

The widespread adaption of parallel computing has been obstructed due to two major reasons: Parallel hardware is still very expensive and there is a lack of powerful parallel programming environments. The first problem has been tackled by the utilization of networked workstations and personal computers (PC) as low-cost parallel computing platforms. In order to attack the second problem, the message passing platforms PVM [6] and MPI [7] have recently been ported to the Windows NT platform [2, 3].

Message passing platforms are an enabling technology for parallel computing, but additional tools are required to develop parallel software efficiently. Of particular interest are integrated development environments (IDE) which support all activities in the development process. Up to now, neither one of the development environments available for UNIX has been ported to Windows NT, nor have any new environments been developed.

In the first part of this paper, we present TRAPPER [11], an interactive development and visualization environment for parallel computing on Windows NT. TRAPPER has been ported to Windows NT, redesigned and improved in the context of the WINPAR[1] project [1]. Several new features have been added

---

[1] The WINPAR (Windows-based Parallel Computing) project is supported by the European Union under ESPRIT IV project No. 23516.

P. Amestoy et al. (Eds.): Euro-Par'99, LNCS 1685, pp. 98–107, 1999.

and the graphical user interface (GUI) has been re-engineered to meet current standards of the Windows platform.

A very important requirement to an IDE on the Windows NT platform is user-friendliness, because demands from users being accustomed to carry out most of their daily work using interactive tools are very high. There exists a variety of development environments for parallel computing, but only few attention has been payed to the aspect of user-friendliness.

To enhance the user-friendliness of TRAPPER, a new component called the *project workspace tool* has been developed. The project workspace tool is the central component of TRAPPER: On the one hand, it offers a structured view on all tools, settings and files belonging to a software project. On the other hand, it serves as a steering and navigation tool which allows to launch tools, to trigger actions like the compilation process or to modify project settings. Moreover, the project workspace tool automatically cares for complete and consistent data and settings and performs a kind of garbage collection for files that are no longer needed. These features allow for an efficient cycle of work and relieve the programmer from many complicated and repetitive tasks.

In the next chapter we give a brief summary of existing development environments for parallel computing. The third chapter gives an introduction into TRAPPER. Chapter 4 describes the project workspace tool in-depth. The last chapter contains the conclusion and outlines future work on the project workspace tool.

2   Related Work

In the following we describe graphical software development environments for message-passing architectures. We refer to [13] for a detailed review of parallel programming tools and environments.

HeNCE (Heterogeneous Network Computing Environment) [4] is an X-based environment designed to assist in developing parallel programs on top of PVM that run on a heterogeneous networks of UNIX computers. HeNCE is composed of integrated graphical tools for creating, compiling, executing, animating and analyzing programs. HeNCE describes the behavioral aspects of the computation via a directed acyclic graph whereas TRAPPER describes the structural aspects of the application.

The GRADE (Graphical Application Development Environment) [8] environment provides tools to construct, execute, debug, monitor, and visualize PVM programs. The structure of the parallel application is specified down to the level of individual communication operations and is hence much more detailed than in TRAPPER. A distributed debugging engine assists the user in debugging programs on distributed memory computer architectures. The monitoring tool TAPE/PVM [10] and the visualization tool PROVE support performance monitoring and visualization of the program behavior. GRADE and TRAPPER are very similar except for that GRADE provides a distributed debugger and has a more elaborated graphical editor.

The EDPEPPS (Environment for the Design and Performance Evaluation of Portable Parallel Software) [5] environment comprises integrated tools for graphical parallel software design and mapping (PVMGraph), monitoring (TAPE/PVM), CPU modeling (cputime), simulation (SES/Workbench) and visualization, animation and analysis of program execution and predicted results (PVMVis). The EDPEPPS tool-set is targeted for a heterogeneous network of workstations and for the PVM parallel programming model. The main feature of this environment is the rapid prototyping approach to parallel software development, since the graphical editor together with the simulation tools allows performance analysis to be done without accessing the target platform.

The *Tool-set* [12] environment for PVM consists of a set of integrated tools which can be divided into automatic tools, e.g. the parallel file system, a checkpoint generator and a load balancer, and interactive tools with a GUI based on OSF/Motif, e.g. a source level debugger, a performance analyzer, a program flow visualizer and a deterministic execution controller. The core of the system is an OMIS (On-line Monitoring Interface Specification) [9] compliant monitoring system (OCM) for the PVM programming library on networks of workstations. *The Tool-set* environment focuses on on-line analysis and debugging of the parallel application and provides powerful and comprehensive facilities for this. It does not cover the graphical specification of a parallel application (along with code generation).

The environments discussed above all support PVM and run on UNIX. TRAPPER differs from them in that it supports both PVM (3.3) and MPI (1.1), and runs on Windows NT. Moreover, TRAPPER has been implemented on top of a portable GUI library and can be ported to UNIX with little effort. The project workspace tool is unique to TRAPPER.

3   The TRAPPER Visual Programming Environment

Figure 1 shows the tools of the TRAPPER development environment. Both, the parallel application and the target hardware, will be specified graphically with the *design manager* and the *graph editor*. The parallel application is represented by a process graph where nodes are processes and edges are message passing communication channels. TRAPPER generates code-skeletons for each process; the user only has to insert code for the process' functionality. The *mapping tool* determines the assignment of the processes to hardware nodes. The *platform settings* comprise paths and flags for compiler, linker, communication library and make tool available on the target hardware. This information is required for *generating* (by a click of a button) makefiles and configuration files which allow to *compile* and *run* the parallel program. The *monitoring settings* decide whether and how the application should be monitored (on-line/off-line). Trace data from the application, the communication platform and the hardware is collected by the *monitoring system*. For debugging and optimization purposes this data can be visualized and analyzed with the *visualization tool*. All tools are embedded in the so-called *project workspace tool*, which will be covered in-depth in chapter 4.

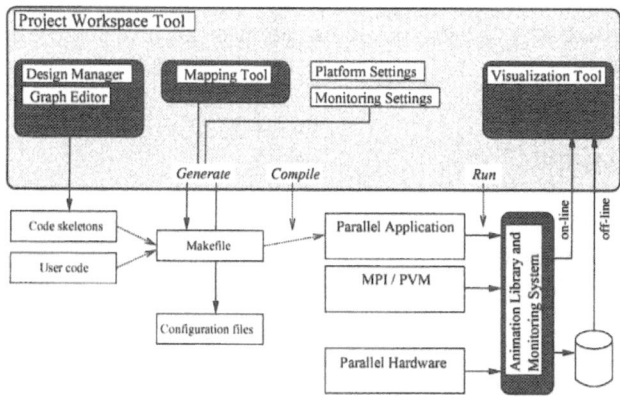

Fig. 1. The TRAPPER architecture

3.1   Designs

The central concept of TRAPPER is a *design*. A design is a graphical specification of a static process net or hardware configurations. Designs are created and modified using the *graph editor*. A design contains rectangles for its subnodes, ports denoting input/output channels and network interfaces, and links for connecting ports to nodes. Busses are used for multiplexing links. They denote group constructs in MPI and PVM program specifications and networks in hardware configurations.

Although the effort needed to specify a parallel system graphically introduces some overhead in early stages of the system development cycle, the graphical specifications gain a high value when the parallel system has to be animated. The node rectangles can be used to display the state of a process or a processor as well as to display contents of process variables and arrays. Links and ports can be highlighted to display communication activities. Since visualization tools are very helpful for debugging and performance tuning, the benefits obtained in later development stages outweigh the initial specification effort.

3.2   Software Development

TRAPPER supports a hybrid software development approach: The parallel structure of the application is described graphically by a process graph, whereas the sequential process code has to be written by the software developer. Each process design has an associated code file. Additional source code files can be associated with the whole application. TRAPPER simplifies coding of a process' code by generating code skeletons, which can later be extended.

3.3 Hardware Design

A graphical hardware design can be composed out of four different types of nodes: hardware clusters, single-processor computers, multi-processor computers and CPUs. These four basic types allow the user an exact and easy description of the hardware.

3.4 The Mapping Tool

The mapping tool performs the assignment of the application processes to nodes of a desired hardware design. TRAPPER provides automatic and manual mapping. The heuristic mapping algorithm computes a partitioning of the process graph with a well distributed computation load and a small communication load between partitions. The computation and communication loads are gained from the attributes of the software and hardware nodes.

3.5 The Monitoring System and the Animation Library

The monitoring system gathers run-time trace information about the parallel application. Trace events can either be stored in a binary trace file and then be analyzed retrospectively (*off-line*), or the events can be analyzed *on-line* while the parallel application is running.

On-line monitoring is realized via portable TCP/IP socket library calls. The benefit of using portable socket library calls is that a Windows NT workstation running TRAPPER can be used as a visualization front-end for a parallel application that runs on an arbitrary remote hardware.

The monitoring system is implemented as a C function library - the *TRAP-PER animation library* - which has to be linked with the application. A language binding for Fortran is available. TRAPPER automatically instruments MPI and PVM applications by wrapping MPI and PVM function calls.

In addition to tracing of communication and process events, the animation library supports tracing of scalar, one-dimensional and two-dimensional program variables.

3.6 The Visualization Tool

The visualization tool supports the developer in understanding, debugging and optimizing his or her application. It displays and animates trace data generated through calls to the animation library. It uses the node rectangles in the graphical software and hardware designs to display state information and views of program variables (fig. 2). The visualization tool also provides time axis diagrams for process states, communication events, communication statistics, scalar program variables, and critical path analysis.

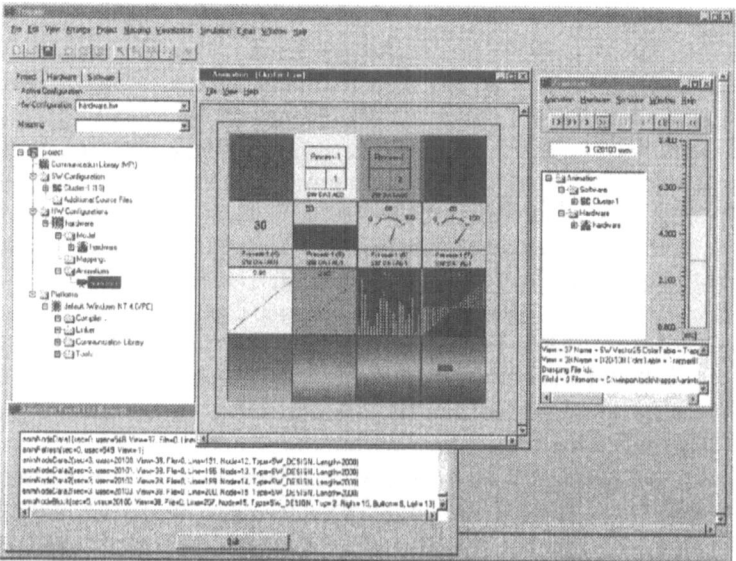

Fig. 2. TRAPPER screen-shot: The project tree (left) and the visualization tool (center)

4   The Project Workspace Tool

4.1   Motivation

Each tool in a development environment works upon data which are generally stored in files. For an environment like TRAPPER there are software and hardware design files, mapping files, source files, makefiles, configuration files for platforms and communication libraries, and animation files. A typical problem is to create an appropriate directory structure and to maintain all files according to this structure. Differentiation of file types is solely accomplished via file suffixes and it is sometimes difficult to bear in mind which files belong together, e.g. which software and hardware model files belong to a mapping. Furthermore the deletion of invalid or obsolete files and data is cumbersome and might be risky because valid files could be removed accidentally.

Another aspect concerns the maintenance of several similar program execution scenarios. For example, the developer tries different mappings or different target platforms to optimize the performance of an application. This could result in a lot of different mappings, configuration files, platform settings and associated animation files which are difficult to manage.

Furthermore, it is very convenient for a developer if the current state of work in the environment (e.g. the tools opened, the settings performed, the windows positioned on the screen, etc.) could be saved and restored later. This allows the developer to continue immediately with his work at the next session instead of configuring everything new.

4.2 Concept

The project workspace tool acts as a shell around the other tools of the TRAPPER environment. It has *structuring*, *visualizing* and *interactive* aspects.

As a fundament we designed a logical structure for the files and settings that are part of the TRAPPER development process. All these files and settings - inherently organized according to this structure - constitute the so-called *project workspace*. For each development project the user will create a separate project workspace. The project workspace tool cares for *consistency* and *completeness* of all elements added to the workspace. Moreover, the project workspace can be made persistent, i.e. the current state of work within the project workspace can be saved to disk and easily restored for the next session.

The visualization aspect of the project workspace tool is realized by a tree gadget[2] GUI - called the *project tree*. Within the project tree each item contained in a project workspace will be displayed as a tree node. The hierarchy of the structure is mirrored by a parent-child relation between the corresponding nodes. Hence all files and settings are ordered visually and can easily be found and accessed by the user. A design/sub-design relation on design level will be mapped by a parent-child relation of the corresponding nodes in the project tree. This technique allows to capture particularly complex designs with many nodes and hierarchy levels. The directory structure of the files contained in the project tree is hidden; the user doesn't have to care where a file is stored. Since folder nodes can be closed or opened the contents can be hidden respective shown allowing the developer to concentrate on the parts he or she currently works upon.

For instant access each node of the project tree provides context sensitive interaction facilities through a context menu.

We will now present the structural aspects of the project workspace and its visualization and interaction facilities within the project tree. At the same time we will discuss a sample application, the *Poisson problem* taken from [7], in order to illustrate the features and benefits of the project workspace tool.

4.3 The Project Workspace and the Project Tree

The general structure of a project workspace is depicted in fig. 3. Figure 4 shows the project tree for the Poisson problem example.

Communication Library. At the beginning of the development process a basic decision concerning the *communication library* has to be made by the developer. This could be either MPI or PVM. The example application which solves the Poisson problem will be implemented on top of MPI.

---

[2] A tree gadget is a GUI element consisting of *folder* nodes and *leaf* nodes arranged as a tree. Each node has an icon and a label. Folder nodes can contain further folder and leaf nodes.

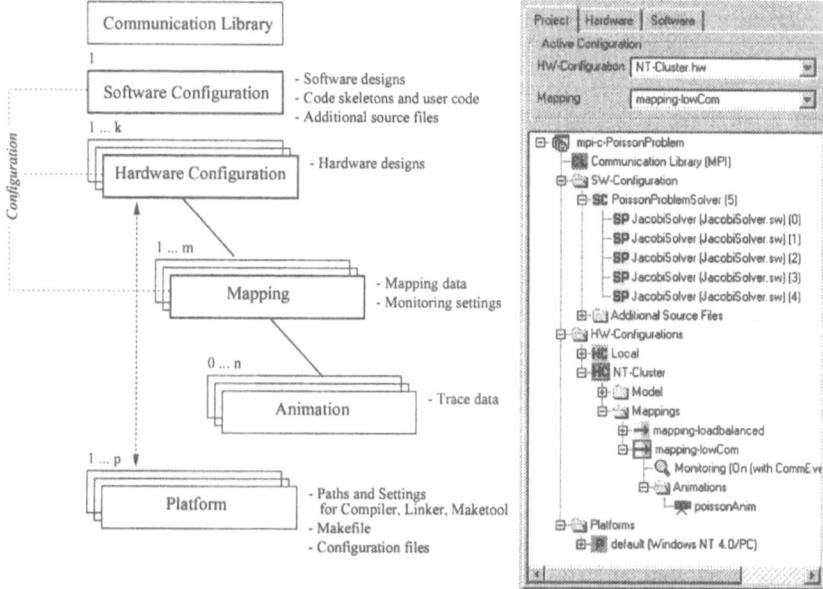

Fig. 3. Structure of the project workspace

Fig. 4. Screen-shot of the project tree

Software/Hardware Configuration and Mapping. The *software configuration* comprises software designs of the parallel application and associated
code skeletons including inserted user code. In addition, it contains source files
which are attached to the application. In our example the software design which
represents the whole parallel application is called PoissonProblemSolver. It is
composed of five JacobiSolver processes, programmed in C.

In order to execute the parallel application on different available hardware
platforms there can be several *hardware configurations*. Each hardware configuration includes hardware designs and an arbitrary number of *mappings*. This
enables the user to evaluate different mapping strategies.

In the above example we consider two different target platforms, named
Local and NT-Cluster. The Local platform contains one single-processor computer and is used for implementation and testing the application. The NT-Cluster hardware consists of three single-processor computer nodes and is used in
the stage of performance optimization. All nodes are attributed with Windows
NT 4.0 as their operating system and Pentium as their cpu. For our application
we focus on mappings for the hardware configuration NT-cluster. We create
two mappings mapping-lowCom and mapping-loadbalanced which concentrate
on minimizing the communication between the processors respective ensuring a
good load balancing.

Configuration. A hardware configuration and a mapping define a unique execution scenario for a parallel application in the project workspace. We denote

such a pair as a *configuration*. The actions *Generate, Compile* and *Run* (fig. 1) are only available when there is at least one configuration. In general there are several configurations contained in a project workspace. Hence there has to be a particular configuration, called *active configuration*, to which these actions refer.

We assume that we want to solve the Poisson problem on the `NT-cluster`. Because communication is a bottleneck for connected PCs we choose the mapping `mapping-lowCom` for our purposes. Hence we have to choose the pair (`NT-cluster`, `mapping-lowCom`) as active configuration.

Since a mapping component identifies a configuration it is used as a container for monitoring settings and *animation* components. The monitoring settings influence whether and how the application generated for this configuration should be monitored. Each animation component contains trace data of an application run made for this configuration. Several trace files can be managed to compare application runs.

In order to obtain trace files we switch on the off-line mode in the monitoring settings of the active configuration (`NT-cluster`, `mapping-lowCom`). The *Generate* action produces a makefile and a MPI process group file. Via the *Compile* action the makefile will be processed by the make tool resulting in an executable which can then be started with the *Run* call. For the program run the monitoring system produces the trace file `poissonAnim`.

Platform. A *platform* is defined by the processor type and the operating system of a hardware design. For each hardware design contained in the project there exists an associated platform dataset which comprises settings and paths for compiler, linker, communication library and make tool. This data is required for the generation of makefiles and configuration files. TRAPPER provides settings for each platform in a *defaults database*. The defaults database eventually has to be adapted once by the system administrator of a target platform; after that the user doesn't have to care about these issues.

Consistency and Completeness. The *consistency* mechanism of the project workspace tool cares for deleting invalid data. For example, if the software configuration has been modified because a process node has been added/removed all existing mappings will become invalid. Hence all mapping data including animation data will be deleted automatically to ensure the consistency of the project workspace.

Completeness checks of required settings will be performed by the mapping tool and the generation process. In case of failure the user is led directly to the concerned dialog.

5   Conclusion

We have described the concept and the implementation of project workspaces for the development of parallel applications. Experiences have shown that the project workspace tool is a suitable tool for speeding up application development. For people who are new to an already running development project, project workspaces can be a valuable aid in discovering content and structure of the project. TRAPPER is therefore an ideal supplementary tool for education in parallel programming.

References

[1] D. Ahr, A. Bäcker, O. Krämer-Fuhrmann, R. Lovas, H. Mierendorff, H. Schwamborn, J. G. Silva, and K. Wolf. WINPAR - Windows based Parallel Computing. In E. H. D'Hollander, G.R. Joubert, F. J. Peters, and U. Trottenberg, editors, *PARALLEL COMPUTING: Fundamentals, Applications and New Directions*, pages 495–502. Elsevier, Amsterdam, The Netherlands, 1998.

[2] A. Alves, L. Silva, J. Carreira, and J. G. Silva. WPVM: Parallel Computing for the People. In *Proceedings of HPCN'95*, pages 582–587, 1995.

[3] Marc Baker. MPI on NT: The current status and Performance of the available environments. In *EuroPVM/MPI'98*, Liverpool, UK, September 1998.

[4] A. Beguelin, J. Dongarra, G. A. Geist, R. Manchek, and K. Moore. HeNCE: A Heterogeneous Network Computing Environment. *Scientific Programming*, 3(1):49–60, 1994.

[5] T. Delaitre, M. J. Zemerly, P. Vekariya, G. R. Justo, J. Bourgeois, F. Schinkmann, F. Spies, S. Randoux, and S. C. Winter. EDPEPPS: A Toolset for the Design and Performance Evaluation of Parallel Applications. In D. Pritchard and J. Reeve, editors, *Euro-Par'98*, pages 113–125, September 1998.

[6] A. Geist, A. Beguelin, J. Dongarra, J. Weicheng, R. Manchek, and V. Sunderam. *PVM: Parallel Virtual Machine - A Users' Guide and Tutorial for Networked Parallel Computing*. MIT Press, 1994.

[7] W. Gropp, E. Lusk, and A. Skjellum. *Using MPI: Portable Parallel Programming with the Message Passing Interface*. MIT Press, 1994.

[8] P. Kacsuk, J. C. Cunha, G. Dózsa, J. Lourenço, T. Fadgyas, and T. Antão. A Graphical Development and Debugging Environment for Parallel Programs. *Parallel Computing*, 22(13):1747–1770, February 1997.

[9] Thomas Ludwig and Roland Wismüller. OMIS 2.0 – A Universal Interface for Monitoring Systems. In M. Bubak, J. Dongarra, and J. Waśniewski, editors, *EuroPVM/MPI'97*, pages 267–276, November 1997.

[10] E. Maillet. TAPE/PVM: An Efficient Performance Monitor for PVM Applications - User Guide. WWW: `ftp://ftp.imag.fr/pub/APACHE/TAPE/`, March 1995.

[11] L. Schäfers, C. Scheidler, and O. Krämer-Fuhrmann. TRAPPER: A Graphical Programming Environment for Parallel Systems. *Future Generations Computer Systems*, 11(4-5):351–361, August 1995.

[12] R. Wismüller, T. Ludwig, A. Bode, R. Borgeest, S. Lamberts, M. Oberhuber, C. Röder, and G. Stellner. The Tool-Set Project: Towards an Integrated Tool Environment for Parallel Programming. In X. De, K. E. Großpietsch, and C. Steigner, editors, *APPT'97*, pages 9–16, Koblenz, Germany, September 1997.

[13] M. J. Zemerly, T. Delaitre, and G. R. Justo. Literature Review (2), EDPEPPS EPSRC Project (GR/K40468) D6.2.2, EDPEPPS/32, Centre for Parallel Computing, University of Westminster, London. WWW: `http://www.cpc.wmin.ac.uk/~edpepps/reports/edpepps32.ps.gz`, July 1997.

PVMbuilder - A Tool for Parallel Programming

Jan B. Pedersen and Alan Wagner

University of British Columbia, Vancouver
matt@cs.ubc.ca, wagner@cs.ubc.ca

Abstract. Message-passing is often used to implement parallel programs to run on workstation clusters. However, writing message-passing programs is a difficult and error prone task. In this paper we describe a graphical interface called PVMbuilder to support the construction of PVM programs. We show the tool produces little overhead in comparison to the hand-written PVM program. PVMbuilder provides a higher level abstract view of the program and supports dynamic process creation along with automatic generation of PVM communication calls.

1   Introduction

Clusters of workstations are a widely available source of computing power. Programs for these systems are typically written in C or Fortran and use libraries such as PVM [3] and MPI [2] to communicate and coordinate the computation. However, it is difficult to reason about processes all executing at the same time, sending and receiving data according to some, often complicated, pattern. Specifying the communication becomes a major difficulty. Communication errors can lead to unpredictable program behaviour such as deadlock, process failure, or incorrect results. Given the inherent problems in debugging message-passing programs, it is important to develop tools that avoid the problem altogether by assisting the programmer to construct correct code in the first place.

PVMbuilder was created to help with the construction of message-passing programs. Our main motivation was to create a tool to assist non-computer scientists to take advantage of cluster computing. The tool supports the following:

1. Automatic generation of communication calls to reduce errors.
2. A high level graph structure that provides a abstract representation of the overall structure of the program.
3. Generation of source code directly from the graph which compiles into C executables, with execution times close to hand-written programs.

PVMbuilder differs from other graphical program builders such as HeNCE [5] and CODE [7] in that it allows explicit message passing thereby supporting a superset of the programs representable in CODE or HeNCE. In addition, PVMbuilder appears to add a problem dependent constant to the runtime of the hand-written code. This constant varies between 6% and 13%.

Graphs are often used by programmers as an aid in constructing a message-passing program, where nodes are used to represent processes (or processors) and

P. Amestoy et al. (Eds.): Euro-Par'99, LNCS 1685, pp. 108–112, 1999.

arcs denote communication or control flow. A natural extension to this process
is to support the creation of these graphs and from there support the automatic
generation of the underlying message-passing program.

As a result the programmer no longer has to manually transcribe the program
and communication patterns into detailed source code instructions, a tedious
and error-prone task. The graph also provides a higher-level specification of the
program which can be used for debugging and program evolution.

2 PVMbuilder

A PVMbuilder program is built by constructing a directed acyclic graph that
reflects the high level control flow of the program and the communication between the various processes. The graph consists of several different node types,
each representing a different task. The resulting graph is then translated into
C-code with PVM communication calls. The programs can then be compiled and
executed via PVM on a cluster of workstations. Figure 1 shows an annotated example of a PVMbuilder graph which uses the seven different node types that can
appear in the graph. The algorithm shown in Fig. 1 uses a master/slave pattern
to solve a hyperbolic differential equation. The slaves are organized into a linear
array of processes with left and right neighbours (except for the endpoints). The
master distributes a one dimensional array to the slaves who compute and communicate the results to their two neighbours within a loop. The master program
is shown on the left in Fig. 1 while the slave program appears on the right.

The following node types are defined:

A BEGIN node marks the beginning of a new process while an END node
marks the end of the process. A CODE node contains sequential program code.
There are three CODE nodes in Fig. 1. There is one in the master, Init, used to
initialize variables, and get command line arguments. The slaves have two: Init,
which again declares and initializes variables, and DoCalculations to perform
the computation.

The SPAWN node is used to create new process groups. The number of new
processes spawned is determined by a function bound to the SPAWN arc. In Fig. 1
there is only a single group of slave processes all executing the same program.
As a result, the master has a single SPAWN node with only one SPAWN arc.

The COMM node denotes communication between two processes. COMM
nodes are connected by a COMM arc that describes exactly what data is sent.
In the example shown in Fig. 1 we find four instances of communication.

LOOPSTART and LOOPEND nodes denote loops containing communication.
Since the slaves in the example execute a number of steps, each containing
communication and computation, a set of LOOPSTART and LOOPEND nodes
are declared. In Fig. 1 the loop condition is shown above the LOOPSTART node.
Figure 1 shows the three different kinds of arcs in a PVMbuilder graph.

PROGRAM arcs connect nodes, thereby describing the control flow.

SPAWN arcs are used to connect SPAWN nodes with the BEGIN nodes of the
slave groups they create. In Fig. 1 the function which determines the number

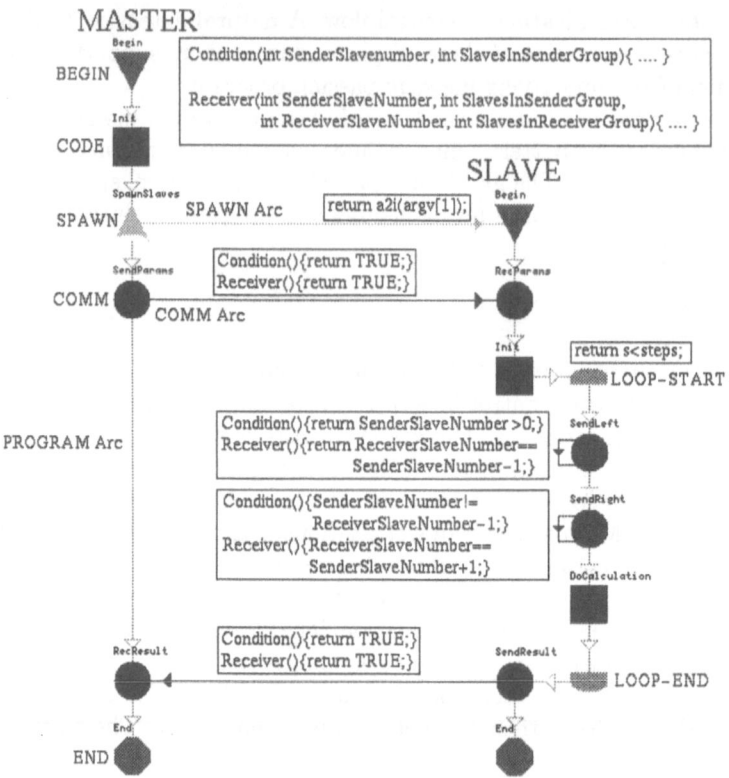

Fig. 1. PVMbuilder graph for the wave equation.

of slave processes to spawn is shown above the SPAWN arc. The COMM arcs connect COMM nodes and denote communication.

It is still necessary to bind program specific information to the appropriate nodes or arcs of a graph. Types, global variables, and functions are bound to BEGIN nodes. Sequential program code is bound to CODE nodes and the loop conditions are bound to LOOPSTART nodes. Figure 1 shows the loop condition above the LOOPSTART node. A runtime evaluable function, which determines the number of slaves to be spawned, is bound to every SPAWN arc.

Bound to every COMM arc are a guard (Condition()) to determine if a send should occur and a predicate (Receiver()) evaluated on every possible sender/receiver pair (COMM arcs in Fig.1 are annotated with these predicates). The data to be sent is described using our own data description language.

The sender and receiver part of the program must both be able to evaluate the two predicates bound to the COMM arc connecting them. The Receiver() predicate is used by the sender to determine if a message should be sent to a receiver, and by the receiver to determine if there is a message. This restriction ensures that all sends are matched with receives, thus preventing communication deadlocks. In [1] we prove that keeping the PVMbuilder graph acyclic is a sufficient condition to avoid several types of communication deadlocks.

The Condition() and Receiver() predicates must return the same result in both the sender and receiver process. Therefore, the predicates can only depend on their arguments and two globally defined variables (mySlaveNumber and myGroupSize). In Fig. 1 we have shown function prototypes for the two predicates. As previously mentioned, we first evaluate the Condition() predicate, and if satisfied, the Receiver() predicate is evaluated for each (SenderSlaveNumber, RecieverSlaveNumber) pair and a message sent whenever it is satisfied.

Once the graph has been constructed and program specific information bound to the nodes and the arcs, PVMbuilder can generate and compile source files. The source files are generated with respect to the control flow of the graph and the appropriate PVM communication calls are inserted. The programs are then compiled and can be executed either as a stand-alone PVM application or from within PVMbuilder. When compiled in PVMbuilder, compilation errors are matched to the line number in the CODE node that generated them.

PVMbuilder also includes a performance monitor that can be used to track and display idle time, computation time and communication time. In addition trace files can be generated. Once debugged and tuned, source code without monitoring calls can be generated.

3   Patterns

Many algorithms share the same underlying graph structure. This gives rise to the notion of program templates or patterns [6] where the same graph structure can be reused to solve different problems.

In PVMbuilder there are two types of patterns that can be reused, graph patterns and communication predicates. A graph pattern in PVMbuilder is simply a graph structure without the problem specific information bound to the nodes and arcs. Communication predicates implicitly define a communication topology (or pattern) which can often be reused. When a predefined communication pattern is chosen PVMbuilder simply inserts the appropriate pre-defined Condition() and Receiver() predicates. The example shown in Fig. 1 is an instance of a master-slave pattern with an underlying linear array communication topology among the slave processes.

4   Related Work

Two notable tools which influenced the design of PVMbuilder were CODE and HeNCE. However, unlike PVMbuilder, CODE and HeNCE are based on a data flow representation of the program and do not support explicit message-passing. Indeed, PVMbuilder was developed as an attempt to extend the data flow model to encompass message-passing. More recently, the authors of HeNCE and CODE have developed the VPE [8] system which also addresses the shortcomings of CODE and HeNCE by including message-passing.

PVMbuilder differs from the previous system in that communication is more abstract in the sense the user does not have to explicitly specify the send/receive primitives. As a result it is possible to check for and avoid communication dead-locks.

PVMbuilder includes both the data flow as the communication between processes as well as the control flow within each process. Therefore, it differs from purely process-based tools such as TRAPPER [4] and the higher level abstraction tools such as Enterprise [9] which represents the program as a template or pattern that abstracts away all of the underlying communication.

5 Conclusion

We have presented a new graphical user interface, PVMbuilder, for developing message-passing programs. It is an integrated environment that includes tools for program creation, reuse, tracing and performance monitoring.

The key feature of PVMbuilder is that programs are expressed in a manner close to how humans think of them: as graphs dealing with both code and data flow. The communication model of PVMbuilder, i.e. the ability to specify what to send and receive at the same time, is a significant improvement over the explicit specification of every send and receive.

In addition, we have shown in [1] that PVMbuilder introduces a relatively low overhead that we believe can be further reduced. And, since PVMbuilder generates code and executables that can be run without the presence of the tool, the number of potential users of the tool, or programs generated from it, is large. Although PVMbuilder generates C code and PVM communication calls it can easily be targeted to other similar languages and communication libraries. For more details see http://www.cs.ubc.ca/~matt/pvmbuilder.html

References

[1] B. B. Blendstrup and J. B. Pedersen. PVMbuilder - et grafisk værktøj til parallel programmering. Master's thesis, Aarhus Universitet, Jan. 1997.
[2] J. Dongarra. MPI: A message passing interface standard. *The International Journal of Supercomputers and High Performance Computing*, 8:165–184, 1994.
[3] A. Geist et al. *PVM: Parallel Virtual Machine. A User's Guide and Tutorial for Networked Parallel Computing*. Prentice Hall, 1994.
[4] F. Heinze et al. Trapper, eliminating performance bottlenecks in a parallel embedded application. *IEEE Concurrency*, pages 28–37, July-September 1997.
[5] J. Dongarra et al. *HeNCE: A Users' Guide. Version 2.0*, http://www.netlib.org/hence edition.
[6] E. Gamma, R. Helm, R. Johnson, and J. Vlissides. *Design Patterns - Elements of Reusable Object-Oriented Software*. Addison-Wesley, 1995.
[7] P. Newton and J. C. Browne. The CODE 2.0 graphical parallel programming language. *Proc. ACM Int. Conf. on Supercomputing*, July 1992.
[8] P. Newton and J. Dongerra. Overview of VPE: A visual environment for message-passing. 1994.
[9] A. Singh, J. Schaeffer, and M. Green. A template-based approach to the generation of distributed applications using a network of workstations. *IEEE Transactions on parallel and distributed systems*, 2(1):52–67, January 1991.

Message-Passing Specification in a *CORBA* Environment

T. Es-sqalli[1], E. Fleury[1], E. Dillon[2], and J. Guyard[1]

[1] LORIA, Campus Scientifique, BP 239
F-54506 VANDŒUVRE-LÈS-NANCY cedex France
[2] Mathematics and Computing Technology
The Boeing Company
P.O. Box 3707 MC 7L-20, sEATTLe, WA 98124-2207, USA

Abstract. Modern technologies including rapid increase in speed and availability of network of supercomputer provides the infrastructure necessary to develop distributed applications and making high performance global computing possible. The performance potential offered is enormous but generally is hard and tedious to obtain due to the complexity of developing applications in such distributed environment. In this paper, we introduce MeDLey which is a specification language of communications for parallel applications. Its aim is to provide to the user an abstract vision of his application in terms of tasks and exchanges between these tasks independently of material architecture and the means of communication used. Based on this specification, the MeDLey compiler generates several levels of implementation for various communication primitives. More precisely, we will detail in this article, a new approach of this language allowing the implementation of the communications based on ideas underlying the Common Object Request Broker Architecture *CORBA*.

1 Overview of MeDLey Language

The increase in speed and availability of network of supercomputers make possible the development of so called high performance "global computing", in which computational and data resources in the network are used to solve large scale problems. The performance potential offered is enormous but generally is hard and tedious to obtain due to the complexity of developing applications in such distributed environment which implies different transport mechanism, several software packages and heterogeneous resources.

To cure this problem, we propose a new language, called MeDLey (*Message Definition Language*) [2]. Its purpose is to describe and specify all exchanges between parallel application tasks. In a MeDLey specification, a distributed application is split into tasks: each MeDLey module is the specification of a task. A task specification in MeDLey mainly contains four parts as shows in figure 1.

Starting from this description, an automatic tool derives an efficient implementation of the exchanges on the communication support chosen by the user in the target language (currently C++). A first approach of MeDLey [3] has been

P. Amestoy et al. (Eds.): Euro-Par'99, LNCS 1685, pp. 113–116, 1999.
© Springer-Verlag Berlin Heidelberg 1999

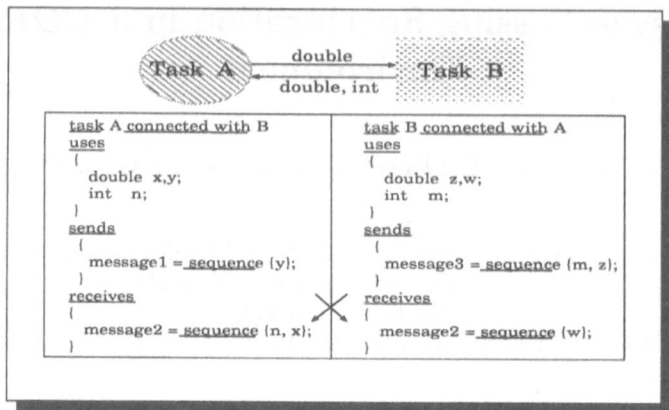

Fig. 1. Example of a MeDLey specification

developed and the implementation of communication was based on standard message passing libraries (*MPI* or *PVM*). It showed us, by experimentation [4], the feasibility and the interest of this language.

Another way to implement exchanges between parallel/distributed application tasks can be done by the modelling tasks by objects in a distributed object system. The communications between these tasks are done via method invocations on these objects. It's the basic idea of the new approach of MeDLey that we will present in this paper.

2   Using MeDLey in a *CORBA* Environment

Motivation   In *CORBA* [5], isolation between client and server, in terms of using network protocol, material infrastructure, programming language and data transport mechanisms, makes its great advantage. Moreover, *CORBA* allows real inter-working and coupling which is not the case of applications developed with more lower level tools/libraries. The success of *CORBA* also stimulates new investigations in its performance for high speed networks and these researches are related to real time applications [6].

Even if at the moment lower-level C and C++ implementations outperform the CORBA (Orbix 2.1 and VisiBroker) implementations, the authors pinpoint [1, 6] the sources of overhead for communication middleware in order to develop scalable and flexible CORBA implementations that can deliver gigabit data rates to applications.

We perform some preliminary tests before launching the development of the communication in MeDLey using *CORBA* and it seems that *CORBA* is able to obtain good results for data exchanges. We are not going to present all the extensive tests here but just a snapshot of the the comparison between two implementations of *CORBA*, called *Orbacus* [7] and *TAO* [6] and the two message-passing library *PVM* and *MPI* between two *Unix* stations connected through a 10 Mbit/s Ethernet networks. The figure 2(a) shows the results of measured data exchange time between two processors for different data sizes (classic ping-pong).

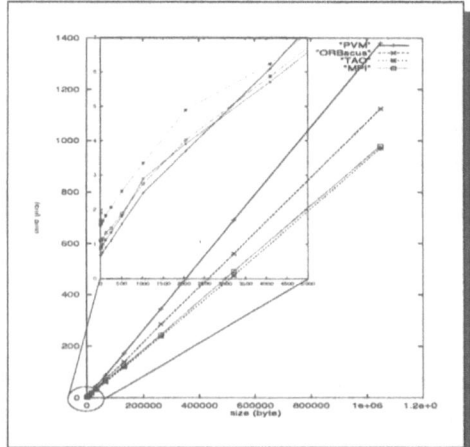

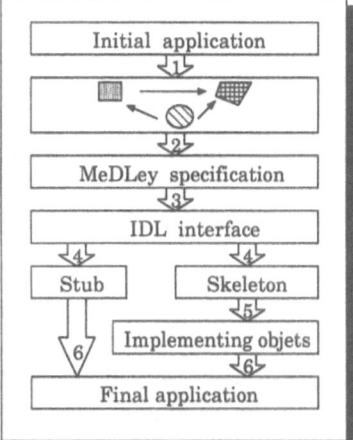

(a) Comparison of communication performances between *Orbacus*, *TAO*, *PVM* and *MPI*

(b) Construction steps of a `MeDLey` application in *CORBA* environment

Fig. 2.

Based on this experimental results, it seems that *CORBA* obtains good communication times. On the level of communication latency, ORBacus is comparable to *PVM* and better than *MPI*. On the other hand, for the great messages, *TAO* is more powerful compared to *PVM*, and slightly better compared to *MPI*.

Taking into account the performance of *CORBA* which is not as bad as the rumour usually claims, and its other advantages described before make *CORBA* to be a possible efficient environment for the development of parallel applications, in particular those using inter-task communications. However, if we consider the large amount of communication modes available and used in parallel applications but also the new intrinsic difficulty introduce by the use of *CORBA* to an unexperimented developer who is more familiar with *MPI* who need to learn and to be aware of how initialise the *ORB*, the OA (Object Adapter), searching the objects and so on..., this task can appear to be tedious and the developer may think it does not worth it. The main goal of the `MeDLey` approach we proposed is to bypass this problem.

Presentation   Our approach relies on modelling the communicating tasks of a parallel application by *CORBA* objects. The communication between these tasks is done via remote invocation methods on these objects. The *ORB* becomes the communication support used within this framework. This approach contains six major steps as described in more details below and it is illustrated by the figure 2(b):

1. structuring the application in communicating tasks;
2. description of the exchanges between these tasks in a MeDLey specification file;
3. mapping of the MeDLey specification to *IDL* interface: such a generation is based on the mapping of the MeDLey types used in the description to their corresponding types in the *IDL* language and on the generation of the communication methods signatures which corresponds to the messages declared in the MeDLey description;
4. mapping of the *IDL* interface to stubs and skeletons: MeDLey runs the *IDL* compiler in order to generate the stub and the skeleton;
5. implementation of the exchanges between tasks: it allows to provide the semantic of the objects which model the parallel application tasks. It is done by implementing the corresponding methods of the exchanges between these tasks;
6. writing the final application using the generated communication functions: at this step, the user holds the key of the parallel program. Indeed, all the necessary methods for the communication are generated. The remaining job, is to develop the computing part of the application, and to call the right communication functions at the right place.

3   Conclusions

This paper described the part of the MeDLey language concerning the specification of the communications for parallel applications in a *CORBA* environment allowing programmers to structure their applications in terms of communicating tasks. We have described the different construction steps of a parallel application based on our approach. The implementation of our approach using *TAO* as a *CORBA* implementation is in progress.

References

[1] J-L. Antoine and P. Chatonnay. Corba : un environnemt d'exécution pour now? *RENPAR10*, pages 149–152, 1998.
[2] E. Dillon. Medley : User's guid. Technical report, CRIN-CNRS/INRIA-Lorraine, February 1997.
[3] E. Dillon, C. Gamboa Dod Santos, and J. Guyard. Medley : An abstract approch to message passing. In *PARA96*, 1996.
[4] T. Es-sqalli, E. Dillon, and J. Guyard. Using medley to resolve the vlasov equation. In *HPCN Europe, Amsterdam, The Netherlands*, April 1999.
[5] J-M. Geib, C. Gransart, and Merle. P. *CORBA, des concepts à la pratique*. InterEditions, 1997.
[6] A. Gokhale, I. Pyarali, C. O'Ryan, V. Kachroo, A. Arulanthu, N. Wang, and D. Schmidt. Design considerations and performance optimizations for real-time ORBs. In *5th USENIX Conference on Object-Oriented Technologies and Systems*, San Diego, CA, May 1999.
[7] OOC. *ORBacus for C++ and Java*. http://www.ooc.com/. Object-Oriented Concepts. Inc, 1997.

Using Preemptive Thread Migration
to Load-Balance Data-Parallel Applications

Gabriel Antoniu and Christian Perez

LIP, ENS Lyon, 46, Allée d'Italie, 69364 Lyon Cedex 07, France
{Gabriel.Antoniu, Christian.Perez}@ens-lyon.fr

Abstract. *Generic* load balancing policies for irregular parallel applications may be efficiently implemented by integrating *preemptive* thread migration into the runtime support. In this context, a delicate issue is to manage pointer validity in a migration-safe way. In [1] we presented an *iso-address* approach to this problem. This paper discusses the impact of the iso-address migration scheme on the runtime of the Adaptor [4] HPF compiler. This runtime (previously modified so as to generate multithreaded code for our PM2 runtime system [3]) now provides a generic support for *dynamic* load balancing, using preemptive thread migration. We report some encouraging results obtained with our system on a HPF flame simulation code, a motivating application of HPF 2.0 [7].

1 Introduction

Load balancing is a key issue for high performance parallel distributed computing. When the application behavior is hardly predictable at compile time, *dynamic* load balancing becomes essential. This can be achieved either by redistributing the data, or by migrating the activities across the available nodes. The former method may be very efficient, but it is generally highly dependent on the application. Besides, data redistribution code has to be explicitly inserted into the application code, at well-defined points so that a good tradeoff between responsiveness to load imbalance and overhead is difficult to obtain. The latter method detects load imbalance and responds by migrating activities from the overloaded nodes to the underloaded ones, provided activity migration is preemptive. In this approach, generic balancing policies may be implemented *outside* the application code and transparently applied to the application activities. This approach may be less efficient than data redistribution if activities are implemented as regular, heavy processes. In this context, threads have proven adequate as a means to accomplish fine-grained load balancing thanks to their ability to be efficiently migrated from one processor to another.

The framework of our study is data-parallel compiling for distributed-memory systems. In the approach we present in [3], the code dedicated to load balancing is separated from the HPF [8] application code. HPF abstract processors are projected onto threads. The load is balanced by transparent thread migration. To achieve this, *preemptive* migration is necessary. However, the migrating threads

P. Amestoy et al. (Eds.): Euro-Par'99, LNCS 1685, pp. 117–124, 1999.

may execute arbitrary codes, in particular they may handle pointers to stack data or to allocated heap data. This is a constraint with respect to preemptive migration, since these pointers may become invalid when the threads get moved to another node. In [1] we propose a solution to this problem, based on an *iso-address* allocation scheme which guarantees that all pointers are migration-safe. This allocator was integrated into our PM2 [12] multithreaded environment, which serves as a runtime support for the Adaptor HPF compiling system [3]. To benefit from the iso-address approach, we have modified the compiler runtime and then tested the whole compiling system using a flame simulation code [7].

The remainder of this paper is organized as follows: in Section 2 we give a brief description of our iso-address allocation scheme. Section 3 shows how the Adaptor HPF compiler runtime was modified to integrate the iso-address allocator. Section 4 describes our experiments with the flame simulation code and presents some performance figures. Section 5 offers a few concluding remarks.

Related Works

Thread migration is provided by several multithreaded systems, the underlying motivations being different from case to case. In Ariadne [10], thread migration is the only way to access remote data. In Millipede [9], it is used to balance the load. In both systems, data are shared. Our study is concerned with a different case, in which data are not shared by multiple threads: each piece of data belongs to some unique thread and has to follow it on migration.

Migrating threads dealing with pointers is a common problem for all systems providing thread migration [6, 9, 10, 15]. Nevertheless, the solutions proposed are not satisfactory with respect to both efficiency and flexibility. In Ariadne [10], thread stacks are relocated at a usually different address on the destination node, such that pointers need to be updated. As shown in [1], several important problems cannot be solved by this approach. In Millipede [9], threads and their data are always relocated at the same virtual addresses on all nodes. Yet, thread creation is expensive, therefore the number of concurrent threads is statically fixed at initialization. In UPVM [5], threads are provided with private heaps whose sizes are fixed at thread creation, so the amount of data that a thread can allocate is limited. Also, UPVM thread creation is expensive, since it is requires a global synchronization. The approach we propose in [1] avoids the static limitations while keeping efficient and provides an appropriate support for building migration-based strategies for dynamic load balancing (see Section 4).

2 Iso-Address Allocation: A Sufficient Condition for Preemptive Migration

Preemptive migration relies on the ability to interrupt a running thread *at an arbitrary point of its instruction stream* and to move its resources (i.e. its stack, descriptor and private data) to another node, where it can resume its execution. A difficulty arises as soon as the migrating thread uses pointers to access data stored in its stack or in the heap of the host process. Two kind of pointers are concerned: *user-level* pointers, explicitly declared in the applications, and *compiler-level* pointers, implicitly generated by the compiler. For instance, frame pointers, used to link stack contexts, belong to this latter class, as well as auxiliary pointers, generated in order to avoid indirections when accessing data. Moreover, pointers may be present in registers when the thread gets interrupted. The migration mechanism must ensure the validity of all these classes of pointers and must guarantee their safe use after migration.

2.1   The Iso-Address Approach

As shown in [1], the best approach to the problem is to design a mechanism which ensures that a thread and its data can always migrate while keeping the same virtual addresses. Our iso-address memory allocator (called `isomalloc`) guarantees that each virtual address range allocated on a node is kept free on any other node, such that iso-address migrations generate no overwriting. After such an operation, the thread may resume its execution without any post-migration processing, since all pointers remain valid.

A simple, but expensive way to ensure this global consistency would be to reserve a virtual address range on all nodes each time an allocation is performed, by means of a global synchronization. A more efficient strategy is proposed by `isomalloc`, allowing each node to locally allocate memory within a set of reserved address ranges without informing the other nodes. A special area in the virtual address space is dedicated to iso-address allocations on all nodes (`iso-address area`). This area is divided into a series of fixed-size ranges called *slots*. Each slot is assigned to a single node and, conversely, each node may *locally allocate* memory into its *globally reserved* slots without informing the other nodes.

2.2   Managing the Iso-Address Area

At any moment, a slot belongs either to a node, or to a thread. Only the slot owner is entitled to use the slot for memory allocation/release (internally carried out through calls to the `mmap`/`munmap` primitives) and to access it in read or write mode. This means that the data stored in a given slot belongs to a unique owner, since *slots are not shared*. At application startup, each slot is assigned to some node according to some arbitrary, user-specified distribution. When a thread gets created, it "buys" a slot from the local node, to store its stack. If the thread dynamically allocates iso-address memory, it gets other slots from the local node. On migration, all these slots automatically migrate along with the thread. Finally, when the thread releases its memory, the corresponding slots are given to the current local node, which may be different from their initial owner.

A particular problem arises when a thread allocates a block larger than the largest iso-address region (i.e. set of contiguous slots) owned by the local node. In this case, the latter may "buy" slots from the other nodes by means of a global negotiation. This operation is obviously expensive and should not occur too frequently. Therefore, a "smart" initial slot distribution (which may be application-dependent) and an appropriate slot size should be chosen, in order to obtain an efficient execution. In our implementation, the slot size is fixed to 16 memory pages, which is large enough to contain the initial resources of a thread (i.e. its stack and its descriptor). Consequently, thread creation is a local operation irrespective of the slot distribution, since a single slot is required for the thread resources.

2.3 Using `isomalloc`

The PM2 programming interface provides two primitives for iso-address allocation/release operations `pm2_isomalloc` and `pm2_isofree`, with the same prototype as the `malloc` and `free` primitives.

Threads should call `pm2_isomalloc` instead of `malloc` to allocate memory for data which must follow the thread on migration. Any area allocated through `pm2_isomalloc` should be released via `pm2_isofree`. These primitives guarantee that the references to the corresponding memory areas are kept valid in case the thread migrates to another node. Consequently, threads may use pointers and still be preemptively migrated. Notice that `pm2_isomalloc` and `malloc` are not incompatible: the `malloc` primitive may still be used to allocate memory for non-migratable data.

3 Impact on the Runtime of a Data-Parallel Compiler

3.1 Integrating `isomalloc` in a HPF Compiler Runtime

As explained in Section 1, we use PM2 as a runtime support for the Adaptor [4] HPF compiler. The compiler runtime was previously modified in order to generate multithreaded code for PM2 [3]: each HPF abstract processor is mapped onto a thread. We have updated the runtime of Adaptor 4.0a, in order to take advantage of the iso-address allocator.

In the runtime version without `isomalloc`, an abstract processor had to explicitly call the migration function when it decided to migrate. The main objective of this function is to pack all data having to follow the migrating abstract processor: *dynamically* allocated data and static data specific to the Adaptor runtime. When arriving at the new node, a symmetric unpack function is invoked. Then, in order to build the environment of the abstract processor as it was on the original node, many calls to the `malloc` function are needed.

In the runtime version with `pm2_isomalloc`, only the static data need to be packed and unpacked. All memory areas dynamically allocated by `pm2_isomalloc` are automatically migrated with the abstract processor. The iso-address allocator has thus allowed a simplification of the runtime, while the migration time for abstract processors remains basically the same.

Preemptive migration has introduced critical sections in the program. The critical sections are located in the communication part of the runtime because that part of the code accesses shared data, such as the list of waiting messages for a thread.

A last point related to preemptive migration is the way the location of an abstract processor is managed. If an abstract processor $ap1$ located on process $p1$ needs to send a message to an abstract processor $ap2$, it has to know on which process $p2$ that abstract processor is located. Our implementation does not need accurate knowledge about abstract processor location. Every message finds its addressee thanks to a message forwarding mechanism. Optimization techniques, such as regular broadcast of abstract processor localization, can be used to avoid a long chain of forwarding.

4 Application Study: A Flame Simulation Code

4.1 Benchmark Description

To evaluate the benefits of preemptive abstract processor migration, we have chosen the flame simulation code (illustrated in Figure 1), which is one of the motivating applications of HPF 2 [7]. The code performs a detailed time-dependent, multi-dimensional simulation of hydrocarbon flames in two phases. The first phase requires communications between neighbor mesh points. The computation load is the same in each point of the mesh. The second phase does not involve communications but the computational cost of each point varies as the computation progresses.

```
!lb$ begin load balancing with work stealing algorithm
     do time = 1, timesteps
C        convection phase
         x(2:NX-1,2:NY-1) = x(2:NX-1,2:NY-1) + F(z(2:NX-1,2:NY-2),
     $         y(1,NX-2),y(3,NX-1),y(2:NX-1,1:NY-2),y(2:NX-1,3:NY))
         y=x
C        Reaction phase
         forall (i=1:NX, j=1:NY) z(i,j) = Adaptive_Solver(x(i,j))
     end do
!lb$ end load balancing
```

Fig. 1. Flame simulation kernel code.

This application is interesting because the two phases have different requirements. The first phase is well suited for a block distribution (it requires neighborhood communication and the computations are regular), whereas the second phase needs an irregular distribution of the data.

4.2 Load Balancing Issues

The block data distribution minimizes communications in the first phase. But, in the second phase, the application suffers the effects of the load imbalance.

The cyclic data distribution has opposite effects compared to the block distribution. The load is quite well balanced in the second phase, but the communications are expensive in the first phase, given that each node, at each iteration, has to send all its data to its 2 (resp. 4) neighboring nodes if one (resp. two) dimension(s) is (are) cyclicly distributed. A more important drawback is the huge waste of memory. To receive data from the neighboring nodes, a node has to allocate 2 (resp. 4) arrays as large as the main array. So, only 33 % (resp. 20 %) of the memory available on each node can be used for local data.

Our approach consists in managing the mapping of HPF abstract processors onto the nodes. Using a block distribution and a regular mapping of abstract processors, best performance can be obtained for the first phase. To handle load imbalance in the second phase, we only have to migrate abstract processor from overloaded nodes to underloaded nodes. Since the load distribution is unknown, abstract processors are migrated according to a work stealing algorithm [2].

4.3   Performance Results

The experiments have been made on a 4-node cluster of 200 MHz PentiumPro PC connected by a Myrinet network [11] and using PM2 on top of the MPI-BIP communication library. The latency in this configuration is 25 μs and the bandwith is 120 MB/s.

We have considered three data sets, which differ in their degree of irregularity for the reaction phase. They range from a regular to a highly irregular computational pattern. For each data set, we first present the execution times obtained with the BLOCK and CYCLIC distributions, using the original Adaptor compiler. Then, we present the times obtained by BLOCK-distributing the data among 64 abstract processors on the 4 nodes, without and with load balancing.

Mode	Regular	Medium irregularity	High irregularity
block	3.69	5.34	9.51
cyclic	5.37	4.34	5.34
block + threads	3.36	3.10	5.32
block + threads + work stealing	3.57	2.85	4.45

Table 1. Time in seconds for different distribution on the flame simulation benchmark on four processors. The grid is 1024x1024 elements.

The experimental results are displayed on Figure 1. We can see that the cyclic distribution incurs severe overhead due to communication in all cases. The block distribution performs well when the computational cost is regular but its performance drops when it becomes irregular. Introducing several abstract processors per node allows communication to be overlapped by computation. It leads to better performance for the block distribution. Load balancing the application by migrating abstract processor according to a work stealing algorithm leads to good performance in all circumstances. When the load is balanced, this strategy does not generate a significant overhead. When the load is not balanced, abstract processor migration improves performances.

4.4 Discussion

Orlando and Perego [14] have studied this application in depth. They hand-coded on top of MPI different solutions. They found that the best performance is obtained when using dynamic load balancing techniques through the SUPPLE support [13]. SUPPLE is a run-time support for the implementation of uniform and nonuniform parallel loops. It adopts a static block distribution but, during the execution, chunks of iterations and associated data may be dynamically migrated towards underloaded processors. Migration is decided at run-time with a work stealing algorithm.

Our approach for this application is very similar. However, there are two main differences. First, our work is fully integrated into an existing HPF compiler, as detailed in [3]. Second, we have separated the execution of the HPF abstract processors from their location. The key point to achieve this is that we embed an HPF abstract processor into a migrantable thread. Thus, we do not have to care about the scheduling of computation and communication, as in the the SUPPLE support. Our approach is not limited to a specific kind of loop. So, we can claim that we have obtained an HPF compiler and runtime that can apply work stealing techniques to *any* HPF program. Nevertheless, the ability to dynamically move abstract processors makes us think that any load balancing strategy could be implemented in our framework.

5 Conclusion and Future Work

This paper studies how preemptive thread migration can be used to load-balance data-parallel applications. The major problem of preemptive thread migration lies in the management of pointers. Our iso-address allocator provides a simple and efficient solution to this problem. Its integration into a multithreaded version of the Adaptor runtime has been easy and has simplified the runtime. Last, the cost of abstract processor migration has not been significantly modified by this new functionnality. To validate the usefulness of preemptive abstract processor migration, we have benchmarked a flame simulation kernel code, which is part of the motivating applications of HPF-2. Previous research [14] working on this application with MPI-based hand-coded program has shown that the cyclic distribution may have good results. Since our solution based on a work stealing strategy leads to globally better performance than both the block and the cyclic distributions, we can conclude that preemptive thread migration is an efficient way to load-balance this kind of HPF applications.

The future developments of this work have two main directions. On one hand, we intend to study the relationship between HPF programs and load balancing strategies. Now that our modified version of Adaptor 6 is operational, we can test HPF 2 distributions like the CYCLIC(N) or the INDIRECT distributions. On the another hand, it seems interesting to study the links between the iso-address allocator and distributed shared memory systems (DSM). The iso-address allocator seems to offer the basic functions to build such a system.

References

[1] G. Antoniu, L. Bougé, and R. Namyst. An efficient and transparent thread migration scheme in the PM2 runtime system. In *Proc. 3rd Workshop on Runtime Systems for Parallel Programming (RTSPP '99)*, Lect. Notes in Computer Science, San Juan, Puerto Rico, April 1999. Springer-Verlag. Workshop held as part of IPPS/SPDP 1999, IEEE/ACM.

[2] Robert D. Blumofe and Charles E. Leiserson. Scheduling multithreaded computations by work stealing. In *35th Annual Symposium on Foundations of Computer Science (FOCS '94)*, pages 356–368, Santa Fe, New Mexico, November 1994.

[3] L. Bougé, P. Hatcher, R. Namyst, and C. Perez. Multithreaded code generation for a HPF data-parallel compiler. In *Proc. 1998 Int. Conf. Parallel Architectures and Compilation Techniques (PACT'98)*, ENST, Paris, France, October 1998.

[4] Th. Brandes. *Adaptor*. Available at http://www.gmd.de/SCAI/lab/adaptor/adaptor_home.html.

[5] J. Casas, R. Konuru, S. W. Otto, R. Prouty, and J. Walpole. Adaptive load migration systems for PVM. In *Proc. Supercomputing '94*, pages 390–399, Washington, D. C., November 1994.

[6] D. Cronk, M. Haines, and P. Mehrotra. Thread migration in the presence of pointers. In *Proc. Mini-track on Multithreaded Systems, 30th Intl Conf. on System Sciences*, Hawaii, January 1997.

[7] HPF Forum. HPF-2 Scope of activities and motivating applications, November 1994. Ver. 0.8.

[8] HPF Forum. *High Performance Fortran Language Specification*. Rice University, Texas, Oct. 1996. Version 2.0.

[9] A. Itzkovitz, A. Schuster, and L. Shalev. Thread migration and its application in distributed shared memory systems. *J. Systems and Software*, 42(1):71–87, July 1998.

[10] E. Mascarenhas and V. Rego. Ariadne: Architecture of a portable threads system supporting mobile processes. *Software: Practice & Experience*, 26(3):327–356, March 1996.

[11] Myricom. Myrinet link and routing specification. Available at http://www.myri.com/myricom/document.html, 1995.

[12] R. Namyst and J.-F. Méhaut. PM2: Parallel multithreaded machine. a computing environment for distributed architectures. In *Parallel Computing (ParCo '95)*, pages 279–285. Elsevier Science Publishers, September 1995.

[13] S. Orlando and R. Perego. SUPPLE: an efficient run-time support for non-uniform parallel loops. Research Report CS-96-17, Dip. di Matematica Applicata ed Informatica, Universita Ca'Foscari di Venezia, December 1996.

[14] S. Orlando and R. Perego. A comparison of implemantation strategies for nonuniform data-parallel computations. *J. Parallel Distrib. Comp.*, 52(2):132–149, March 1998.

[15] B. Weissman, B. Gomes, J. W. Quittek, and M. Holtkamp. Efficient fine-grain thread migration with Active Threads. In *Proceedings of IPPS/SPDP 1998*, Orlando, Florida, March 1998.

FITS—A Light-Weight Integrated Programming Environment[*]

B. Chapman[1], F. Bodin[2], L. Hill[3], J. Merlin[4], G. Viland[3], and F. Wollenweber[5]

[1] Department of Electronics and Computer Science, University of Southampton,
Highfield, Southampton SO17 1BJ, U.K.
bmc@ecs.soton.ac.uk
[2] IRISA, Campus Universitaire de Beaulieu, 35042 Rennes, France
[3] Simulog Sophia Antipolis, Les Taissounieres HB2, Route des Dolines,
06560 Valbonne, France
[4] VCPC, University of Vienna, Liechtensteinstr. 22, 1090 Vienna, Austria
[5] Eumetsat, Am Kavalleriesand 31, 64205 Darmstadt, Germany

Abstract. Few portable programming environments exist to support
the labour-intensive process of application development for parallel systems. Popular stand-alone tools for analysis, restructuring, debugging
and performance optimisation have not been successfully combined to
create integrated development environments.
In the ESPRIT project FITS we have created such a toolset, based
upon commercial and research tools, for parallel application development. Component tools are loosely coupled; with little modification, they
may invoke functions from other components. Thus integration comes at
minimal cost to the vendor, who retains vital product independence. The
FITS interface is publically available and the toolset is easily extensible.

1 Introduction

Integrated software development environments are popular. It is easier to use a
suite of functions which have been crafted to complement each other, than to
learn how to apply a set of individual tools, possibly with different conventions
and terminology and overlapping functionality; moreover, the developer may
then access different kinds of information on a program concurrently.

There have been a number of efforts to build programming environments for
the creation and maintenance of parallel programs. However, the only toolsets
that have been truly successful are those which have been constructed by a single
vendor; these tend to run on a restricted range of platforms or are limited in their
scope. Whilst some individual development tools (e.g. TotalView) have been
remarkably successful, efforts to provide environments based upon a collection
of such tools have not been so. Indeed, the only levels of "integration" that have
been adopted in practice are interaction at the operating system level, e.g. via

---

[*] The work described in this paper was supported by the European Union ESPRIT
project 23502, 'FITS'.

P. Amestoy et al. (Eds.): Euro-Par'99, LNCS 1685, pp. 125–134, 1999.

Unix files, and incorporation within a simple graphical user interface. Thus a tool may benefit from the output of another tool, but no further interaction is implied.

There are sound reasons for this. The complexity of each part of the parallelisation process has led to the specialisation of tools (and their suppliers), which usually handle just one aspect of the parallelisation problem. Thus each potential component of a toolset is developed in isolation from other tools. Each tool has its own internal data structures and programming conventions. The vendors are unlikely to share details of these, let alone adapt existing structures and productisation plans to fit in with those of another product.

In the *Fortran Integrated ToolSet* ('*FITS*') project we have tried to learn from the lessons of the past regarding tool integration. This ESPRIT-funded project has created a parallel programming environment incorporating two existing, separately developed and marketed products, and includes functionality from two research prototypes. In order to permit interoperability, we have sought a light-weight solution which avoids any major re-engineering of the component tools and does not require complex interfaces. As a consequence, continuing product independence is also assured. The result is the FITS toolset, an extensible, portable programming environment for the development and maintenance of parallel applications written in Fortran.

This paper is organised as follows. First we describe the tasks involved in a typical parallelisation project. Then we introduce the components of the FITS toolset and show how they support this process. We outline the functionality of the integrated toolset and give a few examples of the benefits of the FITS tool interaction before describing the mechanisms employed to realise it. We conclude with a summary of related work and future plans.

2   The Manual Parallelisation Process

Coarse-grain parallelisation is a particularly demanding task in terms of the manual labour it involves. A suitable global parallelisation strategy must be devised and implemented, requiring a good understanding of the entire code. Tuning is an intricate process which must be performed separately for distinct architectures, implying that several different parallel program versions may be created and require maintenance. The main activities in the program parallelisation cycle are as follows:

1. **Analysis:** The code is first statically and dynamically analysed in depth in order to fully understand its data usage, to identify computationally intensive regions, and to select a parallelisation strategy.
2. **Restructuring:** The code is cleaned and restructured in preparation for parallelisation and to facilitate subsequent program maintenance. Non-standard constructs are removed to achieve a portable version.
3. **Debugging / verification:** It is debugged and its correctness is verified. The result is now suitable for use as a baseline version for parallelisation.

4. **Parallelisation:** Results of the program analysis and data locality information are used to parallelise the program, either explicitly under the message-passing paradigm or by inserting directives if a higher-level paradigm such as HPF or OpenMP is used.
5. **Debugging / verification:** The parallel program version is debugged and its correctness established.
6. **Tuning:** The performance optimisation cycle begins. The application is run several times on the target architecture. Results are visualised by a performance tool, communication hot spots identified and the corresponding source code re-examined. It may need modification to reduce or re-organise communication. Transformations are applied to optimise the single node performance, in particular to better utilise the memory hierarchy. This process terminates when results are satisfactory.
7. **Maintenance:** The verified parallel version becomes the main production version, and program maintenance becomes the dominant activity.

Many of the individual tasks in the migration process are conceptually simple, but daunting in view of the size of programs and the number of details to be considered. For example, coarse-grain parallelisation, like other large-scale program modifications, requires evaluation of the data and control flow throughout the entire program. Yet a program such as the global weather forecast model IFS [2], developed and used in production mode on a parallel system at the European Centre for Medium-Range Weather Forecasting (ECMWF), consists of over 300,000 lines of code spread throughout many files; there are some 1,400 subroutines. Instances of Fortran storage and sequence association must be identified and, if an impediment to parallelisation, removed. This involves comparing actual and formal arguments at call sites throughout the program, comparing the declarations of common blocks in the entire code, and searching for other forms of implicit and explicit equivalencing of data. But IFS has over 300 different common blocks, involving a large number of variables. Some subroutines have hundreds of arguments in their interface. Codes such as this have often been highly vectorised and may contain proprietary language extensions. Re-engineering of data structures and loops may be needed in order to achieve good performance on a parallel system.

An estimate of the effort required for each of the major steps in the parallelisation of IFS is in general agreement with an early user survey [11]. At ECMWF, about 25% of the effort went into code analysis, and another 25% into cleaning and restructuring the baseline code, to remove vendor-specific code features, reorganise control flow and data structures, and convert it to Fortran 90. A further 20% went into the initial parallelisation. Finally, debugging and performance tuning each took about 15% of the overall effort. Our estimates suggest that the fraction of this effort that is supported by a typical vendor-supplied toolset, likely to consist of a compiler, runtime system, debugger and some kind of performance tool (possibly no more than a profiler), is no more than 30%; in particular, it is unlikely to be of use in the significant initial program analysis and restructuring phases.

3   The FITS Toolset

The FITS project has created an extensible toolset facilitating the development of parallel applications. The tools it contains support almost all of the major activities in parallel application development, as outlined above.

We now describe the main components of the FITS toolset, and conclude this section with an overview of the mechanisms used to integrate them.

3.1   FITS Components

The following tools have been input to the development of the FITS environment:

FORESYS: a Fortran source code restructurer, providing source code display, checking, analysis and transformation, including conversion to Fortran 95.

TSF: a module for the user-driven application of source code transformations.

VAMPIR: a trace generator and performance analysis tool.

ANALYST: a research prototype interactive Fortran program analyser with a sophisticated graphical user interface.

In addition, ideas from another tool, **IDA**, were used to design GIM's displays.

FORESYS [13], from Simulog, is itself an integrated collection of tools to check, analyse and restructure Fortran programs. It accepts Fortran with many extensions, and can transform it into standard-conforming Fortran 77 or Fortran 95 and ensure that it meets quality assurance standards. It 'pretty prints' the source code, using colour and a variety of fonts to make the code structure clearer. It includes an editor, thus permitting code modifications from within the system. It performs a variety of interprocedural checks, and can also display data dependence graphs. This tool is thus particularly useful in steps *1, 2, 4* and *7* of the development process outlined in Section 2.

The TSF module [3] is an extension to FORESYS that provides a set of source code transformations for performance enhancement. Examples are loop transformations such as loop blocking, unrolling, interchanging, etc. These are applied under the user's control, usually in conjunction with steps *2* or *6* in the above description. TSF also provides a scripting mechanism that allows users to write their own transformations and to access program information stored in the FORESYS system. This allows the automation of many of the repetitive tasks that arise at various stages of code parallelisation.

VAMPIR [10], from Pallas, is a tool which supports the program optimisation step by visualising its execution behaviour. It does so by displaying information obtained by tracing a program run. It can gather and present a range of information, including the execution of message-passing constructs, collective communication and procedure calls, and it can present this as system snapshots or animated sequences. It has many measurement options and a range of graphical displays. The user may zoom in on arbitrarily small time intervals of the trace.

ANALYST [5, 6] is an interactive Fortran program analyser. It allows the user to browse through a Fortran code and request various kinds of information about it. Information is presented in both graphical and text formats. The displays are themselves interactive: thus a user may click on an arc of a callgraph to obtain details of the subroutine invocations it represents, or may access the source text of a program unit by clicking on the corresponding node or its name in a list of subprograms. Filter options enable the extraction of data from large graphs. Displayed information is consistently related to the corresponding region of source code; program information is thus provided together with the relevant source text. ANALYST formed the basis for developing the FITS Graphical Interface Module, GIM.

IDA [8] is a command-line-driven tool for interactive program analysis. It is very easy to use and, where its functions are sufficient for the task at hand, is thus a particularly convenient tool. In particular, it can provide a quick textual display of a program's callgraph. It has inspired the provision of similar textual representations of some information in the FITS toolset.

Figure 1 shows how these components are linked together.

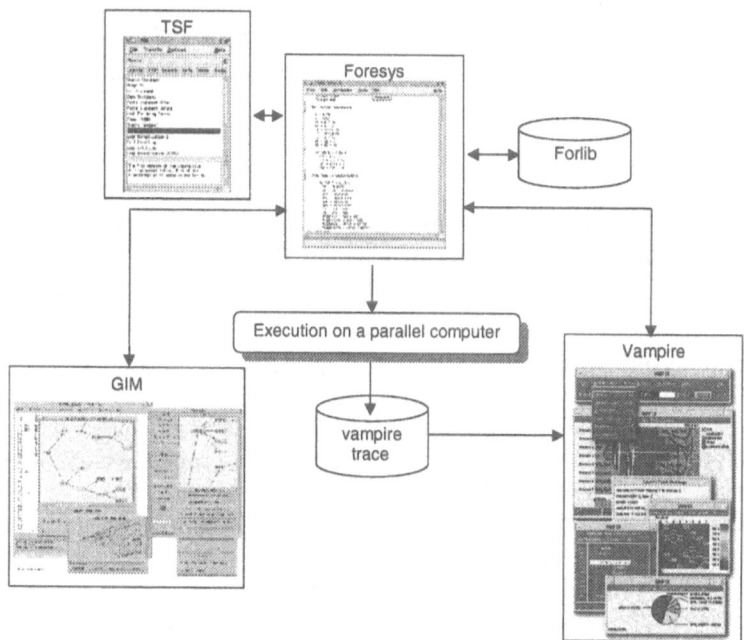

Fig. 1. Overview of the FITS toolset components.

3.2   FITS' Capablilities

As shown in Figure 2, the FITS toolset contains functions to support the majority of tasks in the parallelisation process. It is portable and extensible, permitting

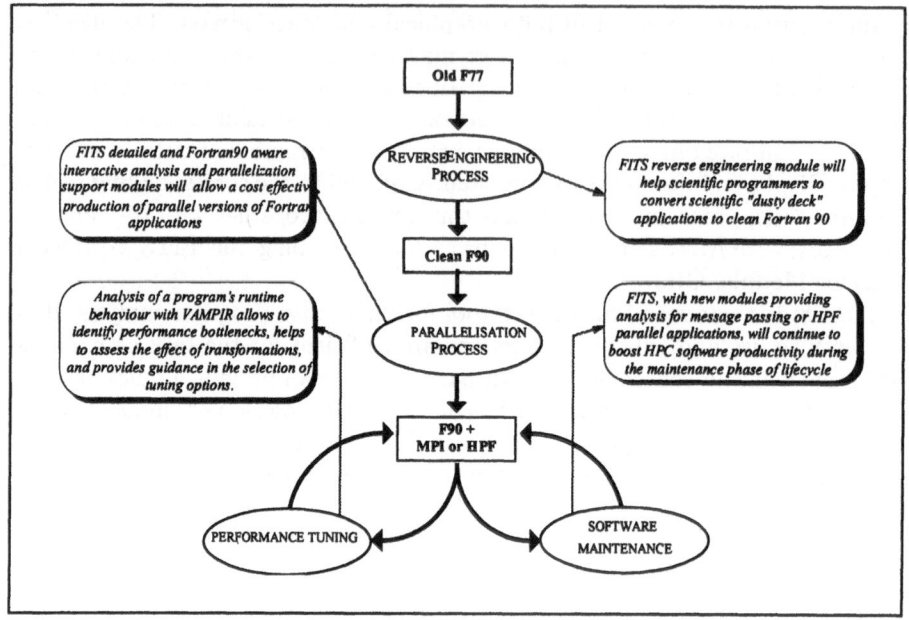

Fig. 2. Application life cycle.

the inclusion of additional tools. Although not exclusively developed for MPI [9], it provides specific functionality to support the development of MPI code.

It provides the following functions to support the parallelisation activities described in Section 2:

Static and dynamic program analysis   (for steps *1* and *6*):
  - Graphical displays of the source program structure, for example its callgraph, USEgraph, and data and control flow. These displays may be manipulated to isolate features of interest, and a graphical element may be linked to a display of the corresponding source text.
  - Display of the data dependence graph for a code region, and details of the dependence relationships it represents.
  - 'Pretty printing' of the source program.
  - Support for analysing a program's runtime behaviour, with the ability to view a source code statement or region together with a display of its execution behaviour.

Code transformations   (for steps *2, 4, 6* and *7*):
  - Code cleaning (e.g. GOTO elimination, etc.) and translation to Fortran 95.
  - Support for inserting MPI constructs into a code.
  - Automatic insertion of instrumentation code to generate a VAMPIR tracefile.
  - An *extensible* set of user-applied transformations to improve the performance of a parallel or sequential program, for example to tune it for the memory hierarchy.

Verification   (for steps *3* and *5*):
- Checking the consistency of parameter passing and common block declarations.
- Static analysis of MPI message-passing in a parallel application.
- Fortran dialect checking.

The combined toolset extends the functionality of the individual tools in several ways. For example, FORESYS can automatically instrument a program so that it generates a tracefile for display by VAMPIR. VAMPIR can then not only visualise the execution behaviour, but it can also invoke FORESYS to display the corresponding location in the source text and/or GIM to display the relevant node in the callgraph, and vice versa. As another example, TSF uses FORESYS to implement program transformations, and these transformations may be invoked from a GIM display.

3.3   Tool Integration Mechanisms

At the heart of the toolset is a database created by FORESYS, called a *For-Lib*. This contains detailed information about the source program, and it also keeps track of other files related to the program such as Makefiles and VAMPIR tracefiles.

Two types of integration are employed within FITS, depending on which tools are coupled.

The displays produced by GIM and the code transformations produced by TSF require a large amount of information to be exchanged with the *ForLib*. Therefore GIM and TSF are *tightly integrated* with FORESYS so that they have a fast and efficient connection to the *ForLib* database.

The coupling between FORESYS and VAMPIR is of a different nature. In this case the tools do not exchange information, but rather each tool can *remotely control* the other. For example, VAMPIR can request that FORESYS provides a source code display, and FORESYS can request that VAMPIR displays some execution events or statistics. Therefore these tools are coupled using a *lightweight integration* mechanism.

The latter mechanism has been designed for ease of use and extensibility. Although it does not provide the close interaction that is possible when a set of tools is designed and constructed by a single organisation, it does permit a variety of useful interactions between them and enhances the functionality of the individual components accordingly. This approach does not require any complex interfaces between the component systems and hence eliminates the need for close coordination of their development.

The general architecture of the FITS integration scheme is shown in Figure 3. Its major components are as follows:

The fits_notify utility: This utility program contains functions for remotely controlling tools in the FITS environment. Each participating tool can invoke fits_notify, e.g. by calling system(), to cause another tool to perform an

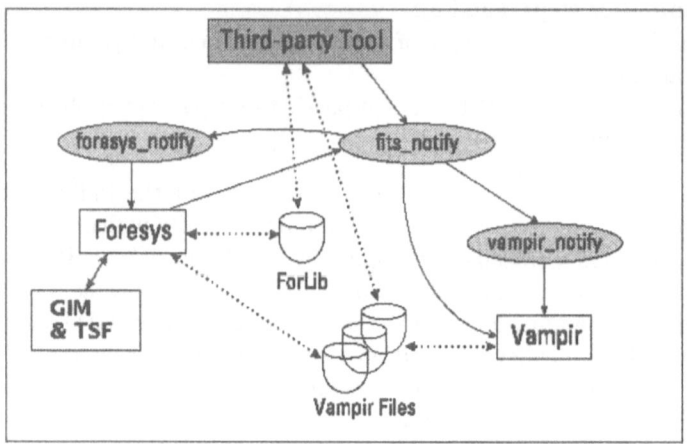

Fig. 3. FITS integration scheme.

action. The protocol is asynchronous; no reply is sent and the calling tool should not depend on the action being executed immediately. The syntax of the call permits the kind of remote action to be specified, thereby implicitly also selecting the target tool, and enables arguments to be given, including such things as a location within a program or trace file.

A set of command-line routines: These enable a tool to interact with the *ForLib*, the database containing source program information.

Interactive routines: These invoke FORESYS functions in order to obtain graphical displays of the source program.

Specified file formats: For example, a specification of the VAMPIR tracefile and configuration file formats.

The realisation depends strongly on the *ForLib* structure. Among the *ForLib*'s publically accessible attributes are: its *name*, which identifies a specific version of an application development; the list of *files* and *include files* comprising the program, with their directory paths; the list of Fortran *program units*; the Fortran *dialect* in which the program is written; a list of *program analysis options* to be applied; and the location of *files related to the program* such as Makefiles and VAMPIR tracefiles.

Operations may be run in batch mode and produce non-graphical output, mainly in text or HTML, or they may be interactive operations that provide graphical output via the displays of FORESYS and GIM. The batch commands enable a user or tool to obtain all public information on the contents of a *ForLib*, and permit its modification, e.g. by adding or removing Fortran files from it. Command-line routines are available to create a *ForLib* and to specify analysis and transformations to be applied to it. The interactive mode allows a third-party tool to request operations such as the display of a call graph, or the display of the source code of a program unit, perhaps with highlighting of a particular location in it such as an executable statement.

Additional interaction possibilities are provided via TSF and via the VAM-PIR configuration file. The latter enables a user or tool to specify options governing the creation and display of tracefiles. The VAMPIR tracefile format is also available as part of the FITS public interface, permitting a tool to read and process the trace information itself. In interactive mode, it is possible to load a tracefile and configuration file, to open certain VAMPIR displays, perhaps related to a particular module, and from there to also request source code views.

4 Related Work

In addition to vendor-supplied parallel program development toolsets, there is a range of product suites from individual vendors which combine several features of the above. These include the Kuck and Associate product line [7] for shared memory parallelisation. The FORGE parallelization tools [1] from Applied Parallel Research enable source code editing and offer control and data flow information, in combination with an HPF compiler. VAST [12] from Pacific Sierra Research helps a user migrate from Fortran 77 to parallel HPF code by automatic code cleanup and translation to Fortran 90 and HPF. It can be augmented by an HPF compiler and by a performance analysis tool, DEEP.

There have been attempts to integrate products from different vendors. The ESPRIT project PPPE (Portable Parallel Programming Environment) [4] enhanced the functionality of a range of programming tools. It began with the ambitious goal of integrating them within the PCTE framework. However, despite the undoubted power of PCTE and its benefits in terms of opportunities for closer tool interaction, the process of integration proved to be too complex and time-consuming for both users and vendors, and this approach was abandoned.

5 Conclusions and Future Work

In this paper we have described the FITS programming environment for the development of parallel Fortran applications. The environment is both portable and extensible. The method for tool interaction was chosen to satisfy the need of software vendors for independence, and to permit the inclusion of additional tools with relative ease, possibly to create customised environments at individual user sites. The realisation has required a central repository of information on a specific application development, the *ForLib*, and a set of functions which permit the asynchronous, remote invocation of functionality provided by component tools. The result is a working environment which will soon be available on the market.

During its development, the toolset was used by two end-user companies in the FITS project, Battelle and QSW, to parallelise two large industrial application codes, namely a CFD code and an electromagnetic scattering simulation. Their feedback provided an important input to the design and development of FITS.

The capabilities of the resulting toolset offer the opportunity to explore new approaches to some parallelisation problems, including introducing interactive guidance for the process of program transformation.

Acknowledgments

We thank our colleagues in the FITS project, especially Markus Egg (VCPC), Yann Mevel (IRISA), Hans-Christian Hoppe and Karl Solchenbach (Pallas), Wolfgang Nagel and Holger Brunst (TU Dresden), Aron Kneer (Battelle), and Ornella Fasolo, Agostino Longo and Carlo Nardone (QSW). We also thank Jerry Yan (NASA Ames) for his valuable comments and suggestions about FITS.

References

[1] Applied Parallel Research: *APR's FORGE 90 parallelization tools for High Performance Fortran*. APR, June 1993

[2] S.R.M. Barros, D. Dent, L. Isaksen, G. Robinson, G. Mozdzynski, and F. Wollenweber, *The IFS model: A parallel production weather code*, Parallel Computing 21, 1621–1638, 1995

[3] F. Bodin, Y. Mevel, and R. Quiniou, *A user level program transformation tool*, Proc. Int. Conf. on Supercomputing, 1998

[4] B. Chapman, T. Brandes, J. Cownie, M. Delves, A. Dunlop, H.-C. Hoppe and D. Pritchard. *The ESPRIT PPPE project: a final report*. Technical report VCPC 05-96, University of Vienna, 1996.
URL: `www.vcpc.univie.ac.at/activities/reports/doc-files/tr_05-96.ps`.

[5] B. Chapman, M. Egg and F. Wollenweber, *ANALYST Version 1.0 User's Guide*, VCPC, April 1996, rev. February 1997

[6] B. Chapman, M. Egg, *ANALYST: Tool Support for the Migration of Fortran Applications to Parallel Systems*, Proc. PDPTA' 97, Las Vegas, June 30–July 3, 1997

[7] Kuck and Associates. *KAP/Pro Toolset for OpenMP*. See `www.kai.com/kpts/`

[8] J.H. Merlin and J.S. Reeve. *IDA—an aid to the parallelisation of Fortran codes*. Technical report, Dept. of Electronics and Computer Science, University of Southampton, Sept. 1995. Software and documentation are available from `www.vcpc.univie.ac.at/information/software/ida/`.

[9] Message Passing Interface Forum. *MPI: A Message-Passing Interface Standard*. Int. Journal Supercomputing Applications, Vol 8 3/4, 1994.

[10] W. Nagel, A. Arnold, M. Weber, H.-C. Hoppe and K Solchenbach. *VAMPIR: Visualisation and analysis of MPI Resources*. Supercomputer 63, 12(1), 69-80, 1996

[11] C.M. Pancake and C. Cook, *What users need in parallel tool support: survey results and analysis*, Proc. Scalable High Performance Computing Conference, 1994

[12] C. Rodden and B. Brode. *VAST/Parallel: automatic parallelisation for SMP systems*. Pacific Sierra Research Corp., 1998.

[13] Simulog SA, *FORESYS, FORtran Engineering SYStem, Reference Manual Version 1.5*, Simulog, Guyancourt, France, 1996

INTERLACE: An Interoperation and Linking Architecture for Computational Engines

Matthew J. Sottile and Allen D. Malony

Department of Computer and Information Science,
University of Oregon, Eugene, Oregon 97403-1202

Abstract. To aid in building high-performance computational environments, INTERLACE offers a framework for linking reusable computational engines in a heterogeneous distributed system. The INTERLACE model provides clients with access to computational servers which interface with "wrapped" computational engines. An API is defined for establishing the server interconnection requirements including data organization and movement, and command invocation. This provides an abstraction layer above the server middleware upon which servers and clients may be constructed. The INTERLACE framework has been demonstrated by building a distributed computational environment with Mat-Lab engines.

1   Introduction

Scientific research that involves the analysis of large data sets, the simulation of complex phenomena, or the execution of detailed computational experiments often requires scientists to develop software that runs on high-performance parallel and distributed computer systems. Unfortunately, these systems can be difficult for a scientist to program and use. The software that is developed tends to be specialized and can be hard to integrate with other tools that are commonly found in the scientist's environment. The reuse and linking of software with components from other projects suffers from the lack of a common interface and interoperation model. More accessible computational programming packages such as MatLab [1], IDL [2], and Mathematica [3] are commonly used for small scale computational tasks. The relative simplicity of the programming system and the wide availability of modules of scientific interest makes these packages both productive and desirable. However, there is minimal support for integrating such tools with high-performance environments to utilize specialized hardware or software libraries. Scientists and other users of high-performance environments could benefit from a framework in which they can both easily create programs utilizing reusable, common tools and scale their applications to larger systems.

In this paper, we propose a framework of this type based on the abstraction of components, or tools, as *computational engines*, and their interoperation as interconnected *computational servers*. A computational server is a persistent program that allows *clients* access to functionality and data provided by the

P. Amestoy et al. (Eds.): Euro-Par'99, LNCS 1685, pp. 135–138, 1999.

computational engines linked to it. Any computational engine that creates and shares data must also be persistent during the application's operation. The server interface provided to clients is abstract, to hide the complexity of the underlying parallel or distributed system. Because the framework supports a computational engine as an abstract component, any tool can be provided as long as its server interface can be created.

2   Architecture

A system view of the INTERLACE[1] framework architecture is shown in Figure 1. The shaded portions of the diagram indicate INTERLACE software components. To make computational engines available to clients through a homogeneous interface, they must be individually wrapped and linked with a computational server. The wrapper acts as a translation and interpretation layer between the INTERLACE environment and the specific engine interface. Given a computational server with a set of available wrapped engines, clients may attach to the public interface of the server and issue data and computation related commands. To provide this layering of interfaces between client, server and engine, INTERLACE specifies two abstract interfaces. The first defines the interface between the wrapper and computational server, while the second defines the interface and protocols for linking individual computational servers and clients.

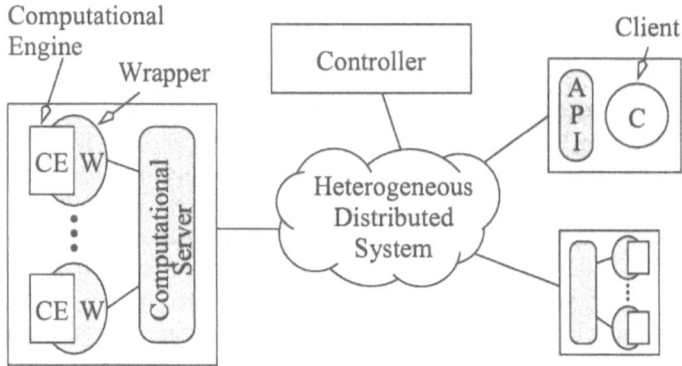

Fig. 1. INTERLACE System Architecture

Before one can assemble an INTERLACE-based environment, the individual servers and wrappers must be created. The wrapping of computational engines is dictated by the engines themselves and is not part of the INTERLACE specification. The interface between the wrapper and the computational server exists in

---

[1]   A complete description of INTERLACE is available as Technical Report CIS-TR 99-06 from the Department of Computer and Information Science at the University of Oregon.

two parts. First, there must be methods available for calling computational functions within the server. Second, if the server will import or export data to other members of an INTERLACE environment, there must be methods for moving data. These functions are then used in the server within the implementation of methods visible to client applications. INTERLACE defines this API as a small set of functions to receive message-based commands, move data, and control the state of the server. The messages and data movement requests are performed by calling appropriate methods in the wrapper, thus hiding the wrapper specific APIs from the client applications.

Once INTERLACE servers are implemented, the environment requires that they are connected in some manner. The method and implementation of such interconnection schemes is independent of INTERLACE. However, the implementation method must provide some form of cross-platform data movement and remote method invocation. The API that the servers present to clients is specified in an abstract manner that does not impose strict requirements on the middleware choice. Current implementations of MatLab-based INTERLACE environments exist in two forms - one built with High Performance C++ (HPC++) [4] for high-performance environments, and one built with the Java Remote Method Invocation (RMI) [5] API for high portability. Servers, data, and computational services are located by a simple identifier, which is mapped to the appropriate URL or other middleware-specific location identifier. Such mappings are maintained at a central location known as the *controller*. All servers and clients are provided the location of the controller, and are then able to dynamically locate and establish connections to each other.

3   Application of INTERLACE with MatLab

We prototyped the INTERLACE framework using servers with MatLab-based engines. The MatLab engine interface provides simple libraries for allowing access to data and functions. Computations are performed by issuing text strings containing MatLab commands to the engine. Data can be moved between the server and engine through both the wrapper and a separate program called from within MatLab, written as a *mex-file*. The mex-file provides an additional MatLab command that allows a MatLab engine access to the INTERLACE environment for moving data or issuing commands to other servers. This also provides means for MatLab to act as a front-end interface to an INTERLACE-based environment.

Our most recent implementations of the INTERLACE middleware layer use Java RMI to increase portability. So that the MatLab frontend could access INTERLACE services, the Java Native Interface (JNI) was used to connect C mex-files and the Java RMI layer. (A similar method was used to connect MatLab using mex-files and HPC++.) At the server interface to the engine wrappers, there is little interpretation since commands are assumed to be basic MatLab string commands. However, a queue of commands can be collected by the server, and when ready, executed in order. Without this support, it would be difficult to create computational scripts that employed multiple servers in an efficient way. For instance, it would allow a user to load large routines (such as an entire MatLab command file, or *.M file*) into a collection of servers, and then let them compute in parallel when ready.

4   Conclusions

The current implementation of an INTERLACE-based system utilizes MatLab engines with either Java or HPC++ middleware. This illustrates the potential of the INTERLACE abstraction above implementation mechanisms. By creating these two implementations, INTERLACE can be shown in high-performance environments that best suit HPC++, and heterogeneous systems that require Java's portability. Continuing work will involve providing wrapper and server interfaces for tools such as IDL and Mathematica. Our primary research interest is in the access and use of heterogeneous engines with a common, homogeneous INTERLACE abstraction layer. Additional research includes wrapper generation techniques and middleware utilization. An important future step will be the ability to integrate two INTERLACE environments implemented with different middleware, such as HPC++ and Java.

Fulfilling the need for a framework for connecting computational engines in a scalable, reusable manner is the central goal of INTERLACE. The abstraction of functionality and data offered by computational engines to the homogeneous INTERLACE API and interconnection scheme provides many advantages. It makes possible dynamic environments in which computational engines may be instantiated as required, while hiding implementation details required to support such behavior. By specifying INTERLACE as an abstract definition of interaction APIs and behaviors, INTERLACE may be implemented using a variety of middleware solutions. Since the computational server framework provided by INTERLACE is an extendible skeleton, the application of this framework to specific tools can be fine-tuned to the user's needs. As a result, INTERLACE is a powerful tool for scientists and others who wish to build a reusable and scalable environment for doing high-performance analysis and research.

References

[1] The Math Works, *MATLAB: High Performance Numeric Computation and Visualization Software Reference Guide* Natick, MA. 1992.
[2] Research Systems, IDL.
    http://www.rsinc.com/.
[3] Wolfram Research, Mathematica.
    http://www.wolfram.com/.
[4] E. Johnson and D. Gannon, HPC++: Experiments with the parallel standard template library. *Technical Report TR-96-51* Indiana University, Department of Computer Science, December 1996.
[5] Javasoft, Java Remote Method Invocation (RMI) Interface.
    http://www.javasoft.com/products/jdk/rmi/index.html.

Multi-protocol Communications and High Speed Networks

Benoît Planquelle[1], Jean-François Méhaut[2], and Nathalie Revol[3]

[1] Université des Sciences et Technologies de Lille, LIFL, France
    Benoit.Planquelle@lifl.fr
[2] École Normale Supérieure de Lyon, LIP, France
    Jean-Francois.Mehaut@ens-lyon.fr
[3] Université des Sciences et Technologies de Lille, ANO, France
    Nathalie.Revol@univ-lille1.fr

Abstract. Due to heterogeneity in modern computing systems, methodological and technological problems arise for the development of parallel applications. Heterogeneity occurs for instance at architecture level, when processing units use different data representation or exhibit various performances. At interconnection level, heterogeneity can be found in the programming interfaces and in the communication performances. In this paper, we focus on problems related to heterogeneity in the frame of interconnected clusters of workstations. In order to help the programmer to overcome these problems, we propose an extension to the multi-threaded programming environment PM^2. It facilitates the development of efficient parallel applications on such systems.

1 Introduction

New technologies for local networks[1] as well as for high speed communications reinforce the interest for parallelism. Architectures built out of PCs interconnected by a high speed network (*clusters*) can be compared to super-computers in terms of performances (with bandwidth beyond a Gigabit per second and latency less than 5 μs). However, legacy and system protocols (TCP/IP) produce significant overhead due to system calls and multiple messages copies. Communication interfaces in user's context (such as BIP, Fast Messages, Active Messages) are actively explored to provide high performance communications for user's programs. Unfortunately, they act at a low level and their use by an average programmer is not straightforward. Higher level interfaces have been developed, such as MPI-FM[5], MPI-BIP or PM^2 HighPerf[2]. But they are still not completely satisfactory: interoperability is lacking. Some environments[4, 3] have been developed to integrate interoperability. In this paper, we describe our multi-cluster approach based on PM^2.

PM^2[6] is a parallel programming environment which is not based upon message-passing procedures: its programming paradigm is to express a parallel program as a set of tasks handled by lightweight processes (threads). These tasks are executed via LRPC (Lightweight Remote Procedure Call). In this work,

P. Amestoy et al. (Eds.): Euro-Par'99, LNCS 1685, pp. 139–143, 1999.

we determined which mechanisms have to be integrated in PM^2 in order to support the simultaneous use of different protocols in the same parallel program. It mainly concerns the polling of several communication interfaces and the data conversions between heterogeneous processors.

This paper is organized as follows: in a first part methodological and technical problems are explained; then solutions to these problems are developed and finally their implementation is presented.

2   The Multi-cluster Approach

We consider a heterogeneous configuration, called a *multi-cluster*, as being structured as a federation of interconnected homogeneous clusters. Let us firstly define a cluster : it is a pool of homogeneous machines connected by the same network and the same communication protocol. We consider that machines are homogeneous when they have similar processors, power, and operating system. Then basic data type have fully compatible representations and thus no data conversion is required for the communications. Inside a cluster, load-balancing is also simplified because the processors exhibit similar performances. Furthermore, the best communication protocol can be employed.

Our objective is to extend the programming environment PM^2 in order to handle this kind of configuration. This concerns the handling of the different machines and the management of the communications between clusters. A multi-cluster is defined as the interconnection of several (homogeneous) clusters. On each cluster, a machine has to deal with the communications from the other clusters, it is referred as the gateway of this cluster. From the network point of view, the clusters are physically connected to Internet via a single gateway. Consequently, we reproduce this connection model in the communication layer of MC-PM2. Usually, the performances achieved by the interconnecting technology between clusters (the network and the communication protocol) are very poor compared to the ones used inside the clusters. Then communications between clusters must be explicit. Since communications occur in PM^2 at LRPC, a new operator has been defined: CLUSTER_LRPC. It has roughly the same semantics as a ASYNC_LRPC[6], the differences being that the thread runs on a machine of another cluster and that this communication is explicitly required by the programmer.

From the user's point of view, which problems arise? He has to take into account that the inter-cluster network is less efficient than the intra-cluster network. To limit the interactions between clusters, the user has to restructure his applications. This can lead to new schemes for sharing data and for communications. For instance, shared data can be replicated on every cluster and a relaxed consistency protocol can be implemented. A paper by H. Bal[7] and his team illustrates how applications have been modified and how gains in performances have been obtained on clusters of PCs/Myrinet interconnected via ATM.

3   Implementation

In this section, we describe some of the main mechanisms that have been added to PM^2 to make it support heterogeneous infrastructures.

Since data conversions take a non negligible part of communications' time, they are deserve to be handled with care. Our choice consists in performing only one conversion at the sending stage. Our data conversion procedures are very simple : since the differences in representations often lie in the order of bytes inside a word, they mainly consist in exchanging bytes in words. When the sender prepares the message, it converts the data into the receiver's internal format, at packing procedures' call. Experimental results are presented on Figure 1: they show that conversions made by MADELEINE ("MAD/TCP Conversion" *minus* "MAD/TCP BY_COPY") are more efficient than XDR, used in PVM("PVM DataDefault" *minus* "PVM DataRaw").

Routing is performed when a communication between two distinct clusters is done. First, the message is automatically routed to the sender's cluster gateway machine, then this machine transfers the message to the gateway in the receiver's cluster and finally the message is transmitted to the receiver's machine. To sum up, our implementation involves at most three communications, two fast intra-cluster communications and one slower inter-cluster communication.

In the context of a homogeneous cluster, message polling is performed by a thread dealing with all messages sent to its parent process. Only that specific thread receives MADELEINE messages, but every thread can emit messages (LRPC). In heterogeneous context (i.e. on gateway machines), several protocols are used since several networks are used. We choose to dedicate specific threads to handle the polling of messages issued by each protocol. The advantage of this approach is to simplify the tuning of polling frequencies: this is achieved by adjusting properly the priority of the polling threads. Moreover, with one independent thread for each protocol, MC-PM^2 can take advantage of Symmetric Multi-Processors (SMP) architectures.

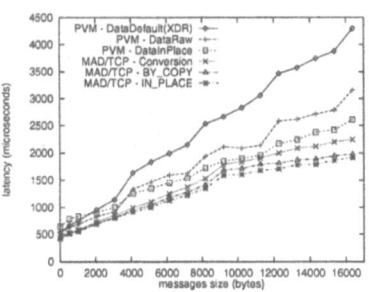

Fig. 1. Ping-pong between two Sun Ultra-SPARC via ATM.

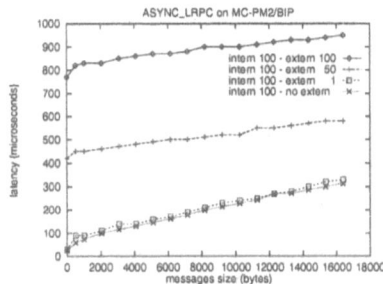

Fig. 2. Performance of LRPC according to extern polling priority on a Myrinet cluster of Pentium Pro 200 MHz.

Since inter-cluster communications are rather slow, they should be avoided by the programmer and intra-clusters communications should be preferred, which implies that: 1/ intra-cluster communications should be privileged by choosing a high frequency, which ensures a good reactivity of the polling thread; 2/ for inter-clusters message polling, a lower priority is assigned. As can be seen on Figure 2, on a fast network such as Myrinet, the frequency of inter-cluster network polling influences dramatically the performance of LRPC on the gateway. When the external polling thread has a priority of 1 and the internal polling thread has a priority of 100, this double polling is not very costly compared to the original PM^2 simple polling method.

4 Conclusion and Future Work

Meta-computing is quickly developing in the United States, thanks to the interconnection of the supercomputers of the main research centers. An analogous approach can be taken by interconnecting clusters of workstations and of PCs. Problems raised by the heterogeneity of these clusters can be solved by techniques tending to limit memory allocations and copies. Due to the heterogeneity in communications, several constraints are also added to the programmer, who should limit the volume of inter-clusters communications, and to the execution support, which should be able to deal with different protocols. In this paper, we have described our extension to PM^2, and we overviewed the following aspects: data conversion and communications routing and polling. Our main goal is to ensure a high level of performance inside clusters. The multi-cluster approach is going to be used for scientific computations (global optimization with interval arithmetic[8]).

References

[1] N. Boden, D. Cohen, R. Feldermann, A. Kulawik, C. Seitz, and W. Su. Myrinet: A Gigabit per second Local Area Network. *IEEE-Micro*, 15-1:29–36, feb 1995. Available from http://www.myri.com/research/publications/Hot.ps.
[2] L. Bougé, JF. Méhaut, and R. Namyst. Madeleine: an efficient and portable communication interface for multithreaded environments. In *Proc. 1998 Int. Conf. Parallel Architectures and Compilation Techniques (PACT'98)*, pages 240–247, ENST, Paris, France, October 1998. IFIP WG 10.3 and IEEE.
[3] M. Brune, J. Gehring, and A. Reinefeld. A lightweight communication interface for parallel programming environments. *Lecture Notes in Computer Science*, 1225:503, 1997.
[4] I. Foster, J. Geisler, C. Kesselman, and S. Tuecke. Managing multiple communication methods in high-performance networked computing systems. *Journal of Parallel and Distributed Computing*, 40:35–48, 1997.
[5] M. Lauria and A. Chien. MPI-FM: High performance MPI on workstation clusters. *Jounal on Parallel and Distributed Computing*, 40(01):4–18, 1997.
[6] R. Namyst, Y. Denneulin, JM. Geib, and JF. Méhaut. Utilisation des processus légers pour le calcul parallèle distribué : l'approche PM^2. *Calculateurs Parallèles, Réseaux et Systèmes répartis*, 10(3):237–258, July 1998.

[7] A. Plaat, H. E. Bal, and R. F.H. Hofman. Sensitivity of parallel applications to large differences in bandwidth and latency in two-layer interconnects. In *Proc of 5th IEEE High Performance Computer Architecture HPCA'99*, pages 244–253, Orlando, USA, January 1999.

[8] N. Revol, Y. Denneulin, J.-F. Méhaut, and B. Planquelle. Parallelization of continuous verified global optimization. In *19th IFIP International Federation for Information Processing TC7 Conference on System Modelling and Optimization*, Cambridge, England, July 1999. to appear.

An Online Algorithm for Dimension-Bound Analysis

Paul A.S. Ward

Shoshin Distributed Systems Group
University of Waterloo, Waterloo Ontario N2L 3G1, Canada

Abstract. The vector-clock size necessary to characterize causality in a distributed computation is bounded by the dimension of the partial order induced by that computation. In theory the dimension can be as large as the number of processes in the computation, but in practice it is much smaller. We present an online algorithm to compute the dimension of a distributed computation. This algorithm requires the computation of the critical pairs of the partial order followed by the creation of extensions of the partial order. This is our next step toward the goal of creating an online vector clock whose size is bounded by the dimension of a distributed computation, not by the number of processes.

1 Motivation

An important problem in distributed systems is monitoring and debugging distributed computations. What makes this problem hard is that events in the computation can be concurrent. The events form a partial order, not a total one. While a process-time diagram which shows this partial order of execution can be useful, it is not possible to display the whole partial order for any non-trivial computation. As a result, distributed debugging systems such as POET [8] need to provide much more than just a drawing. It is necessary to intelligently scroll around the display [14], search for interesting patterns [6], compute differences between subsequent executions of a computation [4], detect race conditions [18], determine appropriate abstractions to provide higher level views [7], and so on. To perform these operations it is frequently necessary to determine event precedence. That is, given two events, it is necessary to be able to efficiently determine if they are ordered or if they are concurrent.

There are several ways in which event precedence may be determined, depending on how the partial order is represented. If we store the partial order as a directed acyclic graph, then precedence determination is a constant time operation. This is because the partial order is transitively closed and so there is an edge between any two events that are ordered. However, the space consumption for this method is unacceptably high. Instead we store the transitive reduction of the partial order. This is space-efficient but requires a (potentially quite slow) search operation on the graph to determine precedence. To compensate for this deficiency a vector clock [2, 10] is associated with each event. If the processes in which the events occur are known, then it is possible to determine precedence with vector clocks in constant time.

The size of a vector clock necessary to capture causality is bounded by the dimension of the partial order induced by the computation. Charron-Bost [1] has shown that the dimension can be as large as the width, and all online vector clocks developed to

P. Amestoy et al. (Eds.): Euro-Par'99, LNCS 1685, pp. 144–153, 1999.

date require a vector with size equal to the number of processes (which forms an upper bound on the width). Since we need to associate such a vector clock with every event in the computation, we are substantially constrained in the number of processes that we can observe. In POET we have found that due to this limitation we can handle at most a few-hundred processes. However, we have already shown [16] that in practice the dimension is much smaller than the number of processes. Our ultimate goal is therefore to develop a vector clock whose size is bounded by dimension, not width, and that vector clock must be implementable in an online algorithm.[1] This paper is a first step in that direction, by developing an online dimension-bound algorithm.

In the remainder of this paper we will specify first the formal model of distributed computation and the terminology necessary to deal with dynamic partial orders. In Sect.3 we will describe the supporting theorems and the online algorithm. We then provide some analysis of the algorithm. Finally we indicate what work remains to be completed to achieve online dimension-bounded vector clock.

2   Background

We use the standard model of distributed systems, initially defined by Lamport [9]: a distributed system is a system comprising multiple sequential processes communicating via message passing. Each sequential process consists of three types of events, send, receive and unary, totally ordered within the process. A *distributed computation* is the partial order formed by the "happened before" relation over the union of all of the events across all of the processes. We will refer to the set of all events with E, the set of events within a given process by E_i (where i uniquely identifies the process) and an individual event by e_i^j (where i identifies the process and j identifies the event's position within the process). Then the Lamport "happened before" relation ($\rightarrow \subseteq E \times E$) is defined as the smallest transitive relation satisfying

1. $e_i^j \rightarrow e_i^l$ if $j < l$
2. $e_i^j \rightarrow e_k^l$ if e_i^j is a send event and e_k^l is the corresponding receive event

Events are concurrent if they are not in the "happened before" relation:

$$e_i^j \parallel e_k^l \Longleftrightarrow e_i^j \nrightarrow e_k^l \wedge e_k^l \nrightarrow e_i^j \tag{1}$$

These definitions presume that the partial order is a static entity. That is, implicit in the "happened before" and "concurrency" relations is a fixed computation to which they refer. In an online monitoring or debugging context, the computation, and hence the partial order, is not fixed. Therefore, we now describe the terminology we use in this paper for dealing with partial orders and changing partial orders.

2.1   Partial-Order Terminology

The basic partial-order terminology is due to Trotter [15], modified as necessary to take into account the dynamic nature of the partial orders that we are dealing with. A *strict*

---

[1] An offline algorithm, the Ore timestamp [11], already exists that is dimension-bounded, not process-bounded.

partial-order (or partially-ordered set, or poset) is a pair (X, P) where X is a finite set[2] and P is an irreflexive,[3] antisymmetric and transitive binary relation on X. An *extension*, (X, Q), of a partial order (X, P) is any partial order that satisfies

$$(x, y) \in P \Rightarrow (x, y) \in Q \tag{2}$$

If Q is a total order, then the extension is called a *linear extension*. A *realizer* of a partial order is any set of linear extensions whose intersection forms the partial order. The *dimension* of a partial order is the cardinality of the smallest possible realizer.

To deal with dynamic partial orders we will develop theorems regarding the relationship between the partial order before and after some element is added to the base set (with resulting changes to the binary relation of the partial order). In other words, the theorems we develop will relate two distinct partial orders. It is therefore necessary to make explicit which partial order the precedence ($\rightarrow$) and concurrency ($\|$) relations refer to. We will do so by subscripting those relations with the base set of the partial order on which the relations hold. We will likewise subscript any other relations or sets we define for partial orders to indicate explicitly the partial order to which the relation or set refers. Subscripts may be omitted if there is only a single partial order under discussion, or if the relation or set is identical for all partial orders in the theorem in question.

We can now continue the discussion of realizers and dimension by recalling the definition of a critical pair.

Definition 1 (critical pair). (x, y) *is a critical pair of partial order* (X, P) *if* $x \|_x y$ *and* $(X, P \cup \{(x, y)\})$ *is a partial order.*

Equivalently

$$(x, y) \in \mathrm{CP}_X \iff (x \|_x y) \wedge \forall_{z \in X} (z \rightarrow_x x \Rightarrow z \rightarrow_x y) \wedge (y \rightarrow_x z \Rightarrow x \rightarrow_x z) \tag{3}$$

where CP_X is the set of all critical pairs of the partial order (X, P). The significance of critical pairs, as regards dimension, is in the following theorem [15]

Theorem 1. *The dimension of a partial order is the cardinality of the smallest possible set of extensions that reverses all of the critical pairs of the partial order.*

A critical pair (x, y) is said to be reversed by a set of extensions if one of the extensions in the set contains $y \rightarrow x$. Note that the extensions need not be linear. Thus the approach we have taken for our algorithm is to incrementally create a set of extensions to the partial order that reverses the critical pairs of the partial order.

Finally, We say that "y is covered by z" or "z covers y" if there is no intermediate element in the partial order between y and z.

$$y <: z \iff y \rightarrow z \wedge (\not\exists_w y \rightarrow w \wedge w \rightarrow z) \tag{4}$$

Before proceeding to describe the algorithm, we will briefly describe some of the work related to our own.

---

[2] Since we are modeling distributed computations, all of the sets will be finite.

[3] If P is reflexive, it is a *partial order* rather than a *strict partial order*. The difference is not an important point in our context. The "happened before" relation, as defined by Lamport [9], is irreflexive, and so we use strict partial orders.

2.2   Related Work

There are essentially two categories of related work. The first group are those who are developing systems for visualizing parallel and distributed systems in a process-time fashion. Such work includes GOLD [12], ParaGraph [5] and our own system, POET [8]. Other than our own, these systems tend to use vector-clocks or variants within the computation, rather than have the information sent to a central server which computes the vector clocks for visualization purposes. These systems tend to take the approach of just using what is presently available, and not being concerned with substantial scalability. The GOLD system uses dependency vectors, developed by Fowler and Zwaenepoel [3], which will be size $O(n)$ by the time they are attached to individual events. ParaGraph has the ability to provide space-time diagrams, but there is no attempt to determine causality. It is merely a visualization based on, possibly badly synchronized, local clocks. It is up to the user to trace the dependencies. Indeed, the authors acknowledge that the size would not scale well beyond 128 processes, as the display becomes too cluttered. POET uses standard Fidge/Mattern vector clocks [2, 10], and so it too requires vectors with size equal to the number of processes.

The second group are those who are trying to reduce the vector-clock size, but in the context of maintaining vector clocks by the processes involved in the computation. The primary work in this area is a technique by Singhal and Kshemkalyani [13]. The problem with the technique offered is that an $O(n^2)$ matrix is required at each process to recover the vector. In the context of a computation, this may be acceptable. In the context of monitoring or debugging, this implies moving from $O(n)$ vector associated with each event to $O(n^2)$ set of data associated with each event.

The closest work to this paper is our previous paper [16] which presented an offline algorithm for computing dimension. That paper presented results which indicated that a significant number of distributed computations have a low dimension. However, the work was offline, where this is an online algorithm.

3   An Online Algorithm

In this section we will describe the algorithms we have developed to bound the dimension of the partial order online. The basic idea is to generate incrementally the critical pairs of the partial order and then create extensions to the partial order that reverse those pairs. To achieve this in an online algorithm, we treat the set of events and their interaction at any given instant as the complete partial order. The arrival of a new event (that is, information pertaining to a previously unknown event) then modifies that partial order. Our algorithm must therefore do three things. It must determine which critical pairs are no longer valid when the new event arrives, which new critical pairs are created by the new event, and it must be able to incrementally reverse these critical pairs into a set of extensions of the partial order. We will now describe our solutions to these problems.

3.1   Computing Critical Pairs Online

To compute the set of critical pairs online we build on the definitions and theorem from our previous paper [16]. First we recall the definitions of leastConcurrent$_E(e)$ and greatestConcurrent$_E(e)$ and their relationship to the critical pairs of a partial order

$$e_l \in \text{leastConcurrent}_E(e) \iff e_l \parallel_E e \wedge \left(\nexists_{e'_l} \, e'_l \parallel_E e \wedge e'_l \to_E e_l \right) \qquad (5)$$

$$e_g \in \text{greatestConcurrent}_E(e) \iff e_g \parallel_E e \wedge \left(\nexists_{e'_g} \, e'_g \parallel_E e \wedge e_g \to_E e'_g \right) \qquad (6)$$

Theorem 2.

$$(x,y) \in \text{CP}_E \iff x \in \text{leastConcurrent}_E(y) \wedge y \in \text{greatestConcurrent}_E(x)$$

These definitions and theorem are sufficient for an offline algorithm where the partial order is fixed. In our online algorithm the set of events that are least or greatest concurrent to a given event may change as new events arrive. As a result, we required two additional theorems: one to indicate which critical pairs were no longer valid in the presence of a new event, and one to indicate which new critical pairs were present as a result of the new event.

In order to be able to efficiently determine the change in the critical-pair set, we must put a restriction on the order in which new events are processed. We require that the events are processed in some linear extension of the partial order. The effect of this condition is that we are now dealing with additions to the partial order of new maximal elements, and of no other kind. This enables the development of our required theorems. The first theorem indicates which critical pairs are no longer valid under the addition to set E of maximal element e.

Theorem 3.

$$(a,b) \in \text{CP}_E \implies (b <: e \wedge a \parallel_{E \cup \{e\}} e) \Leftrightarrow (a,b) \notin \text{CP}_{E \cup \{e\}}$$

(a,b) remains in the set of critical pairs provided b is not covered by e or a is not concurrent with e.

Our second theorem defines the entire set of additional critical pairs that may result from the additional of e.

Theorem 4.

$$(a,b) \in \text{CP}_{E \cup \{e\}} - \text{CP}_E \iff (a = e \vee b = e) \wedge a \in \text{leastConcurrent}_{E \cup \{e\}}(b) \wedge$$
$$b \in \text{greatestConcurrent}_{E \cup \{e\}}(a)$$

The only new critical pairs are those that are formed with the event e as one of the elements of the pair. Furthermore, the combination of these two theorems clearly indicates that if (a,b) is a critical pair but ceases to be after the addition of e, then (a,e) will be a critical pair. We can also observe that in such a case any extension of the partial order that reverses (a,e) must also reverse (a,b) since $b \to e$.

These theorems and observations lead to two possible approaches. We may generate the critical pairs and reverse them without heed to whether or not they are ultimately valid critical pairs, since the extensions we build will reverse any pair that was at one time believed to be a critical pair. This is consistent with the view that the entire set of events received thus far constitutes the whole of the partial order. A drawback with

```
 1: ∀ₑ { // For each new event 'e'          13:     l_C ← leastConcurrent(e)
 2:    ∀_b b <: e {                          14:     ∀_l l ∈ l_C {
 3:       if (a, b) ∈ tentative_CP {          15:         add (l, e) to tentative_CP
 4:          if (a ∥ e) {                     16:     }
 5:             remove (a, b) from tentative_CP   17:   g_C ← greatestConcurrent(e)
 6:          }                                18:     ∀_g g ∈ g_C {
 7:          else if (e is last covering event of b) {   19:       if (e ∈ leastConcurrent(g)) {
 8:             remove (a, b) from tentative_CP   20:           add (e, g) to tentative_CP
 9:             reverse (a, b)                 21:       }
10:          }                                22:     }
11:       }                                   23: }
12:    }
```

Fig. 1. Critical Pair Calculation

this approach is the possible additional cost of such insertions. The alternate approach is to wait to reverse a critical pair until it is certain that the pair will always remain valid. From Theorem 3 we know that a critical pair, (a, b), will always remain valid if it is a critical pair after all events that cover b have been processed. In the context of a distributed computation, this condition is satisfied once the successor event to b in the process and any corresponding receive event (if b is a send event) have been processed. The algorithm shown in Fig.1 implements the second technique. It may be modified to implement the first technique by omitting lines 1–12 and altering lines 15 and 20 to instead reverse the pair, rather than add the pair to a tentative list.

There are three things that must be performed for each new event arrival. First, we must deal with all tentative critical pairs whose second element is covered by the new event. Such pairs cannot be critical pairs if their first element is concurrent with the new event. On the other hand, such pairs must be moved to the permanent set (*i.e.* reversed) if the new event is the last element to be processed that covers the second element of the pair and that new event is not concurrent to the first element of the pair. Second, we must consider all events that are leastConcurrent the the new event. Since the new event is, by requirement, a maximal element, it must be greatestConcurrent to all events that are leastConcurrent to it. As a result, any event leastConcurrent to it forms a tentative critical pair with it. Finally, we must consider events that are greatestConcurrent to the new event. If the new event is leastConcurrent to such events, they too would form a tentative critical pair.

Ignored in this algorithm is what to do at the end of a computation. For the first technique there is no problem. For the second technique, there may be valid critical pairs still in the tentative critical pair list. To deal with this we add a fake unary event to the end of each process. Then for those events, b, covered by the fake unary, which form tentative critical pairs (a, b) we move (a, b) to the permanent set. Note that if b is a send event, we must wait for the corresponding receive event to be processed before adding the fake unary.

Before we proceed to deal with critical pair reversal, we will briefly examine the computation costs. We implemented our algorithm as a client to the POET server. As such we have access to Fidge/Mattern timestamps for each event. Therefore it costs $O(w^2)$ to compute the least concurrent set [16], where w is the width of the partial order (that is, the number of processes in the computation). To compute the greatest concurrent set we observe that, since e is maximal, the only elements that could be greatest concurrent to e must also be maximal. We therefore start with the greatest event we have seen on each process so far. We then remove all events in this set that are not concurrent with e. Finally, we remove from the set any event that precedes some other event in the set. As with the leastConcurrent computation, this will be $O(w^2)$.

Due to the need to compute the least concurrent set of g for each g that is greatest concurrent to e (of which there may be as many as w), the algorithm in Fig.1 is $O(w^3)$. However, we can optimize away the need for this least concurrent of g computation by the following theorem.

Theorem 5.

$$x \parallel y \implies (x \in \text{leastConcurrent}(y) \Leftrightarrow (\forall_z z <: x \Rightarrow z \to y))$$

Thus, rather than compute the complete leastConcurrent set, it is sufficient to check that the events covered by g are predecessors of the new event. Since POET only supports point-to-point communication, the set of events that are immediate successors or predecessors to an event has cardinality less than 3. This means that the cost to compute the critical pairs is $O(w^2)$ per event or $O(w^2 n)$, where n is the total number of events, for the whole distributed computation.

3.2 Incremental Critical Pair Reversal

Building the minimum number of extensions that reverses the critical pairs of a partial order is NP-hard for dimension greater than two [17]. Instead we propose a reasonably efficient algorithm that will not give an optimal solution, but will provide an upper bound on the minimum number of extensions necessary (that is, the dimension). Insofar as the dimension-bound we compute is small, it is a satisfactory tradeoff. To achieve this we developed a two-step algorithm. First we select the desired extension in which we will reverse the current critical pair and then we insert it in that extension.

The first step should not imply that the elements of the critical pair are only inserted into one extension, as per our offline algorithm [16]. In this algorithm all extensions contain all events of the partial order, including the relevant constraints of that partial order. The way we achieve this is to build a transitive reduction DAG of the partial order, with Fidge/Mattern vector clocks, as events arrive. An extension of the partial order will contain some additional constraints on some events, and alter various of the vector clocks. We do this by indexing into the extension for events, and having those events point back to the partial order DAG if the extension has not altered the event.

The second step is, in requirement, identical to that required in our offline algorithm [16]. We can insert a critical pair into an extension if we can add the events of the critical pair to the extension, such that they are reversed, and that the reversal does not cause a cycle. An extension *accepts* a critical pair if the critical pair may be inserted into that extension. An extension *rejects* a critical pair if it does not accept it. Since insertion into an extension may fail, the first step must have a strategy for selecting an alternate extension in which to place the critical pair. The strategy we used is the same greedy one as in our offline algorithm [16].

Extension Insertion For the second step of the algorithm, we had to define a method for inserting critical pairs into an extension that would be acceptably efficient for incremental purposes. The algorithm developed in our previous paper required the examination of every event stored in the extension. While in that paper we merely stored the events that were a part of a critical pair, such extensions could get quite large, and thus unacceptably slow for an incremental algorithm.

The approach we take for this incremental algorithm is to update the vector clocks as a result of the critical pair reversal. We may then use the vector clocks to determine if the insertion of the reversal of the pair would cause a cycle in the extension, and thus must be rejected. Specifically, if we wish to add critical pair (a, b) we first determine if the extension implies $a \rightarrow b$. This may be done using the standard Fidge/Mattern precedence test in constant time. As such, before we change anything, we know whether or not the critical pair will be accepted.

In the event that the critical pair reversal is accepted, we must update the vector clocks of any events affected by this new edge in the DAG. The vector clock update is performed according to standard Fidge/Mattern algorithm. If (a, b) is the critical pair that is being reversed, then the set of events that need to have their vector clock updated is $\{x \mid a \rightarrow x \wedge b \not\rightarrow x\}$. The justification for this set should be clear, since these are precisely those events for which the existence of b was not known in their vector clocks.

Some observations should be made regarding the size of this set. For a given event, it can be as large as $n - 1$. It cannot be as large as $n - 1$ amortized (over critical pair reversals, not events). With some particularly pathological partial orders and linear orderings of same, it can be $O(n)$ amortized, though it appears that in such cases the number of critical pairs would be only a small fraction of n. In practice, it does not appear to be a substantial fraction of n. We therefore use the constant v to represent the average size of this set. The value of v is determined experimentally. Since acceptance is determined in constant time, critical pair reversal will be $O(vw)$ amortized, which reflects v vector clock updates, each with size w.

4 Algorithm Analysis and Experimental Results

We now turn to studying the quality of the algorithm. There are three aspects to this. We wish to study the efficiency of the algorithm, we wish to look at the experimental results, in terms of dimension-bounds produced, and we would like to know how tight those bounds are, given that the algorithm does not produce the optimal bound.

We have already seen that the computation cost to determine the critical pairs is $O(w^2)$ per event, where w is the number of processes. The cost of reversing a critical pair is $O(vw)$. There can be as many as $2w - 2$ tentative critical pairs generated by a new event. The total cost per event then depends on whether we reverse these tentative critical pairs immediately, or wait until it is confirmed that the pair will remain a critical pair under the addition of new events. If we use the first technique, then the total cost per event is $O(w^2 + vw^2)$. If we use the second technique, we observe that at most two critical pairs in the tentative list can be made permanent (and thus must be reversed) for any new event. As such, the cost per event is $O(w^2 + vw)$.

Several comments should be made on this. Clearly the second technique has a lower cost, but how much lower (and whether or not that is significant) depends on two factors. First, it depends on how many tentative critical pairs are generated per permanent critical pair. While we put upper bounds on these numbers analytically, the actual numbers had to be determined experimentally. What we have found is that there are at least 3 and and upwards of 50 tentative critical pairs generated per permanent critical pair, being larger for a partial orders with larger width. Thus, $O(w)$ tentative critical pairs generated per new event does seem to be born out in experiment. The second factor in the cost difference between the techniques is the value v. If this were small, then the cost difference would still be small. Unfortunately this does not appear to be the case, and thus the second technique is clearly preferred from an efficiency perspective.

Ignored in our analysis is the cost of maintaining the list of tentative critical pairs. We ignore the cost of maintaining this list as it may be implemented using any reasonably efficient indexed data structure, which will have a cost much less than $O(w^2)$. The only problem would be if this list started to grow substantially. However, it turns out that this is not the case. This would only happen if the number of critical pairs in the computation exceeded twice the number of events in the computation. While this does occur in some short-lived computations, it is much more common that the number of critical pairs is less than the total number of events.

Our experimental results show little difference in dimension-bound produced from our offline algorithm [16]. In some instances the bound is higher by one, in others lower by one, in most it is the same. We will therefore not repeat the results here, other than to observe that over a range of environments and computations using between 3 and 300 processes, the dimension-bound was 10 or less.

The final question is what is the quality of the dimension-bound produced. This is something that we are currently investigating. We know that the bound we produce will never exceed w as to do so would require more than w mutually conflicting critical pairs. This would imply that the dimension could exceed the width, which is not possible [15]. Beyond this, we do not know how well the dimension-bound produced by the algorithm compares with the actual dimension. We plan to analyze this further.

5   Further Work

There are several areas in which we are actively working. We are attempting to discover the analytical quality of the bounds produced by this algorithm. In addition, we are developing an online variant of the Ore timestamp. Such a timestamp would have a size bounded by dimension, not number of processes. In this regard, we intend to apply our dimension algorithm recursively on the extensions that we develop in the course of this algorithm, as we believe it may give us insight into the problem.

Acknowledgment

The author would like to thank IBM for supporting this work and David Taylor for many useful discussions regarding this work.

References

[1] B. Charron-Bost. Concerning the Size of Logical Clocks in Distributed Systems. *Information Processing Letters*, 39:11–16, July 1991.

[2] Colin Fidge. Fundamentals of Distributed systems Observation. Technical Report 93-15, Software Verification Research Centre, Department of Computer Science, The University of Queensland, St. Lucia, QLD 4072, Australia, November 1993.

[3] Jerry Fowler and Willy Zwaenepoel. Causal Distributed Breakpoints. In *Proceedings of the 10th IEEE International Conference on Distributed Computing Systems*, pages 134–141. IEEE Computer Society Press, 1990.

[4] Jessica Zhi Han. Automatic Comparison of Execution Histories in the Debugging of Distributed Applications. Master's thesis, University of Waterloo, Waterloo, Ontario, 1998.

[5] M. T. Heath and J. A. Etheridge. Visualizing the Performance of Parallel Programs. *IEEE Software*, pages 29–39, September 1991.

[6] Christian E. Jaekl. Event-Predicate Detection in the Debugging of Distributed Applications. Master's thesis, University of Waterloo, Waterloo, Ontario, 1997.

[7] Thomas Kunz. *Abstract Behaviour of Distributed Executions with Applications to Visualization*. PhD thesis, Technische Hochschule Darmstadt, Darmstadt, Germany, 1994.

[8] Thomas Kunz, James P. Black, David J. Taylor, and Twan Basten. POET: Target-System Independent Visualisations of Complex Distributed-Application Executions. *The Computer Journal*, 40(8):499–512, 1997.

[9] Leslie Lamport. Time, Clocks and the Ordering of Events in Distributed Systems. *Communications of the ACM*, 21(7):558–565, 1978.

[10] F. Mattern. Virtual Time and Global States of Distributed Systems. In M. Cosnard et al., editor, *Proceedings of the International Workshop on Parallel and Distributed Algorithms*, pages 215–226, Chateau de Bonas, France, December 1988. Elsevier Science Publishers B. V. (North Holland).

[11] Oystein Ore. *Theory of Graphs*, volume 38. Amer. Math. Soc. Colloq. Publ., Providence, R.I., 1962.

[12] Joseph L. Sharnowski and Betty H. C. Cheng. A Visualization-based Environment for Top-down Debugging of Parallel Programs. In *Proceedings of the 9th International Parallel Processing Symposium*, pages 640–645. IEEE Computer Society Press, 1995.

[13] M. Singhal and A. Kshemkalyani. An Efficient Implementation of Vector Clocks. *Information Processing Letters*, 43:47–52, August 1992.

[14] David J. Taylor. Scrolling Displays of Partially Ordered Execution Histories. In preparation.

[15] William T. Trotter. *Combinatorics and Partially Ordered Sets: Dimension Theory*. Johns Hopkins University Press, Baltimore, MD, 1992.

[16] Paul Ward. An Offline Algorithm for Dimension-Bound Analysis. In *Proceedings of the 1999 International Conference on Parallel Processing*. IEEE Computer Society Press, September 1999.

[17] Mihalis Yannakakis. The Complexity of the Partial Order Dimension Problem. *SIAM Journal on Algebraic and Discrete Methods*, 3(3):351–358, September 1982.

[18] Yuh Ming Yong. Replay and Distributed Breakpoints in an OSF DCE Environment. Master's thesis, University of Waterloo, Waterloo, Ontario, 1995.

Correction of Monitor Intrusion for Testing Nondeterministic MPI-Programs

D. Kranzlmüller[1], J. Chassin de Kergommeaux[2], and Ch. Schaubschläger[1]

[1] GUP Linz, Joh. Kepler University Linz, Austria
[kranzlmueller|schaubschlaeger]@gup.uni-linz.ac.at
[2] ID-IMAG, Grenoble,France
Jacques.Chassin-de-Kergommeaux@imag.fr

Abstract. Software analysis tools apply monitors to retrieve state information about program executions. Unfortunately such observation introduces the probe effect, which means that analysis data are influenced by the monitor's requirements to computing time and space. Additionally, nondeterministic parallel programs may yield different execution patterns due to alterations of event ordering at race conditions. In order to correct these intrusions, two activities are carried out. Firstly, the monitor overhead is removed by recalculating event occurrence times based on measurements of the occurred delay. Secondly, event manipulation and program replay is applied at places, where reordering of events has occurred. The resulting data describes the program's execution without monitoring.

1 Introduction

Performance analysis and error detection rely on observations of program state information that are generated by monitoring tools during execution. Since analysis facilities assume that valid data have been obtained, correctness of the monitor's results is crucial. A main problem is the probe effect [1] which means, that a program's execution is perturbed by observation. The critical issue is the size of the monitor overhead, which consists of the time spent in the instrumented code and the amount of memory needed by the monitor [5].

As a consequence, a primary goal of software monitoring tools must be to generate very small monitor overhead. Yet, even the smallest overhead will introduce an execution delay of the target program, which influences the occurrence time of events. Thus, the data describes events that occurred later in time compared to an execution without monitoring. An additional problem to the displacement of observed events is introduced with nondeterministic behavior. In that case, different executions may be observed in successive program runs, even if the same input data are provided. Furthermore, analysis data may not reflect the same execution as if monitoring were turned off, and errors may be hidden in program branches not taken during the monitored execution.

This paper describes an overhead correction algorithm for the topics mentioned above. It targets on MPI-programs [6], where nondeterminism is introduced at so-called wild card receives. Such receives do not specify the origin of a

P. Amestoy et al. (Eds.): Euro-Par'99, LNCS 1685, pp. 154–158, 1999.

message, but instead provide a special parameter to accept a message from any available process, and different messages may be accepted at a corresponding receive event [8]. In fact, if two or more messages are racing towards the receive, their order may be determined by different processor speeds, scheduling decisions, cache contents, or conflicts on the network. Therefore, it is unpredictable which message may be accepted and different executions of the same program may occur, even if the same input data are provided.

2   Correcting the Monitor Overhead

Perturbation of programs as described above occurs in two areas, time and space. We are concentrating on perturbation of monitor data in time, which can be classified into delay of event occurrence, and alteration of event ordering [5]. An example for both perturbations is displayed in figure 1. Two executions are visualized as space-time diagrams, with the time on the horizontal axis and the processes arranged vertically. The arcs between the process axes are communication events, with the tip of the arc pointing to the receiver.

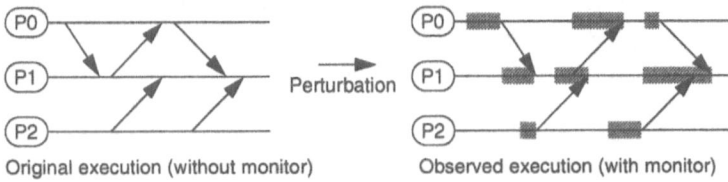

Original execution (without monitor)   Observed execution (with monitor)

Fig. 1. Perturbation of program execution due to monitor overhead

A problem during correction of monitor traces is clock synchronization, which is required because most parallel and distributed systems do not have a global clock. In that case, times on different processors are not comparable, and artifacts, like messages being received before sent, may occur. For synchronization, we are experimenting with three approaches: an existing global clock (e.g. on the IBM SP/2), simple clock synchronization with Lamport's algorithm [3], and extensive clock synchronization with Maillet's method [4].

For the following sections we assume, that either a global clock exists or clock synchronization has been performed. Then the correction algorithm starts by computing the timings of each event. Other approaches ([5], [10]) have already discussed this issue, and our approach for this first step of intrusion correction is similar to [9], which was described for parallel systems based on threads.

In message passing systems the monitor overhead can be identified for non-blocking sends and blocking receives as follows: for the send operation the overhead is generated by monitor functionality before (t_{init}) and after (t_{exit}) the original communication functions (t_{comm}). In addition, a receive operation may be delayed by a certain amount of blocking ($t_{blocking}$). Therefore, the times for a generic send and receive function call are:

$$t_{send} = t_{init} + t_{comm} + t_{exit} \qquad t_{receive} = t_{init} + t_{blocking} + t_{comm} + t_{exit}$$

Hence, the goal of the correction algorithm is to remove the times t_{init} and t_{exit} from t_{send} and $t_{receive}$ for all events in the traces. (For simplification, we assume t_{init} and t_{exit} to be equal for both send and receive events.) A critical point is the blocking time, which can computed based on the assumption that communication time and monitor overhead is fixed. Please note, that communication time means the amount of time spent in the communication function without blocking, which is fixed, because the message is already available. Additionally the monitor overhead is fixed because the activities of the monitor are the same for similar events. These pre-defined portions of the formulae can be evaluated independently from the program, and we apply standard measurements of the SKaMPI benchmarks [7].

After collecting these measurements about the communication functions the correction algorithm is applied to each event in the traces. If an event is a send event, the new event time is computed as

$$t_{event} = t_{prec\_event} - t_{prec\_exit} + t_{send} - t_{init}$$

where t_{prec_event} and t_{prec_exit} denote the times of the preceding event and its exit time, respectively. For receive events the corrections also have to follow Lamport's rule, that messages can only be received after they have been transferred. Thus it is important to determine the transfer time (t_{trans}) for communication between processes, which can be evaluated before the correction algorithm with benchmarks, or based on the times stored in the traces. The latter is possible, since blocking is explicit and receives are waiting until messages are completely available. In that case, the transfer time can be computed based on the times of connected sends and receives as follows:

$$t_{trans} = (t_{recv} - t_{exit}) - (t_{send} - t_{exit})$$

The correction algorithm uses this transfer time to compute the new time for the receive events with the following formula:

$$t_{event} = \max(t_{prec\_event} - t_{prec\_exit} + t_{receive} - t_{init}, t_{send} + t_{transfer})$$

If these formulae are applied to all events in the tracefiles, the resulting trace will contain corrected times as if monitoring would have been turned off. The approach described so far is comparable to conservative approaches [9], which treat nondeterministic events as if they would occur as specified in the traces. However, this need not be the case for obvious reasons and the following section tries to give a solution for that problem.

3   Correcting Nondeterministic Events

The problem of conservative approaches is, that reordering of events is not considered. Yet, if several messages are racing towards a wild card receive, it is unpredictable, which one will succeed. The accepted message is always the message, that arrived first at the receive. Thus, it is important to analyze the reasons that determine the message's arrival date. Besides network transfer time, contention, and scheduling decisions the most obvious prerequisite for the arrival time is the injection time at the sender. From a set of messages being sent to a common receiver, usually the first injected message will be the first to arrive.

However, since the event times of send and receive events are affected by the monitor overhead, it is also possible that the overhead influences the injection time of the messages. It could be, that the global order of send events is scrambled, leading to different arrival times of racing messages. In order to address this problem, the previous approach is extended: For each wild card receive, we determine the racing messages, which are all messages that are possibly accepted at that particular receive [8]. For each of these racing messages we compute the corrected time of the corresponding send event. This produces the following set of send times for n racing messages:

$$T_{send} = \{t_{send(1)}, t_{send(1)}, \ldots, t_{send(n)}\}$$

Based on these send times, we can determine the new order of message arrival by selecting the smallest value. This provides the new value for each wild card receive as:

$$t_{event} = \max(t_{prec\_event} - t_{prec\_exit} + t_{receive} - t_{init}, \min(T_{send} + t_{trans}))$$

At present, we also assume, that the transfer time is equal between all processes, and thus can be determined as above. For varying transfer times calculation has to be performed under consideration of the amount of data and the distance between communicating processes.

After detecting an exchange of event ordering we apply two activities, event manipulation and artificial replay. Both activities have been developed for the MAD environment and serve the purpose of race condition analysis [2]. With event manipulation we can define a different order of message arrival at wild card receives. This is done by specifying the first message from the set of racing messages that arrives at a particular point of exchange (POE). During replay the program is re-executed under the following constraints:

- before the POE: all messages arrive in the order defined by the traces of a previous program run.
- at the POE: the manipulated process waits until the user selected message arrives. Since this is one of the racing messages, its arrival is guaranteed.
- after the POE: the process is executed again without control because the following message transfer is unknown. It is unclear, how the exchanged message order affects the future of the process.

It is important to notice, that only one event manipulation can be evaluated per artificial replay, and only the first message to arrive can be specified. Due to the nondeterministic behavior it is not possible to make any assumptions about the execution after the POE. As a consequence, it may also be necessary to apply event manipulation iteratively, if more than one wild card receive has to be investigated.

4 Conclusion

Monitor intrusion correction is an important issue for tool developers, that want to provide correct and valid data for users. While existing correction algorithms apply conservative approaches, which do not investigate the probability of event reordering, our approach integrates event manipulation and artificial replay. Therefore it is possible to correct the times of each event and additionally the order of racing messages at wild card receives. Furthermore, by specifiying an interval for the arrival time of racing messages, our approach can easily be extended to produce more than one single execution, which is believed to be the program run without monitor, but at set of executions which certainly contains the actual program run.

Acknowledgments The work described in this paper was partially sponsored by the Austrian-French joint-cooperation project Amadee, Austrian contract number 24, French contract number 97005, 1997.

References

[1] J. Gait: A Probe Effect in Concurrent Programs. Software - Practise and Experience, Vol. 16(23), pp. 225–233 (March 1986).
[2] D. Kranzlmüller, S. Grabner, J. Volkert: Debugging with the MAD Environment. Parallel Computing, Vol. 23, No. 1-2, pp. 199–217 (Apr. 1997).
[3] L. Lamport: Time, Clocks, and the Ordering of Events in a Distributed System. Comm. ACM, pp. 558–565 (July 1978).
[4] E. Maillet, C. Tron: On Efficiently Implementing Global Time for Performance Evaluation on Multiprocessor Systems. Journal of Parallel and Distributed Computing, Vol. 28, pp. 84–93 (July 1995).
[5] A. Malony, D. Reed: Models for performance perturbation analysis. Proc. Workshop Parallel Distributed Debugging, ACM SIGPLAN/SIGOPS and Office of Naval Research, (May 1991).
[6] Message Passing Interface Forum: MPI: A Message-Passing Interface Standard - Version 1.1. http://www.mcs.anl.gov/mpi/ (June 1995).
[7] R. Reussner, P. Sanders, L. Prechelt, M. Müller: SKaMPI: A Detailed, Accurate MPI Benchmark. Proc. 5th European PVM/MPI Users' Group Meeting, Springer, Lecture Notes in Computer Science, Vol. 1497, Liverpool, UK, pp. 52–59 (Sept. 1998).
[8] M. Ronsse, D. Kranzlmüller: RoltMP - Replay of Lamport Timestamps for Message Passing Systems. Proc. 6th EUROMICRO Workshop on Parallel and Distributed Processing, University of Madrid, Spain, pp. 87–93, (Jan. 1998).
[9] F. Teodorescu, J. Chassin de Kergommeaux: On Correcting the Intrusion of Tracing Non-deterministic Programs by Software. Proc. EUROPAR'97 Parallel Processing, 3rd Intl. Euro-Par Conference, Springer, Lecture Notes in Computer Science, Vol. 1300, Passau, Germany, pp. 94–101 (Aug. 1997).
[10] W. Wu, R. Gupta, M. Spezialetti: Experimental Evaluation of On-line Techniques for Removing Monitoring Intrusion. Proc. of SPDT'98, SIGMETRICS Symposium on Parallel and Distributed Tools, Welches, Oregon, pp. 30–39, (Aug. 1998).

Improving the Performance of Distributed Shared Memory Environments on Grid Multiprocessors[1]

Dimitris Dimitrelos and Constantine Halatsis

Athens High Performance Computing Laboratory and
Department of Informatics, University of Athens
Panepistimiopolis, Athens 15771, Greece
e-mail: ddimitr@hpcl.uoa.gr, halatsis@di.uoa.gr

Abstract. Distributed Shared Memory (DSM) is a good solution to the scalability, complexity and high cost problems of large scale Shared Memory Multiprocessors, as well as to the difficulty of the programming model problem of the message passing Distributed Memory Multiprocessors. We present a method for improving the performance of Distributed Shared Memory Environments running on grid multiprocessors. The method is based on removing the inherent centralism imposed by X-Y routing that causes congestion in the centre of the grid. Simulation, as well as implementation results on a 1024 processor machine show an improvement of up to 24%.

1. The Environment

Distributed Shared Memory (DSM) allows the execution of programs assuming the Shared Variable programming paradigm in Distributed Memory architecture multiprocessor systems.

All DSM systems must use a coherence protocol, similar to those used by Shared Memory multiprocessors for cache coherence, to keep the shared data consistent at all times. The underlying message passing system makes it unrealistic to use any protocol that assumes broadcast, so directory based protocols are used.

According to directory coherence schemes, each shared object has a directory entry associated with it. In this directory entry the identity of the object's owners, its status (exclusively owned/shared/invalidated) and possibly other information are maintained and updated.

The directory entry of each shared object is kept in some processor. A processor that keeps at least one directory entry is called a *directory handler* (DH). The directory handler maintaining the directory entry of each shared object must be at all times known to every processor in the DSM system, since interaction with the DH is necessary for non-local accesses to shared data.

---

[1] This work was partly supported by the Greek General Secretariat of Research and Technology under the YPER-94 program.

P. Amestoy et al. (Eds.): Euro-Par'99, LNCS 1685, pp. 159-162, 1999.

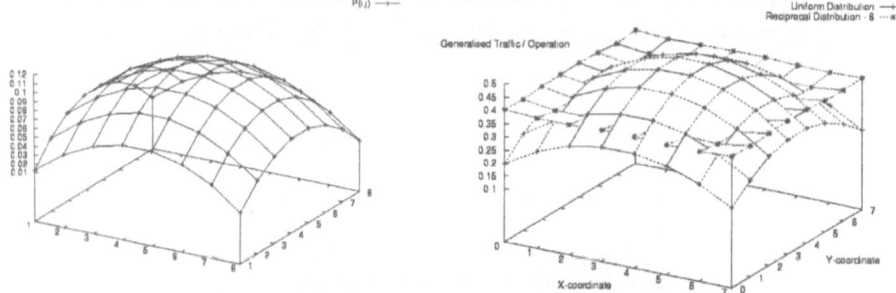

Fig. 1. P(i, j) values for an 8x8 processor grid

Fig. 2. Simulated traffic for uniform vs. reciprocal directory distribution.

The distribution of the directory data is a factor that plays an important role to the performance of the system. *Centralised Directory* schemes use a single processor for maintaining all directory data. According to *Distributed Directory* schemes, the directory data are distributed to more than one processors. According to the *Uniform Distribution* scheme (first introduced by Li in [1] as *Distributed Directory*) the directory entries are *uniformly* distributed among the processors. Even though the Uniform Distribution Directory scheme seems very attractive, its effect on grid multiprocessors using the X-Y routing algorithm for message passing has not been investigated.

2. Reciprocal Directory Distribution

According to the X-Y message routing, a message is propagated from the sender to the receiver first along the X axis until it reaches the column of the receiver, and then along the Y axis. The weak point of X-Y routing is the fact that more messages pass through the centre than through the 'edges' of the grid, on their way to their destination, causing more 'intermediate' traffic to the central processors.

We denote as $P(i, j)$ of processor (i, j) in a MxN processor grid the probability that a random message (not sent from, neither sent to (i, j)) passes through (i, j).

In [2] we formally proved the following theorem:

$$P(i,j) = \frac{-2(i^2 M + j^2 N) + 2iM(N + 1) + 2jN(M + 1) - 3MN - M - N + 1}{(MN - 2)(MN - 1)} \quad (1)$$

The value of P(i, j) among the processors on a 8x8 processor grid is depicted in Figure 1. The X and Y axes correspond to the processor's position in the grid and the Z axis to its P(i, j) value. It is easy to observe the mountain with the peak at the central (or central pair or quadruple) processor.

The Uniform Distribution Scheme for the shared object directory is the natural choice of a DSM system designer, since it provides uniformity in the distribution of data and avoids bottlenecks and hot spots. However, as all DSM operations are

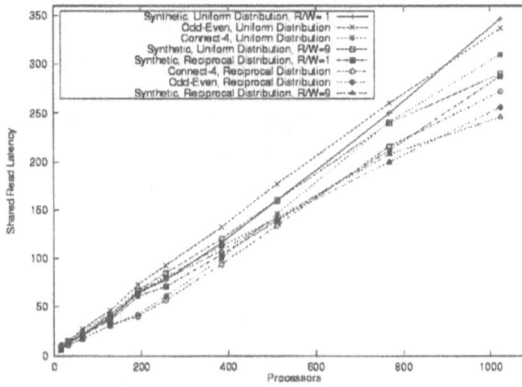

Fig. 3. Benchmark Results: Average Shared Read Times

implemented by explicit message passing between the processors, the centralisation observed at X-Y routing will be passed on to the system and therefore have an impact on the system's performance. It will cause easier congestion to the central parts of the network (that are accepting more traffic), reduce the message bandwidth and effectively reduce the performance of the DSM system.

The idea behind reversing the effects of X-Y centralism is simple. Our simulations revealed that 70%-90% of all messages exchanged in a DSM system start from or end to the Directory Handler. Therefore if we alter the directory distribution in order to move some traffic from the central processors to processors closer to the edges of the grid, we can create an anti-force to the X-Y centrality and equalise the communication load throughout the grid. Intuitively, the distribution should be done inverse proportionately to the 'centrality' of each processor, i.e. the processors residing on the edges of the grid will get a bigger part of the directory (more directory entries) than the processors closer to the centre of the grid.

According to our distribution scheme (*Reciprocal Directory Distribution*) the directory data are distributed proportionately to the sixth power of the average distance between the processor and any other processor. The simulated traffic per operation on an 16x16 processor grid is presented in Figure 2, along with the traffic resulting using the normal uniform directory distribution. The suppression of the centralism caused by X-Y is obvious.

The price that we have to pay for decentralisation is the extra distance that the messages have to travel in order for an operation to complete, as a result from the placement of more directory data in the outer parts of the grid. The increase varies from 3% in the 4x4 case to 9.3% in the 64x64 case.

4. Implementation Results

MaDCoWS[3] (MAssively Distributed COnfigurable Variable Validation System) is a scalable multi-parametric DSM system for 2D grid multiprocessors developed at the Athens High Performance Computing Laboratory and the Dpt. of Informatics of the University of Athens. It implements arbitrary-size variable sharing as well as global semaphore, barrier and read-modify-write operations. MaDCoWS has been developed as a runtime system under the parallel operating environment PARIX[4],

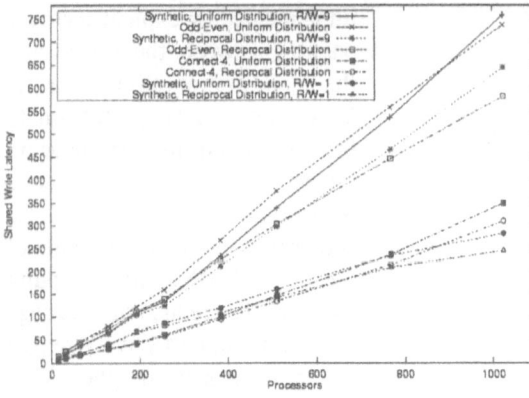

Fig. 4. Benchmark Result: Average Shared Write Times

and has been used, tested and benchmarked on the Parsytec GCel-1024, a 1024 processor machine. The system is currently being ported to the Intel Paragon and to the Parsytec GC-PP.

In order to evaluate the performance and scalability of the reciprocal directory distribution, we run several synthetic benchmarks and two real life applications -two games. The first is a virtual odd-even game played by the processors that exhibits a high degree of sharing, both temporally and spatially. The second application is a connect-4 game played by the system vs. a human.

The average shared read and shared write times for the 3 benchmarks are presented in Figures 3 and 4. In all cases, the Reciprocal Directory Distribution scheme performs better than the Uniform Directory Distribution, improving the DSM system's performance by up to 24% for shared read operations and up to 21% for shared write operations. The scheme performs better in the cases of dense sharing (as in the case of the Odd-Even benchmark), and for large grids. In the case of sparse sharing and very small numbers of processors (<64), a up to 10% increase in the write latency was observed.

Acknowledgments

The authors would like to thank Dr.Alexander Reinefeld and the PC2 centre of the University of Paderborn for allowing us to use their parallel computers for benchmark runs.

References

1. Li, K. 'Shared Virtual Memory on Loosely Coupled Multiprocessors.' PhD thesis, Department of Computer Science, Yale University, September 1986.
2. Dimitrelos, D. and Halatsis, C. 'On the Distribution of Directory Information in a Software Controlled Distributed Shared Memory System'. In Proc. of the Workshop on Parallel Programming and Computation (ZEUS'95), pages 75-89, Linkoping, May 1995.
3. Dimitrelos, D and Halatsis, C. 'MaDCoWS: A Scalable Distributed Shared Memory Environment for Massively Parallel Multiprocessors', in Proc of HPCN'99, pp 784-793
4. Parsytec Computer GmbH 'PARIX 1.2. Software Documentation', March 1993.

Topic 02
Performance Evaluation and Prediction

Jean-Marc Vincent

Local Chair

This topic aims at bringing together people working in the different fields of parallel and distributed systems benchmarking, performance evaluation, and prediction. It addresses methodologies, tools, modelling, of parallel and distributed applications. Especially welcome are contributions that bridge the gap between the fields of analytical modelling, performance measurement, parallel benchmarking and parallel program characterization. 20 papers were submitted in this topic, 8 were selected as regular papers and 6 as short papers.

Performance of Communication Support

Communications appears frequently as a bottleneck for performances. Increasing performances of message passing communication libraries or adapting programming models are two ways to avoid network congestions. The first paper in this session analyses the performances of whormole switching with adaptive routing strategies in a torus network. Analytical solutions and simulation results demonstrate the interest of such planar-adaptive routing. Performances of message passing libraries are crucially related to data distributions. By experimenting on Cray T3E and SGI Origin 2000, the second paper estimates the impact of locality. In the third paper, comparison of libraries on a Fujitsu AP3000 are based on global communication schemes analysis. Using of a software virtual shared memory on a distributed memory architecture improves scalability and portability of parallel programs. This is the aim of VOTE, a communication support, whose performances are analysed in the fourth paper.

Performance of Implementation Strategies

Scheduling strategy is one of the major parameter to increase efficiency of parallel programs. It depends on task decomposition of the program and interleaving of executions. The first paper adresses the issue of performance prediction of parallel program written in a classical language. From source code to the simulation of a synthetic program, the proposed tool offers a guideline to programmers. After decomposing parallel codes as dataflow graphs, near min-cut methods can be used to map the program onto architectures. This is the purpose of the second paper. The third paper studies the impact of non uniform instruction duration on scheduling. Using a probabilistic approach and experiments, sensitivity of schedulers are established.

P. Amestoy et al. (Eds.): Euro-Par'99, LNCS 1685, pp. 163–164, 1999.
© Springer-Verlag Berlin Heidelberg 1999

Data Locality Management

Accessing data needs special mechanisms of cache memory management. Getting a uniform access implies scalability of the architecture. A linear hashing technique, proposed in the first paper, is validated by simulation and experimentation on a real application. In addition, memory hierarchy have a important impact on application behavior. In a numerical parallel applications environment, the second paper analyses and compares performance indexes of caches by modelling and trace driven simulation. The third paper deals with caches in a web server environment and provides a deeper understanding of the interaction between harware caches and operating systems.

Tools for Performance Prediction

Benchmarking approaches offer standardization to performance measurement. The first paper of this session explores the HPFBench suite. Designed for evaluation of HPF compilers, it identifies the bottlenecks in communication primitives. The second paper focuses on the performance at a middleware level and supplies a portable benchmark toolkit. The importance of ORB developments needs a deeper understanding of the performance and scalability tradeoffs. The third paper deals with instrumentation and tracing of concurrent applications.These traces are coupled with an efficient target architecture emulator. The fourth paper adresses the problem of performances of Digital Signal Processing.

Performance Analysis of Wormhole Switching with Adaptive Routing in a Two-Dimensional Torus

M. Colajanni[1], B. Ciciani[2], and F. Quaglia[2]

[1] University of Modena, Italy
colajanni@unimo.it
[2] University of Roma "La Sapienza", Italy
ciciani@dis.uniroma1.it, quaglia@dis.uniroma1.it

Abstract. This paper presents an analytical evaluation of the performance of wormhole switching with adaptive routing in a two-dimensional torus. Our analysis focuses on minimal and fully adaptive routing that allows a message to use any shortest path between source and destination.

1 Introduction

In the wormhole switching technique [2], [11] any message is partitioned into a set of flits. The transmission works as follows: after passing through a channel, the *header flit* tries to get another channel while the *data flits* are transmitted through the already obtained channels. In the case of channel contention, the message flits are stored in the flit buffers of the nodes along the already established path. A channel is released only when the last flit (*tail flit*) of the message is transmitted through it.

If wormhole is combined with adaptive routing, then there is no predefined path from the source to the destination. Since already obtained channels are released only at the end of the message transmission, a mechanism to prevent deadlock must be considered. This is done by using multiple *virtual channels* multiplexed on each physical link and forcing a pre-defined order on the allocation of virtual channels to messages [2], [11]. An adaptive routing policy allowing messages to use only shortest paths is called *minimal*. Moreover, an adaptive routing policy is called *fully adaptive* if a message is allowed to follow any path of the minimal (or non-minimal) class.

In this paper we propose an analytical approach to evaluate the performance of wormhole switching with minimal and fully adaptive routing in a two-dimensional torus. Most previous papers presenting analytical models consider either different switching techniques (such as *circuit-switching* [4] and *virtual-cut-through* [10]) or deterministic routing [1], [3], [5], [6]. To the best of our knowledge, an analytical evaluation of wormhole with adaptive routing is done only in [8]. Such analysis differs from our work as it considers Duato's adaptive routing [7] instead of minimal and fully adaptive routing.

P. Amestoy et al. (Eds.): Euro-Par'99, LNCS 1685, pp. 165–172, 1999.
© Springer-Verlag Berlin Heidelberg 1999

The remainder of the paper is organized as follows. In Section 2 we present the system model. In Section 3 we propose a solution for the estimation of the mean latency time. In Section 4 we validate the analysis with some simulation results.

2   System Model

Each node (i, j) of the two-dimensional torus consists of a *processing element* $P_{i,j}$ and a *router node* $N_{i,j}$ with a controller per each dimension. Nodes are connected through bi-directional full-duplex links[1]. The flit size (bits) is equal to the number of wires B of a link. Therefore, each flit is transmitted in a single cycle link time. This time represents the basic temporal unit of our analysis.

We assume that generation times of messages at each processor are independent and identically distributed in accordance with a Poisson process with rate $1/\tau_{bit}$ (bit/cycle). This assumption and the symmetry of torus topology guarantee that the network is *balanced* [9] that is, we can assume that channels are equally likely to be visited independently of the message destination distribution. As in Dally's approach [5], we assume that the network has no virtual channels.

We consider the transmission of a message as consisting of two consecutive and separate phases: *path-hole* and *data-trail*. During the path-hole phase, the *header flit* builds the path from the source to the destination. All delays due to channel contentions are taken into account in this phase. The data-trail phase models the transmission of data flits along the channels of the selected path. Since we are assuming uniformly chosen destinations, the average length of the path along each dimension is $K = k/4$, where k is the number of nodes of each dimension. We define *average message*, a message which must travel on exactly K channels in each dimension.

Due to network symmetry and bi-directional links, it is possible to partition the torus into four quadrants (north-east, north-west, south-west, south-east), with the same number of nodes, such that the path of an average message belongs entirely to one quadrant. Our analysis focuses on the south-east quadrant. This quadrant is shown in Figure 1, where router nodes (i.e., circles) and channels (i.e., rectangular boxes) are represented through a two-dimensional array notation that identifies their position with respect to the average message's view that is, $N_{1,1}$ is the first router that the average message crosses, $X_{1,1}$ and $Y_{1,1}$ are the first horizontal and vertical channels available to that message, respectively.

Since each message can choose among four directions, the flow considered in the analysis represents 1/4 of the entire flow generated by each node. Let $1/\tau_E = 1/(L\tau_{bit})$ be the message generation rate (messages/cycle) per each node, where $1/\tau_{bit}$ is the emission rate in bit/cycle, and L is the message average length. The average flow considered in the analysis has generation time $\tau = 4\tau_E$.

We can identify three flow streams: (i) the stream β^X denoting messages that use only channels of the dimension X; (ii) the stream β^Y denoting messages that

---

[1] The terms *link* and *channel* are used interchangeably.

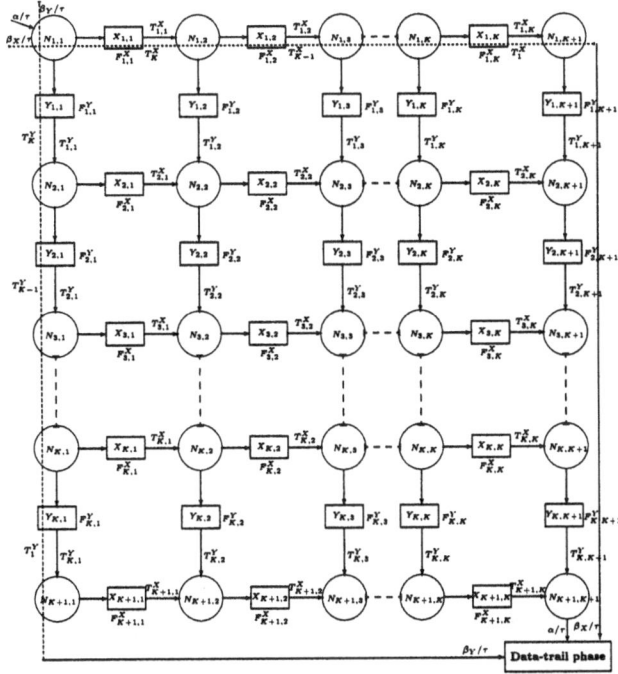

Fig. 1. Communication diagram for the *average message*

use only channels of the dimension Y; (iii) the stream $\alpha = 1 - (\beta^X + \beta^Y)$ denoting messages that can use adaptive path selection along both dimensions. As the traffic is uniform, the value of each stream can be estimated through the ratio between the number of nodes reachable by that stream and the total number of nodes: $\alpha = (k-1)^2/(k^2-1) = (k-1)/(k+1)$, $\beta^X = \beta^Y = (k-1)/(k^2-1) = 1/(k+1)$.

In Figure 1 plain lines denote the paths of the α stream, horizontal dotted lines refer to the β^X stream, and vertical dotted lines refer to the β^Y stream. $F_{i,j}^X$ represents the flow rate of the α stream that uses $X_{i,j}$. $T_{i,j}^X$ denotes the *residual transmission time* from $X_{i,j}$ that is, the mean time that the header flit takes to get from $X_{i,j}$ to the destination. Analogous notation is used for a vertical channel $Y_{i,j}$. A single parameter is sufficient for the identification of the residual transmission time of the streams β^X and β^Y, because they use channels along one dimension only. For example, T_{K-j+1}^X refers to the channel $X_{1,j}$ which is used by β^X, and T_{K-i+1}^Y refers to the channel $Y_{i,1}$ which is used by β^Y. The sub-index denotes the number of router nodes that the β stream has yet to cross.

3   The Analysis

When a message of the α stream reaches a node $N_{i,j}$, it attempts to continue along the dimension X. If the channel $X_{i,j}$ is not available (this happens with probability p_X), then the message tries to continue along the dimension Y. If the channel $Y_{i,j}$ is busy as well (this happens with probability p_Y), the message waits until at least one channel is released. The evaluation of p_X and p_Y is postponed. Let F be the adaptive flow reaching a node, F^X be the horizontal flow exiting that node, and F^Y be the vertical flow exiting that node. The following *bifurcation rule* holds: $F^X = F(1-p_X)/(1-p_X p_Y) = F f^X$ and $F^Y = F p_X (1-p_Y)/(1-p_X p_Y) = F f^Y$ (due to space limitation the proof is omitted). By applying the bifurcation rule to the α stream, we obtain the equations for all the flows of the diagram in Figure 1. We partition the horizontal channels into six classes: the first channel $X_{1,1}$; the first row $X_{1,j}$ but the first channel; the first column $X_{i,1}$ but the first and the last channels; the intermediate channels $X_{i,j}$ but the first row, the first column and the last row; the last channel of the first column $X_{K+1,1}$; the last row $X_{K+1,j}$ but $X_{K+1,1}$. Using the bifurcation rule and Figure 1, we obtain the flow rates associated to each class of channels:

$$F_{1,1}^X = f^X \alpha / \tau \tag{1}$$

$$F_{1,j}^X = F_{1,j-1}^X f^X = (f^X)^j \alpha / \tau \qquad (j = 2, 3, \ldots, K) \tag{2}$$

$$F_{i,1}^X = F_{i-1,1}^Y f^X = (f^Y)^{i-1} f^X \alpha / \tau \qquad (i = 2, 3, \ldots, K) \tag{3}$$

$$F_{i,j}^X = F_{i,j-1}^X f^X + F_{i-1,j}^Y f^X \qquad (i, j = 2, 3, \ldots, K) \tag{4}$$

$$F_{K+1,1}^X = F_{K,1}^Y = (f^Y)^K \alpha / \tau \tag{5}$$

$$F_{K+1,j}^X = F_{K+1,j-1}^X + F_{K,j}^Y \qquad (j = 2, 3, \ldots, K) \tag{6}$$

Similarly, we partition the vertical channels into six classes, with the following flow rates associated to each class:

$$F_{1,1}^Y = f^Y \alpha / \tau \tag{7}$$

$$F_{1,j}^Y = F_{1,j-1}^X f^Y = (f^X)^{j-1} f^Y \alpha / \tau \qquad (j = 2, 3, \ldots, K) \tag{8}$$

$$F_{1,K+1}^Y = F_{1,K}^X = (f^X)^K \alpha / \tau \tag{9}$$

$$F_{i,1}^Y = F_{i-1,1}^Y f^Y = (f^Y)^i \alpha / \tau \qquad (i = 2, 3, \ldots, K) \tag{10}$$

$$F_{i,j}^Y = F_{i,j-1}^X f^Y + F_{i-1,j}^Y f^Y \qquad (i, j = 2, 3, \ldots, K) \tag{11}$$

$$F_{i,K+1}^Y = F_{iK}^X + F_{i-1,K+1}^Y \qquad (i = 2, 3, \ldots, K) \tag{12}$$

The mutual dependences existing among the previous equations can be solved analyzing the flow of each channel starting from $X_{1,1}$, and then following the west-east and north-south direction.

We model each link as an M/G/1 queue with multiple classes of flow indexed $m = 1, 2, \ldots, M$. Assuming that messages of class m arrive with Poissonian rate λ_m and that their mean service time is $\overline{T}_m$, we obtain the following expression for the mean waiting time to get that channel [12, page 276]:

$$\overline{W} = \sum_{m=1}^{M} \frac{\rho_m}{\rho} E[W_m] = \frac{1}{2(1-\rho)} \sum_{m=1}^{M} \lambda_m \left(\overline{T}_m^2 + \sigma_m^2 \right) \tag{13}$$

where σ_m^2 denotes the variance of $\overline{T}_m$, and $\rho = \sum_{m=1}^{M} \lambda_m \overline{T}_m$. The analysis focuses on messages that travel from $N_{1,1}$ to $N_{K+1,K+1}$ following the west-east and north-south directions that is, in the south-east quadrant. Hence, the waiting times of interest are W_{WE} and W_{NE} for the horizontal channels, and W_{NS} and W_{WS} for the vertical channels. We briefly present the steps for the estimation of W_{WE} for a generic horizontal channel $X_{i,j}$. The other waiting times are obtained through analogous considerations. The service time $\overline{T}_m$ (i.e., the link utilization times for the flow of messages of class m) will be evaluated later. To find W_{WE} we have to determine the M classes of flow that use $X_{i,j}$. Depending on the source and destination, that channel may represent the first horizontal channel requested by a message generated at $P_{i,j}$, or a channel of the last row of the diagram in Figure 1, or any horizontal channel in the middle of the same diagram. We have obtained for W_{WE} the following expression where $(S_{i,j}^X)^2 = [(\overline{T}_{i,j}^X)^2 + (\sigma_{i,j}^X)^2]$ (equations for the other waiting times are analogous):

$$W_{WE} = \frac{\sum_{j=1}^{K} F_{K,j}^Y (S_{K+1,j}^X)^2 + \sum_{i=2}^{K} \sum_{j=1}^{K} f^X F_{i-1,j}^Y (S_{i,j}^X)^2 + \frac{\beta^X (S_K^X)^2 + \alpha f^X (S_{1,1}^X)^2}{\tau}}{1 - \rho_{WE}} \tag{14}$$

A message holds a link for a period that we call *link utilization time*. Its value is equal to the residual transmission time of the message minus the number of flits already transmitted through that link. The following equations denote the link utilization time as a function of the link position in the average message transmission:

$$\overline{T}_j^X = T_j^X - j \qquad (j = 1, \ldots, K) \tag{15}$$

$$\overline{T}_i^Y = T_i^Y - i \qquad (i = 1, \ldots, K) \tag{16}$$

$$\overline{T}_{i,j}^X = T_{i,j}^X - (2K - i - j + 2) \qquad (i = 1, \ldots, K+1, j = 1, \ldots, K) \tag{17}$$

$$\overline{T}_{i,j}^Y = T_{i,j}^Y - (2K - i - j + 2) \qquad (i = 1, \ldots, K, j = 1, \ldots, K+1) \tag{18}$$

The general distribution assumed for the link utilization time would require also the specification of the variance of the service time $\overline{T}_{i,j}$, To this purpose, we observe that in wormhole switching the link utilization time is equal to the mean time to transmit the entire message plus the mean waiting time to obtain the remaining links of the path. The former term has a known distribution (that chosen for the message length), whereas the latter term and the covariance between these terms are unknown. We assume null covariance and exponential distribution for the mean waiting time.

The residual transmission time $T_{i,j}^X$ is the average time that the header of a message experiences while traveling from $X_{i,j}$ to the destination. Hence, $T_{latency}$ represents the residual transmission time of a message while traveling from the source node to the destination. To evaluate these residual transmission times, we adopt a backward analysis that moves from the *data trail* node of the diagram in Figure 1 back to the first line of horizontal and vertical channels.

The mean time T_{DT} to complete the *data trail* phase is a known parameter because it corresponds to the transmission time of all flits of an average message.

Therefore, we can write $T_{DT} = L/B$, where B denotes the number of wires per link, and L the average length of the message in flits.

For the messages following adaptive paths we distinguish four classes of horizontal channels: the last channel of the last row $X_{K+1,K}$; the last column $X_{i,K}$ but $X_{K+1,K}$; the last row $X_{K+1,j}$ but $X_{K+1,K}$; the other $X_{i,j}$. These classes have the following residual transmission time:

$$T_{K+1,K}^X = T_{DT} + 1 \tag{19}$$

$$T_{i,K}^X = W_{WS} + T_{i,K+1}^Y + 1 \qquad (i = 1, 2, \ldots, K) \tag{20}$$

$$T_{K+1,j}^X = W_{WE} + T_{K+1,j+1}^X + 1 \qquad (j = 1, 2, \ldots, K-1) \tag{21}$$

$$T_{i,j}^X = (1 - p_X)T_{i,j+1}^X + p_X(1 - p_Y)T_{i,j+1}^Y +$$
$$p_X p_Y \left[r(W_{WE}T_{i,j+1}^X) + (1 - r)(W_{WS}T_{i,j+1}^Y) \right] + 1$$
$$(i = 1, 2, \ldots, K; j = 1, 2, \ldots, K-1) \tag{22}$$

The additional unit term denotes the time to transmit one data flit through a link. The residual transmission time in (22) for the intermediate links is obtained by adding the time values corresponding to the following events multiplied by the probability of their occurrence: the message continues along the dimension X; the message continues along the dimension Y; the message has found both channels busy, and continues when either one becomes free. The equations for the residual transmission time of the vertical channels are obtained in a similar way:

$$T_{K,K+1}^Y = T_{DT} + 1 \tag{23}$$

$$T_{K,j}^Y = W_{NE} + T_{K+1,j}^X + 1 \qquad (j = 1, 2, \ldots, K) \tag{24}$$

$$T_{i,K+1}^Y = W_{NS} + T_{i+1,K+1}^Y + 1 \qquad (i = 1, 2, \ldots, K-1) \tag{25}$$

$$T_{i,j}^Y = (1 - p_X)T_{i+1,j}^X + p_X(1 - p_Y)T_{i+1,j}^Y +$$
$$p_X p_Y \left[s(W_{NS}T_{i+1,j}^X) + (1 - s)(W_{NE}T_{i+1,j}^Y) \right] + 1$$
$$(i = 1, 2, \ldots, K-1; j = 1, 2, \ldots, K) \tag{26}$$

The r and s terms in (22) and (26) are binary variables: if $W_{WS} < W_{WE}$, then $r = 0$, else $r = 1$; if $W_{NE} < W_{NS}$, then $s = 0$, else $s = 1$. The residual transmission times associated to the β^X and β^Y streams are given by:

$$T_1^X = T_1^Y = T_{DT} + 1 \tag{27}$$

$$T_j^X = W_{WE} + T_{j-1}^X + 1 \qquad (j = 2, \ldots, K) \tag{28}$$

$$T_i^Y = W_{NS} + T_{i-1}^Y + 1 \qquad (i = 2, \ldots, K) \tag{29}$$

Equations (27)–(29) are in recursive form. They can be solved through a backward flow analysis starting from T_{DT}, and then following the south-north and west-east direction, alternating horizontal and vertical channels.

The evaluation of the mean latency time requires an estimation of the probability of contention on the vertical and horizontal channels that is, p_X and p_Y.

These probabilities can be obtained as the sum of all flow rates multiplied by time values along horizontal and vertical channels, respectively. In addition to the flows that travel following the west-east direction in X and the north-south direction in Y (as shown by Figure 1), a horizontal channel is used also by the flows that follow the south-north and west-east direction. Analogously, a vertical channel is used also by the flows that follow the north-south and east-west direction. These contributions double the amount of messages on each link and motivate the multiplier terms two in the following equations:

$$p_X = 2 \sum_{i=1}^{K+1} \sum_{j=1}^{K} F_{i,j}^X \overline{T}_{i,j}^X + \frac{2\beta^X}{\tau} \sum_{j=1}^{K} \overline{T}_j^X \tag{30}$$

$$p_Y = 2 \sum_{i=1}^{K} \sum_{j=1}^{K+1} F_{i,j}^Y \overline{T}_{i,j}^Y + \frac{2\beta^Y}{\tau} \sum_{i=1}^{K} \overline{T}_i^Y \tag{31}$$

We evaluate the mean latency time as weighted sum of the residual transmission time values experienced by α, β^X, β^Y that is:

$$T_{latency}(\tau_E) = \alpha T^{\alpha} + \beta^X \left(T_K^X + W_{WE} + W_{NE} \right) + \beta^Y \left(T_K^Y + W_{NS} + W_{WS} \right) \tag{32}$$

where $T^{\alpha} = (1 - p_X)T_{1,1}^X + p_X(1 - p_Y)T_{1,1}^Y + p_X p_Y [v(W_{WE} + W_{NE} + T_{1,1}^X) + (1-v)(W_{NS} + W_{WS} + T_{1,1}^Y)]$. The term v is a binary variable that is, if $(W_{WE} + W_{NE}) < (W_{NS} + W_{WS})$, then $v = 1$, else $v = 0$. The mutual dependency existing among some variables does not permit to achieve a closed formula for the mean latency time. However, a simple backward computation provides any performance value in few iterative steps.

4   Analysis Validation

In this section we validate the analytical model through a discrete event simulator. The Independent Replication Method was used to obtain confidence intervals at 95% level of confidence.

We report values related to four different network dimensions: 4×4, 8×8, 12×12 and 16×16. The average message length is 12 flit, hence we validate the model in the case of: average message length greater than the average path length, average message length equal to the average path length, and average message length smaller than the average path length. The message generation is modeled as a Poisson process with τ_E time cycles per node, and the message destination is an uniformly distributed random variable.

The tables below show the analytical and simulation latency times for the transmission of a message. The results show that the latency evaluated through the model is a good approximation of that obtained through simulation. In particular, the error is under 6% for low and medium arrival rates and under 12% for arrival rates close to the network saturation point.

Gen. rate per node	4 x 4			8 x 8		
	Simulation	Model	Error(%)	Simulation	Model	Error(%)
0.001	13.43	13.65	+1.6	15.55	15.73	+1.1
0.002	13.58	13.70	+0.9	15.96	15.92	-0.3
0.003	13.68	13.75	+0.5	16.27	16.11	-1.0
0.004	13.89	13.81	-0.6	16.81	16.31	-2.9
0.005	14.14	13.87	-1.9	17.10	16.51	-4.6
0.006	14.32	13.93	-2.7	17.66	16.72	-5.3
0.007	14.53	13.98	-3.8	18.15	16.94	-6.6
0.008	14.73	14.04	-4.7	18.65	17.18	-7.9
0.009	14.89	14.10	-5.3	19.14	17.42	-8.9
0.010	15.06	14.17	-5.9	19.52	17.69	-9.4
0.011	15.29	14.23	-6.9	20.12	17.97	-10.7
0.015	16.10	14.50	-9.9	22.18	19.39	-12.6

Gen. rate per node	12 x 12			16 x 16		
	Simulation	Model	Error(%)	Simulation	Model	Error(%)
0.001	17.79	17.90	+0.6	20.07	20.05	+0.1
0.002	18.43	18.31	-0.6	20.99	20.65	-1.6
0.003	19.09	18.76	-1.6	21.85	21.37	-2.2
0.004	19.88	19.27	-3.0	22.82	22.27	-2.5
0.005	20.73	19.87	-4.1	23.99	23.47	-2.2
0.006	21.33	20.58	-3.5	25.06	25.34	+1.1
0.007	22.15	21.40	-3.4	26.27	29.28	+11.4
0.008	22.65	22.52	-0.6	-	-	-
0.009	23.25	24.20	+4.0	-	-	-

References

[1] V.S. Adve, and M.K. Vernon, "Performance analysis of mesh interconnection networks with deterministic routing", *IEEE Trans. on Parallel and Distributed Systems*, vol. 5, Mar. 1994, pp. 225–246.

[2] K.M. Al-Tawil, M. Abd-El-Barr, and F. Ashraf, "A survey and comparison of wormhole routing techniques in mesh networks", *IEEE Network*, Mar.-Apr. 1997, pp. 38–45.

[3] B. Ciciani, M. Colajanni, and C. Paolucci, "An accurate model for the performance analysis of deterministic wormhole routing", Proc. *IEEE 11th Int. Parallel Processing Symposium* (IPPS'97), Geneva, Switzerland, April 1997, pp. 353–359.

[4] M. Colajanni, B. Ciciani, and S. Tucci, "Performance analysis of circuit-switching interconnection networks with deterministic and adaptive routing", *Performance Evaluation*, vol. 34, no. 1, Sept. 1998, pp. 1–26.

[5] W.J. Dally, "Performance analysis of k-ary n-cube interconnection networks", *IEEE Trans. on Computers*, vol. 39, no. 6, June 1990, pp. 775–785.

[6] J.T. Draper, and J. Gosh, "A comprehensive analytical model for wormhole routing in multicomputer systems", *J. Parallel and Distributed Computing*, vol. 23, no. 2, Nov. 1994, pp. 202–214.

[7] J.Duato, "A new theory of deadlock free adaptive routing in wormhole routing networks", *IEEE Trans. on Parallel and Distributed Systems*, vol. 4, no. 12, 1993, pp. 1320–1331.

[8] O. Khaoua, "An analytical model of Duato's fully-adaptive routing algorithm in k-ary n-cubes", Pro. *27th Int. Conference on Parallel Processing*, 1998.

[9] P. Kermani, and L. Kleinrock, "Virtual cut-through: a new computer communication switching technique", *Computer Networks*, vol. 3, 1979, pp. 267–286.

[10] A. Lagman, W.A. Najjar, S. Sur, and P.K. Srimani, "Evaluation of idealized adaptive routing on k-ary n-cubes", Proc. *IEEE Symp. on Parallel and Distributed Processing '94*, Dallas, TX, Dec. 1993, pp. 166–169.

[11] L.M. Ni, and P.K. McKinley, "A survey of wormhole routing techniques in direct networks", *IEEE Computer*, vol. 26, no. 2, Feb. 1993, pp. 62–76.

[12] H. Takagi, *Queueing Analysis - A Foundation of Performance Evaluation*, Elsevier-North-Holland, 1991.

Message Passing Evaluation and Analysis on Cray T3E and SGI Origin 2000 Systems

M. Prieto, D. Espadas, I.M. Llorente, and F. Tirado

Departamento de Arquitectura de Computadores y Automatica
Facultad de Ciencias Fisicas
Universidad Complutense
28040 Madrid, Spain
E-mail: {mpmatias,despadas,llorente,ptirado}@dacya.ucm.es

Abstract. We present the results of different communication tests on some current parallel computers, the Cray T3E and the SGI Origin 2000. The aim of this paper is to study the effect of the local memory use and the communication network exploitation on message sending. For this purpose, we have first designed experiments without network contention to establish the achievable bandwidths. We have then modified this base experiment by increasing the contention of the network and by decreasing the spatial locality properties of the messages. We analyse these results taking into account the underlying architectures and we conclude with some hints for regular applications.

1. Introduction

Since MPI has become the standard for message passing on distributed memory machines, some work has been done on measurement of latency and bandwidth for send/recv operations and other global communications [1][2][3]. It is usually assumed that the data to be communicated is contiguous in memory. This is not always the case in a 2-D or 3-D regular grid domain application. The boundary data are not, in general, contiguous in memory and due to the necessary memory access, the cost of communication depends also on the pattern of data in the user memory that has to be sent. In this case, In this case, variations on classical echo test such as the proposed in [4] and the OCCOMM benchmark [5] provide better information. Indeed, the latter was designed to deal specifically with the exchange of boundaries in a 3-D regular application (an ocean model) and its main goal is to find out the best way of exchanging non-contiguous data using different message passing facilities. It consists of a data exchange between two processes using three different data patterns that can be found in a boundary exchange for three-dimensional decompositions of finite difference grids: contiguous data, contiguous data separated by a constant stride (called single-stride data) and non-unit stride vectors separated by a constant stride (called double-stride data).

We have followed a similar approach to the OCCOMM benchmark. We have extended its results with a detailed analysis of the influence of the underlying architecture and a study of the effect of different measured network loads has been

P. Amestoy et al. (Eds.): Euro-Par'99, LNCS 1685, pp. 173-182, 1999.
© Springer-Verlag Berlin Heidelberg 1999

also done. In addition, at the time of writing this paper, there are no official results for the SGI Origin 2000 on the OCCOMM Web page[5].

This paper is organised as follows. The general components of a simple communication cost model are described in section 2. The essential aspects, network load and the spatial locality analysed in detail on the systems under study, are presented in sections 3 and 4 respectively. Using an example code and based on the previous analysis, the influence in the partitioning of a regular application is presented in Section 5. The paper ends with some conclusions.

2. Communication Cost

Message sending between two tasks located on different processors can be divided into three phases: two of them are where the processors interface with the communication system (the send and receive overhead phases), and a network delay phase, where the data is transmitted between the physical processors. The network delay depends on the number of hops between adjacent network nodes or switches that the message traverses, the rate at which the message data arrives at the destination and the overhead induced by competition of resources with other activities. The interconnection systems of current parallel computers and high-performance networks such as Myricom's Myrinet have taken advantage from the increasing integration density and so the effective bandwidth and latency are now hundreds of times faster than years ago. The peak link bandwidth is growing around 100 percent per year [6]. As the interconnection networks increase their bandwidth, the send and receive overheads are becoming predominant.

Therefore, the degree to which the communication contributes to execution time in a real application depends not only on the amount of communication but also on how the data are structured in the local memories due to their impact on the send and receive overheads and how the network topology is exploited due to contention problems. The following sections analyse in more detail these factors on Cray T3E and Origin 2000 systems.

3. The Cray T3E Message Passing Performance

The T3E-900 system used in this study has 40 DEC Alpha 21164 (DEC Alpha EV5) processors running at 450 MHz. The EV5 contains no board-level cache, but the Alpha 21164 has two levels of caching on-chip: 8 KB first-level instructions and data caches, and a unified, 3-way associative, 96-KB second-level cache [7][8][9].

The local memory is distributed across eight banks, and its bandwidth is enhanced by a set of hardware stream buffers that improve the bandwidth of small-strided memory access patterns and instruction fetching. Each node augments the memory interface of the processor with 640 (512 user and 128 system) external registers (E-registers) [7].

Although it supports shared memory, the natural programming model for the machine is message passing. The library message passing mechanism uses the E-registers to implement transfers, directly from memory to memory. E-registers enhance performance when no locality is available by allowing the on-chip caches to be bypassed. However, if the data to be loaded were in the data cache, then accessing that data via E-registers would be sub-optimal because the cache-backmap would first have to flush the data from data cache to memory [8].

The processors are connected via a 3D-torus network. Its links provide a raw bandwidth of 600 MB/s in each direction with an inter-processor payload bandwidth of 480 MB/s[9]. However, as we have studied previously [10], the effective one-way bandwidth using MPI is smaller due to overhead associated with buffering and with deadlock detection. The maximum achievable bandwidth for large messages is only around half of the peak (around 300 MB/s). The test program employed in that study was slightly different to the code provided by Dongarra to characterise point to point communication [1][2]. Our test use all the processors available in the system where one of them was always the sender processor and the other, which varied, was the receiver. The sender initiated an immediate send followed by an immediate receive and then it waited until both operations had been completed. The receiver began by starting a receive operation and it replied with another message using an immediate send/wait combination. Because this exchange was repeated many times to average results, in order to avoid temporal locality effects, the suppress directive [11] was used to invalidate the entire cache (it forced all entities in the cache to be read from memory) of every iteration. We obtained almost the same asymptotic bandwidth for all the processor pairs.

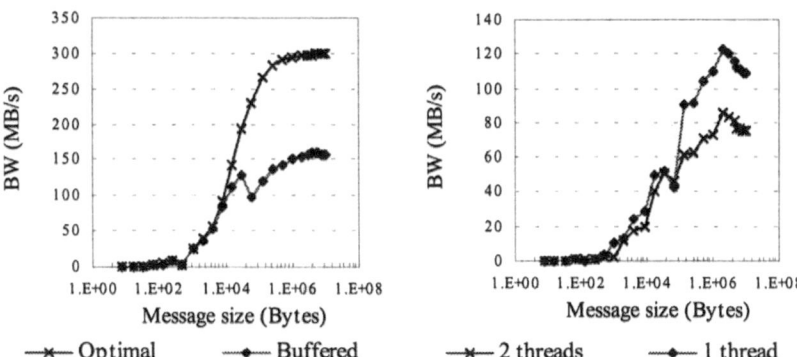

Fig. 1. Message Passing bandwidth in MB/s for contiguous data using buffered and synchronous communication in the CRAY T3E-900 (left-hand chart) and using 1 and 2 threads per node in the SGI Origin 2000 (right-hand chart) respectively.

The maximum bandwidth obtained in these measurements depends on the mode employed for sending and receiving data. Messages can either be buffered by the system or be sent directly in a synchronous way. The environment variable MPI_BUFFER_MAX can be used to control the maximum message size in bytes that will be buffered. The default value of MPI_BUFFER_MAX is no limit, i.e., it is only limited by the total amount of memory available in the system. A large limit is useful if an overlapping of computation with the buffered communication is possible. However it involves additional copies of data and increases the communication overhead. Indeed, the best performance attainable in buffered communications is only half of the maximum (around 160 MB/s). A size of 4099 bytes has been reported to be optimal [3]. This limit solves the trade-off between synchronous and buffered modes, avoiding idle times for small messages and achieving the maximum bandwidth for point-to-point communications (around 300 MB/s) of messages above 2MB. The cache memory used in the buffered mode cause a reduction in the effective bandwidth when the internal buffers do no fit into the second level cache (from message larger than 64 KB in this experiment).

We have extended the previous experiment introducing different network loads. The behaviour of the network under load is complex, e.g. the T3E adaptive routing allows messages to be re-routed around a temporary hot spot, and so we only consider a qualitative description. In the second test, the sender and the receiver exchange their messages as in the previous one, but at the same time, the other processes, which belong to a different MPI communicator, perform a total exchange operation. We have employed the MPI_Alltoall function, a collective operation in which each process sends the same amount of data to every other process in the communicator. The degree of load introduced has been modified by changing the send/receive data count parameter of this function, which sets the number of elements that every process sends to the others. To get the total amount of data that is exchanged during an alltoall operation, one has to multiply this size by N(N-1), where N is the number of processes involved.

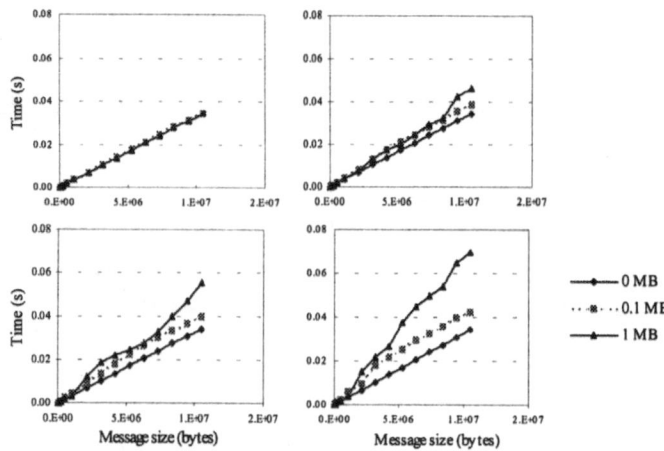

Fig. 2. Transfer times in seconds for varying message sizes and network loads for a point-to-point communication of contiguous data in the Cray T3E-900. The measurement has been taken under different network loads. The two processors involved in the exchange are 0 (top left), 1 (top right), 2 (bottom left) and 3 (bottom right) hops apart respectively.

As figure 2 shows, in a non-contention environment, adding hops does not have a significant effect on the communication bandwidth. However, as the network load increases, the effective bandwidth depends on the network distance between the two processors involved in the exchange for message sizes larger than 1 MB, although the measures are more erratic due to the network contention problems. Thus, 0 and 1 hop distances are advisable for neighbouring processors. An increase of the network load larger than that corresponding to a 1 MB alltoall operation does not produce a significant bandwidth worsening.

The behaviour under a loaded network is different using buffered communication. In this case, the effect of the network load is not so important, though the maximum bandwidth attainable in buffered communications is limited to 160 MB/s.

Figure 3 shows the effect of the spatial data locality. We also use the echo test with only two processors, but we modify the data locality by means of different strides between successive elements of the message. The stride is the number of double precision data between successive elements of the message, so stride-1 represents contiguous data. The measurements have been taking using the MPI datatypes (MPI_Type_vector) instead of the MPI_Pack-Unpack routines. Due to memory constraints the larger message is limited to 256 Kbytes, and although it is not big enough to obtain the asymptotic bandwidth for the stride-1 case, the measures are significant.

As in the first test, the performance depends on the communication mode employed. The bandwidth decreases with the stride using a buffered mode (MPI_MAX_BUFFER unlimited). For 256 KB messages, stride-1 bandwidth is around 5 times larger than stride-32. The extra memory copies needed in this mode explain this behaviour. The local memory is distributed across eight banks, and, when a cacheable load request is missed in the Scache, each bank supplies one word of the 8-word Scache line. For a stride larger than 8, only one word in an Scache line is useful, and thus, from stride 8, the effective bandwidth remains almost constant. We have measured no significant differences in bandwidth with the stream buffers disabled. Using the optimal value of MPI_MAX_BUFFER the performance is improved again.

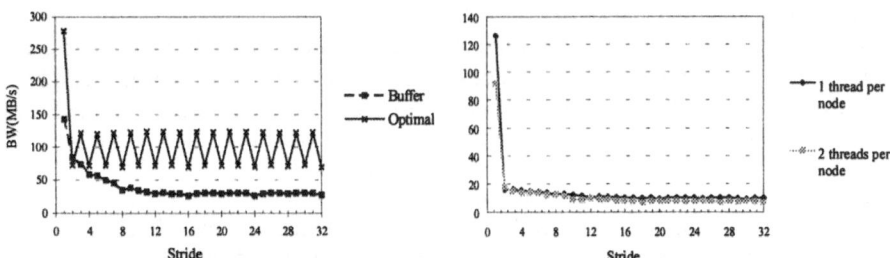

Fig. 3. Bandwidth in MB/s for non-contiguous data using buffered and synchronous communication in the CRAY T3E-900 (left-hand chart) and 1 and 2 thread per node in the SGI Origin 2000 (right-hand chart) respectively. The stride is the number of elements (double precision data) between successive elements of a 256 KB message in the T3E and a 2MB message in the SGI respectively.

Moreover, the bandwidth does not increase proportionally to the stride. The MPI use of E-register access explains this behaviour because the E-registers can be affected by bank conflicts in the local memory system. Strides that are multiples of 2, 4 or 8 are concentrated in 4, 2, or 1 of the eight banks respectively. Consequently, the effective bandwidth for odd strides (around 122 MB/s) is better than for strides that are multiples of 2 (around 72 MB/s), 4 (around 71 MB/s) or 8 (around 68 MB/s).

4. SGI Origin 2000 Message Passing Performance

We repeated these tests on an SGI Origin 2000, a cache-coherent distributed shared memory system. Each node of the Origin contains two processors connected by a system bus (SysAD bus), a portion of the shared main memory on the machine (512 MB in our system), a directory for cache coherence, the Hub (which is the combined communication/coherence controller and network interface) and an I/O interface called Xbow. The system used in this study has the MIPS R10000 running at 250 MHz. Each processor has a 32 Kbyte two-way set-associative primary data cache and a 4-Mbyte two-way set-associative secondary data cache [9][12][13][14].

The peak bandwidths of the bus that connects the two processors and the Hub's connection to memory are 780 MB/s. However the local Memory peak bandwidth is only about 670 MB/s. One important difference between this system and the T3E is that it caches remote data, while the T3E does not. All cache misses, whether to local or remote memory, go through the Hub, which implements the coherence protocol. The Hub's connections to the off-board network router chip and the I/O interface are 1.56 GB/s each. The SGI Origin network is based on a flexible switch called SPIDER, that supports six pairs of unidirectional links, each pair providing over 1.56 GB/s of total bandwidth in the two directions. Two nodes (four processors) are connected to each switch so there are four pairs of links to connect to other routers. Systems with 32 processors are created from a three-dimensional hypercube of switch where only five of the six links on each router are used. Xpress Links are optional cables that connect routers along the diagonal of the cube but the system used in this studied does not have these diagonal connections [9][12][13].

At the user level, due to the cache-coherency protocol and other overheads, the actual effective bandwidth between processors is much lower than the peak. [12][2].

The only different between the ping-pong test used in this system and the T3E one is the process employed to avoid temporal locality effects. Since there is not an SGI instruction similar to the T3E suppress directive, the cache is cleaned by reading a long auxiliary vector that is allocated and freed before each echo iteration.

For the purposes of buffering, MPI messages are grouped into two classes based on length: short (messages with lengths of 64 bytes or less) and long (messages with lengths greater than 64 bytes). The environment variables MPI_BUFS_PER_PROC and MPI_BUFS_HOST set the number of shared and private message buffers (16 KB each) that MPI have to allocate for each process and host respectively. These buffers are used to send long messages, for which one of them has to be different from 0. Unlike the T3E, we have not measured significant differences changing these variables.

Unlike the T3E, Silicon Graphics' MPI implementations use N+1 threads for an N processor job, although the first thread is mostly inactive. The environment variable MPI_DSM_PPM sets the number of MPI threads that can be run on each node. Only values of 1 or 2 are allowed. As figure 1 (right-hand chart) shows, using 1 thread per node, each thread has its own local memory and better performance is obtained since they do not contend for the same SysAD bus.

It is also interesting to note that the measured bandwidth decreases when the message sizes are larger than the L2 cache size. With the information that we have about the message passing implementation of SGI, we cannot be certain of the reason for this fact. Like in the T3E buffered communication mode, the internal message buffers could cause this reduction since, for longer message sizes, they do no fit in the secondary cache and thus, extra main memory accesses are needed.

In this system, like in the T3E, the time required to send a message from one processor to another under an unloaded network varies little with the network distance between them, apart from the case where the two processors share their local main memory where the differences are larger, as we have explained before.

However, as figure 4 shows, the reduction (in percentage) in the effective bandwidth caused by extra network loads is larger than in the T3E. For example, in the worst case (a 1MB alltoall) and for a 2 MB message, the bandwidth is 4 times lower than under an unloaded network when the 2 processors are 3 routes apart. In this case for the 3TE, the reduction is only by a factor of 2. Besides, saturation is reached for a lower network load (corresponding to a 0.1 MB alltoall operation). We suppose that this behaviour could be improved by means of the Xpress Links.

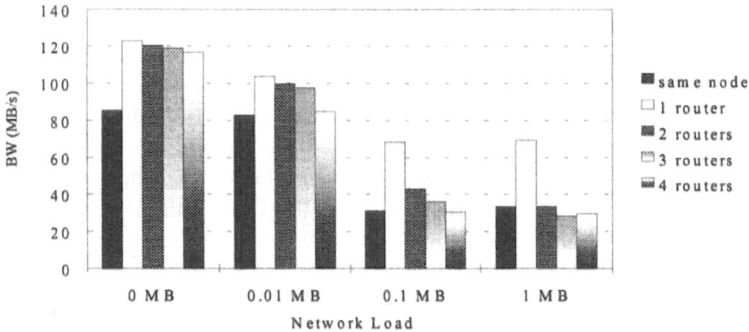

Fig. 4. Bandwidth in MB/s for a 2 MB contiguous message in the SGI Origin 2000. The measurement has been taken under different network loads and the two processors involved in the exchange are on the same node, 1, 2, 3 or 4 routes apart, respectively.

Unlike the T3E, the extra network load also has an effect when the two processes involved in the exchange are located on the same node, due to the cache coherence protocol.

Figure 3 (right-hand chart) shows the effect of the spatial data locality. The behaviour is similar to a buffered mode in the T3E. The second level cache line in this case is 128 Bytes (16 doubles) and thus, the actual bandwidth remains constant from stride 16, between 9 MB/s and 10.2 MB/s using 1 tread per node and between 6.7 MB/s and 8 MB/s using 2 threads.

5. Regular Partitioning

Summing up the results of the previous section, we can say that in the two systems under study, the bandwidth reduction due to non-unit-stride memory access is more important than the reduction due to contention in the network. The message passing performance is better in T3E where a cache coherence protocol is not employed and where it is possible to bypass the second level cache. Non-cache data access is preferred for sending blocks of data that are too large to fit in the cache and for any irregular or strided pattern. How does these conclusions affect real applications?

As a sample problem, we have studied the numerical solution of a regular finite difference application [10]. The code was written in C, so a three dimensional domain is stored in a row-ordered (x,y,z)-array. It can be distributed across a 1D mesh of virtual processors following three possible partitionings: x-direction, y-direction and z-direction.

As is expected, the x and y-direction partitioning were found to be more efficient, because the message data exhibits a better spatial locality. For the 256x256x256 problem size, x- partitioning is found to be around 2 times better than z-partitioning using buffered communications. The difference between contiguous or non-contiguous boundaries is lower using a synchronous mode (around 1.6) as we have measured in the non-unit-stride ping-pong test. However, the performance results are just the opposite: buffered communications are more efficient than a synchronous mode. As in other parallel programs, all the processes of our application communicate at the same time. This can overburden the communication network and cause contention. Obviously, a buffered mode can relieve this problem.

In any case, we should also note that contention could be reduced using correct processor mapping. Although we have employed the process topologies created by the standard `MPI_Cart_create` function, using the option that allows process reordering to get the best performance, MPI on the Cray T3E does not do process reordering. An appropriate algorithm for building better Cartesian topologies has been recently proposed on [15].

In addition, the different E-register usage can cause some discrepancies. The application has some temporal locality and as we have explained in T3E description, if the data to be loaded were in the data cache, then accessing via E-registers involves the invalidation of the corresponding cache line, losing temporal locality advantages.

Although message-passing bandwidth is very important, we should also note that this difference is not only a message passing effect, e.g., partitioning also defines inner loop sizes, which can affect performance. In any case, by means of the MPP Apprentice performance tool we have found that the time spent in the initiation of message sending is 5 times larger in the Z-partitioning simulations using buffered communication.

Equivalent differences in the Origin 2000 can be observed, but are less important than in the T3E case. For the 256-element problem, X partitioning is only 1.3 times better. The large second-level cache of this system, which allows the best exploitation of the temporal locality, influences these results (it is less important the spatial locality).

Taking into account the effect of spatial locality, the choice of an optimal partitioning becomes a trade-off between the reduction of the send and receive overheads (mainly, spatial locality of the data to be exchange) and the efficient exploitation of the interconnection network. The correct processor mapping reduces contention, and thus 3D decompositions are the best choice for the T3E topology. In addition, for a 3D regular application the communication requirements for a process grow proportionally to the size of the boundaries, while computations grow proportionally to the size of its entire partition. However, higher dimensional decompositions require non-contiguous boundaries and they make messages greater in number and hence shorter. Since there is a fixed component of the send and receive overheads, which is associated with initiating or processing a message, it is better to make messages fewer in number, just the opposite.

The left chart of figure 5 compares the different decompositions for our sample application in the Cray T3E. As the number of processors grows, the efficient exploitation of the underlying network becomes important. In the 32-processor simulation an appropriate 2D decomposition solves the trade-off between network exploitation and local memory access. In the larger problem the best 2D decomposition is 5% and 15% better than the 3D and linear decompositions, respectively. In any case, as we have explained before, a better algorithm for building Cartesian topologies could modify the results and its influence will be analysed in a future study. In the SGI Origin 2000 (right chart on figure 9) we have measured similar results, for the 32-processor simulation, the 2D decomposition is 22 % and 8% better than the 1D and 3D decompositions, respectively.

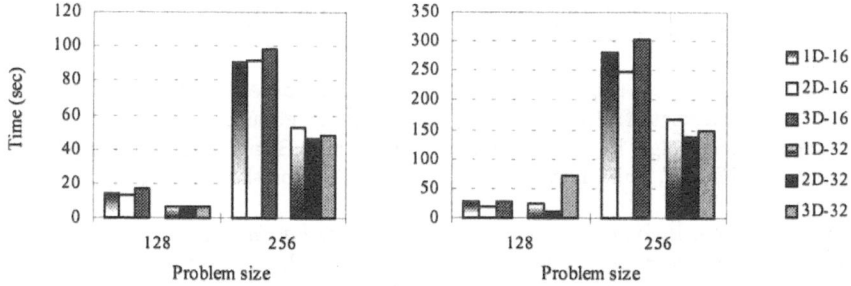

Fig. 5. Different decompositions for our sample program using 16 and 32 processors on the T3E (on the left) and on the SGI Origin 2000 (on the right). The problem size is the number of cells (2 double precision data each) in each dimension.

6. Conclusions

We have first shown how non-unit-stride memory access and network loads affect actual communication bandwidths on two different parallel computers, the Cray T3E and the Origin 2000. In both systems, the bandwidth reduction due to non-unit-stride local memory access is more important than the reduction due to contention in the network. The message passing performance is better in the T3E where a cache coherence protocol is not employed and where it is possible to bypass the second level cache for non-contiguous memory accesses. Secondly, we studied how these architecture properties affect he partitioning of regular domains. Obviously, data partitioning is a trade-off between the improvement of the message data locality and the efficient exploitation of the underlying communication system. Using up to 32 processors, in both systems, an appropriate 2D decomposition, where boundaries with poor spatial locality are not needed, solves that trade-off. The influence of spatial locality is lower in the SGI Origin where the large second level cache allows the best exploitation of the temporal locality. However, we should also note that there are other aspects to the evaluation of a parallel program. A lower-dimensional partitioning program is easier to code, so if we consider implementation cost, 1D partitioning is the best choice. In Addition, it allows the implementation of fast sequential algorithms in the non-partitioned directions as we have explained in [16].

Acknowledgements
This work has been supported by the Spanish research grants TIC 96-1071 and TIC IN96-0510 and the Human Mobility Network CHRX-CT94-0459. We would like to thank Ciemat and CSC (Centro de Supercomputacion Complutese) for providing access to the parallel computers that have been used in this research.

References

[1] J. J. Dongarra, Tony Hey, and Erich Strohmaier. *"Selected Results from the PARKBENCH Benchmark"*, in Proceeding of EuroPar'96 Parallel Processing, Volume II, pages 251--254, August 1996.

[2] Aad J. van der Steen and Ruud van der Pas *"A performance analysis of the SGI Origin 2000"*, in Proceedings of VECPAR 98, pp. 319-332. Porto, Portugal, 1998.

[3] Michael Resch, Holger Berger, Rolf Rabenseifner, Tomas Bönish *"Performance of MPI on the CRAY T3E-512"*, Third European CRAY-SGI MPP Workshop, PARIS (France), Sept. 11 and 12, 1997.

[4] Vladimir Getov, E. Hernández and T. Hey, *"Message-Passing Performance of Parallel Computers"*, in Proceeding of Europar'97, pp. 1009--1016. Passau, Germany, August 1997.

[5] OCCOMM home page: http://www.dl.ac.uk/TCSC/CompEng/OCCOMM.

[6] Shubhendu S. Mukherjee and Mark D. Hill. *"Making Network Interfaces less peripheral"*, pp 70-76 Computer. October 1998

[7] S. L. Scott. *"Synchronization and Communication in the T3E Multiprocessor"*, in Proceeding of the ASPLOS VII, October 1996.

[8] E. Anderson, J. Brooks, C.Grass, S. Scott. *"Performance of the CRAY T3E Multiprocessor"*. in Proceeding of SC97, November 1997.

[9] David Culler, Jaswinder Pal Singh, Annop Gupta. *"Parallel Computer Architecture. A hardware /software approach"* Morgan-Kaufmann Publishers 1998.

[10] M. Prieto, I. M. Llorente, F. Tirado. *"Partitioning of Regular Domains on Modern Parallel Computers"*, in Proceedings of VECPAR 98, pp. 305-318. Porto, Portugal, 1998.

[11] Cray C/C++ Reference Manual, SR-2179 3.0.

[12] J. Laudon and D. Lenoski. *"The SGI Origin: A ccNUMA Highly Scalable Server"*, in Proceeding of ISCA'97.May 1997.

[13] H. J. Wassermann, O. M. Lubeck, F. Bassetti. "Performance Evaluation of the SGI Origin 2000: A Memory-Centric Characterization of LANL ASCI Applications". In Proceeding of the SC97, November 1997.

[14] Silicon Graphics Inc. *"Origin Servers"*, Technical Report, April 1997.

[15] Matthias Müller and Michael M. Resch , *"PE mapping and the congestion problem on the T3E"* in HErmann Lederer and Friedrich Hertweck (Ed.), Proceedings of the Fourth European Cray-SGI MPP Workshop, IPP R/46, Garching/Germany, 1998.

[16] D. Espadas, M. Prieto, I. Martín y F. Tirado *"Parallel Resolution of Alternating-Line Processes by means of Pipelining Techniques"*, in Proceeding of the 7TH Euromicro Workshop on Parallel and Distributed Processing, pp. 289-296. Madeira, Portugal, 1999.

Performance Evaluation and Modeling of the Fujitsu AP3000 Message-Passing Libraries*

Juan Touriño and Ramón Doallo

Dep. of Electronics and Systems, University of A Coruña, Spain
{juan,doallo}@udc.es

Abstract. This paper evaluates, models and compares the performance of the message-passing libraries provided by the Fujitsu AP3000 multi-computer: MPI/AP, PVM/AP and Fujitsu APlib (versions 1.0). Our aim is to characterize the basic communication routines using general models. Several representative parameters and performance metrics help to compare the different primitives and to detect inefficient implementations.

1 Introduction

The performance of the communication primitives of a parallel computer does not only depend on the underlying hardware, but also on their implementation. Users do not know the quality of the message-passing implementations and they can find that the performance of their parallel applications makes worse in other machine or using other message-passing library.

We have focussed on low-level tests to study basic communication primitives on the Fujitsu AP3000 [2]. The AP3000 has UltraSparc-II processors connected via a high-speed communication network (AP-Net) in a two-dimensional torus topology. We have considered point-to-point communications, one-to-all (broadcast) and all-to-one (specifically, a reduction operation). Though more primitives could be analyzed, these ones are the basis for the design of more complex communication patterns in a parallel application.

2 Point-to-Point Communications

The Hockney's model [1] characterizes message latency T:

$$T(n) = \frac{n_{\frac{1}{2}} + n}{Bw_{as}} \tag{1}$$

where n is the message length, Bw_{as} is the asymptotic bandwidth, and $n_{\frac{1}{2}}$ is the half-peak length (n required to obtain $Bw_{as}/2$). Besides, $Bw_{as} = 1/t_b$ and $n_{\frac{1}{2}} = t_s/t_b$, being t_s the startup time and t_b the transfer time per data unit ($T(n) = t_s + t_b n$). The specific performance $\pi_s = 1/t_s$ characterizes short-message performance, while Bw_{as} shows long-message performance. Only blocking primitives are considered: *MPI_Send/Recv*, *pvm_psend/precv* and *Lasend/arecv* (APlib).

---

* This work was supported by the CICYT (Contract TIC96-1125-C03) and by the EU (1FD97-0118-C02). CESGA (Santiago de Compostela, Spain) provided the AP3000

P. Amestoy et al. (Eds.): Euro-Par'99, LNCS 1685, pp. 183–187, 1999.

Table 1. Point-to-point communication parameters and metrics

	MPI/AP	PVM/AP	APlib
t_s (μs)	69	53	46
t_b (μs)	0.0162	0.0162	0.0162
π_s (Kbytes/s)	14.15	18.43	21.23
Bw_{as} (Mbytes/s)	58.87	58.87	58.87
$n_{\frac{1}{2}}$ (bytes)	4260	3272	2840

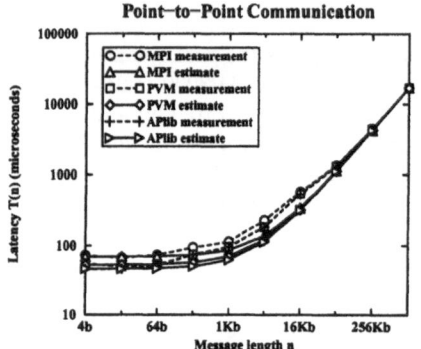 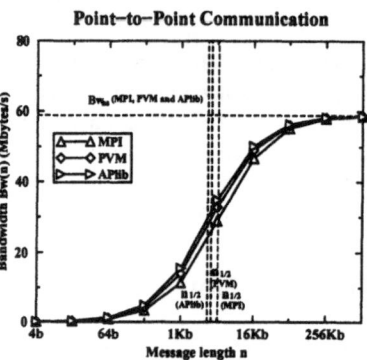

Fig. 1. Latency and bandwidth for point-to-point communications

Table 1 shows the estimated parameters for the three AP libraries. As can be observed, Bw_{as} is the same for the three libraries; therefore, their latencies tend to be similar as the message length increases. Regarding short messages, there are appreciable differences in the startup times. APlib has the lowest t_s and MPI the highest. Consequently, the best $n_{\frac{1}{2}}$ is achieved by APlib. The *pvm_send/recv* routines were also tested: we have estimated $t_s \approx 50\mu s$, but the transfer time increases excessively, $t_b = 0.0263\mu s$, which results in $Bw_{as} = 36.26$ Mbytes/s. It is due to message packing and unpacking operations. Figure 1 depicts in the left-hand graph a comparison between the estimated and measured latencies for the three libraries and different message sizes. The corresponding estimated bandwidth is presented in the right-hand graph.

3   One-to-All Communications: Broadcast

We have applied the model proposed by Xu and Hwang in [3] to characterize one-to-all communications and, more specifically, the broadcast communication:

$$T(n,p) = t_s(p) + \frac{n}{Bw_{as}(p)} \tag{2}$$

where p is the number of processors; $Bw_{as}(p)$ can be also expressed as $1/t_b(p)$ and we can similarly define $n_{\frac{1}{2}}(p) = t_s(p)/t_b(p)$ and $\pi_s(p) = 1/t_s(p)$.

An additional metric is the aggregated (ag.) asymptotic bandwidth Bw_{as}^{ag}, the ratio of the total number of bytes transferred in the collective operation

Table 2. Broadcast parameters and metrics ($k_1=10^6/2^{10}$, $k_2=10^6/2^{20}$)

	MPI/AP	PVM/AP	APlib
$t_s(p)$ (μs)	$69\log_2 p$	$22p$	$46+25(p-2)$
$t_b(p)$ (μs)	$0.0162\log_2 p$	$0.0110p$	$0.0162+0.0110(p-2)$
$\pi_s(p)$ (Kbytes/s)	$k_1/(69\log_2 p)$	$k_1/(22p)$	$k_1/(25p-4)$
$\pi_s^{ag}(p)$ (Kbytes/s)	$k_1(p-1)/(69\log_2 p)$	$k_1(p-1)/(22p)$	$k_1(p-1)/(25p-4)$
π_s^{pag} (Kbytes/s)	43.43	40.69	36.29
$Bw_{as}(p)$ (Mbytes/s)	$k_2 61.73/\log_2 p$	$k_2 90.91/p$	$k_2/(0.0110p-0.0058)$
$Bw_{as}^{ag}(p)$ (Mbytes/s)	$k_2 61.73(p-1)/\log_2 p$	$k_2 90.91(p-1)/p$	$k_2(p-1)/(0.0110p-0.0058)$
Bw_{as}^{pag} (Mbytes/s)	180.63	79.47	83.13
$n_{\frac{1}{2}}(p)=n_{\frac{1}{2}}^{ag}(p)$ (bytes)	4260	2000	$(25p-4)/(0.0110p-0.0058)$
$n_{\frac{1}{2}}^{mag}$ (bytes)	4260	2000	2346

and the time required to perform the operation, as $n \to \infty$. For a broadcast, $Bw_{as}^{ag}(p) = (p-1)Bw_{as}(p)$. Similarly, the ag. specific performance $\pi_s^{ag}(p) = (p-1)\pi_s(p)$ shows the performance of a broadcast for short messages. The ag. half-peak performance $n_{\frac{1}{2}}^{ag}(p)$ can be also defined as n that achieves $Bw_{as}^{ag}(p)/2$ ($n_{\frac{1}{2}}^{ag}(p) = n_{\frac{1}{2}}(p)$). All these measures depend on p. It would be interesting to define peak performance measures to have a global estimate of the behaviour of collective communications. Therefore, we propose the following metrics: the peak ag. bandwidth $Bw_{as}^{pag} = \max_{2 \le p \le p^{max}} Bw_{as}^{ag}(p)$, the peak ag. specific performance $\pi_s^{pag} = \max_{2 \le p \le p^{max}} \pi_s^{ag}(p)$ and the minimum (ag.) half-peak length $n_{\frac{1}{2}}^{mag} = \min_{2 \le p \le p^{max}} n_{\frac{1}{2}}^{ag}(p)$, being p^{max} the maximum p available for users ($p^{max} = 12$ in our machine).

The routines considered in the comparison are: *MPI_Bcast*, *pvm_mcast* and *cbroad* (APlib). The fitting of the components of Eq. 2 and the additional performance metrics are shown in Table 2. Note that, for p=2, in MPI and APlib the numerical values of the model's parameters are the same as the ones of the point-to-point model. Latency is $O(log_2 p)$ in MPI, which reveals that the broadcast in MPI is implemented using a binomial tree-structured approach. In PVM, latency is $O(p)$; it seems that *pvm_mcast* is implemented as a sequence of sends all originating from the root processor. The APlib broadcast is also $O(p)$. Therefore, MPI performance is the best, and it is better as p increases. The results for PVM and APlib are very similar, although $t_b(p)$ is slightly better for the APlib broadcast and the startup time is a bit lower in PVM. Regarding $n_{\frac{1}{2}}(p)$, in MPI and PVM is a constant and in APlib is almost constant, because the complexities of $t_s(p)$ and $t_b(p)$ are the same within each message-passing library.

Figure 2 shows some experimental results for the broadcast routines, by fixing p=8 and n=64 Kb, respectively. The second graph shows that the model is very accurate for PVM and APlib. This graph also reveals that in MPI, for n=64 Kb, the startup time of the model should be a bit higher, although the fitting is acceptable. Figure 3 depicts $\pi_s^{ag}(p)$ and $Bw_{as}^{ag}(p)$. It can be observed that $\pi_s^{ag}(p)$ for MPI is lower than for PVM and APlib because the startup in MPI is

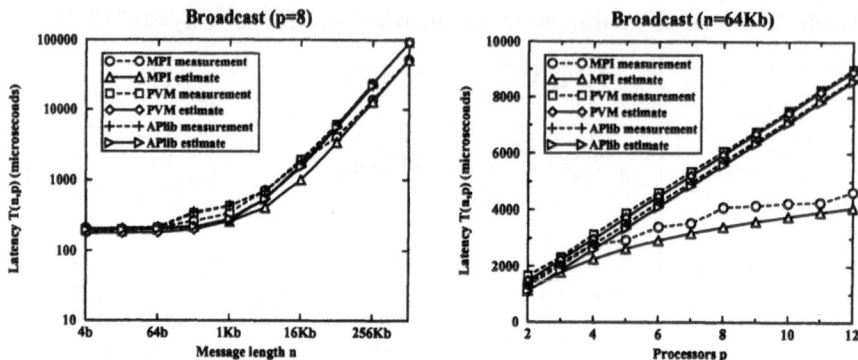

Fig. 2. Latency for broadcast operations: a) p=8, b) n=64 Kb

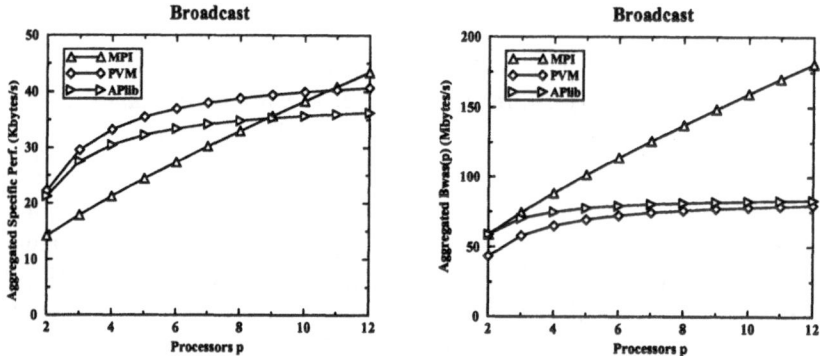

Fig. 3. Broadcast aggregated metrics: a) $\pi_s^{ag}(p)$, b) $Bw_{as}^{ag}(p)$

high for a small number of processors. But, from $p=10$, the $O(log_2 p)$ complexity of t_s leads to a better $\pi_s^{ag}(p)$. This improvement would be more pronounced for a greater number of processors. In the second graph, $Bw_{as}^{ag}(p)$ of MPI is clearly the best due to the log complexity of the broadcast. In PVM and APlib $Bw_{as}^{ag}(p)$ is poor and it tends to be constant as p increases. Consult the corresponding peak values of these functions (π_s^{pag} and Bw_{as}^{pag}) in Table 2.

4   All-to-One Communications: Reduction

The model of Eq. 2 hides a parameter in collective primitives that involve computations (e.g., a reduction): t_c, the cost per byte of the performed computation. We propose an extension of the model, valid for all collective communications:

$$T(n,p) = t_s(p) + \frac{n}{Bw_{as}(p)} + t_c(p)n \qquad (3)$$

Clearly, for a broadcast $t_c(p) = 0$. The metrics defined in Section 3 can be also applied here, and we propose a new metric: the ratio transfer time-computation time $r_{cc}(p) = t_b(p)/t_c(p)$, which provides a view of the weight of the computation factor as opposed to the communication factor in the total latency.

We modeled the sum reduction of doubles in MPI (*MPI_Reduce*) and PVM (*pvm_reduce*). The APlib reduction was not modeled because it only works on single numbers and stores the result in all the processors involved in the reduction; therefore, it is not comparable to MPI and PVM.

We have found that the PVM reduction (and, in general, the group management routines) are poorly implemented. The reduction routine is not robust: it does not work for $p > 6$. Besides, latencies are dominated by very high startup times: for $p=2$, $t_s \approx 5.55ms$ and it seems to be $O(p)$; for $p=3$ and $n=64$Kb, t_s represents $\approx 80\%$ of latency.

Regarding MPI reduction, we have obtained the following results: $t_s(p) = 90log_2p - 15$, $Bw_{as}(p) = 1/(0.0171log_2p + 0.0037)$ and $t_c(p) = 0.0051log_2p - 0.0037$. As expected, MPI reduction is $O(log_2p)$. Additional performance metrics are: $\pi_s^{pag} = 34.92$ Kbytes/s, $Bw_{as}^{pag} = 161.38$ Mbytes/s and $r_{cc}(p) = (0.0171log_2p + 0.0037)/(0.0051log_2p - 0.0037)$. Although $r_{cc}(p)$ varies from 14.86 (for $p=2$) to 4.46 (for $p=12$), it tends to be a constant since $p=4$ ($t_b(p) \approx 5t_c(p)$).

5   Conclusions

The models and metrics used in the previous sections help us to identify design faults in the communication routines and, furthermore, to estimate the performance of parallel programs. Machine vendors should provide the parameters of these models (or, at least, complexities in the case of collective communications) for basic communication routines.

Regarding the AP3000 message-passing libraries, we can conclude that the PVM/AP library (specially, the group routines) is a naive implementation which should be greatly improved. The APlib routines are not robust for long messages (the machine crashes for 1 Mbyte messages), the broadcast implementation is inefficient and the reduction routines only work on single numbers. Besides, APlib is a proprietary library with a small set of primitives compared to MPI.

Currently, MPI/AP is the best choice to program the AP3000. Nevertheless, the performance of the AP3000 hardware is not fully exploited by the MPI/AP library. Message latencies could be reduced by re-designing the low-level communication mechanisms. Hardware improvements, such as the SBus design (the I/O bus which connects the processor and the message controller), could also help to reach this aim.

References

[1] Hockney, R.W.: The Communication Challenge for MPP: Intel Paragon and Meiko CS-2, Parallel Computing **20**(3) (1994) 389–398
[2] Ishihata, H., Takahashi, M., Sato, H.: Hardware of AP3000 Scalar Parallel Server, Fujitsu Sci. Tech. J. **33**(1) (1997) 24–30
[3] Xu, Z., Hwang, K.: Modeling Communication Overhead: MPI and MPL Performance on the IBM SP2, IEEE Parallel & Distributed Technology **4**(1) (1996) 9–23

Improving Communication Support for Parallel Applications*

Joerg Cordsen[1] and Marco Dimas Gubitoso[2]

[1] GMD FIRST — Berlin, Germany
[2] IME-USP — Universidade de São Paulo, São Paulo, Brazil

Abstract. This paper presents the rationale and some implementation aspects of the VOTE communication support system. VOTE supports a global address space by means of a software-implemented virtual shared memory (VSM). VSM consistency maintenance is implemented exploiting a concurrent execution model and avoiding intermediate buffering. The efficiency and scalability of the VOTE solutions is (a) demonstrated by presenting measured results and (b) compared to the predicted execution times of VOTE's performance prediction framework.

1 Introduction

Some systems provide support for both message passing and shared memory communication paradigms. Unanimously, this is done by using an integrated solution where the same memory consistency semantic is applied to both paradigms. This is in contrast to the natural operation models of shared-memory and message-passing communication. Shared-memory communication applies a single reader/single writer (SRSW), or sometimes a multiple reader/single writer scheme (MRSW), operation model to a shared address space and ensures the memory consistency model provided by the memory subsystem. Message-passing communication is based on point-to-point connections of communication partners. This paper presents the VOTE solution of a coexistence of the two communication paradigms keeping hold on their natural operation models.

2 VOTE Communication Support System

VOTE is a communication support system [3] which, to support a scalable and efficient execution of parallel applications, promotes a symbiosis of *architectural transparency* and *efficiency* aspects. Users should not be obliged to know when to use either of the two communication paradigms. Moreover, for reasons of efficiency, a user should any time be able to directly exploit the communication functions to achieve the best possible solution to his or her problem.

---

* This work was in part supported by the Commission of the European Communities, project ITDC-207 and 9124 (SODA), Project FAPESP 99/05040-1 and the German-Brazilian Cooperative Programme in Informatics, project No. 523112/94-7 (PLENA).

P. Amestoy et al. (Eds.): Euro-Par'99, LNCS 1685, pp. 188–192, 1999.

Architectural transparency requires the introduction a global address space, providing a VSM system. At the user-level, this makes shared-memory communication an option even in the absence of shared memory resources. Then, communication is possible via calls to message-passing functions or, alternatively, performed automatically through the on-demand data movement of memory items to processes that request data causing a memory access fault.

VOTE is implemented as part of the PEACE family of operating systems [5]. It exploits a specifically customized kernel of the PEACE family and runs as a library on the parallel MANNA computing system [1]. Parallel programs use VOTE through calls to the application programmer interface (API) which supports Fortran, C, and C++.

VOTE's software architecture distinguishes three functional units implementing the communication support. An *adviser* process serves as a distributed collection of daemon processes and serves as a data repository. The set of *advisers* maintains directory information about the distribution of VSM pages, implements synchronization features and provides mapping functions to group processes for the execution of collective message-passing functions. Each *adviser* is supplemented by a set of threads, called *actor* threads. The purpose of an *actor* is to provide a remotely accessible interface to control the memory management and data transfer functions for a particular address space.

2.1   Shared-Memory Communication

The VSM subsystem of VOTE implements a facility for shared-memory communication in a single global address space. A MRSW driven sequential consistency model is supported, using an invalidation approach to raise memory access faults. VOTE differs from state-of-the-art VSM systems by rejecting the dynamic distributed page ownership strategy [4].

VOTE is based on a fixed distributed scheme and uses a user-directed mapping from VSM address ranges to consistency maintenance processes (*advisers*) and adds a caching mechanism to the consistency maintenance.

Protection Violation	Access Violation			
	cache hit		cache miss	
	read access	write access	read access	write access
$628\mu s^\dagger$	$667\mu s$	$794\mu s^\dagger$	$1670\mu s$	$1396\mu s^\dagger$

† plus logarithmic time complexity for invalidations

Table 1. Times for access fault handling in VOTE

Table 1 summarizes the time an application process faulting on a memory access is suspended from execution. The most simple case is a protection violation requiring $628\mu s$ to resume the application process.

2.2   Message-Passing Communication

Once an application process selects message-passing as its communication paradigm, the memory model is suited to make use of any message-passing communication library. The drawback of most message-passing communication libraries is due to the high latency of message handling. The idea followed by VOTE's implementation of an MPI subset is to support a zero-copy message delivery.

3   Applying the Performance Prediction Framework

This section presents an application case study and the VOTE performance prediction framework. The framework is applied to the case study and its accuracy is shown by a presentation of both predicted and measured performance results.

3.1   Application Case Study

A case study was done with a parallel SPMD-based (single program/multiple data) implementation of a successive overrelaxation with the use of the red&black technique. Data sharing appears between pairs of processes, that is each process communicates with two other processes, the predecessor and successor.

The message-passing version of this program requires four calls to communication functions (exceptionally, the first and last process only performs two calls). A pair of a non-blocking send and a blocking receive function is used to communicate with a neighbor, ensuring synchronization. In comparison to shared-memory communication, programming is less expressive, more error-prone, but has the advantage of a problem-oriented synchronization style.

3.2   Performance Prediction Framework

Performance analysis and prediction are usually done using statistical models and benchmarking. Unfortunately, this approach does not provide a good understanding of the causes and costs of the overheads present in the application [6]. The alternatives are analytical modeling and simulation, the later being less flexible. The drawback of the analytical method is the potential complexity of the generated models.

However, the clear-cut interface and modularization of VOTE and its execution environment allow great simplifications without a large impact on the accuracy of analytical models. Another advantage of VOTE is that there is a relatively small number of overhead classes with enough impact on performance, the most important being communication and contention on the VOTE system services [2].

3.3   Predicted and Measured Results

Finally, this section presents a comparison of the predicted and measured results. Both the shared-memory and the message-passing version of the successive overrelaxation were measured on the parallel MANNA computing system. Measurements were made for up to 16 processors executing the application in a single-tasking SPMD style.

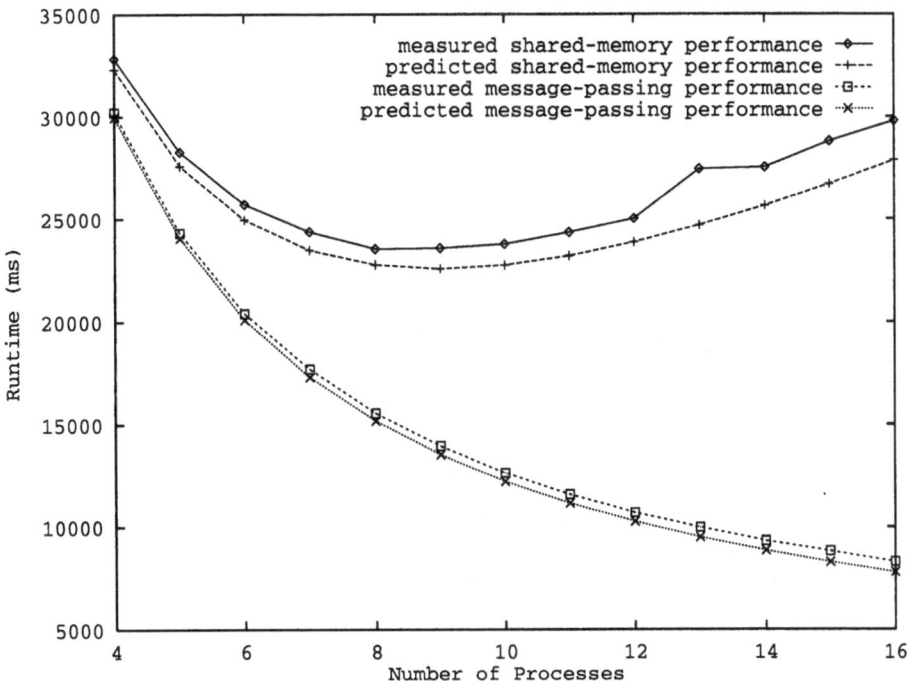

Fig. 1. Comparison of Predicted and Measured Results

Figure 1 shows that the predicted performance results turns out to be at a reasonable level of accuracy. This may be due to the simplicity and regularity of the synchronization and communication of the successive overrelaxation. However, the application of the performance prediction framework to other (yet still simple) parallel kernels [2] support the authors believe that the VOTE performance prediction framework is a viable and yet powerful technique to understand the overheads of parallel programming.

4   Conclusion

VOTE combines two coexistent implementations of the shared-memory and message-passing communication paradigm in a single framework. Consequently, a programmer thus is able to choose the communication paradigm best suited to match the specific application demands and to use the underlying computing system in the most convenient way.

VOTE also supports a performance prediction framework. Modeling the patterns of communication, load unbalance, resource contention and synchronization, it gets possible to find out the overheads of parallel programming. For various uses of the VOTE parallel programming support, it thus gets possible to predict the execution time. In advance, this helps to identify performance bottlenecks and to find the best utilization of the best use of communication and synchronization functions.

References

[1] U. Brüning, W.K. Giloi, W. Schröder-Preikschat, "Latency Hiding in Message Passing Architectures", In *Proceedings of the International Parallel Processing Symposium*, IPPS-8, pp. 704–709, Cancun, Mexico, Apr., 1994.

[2] J. Cordsen, M.D. Gubitoso, "Performance Considerations in a Virtual Shared Memory System", *Arbeitspapiere der GMD*, Nr. 1034, November, 1996.

[3] J. Cordsen, W. Schröder-Preikschat, "On the Coexistence of Shared-Memory and Message-Passing in the Programming of Parallel Applications", *Proceedings of the High-Performance Computing and Networking '97)*, Vienna, Austria, LNCS 1225, pp. 718-727, Springer Verlag, Apr. 1997.

[4] K. Li, "Shared Virtual Memory on Loosely Coupled Multiprocessors", *Ph.D. thesis. RR-492*, Department of Computer Science, Yale University, USA, Sep., 1986.

[5] W. Schröder-Preikschat, "The Logical Design of Parallel Operating Systems", Prentice-Hall, ISBN 0-13-183369-3, 1994.

[6] M. Wagner, Jr., "Modeling performance of parallel programs", *Technical Report 589*, Computer Science Department, The University of Rochester, Rochester, USA, Jun., 1995.

A Performance Estimator for Parallel Programs

Jeff Reeve

The Department of Electronics and Computer Science
The University of Southampton
Southampton SO16 1BJ
UK
jsr@ecs.soton.ac.uk

Abstract. In this paper we describe a Parallel Performance Estimator suitable for the comparative evaluation of parallel algorithms. The Estimator is designed for SPMD programs written in either C or FORTRAN. Simulation is used to produce estimates of execution times for varying numbers of processors and to analyse the communication overheads. Results from the estimator are compared with actual results (obtained on a 16 processor IBM SP2 machine) for an Alternating Direction Implicit (ADI) solver of linear equations and for a Parallel Sort by Regular Sampling (PSRS) sorting program. In both cases the plots of *Execution Time* versus *Number of Processors* are accurate to 20% and show all of the features of the equivalent plots of the measured data.

1   Introduction

Software providers with large production codes have always been concerned with measuring and enhancing the performance of their codes, since a faster, more memory efficient code is commercially advantageous. Even so there has never been a performance prediction tool available, even for sequential codes, that can reliably predict absolute performance. Consequently code optimisation is done by hand for particular codes using particular compilers and running on particular machines.

With parallel codes the problem might seem worse but if one settles for knowing the relative merits of one parallelisation strategy compared with another before embarking on a full scale parallel version of a code then considerable effort can be saved. In view of this our philosophy is to ignore the details of the sequential code and effect a simulation scheme that reproduces the communication pattern of the code on simplified architectures so that the speed of different parallel code versions can be assessed. This approach can also detect deadlock and provide information on the optimal scheduling for small embedded systems.

Our simulator accepts C or FORTRAN 77 source which contains calls to a message passing library. This means that and communication pattern can be accommodated.

Other approaches to the problem of estimating the performance of parallel programs have been for specific problems like data partitioning[3], or have

P. Amestoy et al. (Eds.): Euro-Par'99, LNCS 1685, pp. 193–202, 1999.

taken a statistical approach[10, 4] or have used static information only[5, 6, 8]. Other simulation based approaches that have some commonality with ours are by Kubota et al[14], Aversa et al[2] and Kempf et al[12].

2   Description

The data flow diagram for the Parallel Performance Estimator is shown in Figure 1. Each of the contributing components is explained below.

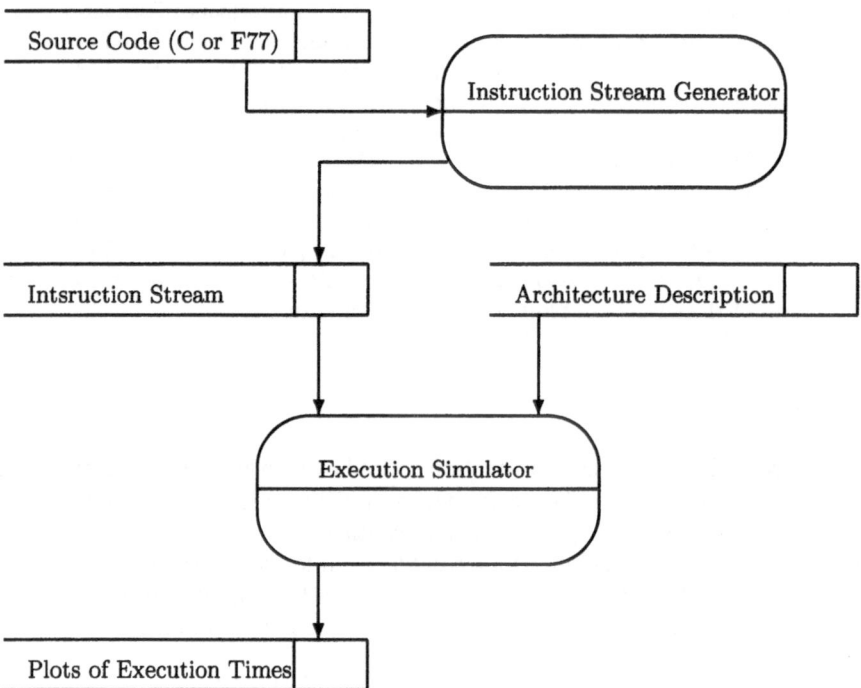

Fig. 1. The Data Flow Structure for the Performance Estimator

The Input Code

The source file input into the code generator takes the form of a F77 or C subroutine with input parameters consisting of the total number of processors (NPROCS) and the processor number (PID).

As this tools is designed to assist with algorithm development and not to analyse existing codes we have restricted the simulated instruction stream to a minimal set, that describe communication and computation events.

These allowed instructions that are accessed through the following subroutine calls are:

1. SEND(PROCESSOR_ID, NO_OF_WORDS)
   This generates a *non-blocking* send instruction of NO_OF_WORDS words to processor PROCESSOR_ID.
2. RECEIVE(PROCESSOR_ID, NO_OF_WORDS)
   This generates a *non-blocking* receive instruction of NO_OF_WORDS words from processor PROCESSOR_ID.
3. BSEND(PROCESSOR_ID, NO_OF_WORDS)
   This generates a send *blocking* instruction of NO_OF_WORDS words to processor PROCESSOR_ID.
4. BRECEIVE(PROCESSOR_ID, NO_OF_WORDS) This generates a *blocking* receive instruction of NO_OF_WORDS words from processor PROCESSOR_ID.
5. WORK(EQUIVALENT_NO_OF_FP_MULTIPLIES)
   This generates a work instruction which takes
   EQUIVALENT_NO_OF _FP_MULTIPLIES times
   TIME_OF_A_FP_MULTIPLY time.
6. WAIT() This generates a wait instruction which causes the simulator to wait for the completion of all outstanding communications events relating to the issueing processor before scheduling any more instructions for that processor.

This minimal set of six instructions is capable of representing almost all of the current flavours of communications schemes. Translation to MPI or PVM is relatively easy. Higher order functionality like *reduction* and *broadcast* routines are implemented using the minimal set using a spanning out tree through the network as the underlying topology for the reduction and distribution of collective data.

For existing codes a tool such as PERFORM[7] could be used to determine the complexity of the algorithm.

The Instruction Stream Generator

The code generator then incorporates the source and executes a driver that calls it successively for values of PID from 0 to NPROCS-1 and records an instruction stream for each of the processors.

The Architecture Description File

The architecture file contains the basic information needed to keep real time in the execution simulation. An example, for our test platform, a 16 processor IBM SP2, gleaned from benchmark results[11], is given below. There are two choices for the network architecture. The type 'bus' is used for single bus multiprocessor architectures such as ethernet systems, and the type 'nobus' is used for arctitectures in which the nodes are effectively equidistant, which is the case for most purpose built MPP systems.

SP2
```
receive latency = 0.24e-4 seconds
send latency = 0.24e-4 seconds
time to send 1 word (8bytes) = 2.39e-7 seconds
time for one multiply = 0.178e-7 seconds
network type = nobus
```

The Execution Simulator

The simulator interprets the instruction stream file, scheduling work and messages in an event queue, when the queue is empty and the instruction stream is exhausted then the program has successfully terminated. If the event queue is empty and the instruction stream isn't empty then the program has deadlocked. Should it occur, this condition is reported and the simulation terminates gracefully.

The initial action of the simulator is to issue a load event for each processor. From this point on the simulator is completely event driven. Each new event is scheduled with a time stamp. The earliest event on the event queue is serviced in turn. The cost of each instruction is determined from the architecture file.

For every architecture it is assumed that the system software is single threaded and offers only one entry point for the send and receive instructions. This means that every send and receive issued generates an event that is stamped with the current time extended by the appropriate send or receive latency time, and it is only when this event is serviced that the corresponding transmission event is put onto a separate queue in readiness for scheduling when both the processors participating in the event are ready and the bus (if appropriate) is ready.

The simulator keeps two queues. One is the event queue ordered in user program execution time, from which events are dequeued and acted on. The other is a queue of potential communications events. When the conditions for a communication are satisfied, ie both participants are ready and the data path is free, then the communication is inserted into the event queue. An outline of the simulator's event descheduler is given below.

```
SWITCH(EVENT_NAME)
    CASE LOAD
        SCHEDULE_NEXT_INSTRUCTION(PROCESS)
    CASE SEND_LATENCY | BSEND_LATENCY |
            RECEIVE_LATENCY | BRECEIVE_LATENCY
        SCHEDULE_NEXT_COMMUNICATION
        SCHEDULE_NEXT_INSTRUCTION(PROCESS)
        ALTER_STATE(PROCESS)
    CASE WORK
        SCHEDULE_NEXT_INSTRUCTION(PROCESS)
    CASE SEND_RECEIVE | BSEND_RECEIVE |
            SEND_BRECEIVE | BSEND_BRECEIVE
        FREE_BUS
```

SCHEDULE_NEXT_INSTRUCTION(SENDING PROCESS)
SCHEDULE_NEXT_INSTRUCTION(RECEIVING PROCESS)
SCHEDULE_NEXT_COMMUNICATION
UNBLOCK - SENDING/RECEIVING PROCESS INSTRUCTION STREAMS

3 Evaluation

We have chosen two exemplar codes to evaluate the Parallel Performance Estimator, the first is an Alternating Direction Implicit (ADI) two dimensional solver of initial value problems[17] and the second is a Parallel Sort by Regular Sampling (PSRS)[16]. In both cases we simulate the performance on our 16 node SP2 and compare the resulting times with actual run times.

The ADI Solver

The ADI solver is rich in communications structure, and behaves quite differently depending on the how the processor array is mapped onto the physical grid.

The ADI solution sequence starts with all of the processors in the first "Y-column", initiating forward elimination in the X direction. When a processor boundary is reached the data is passed to the processor on the right. Once this message is gone then another line of forward elimination in the X direction can be started. The process is repeated for the Y direction. The next step is an edge swap so that each process gets the updated halo data. After this is completed a *reduction* routine is used to determine the termination condition. For this operation the residual from each processor is sent to the bottom left hand corner of the processor array, the worst residual is determined at each node on the gather tree and the result sent to the parent. The root node send the worst residual back down the tree via its own daughters.

```
ADI
    FOR(T=0;T<T;T++)
        UNTIL CONVERGENCE
            FORWARD ELLIMINATE IN THE X DIRECTION
            BACK SUBSTITUTE IN THE X DIRECTION
            SWAP EDGE DATA
            FORWARD ELLIMINATE IN THE Y DIRECTION
            BACK SUBSTITUTE IN THE Y DIRECTION
            SWAP EDGE DATA
            DETERMINE THE CONVERGENCE BY REDUCING THE ERROR
```

The results of this verification are shown in Figures 2 and 3. The two test cases are for a 200×200 system of equations and a 100×100 system. As can be seen from graphs, the predicted timings are within 20% of the actual results, but more encouraging is the consistent accuracy of the prediction over a large number of processors. Figure 3 also shows how the simulator can break up the timings into its constituent parts. The sawtooth structure of the predicted values is due to the fact that the processor array changes configuration. For example for a prime number of processors Np the processors are in a $1 \times Np$ array. In general the processor array is $L_1 \times L_2 = Np$ where $L_1 \leq L_2$ and there is no other pair (L_1', L_2') that satisfies these conditions, for which $L_1' > L_1$.

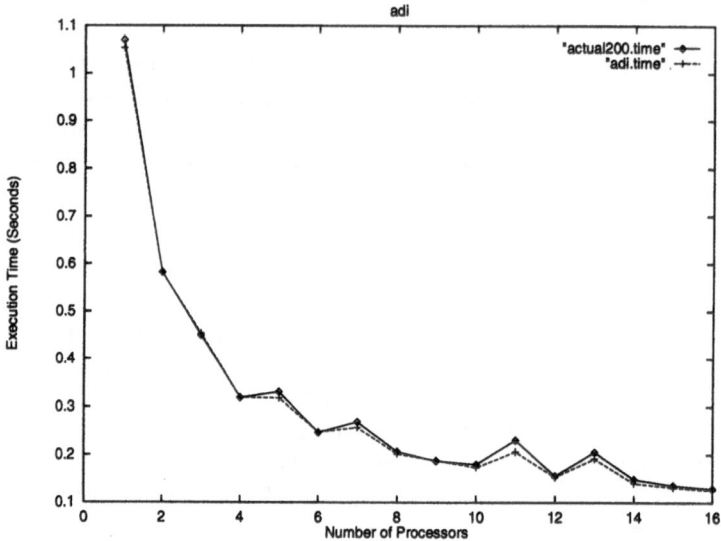

Fig. 2. Actual and Predicted Results Compared for a 200×200 System of Equations Solved by the ADI Method

The effect of the change in processor array is seen by the fact that the most inefficient solution is for a prime number of processors.

Parallel Sorting by Regular Sampling

This is a very effective parallel sorting algorithm developed by Li[15] with its principle attraction being that the number of sorted elements that end up on each processor is bounded and in practise for random initial data the variation in the number of sorted elements per processor is only $2 - 4\%$.

Assume that there are n elements to sort and that the number of processors p divides n without remainder. Initially each processor has $\frac{n}{p}$ of the elements. Each processor uses *quicksort* to order its elements and selects a sample of the elements at indices $1, (n/p^2) + 1, (2n/p^2) + 1, \cdots, (p - 1)(n/p^2) + 1$. These samples are sent to one processor which orders them and selects $p - 1$ pivot values at indices $p + p/2, 2p + p/2, \cdots, (p - 1)p + p/2$. These pivot values are sent to each processor which then divides its list into p parts identified by the pivot values. Now the ith processor keeps its ith partition and sends its jth partition to processor j. Each

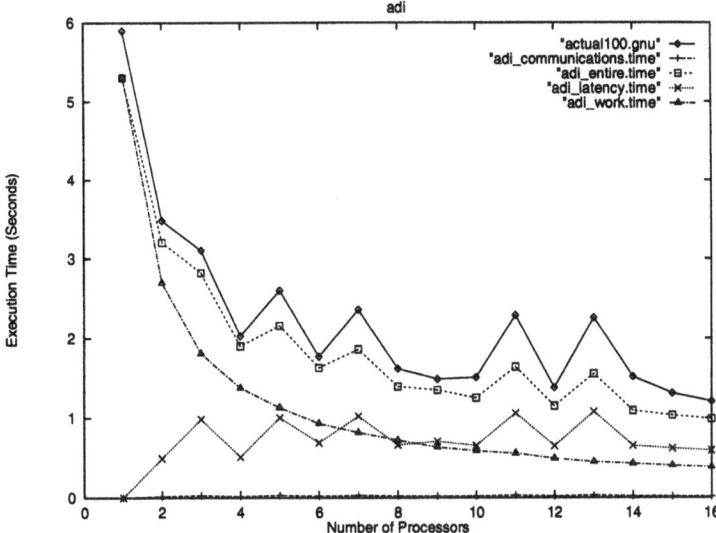

Fig. 3. Analysis of Actual and Predicted Results Compared for a 100×100 System of Equations Solved by the ADI Method

processor then merges its p partitions and the complete set is ordered accross the processors.

The pseudocode for the PSRS algorithm is given below.

```
PSRS
     GENERATE AND DISTRIBUTE THE INITIAL VALUES
     EACH PROCESSOR QUICKSORTS ITS ELEMENTS
     EACH PROCESSOR SELECTS ITS SAMPLES
     THE SAMPLES ARE SENT TO THE MASTER PROCESSOR
     THE MASTER PROCESSOR SELECTS AND BROADCASTS THE PIVOTS
     EACH PROCESSOR PARTITIONS ITS LOCAL LIST
     PARTITIONS ARE DISTRIBUTED AMONG PROCESSORS
     EACH PROCESSOR MERGES ITS LIST
```

The simulation of this sorting method was compared with actual results for data sets of $1,000,000$ and $10,000$ real numbers respectively. The results are shown in Figures 4 and 5 respectively. Again the results are accurate to within 20%.

4   Conclusion

Both sets of test results that we have used show similar trends. While all tests show that the simulation replicates the broad features of the measured data it is

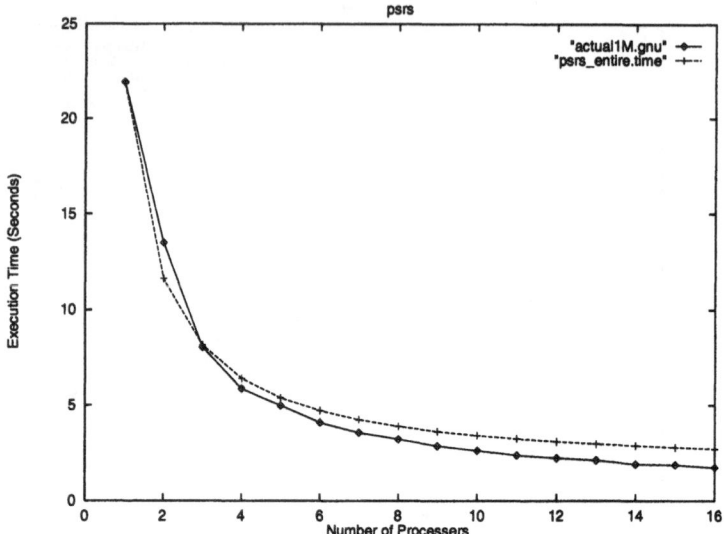

Fig. 4. Actual and Predicted Results Compared for 10^6 Doubles Sorted by PSRS

evident that the simulation is the more accurate when communications is much less important than the work being done. There are we believe two causes for this. The first is that the single processor memory hierachy[18, 9, 1] (cacheing) is becoming important and the second is that the detailed working of the interconnect network is beginning to show. The cache effect can be modelled by introducing structure to the memory access times, but this introduces parameters that are difficult to measure. The network architecture could be simulated[13] but we believe that this effect is secondary compared with the contribution of the memory hierachy. Future work will concentrate on incorporating memorachy hierachy effects without introducing extra parameters.

Finally it is evident from the results of the tests that our Parallel Performance Estimator works effectively for both numerically intensive applications and for applications in which the movement of data dominates.

References

[1] B. Alpern and L. Carter. Towards a model for portable parallel performance: Exposing the memory hierarchy. In A. Hey and J. Ferrante., editors, *Portability and Performance for Parallel Processing.*, pages 21–41, 1994.

[2] R. Aversa, A. Mazzeo, N. Mazzocca, and U. Villano. Developing applications for heterogeneous computing environments using simulation: A case study. *Parallel Computing*, 24(5-6):741–762, 1998.

[3] V. Balasundaram, V. Fox, K. Kennedy, and U. Kremer. A static performance estimator to guide data partitioning decisions. *SIGPLAN Notices*, 26(7):213–223, 1991.

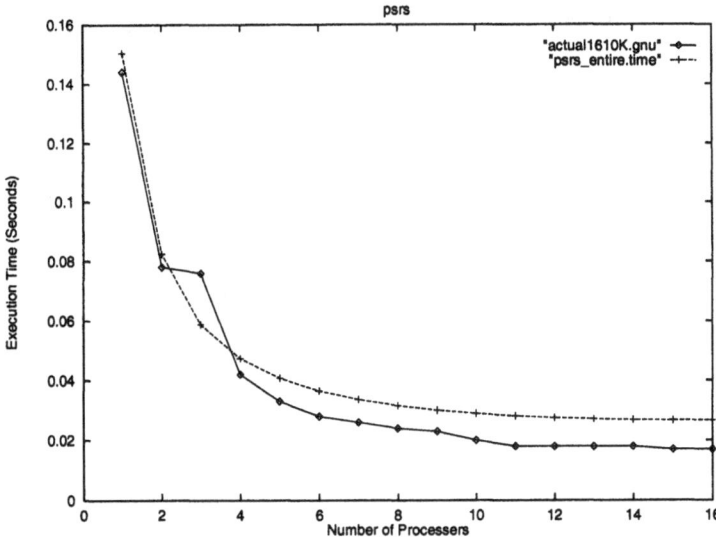

Fig. 5. Actual and Predicted Results Compared for 10^4 Doubles Sorted by PSRS

[4] E. Brewer. High-level optimization via automated statistical modelling. In *Proceedings of the Fifth ACM SIGPLAN Symposium on Principles and Practice of Parallel Programming*, pages 80–91, 1995.

[5] M. Clement and M. Quinn. Architectural scaling and analytical performance prediction. In *Proceedings of the Seventh International Conference on Parallel and Distributed Computing Systems.*, pages 16–21, 1994.

[6] C.N. Dellarocas, E.A. Brewer, A. Colebrook, and W.E. Weihl. Proteus: A high-performance parallel algorithm evaluation system. Technical report, Dept Computer Science, MIT, 1991.

[7] A.N. Dunlop, E. Hernandez, O.Naim, T. Hey, and D. Nicole. A toolkit for optimising parallel performance. *Lecture Notes in Computer Science.*, 919:548–553, 1995.

[8] T. Fahringer. Compile-time estimation of communication costs for data parallel programs. *Journal of Parallel and Distributed Computing*, 39(1):46–65, 1996.

[9] V.S. Getov. Performance characterisation of the cache memory effect. *Supercomputer*, 11:31–49, 1995.

[10] F. Hartleb and V. Mertsiotakis. Bounds for the mean runtime of parallel programs. In *Sixth International Conference on Modelling Techniques and Tools for Parallel Computer Performance Evaluation.*, 1992.

[11] E. Hernandez and T. Hey. Variations on low-level communication benchmarks. *Supercomputer*, 1997.

[12] G. Kempf, A.J. van der Steen, C. Caremoli, and W.Y. Thang. Simulations of scientific programs on parallel architectures. *Lecture Notes In Computer Science*, 1067:536–546, 1996.

[13] O. Kolp. Performance estimation for a parallel system with a heirachical switch network. *Parallel Computing.*, 20:1613–1626, 1994.

[14] K. Kubota, K. Itakura, M. Sato, and T. Boku. Practical simulation of large-scale parallel programs and its performance analysis of the nas parallel benchmarks. *Lecture Notes in Computer Science*, 1470:244–254, 1998.

[15] X. Li, P. Lu, J. Schaeffer, P.S. Wong, and H. Shi. On the versatility of parallel sorting by regular sampling. Technical Report TR91-06, The Department of Computer Science, University of Alberta, March 1992.

[16] M.Quinn. *Parallel Computing: Theory and Practice*, chapter 10. McGraw-Hill, Inc, 1994.

[17] W.H. Press, B.P.Flannery, S.A. Teukolsky, and W.T. Vetterling. *Numerical Recepies*, chapter 17. Cambridge University Press, 1986.

[18] O. Temam, C. Fricker, and W. Jalby. Cache interference phenomena. In *Proceedings of the ACM SIGMETRICS conference on Measurement and Modelling of Computer Systems*, 1994.

Min-Cut Methods for Mapping Dataflow Graphs

Volker Elling[1] and Karsten Schwan[2]

[1] RWTH Aachen, Bärenstraße 5 (Apt. 12), 52064 Aachen, Germany
[2] Georgia Institute of Technology, 801 Atlantic Drive, Atlanta, GA 30332-0280, USA

Abstract. High performance applications and the underlying hardware platforms
are becoming increasingly dynamic; runtime changes in the behavior of both
are likely to result in inappropriate mappings of tasks to parallel machines during application execution. This fact is prompting new research on mapping and
scheduling the dataflow graphs that represent parallel applications. In contrast
to recent research which focuses on critical paths in dataflow graphs, this paper presents new mapping methods that compute near-min-cut partitions of the
dataflow graph. Our methods deliver mappings that are an order of magnitude
more efficient than those of DSC, a state-of-the-art critical-path algorithm, for
sample high performance applications.

1 Introduction

Difficult steps in parallel programming include decomposing a computation into concurrently executable components ('tasks'), assigning them to processors ('mapping')
and determining an order of execution on each processor ('scheduling'). When parallel programs are run on dedicated multiprocessors with known processor speeds, network topologies and bandwidths, programmers can use fixed task-to-processor mappings, develop good mappings by trial-and-error using performance analysis tools to
identify bottlenecks, perhaps based on the output of parallelizing compilers. The difficulties with these approaches are well known. First, many problems, including 'sparse
triangular solves' (see Section 4.3), are too irregular for partitioning by a human. Second, when programs are decomposed by compilers, at a fine grain of parallelism, the
potential number of parallel tasks is too large for manual methods. Third, when using
LAN-connected clusters of workstations (see Figure 1 for the Georgia Tech IHPCL lab)
or entire computational grids (as in Globus [13]), there are irregular network topologies
and changes in network or machine performance due to changes in load or platform
failures. Fourth, task sets may change at runtime either due to the algorithm or because
of computer-user interaction like visualization or steering [19, 15, 8]. Finally, in load
balancing a busy processor occasionally shares tasks with a processor that has become
idle. It is useful to choose the share of the latter processor so that the communication
between both is minimized; this requires min-cut partitioning methods for dataflow
graphs (see below). In all these situations, automatically generated mappings are attractive or even required. For these and other reasons, the ultimate goal of our research is
to develop black-box mapping and scheduling packages that require little or no human
intervention.

P. Amestoy et al. (Eds.): Euro-Par'99, LNCS 1685, pp. 203–212, 1999.

In Section 2, we define the formal problem that is to be solved and present examples for its relevance. Section 3 discusses critical-path vs. min-cut mapping methods. Our algorithmic contributions are presented in Sections 3.3 and 3.4.

Section 4.1 introduces an abstract model of parallel hardware (suited for multiprocessor machines as well as for workstation clusters), which is then used for extending 2-processor mapping methods to an arbitrary number of processors. In Section 4.3, we present two real-world sample problems; simulation results for them are shown, for our two methods as well as DSC, a critical-path algorithm, and 'DSC-spectral', a variant of DSC. For more details about our algorithms and experiments see [9, 10].

2   Formal Problem Definition

Sample Applications. Parallel applications are often modelled using dataflow graphs (also known as 'macro dataflow models', 'task graphs', or 'program dependence graphs'). A dataflow graph is a directed acyclic graph (V, E) consisting of vertices $v \in V$ representing computation steps, connected by directed edges $(v, w) \in E$ which represent data generated by v and used by w. The vertices are weighted with the computational costs $t(v)$, whereas edges are weighted with the amounts of communication $c(v, w)$. We prefer cost measures like 'FLOPS' or 'bytes' rather than 'execution time' resp. 'latency' because the latter depend on where execution/communication take place which is a priori unknown. An example is the graph shown in Figure 1. This graph represents a two-hour iteration step in the Georgia Tech Climate Model (GTCM) which simulates atmospheric ozone depletion. On our UltraSparc-II-cluster, the tasks in the 'Lorenz iteration' execute for about 2 seconds while the chemistry tasks run for about 10 seconds. One execution of the task graph represents 2 hours simulated time; problem size ranges from 1 month (for 'debugging') to 6 years of simulated time. Each graph edge corresponds to 100–400 KB of data. This application constitutes one of the examples addressed by our mapping (and remapping) algorithm. Another example exhibiting finer grain parallelism is presented in Section 4.3 below.

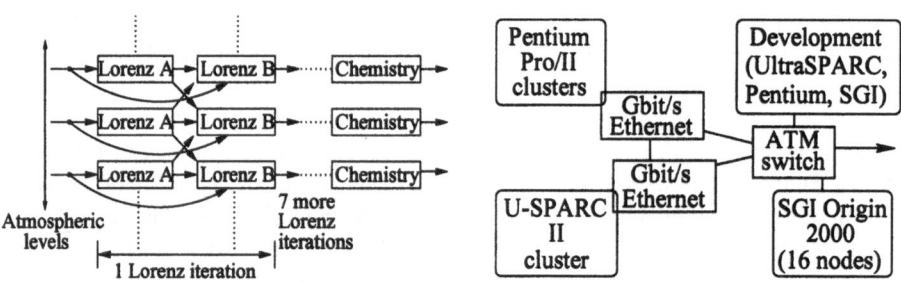

Fig. 1. Left: task graph in the Georgia Tech Climate Model; right: high-performance hardware in the Georgia Tech IHPCL lab

Problem Definition. 'Mapping' is the task of assigning each vertex v to $\pi(v)$, one of P available processors. 'Scheduling' (more precisely, 'local scheduling') is the subsequent stage of determining an order of execution for the vertices on each processor (this order can be strict or advisory). Usually, the objective is to minimize the 'makespan', the time between start of the first and completion of the last task. We define

$$\text{efficiency} = \frac{\sum_{v \in V} \text{load}_v}{\text{makespan} \cdot \sum_{p=1}^{P} \text{speed}_p}$$

where load_v refers to the number of t-units task v takes, and speed_p refers to the speed of processor p in t-units per second. The numerator contains the total workload measured in t-units, the denominator contains the amount of t-units provided by the processors during the makespan.

Many existing mapping/scheduling algorithms are restricted to homogeneous processors connected by a uniform network. For the cluster and grid machines used in our work (see Figure 1), however, we have to deal with processors of different speeds ('weakly heterogeneous multiprocessors') or even different architectures ('strongly heterogeneous multiprocessors' — some processors are well-suited for floating-point computations (MIPS, UltraSPARC) while others were designed for integer tasks (Intel)). The size of on-chip caches is important as well. Some machines (SGI Origin) utilize high performance interconnects, whereas the workstation clusters employ commodity networks like switched Ethernet for interprocessor communication. Finally, processor and network speeds can vary over time due to shared use with other applications. The methods we describe deal with weak heterogeneity, and they are suited for remapping in case of changes in program behavior or resource availability.

The performance of parallel applications is affected by various factors, including CPU performances, amounts of main memory and disk space, network latencies and bandwidths. Some parallel programs are bandwidth-limited in the sense that network congestion causes slowdown. Others are latency-limited: slowdown results from point-to-point delays in frequent, fine-grain communication. This paper considers both latency- and bandwidth-limited programs.

3   Mapping Algorithms

This section first describes a popular mapping algorithm called 'Dominant Sequence Clustering' (DSC). Next, we present new mapping algorithms, called 'spectral mapping' and 'greedy mapping'. In Section 4, these algorithms are shown to be superior to DSC in terms of mapping quality.

3.1   Critical-Path Methods

Critical-path methods require $t(v)$ resp. $c(v, w)$ to be computation resp. communication time. If the sum of $t(v)$ and $c(v, w)$ on a path in the dataflow graph (the 'path length') equals the makespan, the path is called 'critical'. In order to decrease the makespan, the lengths of all critical paths have to be decreased (by assigning v to a faster processor resp. v, w to the same processor or processors connected by a faster link). The most advanced critical-path algorithm known to us is 'Dominant Sequence Clustering' ('DSC', [22]). It is as efficient as older algorithms in minimizing makespan [14]; in addition, it is faster since it recomputes critical paths less often.

Usually, critical-path methods do not try to minimize cut size. DSC forms clusters of tasks; the clusters are assigned to processors blockwise or cyclically. In [22], the use of Bokhari's algorithm [3] for overcoming this limitation is proposed: an undirected, graph is formed, with the clusters as vertices and edges (c, d) weighted with the amount of communication $C(c, d)$ between clusters c and d:

$$C(c, d) := \sum_{v \in c, w \in d} (c(v, w) + c(w, v)).$$

The cut size is reduced by computing small-cut partitions of the cluster graph and assigning them to processors. We propose to partition the graph by spectral bisection rather than Bokhari's algorithm (below, this variant is referred to as 'DSC-spectral').

3.2   Min-Cut Methods

Min-cut methods do not proceed by shortening critical paths; they try to find a mapping $\pi : V \rightarrow \{1, \dots, P\}$ of the task graph with small cut size, and loads approximately proportional to the respective speeds of the processors $p = 1, \dots, P$:

$$\text{cutsize}(\pi) := \sum_{(v,w) \in E, \, \pi(v) \neq \pi(w)} c(v, w), \qquad \text{load}_p(\pi) := \sum_{\pi(v)=p} t(v).$$

Previous Work. To the best of our knowledge, there has been no previous attempt to define explicit min-cut mapping algorithms for arbitrary *dataflow* graphs. However, dataflow graphs often arise from undirected graphs like finite-element grids. There has been progress on min-cut partitioning algorithms for *undirected* graphs ([17, 12, 20, 16]; for a survey see [11]). Unfortunately, these methods cannot be trivially applied to the problem we are solving, since our complex parallel applications typically consist of several coupled subcomponents (e.g., a finite-element elasticity code, a finite-volume gas dynamics code, chemistry and visualisation) that cannot be scheduled easily by partitioning a single physical grid.

In [4], an undirected doubly-weighted graph of communicating long-running processes is partitioned in a Simulated Annealing fashion. [5] proposes a method based on greedy pairwise exchange, for a problem similar to our dataflow graphs. However, this method recomputes the makespan (or a similar objective function) after every exchange which is very expensive.

Min-cut algorithms for directed graphs cannot easily be adapted to dataflow graphs since the partition with smallest cut size might be the *worst* rather than the best mapping (see Figure 2). As evident from the example in the figure, it is important to take the directedness of the graph into account. Toward this end, we define the earliest-start resp. earliest-finish 'times' $\text{est}(v)$ and $\text{eft}(v)$ by

$$\text{est}(v) := \max_{(w,v) \in E} \text{eft}(w), \qquad \text{eft}(v) := \text{est}(v) + t(v)$$

(note that this definition is valid because the graph is acyclic). The nodes are sorted by est and separated into K sets $V_1, \ldots, V_K$ with (almost) equal size so that

$$i \le j, v \in V_i, w \in V_j \quad \Rightarrow \quad \text{est}(v) \le \text{est}(w).$$

Instead of requiring load proportional to speed on each processor, we require proportionality *in each set* V_k:

$$\text{speed}_p \sim \text{load}_{p,k}(\pi) := \sum_{v \in V_k, \, \pi(v)=p} t(v).$$

This means that we have load balance in K 'time intervals' rather than overall. Of course, the exact start and finish times of the tasks are not known in advance; in addition, our definition of est and eft does *not* involve $c(v, w)$. Nevertheless, est and eft are a practical way of estimating the relative execution order of tasks in advance. In our experiments, we compute the longest (here, length = number of vertices) path in the dataflow graph and set K to half its length. Figure 2 shows the improved mapping.

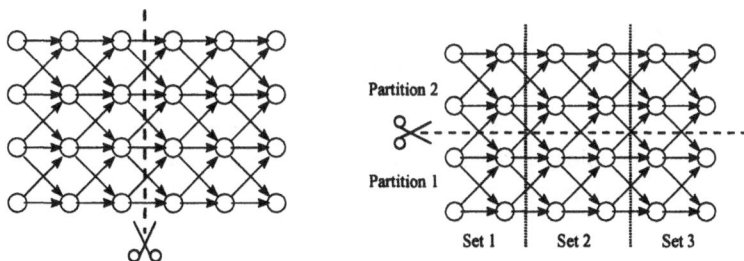

Fig. 2. Left: a bad min-cut mapping; right: improved mapping

3.3 Spectral Mapping

We have adapted spectral bisection[20] to dataflow graphs since it delivers, along with multilevel partitioning[16], the best partitions for undirected graphs. Its disadvantages are low speed and difficult implementation.

For simplicity, we assume that $V = \{1, \ldots, |V|\}$. The 'Laplacian matrix' (compare [7]) $L = (L_{vw})$ is defined by

$$L_{vw} := \begin{cases} -c(v, w), & v \ne w, (v, w) \in E \text{ or } (w, v) \in E \\ \sum_{(v,w) \in E} c(v, w), & v = w \\ 0, & \text{else} \end{cases}$$

It is a positive semidefinite matrix, with $(1, \ldots, 1)$ as eigenvector for the eigenvalue 0. The second smallest eigenvalue is always positive if the graph is connected. The corresponding eigenvector x is called 'Fiedler vector'. It solves

$$\phi(x) := \frac{\sum_{v,w \in E} c(v, w)(x_v - x_w)^2}{\sum_{v \in E} x_v^2} \longrightarrow \min, \quad \text{s.t. } \exists v, w : x_v \ne x_w$$

Note that $\phi(x)$ is small if, for adjacent vertices v, w, the difference $x_v - x_w$ is small. The 'closer' vertices are in the graph, the closer are their x-values. For dataflow graphs, we minimize $\phi(x)$ with respect to the constraints

$$\sum_{v \in V_k} t(v) x_v = 0.$$

This corresponds to finding the smallest eigenvalue and corresponding eigenvector of the operator $P^\top L P$ where P is an orthogonal $n \times (n - K)$ matrix mapping $\mathbb{R}^{n-K}$ into the constraint subspace of $\mathbb{R}^n$. A 'bisection' (2-partition) is obtained by choosing thresholds T_k, $k = 1, \ldots, K$, and setting, for $v \in V_k$,

$$\pi(v) := \begin{cases} 1, & x_v > T_k \\ 0, & \text{else} \end{cases}.$$

Let P_i be the set of processors executing partition i; the T_k are chosen so that

$$\frac{\text{load}_{0,k}(\pi)}{\text{load}_{0,k}(\pi) + \text{load}_{1,k}(\pi)} \approx \alpha := \frac{\sum_{p \in P_0} \text{speed}_p}{\sum_{p \in P_0 \cup P_1} \text{speed}_p}$$

Spectral mapping is much slower than DSC or greedy mapping (see below), but it takes only about 1.7 seconds for a 1000 vertices/4000 edge graph (20 seconds for 10000/40000) on an SGI O2 (R10000 2.6 195 MHz). These times can be improved significantly, as our current implementation is sequential and not very sophisticated; [1, 2] propose multilevel parallelized variants for spectral bisection that achieve an order-of-magnitude performance improvement and can be adapted to spectral mapping.

3.4   Greedy Mapping

A faster but lower-quality method for min-cut bisection is 'greedy mapping'. It corresponds to the greedy bisection methods developed for undirected graphs[17, 12]. At the beginning, an arbitrary initial mapping π is chosen. The 'gain' of a vertex is defined as the decrease in cutsize when this vertex is moved to the other partition ($\pi(v)$ is changed from 0 to 1 or vice versa). The vertices are moved between partitions, in order of decreasing gain. In every iteration, a vertex may be moved only once. A vertex $v \in V_k$ may not be moved if moving it would make

$$\left| \alpha - \frac{\text{load}_{0,k}(\pi)}{\text{load}_{0,k}(\pi) + \text{load}_{1,k}(\pi)} \right|$$

larger than a threshold (for example, 0.07). An iteration finishes when no vertices with nonnegative gain are left. In our experience, there is no improvement after 10–15 iterations.

4   Evaluation

4.1   Modeling Topologies

This section proposes a simple but representative model for real-world networks. A model consists of processors with different speed, each connected to an arbitrary number of buses. Buses themselves can be connected by switches ('zero-speed processors').

In order to apply our bisection algorithms to topologies with more than 2 processors, a hierarchy of 'nodes' is computed. At each step, the bus with highest bandwidth is chosen and, together with the connected processors (atomic nodes), folded into a single parent node. When all buses have been folded, only one node is left. The clustering is undone in reverse order. At each step, one node is unfolded into a bus and the adjacent nodes. The dataflow graph partition corresponding to the parent node is distributed to the children by repeated bisection. α is chosen according to processor performances in each bisection step. Finally, all steps have been undone, and every processor has been assigned a partition of the graph.

As an example, consider Figure 1. Each cluster is (somewhat inaccurately) modeled as a bus to which machines and switches are connected. The p-processor machines can be treated as single-processor machines with p-fold speed, or (for large multiprocessor machines) as a separate bus with p nodes connected to it. Starting with the Gigabit Ethernet links, each link is folded into a node. After this, the ATM links are folded. Our bisection algorithm will try to assign coherent pieces of the dataflow graph to the Gigabit clusters and minimize communication via the slow ATM switch. However, the example also demonstrates the limitations of our simplistic topology treatment: the link 'A' interconnecting the two Gigabit switches might be folded first (before any of the Gigabit-switch-to-processor links is folded). Since it might represent a bottleneck, folding it *last* (i.e. assigning coherent pieces of the dataflow graph to the processors on each side) is more appropriate. A more sophisticated algorithm would consider the cluster topology as an undirected graph and apply spectral bisection to it.

4.2   Local Scheduling

[21] discusses good heuristics for computing local schedules on processors and compares them for randomly generated graphs. Interestingly, taking communication delays into account (as in the 'RCP*' scheme) and neglecting them (in the 'RCP' scheme) does not affect schedule quality. In our experiments we use the RCP scheme (in its non-strict form, i.e. out-of-order execution is allowed if the highest-priority task is not ready).

4.3   Sample Problems

We have chosen two sample problems that are rather complementary and reflect the variety of parallel applications. The first, 'sparse triangular solves', is very fine-grain and latency-limited: each task runs for ≤ 1 μs, each edge corresponds to 12 bytes. The second problem, the Georgia Tech climate model[18] which has already been introduced in Section 2, is a very coarse-grain and usually bandwidth-limited problem. The tasks execute for several seconds; the edges correspond to data in the order of 100 kilobytes. The model runs for a very long time; this is typical for many of the so-called grand-challenge applications.

In our simulations, we use some simplifications: data is sent in packets that have equal size on all buses. The bandwidth of a bus is determined by the number of packets per second. Network interfaces have unlimited memory and perfect knowledge about

$$\begin{pmatrix} 1 & 0 & 0 & 0 & 0 \\ -1 & 2 & 0 & 0 & 0 \\ 2 & 0 & 1 & 0 & 0 \\ 0 & 1 & -1 & 1 & 0 \\ 1 & 3 & 0 & -1 & 1 \end{pmatrix}$$

Fig. 3. Lower triangular matrix and its dataflow graph. Vertices are labeled with the index of the corresponding matrix row.

the other interfaces on the bus. When a bus becomes available, one of the waiting interfaces is chosen randomly with uniform probability. These simplifications are not vital. By varying the packet size it is possible to simulate networks with different latencies.

Sparse Triangular Solves. Our first test problem are sparse triangular solves (STS): solve for x in the linear system $Ax = b$ where A is a lower triangular matrix with few nonzero entries. Task i corresponds to solving for x_i; this requires all x_j with $a_{ij} \neq 0$. The dataflow graph is determined by the sparsity structure of A (see Figure 3). The time for distribution of A, b is neglected (initial data locations depend on the application).

A workstation cluster is appropriate only for very large A; otherwise a multiprocessor machine with good interconnect is mandatory. Results from [6] for a CM-5 implementation of STS with mapping by DSC suggest that DSC is too slow: mapping time exceeds execution time, even if one mapping is reused many times (which is possible in typical applications). Since spectral mapping is slower than DSC, we consider STS as a source for real-world dataflow graphs rather than a practical application. The results shown in Figure 4 were obtained by simulation using the hardware model in Section 4.1, with $t(v) = 1$ μs, $c(v, w) = 12$ Byte (x_i (double), i (long)) and 16 processors connected by a 16 Byte/packet bus. The matrix is 'bcspwr10' from the Harwell-Boeing collection (5300×5300, 8271 below-diagonal nonzeros; available in 'netlib'); other matrices from the 'bcspwr' set yield similar results, as do randomly generated matrices.

Spectral mapping achieves the best results, followed by greedy mapping which offers a fast alternative. DSC-spectral improves DSC performance but cannot compete with the min-cut methods. It is worth noting that for the 'bcspwr05' matrix (443×443, 590 below-diagonal nonzeros), 16 CPUs and an 'infinitely fast' network (bandwidth 10^{30} MB/s), spectral mapping achieves 95 % efficiency while DSC achieves about 51%. Even in this case where communication delays can be neglected, the min-cut algorithm generates better mappings.

Atmospheric Ozone Simulation. In our second application, speedup is bandwidth-limited due to large data items. Our topology consists of two B MB/s buses with 8 equal-speed processors at each, connected by a B MB/s link. This topology represents typical bottleneck situations — in Figure 1, these could occur if a computation is distributed between the nodes in the UltraSPARC cluster and the 16-node SGI Origin. Figure 4 gives results for this problem (with 32 atmospheric levels). In this simulation we used a network packet size of 256 byte; this size is representative for commodity workstation interconnects which are well-suited for GTCM.

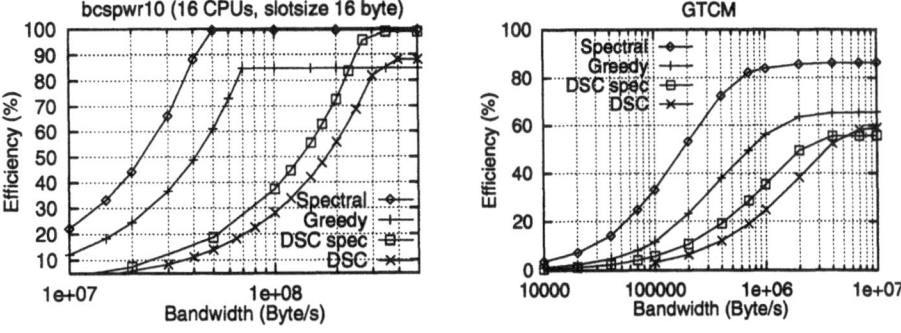

Fig. 4. Simulation results for STS (left) and GTCM (right)

Obviously, small cut size is essential: spectral mapping achieves the same efficiency as DSC for B a factor 10 smaller. Again, greedy mapping qualifies as a fast alternative with fair quality while DSC-spectral is slightly better than DSC without reaching the two min-cut methods.

5 Conclusions

The main contribution of our work is a paradigm for applying undirected-graph min-cut methods to dataflow graph min-cut mapping, by means of the 'time intervals' defined above. We have adapted spectral and greedy bisection to dataflow graph mapping. These methods are applicable to a wide range of processor/network speed ratios. We have demonstrated that, with respect to quality, min-cut mapping methods are slightly better than critical-path methods for fast networks where small cut size does not seem to matter, and that they are clearly superior for slow networks. Mapping speed is traded off for these advantages.

Future work includes the acceleration of spectral mapping in order to make it practical for a wider range of applications, and additional work on greedy spectral mapping in order to assess whether this method performs well for large dataflow graphs (greedy algorithms 'look' at the graph in a 'local' way) — toward this end, multilevel strategies as discussed in [16] for undirected graphs are promising. We have not considered strongly-heterogeneous clusters; for an application that consists, for example, of 'integer' as well as 'floating-point tasks' and runs on a mixed Intel and MIPS CPU cluster, this would lead to serious performance penalties. Finally, it is not clear whether our simplistic topology clustering method is appropriate for all network topologies appearing in practice.

If processor or network performances change during execution, it is necessary to compute a new mapping. For example, GTCM runs for several minutes to several days. During this time, network links and computers can break down and become available again; other users start and stop their own applications on a part of the clusters. It is impossible to adapt to these changes manually because 24-hour operator supervision would be necessary. The only alternative is to develop automatic mapping methods like the ones we describe. Greedy mapping is sufficiently fast for a remapping frequency on the order of 1/second and is easily adapted to take initial data location into account. Spectral mapping is too slow except for long-running applications like our atmospheric simulation.

References

[1] S.T. Barnard. PMRSB: Parallel multilevel recursive spectral bisection. In *Proceedings of Supercomputing*, 1995.

[2] S.T. Barnard and H.D. Simon. A fast multilevel implementation of recursive spectral bisection for partitioning unstructured problems. In *Proc. 6th SIAM Conf. Par. Proc. Sci. Comp.*, pages 711–718, 1993.

[3] S.H. Bokhari. On the mapping problem. *IEEE Trans. Comput.*, C-30(3):207–214, 8/1981.

[4] S.W. Bollinger and S.F. Midkiff. Heuristic technique for processor and link assignment in multicomputers. *IEEE Trans. Comput.*, 40(3):325–333, March 1991.

[5] V. Chaudhary and J.K. Aggarwal. A generalized scheme for mapping parallel algorithms. *IEEE Transactions on Parallel and Distributed Systems*, 4(3):328–346, march 1993.

[6] F.T. Chong, S.D. Sharma, E.A. Brewer, and J. Saltz. Multiprocessor runtime support for fine-grained, irregular DAGs. *Parallel Processing Letters*, 5(4):671–683, December 1995.

[7] Fan R. K. Chung. *Spectral graph theory*. American Mathematical Society, 1997.

[8] G. Eisenhauer and K. Schwan. An object-based infrastructure for program monitoring and steering. In *Proc. 2nd SIGMETRICS Symp. Par. Dist. Tools*, pages 10–20, August 1998.

[9] V. Elling. A spectral method for mapping dataflow graphs. Master's thesis, Georgia Institute of Technology, 1998.

[10] V. Elling and K. Schwan. Min-cut methods for mapping dataflow graphs. Technical Report GIT-CC-99-05, Georgia Institute of Technology, 1999.

[11] U. Elsner. Graph partitioning — a survey. Technical Report SFB393/97-27, TU Chemnitz, SFB "Numerische Simulation auf massiv parallelen Rechnern", Dec. 1997.

[12] C.M. Fiduccia and R.M. Mattheyses. A linear-time heuristic for improving network partitions. In *Proc. 19th IEEE Design Automation Conference*, pages 175–181, 1982.

[13] I. Foster and C. Kesselman. Globus: A metacomputing infrastructure toolkit. *International Journal of Supercomputer Applications*, 11(2):115–128, 1997.

[14] A. Gerasoulis and Tao Yang. A comparison of clustering heuristics for scheduling DAGs on multiprocessors. *J. Par. Dist. Comp., Special Issue on scheduling and load balancing*, 16(4):276–291, 1992.

[15] W. Gu, G. Eisenhauer, K. Schwan, and J. Vetter. Falcon: On-line monitoring for steering parallel programs. *Concurrency: Practice and Experience*, 10(9):699–736, Aug. 1998.

[16] G. Karypis and V. Kumar. A fast and high quality multilevel scheme for partitioning irregular graphs. To appear in SIAM Journal on Scientific Computing.

[17] B.W. Kernighan and S. Lin. An efficient heuristic procedure for partitioning graphs. *Bell System Tech. J.*, 49:291–307, February 1970.

[18] T. Kindler, K. Schwan, D. Silva, M. Trauner, and F. Alyea. Parallelization of spectral models for atmospheric transport processes. *Concurrency: Practice and Experience*, 11/96.

[19] B. Plale, V. Elling, G. Eisenhauer, K. Schwan, D. King, and V. Martin. Realizing distributed computational laboratories. *Int. J. Par. Dist. Sys. Networks, to appear*.

[20] A. Pothen, H.D. Simon, and Kang-Pu Liou. Partitioning sparse matrices with eigenvectors of graphs. *SIAM J. Matrix Anal. Appl.*, 11(3):430–452, July 1990.

[21] Tao Yang and A. Gerasoulis. List scheduling with and without communication. *Parallel Computing Journal*, 19:1321–1344, 1993.

[22] Tao Yang and A. Gerasoulis. DSC: Scheduling parallel tasks on an unbounded number of processors. *IEEE Transactions on Parallel and Distributed Systems*, 5(9):951–967, 1994.

Influence of Variable Time Operations in Static Instruction Scheduling*

Patricia Borensztejn[1], Cristina Barrado[2], and Jesus Labarta[2]

[1] Dep.de Computación, UBA, Argentina. patricia@dc.uba.ar
[2] Dep. de Arquitectura de Computadores, UPC, España

Abstract. Instruction Scheduling is the task of deciding what instruction will be executed at which unit of time. The objective is to extract maximum instruction level parallelism for the code. Compilers designed for VLIW and EPIC architectures do static instruction scheduling in a back-end pass. This pass, known as scheduler, needs to have full knowledge of the execution time of each instruction. But memory access instructions have a variable latency, depending on their locality and the memory hierarchy architecture. The scheduler must assume a constant value, usually the execution time assigned to a hit. At execution a miss may reduce the parallelism because idle cycles may appear before the instructions that need the data. This paper describes a statistic model to evaluate how sensitive are the scheduling algorithms to the variable time operations. We present experimental measures taken over two static scheduling algorithms based on software pipelining.

1 Introduction

Compilers for VLIW, EPIC and superscalar architectures include instruction scheduling as an optimization pass with the objective of minimizing the number of cycles, this is, maximizing parallelism. A special approach that schedules instructions at the loop level is software pipelining [3, 5], where instructions belonging to different iterations of the loop overlap. The *Initiation Interval, II*, is the number of cycles between the initiation of successive iterations. This value is limited by the data dependencies between loop instructions and by the resource constraints of the actual machine. The main goal of different implementations of software pipelining is to achieve a *loop kernel* with the minimum *II*.

The scheduler must have full knowledge of the latencies to schedule instructions. But load and store operations take a variable number of cycles to execute depending on what level of the memory hierarchy holds the values, the buses bandwidths, etc. Schedulers assume a constant value, usually the execution time of a hit. At execution a miss may reduce the parallelism because idle cycles may appear before the instructions that need the data.

This paper presents a statistic model to evaluate the sensitivity of static schedulings to variable time operations (here the memory accesses). Section 2

---

* This work has been supported by the Ministry of Education of Spain under contracts TIC98-0511 and by the MHAOTEU Esprit Project

P. Amestoy et al. (Eds.): Euro-Par'99, LNCS 1685, pp. 213–216, 1999.
© Springer-Verlag Berlin Heidelberg 1999

describes the statistic model proposed to evaluate the II, section 3 presents the statistic measures of two software pipelining methods and the last section summarizes the results and the future work.

2   The Variable Time Model

This section introduces the *Probable II* value, which gives the mean value of the II on the parallel execution. [3] presents how to compute its lower bound:

$$MII = Max(RecMII, ResMII) \tag{1}$$

this is, the II must preserve data dependences that form recurrences (RecMII) and can use only the available resource (ResMII). RecMII can be computed as maximum ratio distance/weight of the edges for all recurrences and ResMII as the maximum ratio used/available of cycles for all resources.

The increase of the latency of an operation (i.e. miss on a load) on the critical path influences more on parallelism than the increase of an operation that is not on the critical path. Our model takes into account all the combinations of misses and hits of the memory access instructions of the loop body.

Given the *dependence graph* of a loop, we define a statistical model where the following conditions are hold: (1) Each memory instruction hits with a probability of P and misses with a probability of $1 - P$, (2) the execution time of a memory instruction is h cycles if it hits and m cycles if it misses, with $h < m$, and (3) each instruction hits or misses with independence of the rest of the memory instructions.

Assuming that L is the number of load/store instructions of a loop ,S_i, then there exists 2^L combinations of hits and misses. We define each combination as an **event**. Each event has a probability of occur, $Prob_{event}$, that can be calculated as in 2. We define the **Probable Initiation Interval**, II_{Prob}, as the mean value of the of the II of each event weighted by the probability of each event.

$$Prob_{event} = \prod_{i=1}^{L} p_i \quad where \; p_i = \begin{cases} P & if \; S_i \;\; hits \\ 1-P & if \; S_i \;\; misses \end{cases} \tag{2}$$

3   Measures of the Influence of Variable Time Operations

In this section we present the experimental measures of the influence of variable time operations (the memory accesses) on the expected speedup of software pipelined loops using two different implementations. The variable time operations are all memory accesses: Latency is 1 cycle on cache hit and 10 cycles on cache miss.

The two software pipelining heuristics applied were **Hypernode Reduction Modulo Scheduling**[4], HRMS, and **Graph Traverse Software Pipelining**[1], GTSP. HRMS is a modulo scheduling based heuristic. Its objective is

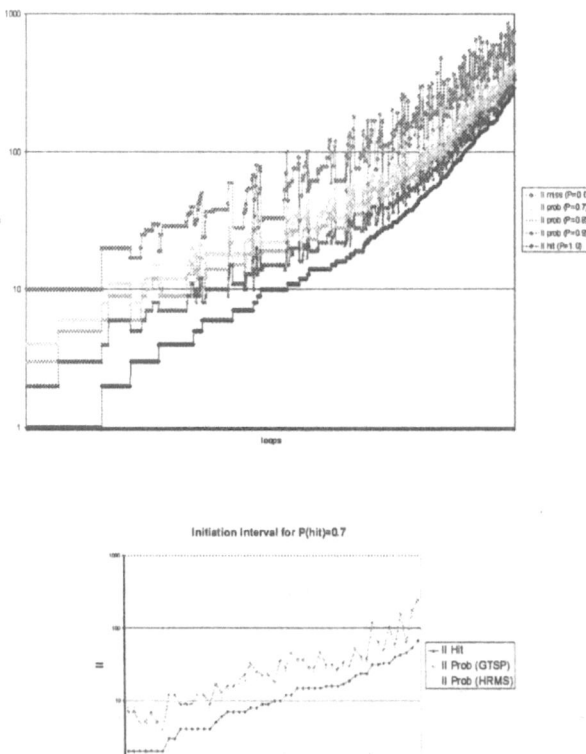

Fig. 1. a) Effects of Variable Latency on II, b) Comparison of two Schedulers

to obtain the minimum II using the minimum number of registers. GTSP is a graph transformation heuristic targeted to minimize the II.

The methodology used first computes the *dependence graphs*, DG, of up to 1258 innermost loops of the Perfect Benchmark Club. Each DG is then software pipelined and represented by a Scheduling Graph, SG. The SG [2] is an extension of the DG where new edges are added to introduce the scheduling restrictions. The variable time model is then applied to each SG and the $IIprob$ evaluated for the given hit probability.

Figure 1-a shows the $IIprob$ of all the loops assuming five hit probabilities and scheduling with HRMS. The plots are sorted in increasing order of the II. The plot labeled $II\ hit$ shows the best II (with a 100% of memory hits). In the other side the plot $II\ miss$ shows the worst II, this is, when every memory access is a miss while the scheduler considered only the hit latency. In between the two bounds the figure shows three more plots, which correspond to intermediate and more realistic situations (hit probabilities of 70%, 80% and 90%). The figure shows that the difference between the plots $P=1.0$ and $P=0.9$ is larger than the differences between the plots $P=0.9$ and $P=0.8$. This shows the importance of the first small variations of the hit probability and that speedup penalties comes out even for the small miss ratios.

Figure 1-b compares the sensitivity of the two heuristics for some loops that gave the same II hit. For a hit probability of 70% the it II prob is, in general, greater for GTSP than for HRMS. Numbers show that one scheduler is more sensitive to variable time operations than the other for loops that *a priori* they were equal. We investigated the cause of such difference and found that GTSP used only one memory unit when possible, while HRMS used most of the time the two memory units.

4 Conclusions

In this paper we show the important influence of variable time operations in the runtime performance of static scheduled loops. We present a statistic model to measure the probable mean parallelism of a loop for a given hit probability. The model shows that the use of the hit latency for memory operations may degrade the application speedup at runtime with a factor equal to the miss latency. Another important outcome is that the performance degradation is not linear. The increment of the II_{Prob} is higher when the hit probability goes from 100% to 90% than when it decrements from 90% to 80%.

We show that static schedules are not equally sensitive to variable time operations. One of the objectives of scheduling heuristics should be to minimize of variable time operations sensitivity. We believe that this could be more important for performance than other considerations like minimizing register pressure.

A direct application of our model could be to introduce it into the compiler in order to help choosing the best scheduling considering their II_{Prob} Apply statistic regression in a future work will help for fast computing the Probable Initiation Interval.

Also the model can be extended to consider different hit probabilities for the loop memory operations. This, together with a static locality analysis, can be very useful to achieve realistic II at compile time.

References

[1] C. Barrado and J. Labarta: "Hamiltonian Recurrence for ILP", EuroPar, 1997.
[2] P. Borensztejn, C. Barrado and J. Labarta: "The Scheduling Graph in the Variable Time Model". Report UPC-DAC-1999-18.
[3] M. Lam: "Software Pipelining: An Effective Scheduling Technique for VLIW machines", Proceedings SIGPLAN'88 PLDI, pp.318-28, 1988.
[4] J. Llosa, M. Valero, E. Ayguadé and A. González: "Hypernode Reduction Modulo Scheduling", Micro-28, pp.350-360, Dec. 1995.
[5] B. Rau: "Iterative Modulo Scheduling: An Algorithm for Software Pipelining Loops", IEEE Micro-27, pp.63-74, Nov. 1994.

Evaluation of LH*LH for a Multicomputer Architecture

Andy D. Pimentel and Louis O. Hertzberger

Dept. of Computer Science, University of Amsterdam
Kruislaan 403, 1098 SJ Amsterdam, The Netherlands

Abstract. Scalable, distributed data structures can provide fast access to large volumes of data. In this paper, we present a simulation study in which the performance behaviour of one of these data structures, called LH*LH, is evaluated for a multicomputer architecture. Our experimental results demonstrate that the access-time to LH*LH can be very small and does not deteriorate for increasing data structure sizes. Furthermore, we also show that parallel access to the LH*LH data structure may speed up client applications. This speed-up is, however, shown to be constrained by a number of effects.

1 Introduction

Modern database applications require fast access to large volumes of data. Sometimes the amount of data is so large than it cannot be efficiently stored or processed by a uniprocessor system. Therefore, a *distributed data structure* can be used that distributes the data over a number of processors within a parallel or distributed system. This is an attractive possibility because the achievements in the field of communication networks for parallel and distributed systems have made remote memory accesses faster than accesses to the local disk [3]. So, even when disregarding the additional processing power of parallel platforms, it has become more efficient to use the main memory of other processors than to use the local disk.

It is highly desirable for a distributed data structure to be scalable. The data structure should not have a theoretical upper limit after which performance degrades (i.e. the time to access data is independent of the number of stored data elements) and it should grow and shrink incrementally rather than reorganising itself totally on a regular basis (e.g. rehashing of all distributed records). For distributed memory multicomputers, a number of Scalable Distributed Data Structures (*SDDS*s) have been proposed [7]. In these distributed storage methods, the processors are divided into *clients* and *servers*. A client manipulates the data by inserting data elements, searching for them or removing them. A server stores a part of the data, called a *bucket*, and receives data-access requests from clients. Generally, there are three ground rules for the implementation of an SDDS in order to realize a high degree of scalability [4]:

- The SDDS should not be addressed using a central directory which forms a bottleneck.
- Each client should have an image of how data is distributed which is as accurate as possible. This image should be improved each time a client makes an "addressing error", i.e. contacts a server which does not contain the required data. The client's state (its current image) is only needed for efficiently locating the remote data; it is not required for the correctness of the SDDS's functionality.

P. Amestoy et al. (Eds.): Euro-Par'99, LNCS 1685, pp. 217–228, 1999.

- If a client has made an addressing error, then the SDDS is responsible for forwarding the client's request to the correct server and for updating the client's image.

For an efficient SDDS, it is essential that the communication needed for data operations
(retrieval, insertion, etc.) is minimised while the amount of data residing at the server
nodes (i.e. the *load factor*) is well balanced. In [7], Litwin et al. propose an SDDS,
called *LH**, which addresses the issue of low communication overhead and balanced
server utilisation. This SDDS is a generalisation of Linear Hashing (LH) [6], which
will be elaborated upon in the next section. For LH*, insertions usually require one
message (from client to server) and three messages in the worst case. Data retrieval
requires one extra message as the requested data has to be returned.

In this paper, we evaluate the performance of a variant of the LH* SDDS, called
LH*LH, which was proposed by Karlsson [4, 5]. For this purpose, we use a simulation
model which is based on the architecture of a Parsytec CC multicomputer. With this
model, we investigate how scalable the LH*LH SDDS actually is and which factors affect its performance. The reason for our interest in the LH*LH SDDS (and the Parsytec
CC) is that LH*LH is considered for using in a parallel version of the Monet database
[1] which will initially be targeted towards the Parsytec CC multicomputer.

The next section explains the concept of Linear Hashing. In Section 3, we discuss
the distributed data structure LH*LH. Section 4 describes the simulation model we have
used in this study. In Section 5, our experimental results are presented. Finally, Section
6 concludes the paper.

2   Linear Hashing

Linear Hashing (LH) [6] is a method to dynamically manage a table of data. More
specifically, it allows the table to grow or shrink in time without suffering from a penalty
with respect to the space utilisation or the access time. The LH table is formed by
$N \times 2^i + n$ buckets, where N is the number of starting buckets ($N \geq 1$ and $n < 2^i$).
The meaning of i and n is explained later on. The buckets in the table are addressed by
means of a pair of hashing functions h_i and h_{i+1}, with $i = 0, 1, 2...$ Each bucket can
contain a predefined number of data elements. The function h_i hashes data keys to one
of the first $N \times 2^i$ buckets in the table. The function h_{i+1} is used to hash data keys to
the remaining buckets. In this paper, we assume that the hash functions are of the form

$$h_i(\text{key}) \rightarrow \text{key} \bmod (N \times 2^i) \tag{1}$$

The LH data structure grows by splitting a bucket into two buckets whenever there is a
collision in one of the buckets, which means that a certain load threshold is exceeded.
Which bucket has to be split is determined by a special pointer, referred to as n. The
actual splitting involves three steps: creating a new bucket, dividing the data elements
over the old and the newly created bucket and updating the pointer n. Dividing the data
elements over the two buckets is done by applying the function h_{i+1} to each element in
the splitting bucket. The n pointer is updated by applying $n = (n + 1) \bmod N \times 2^i$.
Indexing the LH data structure is performed using both h_i and h_{i+1}. Or, formally:

$$\text{index}_{bucket} = h_i(\text{key}) \tag{2}$$
$$\text{if } (\text{index}_{bucket} < n) \text{ then index}_{bucket} = h_{i+1}(\text{key})$$

As the buckets below the n pointer have been split, these buckets should be indexed using h_{i+1} rather than with h_i. When the n pointer wraps around (because of the modulo), i should be incremented.

The process of shrinking is similar to the growing of the LH data structure. Instead of splitting buckets, two buckets are merged whenever the load factor drops below a certain threshold. In this study, we limit our discussion to the splitting within SDDSs.

3   The LH*LH SDDS

The LH* SDDS is a generalisation of Linear Hashing (LH) to a distributed memory parallel system [7]. In this study, we focus on one particular implementation variant of LH*, called LH*LH [4, 5]. The LH*LH data is stored over a number of server processes and can be accessed through dedicated client processes. These clients form the interface between the application and LH*LH. We assume that each server stores *one* LH*LH bucket of data, which implies that a split always requires the addition of an extra server process. Globally, the servers apply the LH* scheme to manage their data, while the servers use traditional LH for their local bucket management. Thus, a server's LH*LH bucket is implemented as a collection of LH buckets. Hence, the name LH*LH. In Figure 1, the concept of LH*LH is illustrated.

As was explained in the previous section, addressing a bucket in LH is done using a key and the two variables i and n (see Equation 2). In LH*LH, the clients address the servers in the same manner. To do so, each client has its own *image* of the values i and n: i' and n' respectively. Because the images i' and n' may not be up to date, clients can address the wrong server. Therefore, the servers need to verify whether or not incoming client requests are correctly addressed, i.e. can be handled by the receiving server. If an incoming client request is incorrectly addressed, then the server forwards the request to the server that is believed to be correct. For this purpose, the server uses a forwarding algorithm [7] for which it is proven that a request is forwarded at most twice before the correct server is found. Each time a request is forwarded, the forwarding server

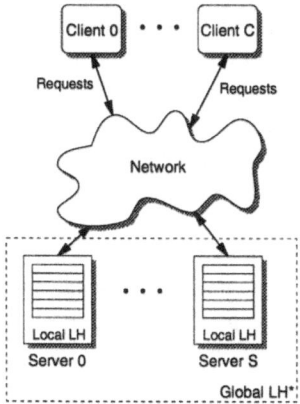

Fig. 1. The LH*LH SDDS.

sends a so-called Image Adjustment Message (IAM) to the requesting client. This IAM contains the server's local notion of i and n and is used to adjust the client's i' and n' in order to get them closer to the global i and n values. As a consequence, future requests will have a higher probability of being addressed correctly.

The splitting of an LH*LH bucket is similar to the splitting of LH buckets. The pointer n is implemented by a special token which is passed from server to server in the same manner as n is updated in LH: it is forwarded in a ring of the servers 0 to $N \times 2^i$, where N is the number of starting servers. When a server holds the n token and its load factor is larger than a particular threshold, the server splits its LH*LH bucket and forwards the n token. Splitting the LH*LH bucket is done by initialising a new server (by sending it a special message) and shipping half of its LH buckets to the new server (remember that the LH*LH bucket is implemented as a collection of LH buckets).

It has been shown in [7] that a splitting threshold which can be dynamically adjusted performs better than a static one. Therefore, LH*LH applies a dynamic threshold T which is based on an estimation of the global load factor [5]:

$$T = M \times V \times \frac{2^i + n}{2^i}$$

where M is a sensitivity parameter and V is the capacity of an LH*LH bucket in number of data elements. Typically, M is set to a value between 0.7 and 0.9. If the number of data elements residing at the server with the n token is larger than the threshold T, then the server splits its data.

4   The Simulation Model

The parallel architecture we focus on in this study is based on that of a Parsytec CC multicomputer. The configuration we have used for our initial experiments consists of 16 PowerPC 604 processors connected in a multistage network with an application-level throughput of 27 MByte/s for point-to-point communication. To model this multicomputer architecture, we have used the Mermaid simulation environment which provides a framework for the performance evaluation of multicomputer architectures [10, 11, 9].

In our model, a multicomputer node can contain either one server or client process. The clients provide the network part of the model with communication requests, which are messages to server processes containing commands that operate on the LH*LH SDDS (e.g. insert, lookup, etc.). In addition, the server processes can also issue communication requests, like when a client request has to be forwarded to another server.

The network model simulates a wormhole-routed network [2] at the flit-level and is configured to model a Mesh of Clos topology [10]. This is the generic topology from which the Parsytec CC's topology has been derived. The Mesh of Clos topology is a mesh network containing clusters of Clos multistage sub-networks. In the case of the Parsytec CC, a Mesh of Clos topology with a 1×1 mesh network is used, i.e. it contains a pure multistage network. For routing, the model uses deterministic XY-routing [8] for the mesh network and a deterministic scheme based on the source node's identity [10] for the multistage subnetworks. Furthermore, messages larger than 4Kb are split up in separate packets of 4Kb each.

We have validated our Parsytec CC network model against the real machine with a suite of small communication benchmarks and found that the error in predicted execution times for these programs is on the average less than 4% [9]. For *some* experiments, we were also able to compare our simulation results with the results from a real LH*LH implementation [4, 5]. Although the measurements for the real implementation were obtained using a multicomputer which is different from the one we have modelled, the comparison between our simulation results and the actual execution results can still give us some insight into the validity of our simulation model. We found that the simulation results correspond closely to the behaviour measured for the real implementation.

5   Experiments

We performed experiments with an application which builds the LH*LH SDDS using a Dutch dictionary of roughly 180,000 words as the data keys. The data elements that are being inserted consist of a key only (they do not have "a data body") unless stated otherwise. Throughout this paper, we use the term *blobsize* when referring to the size of the body of data elements. By default, our architecture model only accounts for the delays that are associated with communication. Computation performed by the clients and servers is not modelled. In other words, the clients and servers are infinitely fast. This should give us an upper bound of the performance of LH*LH when using the modelled network technology. However, we will also show several results of experiments in which we introduce and vary a computational server latency.

The first experiments concern LH*LH's performance on the 16-node Parsytec CC platform. Because the number of available nodes in this platform is rather limited, we needed to calibrate the split-threshold such that the LH*LH SDDS does not grow larger than the available number of nodes. Another result of the limited number of nodes is that the model only simulates up to 5 clients in order to reserve enough nodes for the server part of LH*LH. To overcome these problems, we have also simulated a larger multicomputer platform, of which the results are discussed later in this study. Throughout this paper, we assume that the number of starting servers is one, i.e. $N = 1$.

Figure 2a shows the predicted time it takes to build the LH*LH SDDS on the Parsytec CC when using c clients, where $c = 1, 2, ..., 5$. In the case multiple clients are used, they *concurrently* insert a different part of the dictionary. The data points in Figure 2a correspond to the points in time where LH*LH splits take place. The results show that the build-time scales linearly with the number of insertions. Thus, the insertion latency does not deteriorate for large data structure sizes (it is dominated entirely by the message round trip time). In this respect, Figure 2a clearly indicates that LH*LH is scalable.

The results of Figure 2a also show that the build-time decreases when more clients are used. This is due to the increased parallelism as each client concurrently operates on the LH*LH SDDS. In Figure 2b, the relative effect (i.e. the speedup) of the number of clients is shown. When increasing the number of clients, the obtained speedup scales quite well for the range of clients used.

Another observation that can be made is that the occurrence of splits (the data points in Figure 2a) is not uniformly distributed with respect to the insertions. Most of the splits are clustered in the first 20,000 insertions and near the 180,000 insertions. This

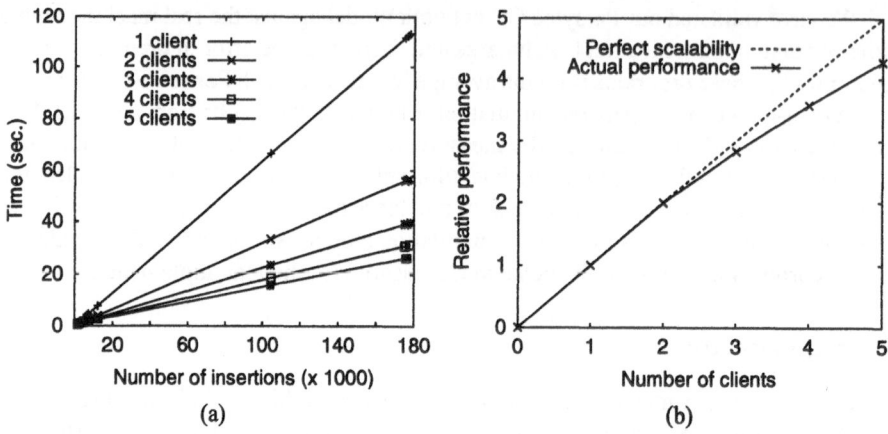

Fig. 2. Building the LH*LH SDDS: absolute performance (a) and scalability (b). In graph (a), the data points correspond to the moments in time where splits take place.

phenomenon is referred to as *cascading splits* [7]. It is caused by the fact that when the load factor on server n (the token-holder) is high enough to trigger a split, the servers that follow server n are often also ready to split. This means that after server n has split and forwarded the n token to the next server, the new token-holder immediately splits as well. As a result, a cascade of splitting servers is formed which terminates whenever a server is encountered with a load factor that is lower than the threshold. Essentially, cascading splits are undesirable as they harm the incremental fashion with which the distributed data structure is reorganised. Litwin et al. [7] have proposed several adjustments to the split threshold function to achieve a more uniform distribution of the splits.

Figure 3a plots the curves of the average time needed for a single insertion, as experienced by the application. The data points again refer to the moments in time where splits occur. Because the data points for the first 20,000 insertions are relatively hard

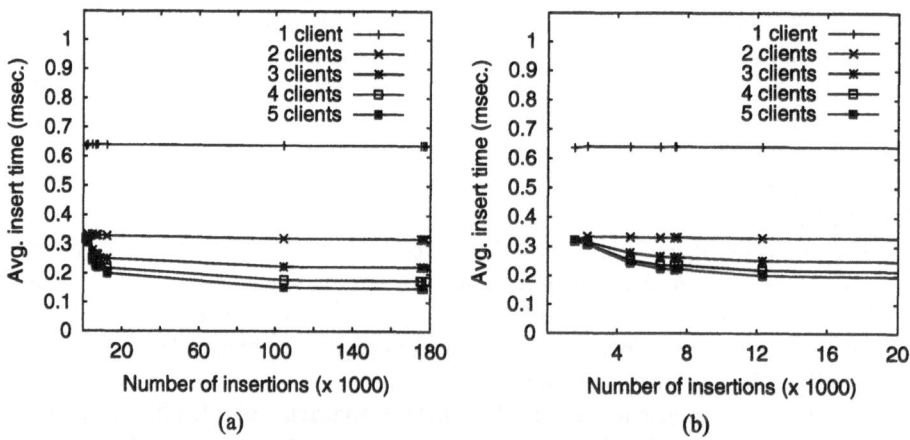

Fig. 3. Average time per insertion in milliseconds.

# Clients	# IAMs	Msg./insertion	Average message overhead
1	14	2.0008	0.04%
2	24	2.0011	0.06%
3	43	2.0012	0.06%
4	48	2.0014	0.07%
5	55	2.0014	0.07%

Table 1. Number of messages required to build the LH*LH SDDS.

to distinguish, Figure 3b zooms in on this particular range. Figure 3a shows that the worst average insertion time (for 1 client) does not exceed 0.65ms. This is about one order of magnitude faster than the typical time to access a disk. However, we should remind the reader that these insertion latencies reflect a lower bound (for the investigated multicomputer architecture) since no computation is modelled at the clients and servers.

As can be expected from Figure 2a, Figure 3 shows that the average insertion time decreases when increasing the number of clients. Additionally, the average insertion time also decreases for an increasing number of insertions. This is especially true during the first few thousands of insertions and for the experiments using 3 or more clients. The reason for this is that a larger number of clients requires the creation of enough servers before the clients are able to effectively exploit parallelism, i.e. allowing the clients to concurrently access the LH*LH SDDS with low server contention. As can be seen in Figure 3b, after about 8,000 insertions, enough server processes have been created to support 5 clients.

In Table 1, the message statistics are shown for building the LH*LH SDDS. For about 180,000 insertions, the maximum number of IAMs (Image Adjustment Messages) that were sent is 55 (for 5 clients). This is only 0.003% with respect to the total number of insertions. The average number of messages per insertion is near the optimum of 2 (needed for the request itself and the acknowledgement). The last column shows the average proportional message overhead (due to IAMs and the forwarding of messages) for a single insertion. These results confirm the statement from Litwin et al. [7] in which they claim that it usually takes one message only to address the correct server.

In Figure 4, the results are shown when experimenting with the blobsize of data elements. Figure 4a plots the curves for the build-time when using a blobsize of 2Kb (the data points again correspond with the server splits). The results show that the build-times are still linear to the number of insertions and that the performance scales properly when adding more clients. In fact, the build-times are not much higher than the ones obtained in the experiment with empty data elements (see Figure 2a). The reason for this is twofold. First, since computational latencies like memory references are not simulated, our simulation model is especially optimistic for data elements with a large blobsize. Second, the modelled communication overhead of the AIX kernel, which is running on the real machine's nodes, is rather high (at least $325\mu s$ per message) and dominates the communication latency. This reduces the effect of the message size on the communication performance.

Figure 4b depicts the average time per insertion as experienced by the application when varying the blobsize from 64 bytes to 8Kb (the latter is building a data structure

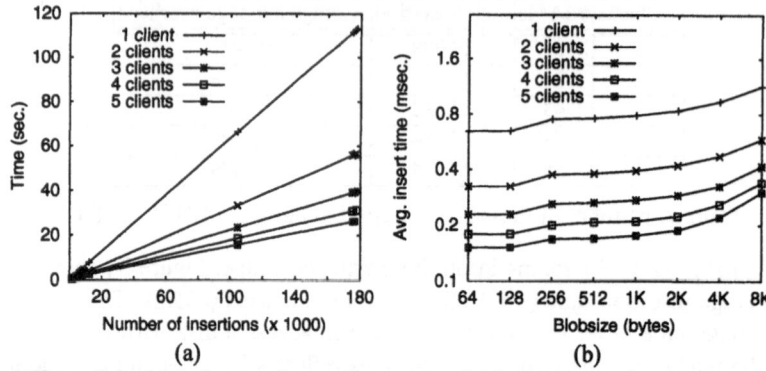

Fig. 4. Effect of the blobsize on the performance: the build-time when using a blobsize of 2Kb (a) and the effect of the blobsize on the average time per insertion (b).

of 1.4 Gbytes). Note that both axes have a logarithmic scale. Two observations can be made from this figure. First, the average insertion time suddenly increases after a blobsize of 128 bytes. This effect is somewhat diminished when using more clients. The reason for the performance anomaly is a peculiarity in the implementation of synchronous communication for the Parsytec CC machine. Messages smaller than 248 bytes can piggy-back on a fast initialisation packet, which sets up the communication between the source and destination nodes. Beyond these 248 bytes, normal data packets have to be created and transmitted, which slow down the communication (e.g. the software overhead is higher). By increasing the number of clients, the effect of the higher communication overhead of big messages (> 248 bytes) is reduced due the parallelisation of the overhead.

A second observation that can be made from Figure 4b is that the curves are relatively flat for blobsizes below the 1Kb. Again, this can be explained by the fact that the large OS overhead dominates the insertion performance for small blobsizes. For insertions with large blobsizes (> 1Kb), the network bandwidth and possibly the network contention become more dominant. To investigate whether or not the network contention plays an important role in the insertion performance, Figure 5 shows the average time that packets were stalled within the network. We should note that the variation of these averages is rather large, which implies that they should be interpreted with care. Nevertheless, the graph gives a good indication of the intensity of the network contention. From Figure 5 can be seen that the contention increases more or less linearly when increasing the blobsize in the case of 1 client. However, when using more clients, the contention starts to scale exponentially. The highest measured average block time equals to $31\,\mu s$ (8Kb blobsize with 5 clients), which is still $1/10^{th}$ of the $325\,\mu s$ software communication overhead for a single message. So, in our experiments, the contention is not high enough to dominate the insertion performance. This suggests that the insertion performance for large blobsizes (see Figure 4b) is dominated by the network bandwidth.

So far, we have assumed that the server (and client) processes are infinitely fast, i.e. computation is not modelled. To investigate the effect of server overhead on the overall performance, we modelled a delay for every incoming insertion on a server (the clients

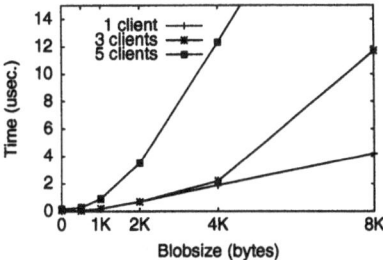

Fig. 5. The average time that packets blocked within the network.

continue to be infinitely fast). During the delay, which takes place after acknowledging the client's request, the server is inactive. Since the server overhead can overlap with the transmission of new client requests, this scheme exploits parallelism for all client configurations (even when 1 client is used). We varied the server overhead from 0 (no overhead) to 50ms per insertion.

Figure 6 depicts the results of the server overhead experiment. In Figure 6a, the effect on the average insertion time is shown. Note that both axes have a logarithmic scale. Figure 6a shows that the insertion latency starts to be seriously affected by the server overhead after a delay of approximately 0.1ms. Beyond a server overhead of 1ms, the insertion latency increases linearly with the server overhead which indicates that the insertion latency is entirely dominated by the server overhead. After this point, the differences between the various client configurations have more or less been disappeared as well, i.e. the curves converge. This implies that the large server overheads reduce the potential parallelism. An important reason for this is that when the server overhead approaches or exceeds the minimum time between two consecutive insertion requests from a single client (which is in our case simply the $325\mu s$ software communication overhead as the clients do not perform any computation), the rate at which a server can process insertion requests becomes lower than the rate at which a single client can produce requests. This means that a server can easily become a bottleneck when adding clients.

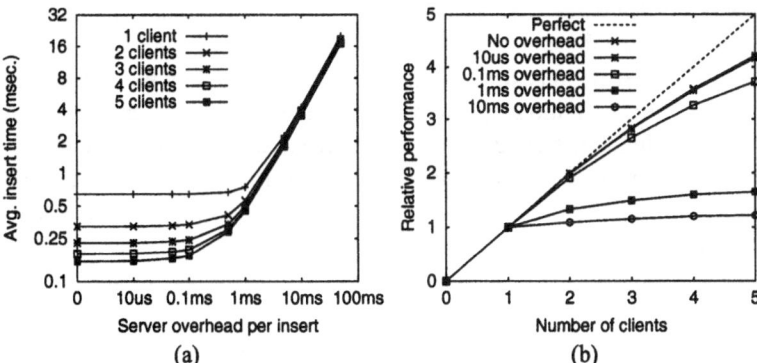

Fig. 6. Effect of the server overhead on the performance: the average time per insertion (a) and the scalability (b).

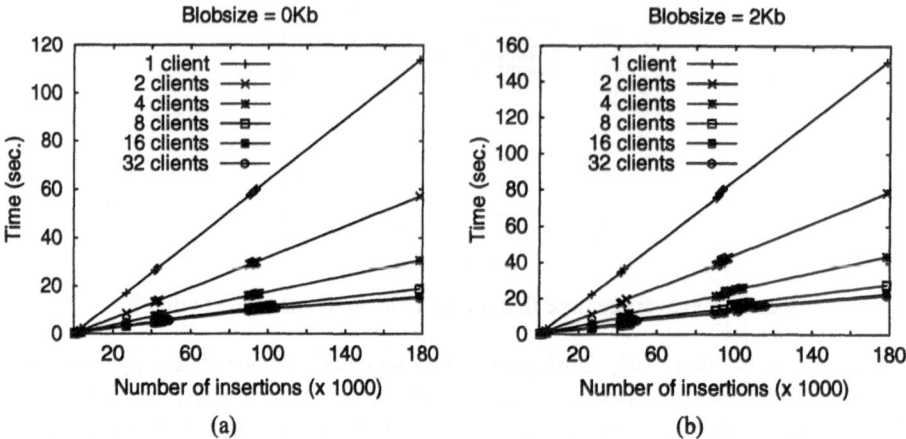

Fig. 7. Build-times for the 64-node Mesh of Clos network using blobsizes of 0 (a) and 2Kb (b).

The reduction of parallelism caused by large server overheads is best illustrated in Figure 6b. This figure shows the speedup for multiple clients when varying the server overhead. It is obvious that while the client-scalability is good for server overheads of 0.1ms and less, the scalability for larger overheads has collapsed completely.

Until now, the experiments have been performed using the simulation model of a 16-node Parsytec CC architecture. This has limited our simulations to allocating a maximum of 5 clients and 11 servers. To investigate how LH*LH behaves for a larger number of clients and servers, we have also simulated a 64-node Mesh of Clos network [10], which connects four multistage clusters of 16 nodes in a 2×2 mesh.

Figure 7 shows the build-times for this Mesh of Clos when using blobsizes of 0 (Figure 7a) and 2Kb (Figure 7b). Again, the data points refer to the moments in time where a split occurs. In these experiments, we have simulated the LH*LH for up to 32 clients. The curves in Figure 7 show the same behaviour as was observed in the experiments with the 16-node Parsytec CC model: the build-time is linear to the number of insertions and decreases with an increasing number of clients. The occurrence of cascading splits is also illustrated by the clusters of points in Figure 7.

In Figure 8, the scalability for several blobsizes is plotted by the curves which are labelled with *normal*. Additionally, the curves labelled with "1/3" and "1/9" show the scalability when the $325\mu s$ communication overhead is reduced by a factor 3 and 9 respectively. A number of observations can be made from Figure 8. First, the *normal* curves indicate that the configurations with blobsizes up to 2Kb scale to roughly 8 clients, whereas the configuration with a blobsize of 8Kb only scales up to about 4 clients. The deterioration of scalability for large blobsizes is due to the increased network contention.

Another, quite interesting, result is that the scalability for the large blobsizes is reduced even more when decreasing the software communication overhead. This effect is also caused by the increase of network contention: the smaller the overhead, the higher the frequency at which the clients inject insertion messages into the network.

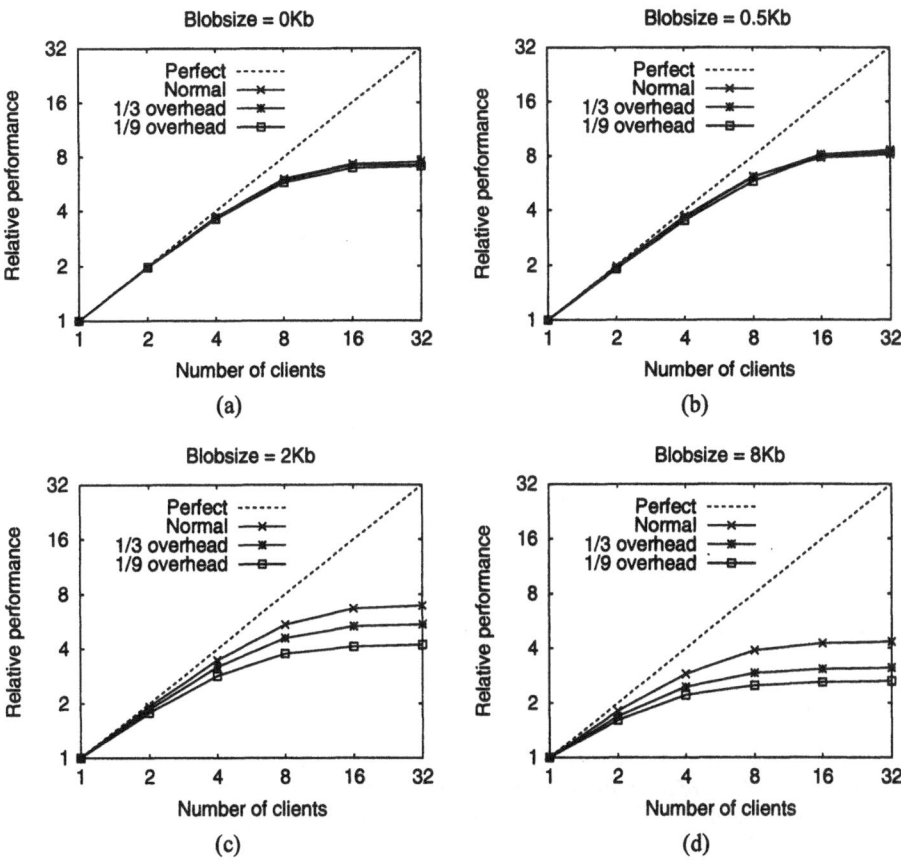

Fig. 8. The scalability for blobsizes of 0 bytes (a), 512 bytes (b), 2Kb (c) and 8Kb (d). The curve labelled with *normal* refers to the results of the original model. The other two curves (1/3 and 1/9) present the results for a modified model in which the software communication overhead is reduced by a factor of 3 and 9 respectively.

Clearly, the higher frequency of insertions with large blobsizes results in more network traffic and thus more contention. So, one can conclude that a part of the measured client-scalability for large blobsizes is due to the high communication overhead of the modelled platform. Apparently, the increase of network traffic in the case of a blobsize of 512 bytes (see Figure 8b) is not severe enough to cause a lot of extra contention.

6   Conclusions

In this study, we evaluated the LH*LH distributed data structure for a multicomputer architecture. Our simulation results show that LH*LH is indeed scalable: the time to insert a data element is independent on the size of the data structure. For the studied multicomputer, the insertion time as experienced by a single client can be an order of magnitude faster than a typical disk access. The insertion time can even be reduced by using multiple clients which concurrently insert data into the distributed data structure. The results indicate that the speedup of insertions scales reasonably well up to about 8 clients for the investigated network architecture and workload.

We have also shown that the scalability of the insertion performance can be affected in more than one way. For instance, the larger the data elements that are inserted, the poorer is the scalability when increasing the number of clients. We found that the performance of insertions with data elements of 512 bytes scales up to 8 clients, whereas the performance of insertions with 8Kb data elements only scales up to 4 clients. Moreover, a large server overhead can seriously hamper the client-scalability of the LH*LH data structure. Our results indicate that it is important to keep the server overhead below the minimum time between two consecutive requests from a single client. This increases the potential parallelism as it ensures that the rate at which a server can process requests is higher than the rate at which a single client can produce requests. Finally, we found that a part of the speedup for multiple clients is due to the parallelisation of software communication overhead (which is quite large for the studied architecture). When the communication is optimised (i.e. the overhead is reduced), lower speedups are achieved.

Acknowledgements

We would like to thank Jonas Karlsson for his feedback regarding our LH*LH model.

References

[1] P. A. Boncz and M. L. Kersten. Monet: An impressionist sketch of an advanced database system. In *Proc. of IEEE BIWIT workshop*, July 1995.

[2] W. J. Dally and C. L. Seitz. The torus routing chip. *Journal of Distributed Computing*, 1(3):187–196, 1986.

[3] J. L. Hennessy and D. A. Patterson. *Computer Architecture, A Quantitative Approach*. Morgan Kaufmann Publishers, Inc., San Mateo, California, 1990.

[4] J. S. Karlsson. A scalable data structure for a parallel data server. Master's thesis, Dept. of Comp. and Inf. Science, Linköping University, Feb. 1997.

[5] J. S. Karlsson, W. Litwin, and T. Risch. LH*lh: A scalable high performance data structure for switched m ulticomputers. In *Advances in Database Technology*, pages 573–591, March 1996.

[6] W. Litwin. Linear hashing: A new tool for file and table addressing. In *Proc. of VLDB*, 1980.

[7] W. Litwin, M-A. Neimat, and D. Schneider. LH*: A scalable, distributed data structure. *ACM Transactions on Database Systems*, 21(4):480–526, Dec. 1996.

[8] L. M. Ni and P. K. McKinley. A survey of wormhole routing techniques in direct networks. *IEEE Computer*, 26:62–76, Feb. 1993.

[9] A. D. Pimentel. *A Computer Architecture Workbench*. PhD thesis, Dept. of Computer Science, University of Amsterdam, Dec. 1998.

[10] A. D. Pimentel and L. O. Hertzberger. Evaluation of a Mesh of Clos wormhole network. In *Proc. of the 3rd Int. Conference on High Performance Computing*, pages 158–164. IEEE Computer Society Press, Dec. 1996.

[11] A. D. Pimentel and L. O. Hertzberger. An architecture workbench for multicomputers. In *Proc. of the 11th Int. Parallel Processing Symposium*, pages 94–99. IEEE Computer Society Press, April 1997.

Set Associative Cache Behavior Optimization*

Ramón Doallo[1], Basilio B. Fraguela[1], and Emilio L. Zapata[2]

[1] Dept. de Electrónica e Sistemas. Univ. da Coruña
Facultade de Informática, Campus de Elviña, s/n
15071 A Coruña, SPAIN
{doallo,basilio}@udc.es
[2] Dept. de Arquitectura de Computadores. Univ. de Málaga
Complejo Tecnológico, Campus de Teatinos. PO Box 4114
E-29080 Málaga, SPAIN
ezapata@ac.uma.es

Abstract. One of the most important issues related to program performance is the memory hierarchy behavior. Programmers try nowadays to optimize this behavior intuitively or using costly techniques such as trace-driven simulations through a trial and error process. A systematic modeling strategy that allows an automated analysis of the memory hierarchy performance is developed in this work. This approach, besides requiring much shorter computation times, can be integrated in a compiler or an optimizing environment. The models consider caches of an arbitrary size, line size and associativity, and as we will show, they have proved a good degree of accuracy and have a wide range of applications. Loop interchange, loop fusion and optimal block size selection are the techniques whose successful application has been driven by the models in this work.

1 Introduction

The increasing gap between processor and main memory cycle times makes more critical the memory hierarchy performance, in which caches play an essential role. The traditional approaches to study and try to improve this performance are based on trial and error processes which give little information about the origins of this behavior, and do not allow to understand the way different program transformations influence it. For example, the most widely used approaches, trace-driven simulations [10] and the use of the hardware built-in counters that some microprocessor architectures implement [11], only provide the number of misses generated. Besides, both approaches require large computation times, especially simulation. On the other hand, built-in counters are limited to the study of the architectures where these devices exist.

General techniques for analyzing cache performance are required in order to propose improvements to the cache configuration or the code structure. A

---

* This work was supported by the Ministry of Education and Science (CICYT) of Spain under project TIC96-1125-C03

P. Amestoy et al. (Eds.): Euro-Par'99, LNCS 1685, pp. 229–238, 1999.

better approach is that of analytical models that extract some of their input parameters from address traces [1], [6], although they still need the generation of the trace each time the program is changed. Finally, there are a few analytical models based directly on the code [9], [4], and most of them are oriented to direct mapped caches. The last one has been implemented in an optimizing environment and later extended to set associative caches in [5]. It is based on the construction of *Cache Miss Equations* (CMEs) suitable for regular access patterns in isolated perfectly nested loops. It does not take into account the probability of hit in the reuse of data structures referenced in previous loops and seems to have heavy computing requirements.

In this work we introduce a systematic strategy to develop probabilistic analytical cache models for codes with regular access patterns, considering set associative caches with LRU replacement policy. Our approach allows the automated generation of equations that provide the number of misses generated in each memory reference and loop. An additional feature of these models, that previous works such as [5] lack, is that they take into account that portions of data structures accessed in previous loops may be in the cache when they are accessed by another set of nested loops, although the worst case hit probability is used. The automatically developed models have been validated using trace-driven simulations of typical code loops, and have proved to be very accurate in most of the cases. As an added benefit, the computing times of the model are very competitive.

The concepts on which our modeling strategy relies are introduced in the following section. The steps for modeling the accesses in regular nested loops are explained in Section 3, while Section 4 is devoted to both validation and usefulness of our approach in driving optimizations based on standard loop transformations. Conclusions and future work are discussed in the last section.

2   Basic Concepts

We classify the misses on a given data structure in a K-way associative cache in two groups: intrinsic misses, those that take place the first time a memory line is referenced, and interference misses, which take whenever a line has been already accessed but it has been replaced since its previous access. If a LRU replacement policy is applied, this means that K or more different lines mapped to the same cache set have been referenced since the last access to the line. In this way, the first time a memory line is accessed, a miss is generated. For the remaining accesses, the miss probability equals the probability that K or more lines associated to its cache set have been accessed since its previous access.

Our model uses the concept of area vector to handle this probability. Given a data structure V, $S_V = S_{V_0}, S_{V_1}, \ldots, S_{V_K}$ is the area vector corresponding to the accesses to V during a portion of the program execution. The i-th position of the vector contains the ratio of cache sets that have received $K - i$ lines of the structure. The first position, S_{V_0}, has a different meaning: it is the ratio of sets that have received K or more lines. Different expressions have been developed that estimate the area vectors as a function of the access pattern of the corresponding data structure [2]. The formulae and algorithms associated to the sequential access and the access to a number of regions of consecutive memory words which have a constant distance between the start of two such regions (access with a constant stride) may be found in [3]. A mechanism for adding the area vectors corresponding to the different data structures referenced between two consecutive accesses to the data structure which is being studied has been developed too.

3 Automatic Code Modeling

We will now present a systematic approach to apply automatically this modeling strategy to codes with regular access patterns. Figure 1 shows the type of code to model. A FORTRAN like notation is used, as this is the language we consider.

```
DO I_Z=1, N_Z, L_Z
  ...
  DO I_1=1, N_1, L_1
    DO I_0=1, N_0, L_0
      A(f_{A1}(I_{A1}), f_{A2}(I_{A2}), ..., f_{AdA}(I_{AdA}))
      ...
      B(f_{B1}(I_{B1}), f_{B2}(I_{B2}), ..., f_{BdB}(I_{BdB}))
      ...
    END DO
    ...
    C(f_{C1}(I_{C1}), f_{C2}(I_{C2}), ..., f_{CdC}(I_{CdC}))
    ...
  END DO
  ...
END DO
```

Fig. 1. General perfectly nested DO loops in a dense code.

The data structures considered are vectors and matrices whose dimensions are indexed by an affine function of the enclosing loop variables. Any of these variables is not allowed to appear in more than one dimension, so that only the two types of regular accesses discussed in the previous section take place. The functions of the indices are of the general form (affine function):

$$f_{Ax}(I_{Ax}) = \Delta_{Ax}I_{Ax} + K_{Ax}, \quad x = 0, 1, \ldots, d_A \tag{1}$$

We obviate the Δ_{Ax} constants in our exposition, as their handling would be analogous to that of the steps L_i of the loops. Their consideration would just require taking into account that the distance between two consecutive points accessed in dimension x is $\Delta_{Ax}S_{Ax}$ instead of S_{Ax}, that is what we shall consider.

3.1   Modeling Perfectly Nested Loops

We will explain here in detail only the modeling of perfectly nested loops with one reference per data structure due to space limitations. When several references are found for the same data structure, the procedure is somewhat different in order to take into account the data reuses that these references may generate. Non perfectly nested loops in which only simple blocks are found between the loops of the nest can be also modeled following this strategy.

Miss Equations First, the equations $F_i(R, p)$ are built. They provide the number of misses on reference R at nesting level i considering a miss probability p in the first access to a line of the referenced matrix. These equations are calculated examining the loops from the inner one containing R to the outer one, in such a way that $F_i(R, p)$ depends on $F_{i-1}(R, p)$ and the value of p for level $i - 1$ is really calculated during the generation of the the formula for level i. The following rules are applied in each level:

1. When the loop variable is one of the used in the indices of the reference, but not the corresponding to the first dimension, the function that provides the number of misses during the complete execution of the loop of level i as a function of probability p is:

$$F_i(R, p) = \frac{N_i}{L_i} F_{i-1}(R, p) \tag{2}$$

   This approach is based on the hypothesis that the first dimension of any matrix is greater or equal to the cache line size. This could not hold for caches with large line sizes and matrices with small dimensions, but the conjunction of both factors gives place to very small miss rates in which the error introduced is small and little advantage can be obtained from the application of the automated model.

2. If the loop variable is not any of those used in the indices of the reference, this is a reuse loop for the reference we are studying. In this case the number of misses in this level is estimated as:

$$F_i(R, p) = F_{i-1}(R, p) + \left(\frac{N_i}{L_i} - 1 \right) F_{i-1}(R, S_0(\mathtt{A}, i, 1)) \tag{3}$$

   where $S(\mathtt{Matrix}, i, n)$ is the interference area vector that stands for the lines that may cause interferences with any of the lines of matrix $\mathtt{Matrix}$ after n iterations of the loop in level i. Function (3) expresses the fact that the first iteration of the loop does not influence the number of misses on matrix $\mathtt{A}$, being this value determined by the probability p, which is calculated externally. Nevertheless, in the following iterations the same regions of the matrix are accessed, which means that the interference area vector in the first accesses to each line in this region is composed by the whole set of elements accessed during one iteration of the reuse loop.

3. If the loop variable is the one used in the indexing of the first dimension and $L_i \geq L_s$, where L_s is the line size, we proceed as in case 1, as each reference will take place on a different line. Otherwise, we have:

$$F_i(R, p) = \frac{N_i}{L_s} F_{i-1}(R, p) + \left(\frac{N_i}{L_i} - \frac{N_i}{L_s}\right) F_{i-1}(R, S_0(\mathtt{A}, i, 1)) \qquad (4)$$

The miss probability in the first access to each referenced line is given by the probability p, externally calculated. The remaining accesses take place on lines referenced in the previous iteration of the loop of level i, so the interference area vector is the corresponding to one iteration of this loop.

In the first level containing the reference $F_{i-1}(R, p)$, the number of misses caused by this reference in the closer lower level, is considered to be p. Once calculated the formula for the outer loop, the number of misses is calculated as $F_z(R, 1)$ (see Figure 1), which assumes that there are no portions of the data structure in the cache when the code execution begins.

Interference Area Vectors Calculation The calculation of the interference area vectors $S(\mathtt{Matrix}, i, n)$ is performed in an automated way by analyzing the references in such a way that:

- If the variable used in the indexing of one dimension I_h, is such that $h < i$, then we assign to this dimension a set of N_h/L_h points with a constant distance of L_h points between each two of them.
- On the other hand, if $h > i$, only one point in this dimension is assigned.
- Finally, if $h = i$, n points with a distance L_i are assigned to the dimension.

There is one exception to this rule. It takes place when the variable associated to the considered dimension I_h belongs to a loop that is a reuse loop for the reference whose number of misses is to be estimated, and there are no non reuse loops for that reference between levels i (not included) and h. In that case only one point of this dimension is considered.

After this analysis, the area of matrix $\mathtt{A}$ affected by reference R during an iteration of loop i would consist of $N_{\mathrm{R}i1}$ regions of one word in the first dimension, with a constant stride between each two regions of $L_{\mathrm{R}i1}$ words. In the second dimension, $N_{\mathrm{R}i2}$ elements would have been accessed, with a constant stride between each two consecutive ones of $L_{\mathrm{R}i2}$ groups of $d_{\mathtt{A}1}$ words (size of the first dimension), and so on. This area could be represented as:

$$\mathcal{R}_{\mathrm{R}i} = ((N_{\mathrm{R}i1}, L_{\mathrm{R}i1}), (N_{\mathrm{R}i2}, L_{\mathrm{R}i2}), \dots, (N_{\mathrm{R}id_\mathtt{A}}, L_{\mathrm{R}id_\mathtt{A}})) \qquad (5)$$

Figure 2 depicts this idea for a bidimensional matrix. This area will typically have the shape of a sequential access or an access to groups of consecutive elements separated by a constant stride, which is what happens in case (b) of the figure (the stride in case (a) is not constant). Both accesses have already been modeled for the calculation of their corresponding cross or auto-interference area vectors (see [3]).

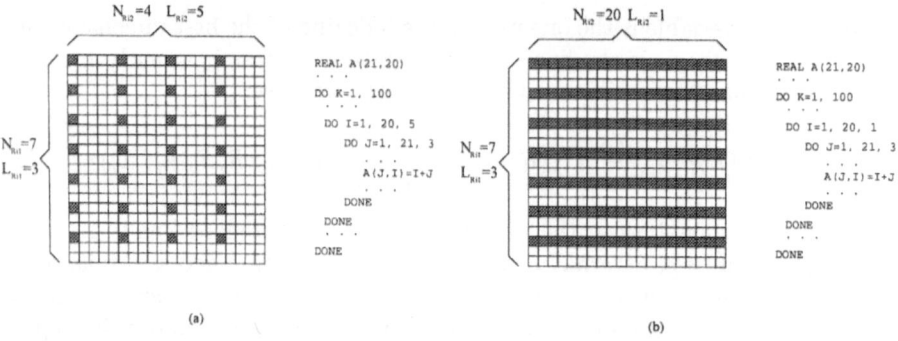

Fig. 2. Areas accessed with strides 5 and 1 for the columns in a bidimensional matrix A during one iteration of the K loop, in the i-th nesting level.

3.2 Imperfectly Nested Loops and Data Reuse

Real applications consist of many sets of loops which may have in each level several other loops, this is, sets of imperfectly nested loops. The data structures are accessed in several of these sets, giving place to a probability of hit in the reuse of their data.

Our model is based on the generation of a miss equation for each reference and loop and the calculation of the miss probability in the access to a line of the considered matrix inside that loop. As we have seen in Section 3.1, when considering an isolated set of perfectly nested loops, the probability of miss in the first access to a line of a data structure is 1. The probability of hit due to accesses in previously completed loops must be considered when imperfectly nested loops are modeled. Our automatic analyzer uses a pessimistic approach to estimate this probability in order to optimize modeling times. The whole regions of the vectors and matrices accessed during the execution of the outer non-reuse loop including a reference to the studied data structure are used to compute the interference area vectors. More accurate approaches have been already developed [2].

4   Validation and Applications

A simple automatic analyzer based in our modeling strategy has been built in order to validate it. Function calls are used to describe the loops, data structures and references to the analyzer, just as the semantic stage of a code analyzer could do. Its predictions are compared with the misses measured using a locally developed simulator whose validity has been checked using dineroIII, a trace-driven cache simulator belonging to the WARTS toolset [7]. We have applied both tools to a series of codes where different program transformations may be applied, being our purpose to validate the automatically developed models and check if they choose the right approach in the different codes. Attention is also paid to the times required to make the choices.

```
DO I=0, N-1
  DO J=0, N-1
    C(I,J)=C(I,J)+A(I,J)*B(I,J)
  ENDDO                                DO I=0, N-1
ENDDO                                    DO J=0, N-1
                                           C(I,J)=C(I,J)+A(I,J)*B(I,J)
DO I=0, N-1                                B(I,J)=B(I,J)+A(I,J)
  DO J=0, N-1                            ENDDO
    B(I,J)=B(I,J)+A(I,J)               ENDDO
  ENDDO
ENDDO

          (a)                                      (b)
```

Fig. 3. Separate (a) and fused (b) loops

4.1 Loop Fusion

Let us consider the codes in Figure 3, where (b) results from the fusion of the two loops in (a). Both the model and the trace-driven simulations recommend the fusion, as Table 1 shows, which also includes the simulation and modeling times for the separate code ((a) in Figure 3). In this table and the following ones, C_s stands for the cache size. In this case, although the model always succeeds in advising the best strategy, important deviations may be observed in the predictions for some combinations of the input parameters. This is due to two facts. On the one hand, the model follows a pessimistic approach when estimating the miss probability in the first access to a line of a vector that has been accessed in a previous loop, as it estimates it as the worst-case probability (the one associated to the full execution of the previous loop). On the other hand, most of the accesses in this code are row-wise, which makes very important the relative positions of the accesses matrices, and the model uses them only to calculate overlapping coefficients, which do not reflect properly their relation with the miss probability.

Table 1. Validation and time data for the codes in Figure 3, where loop fusion is applied

N	C_s	L_s	K	Measured misses(a)	Trace-driven simulation time(a)	Predicted misses(a)	Modeling time(a)	Measured misses(b)	Predicted misses(b)
400	8	8	1	302	0.335	473	0.005	262	324
400	32	8	1	100	0.266	205	0.006	60	138
400	32	8	2	100	0.286	116	0.007	60	76
400	128	8	2	100	0.275	101	0.017	60	61
500	512	16	2	66	0.475	76	0.059	48	48

```
                              DO J2=1, N, BJ
                                DO K2=1, N, BK
                                  DO I=1, N
   DO J=0, N-1                       DO K=K2, K2+BK
     DO I=0, N-1                       RA=A(I,K)
       R=0.0                           DO J=J2, J2+BJ
       DO K=0, N-1                       D(I,J) = D(I,J) + B(K,J) * RA
         R=R+A(I,K)*B(K,J)            ENDDO
       ENDDO                        ENDDO
       D(I,J)=R                   ENDDO
     ENDDO                      ENDDO
   ENDDO                      ENDDO
```

Fig. 4. Dense matrix-dense ma- **Fig. 5.** Dense matrix-dense matrix product
trix product with JIK nesting.   with blocking

4.2 Loop Interchange

Let us consider the dense matrix-dense matrix product in which we use variables
I, J and K to index the three dimensions of the problem. There are 9 versions
of this algorithm depending on the order in which the loops are nested. As an
example, Figure 4 shows the JIK form, where J is the variable for the outer loop
and K is the one used in the inner loop. The model can be used to choose the
optimal form much faster than any other technique.

In this example we have tried the IJK, JIK, JKI and KJI forms, those in
which more sequential accesses take place. The results for some combinations
of the input parameters are shown in Table 2, where trace-driven simulation
results are compared with those of our model. In the following, misses are in
thousands and times in seconds. Results have been obtained in a SGI Origin
200 with R10000 processors at 180 MHz. In this table and the following ones,
N stands for the matrix size, C_s for the cache size in Kwords, L_s for the line
size in words and K for the set size. The table shows both a very good degree
of accuracy, which allows the model to always choose the best loop ordering,
and very competitive modeling times, even comparing them to execution time
(around 0.53 seconds for compilation and 0.45 for execution).

4.3 Optimal Block Sizes

As a final example, let us consider the code in Figure 5, where blocking has
been applied to the dense matrix-dense matrix product with IKJ ordering. Both
trace-driven simulation and modeling have been used to derive the optimal block
sizes for which the dimensions of the block are multiples of 100 for some cache
configurations and values of N. This has been done by trying the different block
sizes, as the model estimates the number of misses, but not the optimum code
parameters. Table 3 shows them in format BJxBK, the number of misses they

Table 2. Validation and time data for several forms of the dense matrix-dense matrix product.

Order	N	C_s	L_s	K	Measured misses	Trace-driven simulation time	Predicted misses	Modeling time
IJK	400	32	8	2	8180	36.735	8181	0.004
JIK	400	32	8	2	8040	35.914	8042	0.004
JKI	400	32	8	2	8040	38.050	8040	0.001
KJI	400	32	8	2	8180	36.899	8180	0.004
IJK	400	32	8	1	8988	35.623	9004	0.006
JIK	400	32	8	1	9502	36.771	9559	0.006
JKI	400	32	8	1	8139	37.195	8172	0.001
KJI	400	32	8	1	8279	37.446	8308	0.006
IJK	400	8	8	1	11528	36.499	11473	0.003
JIK	400	8	8	1	14091	38.290	14110	0.003
JKI	400	8	8	1	8638	37.356	8567	0.001
KJI	400	8	8	1	8779	37.566	8693	0.003

Table 3. Real optimal and predicted optimal block sizes with the number of misses they generate in the trace-driven simulation and time required to derive them.

N	C_s	L_s	K	Optimal block	Measured misses	Trace-driven simulation time	Model-based optimal block	Measured misses	Modeling time
400	8	8	1	100x100	19721	396.07	100x100	19721	0.90
400	32	8	1	100x100	5889	328.72	100x100	5889	0.93
400	128	8	1	100x100	784	306.64	100x100	784	0.73
400	128	8	2	100x100	145	304.03	100x100	145	0.52
200	32	8	2	100x100	70	17.48	100x100	70	0.17
200	128	8	2	200x200	15	16.46	100x100	16	0.24
600	512	16	2	300x600	87	1756.86	300x100	93	2.18

generate in the simulation, and the time required to derive them through simulation and modeling. Although the block recommended by the modeling is not always the optimum one, the difference in the number of misses is very small, and the time required by the simulations is dramatically greater than the one for the modeling. Even the time required to obtain the optimum block by measuring the execution time or the values of the built-in counters is greater. For example, for $N = 400$ it took 26.6 seconds of real time, adding compilations and executions, while the model requires always less than one second.

5   Conclusions and Future Work

A probabilistic analytical modeling strategy has been presented that can be integrated in a compiler or optimizing environment. It is applicable to caches of an arbitrary associativity and provides equations for the number of misses each reference generates in each loop of the program. Hit probability in the reuses of data accesses in previous loops is also considered using a simplified approach which allows non perfectly nested loops better analysis. Our automated modeling approach has been implemented in a simple analyzer in order to validate it. This tool has proved to be very fast and to provide good hints on the cache behavior that allow to choose always optimal or almost optimal program transformations, even when important deviations in the number of predicted misses may raise for certain input parameter combinations.

In this work only affine indexing schemes have been studied, but much more ones appear in real scientific problems (indirections, multiple index affine, etc.). In future we plan to extend the automatization of the modeling process to those indexing schemes. Formulae that estimate the area vectors associated to indirections have already introduced in [2].

We also intend to perform the integration of our automated modeling approach in program optimization frameworks that support it, for example by providing the analysis needed to extract the input parameters from the code (*Access Region Descriptors* [8] could be used as a systematic representation for the access patterns).

References

[1] Buck, D., Singhal, M.: An analytic study of caching in computer systems. J. of Parallel and Distributed Computing. **32(2)** (1996) 205–214

[2] Fraguela, B.B., Doallo, R., Zapata, E.L.: Modeling set associative caches behavior for irregular computations. ACM Performance Evaluation Review (Proc. SIGMETRICS/PERFORMANCE'98) **26(1)** (1998) 192–201

[3] Fraguela, B.B.: Analytical Modeling of the Cache Memories Behavior. Ph. D. Thesis. Dept. de Electrónica e Sistemas, Univ. da Coruña (1999) (in Spanish)

[4] Ghosh, S., Martonosi, M., Malik, S.: Cache miss equations: An analytical representation of cache misses. Proc. 11th ACM Int'l. Conf. on Supercomputing (ICS'97) (1997) 317–324

[5] Ghosh, S., Martonosi, M., Malik, S.: Precise miss analysis for program transformations with caches of arbitrary associativity. ACM SIGPLAN Notices. **33(11)** (1998) 228–239

[6] Jacob, B.L., Chen, P.M., Silverman, S.R., Mudge, T.N.: An analytical model for designing memory hierarchies. IEEE Transactions on Computers. **45(10)** (1996) 1180–1194

[7] Lebeck, A.R., Wood, D.A.: Cache profiling and the SPEC benchmarks: A case study. IEEE Computer. **27(10)** (1994) 15–26

[8] Paek, Y., Hoeflinger, J.P., Padua, D.: Simplification of array access patterns for compiler optimizations. Proc. ACM SIGPLAN'98 Conference on Programming Language Design and Implementation (PLDI). (1998) 60–71

[9] Temam, O., Fricker, C., Jalby, W.: Cache interference phenomena. Proc. SIGMETRICS/PERFORMANCE'94. (1994) 261–271

[10] Uhlig, R.A., Mudge, T.N.: Trace-driven memory simulation: A survey. ACM Computing Surveys. **29(2)** (1997) 128–170

[11] Zagha, M., Larson, B., Turner, S., Itzkowitz, M.: Performance analysis using the MIPS R10000 performance counters. Proc. Supercomputing '96 Conf. (1996) 17–22

A Performance Study of Modern Web Server Applications

Ramesh Radhakrishnan and Lizy Kurian John

Laboratory for Computer Architecture
Electrical and Computer Engineering Department
The University of Texas at Austin, Austin, Texas 78712
{radhakri,ljohn}@ece.utexas.edu

Abstract. Web server programs are one of the most popular computer applications in existence today. Our goal is to study the behavior of modern Web server application programs to understand how they interact with the underlying Web server, hardware and operating system environment. We monitor and evaluate the performance of a Sun UltraSPARC system using hardware performance counters for different workloads. Our workloads include static requests (for HTML files and images) as well as dynamic requests in the form of CGI (Common Gateway Interface) scripts and Servlets. Our studies show that the dynamic workloads have CPIs (Cycles Per Instruction) approximately 20% higher than the static workloads. The major factors that we could attribute to this were higher instruction and data cache miss rates compared to the static workloads, and high external (L2) cache misses.

1 Introduction

In the past few years the World Wide Web [1] has experienced phenomenal growth. Not only are millions browsing the Web, but hundreds of new Web sites are added each day. Yet, despite the increasing number of Web servers in use, little is known about their performance characteristics. Increasing Web server performance means reducing the response time to service a request. This depends on a lot of factors involving the underlying hardware, server software, the interaction between the software and hardware, and the characteristics of the workload. Previous research in Web server performance [2] suggests better hardware system performance to be the most obvious way of improving Web server performance.

Previous studies have shown that Web servers spend around 85% of their cycles executing OS code [3], as compared to 9% by the SPEC95 suite [4]. This suggests that Web server software may well exhibit significantly different characteristics from standard benchmarks like SPEC95 [5]. This, in turn, raises questions about performance of Web servers on current microarchitectures, since many of the recent innovations in microarchitecture have been justified on the basis of increasing SPEC95 performance. Therefore it is important to understand the behavior of modern superscalar processors when running Web server

P. Amestoy et al. (Eds.): Euro-Par'99, LNCS 1685, pp. 239–247, 1999.

workloads. These studies can lead to improved Web server performance by providing insights about microarchitectural design changes or making modifications in server software (to better utilize the hardware resources). Previous research has shown that CPU and memory can form a bottleneck for a heavily loaded server system [6]. However, microarchitectural studies based on Web workloads have not been previously done, to identify the exact nature of these bottlenecks. Efforts have been made to identify the impact of secondary cache size on the performance of the server, however this study was based on static workloads [7].

In this paper, we study the behavior of a popular Web server running on a modern superscalar machine to understand the interaction between the Web server software and hardware. First, we compare Web server performance for static and dynamic workloads. Dynamic requests are excluded in current Web server benchmarks [8] [9] [10], making these benchmarks less representative of real-world workloads. We include dynamic requests in our studies to model a real-world workload. Second, we characterize the performance of a modern superscalar machine while running Web workloads. Previous studies on processor characterization have used technical and commercial workloads, however studies based on Web workloads are lacking.

The remainder of this paper is organized as follows. In the next section we give a brief description of the operation of Web servers. In Section 3, we describe the characteristics of our experimental environment including the key features of the UltraSPARC microarchitecture. We define the metrics of interest in our study and justify our choices in Section 4. We also present our data and analysis in Section 4. We suggest some future work and conclude the paper in Section 5.

2   Background

The World Wide Web (WWW) [1] is a distributed hypertext based information system developed at CERN based on the client-server model. The client forwards the request to the appropriate Web server, which responds with the required information. Till a few years ago, all requests made by the client were static in nature. The function of the server was to load the appropriate file and send it to the client. However, the functionality of current servers has been extended to incorporate dynamic requests, i.e. the server can generate data or information based on the contents of the request. This is possible by running CGI [11] programs or Servlets [12] on the server machine to generate results, which are then sent to the client. To process a dynamic request using CGI the server starts a process upon receiving the request from the client. The process can be a script file, an executable or any number of such processes depending on the request. Thus, CGI enables the server to create information on the fly, which is sent back to the client. CGI programs are executed in real time, to output dynamic information. Servlets are the Java equivalent of CGI programs. To output dynamic information, the server executes Java programs which are run on the JVM (Java Virtual Machine). Upon receiving a request from the client the server invokes the JVM to run Servlets and the generated information is sent back to the client. Although dynamic capabilities add to a Web server's overall capability, there is additional processing involved in the creation of a new process and the subsequent interprocess communication generated.

3   Experimental Setup

3.1   The UltraSPARC-II Microarchitecture

The hardware platform used to characterize the web server running in a UNIX environment was the UltraSPARC-II. We give a summary of the UltraSPARC-II architecture in Table 1. The UltraSPARC-II is a high-performance, superscalar processor implementing a 64-bit RISC architecture [13]. It is capable of sustaining the execution of up to four instructions per cycle. This is supported by a decoupled prefetch and dispatch unit with an instruction buffer. Instructions predicted to be executed are issued in program order to multiple functional units and are executed in parallel. Limited out of order execution is supported for these predictively-issued instructions. Long latency floating point and graphic instructions may complete out of order with respect to independent instructions.

Table 1. The UltraSPARC-II Processor Characteristics

	UltraSPARC System
Model	Sun Ultra5 (SPARC station 5)
Processor	UltraSPARC-II
Memory Size	128 MB
Cache	Split Instruction and Data Cache 16KB, 2-way set associative I Cache 32 byte line size 16KB direct-mapped D Cache, write-through 16-byte sub-blocks per cache line Unified External Cache (Level 2 or L2)
Instruction Issue	Execution of up to 4 instructions per cycle
Instruction Completion	Limited out of order execution
Branch Prediction	Dynamic branch prediction scheme implemented in hardware, based on two-bit history

3.2   Software Environment and Web Server Engine

The main Web server engine used in our study is Apache 1.3.0 [14] [15]. The Apache Web server is widely used, has good performance and is non-proprietary. Apache has built-in capability for handling CGI requests, and can be extended to run Servlets using third-party modules. Apache was run on an UltraSPARC-II and measurements were made using the performance counters that are provided in the processor. We use the Solaris 2.6 Operating System as the software platform to run the server program. In addition to Apache, we use the web server engine *servletrunner* provided with JSDK 2.0. to handle requests for Java Servlets. The JVM that executes the Servlets is the Sun JDK 1.1.6.

The static workload consisted of requests made for an HTML page containing text and a request for an image (GIF file). The dynamic requests consisted of a call to an executable invoked through a CGI [11] interface. A request made for a Java Servlet [12] and formed the other half of the dynamic workload. A dummy

Table 2. Details of the requests made to the server

Web Server Engine	Requested File	Nature of request	File size
Apache 3.0	index.html	static	1622 bytes
Apache 3.0	apache_pb.gif	static	2326 bytes
Apache 3.0	printenv	dynamic	126 bytes
JSDK 2.0	SnoopServlet.class	dynamic	1107 bytes

client, running on the same machine as the server, makes the appropriate client requests. This would minimize any latencies that occur on the network. Since we can specify which processes we want to monitor, we only gather data for the web server daemons and the data collected is minimally affected by the client program. The machine is used exclusively to run the aforementioned experiments, minimizing interference from other processes. The static workload is formed by repeatedly requesting *index.html*, an HTML document, and *apache_pb.gif*, a GIF image. The Perl executable *printenv*, which is invoked using a CGI interface and *SnoopServlet.class*, a Java class file, form the dynamic workload. The details of the different requests made by the dummy client are given in Table 2.

The client program would establish a connection and make 20 requests. A series of such runs are made in order to keep the server busy and heavily loaded. This particular methodology is obviously contrived, but it is simple enough to repeat extensively and serves as a starting point for characterizing how emerging web applications behave.

4   Performance Evaluation

The on-chip counters were accessed using the *Perf-Monitor* [16] package and used to gather statistics to understand the performance of the Web server while servicing different requests. A description of the measurable events and the performance counters can be obtained from the UltraSPARC-II manual [17]. The interesting statistics obtained and our observations based on the data collected are described in this Section.

4.1   Performance of Different Workloads

Table 3 presents the instruction count, cycles executed and CPI (Cycles Per Instruction) for the different workloads. The first interesting observation from Table 3 is that the number of instructions executed are comparable for CGI and static workloads. Even though the server has to create separate processes to handle the dynamic requests, we see that the instruction count is higher for Servlets compared to CGI. This is because the server will invoke the JVM to run the Servlet class file. This causes additional instructions being executed and is reflected in the instruction count. As expected, the CPI is higher while executing

Table 3. Instruction count, cycles executed and CPI for static and dynamic workloads

Static workload	Instr	Cycles	CPI
index(system)	152159904	327325408	2.15
index(user)	11424319	23748612	2.07
index(both)	159348832	350395584	2.13
gif(system)	208139824	433525344	2.08
gif(user)	10934707	21821480	1.88
gif(both)	207659920	442459808	2.04
Dynamic workload	Instr	Cycles	CPI
cgi(system)	141329088	399676160	2.82
cgi(user)	30277060	80925312	2.67
cgi(both)	178319120	482511360	2.70
servlet(system)	151229216	424606304	2.80
servlet(user)	140923376	271369728	1.92
servlet(both)	287816608	686206592	2.38

dynamic requests. This is because new processes have to be created, resulting in context switches and additional overhead. The overhead for creating new processes and calling across these process boundaries deteriorates the performance. This effect is seen in the case of a CGI request, where the CPI is much higher than the static workload although the instructions executed are comparable. In the case of Java Servlets, the CPI is higher than the static workload. However, since the Web server engine used to run this workload, *servletrunner*, is written in Java, it runs the Servlet on the JVM. A higher CPI is observed because the Java environment is interpreted, leading to poor performance as compared to running compiled scripts. Also, *servletrunner* is not a commercial server and its purpose is only to test Servlets before they are deployed in a Web server. Hence, it is not optimized for performance, which is another reason contributing to its high CPI.

Another observation from Table 3 is that a high percentage of cycles are spent in the system mode as compared to the user mode. It is seen that *index* and *cgi* spend more than 80% of cycles executing system code. For Servlets the percentage of cycles spent in system code is lower (around 61%), but is still significant compared to other standard benchmark programs used for measuring system performance.

4.2   Analysis of Pipeline Stalls and Cache Misses

We use the performance counters to measure pipeline stalls and cache misses, which could be factors leading to the higher CPI observed for the server while running the dynamic workload.

Cache Performance Instruction cache misses, data cache misses and external (L2) cache misses were calculated of the different workloads were calculated, and were observed to correspond with the CPI calculated. Figure 1 gives miss rates that were observed for all the three caches. The miss rates for the external cache were similar for the different workloads, with the exception of Servlets which showed a lower external cache miss rate.

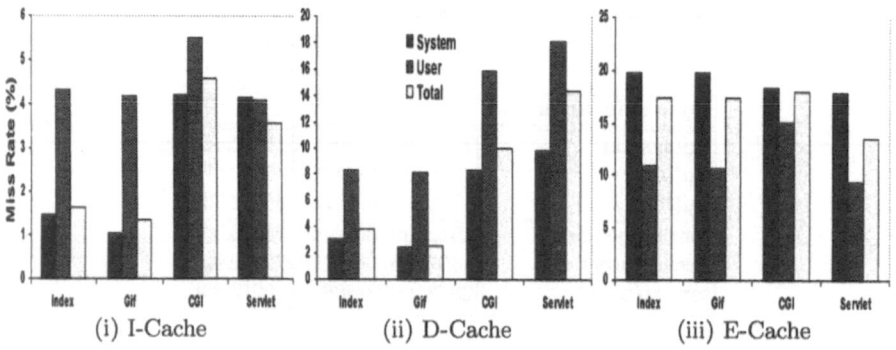

(i) I-Cache   (ii) D-Cache   (iii) E-Cache

Fig. 1. Miss rates for instruction, data and external (Level2) caches

Static workloads were observed to have lower miss rates in both the instruction and data cache, as compared to dynamic workloads. Since the dynamic workloads spawn a new process for handling the request, it brings a new working set into the instruction cache, resulting in higher miss rates. The miss rates for instruction and data caches are higher by a factor of 4 for the dynamic workload, contributing to the higher CPI. The higher cache miss rates seen for the dynamic workload, leads to higher pipeline stalls for the dynamic workload, as seen in the next section.

Pipeline stalls Table 4 gives details about the stalls that occurred in the processor pipeline due to store buffer full conditions, data waiting and stalls in the instruction buffer. The table lists the percentage of total clock cycles that the flag associated with a stall was seen to be true. From this we can estimate the total number of stalls that were caused by each of the different factors. The various stalls are defined as follows:

- I-Buffer Stalls: This is further divided into stalls due to IC miss and branch.
  IC miss: lists the percentage of cycles that the I-buffer is empty from an I-Cache miss. This also includes E-Cache miss processing if an E-Cache miss also occurs.
  branch stalls: lists the percentage of cycles the I-buffer is empty from branch misprediction. Since branch misprediction invalidates any instruction issued after the current instruction, the total number of pipeline bubbles is approximately twice as big as measured from this count.
- Store buffer full: This column gives the percentage of cycles that the first instruction in the group is a store, and the Store buffer can not hold additional stores.
- Load Stall: This is the percentage of cycles that an instruction in the execute depends on an earlier load result that is not yet available.
- Load RAW: This gives the percentage of cycles that there is a load in the execute stage and there is a read-after-write hazard on the oldest outstanding load. This basically means that the load data is being delayed by completion of an earlier store.

Instruction buffer and store buffer stalls cause the processor to stall in the dispatch stage of the pipeline. The static workload shows lower instruction buffer stalls due to an IC miss, as compared to the dynamic workload. However, instruction buffer stalls caused by branch misprediction is higher for static workloads as compared to the dynamic workload. The branch stalls are least for the Servlet, which runs in an interpreted environment and therefore shows better branch behavior.

Stalls in the dispatch stage of the pipeline due to store buffer full conditions were higher by a factor of approximately 5 times for dynamic workloads. Similarly a higher percentage of stalls were observed in the execute and grouping stages of the pipeline for dynamic workloads. These stalls are caused by the processor not being able to execute an instruction because it is waiting for data from a previous load instruction. Load stalls are seen to be high for both static and dynamic workloads.

I-Buffer stalls due to I-Cache misses and stalls due to D-Cache misses contribute to more than 45% of all the stalls in the dynamic workloads. Stalls in the static workload is dominated by branch related I-Buffer stalls and stalls for data (load miss) which together account for more than 75% of all the stalls. Since many of these stalls can overlap, it is extremely difficult to precisely correlate the various stalls with the observed CPI.

5   Conclusions

We studied the behavior of a modern Web server to understand the interaction with the underlying hardware and operating system. We used hardware perfor-

Table 4. Breakdown of stalls in the processor (% of Total cycles)

Different Workloads	I-Buffer Stalls IC miss	branch	Store buffer full	Load Stall	Load RAW
index	8.64	10.72	0.62	16.09	0.65
gif	6.91	13.12	0.49	19.02	0.49
cgi	17.98	6.76	2.47	22.70	1.10
servlet	17.77	5.89	2.97	20.77	0.94

mance counters on the UltraSPARC-II to monitor performance. The behavior of the server program was characterized while servicing static and dynamic requests. The major observations are as follows:

- It was found that the additional functionality required by the dynamic requests (in the form of process creation and inter-process communication) causes the dynamic workload to have a higher CPI. The CPI for dynamic workloads was approximately 20% higher than the static workloads. A strong correlation was observed between the cache miss rates and processor stalls to the measured CPI.
- Dynamic workloads resulted in higher cache misses as compared to the static workloads. The I-Cache misses for static workloads ranged from 1.3% to 1.6%. The I-Cache misses for the dynamic workloads were higher, ranging from 3.6% to 4.5%. We see a similar observation for D-Cache misses also, 2.5% to 4% was the miss rate observed for the static workloads and the dynamic workloads had a higher miss rate of 10% to 14.2%. The cache miss rates are higher for dynamic workloads, since creation of additional processes to service the request brings in a new working set into the cache.
- I-Buffer stalls due to I-Cache misses and stalls due to D-Cache misses contribute to more than 45% of the stalls in the dynamic workloads. Stalls in the static workload is mainly dominated by branch related I-Buffer stalls and stalls for data (load miss) which together account for more than 75% of the stalls.

The server code was seen to spend more than 80% of its cycles executing system code. Benchmark suites like SPEC95 are seen to spend less than 10% of cycles executing system code, hence further studies on modern processor characterization using Web workloads is a significant research area. Including dynamic requests to existing benchmarks is important to model real-world workloads accurately. Further studies on Web workload characterization, taking into account the emerging technologies, is needed to create better benchmarks.

Acknowledgments
This work is supported in part by the National Science Foundation under Grants EIA-9807112 and CCR-9796098, the State of Texas Advanced Technology Program grant #403, and by Dell, Intel, Microsoft and IBM.

References

[1] T. Berners-Lee, R. Calliau, A. Luotonen, H. Nielsen and A. Secret, "The World-Wide Web", *Communications of the ACM*, 37(8), pp. 76-82, August 1993.

[2] E. Tittel, "Measuring Web Server Performance", SunWorld, September 1997, also available at: http://sunsite.uniandes.edu.co/sunworldonline/swol-09-1997/swol-09-webserver.html

[3] B. Chen, Y. Endo, K. Chan, D. Mazieres, A.Dias, M. Seltzer and M. Smith, "The Measured Performance of Personal Computer Operating Systems.", *Proc. of the Fifteenth Symposium on Operating System Principles*, pp. 299-313, 1995.

[4] N. Gloy, C. Young, J. Chen, and M. Smith, "An Analysis of Dynamic Branch Prediction Schemes on System Workloads", *Proc. of ISCA*, May 1996.

[5] The Standard Performance Evaluation Corporation, www.spec.org

[6] J. Almedia, V. Almedia and D. Yates, "Measuring the behavior of a World-Wide Web server", in *Proc. 7th Conf. High Performance Networking*, pp. 57-72, April 1997.

[7] Roy Saharoy, "Performance Impact of Secondary Cache Size on Web Servers", *Second Workshop on Computer Architecture Evaluation Using Commercial Workloads*, Held in Conjunction with HPCA-5, 1999.

[8] G. Trent and M. Sake, "WebStone: The First Generation in HTTP Server Benchmarking", White Paper, Silicon Graphics, February 1995.

[9] SPECweb96 Benchmark, http://www.spec.org/osg/web96/

[10] NetPerf, http://www.cup.hp.com/netperf/NetperfPage.html

[11] M. Felton, CGI: Internet Programming With C++ and C, Prentice Hall Press, 1997.

[12] Phil Inje Chang, "Inside The Java Web Server, An Overview of Java Web Server 1.0, Java Servlets, and the JavaServer Architecture", White Paper, Sun Microsystems, http://java.sun.com/features/1997/aug/jws1.html

[13] K. Normoyle, M. Csoppenzky, A. Tzeng, T. Johnson, C. Furman, and J. Mostoufi, " UltraSPARC-IIi: Expanding the Boundaries of a System on a Chip", IEEE Micro, Vol. 18, No. 2, March/April 1998.

[14] B. Lauri and P. Lauri, "Apache, The Definitive Guide", O'Reilly and Associates, 1997.

[15] The Apache HTTP Server Project, www.apache.org

[16] Magnus Christensson, "Perf-Monitor, A Package for Measuring Low-Level Behavior on the UltraSPARC", Swedish Institute of Computer Science.

[17] The UltraSPARC User's Manual, http://www.sun.com/microelectronics/UltraSPARC-II/

An Evaluation of High Performance Fortran Compilers Using the HPFBench Benchmark Suite

Guohua Jin and Y. Charlie Hu

Department of Computer Science
Rice University
6100 Main Street, MS 132
Houston, TX 77005
{jin, ychu}@cs.rice.edu

Abstract. The High Performance Fortran (HPF) benchmark suite HPF-Bench was designed for evaluating the HPF language and compilers on scalable architectures. The functionality of the benchmarks covers scientific software library functions and application kernels. In this paper, we report on an evaluation of two commercial HPF compilers, namely, *xlhpf* from IBM and *pghpf* from PGI, on an IBM SP2 using the linear algebra subset of the HPFBench benchmarks.

Our evaluation shows that, on a single processor, there is a significant overhead for the codes compiled under the two HPF compilers and their Fortran 90 companions, compared with the sequential versions of the codes compiled using *xlf*. The difference mainly comes from the difference in code segments corresponding to the communications when running in parallel. When running in parallel, codes compiled under *pghpf* achieve from slightly to significantly better speedups than when compiled under *xlhpf*. The difference is mainly from better performance of communications such as *cshift, spread, sum* and *gather/scatter* under *pghpf*.

1 Introduction

High Performance Fortran (HPF) [8] is the first widely supported, efficient, and portable parallel programming language for shared and distributed memory systems. Since the first release of the HPF specification in 1993, a growing number of vendors have made commercially available HPF compilers, with more vendors announcing plans to join the effort. However, there has not been a systematically designed HPF benchmark suite for evaluating the qualities of the HPF compilers, and the HPF compiler vendors have mostly relied on individual application programmers to feed back their experience often with some particular class of applications on a particular type of architecture (see, for example [12]). In [11], we have developed the first comprehensive suite of HPF benchmark codes for evaluating HPF compilers.

The functionality of the HPFBench benchmarks covers linear algebra library functions and application kernels. The motivation for including linear algebra li-

P. Amestoy et al. (Eds.): Euro-Par'99, LNCS 1685, pp. 248–257, 1999.
© Springer-Verlag Berlin Heidelberg 1999

brary functions is for measuring the capability of compilers in compiling the frequently time-dominating computations in many science and engineering applications. The benchmarks were chosen to complement each other, such that a good coverage would be obtained of language constructs and idioms frequently used in scientific applications, and for which high performance is critical for good performance of the entire application. Detailed descriptions of the HPFBench suite can be found in [11]. The source code and the details required for the use of the HPF-Bench suite are covered online at http://www.crpc.rice.edu/HPFF/benchmarks.

In this paper, we use the linear algebra subset of the HPFBench benchmarks to evaluate two commercial HPF compilers, namely, *xlhpf* from IBM and *pghpf* from PGI, on an IBM SP2. We evaluate the functionality, the single–processor performance, parallel performance, and performance of communications of the codes generated by the two compilers. Our evaluation shows that, when running on a single processor, there is a significant overhead for the codes compiled under the two HPF compilers and their Fortran 90 companions, compared with the sequential versions of the codes compiled under *xlf*. The difference mainly comes from the difference in code segments corresponding to the communications when running in parallel. When running in parallel, codes compiled under *pghpf* scale from slightly to significantly better than when compiled under *xlhpf*. The difference is mainly from better performance of communications such as *cshift*, *spread, sum* and *gather/scatter* under *pghpf*.

While numerous benchmarking packages [15, 6, 1, 2, 10] have been developed for measuring supercomputer performance, we are aware of only two that evaluate HPF compilers. The NAS parallel benchmarks [1] were first developed as "paper and pencil" benchmarks that specify the task to be performed and allow the implementor to choose algorithms as well as programming model, though an HPF implementation has been developed recently. The NAS parallel benchmarks consist of only a few benchmarks which are mostly representative of fluid dynamics applications.

The PARKBENCH [10] suite intends to target both message passing and HPF programming models, and to collect actual benchmarks from existing benchmarks suites or from users' submissions. It includes low level benchmarks for measuring basic computer characteristics, kernel benchmarks to test typical scientific subroutines, compact applications to test complete problems, and HPF kernels to test the capabilities of HPF compilers. The HPF kernels comprises several simple, synthetic applications which test implementations of parallel statements, such as FORALL statements, intrinsic functions with different data distributions, and passing distributed arrays between subprograms.

2 The Linear Algebra Subset of the HPFBench

Linear algebra functions frequently appear as the time-dominant computation kernels of large applications, and are often hand-optimized as mathematical library functions by the supercomputer vendors (e.g. ESSL and PESSL [13] from IBM) or by research institutions (e.g. ScaLAPACK [3] from University of Ten-

| Communication | Arrays | | |
Pattern	1–D(BLOCK)	2–D(BLOCK,BLOCK)	3–D(*,BLOCK,BLOCK)
CSHIFT	conj-grad jacobi, FFT-1D	fft-2D, pcr:coef_inst jacobi, pcr:inst_coef	fft-3D
SPREAD		gauss-jordan, jacobi lu:nopivot, lu:pivot qr:factor, qr:solve	matrix-vector
SUM		qr:factor, qr:solve	matrix-vector
MAXLOC		gauss-jordan, lu:pivot	
Scatter		gauss-jordan, fft-2D lu:pivot	fft-3D
Gather		gauss-jordan	

Table 1. Communication pattern of linear algebra kernels (the array dimensions for reduction and broadcast are of source and destination respectively).

nessee, et. al.). These hand-optimized implementations attempt to make efficient use of the underlying system architecture through efficient implementation of interprocessor data motion and management of local memory hierarchy and data paths in each processor. Since these are precisely the issues investigated in modern compiler design for parallel languages and on parallel machine architectures, the subset of the HPFBench benchmark suite is provided to enable testing the performance of compiler generated code against that of any highly optimized library, such as the PESSL and ScaLAPACK, and exploiting potential automatic optimizations in compilers.

The linear algebra library function subset included in the HPFBench suite is comprised of eight routines. Table 1 shows an overview of the data layout for the dominating computations and the communication operations used along with their associated array ranks.

conj-grad uses the Conjugate Gradient method for the solution of a single instance of a tridiagonal system. The tridiagonal system is stored in three 1–D arrays. The tridiagonal matrix–vector multiplication in this method corresponds to a three–point stencil in one dimension. The three–point stencil is implemented using *CSHIFT*s.

fft computes the complex–to–complex Cooley–Tukey FFT [4]. One–, two–, or three–dimensional transforms can be carried out. In the HPFBench benchmark, the twiddle computation is included in the inner loop. It implements the butterfly communication in the FFT as a sequence of *CSHIFT*s with offsets being consecutive powers of two.

gauss-jordan computes the inversion of a matrix via the Gauss–Jordan elimination algorithm with partial pivoting [7]. Pivoting is required if the system is not symmetric positive definite. At each pivoting iteration, this variant of the algorithm subtracts multiples of the pivot row from the rows above as well as below the pivot row. Thus, both the upper and lower triangular matrices are brought to zero. Rather than replacing the original matrix with the identity matrix, this space is used to accumulate the inverse solution.

jacobi uses the Jacobi method [7] to compute the *eigenvalues* of a matrix. *Eigenvectors* are *not* computed within the benchmark. The Jacobi method makes iterative sweeps through the matrix. In each sweep, successive rotations are applied to the matrix to zero out each off–diagonal element. As the sweeps continue, the matrix approaches a diagonal matrix, and the diagonal elements approach the eigenvalues.

lu solves dense system of equations by means of matrix factorization. It includes Gaussian elimination with and without partial pivoting. Load–balance is a well–known issue for LU factorization, and the desired array layout is cyclic.

matrix-vector computes matrix–vector products. Given arrays **x**, **y** and **A** containing *multiple instances* [14] of vectors x and y and matrix A, respectively, it performs the operation $y \leftarrow y + Ax$ for each instance. The version tested in this paper has multiple instances, with the row axis (the axis crossing different rows) of **A** allocated local to the processors, and the other axis of **A** as well as the axes of the other operands spread across the processors. This layout requires communication during the reduction.

pcr solves irreducible tridiagonal system of linear equations $AX = B$ using the *parallel cyclic reduction* method [7, 9]. The code handles multiple instances of the system $AX = B$. The three diagonals representing A have the same shape and are 2–D arrays. One of the two dimensions is the *problem axis* of extent n, i.e., the axis along which the system will be solved. The other dimension is the *instance axis*. For multiple right–hand–sides, B is 3–D. In this case, its first axis represents the right–hand–sides, and is of extent r and is local to a processor. Excluding the first axis, B is of the same shape as each of the arrays for the diagonal A. The HPFBench code tests two situations, with the problem axis being the left and the right parallel axis, denoted as coef_inst and inst_coef, respectively.

qr solves dense linear systems of equations $AX = B$ using Householder transformations [5, 7], commonly referred to as the *QR routines*. With the QR routine, the matrix A is factored into a trapezoidal matrix Q and a triangular matrix R, such that $A = QR$. Then, the solver uses the factors Q and R to calculate the least squares solution to the system $AX = B$. In our experiments, these two phases are studied separately and only the solver is reported. The HPFBench version of the QR routines only supports single–instance computation and performs the Householder transformations *without* column pivoting.

3   Methodology

This paper evaluates two commercial HPF compilers on an IBM SP2. The evaluation therefore focuses on comparing the functionality of the HPF languages supported by the compilers and the performance of the generated codes.

An ideal evaluation of performance of the compiler-generated code is to compare with that of a message–passing version of the same benchmark code, as message–passing codes represent the best performance tuning effort from low level programming. Message passing implementations of LU, QR factorizations and matrix-vector multiply are available from IBM's PESSL library, and in other libraries such as ScaLAPACK. However, the functions in such libraries often use blocked algorithms for better communication aggregation and BLAS performance. A fair comparison with such library implementations thus require sophisticated HPF implementations of the blocked algorithms.

In the absence of message–passing counterparts that implement the same algorithms as the HPFBench codes, we adopt the following two–step methodology in benchmarking different HPF compilers. First, we compare the single-processor performance of the code generated by each HPF compiler versus that of the sequential version of the code compiled under *xlf*. Such a comparison will expose the overhead of the HPF compiler-generated codes. We then measure the speedups of the HPF codes on parallel processors relative to the sequential code. This will provide a notion on how well the codes generated by different HPF compilers scale with the parallelism available in a scalable architecture.

4   Evaluation Results

The evaluation reported here are of two commercial HPF compilers, *pghpf* from the Portland Groups, Inc., and *xlhpf* from IBM. Version 2.2-2 of *pghpf* with compiler switch -O3 -W0,-O4 was used and linked with -Mmpl. *xlhpf* is invoked with compiler switch -O4. To measure the overhead of HPF compilers on a single processor, we also measured code compiled using Fortran 90 compilers, specifically, using *xlf90* -O4 and *pghpf* -Mf90 -O3 -W0,-O4. The later was linked with -Mrpm1 and will be denoted as *pgf90* in the rest of this section.

Our evaluation was performed on an IBM SP2. Each node has a RS6000 POWER2 Super Chip processor running AIX 4.3 at 120Mhz and has 128 Mbytes of main memory. The nodes are communicating through IBM's MPL library on the IBM SP2 high–performance switch network with a peak node–to–node bandwidth of 150 Mbytes/second. All results were collected under dedicated use of the machines.

Due to the space limitation, for each benchmark with several subprograms, we only report on one subprogram. For example, we only report on `fft-2d`, `lu:pivot`, `pcr:inst_coef` and `qr:solve`. Table 2 gives the problem size and the total MFLOPs performed for the problem sizes. The problem sizes are chosen so that the total memory requirement will be around 50 Mbytes.

4.1   Functionality

In comparing the two compilers, we found that *pghpf* supports the full set of library procedures which include gather/scatter operations used to express the gather/scatter communication in `fft` and `gauss-jordan`. On the other hand, *xlhpf* does not support any of the library procedures, and we had to rewrite the gather/scatter communication as `forall` statements. Furthermore, *xlhpf* restricts distributed arrays in a subprogram to have the same number of distributed axes.

Benchmark	Problem Size	Iteration	Mflop
conj-grad	2^{17}	962	3278.37
fft-2d	$2^9 \times 2^8$	1	11.1
gauss-jordan	$2^9 \times 2^9$	1	268.7
jacobi	$2^8 \times 2^8$	6	614.2
lu: pivot	$2^9 \times 2^9$	1	89.5
matrix_vector	$2^8 \times 2^8 \times 2^4$	10	21.0
pcr: inst_coef	$2^9 \times 2^9$, 8 RHS	1	117.4
qr: solve	$2^9 \times 2^9$, 16 RHS	1	872.4

Table 2. Problem size and FLOP count.

Code	xlf		xlf90	xlhpf	pgf90	pghpf
	Time	FLOPrate	Time	Time	Time	Time
	(sec.)	(Mflops/s)	vs. xlf	vs. xlf	vs. xlf	vs. xlf
conj-grad	176.7	18.56	1.07	1.36	1.08	1.17
fft-2D	5.3	2.12	1.44	2.28	1.07	1.08
gauss-jordan	28.3	9.49	1.17	5.48	1.24	1.46
jacobi	54.5	11.27	4.28	6.04	2.20	1.98
lu: pivot	19.8	4.51	1.07	1.08	1.16	1.13
matrix_vector	1.1	18.66	1.00	7.64	2.64	3.09
pcr: inst_coef	25.3	4.64	1.35	3.36	1.22	0.57
qr: solve	75.3	11.58	1.39	2.07	1.53	1.47

Table 3. Single-processor performance of the linear algebra kernels.

4.2 Sequential Performance

Table 3 compares the the performance of the HPF benchmark codes compiled using *xlf90*, *xlhpf*, *pgf90*, and *pghpf*, respectively, versus that of the sequential versions of the same codes compiled using *xlf*, on a single node of the SP2. The table shows that all four HPF or F90 compilers incur significant overhead to the generated code when running on a single processor. More specifically, codes compiled using *xlf90* are up to 4.28 times slower than the sequential codes compiled using *xlf*. Futhermore, codes compiled using *xlhpf* are up to 7.64 times slower than the sequential codes, showing that *xlhpf* generates code with additional overhead. In contrast, codes compiled using *pgf90* and *pghpf* are only up to 3.09 times slower than the sequential codes for all benchmarks except pcr:inst_coef, which is 43% faster than its sequential counterpart. A close look at this benchmark shows that the improvement is due to the default padding (-Moverlap=size:4) by *pghpf* along all parallel dimensions of an array which significantly reduces cache conflict misses in pcr:inst_coef.

To understand the overhead incurred on the codes generated by the four compilers, we measured the time spent on segments of the code that would cause communication when running on parallel processors, and compared them with those of the sequential codes. Table 4 lists the time breakdown for the five versions of each benchmark.

Code	Breakdown	xlf	xlf90	xlhpf	pgf90	pghpf
conj-grad	total	176.66	189.82	240.44	190.45	207.13
	cshift	44.81	47.37	75.15	61.77	65.57
fft-2d	total	5.25	7.54	11.96	5.60	5.67
	scatter	0.10	0.13	2.28	0.93	0.95
	cshift	1.87	3.43	6.08	1.53	1.77
jacobi	total	54.5	233.03	328.98	119.9	107.77
	cshift	14.49	180.60	196.94	58.57	50.37
	spread	7.93	12.62	15.63	14.31	15.08
pcr: inst_coef	total	25.3	34.21	84.59	30.77	14.50
	cshift	2.44	15.44	71.01	2.63	2.59
gauss-jordan	total	28.31	33.18	155.05	35.18	41.36
	spread	4.72	4.80	5.26	4.91	5.06
	maxloc	9.65	14.22	13.58	17.73	18.27
	scatter	0.03	0.03	119.62	0.03	0.09
matrix_vector	total	1.12	1.12	8.37	2.93	3.38
	sum	0.20	0.20	7.45	2.42	2.80
	spread	0.39	0.39	0.41	0.34	0.37
qr: solve	total	75.31	104.56	155.57	115.5	110.51
	sum	6.91	8.44	9.42	15.92	20.62
	spread	11.71	24.70	86.94	30.55	19.86
lu: pivot	total	19.83	21.14	21.34	22.90	22.50
	spread	8.43	9.58	10.04	9.89	9.93
	maxloc	0.03	0.04	0.09	0.07	0.18
	scatter	0.01	0.02	0.15	0.52	0.44

Table 4. Breakdown of single-processor time of the linear algebra kernels.

Table 4 shows that the overhead of the four HPF or F90 compilers occurs in code segments corresponding to *cshift, spread, sum* and *scatter*. First, *cshift* is up to 29 times slower when compiled using *xlhpf* or *xlf90*, and up to 4 times slower when compiled using *pgf90* or *pghpf*, compared with the sequential versions of the codes. This contributed to the longer total time for fft-2D, jacobi, pcr:inst_coef and conj-grad. Second, *scatter* under *xlhpf* is extremely slow and this contributed to the slowdown of gauss-jordan and fft-2D using *xlhpf* when compared to other compilers. Third, *sum* is extremely efficient under *xlf90* and makes matrix-vector under *xlf90* as fast as the sequential code. Lastly, *spread* costs almost the same in all five versions of three benchmarks (lu:pivot, matrix_vector, gauss-jordan), and is up to 7.4 times slower under *xlhpf* and up to 2.6 times slower under *xlf90, pgf90,* or *pghpf,* for jacobi and qr:solve, compared with the sequential codes.

4.3 Parallel Performance

Figure 1 shows the parallel speedups of the benchmarks compiled using *xlhpf* and *pghpf* on up to 16 processors of the SP2, using the performance of the sequential codes as the base. Overall, *pghpf* compiled codes scale from about the same to

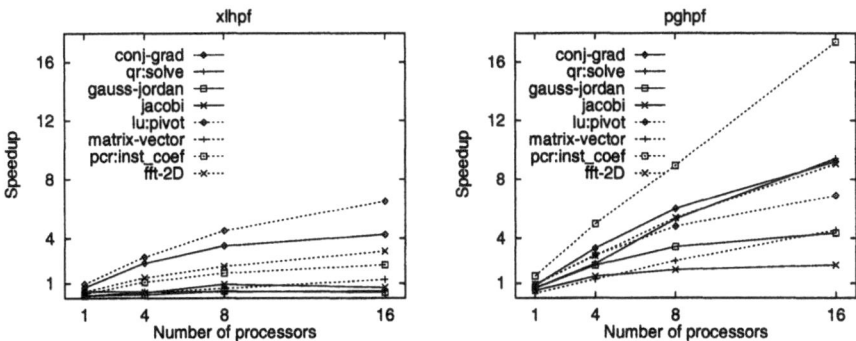

Fig. 1. Speedups of the linear algebra kernels.

significantly better than those compiled with *xlhpf*. When compiled with *xlhpf*, jacobi, gauss-jordan, matrix-vector and qr:solve achieve no speedups on 4, 8, and 16 nodes.

To understand the contributing factors to the performance difference between codes compiled using the two compilers, we further measure, for each benchmark, the time spent in communication for runs on 1, 4, 8, and 16 nodes. Figure 2 shows the measured breakdowns of the total time. The performance difference between the two compilers divides the benchmark codes into two groups. In the first group, lu:pivot shows very close running time under the two compilers. The rest kernels belong to the second group. These codes when compiled using *pghpf* run from slightly to much faster than when compiled using *xlhpf*. The difference is mainly from the difference in *cshift* under the two compilers, as shown by the kernels conj-grad, fft-2D, jacobi, and pcr:inst_coef. However, for gauss-jordan, the *xlhpf* version is much slower than the *pghpf* version because of *gather/scatter*. matrix-vector is slower with the *xlhpf* version than the *pghpf* version because of the difference in *sum*. The reason for the poor performance of qr:solve with the *xlhpf* is from the slowdown in performing *spread*.

5 Conclusion

We have reported a performance evaluation of two commercial HPF compilers, *pghpf* from Portland Groups Inc. and *xlhpf* from IBM, using a subset of the HPFBench benchmarks on the distributed–memory IBM SP2.

We first compare the single–processor performance of the codes when compiled using the two compilers, their Fortran 90 companions versus that of sequential versions of the benchmarks. Our experiments show that the HPF or F90 incur significant overhead to the generated code. Compared to the sequrntial codes, the codes compiled using *xlhpf* are up to 7.64 times slower, and those compiled using *pghpf* up to 3.09 times slower. We further measure the time breakdowns of selected benchmarks and find that the total running time difference largely comes from difference in code segments which would generate communication when running in parallel. Other than *sum* which is highly efficient under *xlf90*,

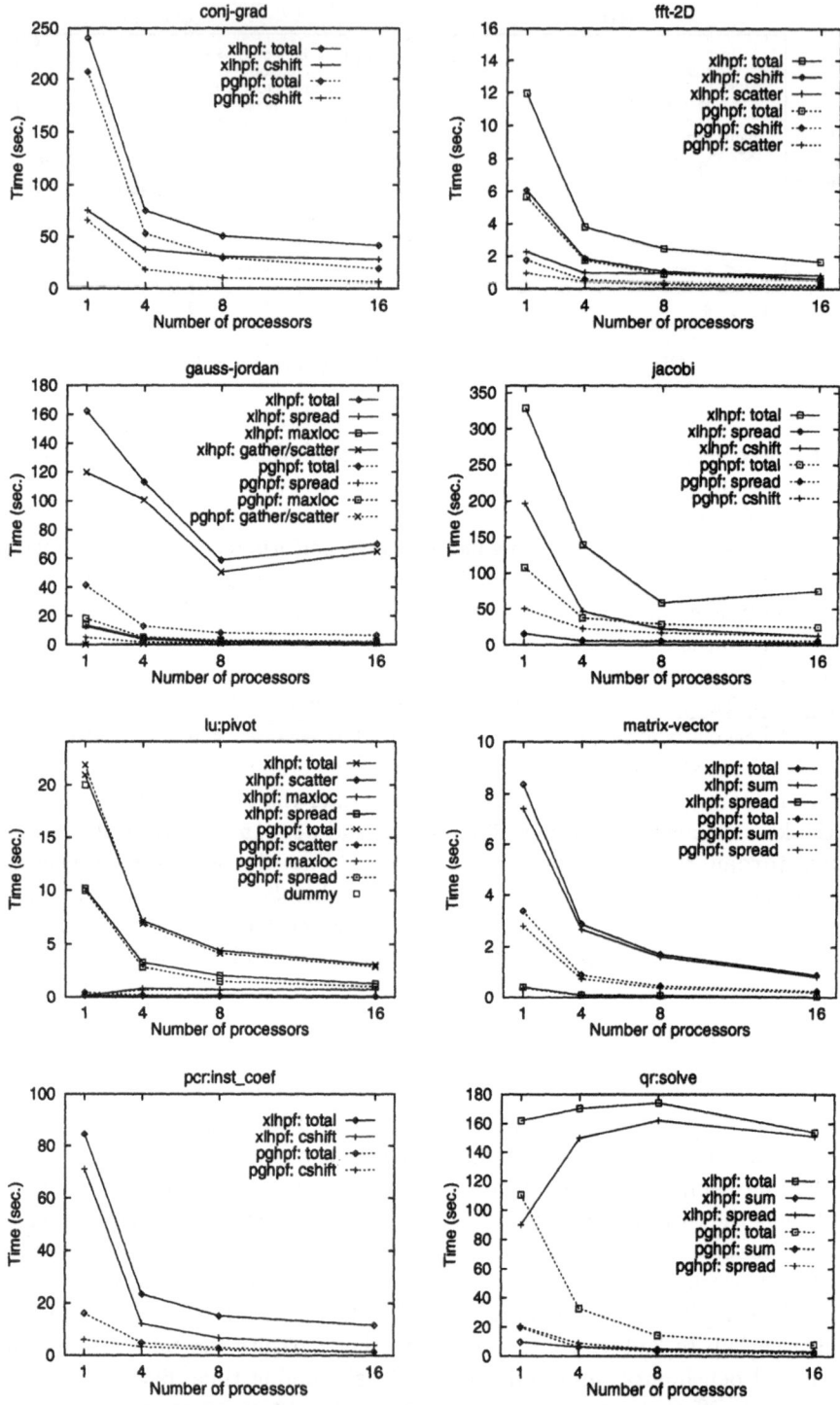

Fig. 2. Total running time and the communication breakdown .

other data movements such as *cshift, spread, gather/scatter* are mostly slower under *xlf90* and *xlhpf* than under *pgf90* and *pghpf*. When running in parallel, codes compiled under *pghpf* show better scalability than when compiled under *xlhpf*. The difference is mainly from better performance of communications such as *cshift, spread, sum* and *gather/scatter* under *pghpf*.

References

[1] D. Bailey, et. al. The NAS parallel benchmarks. Technical Report RNR-94-007, NASA Ames Research Center, Moffett Field, California, March 1994.

[2] M. Berry et. al. The PERFECT Club benchmarks: Effective performance evaluation of supercomputers. *International Journal of Supercomputer Applications*, 3:5 – 40, 1989.

[3] L. S. Blackford, J. Choi, A. Cleary, E. D'Azevedo, J. Demmel, I. Dhillon, J. Dongarra, S. Hammarling, G. Henry, A. Petitet, K. Stanley, D. Walker, and R. C. Whaley. ScaLAPACK: A Linear Algebra Library for Message-Passing Computers. In *SIAM Conference on Parallel Processing*, March 1997.

[4] J. C. Cooley and J. Tukey. An algorithm for the machine computation of complex fourier series. *Math. Comp*, 19:291–301, 1965.

[5] G. Dahlquist, A. Björck, and N. Anderson. *Numerical Methods*. Series in Automatic Computation. Prentice Hall, Inc., Englewood Cliffs, NJ, 1974.

[6] J. J. Dongarra. Performance of various computers using standard linear equations software. Technical Report CS-89-85, University of Tennessee, Department of Computer Science, 1989.

[7] G. Golub and C. vanLoan. *Matrix Computations*. The Johns Hopkins University Press, second edition, 1989.

[8] High Performance Fortran Forum. High Performance Fortran; language specification, version 1.0. *Scientific Programming*, 2(1 - 2):1–170, 1993.

[9] R. W. Hockney and C. Jesshope. *Parallel Computers 2*. Adam Hilger, 1988.

[10] R. Hocney and M. Berry. Public international benchmarks for parallel computers: Parkbench committe report-1. Technical report, Netlib, Oak Ridge National Laboratory, February 1994.

[11] Y. C. Hu, G. Jin, S. L. Johnsson, D. Kehagias, and N. Shalaby. HPFBench: A High Performance Fortran benchmark. Tech. Rep. TR98-322, Computer Science Dept., Rice Univ., 1998. URL: http://www.crpc.rice.edu/HPFF/benchmarks.

[12] Y. C. Hu, S. L. Johnsson, and S.-H. Teng. High Performance Fortran for highly irregular problems. In *Proc. of the 6th ACM SIGPLAN Symposium on Principles and Practice of Parallel Programming*, Las Vegas, Nevada, June 1997.

[13] IBM. *IBM Parallel Engineering and Scientific Subroutine Library Release 2, Guide and Reference*, 1996.

[14] S. L. Johnsson, T. Harris, and K. K. Mathur. Matrix multiplication on the Connection Machine. In *Supercomputing 89*, pages 326–332. ACM, November 1989.

[15] F. McMahon. The Livermore Fortran kernels: A test of numerical performance range. In *Performance Evaluation of Supercomputers*, pages 143 – 186. North Holland, 1988.

Performance Evaluation of Object Oriented Middleware

László Böszörményi*, Andreas Wickner*, and Harald Wolf*

*Institute of Information Technology, University of Klagenfurt
Universitätsstr. 65 – 67, A – 9020 Klagenfurt and
*software design & management GmbH & Co. KG
Thomas-Dehler-Straße 27, D – 81737 München
email: laszlo@itec.uni-klu.ac.at, andreas.wickner@sdm.de, hwolf@edu.uni-klu.ac.at

A method for evaluating several aspects of the performance of object oriented middleware is introduced. Latency, data transfer, parameter marshalling and scalability are considered. A portable benchmark toolkit has been developed to implement the method. A number of actual middleware products have been measured, such as C++ and Java based CORBA implementations, DCOM and Java/RMI. The measurements are evaluated and related to each other.

1   Introduction

Currently, a certain number of de facto and de jure standards of object oriented middleware exist; also a number of implementations are available. Therefore, performance benchmarking is helpful in order to support the decision process at the beginning of the software lifecycle.

Existing measurements are rare and often restricted to a few aspects. Most of them ignore scalability either entirely or they consider the case of many objects on the server side only, ignoring the case of many clients attacking the same server concurrently (see [SG96], [PTB98], [HYI98]). The following research of the performance of a number of actual implementations is based on the most relevant object oriented middleware proposals, such as CORBA [OMG98], DCOM [Mic98] and Java/RMI [Sun98]. The evaluation covers several important performance aspects: latency, data transmission time and scalability.

A key issue in benchmarking are comparisons. Therefore, we have developed a portable benchmark toolkit.

2   Measurement Criteria and Methods

If we want to find out how fast our middleware is, we first have to inspect the latency and the turnaround time in detail (i.e. parameter marshalling/unmarshalling). A simple black box measurement of the turnaround time of parameterless remote method calls is sufficient to inspect the latency. White box measurements were performed with the help of interceptors [OMG98] on remote method calls with parameters of different type and size to measure data transmission time in detail (with measurement points before and after marshalling resp. unmarshalling).

Unfortunately, the standardization of interceptors in CORBA is still in progress. At this time, different manufacturers provide different interceptors (e.g. Orbix Filter, Visibroker interceptors) or do not support interceptors at all (e.g. TAO, JDK 1.2). Moreover, white box measurements are only meaningful if the resolution of the time

P. Amestoy et al. (Eds.): Euro-Par'99, LNCS 1685, pp. 258–261, 1999.
© Springer-Verlag Berlin Heidelberg 1999

measurement is in order of magnitude of microseconds and not milliseconds as usual in Java based systems.

Second, we want to know how the system scales. Special interest has been raised in the question: How does the server cope with a large number of concurrent client requests? It is not trivial to generate many parallel requests in order to evaluate scalability. Because it is hard to handle a large number of client processes (e.g. 1000 clients at once) we preferred to use threads.

3   The Benchmark Toolkit

Since CORBA is a standard and not a product, many implementations exist. The following relevant products have been investigated: Orbix 2.3 (C++) and OrbixWeb 3.0 (Java) from Iona, Visibroker 3.2 (C++ and Java) from Inprise (Borland/Visigenic), TAO 1.0 (C++) from D. Schmidt (Washington University) and JDK 1.2 (Java) from Sun (Beta3). The main focus was on CORBA implementations, because it is a widely used object oriented middleware platform. Moreover, CORBA provides effective support for a portable benchmark toolkit and for white box measurements. Additional measurements with Java/RMI and DCOM were made with separate benchmarks to compare at least basic functions (such as method invocation and object creation). All Java environments but JDK 1.2 included a JIT compiler.

A portable benchmark toolkit was developed by using the new Portable Object Adapter (POA) [SV97]. This enabled enhanced portability of source code and generated code frames (stubs, skeletons) between different CORBA implementations. Unfortunately, the POA is still not widely supported. A wrapper object was developed for the C++ based implementations, in order to hide the differences between different implementations.

The benchmark information is specified and stored in a scenario profile, which contains a list of actions with parameters, such as: number of repetitions, amount of data, measuring offset, etc. The toolkit produces a workload on the system under test according to the profile. A transformation utility computes from the measured data the following indices: response time, marshalling overhead (client, server) and throughput. Minimum, maximum, mean and standard deviation are calculated for all indices.

For the measurements we used good, off-the shelf components. The server used in most cases was a NT host (Dual Pentium II, 233 MHz, 128 MB RAM, Windows NT 4.0, Service Pack-3,Visual C++ 5.0), and the client was a Solaris-Host (Sun Ultra 5,2, UltraSparc 300MHz processors, 256 MB RAM, Solaris 2.5, Sun Workshop C++ Compiler 3.0) connected by a 100 Mbit/sec Ethernet switch. For the DCOM measurements an NT system was used to host the clients as well.

4   Numeric Results

The following section shows a small excerpt from the evaluation. A summary of the entire investigation can be found in [BWW99]. The measurements illustrated in Fig. 1 show the response times for clients invoking remote methods without parameters and results. In each case, client and server are implemented using the same product. The C++ products are about 3 to 4 times faster than the Java based systems. The effect of the JIT compiler seems to be almost negligible in this measurement. The heavy fluctuation in the standard deviation in the Java implementations is caused by the

260 László Böszörményi, Andreas Wickner, and Harald Wolf

garbage collection. DCOM is fast, mainly because it is fully integrated into the Windows NT operating system.

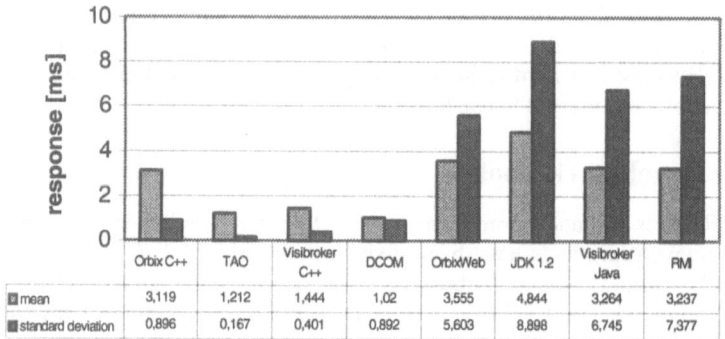

Fig. 1. Method invocation without parameters

The following table shows the response times of transmitting 10 Kbyte of data within a sequence of shorts including the accumulated marshalling and unmarshalling subsections. TAO and DCOM do not support interceptors, thus this comparison contains only Visibroker and Orbix.

	Visibroker	Orbix	Sockets
Server	1017 µs	4058 µs	
Net transfer	2402 µs	2378 µs	2975 µs
Client	843 µs	7301 µs	

Visibroker is fast and has only little overhead compared with pure socket transmissions (43%). Orbix is astonishingly slow, its overhead hardly acceptable (362%). The reason is, unfortunately, unclear. Older Orbix versions were much faster than the current one (Iona has recently shipped Orbix 3.0 which is reported to include fixes for various performance bottlenecks but we have not yet been able to benchmark this version of Orbix.). Java-based products (not shown here) are about 3 to 4 times slower than the C++ based ones in transmitting small sequences, and more than 10 times slower in the case of large sequences. If sequences of complex data types, like nested structures are transmitted, the Java-based systems perform even worse.

Beside investigations on object creation (not shown here) some measurements have been made with special multithreaded CORBA clients (see Fig. 2). The client process is an Orbix-Client on the Solaris workstation that produces heavy load on the server process at the NT-Server. (For technical reasons, TAO, DCOM and Java/RMI were not considered in this test.) Depending on the number of clients the throughput grows to a maximum, then remains at the same level until it slowly decreases. The JDK 1.2 server stops working, when more than 197 client connections are open - that might be a bug in the Beta3 version. Orbix is quite slow, presumably due to heavy lock contention. The server of Visibroker for Java consumes so much memory, that Windows NT stops the server process at 800 open client connections.

5   Conclusions and Future Work

We demonstrated that it is possible to measure and compare several aspects of the performance of different middleware software. A portable benchmark toolkit was developed that can be used repeatedly for the same products and that can also be easily adapted to entirely new products. A number of useful actual measurements and comparisons were presented. The absolute numbers of some products, such as Visibroker, DCOM and TAO are quite promising by showing that the overhead caused by standard middleware is not necessarily high.

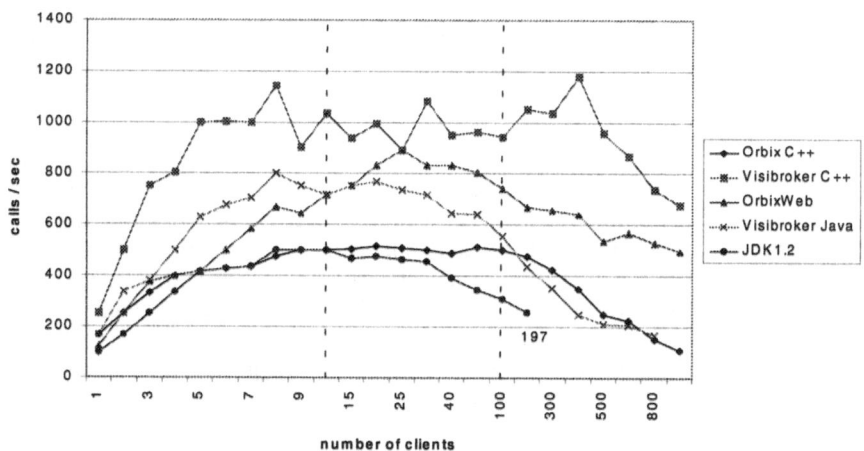

Fig. 2. Throughput with multiple Clients

In future, white box benchmarks could be extended to give more explanations of performance shortages. Emerging benchmark standards of OMG should be included.

6   References

[BWW99]   L. Böszörményi, A. Wickner, H. Wolf: *Performance Evaluation of Object Oriented Middleware – Development of a Benchmarking Toolkit*, Technical Reports of the Inst. of Inf. Technology, University Klagenfurt, TR-9902, http://colossus.itec.uni-klu.ac.at/~laszlo/pubrec.html

[HYI98]   S. Hirano, Y. Ysu, H. Igarashi: *Performance evaluation of Popular Distributed Object Technologies for Java*, ACM Workshop on Java for High-Performance Network Computing, February 1998

[Mic98]   Microsoft Corporation: *DCOM Architecture*, White Paper, 1998

[OMG98]   Object Management Group, Inc.: *The Common Object Request Broker: Architecture and Specification*, Revision 2.2, February 1998

[PTB98]   Fantisek Plasil, Petr Tuma, Adam Buble: *CORBA Comparison Project*, Final Report, Distributed Research Group, Charles University, Prague, June 1998

[SV97]   Douglas C. Schmidt, Steve Vinoski: *Object Adaptors: Concepts and Terminology*, SIGS C++ Report, Vol. 9, No 11., SIGS Publications, 1997

[SG96]   Douglas C. Schmidt, Aniruddha Gokhale: *Measuring the Performance of Communication Middleware on High-Speed Networks*, Proceedings of the SIGCOMM Conference, August 1996

[Sun98]   Sun Microsystems: *Java Remote Method Invocation Specification*, 1998

PopSPY: A PowerPC Instrumentation Tool for Multiprocessor Simulation

C. Limousin, A. Vartanian, and J-L. Béchennec

Équipe Architectures Parallèles
Université de Paris-Sud XI
Laboratoire de Recherche en Informatique, Bât. 490
91405 Orsay Cedex, France
(limousin,alex,jlb)@lri.fr
tel: 33 1 69 15 64 07
fax: 33 1 69 15 65 86

Abstract. PopSPY is an instrumentation tool for the PowerPC microprocessor family that modifies code at the executable level, providing the ability to instrument a wide range of applications including commercial programs. PopSPY instruments sequential, multithreaded and multiprocess applications including their associated dynamic libraries. PopSPY uses a dynamic trace library that either generates execution traces of applications or communicates with an execution driven simulator for accurate multiprocessor simulation.

Topic 02: Performance Evaluation and Prediction

1   Introduction

Trace driven simulation is a commonly used technique for performance evaluation of architectures. The benchmarks are first run on a real architecture (the *host architecture*) where they are monitored to obtain their execution trace[1], which is then injected in the simulator. Traces are valid for applications whose execution doesn't depend on the architecture they are running on: this is true for most of the sequential applications and for parallel applications that statically allocate shared resources, if the spins at the synchronisation points are not recorded [2]. Execution driven simulation [1] solves this problem enforcing the execution order of the synchronization events on the host architecture to be the same as on the target architecture.

Application monitoring to obtain execution traces is commonly done using a code instrumentation technique, which consists in modifying an application by inserting tracing code sequences around memory instructions and basic blocks. The instrumentation may be done at the assembly source level ([2]), at the object-module level ([7]) or at the executable level ([5], [8], [3]), where one doesn't need the application source or object code.

---

[1] The execution trace of an application contains the list of executed instructions, addresses of memory accesses, etc...

P. Amestoy et al. (Eds.): Euro-Par'99, LNCS 1685, pp. 262–265, 1999.
© Springer-Verlag Berlin Heidelberg 1999

We wanted to study the memory hierarchy (that is to simulate a processor and focus on caches) of mainstream computers architectures, namely PowerPC or x86 (multi or monoprocessors). On those platforms, representative benchmarks are mainly commercial application whose code sources or objects are not available. We thus developped PopSPY, a tool for the PowerPC microprocessor that monitors sequential and shared memory multithreaded applications at the executable level[2]. Since we restrict our research with computing-intensive applications, the fact that the operating system is not monitored is not invalidant. PopSPY uses a shared tracing library that either generates the execution trace of the instrumented applications or is plugged to an execution driven simulator. Moreover, dynamic libraries are also monitored. PopSPY runs under MacOS 7.x or later.

Section 2 presents the functionalities of PopSPY. Performance of PopSPY is exposed in section 3.

2   PopSPY Overview

Working scheme The PopSPY code instrumentor annotates the monitored application with function calls to a shared library: when the annotated application executes, these calls generate the execution trace of the application. Since the library is dynamic, one may rewrite the functions for different purposes without reinstrumenting the application. Only the prototypes of the functions are fixed. The library consists in two main functions, `TraceBlock` and `TraceLdSt`. The first one is called in front of each basic block. It receives the calling block number as a parameter. The second one, which calls are added in front of each memory access, receives three parameters: the calling block number, the number of the load-store instruction in this block and a coded description of this instruction (type, used registers, address compute mode, etc...).

Sequential applications traces A sequential application trace is composed of three files: a file (*static file*) generated at instrumentation time containing the application original basic blocks description (size, memory instructions count, etc...); a file (*dynamic file A*) generated by the library at execution time containing the numbers of executed original basic blocks. These numbers are indexes in the static file; a file (*dynamic file B*) generated by the library containing the description of executed memory accesses (kind, address and data read or written).

With the description of each block a simulator obtains the whole trace from the two dynamic files: the bloc number in the dynamic file A gives the instruction block to be read in the application executable and each time a load-store instruction is encountered, it is read in the dynamic file B.

Shared-memory multithreaded applications traces Contrary to sequential applications a multithreaded one is made of several threads that call the library in

---

[2] IDTrace [4] for the i486 is compiler-dependant and does not monitor multithreaded applications.

an interleaved way. The library must generate one trace per thread and thus be able to know which thread called it. The multiprocessor support library for MacOS, Daystar MP, gives this information to the trace library. In a multi-threaded application trace generated by PopSPY, one will found the same static file as in the sequential case, plus a dynamic file A_i and a dynamic file B_i for each thread i. These A_i and B_i files are generated the same way as in the sequential case. However, an other information must be present in the trace in order to correctly simulate a multithreaded application: the multiprocessor events concerning thread i are interleaved with the block numbers in the A_i file. Supported multiprocessor events are the following: thread creation/merging (fork/join), thread termination, locks management (creation, acquisition, release, destruction), Barrier management.

Multiprocess and dynamic libraries traces PopSPY may add at the user choice an identifier to block numbers passed to the trace library. Several applications with different identifiers may then be executed in parallel: their traces are interleaved in the global trace, each traced instruction being tagged with the idenfifier, allowing a simulator to read correctly each traces. A dynamic library is traced the same way: it is given a unique identifier, and its trace is interleaved with the application one.

Execution driven simulation When a simulator is connected to PopSPY via the trace library, the `TraceBlock` and `TraceLdSt` functions send the informations they receive to the simulator. The trace library intercepts the calls from the application to the multithread support library. Whenever an event whose execution order may modify the trace occurs, the trace library freezes the corresponding threads. The simulator executes these events then feed back the trace library with temporal informations. The trace library uses these informations to release the threads in the simulation order.

3   PopSPY Performance

Examples of applications known to run properly after being instrumented by PopSPY are all programs compiled with Metrowerks Codewarrior, and particularly the benchmark suite SPEC95 [6] and the parallel suite SPLASH-2 [9]. Adobe Photoshop 4.0.1 Tryout and Strata StudioPro Blitz 1.75 which are multi-threaded image creation programs are also properly executing after instrumentation.

Measuring the execution of *Adobe Photoshop 4.0.1 Tryout* resizing an image from 72 to 300 dpi, *Strata StudioPro Blitz v1.75* during the Phong rendering of a texture mapped sphere and the five benchmarks *compress*, *go*, *hydro2d*, *mgrid* and *tomcatv* from the SPEC95 suite, executed with the standard data sets on a 90Mhz PowerPC 601 under MacOS 8.0 shows a memory dilation (resp. slow-down) between 3.02 and 6.59 (resp. 2.8 and 16.37), depending of the benchmark and the completeness of instrumentation. As a comparison, Pixie and Goblin have memory dilation of respectively 6 and 4 and slowdowns of 10 and 20.

4   Conclusion

We presented in this paper an instrumentation tool for the PowerPC microprocessors family called PopSPY. This tool instruments applications at the executable level and allowed us to instrument commercial applications like *Adobe
Photoshop Tryout* and *Strata StudioPro Blitz*. PopSPY monitors sequential, multithreaded (such as SPLASH-2) and multiprocess applications, including dynamic libraries. Moreover, the dynamic trace library of PopSPY may be plugged
to an execution-driven simulator.

We saw that the slowdown and memory dilation of PopSPY are in the order
of magnitude of existing executable-level instrumentation tools, making it fully
usable. Moreover, a graphical user interface is provided. An extended version of
this paper can be found at http://www.lri.fr/~limousin/biblio.html.

References

[1] R.C. Covington, S. Madala, V. Mehta, and J.B. Sinclair. The rice parallel process-
    ing testbed. In *Proceedings of the 1988 ACM SIGMETRICS Conf. on measurement
    and Modeling of Computer Systems*, 1988.
[2] S.R. Goldschmidt and J.L. Hennessy. The accuracy of trace-driven simulations of
    multiprocessors. In *Proceedings of the 1993 ACM SIGMETRICS Conference on
    Measuring and Modeling of Computer Systems*, 1993.
[3] J. R. Larus and E. Schnarr. Eel: Machine-independent executable editing. In *SIG-
    PLAN Conference on Programming Language Design and Implementation (PLDI)*,
    June 1995.
[4] J. Pierce and T. Mudge. Idtrace - a tracing tool for i486 simulation. Technical
    report, University of Michigan technical report CSE-TR-203-94, 1994.
[5] M. D. Smith. Tracing with pixie. Technical report, Technical Report, Stanford
    University, 1991.
[6] SPEC. *SPEC Newsletter*, vol. 7, issue 3, 1995.
[7] A. Srivastava and A. Eustace. Atom: A system for building customized program
    analysis tools. In *Proceedings of the SIGPLAN'94 Conference on Programming
    Language Design and Implementation*, 1994.
[8] C. Stephens, B. Cogswell, J. Heinlein, G. Palmer, and J. Shen. Instruction level
    profiling and evaluation of the ibm rs/6000. In *Proceeding of the 18th Annual
    International Symposium on Computer Architecture*, 1991.
[9] S.C. Woo, M. Ohara, E. Torrie, J.P. Singh, and A. Gupta. The splash-2 pro-
    grams: Characterization and methodological considerations. In *Proceedings of the
    22thAnnual International Symposium on Computer Architecture*, 1995.

Performance Evaluation and Benchmarking of Native Signal Processing

Deependra Talla and Lizy K. John

Laboratory for Computer Architecture,
Department of Electrical and Computer Engineering,
The University of Texas at Austin, Austin, TX, USA 78712
deepu@ece.utexas.edu, ljohn@ece.utexas.edu

Abstract. DSP processor growth is phenomenal and continues to grow rapidly, but general-purpose microprocessors have entered the multimedia and signal processing oriented stream by adding DSP functionality to the instruction set and also providing optimized assembly libraries. In this paper, we compare the performance of a general-purpose processor (Pentium II with MMX) versus a DSP processor (TI's C62xx) by evaluating the effectiveness of VLIW style parallelism in the C62xx versus the SIMD parallelism in MMX on the Intel P6 microarchitecture. We also compare the execution speed of reliable, standard, and efficient C code with respect to the signal processing library (from Intel) by benchmarking a suite of DSP algorithms. We observed that the C62xx exhibited a speedup (ratio of execution clock cycles) ranging from 1.3 up to 4.0 over the Pentium II, and the NSP libraries had a speedup ranging from 0.8 to over 10 over the C code.

1 Introduction

Digital Signal Processing (DSP) and multimedia applications are ubiquitous and the demand for such capabilities on general-purpose processors is increasing. Multimedia and signal processing oriented workloads have been projected to be a dominant segment of future workloads [1]. Traditionally, DSP applications were being performed solely on a dedicated DSP processor that is specialized to do the intense number crunching tasks presented by various signal processing algorithms. However, there are several reasons as to why one might want to use a general-purpose processor instead of a DSP processor. The main driving force behind this initiative is the fact that instead of using a DSP processor for signal processing and a general-purpose processor for other tasks; space, cost and overall system complexity can be improved by moving the signal processing onto the general-purpose processor and eliminating the DSP processor. It was also found that in certain applications general-purpose processors outperform DSP processors [2].

General-purpose microprocessors have entered the multimedia and signal processing oriented stream by adding DSP functionality to the instruction set and also providing optimized assembly libraries. Perhaps, one of the earliest and most visible example of this approach to signal processing was Intel's Native Signal Processing (NSP) initiative with the introduction of the MMX (commonly dubbed as MultiMedia

P. Amestoy et al. (Eds.): Euro-Par'99, LNCS 1685, pp. 266-270, 1999.

eXtension) set to the traditional x86 architectures. In addition to MMX instructions, Intel also provides optimized C-callable assembly libraries for signal and image processing [3]. Such media processing capabilities are foreseen to make profound changes in the use of current microprocessors [4]. DSP and media processing applications have been known to be structured and regular. The approach adopted by NSP extensions such as MMX has been to exploit the available data parallelism using SIMD (single-instruction multiple-data) techniques. On the other hand, high performance DSP processors such as the Texas Instrument's TMS320C6201 utilize a VLIW (very long instruction word) architecture.

Previous efforts have analyzed the benefits of multimedia extensions on general-purpose processors [5][6][7]. Performance benefits of MMX instructions over non-MMX code for a Pentium processor (P5 microarchitecture, with MMX) were evaluated in [5]. Effect of Sun's VIS (visual instruction set) on a superscalar processor was measured in [6]. A number of DSP and general-purpose processors have been benchmarked by BDTI [8][9], however details on performance of individual benchmarks are not publicly available. The objectives of this paper are two-fold. We compare the performance of a general-purpose processor versus a DSP processor for several DSP kernels, followed by evaluating the effectiveness of Intel's NSP library by benchmarking a suite of DSP algorithms.

2   Performance Evaluation Methodology

To measure the effectiveness of performing signal processing on general-purpose processors, we chose the highest performance member of the Intel x86 architecture family – P6 microarchitecture (a Pentium II processor with MMX) [10]. It has a three-way superscalar architecture with dynamic execution. As a high performance DSP processor, Texas Instrument's TMS320C6201 – a 32-bit fixed-point, eight-way VLIW was evaluated [11].

DSP performance evaluation is still largely done by evaluation of kernels. We assembled a suite of benchmarks that reflect a majority of the DSP routines used in the real world. Our benchmark suite consists of the following DSP kernels:

- Dot product, autocorrelation, convolution, and vector-matrix operations
- Filtering (FIR – finite impulse response and IIR – infinite impulse response)
- Transforms (FFT – fast Fourier transform, DCT – discrete cosine transform, and DWT – discrete wavelet transform)
- Window functions (Hamming, Hanning, Kaiser, Bartlett, and Blackman)
- Signal generation functions (Gaussian, Triangle, and Tone)

Our performance evaluation utilized both C and assembly code versions of the benchmarks. Assembly code versions of the benchmarks were obtained from Intel and TI libraries, while the C versions were obtained from industry standard sources [12][13]. We used Microsoft Visual C++ version 5.0 and Intel C/C++ compiler for the Pentium II processor and the C62xx compiler for the C6201 processor. Maximum optimization was enabled during compilation in all cases for faster execution times.

While comparing the Pentium II and the C62xx, each of our benchmarks has four versions – Pentium II C code, Pentium II optimized assembly code with MMX, C62xx C code, and C62xx optimized assembly code. While evaluating the NSP li-

braries on the Pentium II processor, each benchmark has the following four versions –
C code using 'float' data type, assembly code using 'float', assembly using 'double'
and assembly using 'short' data types. It is in the 'short' data type that MMX instructions are used to take advantage of data parallelism. We quantified speedup as the
ratio of the execution clock cycles with Pentium II C code as the baseline for the
general-purpose versus DSP processors comparison and C code using 'float' data type
as the baseline for measuring the effectiveness of NSP on the Pentium II processor.

We used VTune [10], performance counters on the Pentium II, and the C62xx
stand-alone simulator [11] for measuring the execution times and profiling the instruction mixes of the algorithms on the Pentium II and C62xx. VTune is Intel's performance analysis tool that can be used to get complete instruction mixes of the code.
The performance counters present in the Intel Pentium line of processors are used to
gather the execution cycle count. Similarly, the stand-alone simulator of the C62xx is
useful for quick simulation of small pieces of code - specifically to gather cycle count
information.

3   Results

Pentium II processor (with MMX) and C62xx are compared utilizing C code and
optimized assembly code for dot product, autocorrelation, FIR, IIR, and FFT. Not all
benchmarks mentioned earlier were evaluated because optimized assembly code for
C62xx was not readily available. Figure 1 below shows the results obtained for the
comparison of the Pentium II and the C62xx processor.

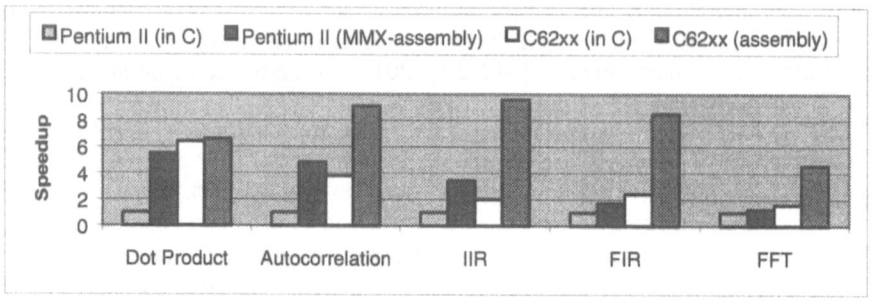

Figure. 1. Comparison of the DSP versus the general-purpose processor

Based on the execution cycle counts, the C62xx DSP is superior to the Pentium II
for the DSP kernels considered. The C62xx has eight functional units (six ALU's and
two load-store units) that allows it to simultaneously operate on multiple data elements. In almost all of the above benchmarks, the maximum instruction bandwidth of
eight is being achieved in the case of the C62xx assembly. However, the compiler-
generated code is not as efficient as the optimized assembly code. In the case of the
Pentium II, the SIMD parallelism is 4 for 16-bit data. Speedup for Pentium II MMX
assembly is over 4 in the case of dot product and autocorrelation. This is because in
the case of C code, the integer multiply takes 10 cycles on a Pentium II, while the
MMX multiply and accumulate operation takes only 3 cycles. Both the dot product

and autocorrelation involve a lot of multiply and accumulates. To make a fair comparison between the two processors, we pre-load the caches in the case of the Pentium II to eliminate cold starts.

An important aspect of the above comparison is that we are measuring the absolute number of execution clock cycles, and not the execution time. The fastest C62xx currently runs at 250 MHz, while the Pentium II processor can run at speeds up to 450 MHz. Other aspects worth mentioning are the price, power consumption, interrupt service latencies, and sizes of the two processors. The C62xx is approximately 1/15 in price, 1/3 in power consumption, 1/10 in interrupt service latency, and 1/100 in volume. Table 1 shows the effectiveness of Intel's NSP libraries. The C version uses a single-precision floating-point data type. The NSP 'float' is single-precision floating-point data, 'double' is double precision floating point, and 'short' is 16-bit fixed-point data using MMX.

Benchmark	C version	NSP float	NSP double	NSP Short
FFT	1	2.57	1.77	2.4
DCT	1	3.51	2.3	7.21
DWT	1	2.3	1.7	1.32
FIR	1	0.81	0.81	0.86
IIR	1	1.7	1.53	2.03
Hamming	1	2.44	1.66	3.84
Hanning	1	2.59	1.72	3.9
Kaiser	1	0.62	0.73	0.42
Bartlett	1	3.04	1.8	5.15
Blackman	1	1.87	1.51	3.38
Gaussian	1	0.99	0.94	0.72
Triangle	1	0.77	0.47	0.61
Tone	1	10.3	10.3	20.75
Vector/Matrix	1	-	-	2.47
Convolution	1	2.29	2.4	-

Table. 1. Speedup for NSP library code over C code

The 'short' data type version of the benchmarks has the potential to use MMX instructions to exploit data parallelism. It is interesting to note that the 'short' data versions for DWT, Gaussian, and Kaiser are slower than the 'float' versions. We believe this is true because that the library code for 'short' does an internal conversion to 'float' as was also observed in [5]. Several benchmarks used MMX instructions to achieve the speedup – vector/matrix (91% of dynamic instructions were MMX related instructions), FIR (10%), IIR (73%), FFT (3%), DCT (57%), Blackman (87%), Bartlett (80%), Hamming (86%), and Hanning (86%). In general as expected, the 'float' version of the benchmarks are faster than the 'double' versions, and the 'short' versions (that use MMX) are faster than all other versions.

4   Conclusions

In this paper, we evaluated the effectiveness of native signal processing on a Pentium II processor. First, we compared several DSP kernels on the Pentium II with respect to the C62xx DSP processor followed by a comparison of efficient C code with Intel's NSP libraries. Our major observations are:

- TI's optimized C62xx assembly code yielded a speedup of 1.2 in the case of the dot product to over 4.0 in the case of the FIR filter over optimized MMX assembly code on a Pentium II processor.
- Compiler generated code for the C62xx showed a speedup ranging from 1.6 to 6.0 over the Pentium II processor.
- NSP libraries are faster than using standard C code for many of the benchmarks. Speedup ranged from 0.8 to over 10.0. Some benchmarks like triangle wave generation and Kaiser windowing function were slower for the NSP libraries over the C code. Such anomalies make it a difficult task to predict if using NSP assembly code is faster than standard C code.

Acknowledgements

This work is supported in part by the State of Texas Advanced Technology Program grant #403, the National Science Foundation under Grants CCR-9796098 and EIA-9807112, and by Dell, Intel, Microsoft, and IBM.

References

1. C.E. Kozyrakis and D.A. Patterson, "A New Direction for Computer Architecture Research", IEEE Computer Magazine, pp. 24-32, Nov. 1998.
2. G. Blalock, "Microprocessors Outperform DSPs 2:1", MicroProcessor Report, vol. 10, no. 17, pp. 1-4, Dec. 1995.
3. Intel, "Performance Library Suite". http://developer.intel.com/vtune/perflibst/index.htm.
4. K. Diefendorff and P.K. Dubey, "How Multimedia Workloads Will Change Processor Design", IEEE Computer Magazine, pp. 43-45, Sep. 1997.
5. R. Bhargava, L. John, B. Evans, and R. Radhakrishnan, "Evaluating MMX Technology Using DSP and Multimedia Applications", IEEE Micro-31, pp. 37-46, Dec. 1998.
6. P. Ranganathan, et. al, "Performance of Image and Video Processing with General-purpose Processors and Media ISA extensions", ISCA-26, pp. 124-135, May 1999.
7. R.B. Lee, "Multimedia Extensions For General-Purpose Processors", Proc. IEEE Workshop on Signal Processing Systems, pp. 9-23, Nov. 1997.
8. G. Blalock, "The BDTIMark: A Measure of DSP Execution Speed", 1997. White Paper from Berkeley Design Technology, Inc. http://www.bdti.com/articles/wtpaper.htm.
9. J. Bier and J. Eyre, "Independent DSP Benchmarking: Methodologies and Latest Results", Proc. ICSPAT, Toronto, Sep. 1998.
10. Intel Literature, P6 architecture developer's manuals. http:/developer.intel.com.
11. TMS320C6x Optimizing C Compiler User's guide, Texas Instruments Inc. Lite. Num. SPRU187B.
12. Siglib version 2.4, Numerix Co Ltd. http://www.numerix.co.uk.
13. P.M. Embree, "C Algorithms for Real-Time DSP", NJ: Prentice Hall, 1995.
14. V. Zivojnovic, J. Martinez, C. Schläger and H. Meyr. "DSPstone: A DSP-Oriented Benchmarking Methodology". Proc. of ICSPAT'94 - Dallas, Oct. 1994.
15. EDN Embedded Microprocessor Benchmark Consortium, http://www.eembc.org.

Topic 03
Scheduling and Load Balancing

Jean-Marc Geib, Bruce Hendrickson, Pierre Manneback, and Jean Roman

Co-chairmen

General Presentation

Mapping a parallel computation onto a parallel computer system is one of the most important questions for the design of efficient parallel algorithms. Especially for irregular data structures the problem of distributing the workload evenly onto parallel computing systems becomes very complex. This workshop will discuss the state of the art in this area.

Besides the discussion of novel techniques for mapping and scheduling irregular dynamic or static computations onto a processor architecture, a number of problems are of special interest for this workshop. One is the development of dynamic load balancing algorithms that adapt themselves to the special characteristics of the underlying heterogeneous parallel or distributed architecture, easing the development in suitable environments of portable applications. Others topics in this uncomplete list of problems of relevance for this workshop concern new sequential or parallel partitioning algorithms, and efficient mapping or scheduling strategies for particular classes of applications.

The papers selected for presentation in the workshop (9 regular papers and 5 short papers) are split into 4 sessions each of which consists of works which cover a wide spectrum of different topics. We would like to thank sincerely the more than 40 referees that assisted us in the reviewing process.

P. Amestoy et al. (Eds.): Euro-Par'99, LNCS 1685, pp. 271–271, 1999.
© Springer-Verlag Berlin Heidelberg 1999

A Polynomial-Time Branching Procedure for the Multiprocessor Scheduling Problem

Ricardo C. Corrêa[1] and Afonso Ferreira[2]

[1] Departamento de Computação, Universidade Fedeal do Ceará,
Campus do Pici, Bloco 910, 60455-760, Fortaleza, CE, Brazil,
correa@lia.ufc.br
[2] CNRS. Projet SLOOP, INRIA Sophia Antipolis, 2004, Route des Lucioles,
06902 Sophia Antipolis Cedex, France

Abstract. We present and analyze a branching procedure suitable for best-first branch-and-bound algorithms for solving *multiprocessor scheduling problems*. The originality of this branching procedure resides mainly in its ability to enumerate all feasible solutions without generating duplicated subproblems. This procedure is shown to be polynomial in time and space complexities.

1   Introduction

Nowadays, most *multiprocessor systems* consist of a set $\mathcal{P} = \{p_1, p_2, \cdots, p_m\}$ of $m > 1$ identical processors with distributed memory and communicating exclusively by message passing through a completely interconnected network. A common feature of many of the algorithms designed for these systems is that they can be described in terms of a set of *modules* to be executed under a number of *precedence constraints*. A precedence constraint between two modules determines that one module must finish its execution *before* the other module starts its execution. In this paper, we consider the following *Multiprocessor Scheduling Problem (MSP)*. Given a *program*, which must be divided in *communicating modules* to be executed in a *given* multiprocessor system under a number of *precedence constraints*, *schedule* these modules to the processors of the multiprocessor system such that the program's execution time (or *makespan*) is *minimized* (for an overview of this optimization problem, see [1, 3, 8] and references therein).

We call *schedule* any solution to the MSP. In general, finding an optimal schedule to an instance of the MSP is computationally hard [3, 4]. The branch-and-bound approach has been used to find exact or approximate solutions for the MSP (for further details, including computational results, see [2, 5, 6]). Briefly speaking, a *branch-and-bound* algorithm searches for an optimal schedule by recursively constructing a search tree, whose root is the initial problem, internal nodes are *subproblems*, and the leaves are schedules. This search tree is built using a *branching procedure* that takes a node as an input and generates its children, such that each child is a subproblem obtained by the scheduling of some tasks that were not yet scheduled in its parent. Although several techniques

P. Amestoy et al. (Eds.): Euro-Par'99, LNCS 1685, pp. 272–279, 1999.
© Springer-Verlag Berlin Heidelberg 1999

can be used to reduce the number nodes of the search tree which are indeed visited during a branch-and-bound search [7], the branching procedure is called many times during a branch-and-bound algorithm. We study in this paper a new such procedure, rather than branch-and-bound algorithms as a whole. As a consequence, we are able to focus our work on the techniques required to speedup this important component of the branch-and-bound method.

The solutions reported in the literature are not satisfactory enough to deal with this central problem. Kasahara and Narita [5] have proposed such a procedure and applied it in a depth-first branch-and-bound algorithm. Unfortunately, their approach generates non-minimal schedules and the height of the search tree depends on the task costs. Thus, it is not suitable for the best-first search context intended to be covered by the branching procedure proposed in this paper.

More recently, another approach, based on state space search techniques, was proposed [2, 6]. In this case, only minimal schedules are generated. However, this branching procedure constructs a search **graph** (not necessarily a tree), which means that each subproblem can be obtained from different sequences of branchings from the initial problem (for more details, see Sections 3.1 and 3.2). For this reason, every new subproblem generated by the branching procedure should be compared with all subproblems previously generated during the search in order to eliminate duplicated versions of identical subproblems. This fact drastically affects the time and space complexities of the branching procedure.

In this paper, we propose a **polynomial time** branching procedure that generates a search **tree** (not a general graph, i.e. each subproblem is generated exactly once). In this search tree, an edge indicates the assignment of a task to a processor, the leaves are feasible schedules, and a path from the root to a leaf defines a total order on the task assignments. In addition, the branching procedure stablishes a set of total orders on the tasks verifying the precedence constraints. If a feasible schedule could be associated to more than one task assignment order (path in the search tree), then only the path which is coherent with one of the total orders stablished by the branching procedure is generated.

This strategy requires relatively small storage space, and the height of the generated search tree depends linearly upon the number of modules to be scheduled. Furthermore, the characteristics of our procedure enables it to be easily plugged into most existing branch-and-bound methods. The branch-and-bound algorithms thus obtained must outperform the original ones in most cases, provided that either the new branching procedure is faster (e.g., when compared to [2, 6]) or the height of the search tree is smaller (e.g., when compared to [5]).

The remainder of the paper is organized as follows. The mathematical definitions used in this paper are presented in Section 2. In Section 3, we describe an enumeration algorithm based on our branching procedure. The proof of correctness and the complexity analysis of our branching procedure are given in Section 4. Finally, concluding remarks are given in Section 5.

2   Definitions and Statements

Each module of the program to be scheduled is called a *task*. The program is described by a (connected) *directed acyclic graph (DAG)*, whose vertices represent the $n > 0$ tasks $T = \{t_1, \cdots, t_n\}$ to be scheduled and edges represent the *precedence relations* between pairs of tasks. An edge (t_{i_1}, t_{i_2}) in the DAG is equivalent to a precedence relation between the tasks t_{i_1} and t_{i_2}, which means that t_{i_1} must finish its execution and send a message to t_{i_2} which must be received before the execution of t_{i_2}. In this case, t_{i_1} is called the *immediate predecessor* of t_{i_2}. The task t_1 is the only one with no immediate predecessors. The execution of any task on any processor and the communication between any pair of tasks costs one time unit each. One may notice that this assumption of identical costs does not yield any loss of generality in terms of the structural issues considered in the rest of the paper.

A *schedule* is a mapping $S_n : T \rightarrow \mathcal{P} \times \mathbb{N}$, which indicates that each task t_i is allocated to be executed on processor $p(i, S_n)$ with rank $r(i, S_n)$ in this processor. The schedules considered are those whose computation of the start time of each task takes into account three conditions, namely: (i) precedence relations; (ii) each processor executes at most one task at a time; and (iii) task preemptions are not allowed. Thus, given a schedule S_n, the computation of the introduction dates $d(i, S_n)$, for all $t_i \in T$, follows the *list heuristic* whose principle is to schedule each task t_i to $p(i, S_n)$ according to its rank $r(i, S_n)$ [3]. In addition, $d(i, S_n)$ assumes the minimum possible value depending on the schedule of its immediate predecessors. We call these (partial) schedules *minimal (partial) schedules*.

Let us define some notation. A *partial schedule* where r tasks, $0 \leq r \leq n$, are scheduled is represented by S_r. An *initial task* is a task, scheduled on some processor p_j in S_r, with rank 1. A schedule S_n is said to be *attainable from* S_r if the tasks that are scheduled in S_r are scheduled in S_n on the same processor and with the same start time as in S_r. The set of *free tasks*, *i.e.*, the non-scheduled tasks of a given S_r whose all immediate predecessors have already been scheduled, is denoted by $FT(S_r)$. In Figure 1, an example of an instance of the MSP is shown, where many of the above parameters are illustrated. The MSP can be stated formally as the search of a minimal schedule S_n that minimizes the makespan $d(n, S_n) + 1$.

3   Enumerating All Minimal Schedules

In this section, we concentrate our discussion on enumerating all minimal schedules in such a way that each partial schedule is enumerated exactly once. Starting from an algorithm that enumerates all minimal schedules at least once, we reach an enumeration algorithm where each partial schedule is visited exactly once, by a refinement on the branching procedure used.

An enumeration algorithm consists of a sequence of *iterations*. Throughout these iterations, the state of the list of partial schedules $\mathcal{L}$ evolves from the initial state $\{S_0\}$ (partial schedule with no tasks scheduled) until it becomes idle.

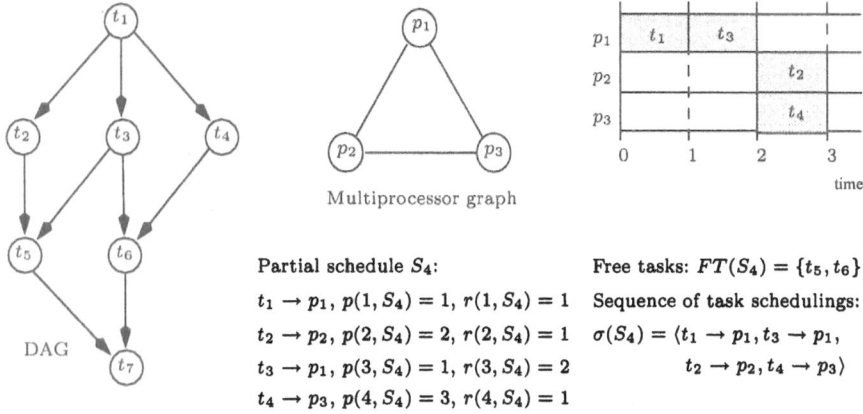

Partial schedule S_4:

$t_1 \to p_1$, $p(1, S_4) = 1$, $r(1, S_4) = 1$

$t_2 \to p_2$, $p(2, S_4) = 2$, $r(2, S_4) = 1$

$t_3 \to p_1$, $p(3, S_4) = 1$, $r(3, S_4) = 2$

$t_4 \to p_3$, $p(4, S_4) = 3$, $r(4, S_4) = 1$

Free tasks: $FT(S_4) = \{t_5, t_6\}$

Sequence of task schedulings:

$\sigma(S_4) = \langle t_1 \to p_1, t_3 \to p_1,$

$t_2 \to p_2, t_4 \to p_3 \rangle$

Fig. 1. Example of a scheduling problem and a partial schedule.

Each iteration starts selecting a partial schedule $S_{\rangle}r$ from the list $\mathcal{L}$. Whenever a partial schedule is selected, it is deleted from $\mathcal{L}$. Then, a branching procedure generates a set of new partial schedules $S_{r+1}^1, S_{r+1}^2, \cdots, S_{r+1}^l$, each of which different from the others and consisting of S_r plus exactly one task $t_i \in FT(S_r)$ scheduled on some processor. Notice that the set of schedules attainable from each new schedule represents a subset of the set of schedules attainable from S_r. Finally, the iteration ends inserting the partial schedules generated into the list $\mathcal{L}$. When a partial schedule S_r is inserted into $\mathcal{L}$, it becomes available for further selection.

3.1   All Minimal Schedules Branching Rule

The following branching rule implements an algorithm that enumerates each minimal schedule at least once, although a number of partial schedules can be enumerated more than once. One could modify the enumeration algorithm to employ an A^* strategy to eliminate the occurrence of duplicated partial schedules, as proposed in [2, 6]. Our approach, to be presented in Section 3.2, is to refine this section's branching procedure, so that it prevents duplicated partial schedules from being generated at all.

As we have already had the opportunity to discuss, each partial schedule S_r, $r \geq 1$, is generated from another partial schedule S_{r-1} by scheduling a free task $t_i \in FT(S_{r-1})$ to a processor p_j. We represent the scheduling of this task by $t_i \to p_j$. The following rule leads the enumeration algorithm to enumerate all minimal schedules at least once (see Lemma 1 in Section 4). The branching procedure adopted in [2, 6] is based on this rule.

Branching rule 1 *Given a partial schedule S_r, $n > r \geq 0$, every partial schedule S_{r+1} consisting of S_r plus $t_i \to p_j$, for all $p_j \in \mathcal{P}$ and $t_i \in FT(S_r)$, is generated.*

Denote by $\Sigma(B_1)$ the multiset of partial schedules enumerated by the enumeration algorithm with a branching procedure based on B_1. This is not necessarily a set because some partial schedules can occur several times in $\Sigma(B_1)$.

3.2   Avoiding Processor Permutations and Duplications

A first drawback of the previous enumeration algorithm is that "equivalent" partial schedules can be enumerated. This equivalence relation is related to processor permutations. A partial schedule S'_r, denoted $S'_r = \Pi(S_r)$, is a *processor permutation* of another partial schedule S_r, $n \geq r \geq 0$, if there is a permutation $\pi : \{1, 2, \cdots, m\} \to \{1, 2, \cdots, m\}$ of the processors such that $p(t_i, S'_r) = \pi(p(t_i, S_r))$ and $r(t_i, S'_r) = r(t_i, S_r)$, for all t_i scheduled in S_r. For the sake of illustration, the first partial schedule in Figure 1 could be modified by exchanging processors p_2 and p_3, which corresponds to a processor permutation where $\pi(2) = 3$ and $\pi(3) = 2$.

Branching rule 1 suffers from a second drawback. Let $\sigma(S_r)$ represent the sequence of task schedulings leading S_0 to a partial schedule S_r, i.e., $\sigma(S_r) = \langle t_{i_1} \to p_{j_1}, \cdots, t_{i_{r-1}} \to p_{j_{r-1}}, t_{i_r} \to p_{j_r} \rangle$. A partial schedule S_r can be generated from two (or more) distinguished partial schedules if S_r can be generated by two (or more) sequences of task schedulings. Recall again Figure 1. Besides the sequence of task schedulings shown in that figure, rule B_1 allows the partial schedule S_4 to be generated with $\langle t_1 \to p_1, t_2 \to p_2, t_4 \to p_3, t_3 \to p_1 \rangle$. The next branching rule eliminates these two anomalies.

Branching rule 2 *Given a partial schedule S_r, $n > r \geq 0$, every partial schedule S_{r+1} consisting of S_r plus $t_i \to p_j$, for all $p_j \in \mathcal{P}$ and $t_i \in FT(S_r)$, is generated if and only if the following conditions hold:*

    i. there is no empty processor $p_{j'}$ in S_r, $j' < j$, where a empty processor is a processor with no task scheduled to it; and

    ii. $r = 0$ or if $l > \max\{k \mid k < r$ and $((i_k, i) \in A$ or $(i_k \to j_r) \in \sigma(S_r))\}$, then $i > i_l$.

Define $\Sigma(B_2)$ as the set of partial schedules enumerated using branching rule B_2. We shall formally see in the next section that $\Sigma(B_2)$ is indeed a set and that all minimal partial schedules are generated nonetheless. Let us illustrate the application of B_2 with an example. Partial schedule S_4 in Figure 1 can be generated with B_2 following the sequence $\sigma(S_4)$ indicated in that figure. However, any sequence starting with $\langle t_1 \to p_1, t_4 \to p_3 \rangle$ is eliminated by condition i, while the sequences starting with $\langle t_1 \to p_1, t_3 \to p_1, t_2 \to p_2 \rangle$ are eliminated by condition ii.

4   Correctness of the Enumeration Algorithms

In this section, we provide a proof of correctness of the enumeration algorithms. The lemma below concerns processor permutations and branching rule 1.

Lemma 1. *For all $n \geq r \geq 0$, every minimal partial schedule S_r is in $\Sigma(B_1)$.*

Proof. This lemma is trivially demonstrated by induction on r. □

In the following lemma, branching rule 2 is used in order to assure that at most one processor permutation of each partial scheduling of $\Sigma(B_1)$ is generated.

Lemma 2. *Let $\Pi_1(S_r)$ and $\Pi_2(S_r)$ be two different processor permutations of some $S_r \in \Sigma(B_1)$. Then $\Pi_1(S_r) \in \Sigma(B_2) \Rightarrow \Pi_2(S_r) \notin \Sigma(B_2)$.*

Proof. The lemma is trivially valid if S_r contains only one non-empty processor. Otherwise, let t_{i_a}, $1 \leq a < r$, be the initial task of a processor p_j and t_{i_b}, $a < b \leq r$, be the initial task of a processor $p_{j'}$, $j < j'$, in S_r. Then $(t_{i_a}, t_{i_b}) \in A$ or $i_a < i_b$. Using this fact, the proof of the lemma is by contradiction. Suppose now that $S_r \in \Sigma(B_1)$ and a processor permutation $\Pi_1(S_r) \in \Sigma(B_2)$. Still, take a processor permutation $\Pi_2(S_r)$, $\Pi_1(S_r) \neq \Pi_2(S_r)$. Let $t_{i_1}, t_{i_2}, \cdots, t_{i_c}$, $c \leq m$, be the initial tasks of all non-empty processors $p_1, p_2, \cdots, p_c$ in S_r. Thus, these are the initial tasks of processors $p_{\pi_1(1)}, p_{\pi_1(2)}, \cdots, p_{\pi_1(c)}$, respectively, in $\Pi_1(S_r)$, and $p_{\pi_2(1)}, p_{\pi_2(2)}, \cdots, p_{\pi_2(c)}$, respectively, in $\Pi_2(S_r)$. It follows that $R(S_r) = R(\Pi_1(S_r)) = R(\Pi_2(S_r))$. Consider two indices d and e, $1 \leq d, e \leq c$, such that $\pi_1(d) < \pi_1(e)$ and $\pi_2(d) > \pi_2(e)$. Such a pair d, e exists since $\Pi_1(S_r) \neq \Pi_2(S_r)$. It steems from the fact mentioned at the begining of the proof that $((t_{i_d}, t_{i_e}) \in A$ or $i_d < i_e)$ and $((t_{i_e}, t_{i_d}) \in A$ or $i_e < i_d)$, which contradicts the assumption that both $\Pi(S_r)$ and $\Pi_2(S_r)$ satisfies branching rule 2. □

The following lemma says that branching rule 2 allows the generation of all minimal partial schedules, as an immediate corollary of the fact that at least one processor permutation of every partial schedule is indeed generated.

Lemma 3. *For each partial schedule $S_r \in \Sigma(B_1)$, there exists a sequence $\sigma(\Pi(S_r))$ such that $\Pi(S_r)$ is a processor permutation of S_r and $\Pi(S_r) \in \Sigma(B_2)$.*

Proof. We demonstrate the lemma by induction on r. Again, the lemma is trivially verified for $r = 0$. Suppose a sequence $\sigma(S_r) = \langle t_{i_1} \rightarrow p_{j_1}, \cdots, t_{i_r} \rightarrow p_{j_r} \rangle$ of task schedulings leading to S_r. If $\sigma(S_r)$ is generated with B_2, then the lemma is proved. Otherwise, let $S_{r-1} \in \Sigma(B_1)$ be a partial schedule obtained with the task schedulings $t_{i_1} \rightarrow p_{j_1}, \cdots, t_{i_{r-1}} \rightarrow p_{j_{r-1}}$ performed in some rank preserving order, that is, an order that preserves, in S_{r-1}, the rank of $t_{i_1}, \cdots, t_{i_{r-1}}$ in S_r. By the induction hypothesis, there exists a processor permutation $\Pi(S_{r-1}) \in \Sigma(B_2)$. Denote S'_r the partial schedule obtained with $\sigma(S'_r) = \langle \sigma(\Pi(S_{r-1})), t_{i_r} \rightarrow p_{\pi(j_r)} \rangle$, where $\sigma(\Pi(S_{r-1}))$ is a sequence of task schedulings generated with B_2 leading to $\Pi(S_{r-1})$. Again, if $S'_r \in \Sigma(B_2)$, then the lemma is proved. Otherwise, we will exhibit another sequence leading to S'_r that is generated.

Rename the tasks scheduled in S'_r such that $\sigma(S'_r) = \langle t_{i'_1} \rightarrow p_{j'_1}, \cdots, t_{i'_r} \rightarrow p_{j'_r} \rangle$. By definition of S'_r, $i'_r = i_r$ and $j'_r = \pi(j_r)$. Since we have assumed that $S'_r \notin \Sigma(B_2)$, it follows that $(t_{i'_{r-1}}, t_{i'_r})$ is not an arc in the DAG. However, by the induction hypothesis, there exists a processor permutation $\Pi(S'_{r-1})$ of a partial schedule S'_{r-1} generated with some rank preserving sequence of the tasks schedulings $t_{i'_1} \to p_{j'_1}, \cdots, t_{i'_{r-2}} \to p_{j'_{r-2}}, t_{i'_r} \to p_{j'_r}$ that is generated by a sequence $\sigma(\Pi(S'_{r-1}))$. This processor permutation Π is chosen in such a way that if $p_{\pi(j_{r-1})}$ is empty, then processors $p_1, p_2, \cdots, p_{\pi(j'_{r-1})}$ are not empty. We can apply the induction hypothesis here because all immediate predecessors of $t_{i'_r}$ are in $\{t_{i'_1}, \cdots, t_{i'_{r-2}}\}$ due to $S_r \in \Sigma(B_1)$ and $(t_{i'_{r-1}}, t_{i'_r}) \notin R(S'_r)$. Certainly, these arguments and $S'_r \in \Sigma(B_2)$ imply that $i'_r < i'_{r-1}$ and $\pi(j'_r) \neq \pi(j'_{r-1})$. Thus, if the sequence $\langle \sigma(\Pi(S'_{r-1})), t_{i'_{r-1}} \to p_{\pi(j'_{r-1})} \rangle$ is generated, then it is a processor permutation of S_r. Otherwise, we can follow recursively the same arguments until finding a sequence $\sigma(\Pi(S'_q))$, for some $l < q \leq r-1$, such that a processor permutation $\langle \sigma(\Pi(S'_{q-1})), t_{i_r} \to p_{\pi(j_r)}, t_{i'_q} \to p_{\pi(j'_q)}, \cdots, t_{i'_{r-1}} \to p_{\pi(j'_{r-1})} \rangle$ of S_r is generated. □

In what follows, we analyze the role played by B_2 in avoiding duplications.

Lemma 4. $\Sigma(B_2)$ *is a set.*

Proof. By contradiction, let S_r be a partial schedule with $r > 0$ tasks scheduled such that B_2 induces two disjoint sequences $\sigma(S_r)$ and $\sigma'(S_r)$ leading from S_0 to S_r. If $r \leq 2$, a contradiction can be easily found. Assume $r > 2$ and let S_q, $r > q \geq 0$, be the most recent partial schedule in the intersection of $\sigma(S_r)$ and $\sigma'(S_r)$, and $\sigma(S_q)$ be its sequence of task schedulings. Denote by $t_i \to p_j$ the task scheduling performed from S_q in $\sigma(S_r)$. Still, denote by $t_{i'} \to p_{j'}$ the task schedulings from S_q in $\sigma'(S_r)$. Clearly, $i \neq i'$ and $j \neq j'$. It follows that $(t_i \to p_j) \in \sigma'(S_r)$ and $(t_{i'} \to p_{j'}) \in \sigma(S_r)$, since both $\sigma(S_r)$ and $\sigma'(S_r)$ lead S_q to S_r. Define S_{q_1} to be the partial schedule generated by $t_{i'} \to p_{j'}$ in $\sigma(S_r)$, and S_{q_2} to be the partial schedule generated by $t_i \to p_j$ in $\sigma'(S_r)$. We observe that either $t_{i'} \to p_{j'}$ occuring after $t_i \to p_j$ or $t_i \to p_j$ occuring after $t_{i'} \to p_{j'}$ violate condition ii of branching rule 2 in $\sigma(S_r)$ or $\sigma'(S_r)$, respectively, which is a contradiction. □

Here is the theorem that follows directly from Lemmas 2, 3 and 4.

Theorem 1. *If S_n is a minimal schedule, then $S_n \in \Sigma(B_2)$, and $\Sigma(B_2)$ is minimal with respect to this property.*

The time complexity of a branching procedure based on B_2 is determined by four components. First, examining the tasks in $FT(S_r)$ takes $O(n)$ time, while scheduling a free task to each one of the processors takes $O(m)$ time. Setting the new free tasks has time complexity $O(n)$. Finally, testing condition ii of branching rule 2 takes $O(n)$ time. Besides the $O(n^2)$ storage space required for representing the DAG, the storage space required by each new partial schedule generated with BRANCH_PARTIAL_SCHEDULE is $O(n)$ and corresponds to $FT(S_{r+1})$. This proves the following theorem.

Theorem 2. *The time complexity of BRANCH_PARTIAL_SCHEDULE is $O(mn^3)$ and requires $O(mn^2)$ storage space.*

5   Concluding Remarks

In this paper, we proposed a polynomial time and storage space branching procedure for the multiprocessor scheduling problem. This procedure can be implemented in a branch-and-bound algorithm, and it is suitable for this purpose either in case of unit task costs (this case was considered in the previous sections for the sole sake of simplicity) or when the task costs are arbitrary (which means that different tasks can be assigned different costs), implying in both cases an associated search tree whose height is equal to the number of tasks in the MSP instance. A best-first branch-and-bound algorithm using the branching procedure proposed in this paper will be effective provided that tight lower and upper bounds procedures are available. In addition, this new branching procedure is more adapted to parallel implementations of branch-and-bound algorithms than other branching procedures which generate each subproblem more than a once during the search because checking for duplicated subproblems in parallel slowdown the parallel algorithm significantly.

Acknowledgments We are grateful to anonymous referees for their useful suggestions. *Ricardo Corrêa* is partially supported by the brazilian agency CNPq. *Afonso Ferreira* is partially supported by a Région Rhône Alpes International Research Fellowship and NSERC.

References

[1] T. Casavant and J. Kuhl. A taxonomy of scheduling in general-purpose distributed computing systems. *IEEE Trans. on Software Engineering*, 14(2), 1988.

[2] P.C. Chang and Y.S. Jiang. A State-Space Search Approach for Parallel Processor Scheduling Problems with Arbitrary Precedence Relations. *European Journal of Operational Research*, 77:208–223, 1994.

[3] M. Cosnard and D. Trystram. *Parallel Algorithms and Architectures*. International Thomson Computer Press, 1995.

[4] M. Garey and D. Johnson. *Computers and Intractability: A Guide to the Theory of NP-Completeness*. W. F. Freeman, 1979.

[5] H. Kasahara and S. Narita. Practical multiprocessor scheduling algorithms for efficient parallel processing. *IEEE Trans. on Computers*, C-33(11):1023–1029, 1984.

[6] Y.-K. Kwok and I. Ahmad. Otimal and near-optimal allocation of precedence-constrained tasks to parallel processors: defying the high complexity using effective search techniques. In *Proceedings Int. Conf. Parallel Processing*, 1998.

[7] L. Mitten. Branch-and-bound methods: General formulation and properties. *Operations Research*, 18:24–34, 1970. Errata in *Operations Research*, 19:550, 1971.

[8] M. Norman and P. Thanisch. Models of machines and computations for mapping in multicomputers. *ACM Computer Surveys*, 25(9):263–302, Sep 1993.

Optimal and Alternating-Direction Load Balancing Schemes

Robert Elsässer[1,*], Andreas Frommer[2], Burkhard Monien[1,*], and
Robert Preis[1,*]

[1] Department of Mathematics and Computer Science
University of Paderborn, D-33102 Paderborn, Germany
{elsa,bm,robsy}@uni-paderborn.de
[2] Department of Mathematics and Institute for Applied Computer Science
University of Wuppertal, D-42097 Wuppertal, Germany
frommer@math.uni-wuppertal.de

Abstract. We discuss iterative nearest neighbor load balancing schemes
on processor networks which are represented by a cartesian product
of graphs like e.g. tori or hypercubes. By the use of the Alternating-
Direction Loadbalancing scheme, the number of load balance iterations
decreases by a factor of 2 for this type of graphs. The resulting flow is
analyzed theoretically and it can be very high for certain cases. Therefore, we furthermore present the Mixed-Direction scheme which needs
the same number of iterations but results in a much smaller flow.

Apart from that, we present a simple optimal diffusion scheme for general
graphs which calculates a minimal balancing flow in the l_2 norm. The
scheme is based on the spectrum of the graph representing the network
and needs only $m-1$ iterations in order to balance the load with m being
the number of distinct eigenvalues.

1 Introduction

We consider the load balancing problem in a synchronous, distributed processor
network. Each node of the network has a computational load and, if it is not
equally distributed, it will have to be balanced by moving parts of the load via
communication links between two processors. We model the processor network
by an undirected, connected graph $G = (V, E)$ in which node $v_i \in V$ contains
a computational load of w_i. We want to determine a schedule in order to move
load across edges so that finally the weight on each node is approximately equal.
In each step a node is allowed to move any size of load to each of its neighbors
in G. Communication is only allowed along the edges of G.

This problem models load balancing in parallel adaptive finite element simulations where a geometric space, which is discretized using a mesh, is partitioned
into sub-regions. Besides, the computation proceeds on mesh elements in each

---

* Supported by European Union ESPRIT LTR Project 20244 (ALCOM-IT) and German Research Foundation (DFG) Project SFB-376.

P. Amestoy et al. (Eds.): Euro-Par'99, LNCS 1685, pp. 280–290, 1999.
© Springer-Verlag Berlin Heidelberg 1999

sub-region independently (see e.g. [4, 5]). Here we associate a node with a mesh region, an edge with the geometric adjacency between two regions, and load with the number of mesh elements in each region. As the computation proceeds, the mesh gets refined or coarsened depending on the characteristics of the problem and the size of the sub–regions (numbers of elements) has to be balanced. Since elements have to reside in their geometric adjacency, they can only be moved between adjacent mesh regions, i.e. via edges of the graph [5]. In this paper we consider the calculation of the load balancing flow. In a further step, the moved load elements have to be chosen and moved to their new destination.

Scalable algorithms for our load balancing problem iteratively balance the load of a node with its neighbors until the whole network is globally balanced. The class of local iterative load balancing algorithms distinguishes between *diffusion* [2] and *dimension exchange* [2, 13] iterations. Diffusion algorithms assume that a node of the graph can send and receive messages to/from all its neighbors simultaneously, whereas dimension exchange iteratively only balances with one neighbor after the other. The *quality* of a balancing algorithm can be measured in terms of numbers of iterations that are required in order to achieve a balanced state and in terms of the amount of load moved over the edges of the graph. The earliest local method is the diffusive *first order scheme* (FOS) by [2]. It lacks in performance because of its slow convergence. With the idea of over-relaxation FOS can be sped up by an order of magnitude [6] (*second order scheme*, SOS). Diffusive algorithms have gained new attention in the last couple of years [4, 5, 7, 8, 10, 12, 13] and it has been shown that *all* local iterative diffusion algorithms determine the unique l_2-minimal flow [3].

In Sect. 3, we discuss a combination of the diffusion and the dimension-exchange model which we call the *Alternating Direction Iterative* (ADI) model. It can be used for cartesian products of graphs like tori or hypercubes, which often occur as processor networks. Here, in each iteration, a processor first communicates with its neighbors in one and then in the other direction. Thus, it communicates once per iteration with each neighbor. We show that for these graphs the upper bound on the number of load balance steps will be halved. As a drawback, we can prove that the resulting flow is not minimal in the l_2-norm and can be very large if optimal parameters for the number of iterations are used. As a (partial) remedy to this problem we present the *Mixed Direction Iterative* (MDI) scheme which needs the same number of iterations but results in a much smaller flow.

[3] introduced an optimal load balancing scheme based on the spectrum of the graph. It only needs $m - 1$ iterations with m being the number of distinct eigenvalues. This scheme keeps the load-differences small from step to step, i.e. it is numerically stable. In Sect. 4, we present a much simpler optimal scheme OPT which also needs only $m-1$ iterations. It might trap into numerical instable conditions, but we present rules of how to avoid those. The calculation of the spectrum for arbitrary large graphs is very time-consuming, but it is known for many classes of graphs and can efficiently be computed for small graphs. In Sect. 5 experiments are presented to underline the theoretical observations.

2   Definitions

Let $G = (V, E)$ be a connected, undirected graph with $|V| = n$ nodes and $|E| = N$ edges. Let $w_i \in I\!R$ be the *load* of node $v_i \in V$ and $w \in I\!R^n$ be the vector of load values. $\overline{w} := \frac{1}{n}(\sum_{i=1}^{n} w_i)(1, \ldots, 1)$ denotes the vector of the average load. Define $A \in \{-1, 0, 1\}^{n \times N}$ to be the node-edge *incidence matrix* of G. Every column has exactly two non-zero entries "1" and "−1" for the two nodes incident to the corresponding edge. The signs of these non-zeros implicitly define directions for the edges of G which will later on be used to express the direction of the flow. Let $B \in \{0, 1\}^{n \times n}$ be the *adjacency matrix* of G. As G is undirected, B is symmetric. Column/row i of B contains 1's at the positions of all neighbors of v_i. The *Laplacian* $L \in \mathbb{Z}^{n \times n}$ of G is defined as $L := D - B$, with $D \in I\!N^{n \times n}$ containing the node degrees as diagonal entries and 0 elsewhere. Obviously, $L = AA^T$. Let $x \in I\!R^N$ be a flow on the edges of G. x is called a *balancing flow* on G iff $Ax = w - \overline{w}$. Among all possible flows we are interested in balancing flows with minimal l_2-norm defined as $\|x\|_2 = (\sum_{i=1}^{N} x_i^2)^{1/2}$. Let us also define the l_1-norm $\|x\|_1 = \sum_{i=1}^{N} x_i$ and the l_∞-norm $\|x\|_\infty = \max_{i=1}^{N} x_i$.

We consider the following *local iterative load balancing algorithm* which performs iterations on the nodes of G with communication between adjacent nodes only. This method performs on each node $v_i \in V$ the iteration

$$\forall e = \{v_i, v_j\} \in E : \quad y_e^{k-1} = \alpha(w_i^{k-1} - w_j^{k-1}); \quad x_e^k = x_e^{k-1} + y_e^{k-1};$$

$$\text{and } w_i^k = w_i^{k-1} - \sum_{e=\{v_i, v_j\} \in E} y_e^{k-1} \tag{1}$$

Here, y_e^k is the amount of load sent via edge e in step k with α being a parameter for the fraction of the load difference, x_e^k is the total load sent via edge e until iteration k and w_i^k is the load of node i after the k-th iteration. The iteration (1) converges to the average load $\overline{w}$ [2]. This scheme is known as the *diffusion algorithm*. Equation (1) can be written in a matrix notation as $w^k = Mw^{k-1}$ with $M = I - \alpha L \in I\!R^{n \times n}$. M contains α at position (i, j) of edge $e = (v_i, v_j)$, $1 - \sum_{e=\{v_i, v_j\} \in E} \alpha$ at diagonal entry i, and 0 elsewhere. Now let $\lambda_i(L)$, $1 \leq i \leq m$ be the distinct eigenvalues of the Laplacian L in increasing order. It is known that $\lambda_1(L) = 0$ is simple with eigenvector $(1, \ldots, 1)$ and $\lambda_m(L) \leq 2 \cdot deg_{max}$ (with deg_{max} being the maximum degree of all nodes) [1]. Then M has the eigenvalues $\mu_i = 1 - \alpha \lambda_i$ and α has to be chosen such that $\mu_m > -1$. Note that $\mu_1 = 1$ is a simple eigenvalue to the eigenvector $(1, 1, \ldots, 1)$. Such a matrix M is called *diffusion matrix*. We denote by $\gamma = \max\{|\mu_2|, |\mu_m|\} < 1$ the second largest eigenvalue of M according to absolute values and call it the *diffusion norm* of M.

Let $w^0 = \sum_{i=1}^{m} z_i$ be represented as a sum of (not necessarily normalized) eigenvectors z_i of M with $Mz_i = \mu_i z_i$, $i = 1, \ldots, m$. Then $\overline{w} = z_1$ [3].

Definition 1. *A polynomial based* load balancing scheme *is any scheme for which the work load w^k in step k can be expressed in the form $w^k = p_k(M)w^0$ where $p_k \in \overline{\Pi}_k$. Here, $\overline{\Pi}_k$ denotes the set of all polynomials p of degree $\deg(p) \leq k$ satisfying the constraint $p(1) = 1$.*

Condition $p_k(1) = 1$ implies that all row sums in matrix $p_k(M)$ are equal to 1. For a polynomial based scheme to be feasible practically, it must be possible to rewrite it as an update process where w^k is computed from w^{k-1} (and maybe some previous iterates) involving *one* multiplication with M only. This means that the polynomials p_k have to satisfy some kind of short recurrence relation. The convergence of a polynomial based scheme depends on whether (and how fast) the 'error' $e^k = w^k - \overline{w}$ tends to zero. These errors e^k satisfy $e^k = p_k(M)e^0$ which yields $\|e^k\|_2 \leq \max_{i=2}^m |p_k(\mu_i)| \cdot \|e^0\|_2$ $k = 0, 1, \ldots$, where $e^0 = \sum_{i=2}^m z_i$ [3].

As an example, take the first order-scheme (FOS) [2], where we have $p_k(t) = t^k$. These polynomials satisfy the simple short recurrence $p_k(t) = t \cdot p_{k-1}(t)$, $k = 1, 2, \ldots$, so that we get $w^k = Mw^{k-1}$, $k = 1, 2, \ldots$. Now $|p_k(\mu_i)| = |\mu_i^k| \leq \gamma^k$ for $i = 2, \ldots, m$ and therefore

$$\|e^k\|_2 \leq \gamma^k \cdot \|e^0\|_2 .$$

Here, $\gamma(M) = \max\{|1-\alpha\lambda_2(L)|, |1-\alpha\lambda_m(L)|\}$ and the minimum of γ is achieved for the optimal value $\alpha = \frac{2}{\lambda_2(L)+\lambda_m(L)}$. Then we have $\gamma = \frac{\lambda_m(L)-\lambda_2(L)}{\lambda_m(L)+\lambda_2(L)} = \frac{1-\rho}{1+\rho}$ with $\rho = \frac{\lambda_2(L)}{\lambda_m(L)}$ called the condition number of the Laplace matrix L.

The flow characterized by FOS is l_2 minimal [3]. The solution to the l_2 minimization problem "*minimize* $\|x\|_2$ *over all* x *with* $Ax = w^0 - \overline{w}$" is given by $x = A^T z$, where $Lz = w^0 - \overline{w}$ [3],[8]. Now z can be expressed through the eigenvalues λ_i and eigenvectors z_i of L as $z = \sum_{i=2}^m \frac{1}{\lambda_i} z_i$ where $w^0 = \sum_{i=1}^m z_i$ and $w^0 - \overline{w} = \sum_{i=2}^m z_i$. As it has been shown in [3], for a polynomial based load balancing scheme with $w^k = p_k(M)w^0$, the corresponding l_2-minimal flow x^k (such that $w^0 - w^k = Ax^k$) is given by $x^k = q_{k-1}(L)w^0$, the polynomial q_{k-1} being defined by $p_k(1 - \alpha t) = 1 + tq_{k-1}(t)$. The flows x^k converge to x, the l_2-minimal balancing flow. It was also shown in [3] that x^k can be very easily computed along with w^k if the iteration relies on a 3-term recurrence relation (as is the case for FOS, SOS and the OPT scheme of Sect.4).

3   Alternating Direction Iterative Schemes

Cartesian products of lines and paths have been discussed previously in [13], but no analysis of the quality of the flow have been made. We generalize the problem to the product of arbitrary graphs, show that one can reduce the number of loadbalance iterations by alternating the direction of the loadbalancing and analyze the resulting flow. We propose a new Mixed-Direction scheme that results in a better flow than the classical Alternating-Direction scheme.

Let G be the cartesian product of two graphs G' and G'' with vertex sets V' and V'', edge sets E' and E'' and the spectra $\lambda_1(G'), \ldots, \lambda_p(G')$ and $\lambda_1(G'')$, $\ldots, \lambda_q(G'')$, where we denote with $p = |V'|$ and $q = |V''|$. Now G has the vertex set $V = V' \times V''$ and the edge set $E = \{((u_i', u_j''), (v_i', v_j'')) \mid$ where $u_i' = v_i'$ and $(u_j'', v_j'') \in E''$ or $u_j'' = v_j''$ and $(u_i', v_i') \in E'\}$. Any eigenvalue $\lambda(G)$ of G is of the form $\lambda(G) = \lambda_i(G') + \lambda_j(G'')$ with $\lambda_i(G')$ eigenvalue of G' $i \in \{1, \ldots, p\}$ and $\lambda_j(G'')$ eigenvalue of G'' $j \in \{1, \ldots, q\}$. (The same relation also holds for the eigenvalues of the adjacency matrices.) We denote with I_k the identity matrix of order k. Now the adjacency matrix B_G of G is $B_G = I_q \otimes B_{G'} + B_{G''} \otimes I_p$ where $\otimes$ denotes the tensor product. Obviously, the matrices $I_q \otimes B_{G'}$ and $B_{G''} \otimes I_p$ have a common system of eigenvectors.

Now we apply the *alternating direction iterative* (ADI) load balancing strategy on G, which is similar to the alternating direction method for solving linear systems [11]. In an iteration, firstly balance *among one component* of the cartesian product (for example G') using FOS, and secondly *among the other component* (in our case G'') using FOS. We denote with $M' = 1 - \alpha' L'$ and $M'' = 1 - \alpha'' L''$ the diffusion matrices of G' and G''. The diffusion scheme ADI-FOS then has the form

$$\forall e = \{(u_i', u_j''), (u_i', v_l'')\} \in E : \; y_e^{k-1} = \alpha''(w_{(u_i', u_j'')}^{k-1} - w_{(u_i', v_l'')}^{k-1}); \text{ and} \quad (2)$$

$$\forall e = \{(u_i', u_j''), (v_l', u_j'')\} \in E : \; y_e^{k} = \alpha'(w_{(u_i', u_j'')}^{k} - w_{(v_l', u_j'')}^{k}); \quad (3)$$

Then we have $w^{i+1} = (M'' \otimes I_p)(I_q \otimes M')w^i = (M'' \otimes M')w^i$.

Theorem 1. *Let G be the cartesian product of two connected graphs G' and G''. Denote L, L', L'' the corresponding Laplacians and $M = I - \alpha L, M' = I - \alpha' L', M'' = I - \alpha'' L''$ the diffusion matrices for the FOS with optimal parameters $\alpha, \alpha', \alpha''$. Moreover, let γ_M and $\gamma_{M'' \otimes M'}$ be the diffusion norms. Then $\gamma_{M'' \otimes M'} < \gamma_M$. If G' and G'' are isomorphic, then $\gamma_{M'' M'} < \gamma_M^2$, i.e. FOS combined with the ADI scheme only needs half of the number of iterations in order to guarantee the same upper bound on the error as FOS.*

Proof. Clearly, $\gamma_{M' \otimes M''} = \max\{\gamma_{M'}, \gamma_{M''}\}$ whereas $\gamma_M = 1 - \frac{2\lambda_2(L)}{\lambda_{pq}(L) + \lambda_2(L)}$. Using $\lambda_2(L) = \min\{\lambda_2(L'), \lambda_2(L'')\}$ and $\lambda_{pq}(L) = \lambda_p(L') + \lambda_q(L'')$, one then concludes $\gamma_{M'} \le \gamma_M$ and $\gamma_{M''} \le \gamma_M$. If G' and G'' are isomorphic, then $\frac{\lambda_p(L') - \lambda_2(L')}{\lambda_p(L') + \lambda_2(L')} < \left(\frac{\lambda_p(L') + \lambda_p(L') - \lambda_2(L')}{\lambda_p(L') + \lambda_p(L') + \lambda_2(L')} \right)^2$. □

The flow calculated by this scheme is not minimal in the l_2- norm and we study the flow in the following. Denote by $z_{i,j}$ the common eigenvectors of the matrices $I_q \otimes B_{G'}$ and $B_{G''} \otimes I_p$ and by A' and A'' the edge-incidence matrices of the graphs G' and G''. We assume that the graphs G' and G'' are isomorphic with common eigenvalues μ_i. Now let $A_1 = I \otimes A'^T$ and $A_2 = A''^T \otimes I$. Let w_1^k and w_2^k be the load after k iterations after the balancing in one dimension and after balancing in both dimensions. Let F_1^{k+1} and F_2^{k+1} be the flows in iteration $k+1$ among the first and the second dimension with $F_1^{k+1} = \alpha A_1^T w_2^k$ and $F_2^{k+1} = \alpha A_2^T w_1^{k+1}$. For the total flow in these directions we get

$$F_2 = \sum_{k=0}^{\infty} F_2^{k+1} = \alpha \sum_{k=0}^{\infty} A_2^T \left(\sum_{i,j=1}^{|V'|} \mu_i(\mu_i\mu_j)^k z_{i,j} \right)$$

$$= \alpha \sum_{i,j=1}^{|V'|} \mu_i \sum_{k=0}^{\infty} (\mu_i\mu_j)^k A_2^T z_{i,j} = \sum_{i,j=1}^{|V'|} \frac{1 - \alpha\lambda_i}{\lambda_i + \lambda_j - \alpha\lambda_i\lambda_j} A_2^T z_{i,j}$$

$$F_1 = \sum_{k=0}^{\infty} F_1^{k+1} = \alpha \sum_{k=0}^{\infty} A_1^T \left(\sum_{i,j=1}^{|V'|} (\mu_i \mu_j)^k z_{i,j} \right)$$

$$= \alpha \sum_{i,j=1}^{|V'|} \sum_{k=0}^{\infty} (\mu_i \mu_j)^k A_1^T z_{i,j} = \sum_{i,j=1}^{|V'|} \frac{1}{\lambda_i + \lambda_j - \alpha \lambda_i \lambda_j} A_1^T z_{i,j}.$$

We observe, that the quality of the flow depends on the value of α. The result of the classical diffusion method is a l_2-minimal flow. In this case we get

$$F = \sum_{k=0}^{\infty} \alpha A^T w^k = \alpha \sum_{l=1}^{|V|} \sum_{k=0}^{\infty} \mu_l^k (M) A^T z_l' = \sum_{i,j=1}^{|V'|} \frac{1}{\lambda_i + \lambda_j} A^T z_{i,j},$$

with z_l' being the eigenvectors of the diffusion matrix M to the eigenvalues $\mu_l(M)$ which are identical to the eigenvectors $z_{i,j}$ of the Laplacian L with the eigenvalues $\lambda_i + \lambda_j$. We observe that the l_2-minimal flow does not depend on α whereby the flow of the ADI-method will only converge to the l_2-minimal flow if α converges to zero. This implies a larger number of iterations.

To give an example, the $n \times n$ torus with an initial load of $w^0 = K z_{1,1} + K z_{n,n}$, for which the optimal value for α converges to $\frac{1}{2}$ with an increasing n. Therefore, the flows F_1 and F_2 for each edge are also increasing with increasing n. It turns out that with this scheme and this initial load, the load is only slightly balanced in each iteration, whereas a large amount of load is added on all circles of length 4. The result is a final flow with many heavy-weighted circles.

Therefore, we construct a new *mixed direction iterative* (MDI) scheme. In each iteration with an even number we first balance *among component 1* and then *among component 2*, whereas in each iteration with an odd number e first balance *among component 2* and then *among component 1*, Thus, the order of the components for each sub-iteration is changed for each iteration. With similar arguments as before, the diffusion norm is the same as for ADI, but we get the flows

$$F_1 = \sum_{i,j=1}^{|V'|} \frac{1 + \mu_i \mu_j^2}{1 - \mu_i^2 \mu_j^2} A_1^T z_{i,j} \quad \text{and} \quad F_2 = \sum_{i,j=1}^{|V'|} \frac{\mu_i + \mu_i \mu_j}{1 - \mu_i^2 \mu_j^2} A_2^T z_{i,j}.$$

Now, for the same load w^0 as before ($w^0 = F z_{1,1} + F z_{n,n}$), we can observe that, if n increases, the flows F_1 and F_2 will be bounded. Although the flow is not necessarily bounded for other initial load distributions, MDI generally calculates a smaller flow than ADI. This fact can also be seen in the experiments of Sect. 5.

4   Optimal Schemes

An optimal scheme OPS has been introduced in [3]. We present another optimal scheme OPT with the same properties but with a much simpler construction. As it was shown in Sect. 2, the error e^k of any polynomial based scheme satisfies $e^k = p_k(M)(\sum_{i=2}^{m} z_i) = \sum_{i=2}^{m} p_k(M) z_i = \sum_{i=2}^{m} p_k(\mu_i) z_i$. Now we apply a variant of the local iterative algorithm (1) where the parameter α varies from step to step. More precisely, given an arbitrary numbering λ_k, $1 \leq k \leq m-1$, for the non-zero eigenvalues of L, we take $\alpha = \alpha_k = \frac{1}{\lambda_k}$ to get

$$\forall e = \{v_i, v_j\} \in E : \; y_e^{k-1} = \frac{1}{\lambda_k}(w_i^{k-1} - w_j^{k-1}); \quad x_e^k = x_e^{k-1} + y_e^{k-1};$$

$$\text{and } w_i^k = w_i^{k-1} - \sum_{e=\{v_i, v_j\} \in E} y_e^{k-1} \tag{4}$$

In each iteration k, a node i adds a flow of $\frac{1}{\lambda_k}(w_i^{k-1} - w_j^{k-1})$ to the flow over edge $\{i, j\}$, choosing a different eigenvalue for each iteration. We obtain

$$w^k = \left(I - \frac{1}{\lambda_{k+1}}L\right)w^{k-1} \qquad \text{and} \qquad e^{m-1} = \prod_{i=2}^{m}\left(I - \frac{1}{\lambda_i}L\right)\sum_{j=2}^{m}z_j = 0,$$

for the weight w^k after k iterations and the final error e^{m-1}. It shows that the load is perfectly balanced after $m-1$ iterations and the flow is l_2-minimal as stated in Sect. 2. For each iteration a different eigenvalue of L is used, eliminating one eigenvector at a time. The order of the eigenvalues may be arbitrary, however, it may influence the numerical stability of the calculation. Even with a favorable order, the new scheme will generally be more sensitive to rounding errors than the OPS scheme from [3]. This issue will be discussed in Sect. 5. No further parameters besides the eigenvalues have to be used for this scheme and it is a much simpler optimal scheme than the one in [3].

Consider for example the k-dimensional hypercube with an initial load of $w^0 = (K, 0, \ldots, 0)$. The dimension exchange strategy is well-known for load balancing on the hypercube, where in each iteration only the balancing in one dimension is allowed and a node equalizes its load with the load of its neighbor. Thus, error e^k decreases to zero after k steps and the l_2-norm of the resulting flow is $F_{de} = K\sqrt{\frac{2^k - 1}{2^{k+1}}}$. The k-dim. hypercube has $k+1$ distinct eigenvalues $0, 2, 4, \ldots, 2k$ and in the algorithm of the OPT scheme the new flow y_e^t over edge $e = \{v_i, v_j\}$ in iteration t is simply $y_e^t = \frac{1}{2t}(w_i^t - w_j^t)$. OPT calculates the l_2-minimal flow $F_{l_2} = \sqrt{\sum_{i=0}^{d-1}\left(\frac{K - \frac{K}{2^k}(\sum_{j=0}^{i}\binom{d}{j})}{\binom{d}{i}(d-i)}\right)^2}$, much smaller than F_{de}.

The main disadvantage of the OPT scheme is the fact that it requires all eigenvalues of the graph. Although their calculation can be very time-consuming for large graphs, they can easily be computed for small processor networks and they are known for many classes of graphs like e.g. paths, cycles, stars, hypercubes, grids or tori, which often occur as processor networks.

In Sect. 3 we have introduced ADI-FOS, now ADI-OPT is introduced.

Theorem 2. *Let G be a graph and $G \times G$ the cartesian product of it. The applying of the OPT scheme (4) for each direction of ADI gives a loadbalancing scheme ADI-OPT which only needs $m-1$ iterations for the loadbalancing on $G \times G$, where m is the number of distinct eigenvalues of the Laplacian of G.*

Proof. Applying OPT for $G \times G$ with $G = (V, E)$ we get $w^i = (I - \frac{1}{\lambda_i}L')w^{i-1}$, $1 \leq i \leq \frac{m(m+1)}{2} - 1$ and λ_i is a nonzero eigenvalue of $L' = I \otimes L + L \otimes I$. By combining OPT with ADI we obtain $w^i = (I - \frac{1}{\lambda_{i+1}}I \otimes L)(I - \frac{1}{\lambda_{i+1}}L \otimes I)w^{i-1}$, where λ_{i+1} is a nonzero eigenvalue of $I \otimes L$ and $L \otimes I$. Then

$$e^{m-1} = \prod_{i=2}^{m}(I - \frac{1}{\lambda_i}I \otimes L)(I - \frac{1}{\lambda_i}L \otimes I)e^0 = \prod_{i=2}^{m}(I - \frac{1}{\lambda_i}I \otimes L)(I - \frac{1}{\lambda_i}L \otimes I)\sum_{k,j \neq 1} z_{k,j} = 0,$$

where $z_{i,j}$ are the corresponding eigenvectors of $L' = I \otimes L + L \otimes I$.   □

The Laplacian matrix of $G \times G$ can have up to $\frac{m(m+1)}{2}$ distinct eigenvalues, whereas the new ADI-OPT scheme uses only $m - 1$ distinct eigenvalues of the Laplacian of G to eliminate all eigenvectors and to reduce the error to zero.

5 Experiments

Firstly, we perform experiments to find out in which order the $m - 1$ distinct non-zero eigenvalues $\lambda_2 < \lambda_3 < ... < \lambda_m$ in the OPT scheme are to be used in the iterations. Although the load after $m - 1$ iterations is, in theory, independent of the afore chosen order, one may trap in numerical instable conditions with some orders. The load balancing is performed on a path of 32 vertices ($m = 32$), for which we observed the most difficult numerical problems, and the results are presented in Fig. 1. It is stopped after $m - 1 = 31$ iterations. In the initial load distribution RAN, $100 * |V|$ load elements are randomly distributed among the $|V|$ nodes (different random distributions exhibited the same behavior). With an increasing order of eigenvalues ($\lambda_2, \lambda_3, \lambda_4, ...$), the error be-

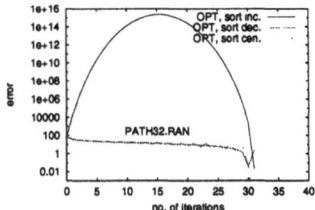

Fig. 1. Optimal OPT scheme on a path of 32 nodes with sorting of the eigenvalues in increasing, decreasing and center-started order. The initial load distribution is random.

comes very high after a few iterations and the final error is about 0.01, due to numerical problems with the high numbers. With a decreasing order (λ_m, λ_{m-1}, λ_{m-2}, ...), the error decreases monotonously until the last iteration, where it increases to a final error of about 1. The best behavior can be observed with the center-started order $(\lambda_{\frac{m}{2}}, \lambda_{\frac{m}{2}-1}, \lambda_{\frac{m}{2}+1}, \lambda_{\frac{m}{2}-2}, \lambda_{\frac{m}{2}+2}, \ldots$ for even m and $\lambda_{\frac{m-1}{2}}, \lambda_{\frac{m-1}{2}+1}, \lambda_{\frac{m-1}{2}-1}, \lambda_{\frac{m-1}{2}+2}, \lambda_{\frac{m-1}{2}-2}, \ldots$ for odd m). Although the error alternately increases/decreases due to the changing order of high and low eigenvalues, the final error is almost zero and we use this very robust order for the following experiments. An even better order can be the use of Leja Points [9].

We compare the new optimal scheme OPT with the First-Order scheme (FOS) and the optimal scheme OPS in [3] on a hypercube of dimension 6 ($m = 7$) and an 8×8 torus ($m = 13$) in Fig. 2. Apart from the RAN initial distribution we use PEAK, in which one node has a load of $100 * |V|$ and the others 0. The balancing stops after t iterations when the error $\|w^t - \overline{w}\|_2$ is less than 10^{-6}. Both optimal schemes behave similarly to FOS for the first iterations, but then they suddenly drop down to 0 after $m - 1$ iterations. OPS and OPT show the same behavior, but our new optimal scheme OPT has a much simpler construction.

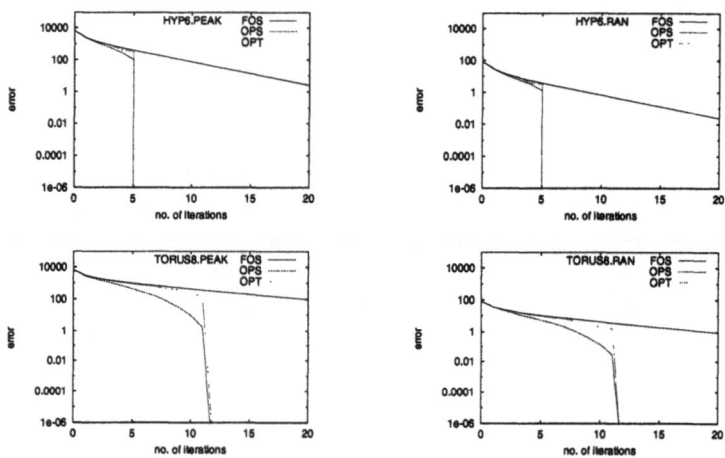

Fig. 2. Iterative Diffusion Loadbalancing on a hypercube of dimension 6 with initial PEAK (upper left) and RAN (upper right) load distribution and on an 8×8 Torus with initial PEAK (lower left) and RAN (lower right) load distribution.

The 16×16 torus is the cartesian product of a cycle with 16 vertices. Tab. 1 shows how the flow depends on α for the ADI-FOS scheme with initial load PEAK. It confirms the theoretical analyses that with decreasing values of α the flow tends to the l_2-minimal flow, but also the number of iterations increases.

Tab. 2 compares the FOS and OPT schemes with the ADI and MDI modifications. Both the FOS and OPT schemes compute the same l_2-minimal flow (Sect. 2), whereas the flow in the ADI and MDI case may differ from each other. As proven before, neither of their flows is minimal in the l_2 norm and they are also worse with respect to the l_1 and l_∞ norms of our test cases. Furthermore, the flow of ADI-OPT is much smaller than of ADI-FOS in all norms and it is

Table 1. Flow for ADI-FOS scheme with different values for α on a 16×16 torus.

	$\alpha = 0.49$	$\alpha_{opt} = 0.4817$	$\alpha = 0.4$	$\alpha = 0.2$	$\alpha = 0.1$	$\alpha = 0.01$	l_2-min. flow
l_1	627200	355082.51	207242.41	204800	204800	204800	204800
l_2	52500.69	39311.89	21361.38	18188.09	17967.10	17919	17918.62
l_∞	19066.10	16743.38	10828.53	7637.55	6908.24	6422.10	6375
iter.	528	291	349	708	1427	14366	

only a small fraction larger than the l_2-minimal flow. The flow of MDI improves over ADI in all test cases. The number of iterations for ADI and MDI are the same and are always smaller than the examples without ADI or MDI. As proven before, the upper bound on the number of iterations for FOS is twice as high as for ADI-FOS. In the experiments ADI-FOS halves the number of iterations for the RAN case and almost halves them for the PEAK case. For OPT, the number of iterations is $m - 1$ with $m = 41$ for the 16×16 torus and $m = 9$ for the cycle of 16 vertices (the product being the 16×16 torus). Thus, the number of iterations will reduce from 40 to 8 if the ADI or MDI schemes are used.

Table 2. Flow of ADI/MDI for PEAK(left) and RAN(right) loads on a 16×16 torus.

			ADI-		MDI-				ADI-		MDI-	
	FOS	OPT	FOS	OPT	FOS	OPT	FOS	OPT	FOS	OPT	FOS	OPT
l_1	204800	204800	355083	204800	205663	204800	2581	2581	4123	2727	3048	2716
l_2	17919	17919	39312	20235	23700	19993	142	142	228	152	171	152
l_∞	6375	6375	16743	9927	12185	9751	19.4	19.4	29.9	19.7	22.3	19.7
iter.	578	40	291	8	291	8	458	40	229	8	229	8

Acknowledgment. We thank Holger Arndt from Wuppertal for his suggestions improving an earlier version of this manuscript.

References

[1] D.M. Cvetkovic, M. Doob, and H. Sachs. *Spectra of Graphs.* Johann Ambrosius Barth, 3rd edition, 1995.

[2] G. Cybenko. Load balancing for distributed memory multiprocessors. *Journal of Parallel and Distributed Computing*, 7:279–301, 1989.

[3] R. Diekmann, A. Frommer, and B. Monien. Efficient schemes for nearest neighbor load balancing. In G. Bilardi et al., editor, *European Symposium an Algorithms*, LNCS 1461, pages 429–440. Springer, 1998.

[4] R. Diekmann, D. Meyer, and B. Monien. Parallel decomposition of unstructured FEM-meshes. *Concurrency: Practice and Experience*, 10(1):53–72, 1998.

[5] R. Diekmann, F. Schlimbach, and C. Walshaw. Quality balancing for parallel adaptive FEM. In *IRREGULAR'98*, LNCS 1457, pages 170–181. Springer, 1998.

[6] B. Ghosh, S. Muthukrishnan, and M.H. Schultz. First and second order diffusive methods for rapid, coarse, distributed load balancing. In *SPAA*, pages 72–81, 1996.

[7] Y.F. Hu and R.J. Blake. An improved diffusion algorithm for dynamic load balancing. *Parallel Computing*, 25:417–444, 1999.

[8] Y.F. Hu, R.J. Blake, and D.R. Emerson. An optimal migration algorithm for dynamic load balancing. *Concurrency: Prac. and Exp.*, 10(6):467–483, 1998.

[9] L. Reichel. The application of leja points to richardson iteration and polynomial preconditioning. *Linear Algebra and its Applications*, 154-156:389–414, 1991.

[10] K. Schloegel, G. Karypis, and V. Kumar. Parallel multilevel diffusion schemes for repartitioning of adaptive meshes. In *Proc. EuroPar'97*, LNCS. Springer, 1997.

[11] R.S. Varga. *Matrix Iterative Analysis*. Prentice-Hall, 1962.

[12] C. Walshaw, M. Cross, and M. Everett. Dynamic load-balancing for parallel adaptive unstructured meshes. In *SIAM Conf. on Parallel Proc. for Sci. Comp.*, 1997.

[13] C. Xu and F.C.M. Lau. *Load Balancing in Parallel Computers*. Kluwer, 1997.

Process Mapping Given by Processor and Network Dynamic Load Prediction

Jean-Marie Garcia, David Gauchard, Thierry Monteil, and Olivier Brun

LAAS-CNRS, 7 avenue du Colonel Roche 31077 Toulouse, France
tél : 05 61 33 69 13, fax : 05 61 33 69 69,
jmg@laas.fr

keywords: task allocation, observation, prediction, Round-Robin, network of workstations

1 Introduction

This article describes a process mapping algorithm for a virtual parallel host. The algorithm is a part of a mapping and virtual machine resource watcher tool: the Network-Analyser. This system collects information about the states of the hosts (CPU, memory, network traffic, etc), saves some of this information, and gives the ability to use different mapping algorithms. All of this is accessible by a graphical interface and an API [1]. The algorithm uses both the current and the average load of the hosts and the network, so that it is able to estimate the execution time of the tasks. The model of the processor and the network operation is based on a "round robin" model. Next, a differential equation describes the transient behaviour of the queues. The running time of regular parallel applications can be estimated by the integration of that equation, taking into account the stochastic load of the hosts and the networks. This algorithm is compared to other algorithms, by making event simulations.

2 Load Prediction Mapping Algorithm

2.1 Model of a Loaded Processor

The model of a loaded processor can be made using a "round robin" scheme [2]. Let's assume that Q is the time quantum, λ the poisson arrival rate, σ the leaving probability, and $\mu = (1 - \sigma)/Q$. This system can be described using Markov chain theory. Even though this model is far simpler than systems like UNIX, it gives a good description of the behaviour of the hosts. Assuming that $P_i(t)$ is the probability to have i processes in the system at time t, and $X(t)$, $\dot{X}(t)$ are the expectancy and its variation at time t, we get the equation (1).

$$\dot{X}(t) = \lambda - \mu(1 - P_0(t)) \tag{1}$$

If $t \to \infty$ in equation (2.1), it can be seen that we get the stationary case. Then we approximate $(1 - P_0(t))$ using the expression $\frac{X(t)}{1+X(t)}$. Thus we obtain an autonomous equation (2.2) in which $X_a(t)$ is the approximation of $X(t)$.

P. Amestoy et al. (Eds.): Euro-Par'99, LNCS 1685, pp. 291–294, 1999.
© Springer-Verlag Berlin Heidelberg 1999

$$1 - P_0(t \to \infty) = \rho = \frac{X(t \to \infty)}{1 + X(t \to \infty)}, \frac{\partial X_a(t)}{\partial t} = \lambda - \mu \frac{X_a(t)}{1 + X_a(t)} \quad (2)$$

For a real host, coefficients λ and μ are time-dependant. We observed our laboratory's hosts for several months. So that we have been able to distinguish two different types of behaviour according to the hosts: we use the current value to predict the near future behaviour:

- special: the current load of the host is very different from its average load for this day of the week, or the host has a constant load (perhaps no load).
- cyclic: the current load is close to the average load for this particular week day and time: we use an estimator that will smoothly forget the current value and which will approach the average values of the last few weeks.

2.2   Execution on Processor Time Prediction

We wish to predict the deterministic process execution time in the stochastic model. Let's say that we have a process to run at date t_0 that needs t_c processor units, and there are $X(t_0)$ processes in the system. We would like to find out the exit time expectancy $E(t_f)$. UNIX has a time shared policy, and for a long enough process, the UNIX mechanisms (scheduler, priority, ...) have a light influence. So that we can write the relation (3.1). It can be more general for N processes that verify $t_c^1 \leq t_c^2 \leq \ldots \leq t_c^N$ (equation (3.2)[1]) ($t_f^0 = t_0$).

$$t_c = \int_{t_0}^{E(t_f)} \frac{1}{1 + X(t)} dt, t_c^j = \sum_{i=0}^{j} \int_{E(t_f^{i-1})}^{E(t_f^i)} \frac{1}{(1 + X(t))(N - i + 1)} dt \quad (3)$$

2.3   Network Modelisation

Network packets may have a variable size, and are serialized into a wire. This allows us to model the network as in the above scheme. In this case, the time given by the server (network) to the client (emitted message) to process a packet is the time needed to transmit the packet over the wire. We noticed with the Network-Analyser that the network behaviour is nearly the same as the cpu behaviour. Thus we are able to use the same prediction method for the network.

2.4   Task Mapping over the Network

We use an application model in which there is a rendez-vous between all the tasks at each calculus step (figure 1). Let's assume that N is the number of tasks, K the number of steps, M the number of processors, $t_i(k)$ the date when the step k ends, $t_c^n(k)$ the time used by the calculus step of the task n during the step k, $t_r^n(k)$ the time needed by the communication step of the task n during the step k, and $t_s^n(k)$ the time needed to synchronize between the task n and with the others. Mapping the task in the optimal way can be done if we know

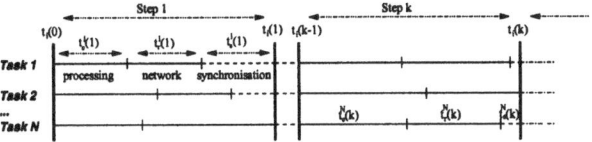

Fig. 1. Application model

the optimal mapping function $\hat{P}$, where $P : N \longrightarrow M, n \longmapsto m$ is a function which associates a task n with a host m. Let φ be the ensemble of the mapping function P. At step k, we have:

$$t_c^n(k) + t_r^n(k) + t_s^n(k) = t_i(k+1) - t_i(k) \tag{4}$$

For a given task n of the application, we look for $\hat{P}$ such that the total time of the application is a minimum:

$$\min_{P \in \varphi} \sum_{k=1}^{K} E(t_c^n(k) + t_r^n(k) + t_s^n(k)) \tag{5}$$

Using the results of the equation (3), we can evaluate $E(t_c^n(k))$ and $E(t_r^n(k))$. To find $t_s^n(k)$, one can use the following two relations ((6.1) and (6.2)).

$$t_i(k) = \max_{n \in 1...N} t_c^n(k) + t_r^n(k), t_s^n(k) = t_i(k) - (t_c^n(k) + t_r^n(k)) \tag{6}$$

For speed reasons, we use an heuristic that comes from the LPT algorithm [3] to determine an approximate value of $\hat{P}$: Tasks are mapped one-by-one. Ghost communications with still non mapped tasks are assumed to take into account the network even from the very beginning of the algorithm. These are mapped on the fastest links connected to the processor which has the current task.

3   Simulations

3.1   System Description

The network on which we did our simulations consists of two clusters of four processors each. The hosts of the first cluster are interconnected by a 10Mbits/s link, while the others connected to both the first link, and a 1Gbits/s link. Each application is a group of six tasks each having five calculus steps, and communications between them are all-to-all. The type of the algorithms we tested are random mapping, cyclic mapping, minimum run queue mapping, processor load prediction, and network and processor load prediction. We use granularity 20 for 5 times faster processors on the fast network, and granularity 2 for homogoneous processors.

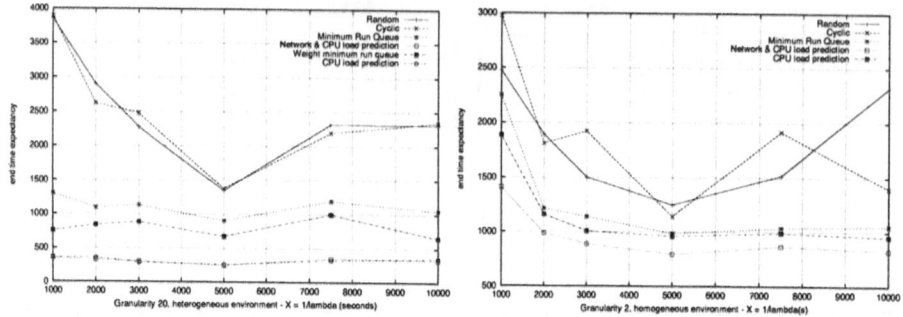

Fig. 2. Simulation results

Mapping policies become critical when the whole system is not well balanced, which is frequent. Figure 2 (granularity 20) shows that the applications with large granularity are badly mapped with the blind algorithms. Though, policies taking into account the current state of the system are much more efficient (closed loop policy). The proposed algorithm (and its simpler version not taking into account the network) has a better behaviour. Generally, one can say that the minimum run queue algorithm is a correct heuristic and very easy to implement. The network and processor load prediction algorithm is more complex to implement but offers a better guarantee for efficient use of the resources.

4   Conclusion

Sytems like the Network-Analyser are essential for an efficient parallelisation. The mapping algorithms presented in this article show their superiority compared to the classical algorithm we usually use - thanks to their precision over the execution time. The main point is the use of time in the decision model, assuming the distributed system as a stochastic system; the proposed algorithm takes into account heterogeneous communication networks.

References

[1] T. Monteil : Etude de Nouvelles Approches pour les Communications, l'Observation et le Placement de Tâches dans l'Environnement de Programmation Parallèle LANDA. Thèse, LAAS-CNRS, France, novembre 1996.
[2] E.G. Coffman, P.J Denning : Operationg Systems Theory. Prentice-Hall series in automatic computation, 1973.
[3] R.L. Graham : Bounds on Multiprocessing Timing Anomalies. SIAM J. Appl. Math.. Vol 17, No. 2, mars 1969.

Ordering Unsymmetric Matrices into Bordered Block Diagonal Form for Parallel Processing

Y.F. Hu, K.C.F. Maguire, and R.J. Blake

CLRC Daresbury Laboratory, Daresbury, Warrington WA4 4AD, United Kingdom

Abstract. Large, sparse, unsymmetric systems of linear equations appear frequently in areas such as chemical engineering. One way of speeding up the solution of these linear systems is to solve them in parallel by reordering the unsymmetric matrix into a bordered block-diagonal (BBD) form. A multilevel ordering algorithm is presented in this paper. Numerical results of the algorithm are given to demonstrate that this algorithm gives better orderings than existing algorithms,

1 Introduction

Large, sparse, unsymmetric systems of linear equations appear frequently in areas such as chemical engineering.

One way of speeding up the solution of these linear systems is to solve them in parallel. For example, frontal and multi-frontal linear solvers have been used to better exploit fine-grained parallelism in vector and shared memory parallel architectures (e.g., [1],[19]). However frontal solvers are more suitable for shared memory architectures. In order to use state of the art high performance computers, which are mostly of a distributed memory architecture, algorithms that exploit coarse-grain parallelism are required. For symmetric positive definite systems, such parallel algorithms have been proposed [8].

It is possible to achieve coarse-grain parallelism for the solution of unsymmetric systems, by reordering the unsymmetric matrix into a bordered block-diagonal form, and then factorizing the diagonal blocks independently [17, 6, 16, 15]. The quality of such a reordering affects the parallel efficiency directly [16, 15]. This quality is measured by the size of the border (the net-cut) and the load balance of the diagonal blocks. Since in areas such as chemical process simulation, systems of the same structure are often solved repeatedly, effort in seeking a good quality reordering can have long term pay-back [16, 15].

This paper introduces an efficient, high quality algorithm for ordering unsymmetric matrices into BBD form. There have been much work in this area. The MNC algorithm [6] reorders with row and column exchanges, while maintaining a zero-free diagonal. It is essentially an application of the Kernighan-Lin algorithm (KL) [14] to a bipartite graph representation of the unsymmetric matrix. A simple algorithm designed for structurely symmetric matrices [5] has also been applied to the symmetrilized form of unsymmetric matrices, but the results were not satisfactory [16, 15].

P. Amestoy et al. (Eds.): Euro-Par'99, LNCS 1685, pp. 295–302, 1999.

Camarda and Stadtherr [3] simplified the MNC algorithm by abandoning the need to maintain a zero-free diagonal. The resulting algorithm, GPA-SUM, which reorders through row exchanges only, was demonstrated to give matrices in BBD form with smaller border size, more diagonal blocks and better load balance.

Both the MNC and GPA-SUM algorithms are based on the KL algorithm, which is known to be a greedy local optimization algorithm. The quality of ordering depends strongly on the quality of the initial ordering [11]. To overcome this limit of the KL algorithm, in the area of undirected graph partitioning, multilevel techniques, which provides a "global view" of the problem, have been combined with the algorithm to form efficient and high quality partitioning algorithms [2, 9, 13, 18].

These graph partitioning algorithms may be used for ordering symmetric matrices into BBD form, however it is noted that graph partitioning algorithms seek to minimize the edge-cut of the graphs, rather than the net-cut of the matrix, a more important measure that is directly linked with the border size. These algorithms can not however be used directly to order an unsymmetric matrix A into BBD form, unless the matrix is make symmetric by considering $A + A^T$. However ordering with this method has been found to give large border size [15, 16, 6]. Therefore algorithms specifically designed for the ordering of unsymmetric matrices are necessary. This paper reports our work in the design and implementation of a multilevel ordering algorithm for unsymmetric matrices. The work is ongoing and will be reported further in [12].

There has been no reported application of the multilevel approach, so far as the authors are aware, to the problem of minimizing the net-cut in the ordering of unsymmetric matrices into BBD form. Two related works are [4] and [10], which were brought to the authors' attention after the completion of this work, and are motivated by the problem of minimizing the communication time in parallel sparse matrix-vector multiplications. Either a hyper-graph [4] or a bipartite graph [10] representation was used as the model for the problem, and in both cases edge-cut was minimized. This differs from our motivation of reordering unsymmetric matrices into BBD form. The techniques used in this work, based on working directly with the matrix, the row connectivity graph, heavy edge collapsing and the Galerkin operator, are also distinct.

In section 2 the problem of ordering unsymmetric matrices into BBD form is defined. Section 3 presents the implementation of the KL algorithm for this problem, and Section 4 the multilevel implementation. In Section 5 our algorithm is compared with existing algorithms. Finally Section 6 discusses future work.

2   The Ordering Problem

The aim of the ordering algorithm is to order a general unsymmetric matrix A into the following (single) bordered block-diagonal form

$$
\begin{pmatrix}
A_{11} & & & & S_1 \\
& A_{22} & & & S_2 \\
& & \ddots & & \vdots \\
& & & A_{pp} & S_p
\end{pmatrix}.
\tag{1}
$$

where A_{ii} are $m_i \times n_i$ matrices, in general rectangular with $m_i \geq n_i$. Such a matrix can be then factorized in parallel, details of which can be found in [16, 15].

Using p processors, each processor i factorizes A_{ii} in parallel, allowing row or column pivots within each block. Those entries of the diagonal blocks that can not be eliminated are combined with the border elements to form a double bordered block-diagonal matrix

$$
\begin{pmatrix}
L_1 U_1 & & & & U_1' \\
& L_2 U_2 & & & U_2' \\
& & \ddots & & \vdots \\
& & & L_p U_p & U_p' \\
L_1' & L_2' & \cdots & L_p' & F
\end{pmatrix}.
\tag{2}
$$

Here F is the interface matrix containing contributions from the factorization of each diagonal block. The interface problem is then solved, after that the rest of the solution can be found via back-substitution, which again can be performed in parallel.

The factorization of the interface matrix F forms a sequential bottleneck of the solution process. It is therefore essential that the size of this interface matrix is kept to the minimum. The size of F is related directly to the size of the border of the BBD form (1). Moreover, to achieve load balance in the parallel factorization, diagonal blocks are preferably of similar size.

A *net* [6] is defined as the set of all row indices of the nonzero entries in a column. For example, the net corresponding to column 1 of the matrix in Figure 1 is $\{3, 6, 8\}$.

A *net* is *cut* by a partition of the matrix, if two of its row indices belong to different parts of the partition. The *net-cut* of a partition is defined as the number of nets that are cut by that partition. For example in Figure 1, every net is cut by the partition (shown by the broken line), so the net-cut is eight. The purpose of the reordering is to minimize this net-cut. For example, Figure 2 shown the same matrix after the reordering, the matrix is now in a BBD form, with the border marked by the letter "B". The row ordering is shown to the left and the column ordering is shown above the matrix. The net-cut of the partition of the reordered matrix is one.

The *gain* associated with moving a row of a partitioned matrix to another partition is the reduction in the net-cut that will result after such a move. The gain is negative if such a move increases the net-cut. The gain associated with each row of the partitioned matrices in Figure 1 and Figure 2 are shown to the right of the matrices.

	1	2	3	4	5	6	7	8	gain
1		×	×	×		×		×	0
2			×						0
3	×		×		×		×		3
4		×		×		×		×	0
5		×	×	×		×		×	1
6	×				×		×		0
7			×	×		×		×	0
8	×		×		×		×		0

Fig. 1. An 8×8 matrix

	2	4	6	8	1	5	7	3	gain
1	×	×	×	×				B	−4
4	×	×	×	×					−4
5	×	×	×	×				B	−4
7	×	×	×					B	−3
2								B	0
3					×	×	×	B	−3
6					×	×	×		−3
8					×	×	×	B	−3

Fig. 2. The matrix of Figure 1 after reordering

Two rows of a matrix A are *connected* to each other, if they share one or more column index. The *row connectivity graph* G_A of a matrix A is a graph consisting of the row indices of A as vertices. Two vertices i and j are connected by an edge if rows i and j of matrix A are connected. The number of shared column indices between row i and row j of matrix A is called the *edge weight* of the edge (i,j) of graph G_A, and the number of nonzero entries in row i of matrix A is called the *vertex size* of vertex I of graph G_A

Let $\bar{A}$ be the matrix A but with all nonzero entries set to 1. It can be proved that the edge weights of G_A are given by the off-diagonal entries of $\bar{A}\bar{A}^T$. The vertex sizes, on the other hand, are given by the diagonal entry of the same matrix.

3 Implementation of the KL Algorithm

The KL (Kernighan-Lin) algorithm [14] is an iterative algorithm. Starting from a initial bisection, the gain of each vertex is calculated. At each inner iteration, the vertex which has the highest gain is moved from the partition in surplus to the partition in deficit. This vertex is locked and gains updated. The procedure is repeated until all vertices are locked. The iterative procedure is restarted from the best bisection so far, and terminated if no improvement in the reduction of net-cut is achieved. This algorithm is usually implemented using the data structure suggested in [7].

For efficient implementation, we follow the practice in undirected graph partitioning [13, 18] in that only rows on the partitioning boundary are allowed to move, and only gains of those rows that are connected to a "just-moved" row are updated.

4   The Multilevel Implementation

There are three phases for the multilevel implementation, namely, coarsening, partitioning and prolongation.

The coarsening is done using the popular heavy edge-collapsing method (e.g.,[13]). Row vertices are randomly visited. For each of these vertices, the unmatched edge (of the row connectivity graph) from the vertex with the highest edge weight is selected. The corresponding rows of the matrix are combined to give a row of the coarsened matrix. This approach ensures that the net-cut on the coarsest matrix equals that of the (finest) original matrix, provided that the partitioning of the coarsest matrix is injected to the finer matrices without any refinement.

The partitioning of the coarsest matrix is done using the KL algorithm, starting with the natural bisection of the matrix.

In the prolongation phase, the partitioning of a coarse matrix is injected to the fine matrix, and refined using the KL algorithm.

Thus essentially the multilevel process takes a multigrid like approach, with the coarse matrix given by a Galerkin like operator applied on the fine matrix.

As an example, the above multilevel algorithm is applied to the west0381 from the Harwell-Boeing collection and demonstrated in Figure 3. Here the matrix is coarsened three times, on the left side of the figure, to give the coarsest matrix (bottom left). This coarsest matrix is partitioned and reordered (bottom right), and prolongated and refined to give the final matrix of BBD form (top right), with 2 diagonal blocks and a net-cut of 51 (13.39% of the matrix size).

5   Comparison with Existing Algorithms

The software which implements the multilevel approach has been named MONET (Matrix Ordering for minimal NET-cut). It is available on request.

MONET has been compared with other existing algorithms on a suite of unsymmetric test matrices from the Harwell-Boeing Collection (http://www.dci.clrc.ac.uk/Activity/SparseMatrices) and the University of Florida Sparse Matrices Collection (http://www.cise.ufl.edu/~davis).

Two measures are used for the quality of the ordering. The first measure is *net-cut*, expressed here in percentage term relative to the size of the matrix.

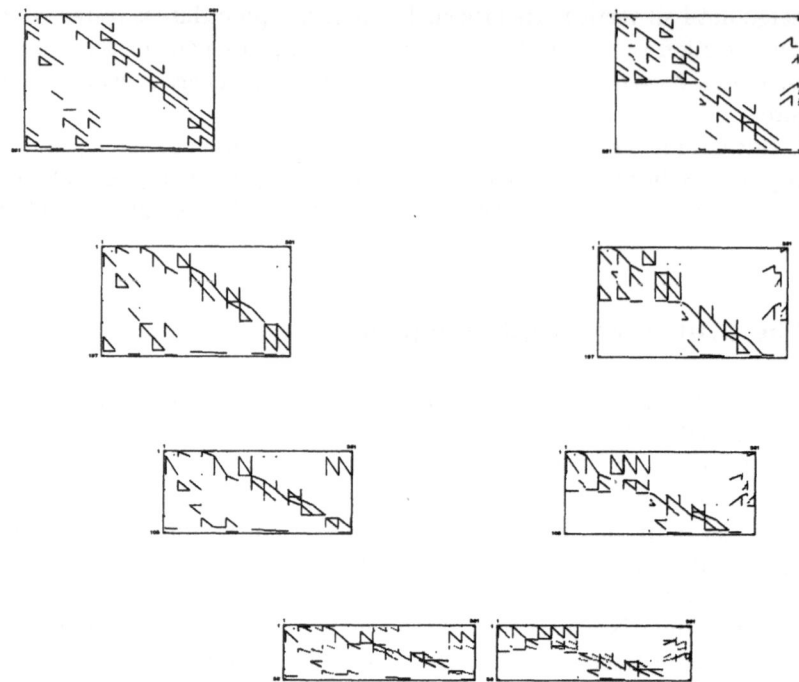

Fig. 3. The multilevel algorithm applied to the west0381 matrix

The second measure is *load imbalance*, which is defined as the size of the largest diagonal block divided by the average block size, again expressed as a percentage. Low net-cut and small load imbalance give good ordering.

Table 1 compares MONET with two existing algorithms, GPA-SUM [3] and MNC [6]. The lowest net-cut is highlighted. The net-cut and load imbalance figures for GPA-SUM and MNC are taken from [3]. It can been seen that MONET gives much less load imbalance. In terms of net-cut, MONET is better than both GPA-SUM and MNC for most of the test cases. This is particularly significant on the some of the matrices from the Harwell-Boeing Collections (lns_3937, sherman5, west2021, west0989). Notice that the number of blocks in Table 1 are relatively small. This is because results for larger number of blocks for the GPA-SUM and MNC algorithms are not available to the author. However MONET itself can work with large number of blocks.

6   Future Work

In this paper a multilevel ordering algorithm for BBD form is presented. The MONET algorithm was demonstrated to give better quality ordering compared with existing algorithms, in terms of lower net-cut and less load imbalance.

Table 1. Comparing MONET with GPA-SUM and MNC. The entries give number of blocks/net-cut (%)/load imbalance (%)

problem	matrix size	MONET	GPA-SUM	MNC
lhr17	17576	4/**4.48**/0.86		4/6.90/9.3
lhr14	14270	8/**5.31**/2.70	8/9.01/6.0	6/5.40/58.8
lhr11	10964	8/**7.45**/0.84	8/9.87/6.9	10/9.64/148.8
lhr10	10672	8/**7.19**/0.07	8/9.59/6.0	10/9.65/151.6
lhr07	7337	16/19.38/2.06	16/**18.1**/11.0	-
hydr1	5308	4/**2.00**/8.59	-	4/3.43/8.8
lhr04	4101	4/7.75/0.07	4/7.85/3.4	3/**6.91**/49.6
lns_3937	3937	4/**9.98**/0.08	4/55.2/10.2	6/18.0/72.2
sherman5	3312	4/**7.88**/0.00	4/41.8/16.7	4/22.7/105.4
west2021	2021	2/**5.54**/0.05	2/15.0/15.3	3/9.25/70.6
west1505	1505	2/**2.06**/0.07	2/15.1/1.0	3/9.17/62.7
gre_1107	1107	2/12.65/0.09	2/34.8/3.5	2/**11.8**/5.9
west0989	989	2/**5.66**/0.1	2/15.5/1.1	3/10.4/59.6
bp_1000	822	2/15.45/0.00	2/20.2/6.3	2/**14.1**/19.0

While it is possible to reorder an unsymmetric matrix into the BBD form with a preset number of diagonal blocks, it is not possible to know in advance what is the largest possible number of blocks that gives a net-cut not exceeding a certain percentage of the matrix size, assuming the underlying physical structure of the application imply a bordered block diagonal structure that is not directly reflected in the original form of the matrix. Thus for the moment the optimal number of blocks can only be derived by trial and error. This remains a question to be answered by further research. Work to combine the ordering algorithm of this paper with a direct sparse solver to form a parallel direct solver or preconditioner is also underway.

References

[1] Amestoy P. R., Duff I. S.: Vectorization of a multiprocessor multifrontal code. International Journal of Supercomputer Applications and High Performance Computing **3** (1989) 41-59

[2] Barnard S. T., Simon H. D.: Fast multilevel implementation of recursive spectral bisection for partitioning unstructured problems. *Concurrency: Practice and Experience* **6** (1994) 101-117.

[3] Camarda K. V., Stadtherr M. A.: Matrix ordering strategies for process engineering: graph partitioning algorithms for parallel computation, Computers & Chemical Engineering, to appear.

[4] Çatalyürek U. V., Aykanat C.: Decomposing Irregularly Sparse Matrices for Parallel Matrix-Vector Multiplication. Lecture Notes in Computer Science 1117 (1996) 75-86.

[5] Choi H., Szyld D. B.: Application of threshold partitioning of sparse matrices to Markov chains. Technical Report 96-21, Dept. of Mathematics, Temple Univ., Philadephia, PA. (1996). Available from http://www.math.temple.edu/~szyld.

[6] Coon A. B., M. A. Stadtherr M. A.: Generalized block-tridiagonal matrix orderings for parallel computation in-process flowsheeting. Computers & Chemical Engineering **19** (1995) 787-805.

[7] Fiduccia C. M., Mattheyses R. M.: A linear time heuristic for improving network partitions. in Proc. 19th ACM-IEEE Design Automation Conf., Las Vegas, ACM (1982).

[8] Gupta A., Karypis G., Kumar V.: Highly scalable parallel algorithms for sparse matrix factorization. IEEE Transactions on Parallel and Distributed Systems **5** (1997) 502-520.

[9] Hendrickson B., Leland R.: A multilevel algorithm for partitioning graphs. Technical Report SAND93-1301, Sandia National Laboratories, Allbuquerque, NM (1993).

[10] Hendrickson B., Kolda T. G.: Partitioning Rectangular and Structurally Non-symmetric Sparse Matrices for Parallel Processing, submitted to SIAM Journal of Scientific Computing, available from http://www.cs.sandia.gov/~bahendr/papers.html

[11] Hu Y. F., Blake R. J.: Numerical experiences with partitioning of unstructured meshes. Parallel Computing **20** (1994) 815-829.

[12] Hu Y. F., Maguire K. C. M., Blake R. J.: A multilevel unsymmetric matrix ordering algorithm for parallel process simulation, in prepartion.

[13] Karypis G., Kumar V.: A fast and high quality multilevel scheme for partitioning irregular graphs. SIAM Journal on Scientific Computing **20** (1999) 359-392.

[14] KERNIGHAN B. W., LIN S.: An efficient heuristic procedure for partitioning graphs. Bell Systems Tech. J. **49** (1970), pp. 291-308.

[15] Mallya J. U., Zitney S. E., Choudhary S., Stadtherr M. A.: A parallel frontal solver for large-scale process simulation and optimization. AICHE JOURNAL **43** (1997) 1032-1040.

[16] Mallya J. U., Zitney S. E., Choudhary S., Stadtherr M. A.: A parallel block frontal solver for large scale process simulation: Reordering effects. Computers & Chemical Engineering **21** (1997) s439-s444.

[17] Vegeais J. A., Stadtherr M. A.: Parallel processing strategies for chemical process flowsheeting. AICHE Journal **38** (1992) 1399-1407.

[18] Walshaw C., Cross M., Everett M. G.: Parallel dynamic graph partitioning for adaptive unstructured meshes. Journal of Parallel and Distributed Computing **47** (1997) 102-108.

[19] Zitney S. E., Mallya J., Davis T. A., Stadtherr M. A.: Multifrontal vs frontal techniques for chemical process simulation on supercomputers Computers & Chemical Engineering **20** (1996) 641-646.

Dynamic Load Balancing for Ocean Circulation Model with Adaptive Meshing

Eric Blayo, Laurent Debreu, Grégory Mounié, and Denis Trystram

LMC-IMAG, BP 53, 38041 Grenoble Cedex 9, France
tel: (33) (0)476514500, fax: (33) (0)476631263
{name}@imag.fr

Abstract. This paper reports the parallel implementation of adaptive mesh refinement within finite difference ocean circulation models. The implementation is based on the model of *Malleable Tasks* with inefficiency factor which allows a simple expression of the different levels of parallelism with a good efficiency. Our goal within this work was to validate this approach on an actual application. For that, we have implemented a load-balancing strategy based on the well-known level-by-level mapping. Preliminary experiments are discussed at the end of the paper.

Keywords: Load Balancing - Malleable Tasks - Adaptive Mesh Refinement - Ocean Modeling

1 Introduction

Numerical modeling of the ocean circulation started in the sixties and was continuously developed since that time. Today the motivation for ocean modeling is mainly twofold: The first goal is climate study (prediction of the evolution of the climate, at ranges from a few months to a few years), while the second motivation is operational oceanography, i.e. near real time forecast of the "oceanic weather", in a way similar to operational meteorology.

A major practical problem within ocean general circulation models (OGCMs) is their very large computational cost. These models are run on vector and/or parallel supercomputers (Cray C90 or T3E, Fujitsu VPP ...), where a usual simulation requires several hundred or thousand hours of CPU-time.

In this context, it is clear that adaptive meshing could be of real interest for ocean modelers. It could reduce the computational cost of models by taking advantage of the spatial heterogeneity of oceanic flows and thus using a fine mesh only where and when necessary. Moreover, this would allow local zooms on limited areas of particular interest, e.g. for operational purposes. The finite element technique permits of course the use of such a non-uniform adaptive grid over the computational domain. However, this approach is mostly reserved, in the field of ocean modeling, to coastal or tidal models, and all major OGCMs use finite difference methods on structured grids. Starting from this consideration, Blayo and Debreu [2] are currently developing a Fortran 90 software package, which

P. Amestoy et al. (Eds.): Euro-Par'99, LNCS 1685, pp. 303–312, 1999.

will furnish any finite difference ocean model the capability of adaptive mesh-ing and local zooming. The method is based on the adaptive mesh refinement (AMR) algorithm proposed by Berger and Oliger [1], which features a hierar-chy of grids at different resolutions, with dynamic interactions. Since the mesh refinement is adaptive over the time step, the number of grids as well as their sizes and resolutions vary during the simulation. Thus, the computational load is also varying in time, and a dynamic load balancing is necessary to implement efficiently this AMR method on a parallel computer.

This paper will discuss the design, implementation and performance of a load balancing package and its integration into a code for large scale simulations of ocean circulation. It is organized as follows: in section 2, we describe some features of ocean models useful for the rest of the paper and present the AMR method. The model of Malleable Tasks with inefficiency factor is presented in section 3. Then, an adaptation of the level-by-level load balancing algorithm is detailed and analyzed in section 4. Some numerical experiments are reported in section 5, in order to illustrate the behavior of the algorithm on a parallel machine. Finally, some conclusions and perspectives are discussed.

2   Ocean Models and Adaptive Mesh Refinement

2.1   About Ocean Models

The basic relations describing the ocean circulation are equations for conserva-tion of momentum (Navier-Stokes equations), mass (continuity equation), heat and salt, and an equation of state. The addition of the Boussinesq's approxima-tion (variations of density are small) and the hydrostatic approximation leads to the so-called "primitive equations" (PE), which are solved by most OGCMs, using finite difference techniques. Simpler models of ocean circulation can also be derived by making additional assumptions.

Thus, an ocean model can be written in a symbolic way as: $\frac{\partial X}{\partial t} = F(X)$ where t is the time, X is the state variable and F is a non-linear operator. The time discretization scheme is generally explicit, which leads in most cases to discretized equations of the form $X(t + \delta t) = G(X(t), X(t - \delta t))$ where δt is the discretization time step. However, some models use also an implicit scheme for one of the equations (equation for barotropic – or depth-averaged – motion), which implies in that case to solve at each time step a linear system for $X(t+\delta t)$.

The parallelization of ocean models is performed by domain decomposition techniques : the geographical domain is divided into subdomains, each of them being affected to a processor (e.g. [11] [3] [12]). An important point for the purpose of this work is to emphasize that ocean models are *regular* applications, in the sense that the volume of computations can be estimated quite precisely as a function of the grid size and the number of processors (at least, it can be measured by adequate benchmarks).

2.2 Adaptive Mesh Refinement (AMR)

The basic principle of AMR methods consists in locally refining or coarsening the computation mesh, according to some mathematical or physical criteria (like error estimates or eddy activity). Such techniques are widely used with finite element codes, but rather rarely with finite differences because refinements lead to non-homogeneous grids and thus complicate the handling of the code. However, Berger and Oliger [1] proposed an AMR algorithm which avoids this drawback, by considering a locally multigrid approach. In their method, the refinement is not performed on a unique non-homogeneous grid, but on a hierarchy of grids, i.e. a set of homogeneous embedded grids of increasing resolutions, and interacting among themselves. Without going deeply into more details (the reader can refer for example to [1] or [2]), the principle of the algorithm is as follows. Consider a hierarchy of grids, like in Figure 1: it consists in a root (or level-0) grid covering the entire domain of computation with coarse resolution Δh_0 and coarse time step Δt_0, and a number of subgrids (or level-1 grids) with a finer resolution $\Delta h_1 = \Delta h_0/r$ and a finer time step $\Delta t_1 = \Delta t_0/r$ focused only on some subdomains (r is an integer called the refinement ratio). This structure is recursive, in the sense that any level-l grid can contain finer level-$(l+1)$ subgrids, with $\Delta h_{l+1} = \Delta h_l/r$ and $\Delta t_{l+1} = \Delta t_l/r$ (of course until a maximum level $l_{\max}$).

Time integration is performed recursively starting from the root grid. Any level-l grid is advanced forward one time step Δt_l. The solutions at time t and $t + \Delta t_l$ are used to provide initial and boundary conditions for the integration of the level-$(l + 1)$ subgrids. These subgrids can then be advanced forward r time steps Δt_{l+1}, to provide a more precise solution at time $t + r\Delta t_{l+1} = t + \Delta t_l$ on the regions covered by the subgrids. These solutions at level $(l+1)$ are then used to improve the solution at level l, via an update procedure.

The relevance of the grid hierarchy is checked regularly every N coarse time steps Δt_0. A criterion is evaluated at every grid point to determine whether the local accuracy of the solution seems sufficient or not. Subgrids can then be created, resized or removed.

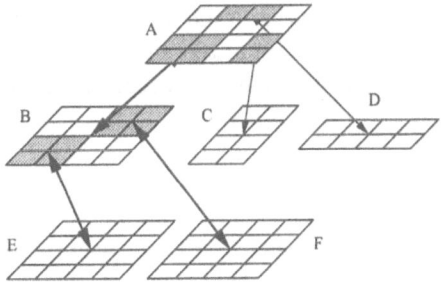

Fig. 1. Example of grid hierarchy

A major interest of this AMR method is that it can be used without modifying the model. The model can be seen like a black-box (corresponding to the Step routine in the previous algorithm) needing only configuration parameters (domain of computation, mesh size Δh, time step Δt, initial conditions, boundary conditions) to be run. This is one of the key ideas of the package that we are presently developing.

In the parallel version of our package, the model will still be used as a black-box like in the sequential case, but with an additional parameter: the number of processors used to run the model. Since the different grids at a given level l can be run simultaneously, they will be affected to different groups of processors. The problem is then to determine, for a given hierarchy, which is the best grids-to-processors correspondence for computational efficiency. These assignments must be determined at every regridding step, because of the evolution of the grid hierarchy during the simulation. The model of Malleable Tasks presented in the next section is an efficient tool for solving such problems.

3   Parallel Implementation

3.1   Malleable Tasks

Since the eighties, many works have been developed for parallelizing actual large scale applications like the oceanographic simulation considered in this paper. The solution of the scheduling problem (in its larger acception) is a central question for designing high-performance parallelization. It corresponds to finding a date and a processor location for the execution of each task of the parallel program. Among the various possible approaches, the most commonly used is to consider the tasks of the program at the finest level of granularity and apply some adequate clustering heuristics for reducing the relative communication overhead [6]. The main drawback of such an approach is that communications are taken into account explicitly (they are expressed assuming a model of the underlying architecture of the system which is very difficult to establish and often far from the actual asynchronous and non-deterministic behavior of the parallel execution). It is well-known that the introduction of explicit communications into the scheduling algorithms renders the problem harder than without communication [4]. Recently, a new computational model has been proposed [5]. *Malleable tasks* (denoted MT in short) are computational units which may be themselves executed in parallel. It allows to take into account implicitly the influence of communications.

3.2   Motivation for Using MT

There are several reasons for introducing MT as a way to design efficient parallel applications. Let us summarize them below:

- MT allow to hide the complex behavior of the execution of a subset of tasks by using a parameter (the inefficiency factor μ discussed later) to abstract the overhead due to the management of the parallelism (communications, idle times due to internal precedence constraints, etc..).

- MT reflect the structure of some actual parallel applications because they unify the usual single processor tasks and tasks which may require more than one processor for their execution. The structural hierarchy is then included into the model and the user does not specify when to use one model or another.
- MT simplify the expression of the problems in the sense that the user does not have to consider explicitly the communications. Thus, the same code may be used on several parallel platforms (only the μ factors will change), leading to a better portability.

The idea behind this model is to introduce a parameter which will implicitly give the communication overhead. An average behavior has to be determined using the *inefficiency* factor that will be now discussed in more details.

3.3   Inefficiency Factor: Definition and Properties

The efficiency of the execution of a parallel application depends of many factors like the algorithm, the data size, the data partitioning between available processing units, communication volume, network latency, topology, throughput, etc.. Ideally, a parallel system with m processors could complete the execution of an application m times faster than a single processor. However, an actual m processors system does not achieve such speedup due to some overhead introduced by inter-processor communication (which is slow compared to basic processor computation) and synchronization between the tasks. The ratio between achieved speedup and theoretical speedup depends on the number m of processors and the size N of the application. It will be denoted by $\mu(m, N)$ and called *inefficiency factor*. We give below the definition using a geometric interpretation.

Definition 1. *The inefficiency factor is the expansion of the computational area while using m processors: $t_m = \mu(m, N)\dfrac{t_1}{m}$ where t_1 (resp. t_m) is the time required to complete a malleable task of size N on one processor (resp. on m processors).*

Generally, the inefficiency factor increases with the number of processors and decreases with the size of the parallel task (at least until a certain threshold). The general shape of μ function may be divided into three zones of consecutive intervals of number of processors: (I) corresponds to the start-up overhead due to the management of the parallelism, where the efficiency may be quite bad. (II) is the region where the addition of extra processors cost almost nothing, speed-ups here are generally linear. (III) corresponds to a degradation due to the lack of parallelism (for instance, for too small MT). Some properties of μ are presented below. They will be useful for bounding the time in the load-balancing heuristics.

Property 1. $\mu(q,.)$ is an increasing function of q

Proof. Adding a processor generally involves an extra cost for managing communications and synchronizations between this processor and the others. This comes directly from the Brent's Lemma [9]. □

Property 2. Given a specific m processor system and a particular application, $\mu(1,N) = 1$ and $\mu(q+1,N)\dfrac{t_1}{q+1} < \mu(q,N)\dfrac{t_1}{q}$ $\forall q, 1 \le q \le m-1$.

Proof. These relations are usual hypotheses in parallel processing. The first part comes from the sequential execution of a task. For the second relation, just remark that it is useless to solve the application with one more processor if t_{q+1} is greater than t_q. □

A direct corollary of this last property ensures that $\dfrac{\mu(q,.)}{q}$ is a decreasing function of q.

4 Load-Balancing Issues

4.1 Definition and Context

A parallel application is usually described as a set of *tasks* (which are computational units) plus their interactions which are represented as a graph, called the *precedence task graph* [9]. For the specific oceanographic application we are considering in this paper, the tasks correspond to solve the equations on a mesh at a given time step. The parameters of a task are the size of the mesh (n_1 and n_2 plus the height along the vertical axis, at each time step k). This graph is not known in advance, it evolves in time and a mesh refinement leads to introduce a new series of tasks in the graph. In this sense, the problem is dynamic.

According to most studies in parallel processing, we are interested in minimizing the total execution time. It is influenced by two contradictory criteria, namely the idle time (or load-balance) which decreases as the size of the grain decreases and the communication overhead which grows with the number of processors. The load-balancing problem corresponds formally to determine an application which associates to each (malleable) task a sub-set of the processors. The resulting scheduling problem (defined as determining the date of which each task will start its execution, synchronously on the sub-set of processors) should add some constraints; namely, a task is allocated only once and should not be preempted. One contribution of this work is to introduce the influence of the inefficiency factor into load-balancing policies.

The well-known *gang scheduling* policy corresponds to allocate the total resources of the parallel system for executing each task [7]. Gang will be used as a basis for comparison.

4.2   Level-by-Level Mapping

The well-known level-by-level load-balancing policy can also be adapted to MT with inefficiency. A complete presentation of the basic heuristic can be found in [8]. It was proposed for a classical model of tasks with no communication delays. The basic idea of the level-by-level mapping is to distribute the tasks level by level in the precedence task graph. The level i is defined as the sub-set of independent tasks (not related by the precedence constraints) at distance i from the root of the corresponding not-weighted task graph. In other words, this policy may be interpreted as a series of successive synchronized Macro-Gang schedulings with sets of independent malleable tasks.

5   Experiments

5.1   The Ocean Model

The ocean model used for this study is a simple quasi-geostrophic box model, since our intention is to validate the MT approach and to design a good scheduling heuristic before running the simulation on the operational models. This type of model has been widely used in the ocean modeling community, and is known as a simple prototype of eddy-active large scale circulation in the mid-latitudes. Blayo and Debreu [2] already implemented the AMR method in such a multi-layered model, and demonstrated its interest.

In the present study, we use a barotropic (i.e. one layer) version of this model (see for instance [10] for a detailed description of such a model). Its governing equation can be written as:

$$\frac{\partial \Delta \psi}{\partial t} + J(\psi, \Delta \psi) + \beta \frac{\partial \psi}{\partial x} = \mathrm{curl}\tau - r\Delta\psi + A\Delta^2\psi \tag{1}$$

where ψ is the streamfunction, J is the Jacobian operator, β is the meridional gradient of the Coriolis parameter at the mid-latitude of the domain, τ is the horizontal wind stress at the surface of the ocean, r is a bottom friction coefficient, and A is a lateral viscosity coefficient.

5.2   Evaluation of Inefficiency

The inefficiency factors μ correspond to the penalty coefficients of the parallel time due to the overhead coming from the management of the parallelism. The parallel code was benchmarked in order to determine empirically an expression of the inefficiency factors for including them into the load-balancing policies. The μ is measured using single grid parallel computation. The variation of these curves in regard to the number of processors confirms the same global behaviour. Two aspects were studied:

– The variation of μ versus the grid size (cf Fig. 2) for a given number of proc. Note that for other number of processors, the general shape of the curves remains the same. The greater the number of processors, the greater the number of anomalies.

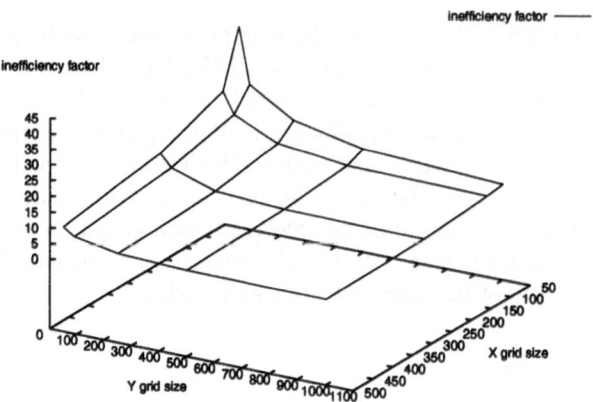

Fig. 2. Inefficiency factor as a function of the number of gridpoints in $x-$ and $y-$directions for 15 processors

- Figure 3 depicts the typical behaviour of μ versus the number of processor for a small and a large grid. This behaviour shows clearly some limitations of speedup due to inefficiency for small grids (speed down) and a linear speedup of larger problems.

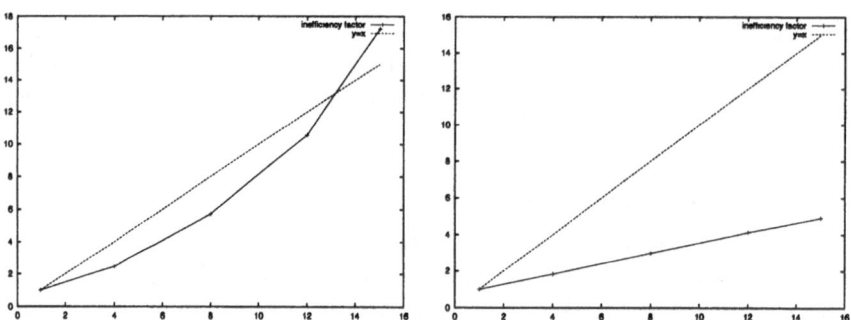

Fig. 3. Inefficiency factor as a function of the number of processors. Left panel: with anomaly (small grid size); right panel: without anomaly (large grid size). The reference curve $y = x$ is also plotted.

5.3   Experimental Results

We are currently developing numerical experiments on an IBM-SP2 parallel machine with 16 nodes connected on a fast switch.

Since the oceanographic model considered here is rather simple, the present study should be interpreted as a preliminary study for comparing the load-balancing policies and demonstrating the interest of using MT model for actual applications.

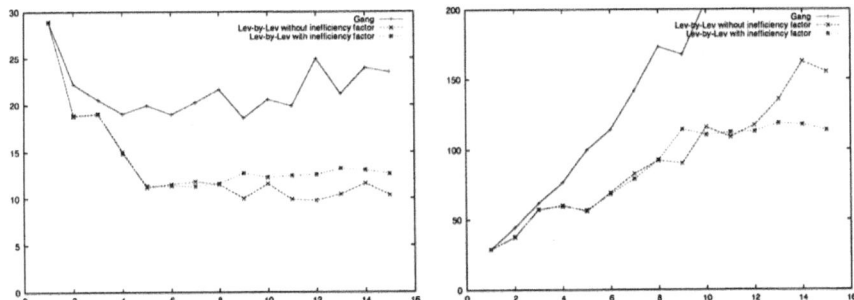

Fig. 4. Time and Work achieved by Gang, Level-by-Level with and without inefficiency factor scheduling algorithms versus the number of processors.

The tests reported in Fig. 4 compare speedup and efficiency of the three load-balancing policies: gang, level-by-level with and without inefficiency factor for multiple runs of a test case.

On a small number of processors the difference between algorithms is small. In fact, as the load imbalance produces large overhad, the gang scheduling perform better than Level by Level algorithms in some cases.

On a greater number of processor, the gang scheduling induces a large communication overhead.

The two Level by Level algorithms (with and without inefficiency factor) seem quite close in appearance on left panel of Fig. 4, namely in achieved makespan. However the Level by Level algorithm with inefficiency factor never used more than 9 processors while it achieves almost the same time. The amount of computation reported in Fig. 4 show clearly that inefficiency factor allows to bound it without large performance penalty.

The good behavior of the Level-by-Level scheduling algorithm with inefficiency factor allows us to expect good results for future works using a large number of processors, for example on a 128 processors Cray T3E.

Note, since the adaptive algorithm use the oceanographic model as black boxes, it is intrinsically synchronous; communications between MT only occur before and after computation. Thus the makespan is quite sensitive to the synchronization overhead between MT. This is especially true for the Gang scheduling and partially explains its poor efficiency. Gang scheduling is also highly sensitive to perturbations occurring in multi-user environment.

6 Conclusion and Future Works

We have presented in this work a new way to parallelize actual applications. It is based on the model of Malleable Tasks with inefficiency factor. We have discussed some fundamental properties of MT and showed how they are suitable for designing load-balancing heuristics. Some adaptations of well-known heuristics have been done, and preliminary experiments on a test problem coming from oceanography are currently carried out to show the feasibility and potential of MT model.

Other short term perspectives are to develop other load-balancing heuristics, and implement them within operational models of ocean circulation.

References

[1] Berger M. and J. Oliger, 1984: Adaptive mesh refinement for hyperbolic partial differential equations. *J. Comp. Phys.*, **53**, 484-512.

[2] Blayo E. and L. Debreu, 1998: Adaptive mesh refinement for finite difference ocean models: first experiments. *J. Phys. Oceanogr.*, **29**, 1239-1250.

[3] Bleck R., S. Dean, M. O'Keefe and A. Sawdey, 1995: A comparison of data-parallel and message passing versions of the Miami Isopycnic Coordinate Ocean Model (MICOM). *Paral. Comp.*, **21**, 1695-1720.

[4] Hoogeveen J., Lenstra J.-K., and Veltman B., 1994: Three, four, five, six, or the complexity of scheduling with communication delays. *Operations Research Letters*, **16**, 129–137

[5] Turek J., Wolf J. and Yu, P., 1992: Approximate algorithms for scheduling parallelizable tasks. *4th Annual ACM Symposium on Parallel Algorithms and Architectures*, 323-332

[6] Gerasoulis A. and Yang T., 1994: DSC : Scheduling Parallel Tasks on an Unbounded Number of Processors. *IEEE Transaction on Parallel and Distributed Systems*, **5**,951-967

[7] Scherson I.D., Subramanian S.D., Reis V. L. M. and Campos L. M., 1996: Scheduling computationally intensive data parallel programs. *Placement dynamique et répartition de charge : application aux systèmes parallèles et répartis (École Française de Parallélisme, Réseaux et Système)*, Inria, 107-129

[8] Kumar V., Grama A., Gupta A. and Karypis G., 1994: Introduction to Parallel Computing: Design and Analysis of Algorithms, *Benjamin/Cummings*

[9] Cosnard M. and Trystram D., 1993: Algorithmes et architectures parallèles. InterEditions, collection IIA.

[10] Pedlosky J., 1987: Geophysical fluid dynamics. Springer-Verlag, 710.

[11] Smith R.D., J.K. Dukowicz and R.C. Malone, 1992: Parallel ocean general circulation modeling. *Physica D*, **60**, 38-61.

[12] Wallcraft A.J. and D.R. Moore, 1997: The NRL layered ocean model. *Paral. Comp.*, **23**, 2227-2242.

DRAMA: A Library for Parallel Dynamic Load Balancing of Finite Element Applications*

Bart Maerten[1], Dirk Roose[1], Achim Basermann[2], Jochen Fingberg[2], and Guy Lonsdale[2]

[1] Dept. of Computer Science, K.U.Leuven, Heverlee-Leuven, Belgium
(bart.maerten,dirk.roose)@cs.kuleuven.ac.be
[2] C&C Research Laboratories, NEC Europe Ltd., Sankt Augustin, Germany
(basermann,fingberg,lonsdale)@ccrl-nece.technopark.gmd.de

Abstract. We describe a software library for dynamic load balancing of finite element codes. The application code has to provide the current distributed mesh and information on the calculation and communication requirements, and receives from the library all necessary information to re-allocate the application data. The library computes a new partitioning, either via direct mesh migration or via parallel graph re-partitioning, by interfacing to the ParMetis or Jostle package. We describe the functionality of the DRAMA library and we present some results.

1 Introduction

The aim of the Esprit project DRAMA, **D**ynamic **Re**-**A**llocation of **M**eshes for parallel finite element **A**pplications, is to develop a library with dynamic mesh re-partitioning algorithms for parallel unstructured finite element applications with changing work load and communication requirements. Although the aim is to develop a general purpose library, emphasis lies on two industrially relevant application codes, i.e., PAM-CRASH/PAM-STAMP and FORGE3. FORGE3 applies rather frequently refinement/de-refinement techniques together with 3D re-meshing. PAM-CRASH has varying calculation and communication costs due to (dynamically changing) contacts between different parts of the mesh. PAM-STAMP uses adaptive (de-)refinement techniques without re-meshing.

2 Mesh Re-partitioning Methods

Compared to (static) mesh partitioning, which is typically performed as a pre-processing phase, (dynamic) mesh re-partitioning strategies must satisfy the following *additional* requirements: *it must be fast and parallel, take the current partitioning into account, and interact with the application program*. The input

---

* DRAMA is supported by the EC under ESPRIT IV, Long Term Research Project No. 24953. More information can be found on WWW at http://www.cs.kuleuven.ac.be/cwis/research/natw/DRAMA.html

P. Amestoy et al. (Eds.): Euro-Par'99, LNCS 1685, pp. 313–316, 1999.

from the application program consists of the current distributed mesh, calculation and communication cost parameters and timings [2]. Based on the output of the mesh re-partitioner, i.e., the newly computed partitioning and the relation to the old partitioning, the application must decide whether the gain of the re-distribution will outweigh the cost of the data migration and if so, migrate the application data according to the new partitioning.

The DRAMA library contains two types of mesh re-partitioning techniques:
a) mesh-migration in combination with a parallel re-meshing technique [3]
b) global mesh re-partitioning by parallel re-partitioning a suitable graph representation using a multilevel algorithm, by calling existing libraries, i.e., ParMetis [4] and Jostle [5].

Further, DRAMA also contains a simple geometric partitioner. In this paper we only discuss the DRAMA mesh re-partitioning via graph re-partitioning.

3   DRAMA Cost Function and Its Graph Representation

The DRAMA cost function [1] models the execution time of parallel finite element calculations, taking into account changing work load and communication requirements and heterogeneous machine architectures. Full details of the DRAMA cost model are described in [1]. Here we only mention that the cost function takes into account that part of the operations within an FEM-code is performed element-wise, e.g., the assembly phase, while other operations are node-based, e.g., solving linear systems. Further, even when the calculations are done element-wise, communication is frequently carried out using node lists. Therefore, the computational cost consists of element-based and node-based contributions, while the communication cost can consist of element-element, node-node, and element-node based contributions.

In order to call one of the graph re-partitioning libraries (ParMetis, Jostle) a (distributed) weighted graph representation of the mesh and the calculation and communication requirements must be constructed. Graph vertex weights correspond to calculation costs and edge weights correspond to potential communication costs. Within the DRAMA library, the user can choose between an element-based graph (dual graph), a nodal-based graph or a combined element-node graph. In the first case, only element-based calculation and communication costs are represented. In the second case, only node-based calculation and communication costs are represented. Hence, by selecting a specific graph, the user chooses how accurately the various costs of the application are represented.

The output of graph partitioners like ParMetis and Jostle is just an array indicating for each graph vertex to which processor/partition it should be migrated to. Note that when the dual graph is partitioned, the output of the graph partitioner only gives a new distribution for the elements of the mesh, and a new distribution for the nodes still has to be determined.

Before the application program can perform the actual data migration, relations between the old and the new partitioning of the mesh, and a new numbering of elements and nodes must be determined. These tasks are performed within the DRAMA library and the new mesh distribution together with the relations between the old and the new mesh are given back to the application program. The latter relations are necessary because the DRAMA library migrates only the mesh description [2]. The application data (temperature, velocity, etc.) associated with the mesh have to be migrated to their new location by the application code, since the DRAMA library has no knowledge about the data structures used in the application.

4 Results

We present some results obtained with the graph based mesh repartitioning methods in DRAMA. As a first test to evaluate the DRAMA library we use a typical mesh for the FORGE3 application code, consisting of 231846 tetrahedral elements and 49666 nodes. This mesh is partitioned into 4 heavily unbalanced subdomains: about half of the mesh resides on one processor.

Each time step of FORGE3 consists of a varying number of iterations of an implicit iterative solver (on average about 2000). The implicit solver consists of both element-based and node-based operations. The element-based costs are associated with the construction of the matrix, which can be neglected compared to the nodal-based costs, associated with the iterative solution procedure.

The dual and nodal graph representations were passed to ParMetis and Jostle. Allowed load imbalances for ParMetis were 1%, 2%, and 5% imbalance. Jostle has been called with local and global matching options. The element-node partitioning is completed based on the graph re-partitioner results. The re-partitioning results are reported in Table 1.

Before and after re-partitioning, we have computed the predicted execution time for one iteration of the implicit solver by evaluating the DRAMA cost function $F^c = \max_{i=0...p-1} F_i^c$, where F_i^c represents *all* calculation and communication costs associated with processor i (also those costs which are not represented in the graph). Also the average$_{i=0...p-1} F_i^c$ and the load imbalance $\lambda = \frac{\text{maximum}_{i=0...p-1}(F_i^c)}{\text{average}_{i=0...p-1}(F_i^c)}$ are presented, together with the number of elements and nodes moved and the re-partitioning time on the NEC Cenju-4 (using 4 processors). The re-partitioning time is the time needed by the DRAMA library. The fraction of the re-partitioning time used by ParMetis depends on the graph representation: 25% when using the nodal graph and 40% when using the dual graph. Jostle is slower than ParMetis but its resulting partitioning is slightly better. Re-allocation costs are not considered here.

The results show that excellent re-partitioning is obtained (within the allowed imbalance tolerance), for both the nodal and the dual graph, despite the fact that some calculation costs and the communication latency are neglected in the graph representations. The nodal graph clearly leads to a good solution in less time and with less memory usage. Further, the re-partitioning time ($\approx 4.3s$) is reasonable compared to the time between two calls to the DRAMA library. Indeed, one integration time step (≈ 2000 iterations of implicit solver) requires

Table 1. Overview of results

dual graph						
		ParMetis			Jostle	
	initial	after (1%)	after (2%)	after (5%)	local	global
$F^c = \max F_i^c$	4.886e-02	2.861e-02	2.821e-02	2.822e-02	2.795e-02	2.790e-02
average F_i^c	2.686e-02	2.795e-02	2.748e-02	2.737e-02	2.689e-02	2.657e-02
imbalance λ	81.92%	2.33%	2.64%	3.12%	3.97%	4.99%
elements moved	-	60037	56725	52331	118573	101891
nodes moved	-	11961	11276	10397	24913	21515
re-partitioning time	-	10.92	10.76	10.74	16.90	16.99
memory used		40MB required for largest domain				

nodal graph						
		ParMetis			Jostle	
	initial	after (1%)	after (2%)	after (5%)	local	global
$F^c = \max F_i^c$	4.886e-02	2.836e-02	2.821e-02	2.877e-02	2.696e-02	2.693e-02
average F_i^c	2.686e-02	2.792e-02	2.757e-02	2.755e-02	2.690e-02	2.684e-02
imbalance λ	81.92%	1.57%	2.32%	4.44%	0.25%	0.31%
elements moved	-	57335	56869	54910	98553	99052
nodes moved	-	11236	11157	10717	20874	20981
re-partitioning time	-	4.32	4.34	4.34	6.35	6.18
memory used		25MB required for largest domain				

about 56 seconds and re-partitioning is typically done every 10 to 20 time steps. Hence, the predicted execution times based on the DRAMA cost function are:

with unbalanced mesh: time $= 20 \cdot 2000 \cdot 4.886 \cdot 10^{-2} \qquad = 1954.4$ s

with balanced mesh by re-partitioning the nodal graph every 20 time steps by

calling ParMetis: $\qquad$ time $= 20 \cdot 2000 \cdot 2.821 \cdot 10^{-2} + 4.34 = 1132.7$ s

calling PJostle: $\qquad$ time $= 20 \cdot 2000 \cdot 2.693 \cdot 10^{-2} + 6.18 = 1083.4$ s

References

[1] A. Basermann, J. Fingberg, and G. Lonsdale, *Final DRAMA Cost Model*, DRAMA Project Deliverable D1.1b, available at http://www.cs.kuleuven.ac.be/cwis/research/natw/DRAMA.html, 1999.

[2] A. Basermann, J. Fingberg, and G. Lonsdale, *The DRAMA Library Interface Definition*, DRAMA Project Deliverable D1.2c, available at http://www.cs.kuleuven.ac.be/cwis/research/natw/DRAMA.html, 1999.

[3] T. Coupez, *Parallel adaptive remeshing in 3D moving mesh finite element*, in Numerical Grid Generation in Computational Field Simulation, B. Soni et al., eds., Mississipi University, 1 (1996), pp 783–792,

[4] G. Karypis, K. Schloegel, and V. Kumar, *ParMetis: Parallel Graph Partitioning and Sparse Matrix Ordering Library, version 2.0*, Technical Report, Dept. of Computer Science, University of Minnesota.

[5] W. Walshaw, M. Cross, and M. Everett, *Dynamic load-balancing for parallel adaptive unstructured meshes.*, in Parallel Processing for Scientific Computing, M. Heath et al., eds., SIAM, Philadelphia, 1997.

Job Scheduling in a Multi-layer Vision System

M. Fikret Ercan[1], Ceyda Oğuz[2], and Yu-Fai Fung[1]

[1] Department of Electrical Engineering,
The Hong Kong Polytechnic University, Hong Kong SAR
{eefikret, eeyffung}@ee.polyu.edu.hk
[2] Department of Management,
The Hong Kong Polytechnic University, Hong Kong SAR
msceyda@polyu.edu.hk

Abstract. We consider job scheduling in a multi-layer vision system. We model this problem as scheduling a number of jobs, which are made of multiprocessor tasks, with arbitrary processing times and arbitrary processor requirements in a two-layer system. Our objective is to minimize the makespan. We have developed several heuristic algorithms that include simple sequencing and scheduling rules. The computational experiments show that three of these heuristic algorithms are efficient.

1 Introduction

A multi-layer vision system provides both spatial and temporal parallelism for solving integrated vision problems [1, 6]. However, its performance does not only depend on the performance of its components but also efficient utilization of them, which can be achieved by effective handling of data partitioning, task mapping, and job scheduling. Although data partitioning and task mapping are well studied in the literature, the job scheduling aspect is usually ignored [3].

In this paper, job scheduling on a multi-layer computer vision architecture is considered. In this architecture, a typical integrated vision algorithm employs more than one feature extracting algorithm to strengthen the evidence about object instances. This requires multiple low and intermediate level algorithm pairs to be performed on a single image frame. More specifically, every incoming image frame first initiates a number of algorithms at the first layer, and then the output of these algorithms for that image is passed on to the second layer where another set of algorithms is initiated, and so on. Due to data dependency between low, intermediate, and high level algorithms, parallelism is exploited by executing them at the dedicated layers of a multi-layer system creating a pipelining effect for continuous image frames. On the other hand, spatial parallelism is exploited by decomposing an operation to the processors of a layer.

We can model the above problem as follows: There is a set $\mathcal{J}$ of n independent and simultaneously available jobs (algorithm pairs) to be processed in a two-stage flow-shop where stage j has m_j identical parallel processors ($j = 1, 2$). Each job $J_i \in \mathcal{J}$ has two multiprocessor tasks (MPTs), namely $(i, 1)$ and $(i, 2)$.

P. Amestoy et al. (Eds.): Euro-Par'99, LNCS 1685, pp. 317–321, 1999.

$MPT(i, j)$ should be processed on $size_{ij}$ processors (processor requirement) simultaneously at stage j for a period of p_{ij} (processing time) without interruption ($i = 1, 2, \ldots, n$ and $j = 1, 2$). Jobs flow through from stage 1 to stage 2 by utilizing any of the processors while satisfying the flow-shop and the MPT constraints. The objective is to find an optimal schedule for the jobs so as to minimize the maximum completion time of all jobs, i.e. the makespan, C_{max}. In this scheduling problem both the processing times and the processor requirements are assumed to be known in advance. Although these parameters for some vision algorithms cannot be known until run-time, this problem can be overcome by obtaining them from the results of previously processed image frame. Our experiments show that this is a reasonable approximation. Lastly, although this model includes only two layers, it can be easily extended to include more layers.

2   Heuristic Algorithms

As the above scheduling problem is known to be NP-hard [4] and because a quick and good solution is required from a practical point of view, efficient heuristic algorithms are developed to derive approximate solutions. In developing the heuristic algorithms, we considered different rules from the literature and modified them to incorporate the processing times and the processor requirements of the tasks. This resulted in 17 heuristic algorithms. An extensive computational analysis on the average performance of these heuristic algorithms showed that the following three heuristic algorithms outperform the others significantly.
Heuristic 1 ($H1$):

1. Find a sequence S by applying Johnson's Algorithm [2] while assuming that $size_{ij} = m_j = 1$ ($i = 1, 2, \ldots, n$ and $j = 1, 2$).
2. Given the sequence S of the jobs, construct a schedule at stage 1 by assigning the first unassigned job J_i in S to the earliest time slot where at least $size_{i1}$ processors are available ($i = 1, 2, \ldots, n$).
3. As the jobs are processed and finished at stage 1 in order, schedule the jobs to the earliest time slot at stage 2 so that neither processor requirements nor flow-shop constraints are violated.

Heuristic 2 ($H2$):

1. First, sort the jobs in non-increasing order of stage 2 processor requirements. Then, sort each group of jobs requiring the same number of processors in non-increasing order of their stage 2 processing times. Call this sequence S.
2. Given the sequence S of the jobs, construct a schedule at stage 2 by assigning the first unassigned job J_i in S to the earliest time slot where at least $size_{i2}$ processors are available ($i = 1, 2, \ldots, n$).
3. Let T denote the completion time of the schedule obtained in Step 2. Starting from T, schedule the jobs backward at stage 1 to the latest time slot so that neither processor requirements nor flow-shop constraints are violated. If necessary, shift the schedule to the right.

Heuristic 3 ($H3$):

1. Same as in $H1$.
2. Same as in $H2$.
3. Same as in $H2$.

Algorithm H1 benefits from the machine environment of the problem. Since Johnson's algorithm was developed for two-stage flow-shop problem with classical job definitions, it might do well for the same machine environment with MPTs. In H2, two simple rules, namely the Longest Processing Time and its modified version, the Largest Processor Requirement, are combined to benefit from the structure of the jobs so as to minimize the idle time at both stages. The backward scheduling applied at stage 1 assumes that the idle time will be minimized at stage 1 if the sequence is kept as close as to that of stage 2 starting from the end of the schedule. H3 combines the benefits of H1 and H2.

3   Computational Study and an Application

For every $MPT(i,j)$ of job J_i at stage j, $size_{ij}$ and p_{ij} were generated from a uniform distribution over $[1, m_j]$ and $[1, 20]$, respectively ($i = 1, 2, \ldots, n$ and $j = 1, 2$). The number of jobs was selected so that $n = 50, 100, 150$. The number of processors was chosen as $m_1 = 2m_2 = 2^k$ and $m_1 = m_2 = 2^k$ with $k = 1, 2, 3, 4$. The first design corresponds to the actual design of the computer vision system considered in this study [1]. In each design, for every combination of n and $(m_1; m_2)$, 25 problems were generated.

In Table 1, the Average Percentage Deviation (APD) of each heuristic solution from the lower bound is given. This deviation is defined as $((C_{max}(Hl) - LB)/LB) \times 100$ where $C_{max}(Hl)$ denotes the C_{max} obtained by heuristic Hl, $l = 1, 2, 3$, and LB is the minimum of four lower bounds used. These four lower bounds are presented in [5]. For comparison, we also included APD results for the First Come First Serve (FCFS) heuristic which is commonly used in literature. The computational results demonstrate that all three of the heuristic algorithms are very effective. Although, none of the three proposed heuristics outperforms the others, it is observed that $H1$ performs better than $H2$ and $H3$ in most cases. We also observe that if a large number of processors is used, the FCFS may outperform $H2$ and $H3$. It can be seen from Table 1 that APD seems to decrease as n increases for each heuristic algorithm. This result is expected because as n increases, the lower bound becomes more effective. Lastly, the time complexity of all the heuristic algorithms is $O(n \log n)$ due to the sorting step.

Now, we report the results of applying our heuristic algorithms to a multi-layer image interpretation system [1]. In this system, a real-time aerial image interpretation algorithm is performed in which continuous image frames are processed by a series of algorithms. Due to real-time processing limitations, synchronizing the computational throughput with the image input rate is important so that overall run time for an image frame should be as small as possible. The computation in this application is partitioned as low, intermediate and high level

320   M. Fikret Ercan, Ceyda Oğuz, and Yu-Fai Fung

and distributed to a three layered system. There are three independent feature-extraction and feature-grouping task pairs that are performed at the first and the second layers of the system, which will be scheduled by the proposed heuristic algorithms. A histogram computation ($T1$) is performed at the low-level and the analysis of object segments ($T'1$) is conducted by referring to the intensity map obtained by $T1$ at the intermediate level. Similarly, a line detection algorithm ($T2$) is performed to find the linear-line features and a grouping operation ($T'2$) is performed on the extracted line segments. A right angle corner detection algorithm ($T3$) is performed first and a grouping of right angle corners ($T'3$) is performed later to generate a list of possible rectangle candidates.

Table 1. The APD results for various machine and job parameters.

			$m_1 = 2m_2$				$m_1 = m_2$		
n	k	FCFS	H1	H2	H3	FCFS	H1	H2	H3
50	1	2.56	0	1.68	0	8,73	3.41	3.96	3.75
	2	9.54	4.19	5.21	3.35	11.73	6.83	7.88	9.08
	3	12.72	8.22	9.67	12.71	13.69	9.18	11.25	14.36
	4	16.84	11.71	14.88	17.16	16.96	11.98	18.96	19.26
100	1	1	0	1.01	0	5.8	2.33	1.87	2.29
	2	4.14	1.93	1.28	1.01	8.66	5.39	5.99	5.35
	3	8.95	5.17	4.71	5.8	11.59	7.75	10.38	8.58
	4	11.15	8.3	9.18	11.58	13.51	10.3	15.22	14.38
150	1	0.49	0	0.68	0	4.43	1.9	1.87	1.72
	2	3.42	1.46	0.93	0.45	7.28	4.01	3.89	4.67
	3	8.32	5.1	3.51	5.51	10.24	7.34	7.51	8.99
	4	10.18	7.41	9.48	10.29	11.18	9.26	11.58	13.04

We execute our scheduling algorithms at the controller unit, which is developed particularly for task initiation, processor allocation, and inter-layer data transmission. The controller at each layer assigns tasks from the scheduling list by considering availability of the processors. In Table 2, processing time of each task is presented. It should be noted that these figures do not include any control overhead or interlayer communication cost since they were insignificant compared to the execution times.

From Table 2, the optimal makespan for these three jobs will be 364.677 msec for $k = 1$, and 195.798 msec for $k = 2$ by scheduling them in $T3$-$T'3$, $T1$-$T'1$, and $T2$-$T'2$ order. Total execution time of the example problem is found as 384.17, 382.39, and 386.27 by $H1$, $H2$, and $H3$, respectively, for $k = 1$. Similarly, the results are 237.12, 243.89, and 239.01 for $H1$, $H2$, and $H3$, respectively, when $k = 2$. These show that the best approximation to the optimal solution is obtained by heuristics $H1$ and $H2$. Total execution time is 394.048 msec for $k = 1$ and 240.232 msec for $k = 2$ if FCFS order is used. The outcome of

Table 2. Experimental processing times of vision tasks on prototype system (msec.)

Task	m_1		Task	m_2	
	2	4		1	2
$T1$	79.46	43.07	$T'1$	238.02	127.12
$T2$	266.42	137.17	$T'2$	15.71	13.002
$T3$	3.07	2.546	$T'3$	18.03	10.74

this application confirms our computational results. We can also conclude that more significant improvements can be expected for more jobs with a variety of processing times and processor requirements.

4 Conclusions

In this paper, job scheduling problem of vision algorithms in a multi-layer multiprocessor environment is considered. Three heuristic algorithms are presented for the solution of the problem and their average performance is analyzed. These scheduling algorithms are also applied to a prototype system. Overall, the results show that all three proposed heuristic algorithms are very effective and can be used for problems with a large number of jobs and processors.

Acknowledgment

The work described in this paper was partially supported by a grant from The Hong Kong Polytechnic University (Project No. 350-271) for the first author and by a grant from The Hong Kong Polytechnic University (Project No. G-S551) for the second and the third authors.

References

[1] Ercan, M.F., Fung, Y.F., Demokan, M.S.: A Parallel Architecture for Computer Vision. Proc. of ISSPR98, Vol. 1 (1998) 253–258
[2] Johnson, S.M.: Optimal Two and Three-Stage Production Schedules with Setup Times Included. Naval Res. Logist. Q. 1 (1954) 61–68
[3] Krueger, P., Lai, T.H., Dixit-Radiya, V.A.,: Job Scheduling is More Important Than Processor Allocation for Hypercube Computers. IEEE Trans. on Parallel and Dist. Sys. 5 (1994) 488–497
[4] Lloyd, E.L.: Concurrent Task Systems. Opns Res. 29 (1981) 189–201
[5] Oğuz, C., Ercan, M.F., Fung, Y.F.: Heuristic Algorithms for Scheduling Multi Layer Computer Systems. Proc. of IEEE ICIPS'97 (1997) 1347–1350
[6] Weems, C.C., Riseman, E.M., Hanson, A.R.: Image Understanding Architecture: Exploiting Potential Parallelism in Machine Vision. IEEE Computer. 25 (1992) 65–68

A New Algorithm for Multi-objective Graph Partitioning*

Kirk Schloegel, George Karypis, and Vipin Kumar

[1] Department of Computer Science and Engineering,
University of Minnesota
[2] Army HPC Research Center,
Minneapolis, Minnesota
(kirk, karypis, kumar)@cs.umn.edu

Abstract. Recently, a number of graph partitioning applications have emerged with additional requirements that the traditional graph partitioning model alone cannot effectively handle. One such class of problems is those in which multiple objectives, each of which can be modeled as a sum of weights of the edges of a graph, must be simultaneously optimized. This class of problems can be solved utilizing a multi-objective graph partitioning algorithm. We present a new formulation of the multi-objective graph partitioning problem and describe an algorithm that computes partitionings with respect to this formulation. We explain how this algorithm provides the user with a fine-tuned control of the tradeoffs among the objectives, results in predictable partitionings, and is able to handle both similar and dissimilar objectives. We show that this algorithm is better able to find a good tradeoff among the objectives than partitioning with respect to a single objective only. Finally, we show that by modifying the input preference vector, the multi-objective graph partitioning algorithm is able to gracefully tradeoff decreases in one objective for increases in the others.

1 Introduction

Graph partitioning is an important and well-understood problem. Here, the goal is to find a partitioning of a graph into k disjoint subdomains subject to the constraint that each subdomain has a roughly equal number of vertices, and with the objective to minimize the number of edges that are cut by the partitioning

* This work was supported by NSF CCR-9423082, by Army Research Office contracts DA/DAAG55-98-1-0441 and DA/DAAH04-95-1-0244, by Army High Performance Computing Research Center cooperative agreement number DAAH04-95-2-0003/contract number DAAH04-95-C-0008, the content of which does not necessarily reflect the position or the policy of the government, and no official endorsement should be inferred. Additional support was provided by the IBM Partnership Award, and by the IBM SUR equipment grant. Access to computing facilities was provided by AHPCRC, Minnesota Supercomputer Institute. Related papers are available via WWW at URL: http://www-users.cs.umn.edu/~karypis

P. Amestoy et al. (Eds.): Euro-Par'99, LNCS 1685, pp. 322–331, 1999.
© Springer-Verlag Berlin Heidelberg 1999

(referred to as the *edge-cut*). There are many applications for this problem from a wide range of diverse domains. Some examples are the parallelization of numerical simulations, computation of fill-reducing orderings of sparse matrices, the efficient fragmentation of databases, and the partitioning of VLSI circuits. The key characteristic of these applications is that they require the satisfaction of a single balance constraint along with the optimization of a single objective.

Recently, a number of applications have emerged with requirements that the traditional (i.e., single-objective) graph partitioning model cannot effectively handle. Specifically, there is a need to produce partitionings that optimize multiple objectives simultaneously. For example, a number of preconditioners have been developed that are focused on the subdomains assigned to each processor and ignore any intra-subdomain interactions (eg., block diagonal preconditioners and local ILU preconditioners). In these preconditioners, a block diagonal matrix is constructed by ignoring any intra-domain interactions, a preconditioner of each block is constructed independently, and then they are used to precondition the global linear system. The use of a graph partitioning algorithm to obtain the initial domain decomposition ensures that the number of non-zeros that are ignored in the preconditioning matrix is relatively small. However, the traditional partitioning problem does not allow us to control both the number as well as the magnitude of these ignored non-zeros. Ideally, we would like to obtain a decomposition that minimizes both the number of intra-domain interactions (reducing the communication overhead) and the numerical magnitude of these interactions (potentially leading to a better preconditioner).

Another example is the problem of minimizing the overall communications of parallel multi-phase computations. Multi-phase computations consist of m distinct computational phases, each separated by an explicit synchronization step. In general, the amount of interaction required between the elements of the mesh is different for each phase. Therefore, it is necessary to take the communications requirements of each phase into account in order to be able to accurately minimize the overall communications.

A natural extension to the graph partitioning problem is to assign a weight vector, w^e, of size m to each edge and a scalar weight, w^v, to each vertex. Now the k-way partitioning problem becomes that of partitioning the vertices of the graph into k disjoint subdomains such that each subdomain has a roughly equal amount of vertex weight and at the same time each one of the m objectives is optimized. Karypis and Kumar refer to this formulation as the *multi-objective graph partitioning problem* in [3]. Both the problems of minimizing the number and magnitude of the non-zero elements ignored by block-diagonal preconditioners and minimizing the overall communication of parallel multi-phase computations can be modeled as multi-objective partitioning problems.

In this paper, we present a formulation for the multi-objective graph partitioning problem, as well as an algorithm for computing multi-objective partitionings with respect to this formulation. We explain how this scheme is able to be tuned by a user-supplied preference vector in order to control the tradeoffs among the different objectives in the computed partitionings, how this scheme results in predictable partitionings based on these inputs, and how this scheme is able to handle both similar and dissimilar objectives. We show that this multi-objective graph partitioner is better able to balance the tradeoffs of different objectives than partitioning with respect to a single objective only. We also show that by modifying the input preference vector, the multi-objective graph partitioning algorithm is able to gracefully tradeoff decreases in one objective for increases in the others.

2 Challenges in Multi-objective Graph Partitioning

One of the real difficulties in performing multi-objective optimization is that no single optimal solution exists. Instead, an optimal solution exists for each objective in the solution space. Furthermore, finding an optimal solution for one objective may require accepting a poor solution for the other objectives [4]. The result is that the definition of a good solution becomes ambiguous. This being the case, before a multi-objective graph partitioning algorithm can be developed, it is first necessary to develop a formulation that allows the user to disambiguate the definition of a good solution. This formulation should satisfy the following criteria.

(a) It should allow fine-tuned control of the tradeoffs among the objectives. That is, the user should be able to precisely control the amount that one or more objectives may be increased in order to decrease the other objectives when these objectives are in conflict with one another.
(b) The produced partitionings should be predictable and intuitive based on the user's inputs. That is, the output partitioning should correspond to the user's notion of how the tradeoffs should interact with each other.
(c) The formulation should be able to handle objectives that correspond to quantities that are both of similar as well as of different types. Consider the example of minimizing the overall communications of a multi-phase computation. Here, all of the objectives represent similar quantities (i. e., communications overhead). However, other applications exist whose objectives are quite dissimilar in nature. The preconditioner application is an example. Here, both the number and the magnitude of the non-zero elements off of the diagonal are to be minimized. Minimizing the number of non-zero elements off of the diagonal will reduce the communications required per iteration of the computation, while minimizing the magnitude of these elements will improve its convergence rate. These are quite dissimilar in nature.

Two straightforward means of disambiguating the definition of a good multi-objective solution are (i) to prioritize the objectives, and (ii) to combine the objectives into a single objective. We next discuss each of these in the context of the above formulation criteria.

Priority-based Formulation. The definition of a good multi-objective solution is ambiguous when the relationship between the objectives is unclear. One simple way of defining this relationship is to list the objectives in order of priority [5]. Therefore, one possible formulation of the multi-objective graph partitioning problem is to allow the user to assign a priority ranging from one to m to each of the m objectives. Now, the multi-objective partitioning problem becomes that of computing a k-way partitioning such that it simultaneously optimizes all m objectives, giving preference to the objectives with higher priorities.

This formulation is able to handle objectives of different types as the relationship between the objectives is well-described and the objectives are handled separately. It will also result in predictable partitionings, in that the highest priority objective will be minimized with the rest of the objectives minimized to the amount possible without increasing higher priority objectives. However, this formulation does not provide the user with a fine-tuned control of the tradeoffs among the objectives. For example, consider a two-objective graph from which three partitionings are computed. The first has an edge-cut of $(1.0, 100.0)$, the second has an edge-cut of $(1.01, 2.0)$, and the third has an edge-cut of $(500.0, 1.0)$. Typically, the second of these would be preferred, since compared to the first partitioning, a 1% increase in the edge-cut of the first objective will result in a 98% decrease in the edge-cut of the second objective. Also, compared to the third partitioning, a 100% increase in the second objective will yield a 99.8% decrease in the first objective. However, under the priority-based formulation, the user is unable to supply a priority list that will produce the second partitioning. This is because partitionings exists with better edge-cuts for each of the two objectives. From this example, we see that the priority-based formulation is unable to allow the user fine-tuned control of the tradeoffs between the objectives.

Combination-based Formulation. While prioritizing the objectives describes their relationship well enough to disambiguate the definition of a good solution, it is not powerful enough to allow the user fine-tuned control over the tradeoffs among the objectives. One simple approach that has been used in other domains is to combine the objectives into a single objective and then to use a single objective optimization technique [1, 5]. This is done by taking the sum of the elements of the objective vector weighted by a *preference* vector, p. A formulation of the multi-objective graph partitioning problem that uses this approach is to assign to each edge a scalar combined weight, w^c, such that

$$w^c = \sum_{i=1}^{m} w_i^e p_i. \tag{1}$$

The resulting graph can then be partitioned utilizing an existing (single-objective) graph partitioning algorithm. For example, if we would like to scalarize the edge weights of a three-objective problem and want to give the second objective five times the weight of the first and third objectives, we could use a preference vector of $(1, 5, 1)$. An edge with a weight vector of $(2, 2, 1)$, for example, would then be assigned a combined weight of $2+10+1=13$.

Unlike the priority-based formulation, this one allows a fine-tuned control of the tradeoffs among the objectives, since the objectives are weighted and not merely ordered. However, this formulation cannot handle dissimilar objectives. This is because it requires that the objectives be combined (by means of a weighted sum). For dissimilar objectives (such as the number and magnitude of the ignored elements off of the diagonal), this combination can be meaningless.

One possible solution is to divide each edge weight by the average edge weight for the corresponding objective in an attempt to normalize all of the objectives. Normalization will help us to combine dissimilar objectives in a more meaningful way. However, this solution fails the predictability criteria. Consider the example two-edge-weight graph depicted in Figure 1(a). In the center of this graph is a clique composed of edges with edge weight vectors of (10000, 1), while the rest of the edges have vectors of (1, 1). (Note, not all of the edges have their weights marked in the figure.) In this example, the first edge weight represents the degree to which we would like to have the vertices incident on an edge to be in the same subdomain. The second edge weight represents the interaction between vertices as normal. Therefore, in partitioning this graph, we have two objectives. (i) We would like the vertices of the clique to be in the same subdomain. (ii) We would like to minimize the edge-cut. The intuitive meaning of this graph is that the clique should be split up only if doing so reduces the edge-cut by tens of thousands. Figure 1(b) gives the new edge weights after normalization by the average edge weight of each objective. Here we see that the first edge weights of the clique edges have been scaled down to about five, the first edge weights of the non-clique edges have been scaled down to about zero, and the second edge weights of all edges have remained at one. Now, consider what will happen if we input different preference vectors and partition the graph. If the preference vector gives each objective equal weight (i.e., a preference vector of (1, 1)), then the clique will not be split during partitioning. This is as expected, since the clique edges in the input graph have very high edge weights for the first objective. Also, if we favor the first edge weight to any extent (i.e., supply a preference vector between (2, 1) and (1000000, 1)), this trend continues. However, if we favor the second edge weight only moderately (eg., a preference vector of (1, 6)), then the optimal partitioning will split the clique. This is because normalizing by the average edge weight has caused the first edge weights of the clique edges to be scaled down considerably compared to the second edge weights. The result is that we lose the intuitive meaning that the first edge weights had in Figure 1(a).

3   A Multi-objective Graph Partitioning Algorithm

As discussed in the previous section, existing formulations of multi-objective optimization problems, including that of multi-objective graph partitioning, do not fully address the requirements of the multi-objective graph partitioning problem. In this section, we present a novel new formulation that allows the user to control of the tradeoffs among the different objectives, produces predictable partitionings, and can handle objectives of both similar as well as dissimilar type.

Our formulation is based on the intuitive notion of what constitutes a good multi-objective solution. Quite often, a natural way of evaluating the quality of

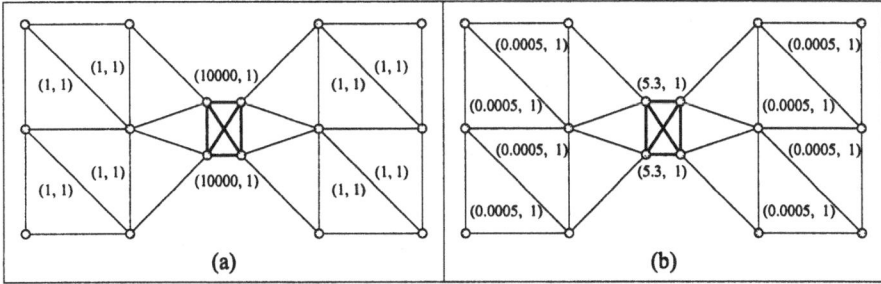

Fig. 1. This is a two-edge-weight graph. The edges of the clique in the center of the graph are weighted (10000, 1) in (a) and (5.3, 1) in (b). The other edges are weighted (1, 1) in (a) and (.0005, 1) in (b). The edge weights in (b) are computed by normalizing the edge weights in (a) by the average edge weights of the corresponding objective.

a multi-objective solution is to look at how close it is to the optimal solutions for each individual objective. For example, consider a graph with two objectives, let $P_{1,2}$ be a multi-objective partitioning for this graph, and let C_1 and C_2 be the edge-cuts induced by this partitioning for the first and second objectives, respectively. Also, let P_1 be the optimal partitioning of the graph with respect to only the first objective, and let C_1^o be the corresponding edge-cut. Finally, let P_2 be the optimal partitioning of the graph with respect to only the second objective, and let C_2^o be the corresponding edge-cut. Given these definitions, we can easily determine whether or not $P_{1,2}$ is a good multi-objective partitioning by comparing C_1 against C_1^o and C_2 against C_2^o. In particular, if C_1 is very close to C_1^o and C_2 is very close to C_2^o, then the multi-objective partitioning is very good. In general, if the ratios C_1/C_1^o and C_2/C_2^o are both close to one, then the solution is considered to be good.

Using this intuitive notion of the quality of a multi-objective partitioning, we can define a scalar combined edge-cut metric, c^c, to be equal to

$$c^c = \sum_{i=1}^{m} \frac{c_i}{c_i^o} \qquad (2)$$

where c_i is equal to the actual edge-cut of the ith objective, and c_i^o is equal to the optimal edge-cut of the ith objective. We can augment this definition with the inclusion of a preference vector, p. So Equation 2 becomes

$$c^c = \sum_{i=1}^{l} p_i \frac{c_i}{c_i^o}. \qquad (3)$$

Equation 3 then becomes our single optimization objective. In essence, minimizing this metric attempts to compute a k-way partitioning such that the edge-cuts with respect to each objective are not far away from the optimal. The distance that each edge-cut is allowed to stray from the optimal is determined by the preference vector. A preference vector of (1, 5) for example, indicates that we need to move at least five units closer to the optimal edge-cut of the second objective for each one unit we move away from the optimal edge-cut of the first objective. Conversely, we can move one unit closer to the optimal edge-cut of the first objective if it moves us away from the optimal edge-cut of the second objective by five or less. In this way, the preference vector can be used to traverse the area between the optimal solutions points of each objective. This results in predictable partitionings based on the user-suppled preference vector as well as fined-tuned control of the tradeoffs among the objectives. Finally, since dividing by the optimal edge-cuts in Equation 3 normalizes the objectives, we can combine similar and dissimilar objectives in this way.

If we further expand the c_i term of Equation 3, it becomes

$$= \sum_{i=1}^{m} p_i \frac{\sum_{j \epsilon cut} w_i^{e_j}}{c_i^o} \tag{4}$$

where the term $\sum_{j \epsilon cut} w_i^{e_j}$ represents the sum of the ith weights of the edges cut by the partitioning. By manipulating this equation, we get the following.

$$= \sum_{j \epsilon cut} \left(\sum_{i=1}^{m} \frac{p_i w_i^{e_j}}{c_i^o} \right) \tag{5}$$

The term $\sum_{i=1}^{m} \frac{p_i w_i^{e_j}}{c_i^o}$ in Equation 5 gives the scalar combined weight of an edge (defined in Equation 1) with each weight normalized by the optimal edge-cut of the corresponding objective. Since Equations 3 and 5 are equal, we have shown that minimizing the normalized combined weights of the edges cut by the partitioning minimizes the combined edge-cut metric from Equation 3. That is, the problem of computing a k-way partitioning that optimizes Equation 3 is identical to solving a single-objective partitioning problem with this proper assignment of edge weights.

We have developed a multi-objective partitioning algorithm that is able to compute partitionings with respect to this new formulation. In our algorithm, we utilize the k-way graph partitioning algorithm in MⅇⅡS [2] to compute a partitioning for each of the m objectives separately. We record the best edge-cuts obtained for each objective. Next, our scheme assigns to each edge a combined weight equal to the sum of the each edge weight vector normalized by the best edge-cuts and weighted by the preference vector. Finally, we utilize MⅇⅡS to compute a partitioning with respect to these new combined edge weights.

4   Experimental Results

The experiments in this section were performed using the graph, MDUAL2. It is a dual of a 3D finite element mesh and has 988,605 vertices and 1,947,069 edges. We used this graph to construct two types of 2- and 4-objective graphs. For the first type, we randomly select an integer between one and 100 for each edge weight element. For the second type, we obtained the edge weights in the following manner. For the first objective, we computed a set of nine different 7-way partitionings utilizing the k-way graph partitioning algorithm in MEIIS and kept a record of all of the edges that were ever cut by any one of these partitionings. Then, we set the first edge weight element of each edge to be one if this edge had ever been cut, and five otherwise. Next, we computed a set of eight different 11-way partitionings utilizing the k-way graph partitioning algorithm in MEIIS and set the second edge weight elements to one if the edge had been cut by any one of these partitionings and 15 otherwise. For the 4-objective graphs we then similarly computed a set of seven different 13-way partitionings and set the third edge weight elements to either one or 45 depending on whether the corresponding edge had been cut. Finally for the 4-objective graphs, we computed a set of six 17-way partitionings and similarly set the fourth edge weight elements to either one or 135.

We generated the Type 1 problems to evaluate our multi-objective graph partitioning algorithm on randomly generated problems. We generated the Type 2 problems to evaluate our algorithm on some particularly difficult problems. That is, each graph should have a small number of good partitionings for each objective. However, our strategy of basing these good partitionings on 7-, 11-, 13-, and 17-way partitionings was designed to help ensure that the good partitionings of each objective do not significantly overlap. Therefore, computing a single partitioning that simultaneously minimizes each of these objectives is difficult.

Qualitative Evaluation of the Multi-objective Graph Partitioner. Figure 2 compares the results obtained by the multi-objective graph partitioner with those obtained by partitioning with respect to each of the objectives alone with the single objective graph partitioner implement in MEIIS. Specifically, we show the results of 2- and 4-objective 64-way partitionings of Type 1 and 2 problems. In Figure 2(a) we give the results of partitioning the 2-objective Type 1 problem. In (b) we give the results of partitioning the 2-objective Type 2 problem. In (c) we give the results of partitioning the 4-objective Type 1 problem. Finally, in (d) we give the results of partitioning the 4-objective Type 2 problem. All of the edge-cuts are normalized by those obtained by partitioning with respect to a single objective only. Therefore, they give an indication of how far away is each objective from the optimal edge-cut.

Figure 2 shows that our multi-objective algorithm is able to compute partitionings such that a good tradeoff is found among the edge-cuts of all of the objectives. Partitioning with respect to a single objective obtains good edge-cut results for only a single objective. All of the other objectives are worse than those obtained by the multi-objective algorithm.

Control of the Tradeoffs by the Preference Vector. Figure 3 demonstrates the ability of our multi-objective graph partitioner to allow fine-tuned control of the tradeoffs of the objectives given a user-suppled preference vector. Specifically, it gives the results of a number of preference vectors for 4-objective 64-way partitionings of Type 2 problems. Here, three of the four elements of the preference

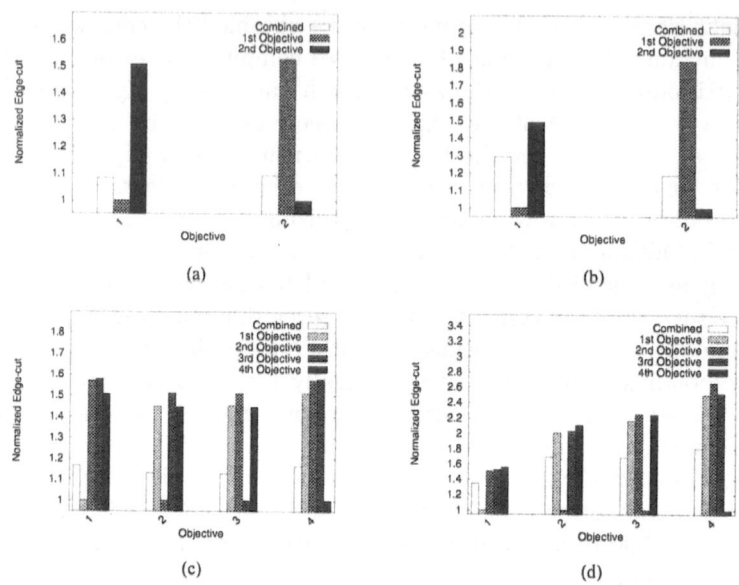

Fig. 2. This figure compares the normalized edge-cut results obtained by the multi-objective graph partitioner and those obtained by partitioning with respect to each of the objectives alone for a 2-objective Type 1 problem (a), a 2-objective Type 2 problem (b), a 4-objective Type 1 problem (c), and a 4-objective Type 2 problem (d).

vector are set to one, while the fourth is set to a value x. This value is plotted on the x-coordinate. The y-coordinate gives the values of the edge-cuts of each of the four objectives obtained by our multi-objective algorithm. So, as we move in the direction of positive infinity on the x-axis, a single objective is increasingly favored. The preference vector used in (a) is $(x, 1, 1, 1)$, (b) is $(1, x, 1, 1)$, (c) is $(1, 1, x, 1)$, and (d) is $(1, 1, 1, x)$. All of the edge-cuts are normalized by those obtained by partitioning with respect to a single objective only.

Figure 3 shows that by increasing the values of one of the elements of the preference vector, it is possible to gracefully tradeoff one objective for the others with the multi-objective partitioner. We see that in each result, the value at $x = 1$ is a good tradeoff among the four objectives. As x is increased, the edge-cut of the corresponding objective approaches that of the partitioning with respect to that objective only. The edge-cuts of the other objectives increase gracefully.

5   Conclusion

We have described a new formulation for the multi-objective graph partitioning problem and an algorithm that computes multi-objective partitionings with respect to this formulation. We have shown that this algorithm is able to minimize the edge-cut of a multi-objective problem given a user-supplied preference vector. We have shown that this algorithm provides the user with a fine-tuned

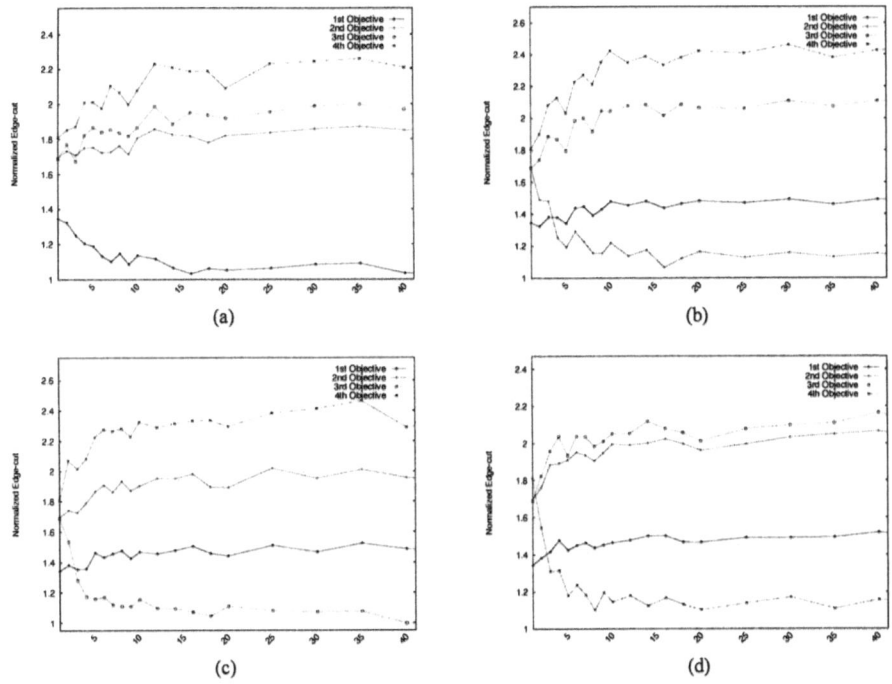

Fig. 3. This figure gives the normalized edge-cut results for 4-objective 64-way partitionings of Type 2 problems. The preference vector used in (a) is $(x, 1, 1, 1)$, (b) is $(1, x, 1, 1)$, (c) is $(1, 1, x, 1)$, and (d) is $(1, 1, 1, x)$.

control of the tradeoffs among the objectives, results in intuitively predictable partitionings, and is able to handle both similar and dissimilar objectives. Finally, we have shown that this algorithm is better able to find a good tradeoff between the objectives than partitioning with respect to a single objective only.

References

[1] P. Fishburn. *Decision and Value Theory*. J. Wiley & Sons, New York, 1964.
[2] G. Karypis and V. Kumar. MEIIS: A software package for partitioning unstructured graphs, partitioning meshes, and computing fill-reducing orderings of sparse matrices, version 4.0. 1998.
[3] G. Karypis and V. Kumar. Multilevel algorithms for multi-constraint graph partitioning. Technical Report TR 98-019, Dept. of Computer Science, Univ. of Minnesota, 1998.
[4] M. Makowski. Methodology and a modular tool for multiple criteria analysis of lp models. Technical Report WP-94-102, IIASA, 1994.
[5] P. Yu. *Multiple-Criteria Decision Making: Concepts, Techniques, and Extensions*. Plenum Press, New York, 1985.

Scheduling Iterative Programs onto LogP-Machine

Welf Löwe and Wolf Zimmermann

Institut für Programmstrukturen und Datenorganisation, Universität Karlsruhe,
76128 Karlsruhe, Germany,
{loewe|zimmer}@ipd.info.uni-karlsruhe.de

Abstract. Usually, scheduling algorithms are designed for task-graphs. Task-graphs model oblivious algorithms, but not iterative algorithms where the number of iterations is unknown (e.g. while-loops). We generalize scheduling techniques known for oblivious algorithms to iterative algorithms. We prove bounds for the execution time of such algorithms in terms of the optimum.

1 Introduction

The communication behavior of many data-parallel programs depends only on the size of their inputs. If this size is known in advance, it may be exploited for translation and optimization to improve the efficiency of the generated code [13]. In contrast to general compiling techniques using data-dependency analysis [12, 10, 5, 11], all synchronization barriers can be removed and data as well as processes can be distributed automatically. Hence, programmers may focus on the inherent parallelism of the problems and ignore the parameters of target machines. More specifically, programmers can use a synchronous, shared memory programming model. Neither data alignment nor mapping of processes onto processors is required in the source code[1]. The LogP-machine [1] assumes a cost model reflecting latency of point-to-point-communication in the network, overhead of communication on processors themselves, and the network bandwidth. These communication costs are modeled with parameters Latency, $o$verhead, and $g$ap. The gap is the inverse bandwidth of the network per processor. In addition to L, o, and g, parameter P describes the number of processors. The parameters have been determined for several parallel computers [1, 2, 3, 4]. These works confirmed all LogP-based runtime predictions.

However, among those problems that are not oblivious, we identify another class of high practical interest: problems that iterate an oblivious program where the number of iterations is determined at run time. We call those problem *iteratively oblivious*. The Jacobi-Iterations for solving linear equation systems or solvers based on the method of conjugated gradients (CG-solvers), e.g., are iteratively oblivious. The present work shows how to compile those programs to

---

[1] Equivalent to the PRAM-machine model, see [6].

P. Amestoy et al. (Eds.): Euro-Par'99, LNCS 1685, pp. 332–339, 1999.
© Springer-Verlag Berlin Heidelberg 1999

LogP-architectures, how to optimize them on a specific instance of the LogP-machine, and proves time bounds for their execution in terms of the theoretical optimum.

The paper is organized as follows: Section 2 introduces the basic definitions. Section 3 defines translations of iterative oblivious programs to LogP architectures. Section 4 proves time bounds of the resulting programs.

2   Basic Definitions

2.1   Oblivious and Iterative Oblivious Programs

First, we define the relation between PRAM-programs $\mathcal{P}$ and task-graphs $G_x = (V_x, E_x, \tau_x)$ for inputs x. The basis is a PRAM-model with an arbitrary number of processors with local memory, a global memory and a global clock. The task-graph of a PRAM-program on input x is basically the data-flow graph w.r.t. input x. More precisely, we partition the program on each processor into phases with the following structure: (1) read data from global memory, (2) perform computation without access of global memory, and (3) write data to global memory. Each of these phases becomes a task whose weight is the execution time of its computation. The task-graph G_x has an edge (v, w) iff phase v writes a value into the global memory that is read by w, i.e. there is data-flow from v to w via the global memory. The execution time $T_{\mathcal{P}}(x)$ of PRAM-program $\mathcal{P}$ on input x is simply the weight of the longest weighted path of G_x. The work $W_{\mathcal{P}}(x)$ is the sum of all task weights. A PRAM-program $\mathcal{P}$ is *oblivious* iff $G_x = G_y$ for any two valid inputs x and y of the same size. For simplicity we assume that we operate only on arrays of floating point numbers and the size is measured by the number of array elements. For oblivious programs, we write G_n for the task-graph for inputs of size n. A PRAM-program $\mathcal{P}$ is *iteratively oblivious* iff (i) it is oblivious, (ii) a sequential composition of iteratively oblivious programs, or (iii) a loop over iteratively oblivious programs. We assume that the termination condition of a loop belongs to the body of the loop and the result of its evaluation is written into global memory. The class of iteratively oblivious programs is a generalization of oblivious programs.

2.2   LogP-Schedules

For scheduling task-graphs to LogP-programs, we have to define when and where computations, send-operations and receive operations have to be executed. A *LogP*-schedule for a task-graph $G = (V, E, \tau)$ is a mapping $s : Op(V) \to 2^{N \times N}$ compatible to precedence constraints defined by G and the LogP-model. $(i, t) \in s(o)$ means that operation o is scheduled on processor P_i for time t. The *execution time* $T(s)$ is the completion time of the last task computed by s. An *input task* is a task containing an input value, an *output task* is a task computing an output value.

In this paper, we discuss three scheduling strategies: The *naive transformation* simply maps every task onto a single processor and assumes that there

Algorithm 1

Process *for task* v *with predecessors* $u_1, \ldots, u_n$, *successors* $w_1, \ldots, w_m$, *and computation* $\Phi(u_1, \ldots, u_n)$

(1)   **process** P_v
(2)   **do** n *times:* $m := recv;\ a_{m.src} := m.val;$
(3) $res := \Phi(a_{u_1}, \ldots, a_{u_n});$
(4) $send(res, v, w_1), \ldots, send(res, v, w_m)$
(5)   **end**

---

are enough processors available. Every task v is implemented by a process P_v which first receives messages from its predecessors containing the operands, then performs its computation, and finally sends the results to its successors. The messages contain a value, their source, and their destination. Algorithm 1 sketches the transformation.

A *linear clustering* computes a vertex-disjoint path cover of G (not necessarily a minimal one) and maps each path onto a separate processor. It also assumes that there are enough processors available.

The last scheduling strategy uses a layering of the task graph G, i.e. a partition λ of the vertices of G such that $\Lambda_0 = \{v : idg_v = 0\}$ and $\Lambda_{i+1} = \{v : PRED_v \subseteq \Lambda_{\leq i} \wedge v \notin \Lambda_{\leq i}\}$ where $\Lambda_{\leq i} = \Lambda_0 \cup \cdots \cup \Lambda_i$, $PRED_v$ is the set of direct predecessors of vertex v, and idg_v is the *indegree* of v. *Brent-clustering* schedules each layer separately onto P processors. The upper bound $L_{L,o,g}^{\max}(u,v)$ of the total communication costs between two vertices u and v is $(L + 2o + (odg_u + idg_v - 2) \times \max[o, g])$, cf.[13], where odg_u is the outdegree of u. We define the notion of granularity of a task graph G which relates computation costs to communication costs, for a task v it is $\gamma_{L,o,g}(v) = \min\limits_{u \in PRED_v} \tau_u / \max\limits_{u \in PRED_v} L_{L,o,g}^{\max}(u,v)$ The granularity of G is defined as $\gamma_{L,o,g}(G) = \min\limits_{v \in G} \gamma_{L,o,g}(v)$. Note that the maximum communication costs and, hence, the granularty of a task-graph depend on the communication parameters L, o and g of the target machines. We summarizes the upper bounds on the execution time for scheduling task graphs – naive transformation: $(1 + 1/\gamma_{L,o,g}(G))$ cf. [13], linear clustering: $(1 + 1/\gamma_{L,o,g}(G)$ cf.[9], Brent-clustering: $(1 + 1/\gamma_{L,o,g}(G)(W(G)/P + T(G))$ cf. [8].

3   Implementing Iterative Oblivious Programs

Oblivious programs can be easily scheduled to LogP-machines if the input size is known: compute the task graph $G(n)$ for inputs of size n and schedule it according to one of the scheduling strategies for LogP-machines. This is not possible for iterative oblivious programs since it is unknown how often the loop bodies in the program are executed. The basic idea is to perform a synchronization after a loop iteration. The oblivious loop bodies are scheduled according to a scheduling strategy for oblivious programs. Furthermore, it might be necessary to communicate results from the end of one loop iteration to the next or to the tasks following the loop. The transformation assume that the input size n is known. We first discuss the naive transformation and then demonstrate the generalization to linear clusterings and Brent-clusterings.

3.1 Naive Transformation of Iterative Oblivious Programs

The naive transformations of iterative oblivious program $\mathcal{P}$ for inputs of size n to LogP-programs works in two phases: The first phase computes naive transformations for the oblivious parts recursively over the structure of iterative oblivious programs, the second phase connects the different parts. The tasks determined by Phase 1 has the following property:

Lemma 1. *Every loop in an iteratively oblivious program, has a unique task computing its termination condition.*

Proof. Since the termination condition of a loop computes a boolean value, it is computed by a single task. Naive transformations of oblivious programs are non-redundant. Thus the task computing the termination condition is uniquely determined.

We identify loops with the tasks computing their termination condition. Phase 2 transforms the processes computed by Phase 1 for implementing $\mathcal{P}$. The processes computing tasks not belonging to a loop require no further transformation, because the preceding and succeeding tasks are always known are executed once. The tasks belonging to a loop can be executed more than once. The source of messages received by input tasks of a loop body depends on whether the loop is entered or a new iteration is started. However, the sources of the messages determine the operand of the computation of an input task. This mapping is implemented by the table *opd* and can also be defined for all tasks. Output tasks of loop bodies require a special treatment. Their sending messages can be classified as follows: the destination of the message is definitely a successor in the loop body (*oblivious messages*), the destination of the message depends on whether the loop is terminated or not (*non-oblivious messages*).

Because loops can be nested, an output task of an inner loop can also be an output task of an outer loop. Let $LOOPS_v$ be loops where v is an output tasks and $OUTER_v$ be the outermost loop in $LOOPS_v$. Then, every non-oblivious message can be classified according to their potential destination w.r.t. a loop in $LOOPS_v$: $FALSE_{v,l}$ is the set of potential successors if loop l does not terminate. $TRUE_{v,l}$ is the set of potential successors of the enclosing loop, if loop l does terminate. $DEFER_{v,l}$ is the set of potential successors outside of the enclosing loop. Obviously, it holds $DEFER_{v,OUTER_v} = \emptyset$. Thus, an output vertex v receives messages on termination of the loops $LOOPS_v$ and sends its messages to either $TRUE_{v,l}$ or $FALSE_{v,l}$ (lines (6)–(13)). Algorithm 2 summarizes this discussion. If the task is also an output task on the whole program, then the

Algorithm 2

Process *for task v belonging to a loop with oblivious successors $w_1, \ldots, w_m$, and computation Φ with n operands*

```
(1)        Process P_v
(2)           loop
(3)              do n times: m := recv; arg_{opd_{m.src}} := m.val;
(4)              res := Φ(arg_1, ..., arg_n)
(5)              send(res, v, w_1), ..., send(res, v, w_m);
                 -- lines (6)-(13) only if v is output vertex
(6)              term := recv_term_message; l := term.src;
(7)              while term.terminated ∧ l ≠ OUTER_v do
(8)                 for w ∈ TRUE_{v,l} do send(res, w);
(9)                 term := recv_term_message; l := term.src;
(10)             end;
(11)             if term.terminated then
(12)                for w ∈ TRUE_{v,l} do send(res, w);
(13)             else for w ∈ FALSE_{v,l} do send(res, w)
(14)          end
(15)       end
```

Algorithm 3

Process *for task v computing the termination condition:*

```
(1)        process P_v
(2)           loop
(3)              do n times: m := recv; arg_{opd_{m.src}} := m.val;
(4)              res := Φ(arg_1, ..., arg_n)
(5)              broadcast(res, OUT_v);
(6)           end;
(7)        end;
```

process is terminated. For simplicity we assume that the termination messages arrive in order of their broadcast. Algorithm 3 shows the processes computing a termination condition. It must be broadcasted to all output vertices of the loop.

Theorem 1. *The naive transformation on the LogP-Machine is correct.*

Proof. (-sketch:) Let $\mathcal{P}$ be an iteratively oblivious PRAM-Program and $[\![\mathcal{P}]\!]$ be the LogP-program obtained from $\mathcal{P}$ by the naive transformation. We prove the correctness of the naive transformation by induction on the structure of $\mathcal{P}$.
CASE 1: $\mathcal{P}$ is oblivious. The correctness follows directly from the results of [13].

CASE 2 $\mathcal{P} \equiv \mathcal{P}_1; \mathcal{P}_2$. Suppose the naive transformation is correct for $\mathcal{P}_1$ and $\mathcal{P}_2$. Then the output tasks of $[\![\mathcal{P}_1]\!]$ send their results to the input tasks of $[\![\mathcal{P}_2]\!]$. If these messages are sent to the correct destination, then the naive transformation is also correct for $\mathcal{P}$, because the mapping *opd* assigns the operands correctly.

If $\mathcal{P}_1$ is not a loop, then the messages sent by the output vertices of $[\![\mathcal{P}_1]\!]$ are oblivious. Thus the messages of the output vertices are sent to the correct destination. Let $\mathcal{P}_1$ be a loop and v be an output task v. Since the termination conditions of all loops in $LOOPS_v$ must yield true, every input task w that is a potential successor of v satisfies $w \in DEFER_{v,l}$ for all $l \in LOOPS_v \setminus OUTER_v$, or $w \in TRUE_{v,OUTER_v}$ where the sets are constructed w.r.t. $\mathcal{P}$. By construction of these sets, the messages are sent to the correct destination (or are output vertices of $\mathcal{P}$).

CASE 3 $\mathcal{P}$ is a loop with body $\mathcal{B}$. This case requires a proof by induction on the number of iterations of $\mathcal{P}$. The induction step can be performed with similar arguments as in Case 2.

Linear schedules and Brent-schedules can be implemented similarly. The first phase determines the processes obtained by linear clustering or Brent-clustering. The receiving phases and sending phases for task v is the same as in Algorithm 2 and occur also immediately before and after the computation performed by task v. In particular, if v is an output task, then non-oblivious messages are sent. If process corresponding to a cluster is in a loop body, then it is iterated as in Algorithm 2. We denote this transformations by $[\![\cdot]\!]_L$ and $[\![\cdot]\!]_B$, respectively.

4   Execution Time

The estimation on the execution time is done by two steps. First, we estimate the execution time if the broadcast performed per loop iteration can be executed in zero time. Then, we add the factor lost by this broadcast. For both cases,

Let $\mathcal{P}$ be an iterative oblivious PRAM-program, x be an input of size n, and $\mathcal{P}'$ obtained by one of the transformations of Section 3. $TIME_{\mathcal{P}'}(x)$ denotes the execution time of a LogP-program $\mathcal{P}'$ on input x. Suppose we monitor the execution of $\mathcal{P}'$ on x. Then, we obtain a clustering Cl of G_x (each iteration of a loop body defines a separate cluster). We say that $\mathcal{P}'$ *corresponds* to the clustering Cl of G_x.

Lemma 2 (Transfer Lemma). *The following correspondences hold for $\mathcal{P}$ and inputs x of size n:*

- *If $\mathcal{P}' = [\![\mathcal{P}]\!](n)$ then $\mathcal{P}'$ corresponds to the naive transformation of G_x and $TIME_{\mathcal{P}'}(x) \leq (1 + 1/\gamma_{L,o,g}(G_x))T(G_x)$.*
- *If $\mathcal{P}' = [\![\mathcal{P}]\!]_L(n)$ then $\mathcal{P}'$ corresponds to a linear clustering of G_x and $TIME_{\mathcal{P}'}(x) \leq (1 + 1/\gamma_{L,o,g}(G_x))T(G_x)$.*
- *If $\mathcal{P}' = [\![\mathcal{P}]\!]_B(n)$ then $\mathcal{P}'$ corresponds to a Brent-clustering of G_x and $TIME_{\mathcal{P}'}(x) \leq (1 + 1/\gamma_{L,o,g}(G_x)) \times (W(G_x)/P + T(G_x))$.*

where the execution times to not count for the broadcasts.

We now add the broadcast times. Broadcasting on the LogP machine has been studied by [7] who showed that an greedy implementation (i.e. every processor sends as fast as possible the broadcast item as soon as it receives the item) is optimal. Provided that the processor initializing the broadcast is known in advance (as it is true in our case) all send and receive operations can be determined statically for an instance of the LogP machine, i.e. optimal broadcasting is oblivious. Let $B_{L,o,g}(P)$ be the execution time for a broadcast to P processors.

We relate now the execution time of a single loop iteration on the PRAM with the costs for broadcasting the termination condition to its output tasks. Let $\mathcal{P}$ be an iteratively oblivious PRAM-program. The *degree of obliviousness* of a loop l of $\mathcal{P}$ for inputs of size n, is defined by $\rho_{L,o,g}(l,n) = T(b)/(T(b) + B_{L,o,g}(n))$. where b is the body of l. Let $LOOPS(\mathcal{P})$ be the set of loops of $\mathcal{P}$ The *degree of obliviousness* of $\mathcal{P}$ is defined by $\rho_{L,o,g}(\mathcal{P},n) = \min_{l \in LOOPS(\mathcal{P})} \rho_{L,o,g}(l,n)$. The time bounds of Lemma 2 are delayed by at most a factor of $1/\rho_{L,o,g}(\mathcal{P},n)$. We define the notion of *granularity* of an iteratively oblivious PRAM-program $\mathcal{P}$ for all inputs of size n by $\gamma_{L,o,g}(\mathcal{P},n) = \min_{x \in I_n} \gamma_{L,o,g}(G_x)$ where I_n is the set of all inputs of size n. We obtain directly our final result:

Theorem 2 (Execution Time on Iteratively Oblivious Programs). *Let $\mathcal{P}$ be an iteratively oblivious PRAM-program. Then*

$$TIME_{[\mathcal{P}]}(n) \leq \left(1 + \frac{1}{\gamma_{L,o,g}(\mathcal{P},n)}\right) \frac{T(\mathcal{P})}{\rho_{L,o,g}(\mathcal{P},n)}$$

$$TIME_{[\mathcal{P}]_L}(n) \leq \left(1 + \frac{1}{\gamma_{L,o,g}(\mathcal{P},n)}\right) \frac{T(\mathcal{P})}{\rho_{L,o,g}(\mathcal{P},n)}$$

$$TIME_{[\mathcal{P}]_B}(n) \leq \left(1 + \frac{1}{\gamma_{L,o,g}(\mathcal{P},n)}\right) \left(\frac{W(\mathcal{P})}{P} + T(\mathcal{P})\right) \rho_{L,o,g}(\mathcal{P},n)^{-1}$$

5   Conclusions

We extended the application of scheduling algorithms from task-graphs to iterative programs. The main idea is to apply standard scheduling algorithms (linear scheduling, Brent-scheduling) to the parts of the algorithm that can be modeled as task-graphs and to synchronize loop iterations by barrier synchronization. We showed that the well-known approximation factors for linear scheduling and Brent-scheduling for LogP-machines are divided by a *degree of obliviousness* $\rho_{L,o,g}$. This degree of obliviousness models how close an iterative algorithm comes to an oblivious algorithm (oblivious programs have a degree of obliviousness $\rho_{L,o,g} = 1$).

Although we can apply all scheduling algorithms for task graphs to iterative programs, we cannot generalize the above results on the time bounds to these algorithms. The problem occurs when the algorithm produces essentially different schedules for unrolled loops and composed loop bodies, i.e. the Transfer Lemma 2 is not satisfied.

The technique of loop synchronization might be useful for implementing scheduling algorithms in compilers. Instead of unrolling (oblivious) loops completely, it might be useful to unroll loops just a fixed number of times. In this case, we have the same situation as with iterative oblivious algorithms.

References

[1] D. Culler, R. Karp, D. Patterson, A. Sahay, K. E. Schauser, E. Santos, R. Subramonian, and T. von Eicken. LogP: Towards a realistic model of parallel computation. In *4th ACM SIGPLAN Symposium on Principles and Practice of Parallel Programming (PPOPP 93)*, pages 1–12, 1993. published in: SIGPLAN Notices (28) 7.

[2] D. Culler, R. Karp, D. Patterson, A. Sahay, K. E. Schauser, E. Santos, R. Subramonian, and T. von Eicken. LogP: A practical model of parallel computation. *Communications of the ACM*, 39(11):78–85, 1996.

[3] B. Di Martino and G. Ianello. Parallelization of non-simultaneous iterative methods for systems of linear equations. In *Parallel Processing: CONPAR 94 - VAPP VI*, volume 854 of *Lecture Notes in Computer Science*, pages 253–264. Springer, 1994.

[4] Jörn Eisenbiegler, Welf Löwe, and Andreas Wehrenpfennig. On the optimization by redundancy using an extended LogP model. In *International Conference on Advances in Parallel and Distributed Computing (APDC'97)*, pages 149–155. IEEE Computer Society Press, 1997.

[5] I. Foster. *Design and Building Parallel Programs - Concepts and Tools for Parallel Software Engeneering*. Addison–Wesley, 1995.

[6] R. M. Karp and V. Ramachandran. Parallel algorithms for shared memory machines. In J. van Leeuwen, editor, *Handbook of Theoretical Computer Science Vol. A*, pages 871–941. MIT-Press, 1990.

[7] R.M. Karp, A. Sahay, E.E. Santos, and K.E. Schauser. Optimal broadcast and summation in the LogP model. *ACM-Symposium on Parallel Algorithms and Architectures*, 1993.

[8] W. Lwe, J. Eisenbiegler, and W. Zimmermann. Optimizing parallel programs on machines with fast communication. In *9. International Conference on Parallel and Distributed Computing Systems*, pages 100–103, 1996.

[9] Welf Löwe, Wolf Zimmermann, and Jörn Eisenbiegler. On linear schedules for task graphs for generalized LogP-machines. In *Europar'97: Parallel Processing*, volume 1300 of *Lecture Notes in Computer Science*, pages 895–904, 1997.

[10] M. Philippsen. *Optimierungstechniken zur bersetzung paralleler Programmiersprachen*. PhD thesis, Universitt Karlsruhe, 1994. VDI-Verlag GmbH, Dsseldorf, 1994, VDI Fortschritt-Berichte 292, Reihe 10: Informatik.

[11] M. Wolfe. *High Performance Compilers for Parallel Computing*. Addison–Wesley, 1995.

[12] H. Zima and B. Chapman. *Supercompilers for Parallel and Vector Computing*. ACM–Press, NY, 1990.

[13] W. Zimmermann and W. Löwe. An approach to machine-independent parallel programming. In *Parallel Processing: CONPAR 94 - VAPP VI*, volume 854 of *Lecture Notes in Computer Science*, pages 277–288. Springer, 1994.

Scheduling Arbitrary Task Graphs on LogP Machines

Cristina Boeres, Aline Nascimento*, and Vinod E.F. Rebello

Instituto de Computação, Universidade Federal Fluminense (UFF)
Niterói, RJ, Brazil
{boeres,aline,vefr}@pgcc.uff.br

Abstract. While the problem of scheduling weighted arbitrary *DAGs* under the *delay model* has been studied extensively, comparatively little work exists for this problem under a more realistic model such as *LogP*. This paper investigates the similarities and differences between task clustering algorithms for the delay and *LogP* models. The principles behind three new algorithms for tackling the scheduling problem under the *LogP* model are described. The quality of the schedules produced by the algorithms are compared with good delay model-based algorithms and a previously existing *LogP* strategy.

1 Introduction

Since tackling the scheduling problem in an efficient manner is imperative for achieving high performance from message-passing parallel computer systems, it continues to be a focus of great attention in the research community. Until recently, the standard communication model in the scheduling community has been the *delay model*, where the sole architectural parameter for the communication system is the *latency* or *delay*, i.e. the transit time for each message word [12]. However, the dominant cost of communication in today's architectures is that of crossing the network boundary. This is a cost which cannot be modelled as a latency and therefore new classes of scheduling heuristics are required to generate efficient schedules for realistic abstractions of today's parallel computers [3]. This has motivated the design of new parallel programming models, e.g. *LogP* [4]. The *LogP model* is an MIMD message-passing model with four architectural parameters: the latency, L; the overhead, o, the CPU penalty incurred by a communication action; the gap, g, a lower bound on the interval between two successive communications; and the number of processors, P.

In this paper, the term *clustering algorithm* refers specifically to the class of algorithms which initially consider each task as a cluster and then merge clusters (tasks) if the completion time can be reduced. A later stage allocates the clusters to processors to generate the final schedule. A number of clustering algorithms

---

* The author was supported by a postgraduate scholarship from Fundação de Amparo à Pesquisa do Estado do Rio de Janeiro (FAPERJ) and CAPES, Ministério de Educação e Cultura (MEC), Brazil.

P. Amestoy et al. (Eds.): Euro-Par'99, LNCS 1685, pp. 340–349, 1999.
© Springer-Verlag Berlin Heidelberg 1999

exist for scheduling tasks under the delay model, well known examples include: *DSC* [13]; *PY* [12]; and an algorithm by Palis et al. [11] which we denote by *PLW*. It has been proved that these algorithms generate schedules within at least a factor of two of the optimal for arbitrary *DAGs*.

In comparison, identifying scheduling strategies that consider *LogP*-type characteristics is difficult. Only recently have such algorithms appeared in the literature. To the best of our knowledge, all of these algorithms are limited to specific types of *DAGs* or restricted *LogP* models. Kort and Trystram [8] presented an optimal polynomial algorithm for scheduling *fork* graphs on an unbounded number of processors considering $g = o$ and equal sized messages. Zimmerman et al. [14] extended the work of linear clustering [7, 13] to propose a clustering algorithm which produces optimal k-linear schedules for tree-like graphs when $o \geq g$. Boeres and Rebello [2, 3] proposed a task replication strategy for scheduling arbitrary UET-UDC (unit execution tasks, unit data communication) *DAGs* on machines with a bounded number of processors under both the *LogP* (also when $o \geq g$) and *BSP* models.

The purpose of this work is to study, from first principles, the problem of scheduling arbitrary task graphs under the *LogP* model using task replication-based clustering algorithms, by generalising the principles and assumptions used by existing delay model-based algorithms. The following section introduces the terminology and definitions used throughout the paper and outlines the influence of the *LogP* parameters on the scheduling problem. The issues that need to be addressed in the design of a clustering-based scheduling algorithm for the *LogP* model are discussed in Sect. 3. In Sect. 4, three replication-based task clustering algorithms are briefly presented for scheduling general arbitrary weighted *DAGs* on *LogP* machines (although an unbounded number of processors is assumed). Section 5 compares the makespans produced by the new algorithms against other well known clustering algorithms under both delay and *LogP* model conditions. Section 6 draws some conclusions and outlines our future work.

2   Scheduling under the Delay and *LogP* Models

A parallel application is often represented by a *directed acyclic graph* or *DAG* $G = (V, E, \varepsilon, \omega)$, where: the set of vertices, V, represent *tasks*; E the precedence relation among them; $\varepsilon(v_i)$ is the execution cost of task $v_i \in V$; and $\omega(v_i, v_j)$ is the weight associated to the edge $(v_i, v_j) \in E$ representing the amount of data units transmitted from task v_i to v_j. Two dependent tasks (i.e. tasks v_i and v_j for which $\exists (v_i, v_j) \in E$) are often grouped together to avoid incurring the cost of communicating. If a task has a successor in more than one cluster it may be advantageous to *duplicate* the ancestor task and place a copy of it in each of the clusters containing the successor. In any case, with or without task duplication, the task clustering problem is NP-hard [12].

This paper associates independent overhead parameters with the *sending* and *receiving* of a message, denoted by λ_s and λ_r, respectively (i.e. $o = \frac{\lambda_s + \lambda_r}{2}$). For the duration of each overhead the processor is effectively *blocked* unable to execute other tasks in V or even to send or receive other messages. Consequently, any scheduling algorithm must, in some sense, view these sending and receiving overheads also as "tasks" to be executed by processors. The rate of executing communication events may be further slowed due to the network (or the network interface) capacity modeled by the gap g. The execution order of the overhead tasks is important since two or more messages can no longer be sent simultaneously as in the delay model. Throughout this paper, the term *delay model conditions* is used to refer to situation where the parameters λ_s, λ_r and g are zero and the term *LogP conditions* used when otherwise.

In [12], the *earliest start time* of a task v_i, $e(v_i)$, is defined as a *lower bound* on the optimal or earliest possible time that task v_i can start executing. As such it is not always possible to schedule task v_i at this time [12]. In order to avoid confusion, we introduce the term $e_s(v_i)$, the *earliest schedule time*, to be the earliest time that task v_i can start executing on any processor. Thus, the completion time or *makespan* of the optimal schedule for G is $\max_{v_i \in V}\{e_s(v_i)+\varepsilon(v_i)\}$.

Many scheduling heuristics prioritise each task in the graph according to some function of the perceived computation and communication costs along all paths to that task. The costliest path to a task v_i is often referred to as the *critical path* of task v_i. Possible candidate functions include both $e(v_i)$ and $e_s(v_i)$. One must be aware that the critical path of a task v_i may differ depending on the cost function chosen. Scheduling heuristics which use some form of critical path analysis need to define how to interpret and incorporate the *LogP* parameters when calculating the costs of paths in the *DAG*. One of the difficulties being that these are communication costs paid for by computation time.

The *LogP* model is not an extension of the delay model but rather a generalisation, i.e. the delay model is a specific instance of the *LogP* one. Therefore, it is imperative to understand the underlying principles and identify the assumptions which can be adopted in *LogP* clustering algorithms.

3   Cluster-Based Scheduling for the *LogP* Model

Clustering algorithms typically employ two stages: the first determines which tasks should be included and their order within a cluster; the second, identifies the clusters required to implement the input *DAG G* mapping each to a processor. These algorithms aim to obtain an optimal schedule for a *DAG G* by attempting to construct a cluster $C(v_i)$ for each task v_i so that v_i starts executing at the earliest time possible (i.e. $e_s(v_i)$). In algorithms which exploit task duplication to achieve better makespans, $C(v_i) \forall v_i \in V$ will contain the *owner* task v_i and, determined by some cost function, *copies* of some of its ancestor tasks.

The iterative nature of creating a cluster can be generalised to the form shown in Algorithm 1. In lines 2 and 5, the list of candidate tasks (*lcands*) for inclusion into a cluster consists of those tasks which become immediate ancestors of the cluster, i.e. $iancs(C(v_i)) = \{u \mid (u,t) \in E \wedge u \notin C(v_i) \wedge t \in C(v_i)\}$.

Algorithm 1 : *mechanics-of-clustering* $(C(v_i))$;

1 $C(v_i) = \{v_i\}$;   /* Algorithm to create a cluster for task v_i */
2 $lcands = iancs(C(v_i))$;
3 while *progress condition* do
4   let $cand \in lcands$; $C(v_i) = C(v_i) \cup \{cand\}$;
5      $lcands = iancs(C(v_i))$;

The goodness of a clustering algorithm (and the quality of the results) depend on four crucial design issues: (i) calculating the makespan; (ii) the ordering of tasks within the cluster; (iii) the *progress condition* (line 3); and (iv) the choice of candidate task for inclusion into the cluster (line 4).

The makespan of a cluster $C(v_i)$ can be determined by the costliest path to task v_i (critical path). Effectively, there are only two types of path: the path due to *computation* in the cluster; and the paths formed by the immediate ancestor tasks of the cluster. The *cluster path cost*, $m(C(v_i))$, is the earliest time that v_i can start due to cluster task computation, including receive overheads but ignoring any idle periods where a processor would have to wait for the data sent by tasks $\in iancs(C(v_i))$ to arrive at $C(v_i)$. The order in which the task are executed in the cluster has a significant influence on this cost (finding the best ordering is a NP-complete problem [6]). The *ancestor path cost*, $c(u, t, v_i)$ is the earliest time that v_i can start due solely to the path from task u, an immediate ancestor of task $t \in C(v_i)$, to v_i. The ancestor path with the largest cost is known as the *critical ancestor path*

The *progress condition* determines when the cluster is complete. Commonly this condition is often derived from a comparison of the makespan before and after the inclusion of the chosen candidate task. This can be simplified to compare the two critical paths if an appropriate cost function is used. A candidate task is added to the cluster only if its inclusion does not increase the makespan. Otherwise, the creation process stops since the algorithm assumes that the makespan can be reduced no further, i.e. that the makespan is a monotonically decreasing function, up to its global minima, with respect to the inserted tasks. This assumption is based on the fourth issue – every choice of candidate has to be the best one. The list *lcands* contains the best candidate since the inclusion of a non-immediate ancestor can only increase the makespan of the cluster. But which of the tasks is the best choice?

Under the delay model, the inclusion of a task has two effects: it will increase the cluster path cost at the expense of removing the critical ancestor path; and introduce a new ancestor path for each immediate ancestor of the inserted task not already in the cluster. Because the cluster path cost increases monotonically under this model, the progression condition can be simplified to a combination of the following two conditions, the algorithm is allowed to progress while: the cluster path is not the critical path (Non-Critical Path Detection (NCPD)); and inserting the chosen task does not increase the makespan (Worth-While Detection (WWD)). The makespan of a cluster can only be reduced if the cost of the critical path is reduced. If the critical ancestor path is the critical path of the cluster, the critical ancestor task u is the only candidate whose inclusion into the cluster has the potential to diminish the makespan. While selecting this task is on the whole a good choice, it is not always the best. The critical path cost could increase the makespan sufficiently to cause the algorithm to stop prematurely, at a local optimum for the makespan, because the inserted task u has a high communication cost with one of its ancestor (a cost not reflected by the chosen cost function value for u).

The effects of including a candidate task under $LogP$ conditions are similar except that the cluster path cost may not necessarily increase. Communication overheads need to be treated like tasks since they are costs incurred on a processor even though they are associated with ancestor paths. This means that an immediate ancestor path can affect the cluster path cost and the cost of the other ancestor paths of the cluster. Not only can the inclusion of a non-critical ancestor path task reduce the makespan but improvements to the makespan may be achieved by progressing beyond the NCPD point through the removal of the receive overheads incurred in the cluster. In addition to dependencies between program tasks, the problem of ordering tasks within a cluster now has to address the dependencies of these overhead "tasks" to their program tasks and their network restrictions (i.e. the effects of parameter g).

All in all, the quality of makespans produced by a clustering algorithm depends on the quality of the implementation of each of the above design issues. These issue are not independent, but interrelated by the multiple role played by the cost function. How paths are costed is important since the same cost function is often used: to calculate the makespan and thus affecting the progress condition; and to order tasks within a cluster. Furthermore, the cost function may also be the basis for choosing a candidate task. If a design issue cannot be addressed perfectly, the implementation of the other issues could try to compensate for it. For example, if the best candidate is not always chosen, the makespan will not be a monotonically decreasing function. Therefore, progress condition must be designed in such way that will allow the algorithm to escape local minima but not to execute unnecessarily creating a poor cluster.

3.1   Adopted Properties for $LogP$ Scheduling

In order to schedule tasks under a $LogP$-type model, we propose that clustering algorithms apply some *scheduling restrictions* with regard to the properties of clusters. In the first restriction, only the owner task of a cluster may send data to a task in another cluster. This restriction does not adversely affect the makespan since, by definition of $e_s(v_i)$, an owner task will not start execution later than any copy of itself. Also, if a non-owner task t in cluster $C(v_i)$ were to communicate with a task in another cluster, the processor to which $C(v_i)$ is allocated would incur a send overhead after the execution of t which in turn might force the owner v_i to start at a time later than its earliest possible.

The second restriction specifies that each cluster can have only one successor. If two or more clusters, or two or more tasks in the same cluster, share the same ancestor cluster then a unique copy of the ancestor cluster is assigned to each successor irrespective of whether or not the data being sent to each successor is identical. This removes the need to incur the multiple send gap costs and send overheads which may unnecessarily delay the execution of successor clusters.

The third restriction, currently a temporary one, assumes that receive overheads are executed immediately before the receiving task even though they have to be scheduled at intervals of atleast g. Future work will investigate the merits of treating these overheads just like program tasks and allowing them to be scheduled in any valid order within the cluster. The fourth restriction is related to the third and to the behaviour of the network interface. This restriction assumes that the receive overheads, scheduled before their receiving task, are ordered by the arrival time of their respective messages. It is assumed that the network interface implements a queue to receive messages.

These restrictions are new in the sense that they do not need to be applied by scheduling strategies which are used exclusively under delay model conditions. The purpose of these restrictions is to aid in the minimisation of the makespan, though they do incur the penalty of increasing the number of clusters required and thus the number of processors needed. Where the number of processors used is viewed as a reflection on the cost of implementing the schedule, post-pass optimisation can be applied in a following stage of the algorithm to relax the restrictions and remove clusters which become redundant (i.e. clusters whose removal will not increase the makespan).

4   Three New Scheduling Algorithms

This section discusses the design of three clustering algorithms, focusing on the cluster creation stage. These algorithms are fundamentally identical but differ in the manner they address the design issues discussed in Sect. 3. The second stages of these algorithms, which are identical, determine which of the generated clusters are needed to implement the schedule.

Unlike traditional delay model approaches which tend to use the earliest start time $e(v_i)$ as the basis for the cost function, the $LogP$ algorithms presented here use the earliest schedule time $e_s(v_i)$. Using $e_s(v_i)$ should give better quality results since this cost represents what is achievable (remember that $e(v_i)$ is a lower bound and that cost is an important factor in the mechanics of cluster algorithms). On the other hand, since finding $e_s(v_i)$ is much more difficult, these algorithms in fact create the cluster $C(v_i)$ to find a good approximation, $s(v_i)$, to $e_s(v_i)$ which is then used to create the clusters of task v_i's successors. $s(v_i)$, therefore, is the *scheduled time* or *start time* of task v_i in its own cluster $C(v_i)$.

In the first algorithm, $BNR2$ [10], tasks are not ordered in non-decreasing order of their earliest start time (as in most delay model approaches) nor by their scheduled time but rather in an order determined by the sequence in which tasks were included in the cluster and the position of their immediate successor task at the time of their inclusion. The *ancestor path cost*, $c(u, t, v_i)$, for immediate ancestor task u of task t in $C(v_i)$ is the sum of: the execution cost of $C(u)$

$(s(u) + \varepsilon(u))$; one sending overhead (λ_s) (due to the second $LogP$ scheduling restriction); the communication delay $(\tau \times \omega(u,t))$; the receiving overhead (λ_r); and the computation time of tasks t to v_i, which includes receive overheads of edges arriving after the edge (u,t) and takes into account g and restrictions 3 and 4. The critical ancestor task is the task $u \in iancs(C(v_i))$ for which $c(u,t,v_i)$ is the largest.

If critical ancestor path cost, $c(u,t,v_i)$, is greater than the cluster path cost then task u becomes the chosen candidate for inclusion into cluster $C(v_i)$ (the NCPD point of the progress condition). Before committing u to $C(v_i)$, $BNR2$ compares this critical ancestor path cost with the cluster path cost of $\{C(v_i) \cup \{u\}\}$ (the WWD point of the progress condition). Note this detection point does not actually check for an non-increasing makespan. This less strict condition allows the algorithm to proceed even if the makespan increases, permitting $BNR2$ to escape some local makespan minimas as long as the NCPD condition holds. One drawback is that the algorithm can end up generating a worse final makespan. A solution might be to keep a record of the best makespan found so far [10].

The second algorithm, $BNR2_s$, is a simplification of the previous algorithm. Only candidates from the set of *immediate ancestors* of task v_i are considered for inclusion into $C(v_i)$. But the inclusion of an ancestor task u implies that all tasks in $C(u)$ should be included in $C(v_i)$. Also, the immediate ancestor clusters $C(w)$ of $C(u)$ become immediate ancestors of cluster $C(v_i)$ unless owner task w is already in $C(v_i)$. The mechanics of the algorithm are almost the same as $BNR2$: the candidate chosen is still the task on the critical ancestor path; however, the progress condition compares the two makespans (before and after the candidate's inclusion); and the tasks in the cluster are ordered in nondecreasing order of their scheduled time.

The third algorithm, $BNR2_h$, is a hybrid of the first two. As in $BNR2_s$, the cluster tasks are ordered in nondecreasing order of their scheduled time and the critical ancestor path task is the chosen candidate. Initially the algorithm behaves like $BNR2_s$, inserting into $C(v_i)$ copies of all tasks in $C(u)$ (u being the first chosen immediate ancestor) and inheriting copies of the immediate ancestor clusters $C(w)$ of $C(u)$. After this cluster insertion, the algorithm's behaviour switches to one similar to $BNR2$ where the cluster is now grown a task at a time. Note that the initial inherited ancestor tasks w are never considered as candidates for inclusion. The main difference from $BNR2_h$ is the progress condition. This algorithm is allowed to continue while the *weight* of the cluster (this being the sum of the weights of the program tasks in the cluster) is less than both its makespan and best makespan found so far. This is a monotonically increasing lower bound on the makespan of $C(v_i)$. The progress condition thus allows the algorithm to speculatively consider possible worthwhile cluster schedules even though the makespan may increase. When the algorithm stops, the best makespan found will be the global minima (given the tasks inserted and ordering adopted). This is not necessarily the optimal makespan since this depends on addressing the design issues (outlined in Sect. 3) perfectly.

5   Results

The results produced by the three algorithms have been compared with the clustering heuristics *PLW* and *DSC* under delay model conditions, and *MSA* under *LogP* conditions using a benchmark suite of UET-UDC *DAG*s which includes out-trees (Ot), in-trees (It), diamond *DAG*s (D) and a set of randomly generated *DAG*s (Ra) as well as irregular weighted *DAG*s (Ir) taken from the literature. A sample of the results are presented in the following tables, where the number of tasks in the respective graphs is shown as a subscript. For fairness, the value of the makespan M is the execution time for the schedule obtained from a *parallel machine simulator* rather than the value calculated by the heuristics. The number of processors required by the schedule is P. C, the number of times the progression condition is checked, gives an idea of the cost to create the schedule.

Although the scheduling restrictions cause the *LogP* strategies to appear greedy with respect to the number of processors required, it is important to show that the makespans produced and processors used are comparable to those of delay model approaches under their model (Table 1). Scheduling approaches such as *DSC* and *PLW* have already been shown to generate good results (optimal or near-optimal) for various types of *DAG*s. Table 2 compares the makespans generated for UET-UDC *DAG*s with those of *MSA* (assuming $\lambda_r \geq g$).

Table 1. Results for a variety of graphs under the delay model.

DAG	τ	DSC		PLW		MSA		BNR2			BNR2$_s$			BNR2$_h$		
		P	M	P	M	P	M	P	M	C	P	M	C	P	M	C
Ir_{41} [7]	1	13	16	22	16	10	16	10	15	120	17	15	67	16	15	71
Ir_{41}	2	9	21	20	18	10	18	13	17	283	12	18	68	13	17	82
Ir_{41}	4	8	31	19	27	10	23	11	22	305	9	24	67	12	21	93
Ir_{41}	8	3	44	19	39	5	27	10	27	381	5	30	69	9	26	98
Ra_{80}	4	27	25	41	31	48	22	31	18	294	55	19	111	57	19	125
Ra_{186}	4	66	22	87	19	52	19	91	14	642	117	14	225	123	14	233
It_{511}	4	98	39	167	27	85	25	134	25	1357	48	37	510	85	28	797
Ot_{511}	4	146	27	264	13	256	9	256	9	4096	256	9	510	256	9	510
Di_{400}	4	27	146	165	140	37	116	31	98	4983	289	112	760	267	104	736

DAG	DSC		PLW		BNR2		BNR2$_s$		BNR2$_h$	
	P	M	P	M	P	M	P	M	P	M
Ir_{7a} [5]	2	9	3	8	3	8	3	8	3	8
Ir_{7b} [13]	3	8	5	12	3	8	3	8.5	3	8
Ir_{10} [11]	3	30	4	27	3	26	4	26	4	26
Ir_{13} [1]	7	301	8	275†	8	246	9	246	9	246
Ir_{18} [9]	5	550†	8	480	9	370	8	400	9	380

†[9] reports a makespan of 460 on 6 processors.

The results produced by the three cluster algorithms $BNR2$, $BNR2_s$ and $BNR2_h$ are, in the majority of cases, better than those of *DSC*, *PLW* and *MSA* under their respective models. $BNR2_s$ on the whole produces results only slightly worse than the other two in spite of its simplicitic approach (which is reflected by its C value). Under the delay model, we also analysed the effect of task ordering on the makespans produced by $BNR2_s$ and $BNR2_h$. Only in two cases for $BNR2_h$, does optimising the order improve the makespan (becoming 25 for It_{511} and 370 for Ir_{18}). The ease by which $BNR2_s$ and $BNR2_h$ find

Table 2. Results considering various LogP parameters values.

DAG	λ_s	λ_r	τ	MSA Prs	M	BNR2 Prs	M	C	BNR2$_s$ Prs	M	C	BNR2$_h$ Prs	M	C	λ_s	λ_r	τ	MSA Prs	M	BNR2 Prs	M	C	BNR2$_s$ Prs	M	C	BNR2$_h$ Prs	M	C
Ir_{41}	1	1	1	8	30	34	24	225	14	23	68	15	23	114	2	1	8	6	37	11	32	396	4	32	69	4	32	110
Ir_{41}	1	1	2	10	32	18	24	313	14	25	67	14	25	122	2	2	1	6	35	21	32	267	11	30	69	11	30	142
Ir_{41}	1	1	4	3	33	12	27	332	9	29	68	11	28	116	2	2	2	6	36	15	32	300	11	31	67	9	31	132
Ir_{41}	1	1	8	5	36	11	31	394	4	32	69	5	31	112	4	0	1	6	27	14	23	354	12	25	68	15	24	86
Ir_{41}	2	0	1	10	25	31	19	276	14	21	69	14	21	86	4	0	2	6	28	16	24	344	9	27	69	15	25	97
Ir_{41}	2	0	2	6	25	26	22	305	14	24	67	13	22	93	4	0	4	6	30	14	27	381	5	30	69	7	29	98
Ir_{41}	2	0	4	6	26	16	24	344	9	27	68	15	25	97	4	0	8	6	34	10	31	374	3	33	69	4	33	108
Ir_{41}	2	0	8	6	30	10	29	368	4	31	69	5	31	104	4	1	1	6	34	12	27	332	9	29	68	11	28	116
Ir_{41}	2	1	1	10	33	18	24	313	14	25	67	14	25	122	4	1	2	6	34	13	29	335	8	31	69	6	28	119
Ir_{41}	2	1	2	10	34	15	25	326	11	27	67	14	27	135	4	1	4	5	35	12	30	384	4	32	69	5	30	110
Ir_{41}	2	1	4	6	34	13	29	335	8	31	69	6	28	119	4	4	1	1	41	15	49	316	2	40	69	2	40	113

DAG	λ_s	λ_r	τ	MSA P	M	BNR2 P	M	C	BNR2$_s$ P	M	C	BNR2$_h$ P	M	C
Ra_{80}	4	2	2	25	56	146	46	238	150	60	121	113	60	329
Ra_{186}	4	2	2	78	47	158	36	660	134	38	232	102	38	416
It_{511}	4	2	2	85	45	85	45	1545	30	58	510	64	46	1106
Ot_{511}	4	2	2	256	9	256	9	4096	256	9	510	256	9	510

optimal schedules for out-trees is reflected by closeness of C to the number of tasks in the graph.

6 Conclusions and Future Work

We present three clustering algorithms for scheduling arbitrary task graphs with arbitrary costs, under the *LogP* model, onto an unbounded number of processors. To the best of our knowledge, no other general *LogP* scheduling approach exists in the literature. This work is not based on extending an existing delay model algorithm to a *LogP* one. Instead, these algorithms were design from first principles to gain a better understanding of the mechanics of *LogP* scheduling. Fundamental differences exist between these algorithms and their nearest delay model relatives (e.g. *PLW*), due in part to interaction of the *LogP* model parameters on the design issues and to the assumptions used to simplify the problem under the delay model.

Based on the results obtained so far, initial versions of these algorithms compare favourably against traditional clustering-based scheduling heuristics such as *DSC* and *PLW* which are dedicated exclusively to the delay model. For UET-UDC *DAG*s, all three algorithms perform surprising well compared to *MSA* which exploits the benefits of bundling messages to reduce the number of overheads incurred. Results of further experiments using graphs with a more varied range of granularities and connectivities are needed to complete the practical evaluation of these algorithms.

Future work will primarily focus on studying the effects of relaxing the scheduling restrictions and of various design issue decisions e.g. on what basis should program and overhead tasks be ordered within a cluster [10], are there alternative cost functions better able to choose candidate tasks?

References

[1] I. Ahmad and Y-K Kwok. A new approach to scheduling parallel programs using task duplication. In K.C. Tai, editor, *International Conference on Parallel Processing*, volume 2, pages 47–51, Aug 1994.

[2] C. Boeres and V. E. F. Rebello. A versatile cost modelling approach for multicomputer task scheduling. *Parallel Computing*, 25(1):63–86, 1999.

[3] C. Boeres, V.E.F. Rebello, and D. Skillicorn. Static scheduling using task replication for LogP and BSP models. In D. Pritchard and J. Reeve, editors, *The Proceedings of the 4th International Euro-Par Conference on Parallel Processing (Euro-Par'98)*, LNCS 1470, pages 337–346, Southampton, UK, September 1998. Springer-Verlag.

[4] D. Culler, R. Karp, D. Patterson, A. Sahay, K.E. Schauser, E. Santos, R. Subramonian, and T. von Eicken. LogP: Towards a realistic model of parallel computation. In *Proceedings of the 4th ACM SIGPLAN Symposium on Principles and Practice of Parallel Programming*, San Diego, CA, May 1993.

[5] A. Gerasoulis and T. Yang. A comparison of clustering heuristics for scheduling directed acyclic graphs on multiprocessors. *Journal of Parallel and Distributed Computing*, 16:276–291, 1992.

[6] A. Gerasoulis and T. Yang. List scheduling with and without communication. *Parallel Computing*, 19:1321–1344, 1993.

[7] S.J. Kim and J.C. Browne. A general approach to mapping of parallel computations upon multiprocessor architectures. In *Proceedings of the 3rd International Conference on Parallel Processing*, pages 1–8, 1988.

[8] I. Kort and D. Trystram. Scheduling fork graphs under LogP with an unbounded number of processors. In D. Pritchard and J. Reeve, editors, *The Proceedings of the 4th International Euro-Par Conference on Parallel Processing (Euro-Par'98)*, LNCS 1470, pages 940–943, Southampton, UK, September 1998. Springer-Verlag.

[9] Y. K. Kwok and I. Ahmad. Dynamic critical-path scheduling: An effective technique for allocating tasks graphs to multiprocessors. *IEEE Transactions on Parallel and Distributed Systems*, 7(5):505–521, May 1996.

[10] A. Nascimento. Aglomeração de tarefas em arquiteturas paralelas com memória distribuída. Master's thesis, Instituto de Computação, Universidade Federal Fluminense, Brazil, 1999. (In Portuguese).

[11] M.A. Palis, J.-C Liou, and D.S.L. Wei. Task clustering and scheduling for distributed memory parallel architectures. *IEEE Transactions on Parallel and Distributed Systems*, 7(1):46–55, January 1996.

[12] C.H. Papadimitriou and M. Yannakakis. Towards an architecture-independent analysis of parallel algorithms. *SIAM J. Comput.*, 19:322–328, 1990.

[13] T. Yang and A. Gerasoulis. DSC: Scheduling parallel tasks on an unbounded number of processors. *IEEE Transactions on Parallel and Distributed Systems*, 5(9):951–967, 1994.

[14] W. Zimmermann, M. Middendorf, and W. Lowe. On optimal k-linear scheduling of tree-like task graph on LogP-machines. In D. Pritchard and J. Reeve, editors, *The Proceedings of the 4th International Euro-Par Conference on Parallel Processing (Euro-Par'98)*, LNCS 1470, pages 328–336, Southampton, UK, September 1998. Springer-Verlag.

Scheduling with Communication Delays and On-Line Disturbances

Aziz Moukrim[1], Eric Sanlaville[1], and Frédéric Guinand[2]

[1] LIMOS, Université de Clermont-II, 63177 Aubière Cedex France
{moukrim, sanlavil}@ucfma.univ-bpclermont.fr
[2] LIH, Université du Havre, BP 540 76058 Le Havre Cedex France
guinand@fst.univ-lehavre.fr

Abstract. This paper considers the problem of scheduling tasks on multiprocessors. Two tasks linked by a precedence constraint and executed by two different processors must communicate. The resulting delay depends on the tasks and on the processor network. In our model an estimation of the delay is known at compile time; but disturbances due to network contention, link failures,... may occur at execution time. Algorithms computing separately the processor assignment and the sequencing on each processor are considered. We propose a partially on-line scheduling algorithm based on critical paths to cope with the possible disturbances. Some theoretical results and an experimental study show the interest of this approach compared with fully on-line scheduling.

1   Introduction

With the growing importance of Parallel Computing issues, the classical problem of scheduling n tasks subject to precedence constraints on m processors in the minimum amount of time attracted renewed attention. Several models including communication delays for both shared and distributed memory multiprocessor systems have been proposed and widely studied [1, 8]. If two tasks T_i and T_j are executed by two different processors, there is a delay between the end of T_i and the beginning of T_j due to data transfer between the two processors. The problem is NP-complete even for unit execution and communication times (the UECT problem), on an arbitrary number of processors or on an unlimited number (see the pioneering work of Rayward Smith [9], or the survey of Chrétienne and Picouleau [2]).

In many models the communication delays only depend on the source and destination tasks, not on the communication network. The general assumption is that this network is fully connected, and that the lengths of the links are equal ([3, 10]). This is rarely the case for a real machine: the network topology and the contention of the communication links, may largely influence the delays, not to speak of communication failures. Some scheduling methods take into account the topology as in [5]. It is also possible to use more precise models (see [7, 11] for simultaneous scheduling and routing) but the performance analysis is then

P. Amestoy et al. (Eds.): Euro-Par'99, LNCS 1685, pp. 350–357, 1999.

very difficult. In general, building an accurate model of the network is much complicated and entails a very intricated optimization problem.

This paper proposes the following approach: an estimation of the communication delays is known at compile time, allowing to compute a schedule. But the actual delays are not known before the execution. Building the complete schedule before the execution is then inadvisable. Conversely, postponing it to the execution time proves unsatisfactory, as we are then condemned to somewhat myopic algorithms. The method presented is a trade-off between these two approaches.

The paper is organized as follows. In section 2, the model is stated precisely, and the different algorithmic approaches are presented. The choice of two phase methods, processor assignment then sequencing, is justified. Section 3 presents a new partially on-line sequencing policy adapted to on-line disturbances. Some theoretical results for special cases are presented in section 4, and an experimental comparison is conducted and interpreted in section 5: for a fixed assignment, our approach is compared with on-line scheduling.

2 Preliminaries

Model and Definitions We consider the schedule of n tasks $T_1, \ldots, T_n$ subject to precedence relations denoted $T_i \prec T_j$, on m identical processors. Preemption (the execution of a task may be interrupted) and task duplication are not allowed. One processor may not execute more than one task at a time but can perform computations while receiving or sending data. The duration of task T_i is p_i. If two tasks T_i and T_j verify $T_i \prec T_j$ and are executed on two different processors, there is a minimum delay between the end of T_i and the beginning of T_j. The communication delay between tasks executed on the same processor is neglected. At compile time, the communication delays are estimated as $\tilde{c}_{ij}$ time units. It is expected however (see section 5) that these estimations are well correlated with the actual values c_{ij} (communication delays at execution time). The goal is to minimize the maximum task completion time, or *makespan*.

A schedule is composed of the *assignment* of each task to one processor and of the *sequencing* (or ordering) of the tasks assigned to one processor. A scheduling algorithm provides a complete schedule from a given task graph and a given processor network. We shall distinguish between *off-line* algorithms and *on-line* algorithms. An algorithm is off-line if the complete schedule is determined before the execution begins. An algorithm is on-line if the schedule is built during the execution; the set of rules allowing to build the final schedule is then called a policy. In our model, the communication delays c_{ij} are only known at execution time, once the data transfer is completed. In order to take into account the estimated delays a trade-off between off-line and on-line scheduling consists in a *partially on-line scheduling*, that is, after some off-line processing, the schedule is built on-line.

List-Scheduling approaches Without communication delay the so called List Schedules (LS) are often used as they provide good average performances, even as the performance bounds are poor. Remember LS schedules are obtained from a complete priority order of the tasks. In most cases the choice is based upon Critical Paths (CP) computation. When a processor is idle, the ready task (all its predecessors are already executed) of top priority is executed on this processor. A way to tackle scheduling with communication delays is to adapt list scheduling. Then the concept of ready task must be precised. A task is *ready on processor* π at time t if it can be immediately executed on that processor at that time (all data from its predecessors have arrived to π). This extension is called *ETF* for Earliest Task First scheduling, following the notation of Hwang et al [5].

Clustering approaches Another proposed method is the clustering of tasks to build a pre-assignment [3, 10]. The idea is to cluster tasks between which the communication delays would be high. Initially, each task forms a cluster. At each step two clusters are merged, until another merging would increase the makespan. When the number of processors m is limited, the merging must continue until the number of clusters is less than m. The sequencing for each processor is then obtained by applying the CP rule.

Limits of these approaches The above methods suppose the complete knowledge of the communication delays. Now if unknown disturbances modify these durations, one may question their efficiency. ETF schedules might be computed fully on-line, at least for simple priority rules derived from the critical path. The first drawback is the difficulty to build good priority lists. Moreover, fully on-line policies meet additional problems: if the communication delays are not known precisely at compile time, the ready time of a task is not known before the communication is achieved. But if this task is not assigned yet, the information relevant to that task should be sent to all processors, to guaranty its earliest starting time. This is impossible in practice.

The clustering approach would better fit our needs since the assignment might be computed off-line, and the sequencing on-line. But it suffers from other limitations. The merging process will result in large makespans when m is small. It is also poorly suited to different distances between processors.

It follows from the above discussion that the best idea is to compute the assignment off-line by one method or another (anyway, finding the optimal assignment is known to be NP-complete [10]). Then the sequencing is computed on-line so that the impact of communication delay disturbances is minimized. The on-line computation should be very fast to remain negligible with regard to the processing times and communication delays themselves. But it can be distributed, thus avoiding fastidious information exchanges between any processor and some "master" processor. Note that finding the optimal schedule when the assignment is fixed is NP-hard even for 2 processors and $UECT$ hypotheses [2].

3   An Algorithm for Scheduling with On-Line Disturbances

New approach based on partially on-line sequencing For a fixed assignment, the natural way to deal with on-line disturbances on communication delays is to apply a fully on-line sequencing policy based on ETF. After all communication delays between tasks executed on a same processor are zeroed, relative priorities between tasks may be computed much more accurately. It is expected however that this approach will result in bad choices. Suppose a communication delay is a bit larger than expected, so that at time t a task with high priority, say T_i, is not yet available. A fully on-line policy will not wait. It will instead, schedule some ready task T_j that might have much smaller priority. If T_i is ready for processing at $t + \epsilon$, for ϵ arbitrarily small, its execution will nonetheless be postponed until the end of T_j. We propose the following general approach (the choices for each step are detailed next).

Step 1 compute an off-line schedule based on the $\tilde{c}_{ij}$'s.

Step 2 compute a partial order $\prec_p$ including $\prec$, by adding precedences between tasks assigned to a same processor.

Step 3 at execution time, use some ETF policy to get a complete schedule considering assignment of step 1 and partial order of step 2.

Detailed partially on-line algorithm   The schedule of **step 1** is built by an ETF algorithm, based on Critical Path priority. We said that Critical Path based policies seemed best suited, as shown by empirical tests with known communication delays [12], and also by bound results (see [5], and [4] for a study of the extension of the Coffman–Graham priority rule to communication delays). These empirical tests, and ours, show that the best priority rule is the following: for each task T_i compute $L^*(i)$, the longest path to a final task, including processing times and communication delays (between tasks to be processed on different processors), the processing time of the final task but **not** the processing time of T_i. The priority of T_i is proportional with $L^*(i)$. The heuristic that sequences ready tasks using the above priority is called RCP^* in [12] (RCP if the processing time of T_i is included).

The partial order of **step 2** is obtained as follows. Suppose two tasks T_i and T_j are assigned to the same processor. If the two following conditions are respected:

1. T_i has larger priority than T_j for RCP^*
2. T_i is sequenced before T_j in the schedule of step 1,

then a precedence relation is added from T_i to T_j. This will avoid that a small disturbance leads to execute T_j before T_i at execution time.

In **step 3**, RCP^* is again used as sequencing on-line policy to get the complete schedule.

The resulting algorithm is called $PRCP^*$ for Partially on-line sequencing with RCP^*. The algorithm for which the sequencing is obtained Fully on-line by RCP^* is denoted by $FRCP^*$.

Example   Figures 1 and 2 show how our approach can outperform both fully off-line and fully on-line sequencing computations. A set of 11 unitary tasks is to be scheduled on two processors. All estimated communication delays are 1. Schedule a) is obtained at compile time by RCP^*. The resulting assignment is kept for scheduling with the actual communication delays. Schedule b) supposes the sequencing is also fixed, regardless of on-line disturbances (fully off-line

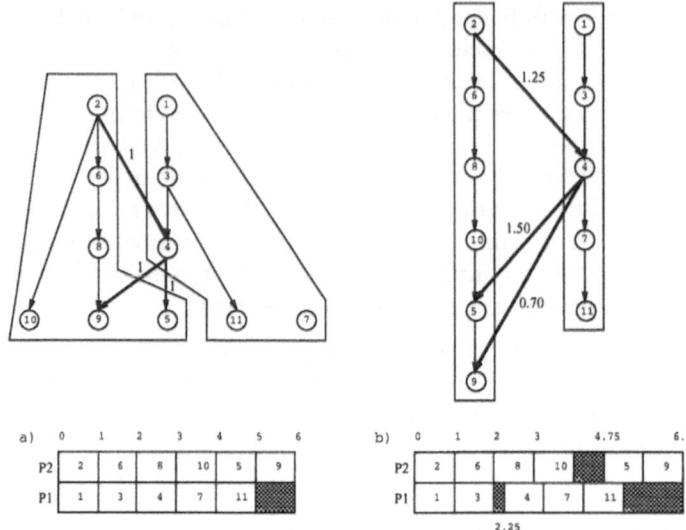

Fig. 1. RCP^* schedule at compile time and associated completely off-line schedule

schedule). Schedule c) is obtained by $PRCP^*$, and schedule d) by $FRCP^*$. After zeroing the internal communication delays, only $\tilde{c}_{24}$, $\tilde{c}_{45}$, and $\tilde{c}_{49}$ remain (bold arcs). The following precedences are added by our algorithm: for processor P_1, $T_4 \prec_p T_7$, $T_4 \prec_p T_{11}$, and for processor P_2, $T_8 \prec_p T_5$, and $T_8 \prec_p T_{10}$ (dashed arcs). Consider now the following actual communication delays: $c_{24} = 1.25$, $c_{45} = 1.50$, and $c_{49} = 0.70$. At time 2, task T_4 is not ready for processing, but T_7 is. $FRCP^*$ executes T_7 whereas $PRCP^*$, due to the partial order $\prec_p$, waits for T_4. At time 4 for $PRCP^*$, processor P_1 is available, and T_9 is ready. No additional precedence was added between T_5 and T_9 as they have the same priority, hence T_9 is executed and then T_5 is ready, so that P_2 has no idle time. For $FRCP^*$, exchanging T_7 and T_4 results in idleness for P_1, as T_4 is indeed critical.

In what follows we consider the merits of the different strategies for sequencing, once the assignment has been fixed. Hence optimality is to be understood within that framework: minimum makespan among all sequencing policies, for a fixed assignment. There is little hope to obtain theoretical results (optimality, bounds) for anything but very special cases. Two of them are presented in section 4.

4 Optimality of $PRCP^*$ for Fork and Join Graphs

Fork graphs In that case one task T_1 is the immediate predecessor of all the others, which are final tasks. For fixed communication delays, RCP or RCP^* find the minimum makespan (see [12]).

Adding disturbances do not much complicate. After the assignment a group of final tasks is to be sequenced on the same processor as T_1, say P_1. All com-

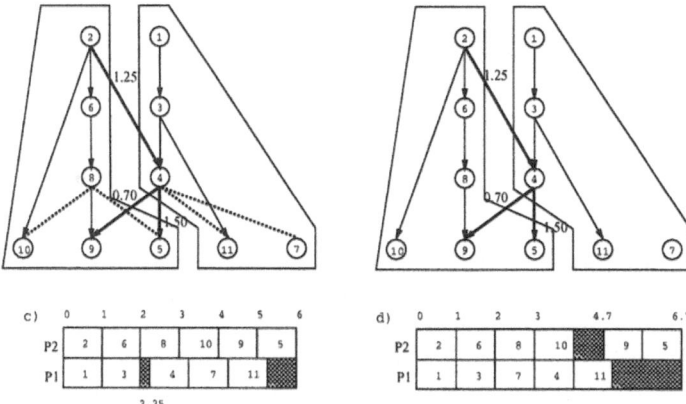

Fig. 2. $PRCP^*$ and $FRCP^*$ schedules

munication delays are zeroed, hence the completion time on P_1 is the sum of the completion times of all these tasks, and is independent of the chosen policy. Consider now a group of final tasks executed on another processor. Each has a ready time (or release date), $r_i = p_1 + c_{1i}$, and a processing time p_i. Minimizing the completion time on this processor is equivalent to minimizing the makespan of a set of independent tasks subject to different release dates on one machine. This easy problem may be solved by sequencing the tasks by increasing release dates and processing them as soon as possible in that order. If a fully on-line policy is used, it will do precisely that, whatever priority is given to the tasks. On the other hand, a partially off-line algorithm may add some precedence relations between these tasks. This may lead to unwanted idleness. Indeed, using RCP may enforce a precedence between T_i and T_j if $\tilde{c}_{1i} \leq \tilde{c}_{1j}$, and $p_i > p_j$. During the execution you may have $c_{1i} > c_{1j}$, and the processor may remain idle, waiting for T_i to be ready. However with RCP^*s, all priorities of the final tasks are equal, hence no precedence will be added, which guaranty optimality.

Theorem 1. *A $PRCP^*$ sequencing is optimal for a fixed assignment when the task graph is a fork graph.*

Join graphs RCP^* is optimal for join task graph (simply reverse the arcs of a fork graph) and fixed communication delays, whereas RCP is optimal only if all tasks have the same duration (see [12]). This is no longer true if the communication delays vary, as in that case there is no optimal policy: the sequencing of the initial tasks must be done using the $\tilde{c}_{in}$'s. During the execution the respective order of these communication delays may change, thus implying a task exchange to keep optimality; but of course the initial tasks are already processed.

However if some local monotonic property is verified by expected and actual communication delays (for instance, when disturbances depend only on the source and target processors), $PRCP^*$ is optimal. Indeed consider the following property on the disturbances (task T_n is final):

$(T_i$ and T_j are assigned to the same processor and $\tilde{c}_{in} \leq \tilde{c}_{jn}) \Rightarrow c_{in} \leq c_{jn}$

		LCT			SCT			UECT		
n	m	Mean	nb	Max	Mean	nb	Max	Mean	nb	Max
50	3	0.93	28	5.65	0.07	4	2.2	1.83	50	9.42
50	4	0.59	28	5.00	1.71	38	11.9*	2.64	50	9.13
50	10	0.44	22	4.83	0.11	12	1.25	0.23	6	4.40
100	3	1.35	68	5.17	0.04	16	1.50	1.17	50	8.01
100	4	0.77	48	7.10	2.16	66	7.79	2.26	76	5.33
100	10	0.24	20	2.61	0.47	34	3.04	0.87	38	4.91
150	3	0.75	54	5.47	0.34	48	1.39	0.99	60	4.87
150	4	0.85	52	3.44	2.45	86	6.57	1.88	90	4.36
150	10	0.18	16	1.72	0.25	20	1.69	0.47	24	5.33
200	3	0.53	66	4.32	0.14	34	1.57	1.01	72	3.83
200	4	0.85	62	3.37	1.23	92	5.38	2.13	84	5.01
200	10	0.39	36	2.73	0.63	48	3.90	0.90	62	3.20
250	3	1.14	68	5.01	0.28	64	1.00	0.46	44	5.19
250	4	0.37	56	3.03	1.73	92	4.68	2.15	96*	5.13
250	10	0.69	58	3.29	0.34	46	1.29	0.98	76	4.23

Table 1. Results for wide task graphs, 3, 4 and 10 processors

Theorem 2. *If the communication delays respect the local monotonic property, then $PRCP^*$ sequencing is optimal for a fixed assignment*

Proof. Any sequencing respecting the non increasing order of the c_{in}'s on all processors is optimal, as the induced release date for T_n is then minimized. $PRCP^*$ respects this order for the $\tilde{c}_{ij}$'s, hence for the c_{in}'s because of the property. □

The trees are a natural extension of fork and join graphs. However the problem is already NP-complete in the UECT case and an arbitrary number of processors (see [6]).

5   Experimental Results

In this section, $PRCP^*$ and $FRCP^*$ are compared. The fully off-line approach is not considered, as it leaves no way to cope with disturbances. If the $\tilde{c}_{ij}$'s are poor approximations of the actual communication delays, the policy of choosing any ready task for execution might prove as well as another. Hence we admitted the following assumptions: c_{ij} is obtained as $c_{ij} = \tilde{c}_{ij} + \pi_{ij}$, where π_{ij}, the disturbance, is null in fifty percent of the cases. When non zero, it is positive three times more often than negative. Finally, its size is chosen at random and uniformly between 0 and 0.5. 500 task graphs were randomly generated, half relatively wide with respect to the size of sets of independent tasks (wide graphs), half strongly connected, thus having small antichains and long chains. The results for the first case are presented. The number of vertices varies from 50 to 250. The durations are generated three times for a given graph to obtain LCT, SCT and UECT durations (Large and Small *estimated* Communications Times, and

Unit Execution and estimated Communication Times, respectively). In the first case p_i is taken uniformly in $[1,5]$ and $\tilde{c}_{ij}$ in $[5,10]$, in the second case it is the opposite, in the third all durations are set to 1. For each graph, 5 duration sets of each type are tested. The table displays the mean percentage of improvement of $PRCP^*$ with respect to $FRCP^*$, the number of times (in percentage) $PRCP^*$ was better, and the maximum improvement ratio obtained by $PRCP^*$.

$PRCP^*$ is always better in average. The mean improvements are significant, as the assignment is the same for both algorithms. Indeed an improvement of 5% is frequent and might be worth the trouble. The results are logically more significant in the SCT and UECT cases. In the case of strongly connected graphs, the differences are less significant, as often there is only one ready task at a time per processor. But when there are differences they are in favor of $PRCP^*$.

References

[1] BAMPIS E., Guinand F., Trystram D., *Some Models for Scheduling Parallel Programs with Communication Delays*, Discrete Applied Mathematics, 51, pp. 5–24, 1997.

[2] CHRÉTIENNE Ph., Picouleau C., *Scheduling with communication delays: a survey*, in Scheduling Theory and its Applications, P. Chrétienne, E.G. Coffman, J.K. Lenstra, Z. Liu (Eds), John Wiley Ltd 1995.

[3] GERASOULIS A., Yang T., *A Comparison of Clustering Heuristics for Scheduling DAGs on Multiprocessors*, J. of Parallel and Distributed Computing, 16, pp. 276–291, 1992.

[4] HANEN C, Munier A, *Performance of Coffman Graham schedule in the presence of unit communication delays*, Discrete Applied Mathematics, 81, pp. 93–108, 1998.

[5] HWANG J.J., Chow Y.C., Anger F.D., Lee C.Y., *Scheduling precedence graphs in systems with interprocessor communication times*, SIAM J. Comput., 18(2), pp. 244–257, 1989.

[6] LENSTRA J.K., Veldhorst, M., Veltman B., *The complexity of scheduling trees with communication delays*, J. of Algorithms 20, pp. 157–173, 1996.

[7] MOUKRIM A., Quilliot A., *Scheduling with communication delays and data routing in Message Passing Architectures*, LNCS, vol. 1388, pp. 438–451, 1998.

[8] PAPADIMITRIOU C.H., Yannakakis M., *Towards an Architecture-Independent Analysis of Parallel Algorithms*, SIAM J. Comput., 19(2), pp. 322–328, 1990.

[9] RAYWARD-SMITH V.J., *UET scheduling with interprocessor communication delays*, Discrete Applied Mathematics, 18, pp. 55–71, 1986.

[10] SARKAR V., Partitioning and Scheduling Parallel Programs for Execution on Multiprocessors, The MIT Press, 1989.

[11] SIH G.C., Lee E.A., *A compile-time scheduling heuristic for interconnection-constrained heterogeneous processor architectures*, IEEE Trans. on Parallel and Distributed Systems, 4, pp. 279–301, 1993.

[12] YANG T., Gerasoulis A., *List scheduling with and without communication delay*, Parallel Computing, 19, pp 1321–1344, 1993.

Scheduling User-Level Threads on Distributed Shared-Memory Multiprocessors*

Eleftherios D. Polychronopoulos[1] and Theodore S. Papatheodorou[1]

High Performance Computing Information Systems Laboratory
Department of Computer Engineering and Informatics
University of Patras, Rio 26 500, Patras, GREECE
{edp,tsp}@hpclab.ceid.upatras.gr

Abstract. In this paper we present Dynamic Bisectioning or DBS, a simple but powerful comprehensive scheduling policy for user-level threads, which unifies the exploitation of (multidimensional) loop and nested functional (or task) parallelism. Unlike other schemes that have been proposed and used thus far, DBS is not constrained to scheduling DAGs or singly nested parallel loops. Rather, our policy encompasses the most general type of parallel program model that allows arbitrary mix of nested loops and nested DAGs (directed acyclic task-graphs) or any combination of the above. DBS employs a simple but powerful two-level dynamic policy which is adaptive and sensitive to the type and amount of parallelism at hand. On one extreme DBS approximates static scheduling, hence facilitating locality of data, while at the other extreme it resorts to dynamic thread migration in order to balance uneven loads. Even the latter is done in a controlled way so as to minimize network latency.

1 Introduction

Thread scheduling is a problem that has a significant impact on the performance of parallel computer systems, and has been an active area of research during the last few years. With the proliferation of NUMA parallel computers efficient scheduling at user-level has become more important than ever before. This is particularly true given the fact that these machines are used not only as powerful performance boosters, but also as multi-user systems that execute a multitude of processes of different types at any given time. Thus far user-level scheduling policies have focused on either simple queueing models that handle scalar threads, or fine-tuned heuristic algorithms that try to optimize execution of (mostly singly) nested parallel loops. In this paper we present a novel comprehensive scheduling policy that targets the most general scenario of parallel applications: processes that are composed of nested parallel loops, nested parallel regions with scalar threads, and arbitrary combinations of the former two. Unlike other similar schemes, the proposed **D**ynamic **B**isectioning **S**scheduling (DBS) environment, accommodates adaptivity of the underlying hardware as well as of applications: processors can be added or preempted in the middle of program execution without affecting load balancing. This work was also part of an earlier collaboration with the SGI OS kernel development group and influenced the kernel scheduling algorithms used in IRIX [Bit95].

---

* This work was supported by the NANOS ESPRIT Project (E21907).

DBS has also been implemented as the user-level scheduling policy of the Illinois-Intel Multithreading Library [Girk98]. In this paper, however, we discuss an independent implementation that does not take advantage of specific kernel support, and it can thus be used on any distributed shared-memory architecture.

DBS was implemented on the Nano-Threads Programming Model (NPM) as part of NANOS, a large-scale project involving the development of a runtime library for user-level threads, OS support, and a compiler that automatically generates threads using the library interface [Nano97a, Nano98b]. The nano-Threads environment aims at delivering optimized lightweight threads, suitable for both fine-grain parallel processing and event-driven applications [Niko98]. An important feature of nano-Threads is that it provides for coordinated and controlled communication between the user-level and the kernel-level schedulers. In this work we used "NthLib", a nano-Threads Library [Mart96] built on top of QuickThreads for the implementation and performance evaluation of the proposed scheduling policy.

Most well known scheduling policies address simple loop parallelism, and although there has been evidence [More95], that functional parallelism can be present in substantial amounts in certain applications, little has been done to address the simultaneous exploitation of functional or irregular and loop parallelism; both in terms of compiler support and scheduling policies. Although classical DAG scheduling (list scheduling) heuristics can be used for simple thread models, they address neither loop parallelism nor nested functional parallelism (which, in general, invalidate the "acyclic" nature of DAGs). Unlike functional parallelism, loop-level parallelism and loop scheduling has been studied extensively. Some of the most well-known such policies are briefly outlined in Section 3 and include *Static scheduling*, as well as dynamic policies such as *Affinity Scheduling* and *Guided Self-scheduling*. Similar approach for scheduling user-level threads using work stealing mechanism for balancing the workload, is implemented in the Cilk multithreading runtime system [Blum95]. However, due to the fact that the philosophy of the Cilk environment is different than that of Nanos the two approaches differ both in their implementation as well as in their target. While the Cilk scheduler, schedules fully strict directed DACs the scheduler of Nanos, encompasses the most general type of parallel program model that allows arbitrary mix of nested loops and nested DAGs.

DBS is implemented for the NPM on a Silicon Graphics Origin 2000 multiprocessor system. A benchmark suite of one synthetic application and five application kernels, described in Section 4, was used to compare four representative user-level scheduling policies. The results indicate that generally DBS is superior, as it presents performance gains ranging between 10% and 80%. These results also demonstrate the importance of functional parallelism which is not exploited by the other scheduling policies. As a consequence enhanced versions, integrating exploitation of functional parallelism, of the four policies were created and the same set of experiments were again tested. These results demonstrate the importance of the exploitation of functional parallelism, as well as the superiority of DBS compared to other scheduling policies and the affinity exploitation gain.

The rest of the paper is organized as follows: In Section 2, the nano-Threads environment is outlined. DBS is described in Section 3. An outline of known policies used for the experiments and the evaluation results are presented in Section 4. Finally a brief conclusion is provided in Section 5.

2   The Nano-Threads Environment

The integrated compilation and execution environment of the NPM as described in [Nano98b], consists of three main components :

The compiler support in NPM (based on Parafrase2) performs fine-grained multilevel parallelization and exploits both functional and loop parallelism from standard applications written in high level languages such as C and FORTRAN. Applications are decomposed in multiple levels of parallel tasks. The compiler uses the Hierarchical Task Graph (HTG) structure to represent the application internally. Management of multiple levels of parallel sections and loops allows the extraction of all useful parallelism contained in the application. At the same time, the execution environment combines multiprocessing and multiprogramming efficiently on general purpose shared-memory multiprocessors, in order to attain high global resource utilization.

During the execution of an application, nano-Threads are the entities offered by the user-level execution environment to instantiate the nodes of the HTG, with each nano-Thread corresponding to a node. The user-level execution environment controls the creation of nano-Threads at run-time, ensuring that the generated parallelism matches the number of processors allocated by the operating system to the application. The overhead of the run-time library is low enough to make the management of parallelism affordable. In other words, the nano-Threads environment implements dynamic program adaptability to the available resources by adjusting the granularity of the generated parallelism.

The user-level execution environment includes a ready queue which contains nano-Threads submitted for execution. Processors execute a scheduling loop, where they continuously pick up work from the ready queue until the program terminates.

The last component of NPM is the operating system, which distributes physical processors among running applications. The resulting environment is multi-user and multi-programmed, allowing each user to run parallel and sequential applications. The operating system offers *virtual processors* to applications, as the kernel abstraction of physical processors on which applications can execute in parallel. Virtual processors provide user-level contexts for the execution of nano-Threads. In the remaining of the paper the terms "virtual processor" and "kernel thread" are used interchangeably.

The main scheduling objective of NPM is that both application scheduling (that is, nano-Threads to virtual processors mapping) and virtual processor scheduling (virtual to physical processors mapping) must be tightly coordinated in order to achieve high performance. In addition, the kernel-level scheduler [Poly98] must avoid any interference with the user-level scheduler, in order to keep runtime scheduling overhead low.

3   The DBS Algorithm

Dynamic Bisectioning Scheduling is designed for multithreading programming environments exploiting loop and functional parallelism. The NUMA property of the targeting architectural platform led us to incorporate a central queue and per processor local queues for facilitating scheduling policies in order to balance the workload. The centralized queue is accessible by all processors. All tasks generated by the compiler of the NthLib, are enqueued in the tail of the central queue. All processors start searching for work, by visiting first the head of the central queue. Since the queue is centralized the enqueueing and dequeuing operations are rather trivial:

- *Enqueueing for central queue*: all generated tasks are enqueued in the tail of the central queue.
- *Dequeuing for central queue* : a processor always dequeues the task descriptor at the head of the queue. If the descriptor denotes regular (serial) task then it dequeues the task and executes. If it denotes a loop (a compound task), then two possibilities exist:
  1. All iterations of the loop have been issued to other processors. In this case the processor searches to dequeue another thread.
  2. There are N iterations left in the queue that are not yet scheduled ($N > 0$). Then the processor dispatches N/P iterations and changes the index of the loop in the descriptor, leaving the remaining $N - (N/P)$ iterations to other processors. Here P denotes the total number of processors allocated to the specific application at this moment.

The per processor local queues are located in shared memory, hence they can be accessed by all processors working in the specific application. However, a per processor local queue is accessed faster by the "owner" processor since it resides in memory location which is local to the owner. For local queues the enqueueing and dequeuing policies work as follows:

- Enqueueing: Tasks that have been enabled by the processor either at the central or the local queue can be enqueued in the head of the local queue or in the head of another processor's local queue, but never in the central queue.
- Dequeuing: A processor first searches for a task to dequeue in the head of its local queue. There are two cases:
  1. The processor finds a task (local queue not empty). Then two sub-cases are distinguished:
     (a) if this is a serial task then it dequeues the task and executes, as with the central queue case.
     (b) if however this is a compound task then it dispatches and executes $1/P$ of the remaining iterations, leaving the remaining iterations in the local queue.
  2. The local queue is empty, then the processor first searches the head of the central queue for tasks. If it succeeds then it follows the dequeuing process for the centralized queue. If the central queue is empty the processor searches the local queues of other processors for work to steal, as described below. If all queues are empty the processor is self-released. If it finds a non empty local queue it bisects the load in that queue, i.e. it steals half of the tasks present and enqueues the stolen tasks in the head of its local queue.

There are several ways for a processor to search for work in other local queues and some of these ways need further investigation. One basic task is to exploit proximity, in the sense that it is better for a processor to steal work from the local queue of a neighboring processor who also works in the same application. This would reduce memory page faults and cache misses. Another significant factor is reducing the search overhead, thus it is important to use a process that it is as simple as possible. One such policy, that we use in this work, is as follows. We define a bit vector V, the size of which is equal to the total number of physical processors and such that at any moment $V(j) = 1$ if and only if the local queue of processor j is nonempty. If processor j finds its local queue empty immediately sets $V(j) = 0$. If this processor is newly assigned to the application or if it steals work from other local queues or dispatches work from the central queue, it sets $V(j) = 1$. Processor j uses V when it needs to search other local queues, in order to detect which of them are nonempty from the fact that the corresponding coordinates of V have value equal to 1. The examination of these coordinates may take place in several ways. In the case implemented here the processor examines the coordinates of V in a circular fashion.

4 Experimental Evaluation

In this section we discuss our experimentation frame work and several performance measurements obtained on a dedicated sixteen processor SGI Origin 2000 comparing the proposed scheduling policy to the most widely used heuristic algorithms for loop parallelism. Execution time is used as the metric for comparison between different scheduling policies.

4.1 Framework

In our experiments we used an application suite consisting of five application codes that have been used previously by other researchers for similar comparisons [More95]. These are two "real life", complete applications, namely *Computational Fluid Dynamics (CFD)* and *Molecular Dynamics (MDJ)*, a *Complex Matrix Multiply (CMM)* and two kernels that are found in several large applications, namely *LU Decomposition (LU) and Adjoint Convolution (AC)*. We have incorporated the kernels of LU and AC in two synthetic benchmarks in order to achieve additional functional parallelism. The two resulting benchmarks are *Synthetic LU Decomposition (SLU)* and *Synthetic Adjoint Convolution (SAC)*. As we have already mention, we are also using a representative synthetic application, *Multi Triangular Array (MTA)*. The kernel of MTA, is an upper triangular array NxN with dimension N=512, which is split in N columns. Each column can only be executed serially, due to a recursive computation among the elements of the column. There are no dependencies between the columns of the array. Therefore, each column will be executed serially and all columns of the array in parallel.

In the first set of experiments the effect of introducing the DBS policy is first studied through comparison with three well established scheduling policies. These are *Static scheduling (STATIC)*, *Affinity Scheduling (AFS)* and *Guided Self-Scheduling (GSS)* and are outlined later in this section. It is worth noting that, as opposed to DBS, all known techniques work with simple loop parallelism only, while DBS exploits multi-level loop parallelism and functional parallelism as well. Thus, in order to supply the other methods with equivalent functionalities we created enhanced versions of AFS, GSS and STATIC scheduling that can also exploit functional parallelism.

For our experiments, we implemented some of the most common scheduling policies employed by most of the existing commercial systems. The scheduling mechanisms used in this paper for our comparisons are outlined in the following paragraphs.

Static Scheduling (STATIC). Under static scheduling loop partitioning is done at compile time and allocation is predetermined during run-time. Static scheduling incurs minimal run-time synchronization overhead, but it suffers on load balancing with irregular applications. In static scheduling, load imbalance can occur either in case where all iterations of a parallel loop do not take the same amount of time to execute or in the case where processors execute iterations at different points in time. With irregular applications, this policy results in low system utilization and long execution times.

Guided Self-Scheduling (GSS) [Poly87], is a dynamic policy that adapts the size of iteration parcels of a loop dispatched to each processor at run-time. Guided self-scheduling allocates large chunks of iterations at the beginning of the loop in order to reduce synchronization overhead (and promote locality), while it allocates small chunks towards the end of the loop so as to balance the workload among the executing processors. At each step, GSS allocates to each processor $1/P$ of the remaining loop iterations, where P is the total number of processors. In GSS all processors finish within one iteration of each other and use a minimal number of synchronization operations, assuming that all iterations of the parallelized loop need the same amount of time to complete. One major drawback in GSS is the excessive attempt of processors to access the work-queue towards the end of the loop, competing for loop iterations in the head of the work-queue. Another drawback is that, in the case of unbalanced parallel loops such as triangular loops, the first few processors will end up taking most of the work of the loop, thus resulting in load imbalance.

Affinity Scheduling (AFS) [Mark92], is also a dynamic scheduling policy based in the observation that for many parallel applications the time spent to bring data in local memory is a significant source of overhead, ranging from 30–60% of the total execution time. AFS attempts to combine the benefits of static scheduling while resorting to dynamic load balancing when processors complete their assigned tasks. This is achieved by partitioning and allocating chunks of N/P iterations to the work queue of each processor. Each processor dispatches and executes $1/P$ iterations of the total iterations enqueued in its local queue. Also for purposes of load balancing, an idle processor examines the work queue of all processors participating in the execution of a loop, and steals $1/P$ iterations from the most loaded queue. Overhead is introduced in keeping track and updating processor loads, but most importantly, from the potential repetitive locking of the same local queue by several processors simultaneously.

4.2 Results and Discussion

The above scheduling policies represent the state-of-the-art in loop scheduling used by research prototypes and commercial systems. However, although recent research activities have focused on tackling the problem of irregular parallelism (task or functional

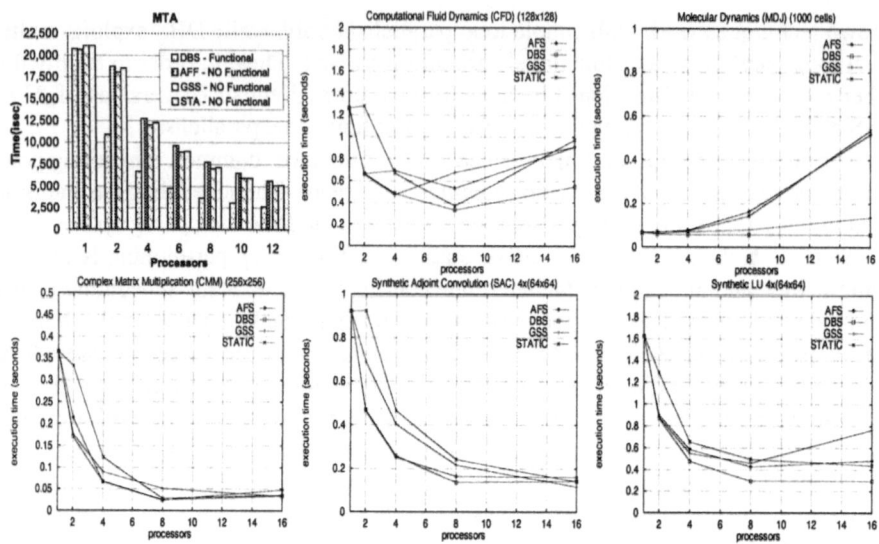

Fig. 1. Execution times with Loop & Functional Parallelism for DBS and Loop only for AFS, GSS, and STATIC. In the first graph the execution time of the synthetic MTA application is depicted.

parallelism), at present there exists no commercial implementation that supports a comprehensive parallel execution model: multidimensional parallelism (arbitrarily nested parallel loops), nested task parallelism (nested cobegin/coend parallel regions) or arbitrary combinations of the two. DBS, on the other hand, addresses such a comprehensive model as does our implementation of the NthLib runtime library. This fact gives an off-the-start advantage to DBS due to the ability of the latter to exploit general task parallelism. Furthermore, due to the 2-level queue model employed by DBS, the latter may involve more run-time overhead than necessary in computations where parallelism is confined to one-dimensional parallel loops. In order to alleviate the limited ability of GSS, AFS, and STATIC, to exploit functional parallelism we implemented enhanced versions of the above which enable task-level parallelism exploitation.

Our performance studies focused on the execution time as well as the aggregate cache misses as an approximate estimate of locality behavior for each of the algorithms studied. The first graph of Figure 1 shows the raw execution times of the four different scheduling policies for the synthetic application, MTA. In the same figure the raw execution time for each of the five benchmarks and each of the four scheduling policies, for execution on 2, 4, 8, 12 and 16 processor configurations of the Origin 2000. In this case, DBS has the advantage of exploiting task-level parallelism unlike the other three (predominantly loop) policies; both CFD and MDJ codes expose small amounts of functional parallelism. In all five cases DBS outperformed AFS, GSS, and even static scheduling. In the case of CFD, DBS significant improvement in execution time approaching a speedup of 2 compared to static, and a 40% improvement over AFS, while for the MDJ code DBS outperformed AFS and static scheduling by a factor of approximately 3, and GSS by 58%.

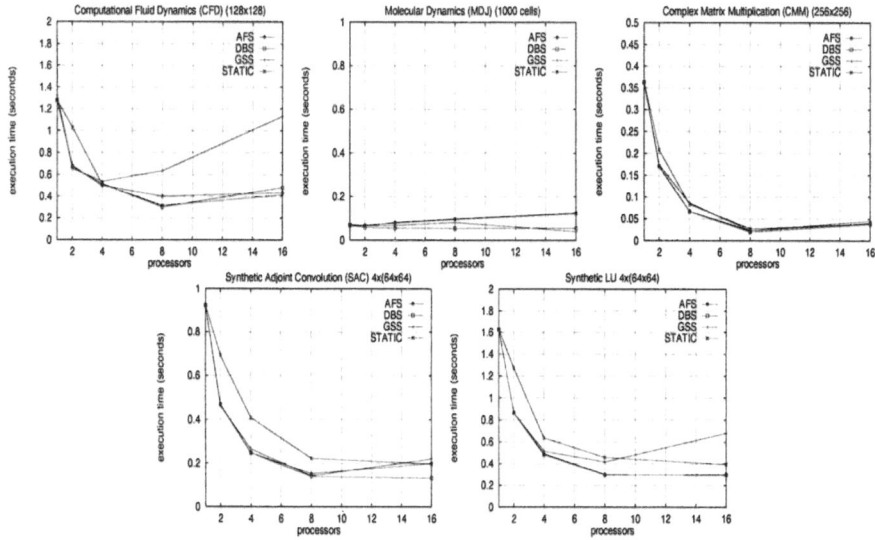

Fig. 2. Execution times for DBS, AFS, GSS and STATIC with Loop and Functional Parallelism

The case of CMM shows a border-case with a four-way functional parallelism and regular loop parallelism which is ideal for an even-handed comparison of the scheduling algorithms. As expected all four policies performed within a small range from each other for 16 processors. However, for a small number of processors DBS, AFS, and GSS outperform static; although somewhat counter-intuitive, this is due to the fact that both DBS and AFS take full advantage of static scheduling by equi-partitioning parallel loops among the available processors in their first phase of scheduling. Unlike static, however, the dynamic schemes have the ability to balance the load when imbalances arise due to false sharing, asynchronous processor interrupts and context switches etc. As the plots indicate, the overhead involved in load balancing is low and more than justifies the benefits of balanced loads.

An important observation from Figure 1 is that DBS scales well with machine size and for all five benchmarks, albeit the number of processors used was up to 16. Nevertheless, this is not the case with any of the other three policies. Improved scalability offered by DBS is due, in part, to the additional (functional) parallelism exploited by DBS as well as its aggressive on-demand load balancing.

Figure 2 illustrates the results for the same set of experiments with the following important difference: in order to enable functional parallelism exploitation by AFS, GSS, and static policies, we encoded functional parallelism (parallel regions) in the form of parallel loops with each loop iteration corresponding to each of the scalar tasks. All three policies were thus successful at distributing different tasks of a parallel region to different processors. This resulted in a noticeable improvement of AFS for the CFD and LU benchmarks and to a lesser degree for the SAC code. However, overall the behavior of all five benchmarks in terms of both scalability and absolute performance

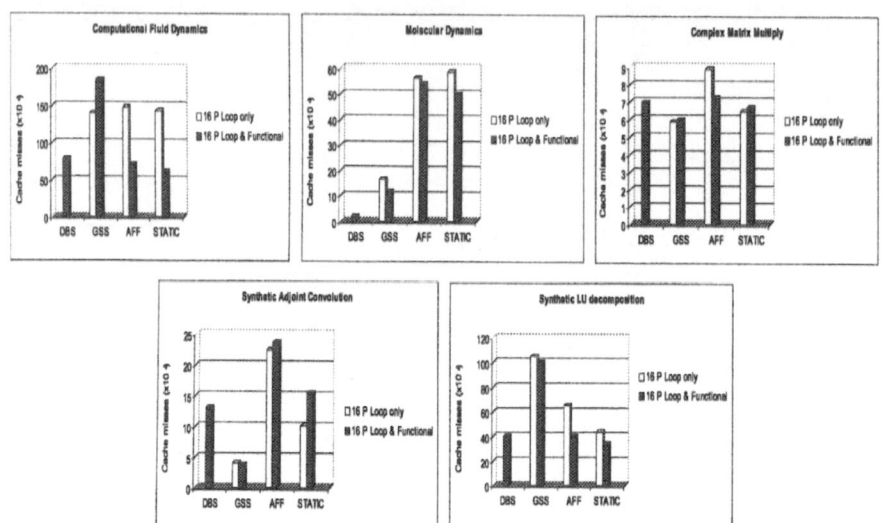

Fig. 3. Aggregate Cache Misses, DBS with Loop & Functional Parallelism, AFS, GSS and STATIC with Loop only and Loop & Functional Parallelism for 16 pro cessors

remained the same. Nevertheless, uniform, stable behavior by DBS is also apparent in Figure 2.

In order to measure a first-order-magnitude effect of locality of data, we profiled the per-processor cache misses for all four algorithms and for each of the five benchmarks. Figure 3 shows measurements of cache misses for all methods. It follows that in the case of CFD and MDJ, i.e. the applications for which loop parallelism is not dominant, DBS presents a very sizeable reduction in the number of cache misses, up to about 45% in the CFD case, as compared with the best of the remaining policies (AFS) and up to about 85% in the MDJ case as compared to the best of the remaining policies which is GSS. In other cases DBS incurs approximately the smallest number of cache misses along with GSS and STATIC (synthetic LU, competing with STATIC), third and second (SAC, competing mainly with GSS) and third (CMM, in close competition with STATIC and GSS).

At first glance, these measurements may appear counter-intuitive, since one would expect the smallest number of cache misses (best locality of data) for the STATIC approach, and the worst cache behavior from fully dynamic schemes. The measured results are due to the synergy of two factors. First, static approaches (or dynamic approaches such as AFS that attempt to exploit the benefits of static scheduling off-start) are more amenable to false sharing. Due to the large cache block in the Origin 2000, a high cache miss rate due to false sharing is exacerbated. Secondly, unlike the other schemes, DBS favors locality of data by giving execution priority to locally-spawned tasks which are executed entirely by the processors that fire them - such tasks (see Figure 2) are never queued in the global queue.

Our measurements make a strong case for functional parallelism which, although not in abundance in the majority of applications, is found in substantial amounts in real-world applications such as MDJ and CFD codes. The results indicate that DBS not only outperforms common loop scheduling schemes, but scales much better than any of these heuristics. As parallel machines proliferate toward the low-end spectrum, task-level parallelism exploitation (common in commercial applications) will become increasingly important.

5 Conclusions and Future-Work

Dynamic Bisectioning Scheduling was introduced as a novel, comprehensive scheduling policy for user-level thread scheduling. The proposed 2-level policy has embedded properties that promote locality of data and on-demand load balancing. Experiments with real-world applications and application kernels indicate that DBS achieves significant performance improvements compared to other common scheduling policies used as default schedulers in today's high-end parallel computers. Moreover, our experiments indicate that DBS is a cache-friendly dynamic scheduler less susceptible to false sharing than static or other dynamic approaches. Our future work is oriented towards further investigation of the exploitation of functional parallelism by conducting more experiments in a well defined and comprehensible application suite.

6 Acknowledgements

We would like to thank Constantine Polychronopoulos, Dimitrios Nikolopoulos and the people in the European Center for Parallelism in Barcelona (CEPBA).

References

[Girk98]   M. Girkar, M Haghighat, P. Grey, H. Saito, N. Stavrakos, C. Polychronopoulos, *"Illinois-Intel Multithreading Library: Multithreading Support for Intel Architecture Based Multiprocessor Systems"*, Intel Technology Journal, 1st Quarter'98,

[Blum95]   Robert D. Blumofe, Christopher F. Joerg, Bradley C. Kuszmaul, Charles E. Leiserson, Keith H. Randall, and Yuli Zhou, *"Cilk: An Efficient Multithreaded Runtime System,"*, In 5th ACM SIGPLAN Symposium on Principles and Practice of Parallel Programming (PPOPP '95), July 19-21, 1995, Santa Barbara, California.

[Mark92]   E. P. Markatos and T. J. LeBlanc. *"Using processor affinity in loop scheduling an shred-memory multiprocessors."*, In Supercomputing '92, pages 104-113, November 1992.

[Mart96]   X. Martorell, J. Labarta, N. Navarro and E. Ayguade, *"A Library Implementation of the Nano-Threads Programming Model"*, In 2nd International Euro-Par Conference, pp. 644-649, Lyon, France, August 1996.

[More95]   J.E. Moreira, *"On the Implementation and Effectiveness of Autoscheduling for Shared-Memory Multiprocessors"*, PhD. thesis, Department of Electrical and Computer Engineering, Univ. of Illinois at Urbana-Champaign, 1995.

[Nano97a]   ESPRIT Nanos Project. *"Nano-Threads: Programming Model Specification"*, Project Deliverable M1D1, http://www.ac.upc.es/NANOS/NANOS-delm1d1.ps, July 1997

[Nano98b] ESPRIT Nanos Project. *"n-RTL Implementation"*, Project Deliverable M2D2, http://www.ac.upc.es/NANOS/NANOS-delm2d2.ps, April 1998

[Niko98]   D. S. Nikolopoulos, E. D. Polychronopoulos and T. S. Papatheodorou, , *"Efficient Runtime Thread Management for the Nano-Threads Programming Model"*, Proc. of the Second IEEE IPPS/SPDP Workshop on Runtime Systems for Parallel Programming, LNCS Vol. 1388, pp. 183–194, Orlando, FL, April 1998.

[Poly87]   Polychronopoulos C. D. and D. J. Kuck. *"Guided Self-Scheduling: A practical scheduling scheme for Parallel Supercomputers"*, IEEE Transactions on Computers. December 1987, pp. 1425-1439.

[Bit95]   N. Bitar, *"personal communication"*, 1995 through 1998.

[Poly98]   E. Polychronopoulos, X. Martorell, D. Nikolopoulos, J. Labarta, T. Papatheodorou and N. Navarro, *"Kernel-Level Scheduling for the Nano-Threads Programming Model"*, 12th ACM International Conference on Supercomputing, Melbourne, Australia 1988.

Using Duplication for the Multiprocessor Scheduling Problem with Hierarchical Communications

Evripidis Bampis[1], Rodolphe Giroudeau[1] and Jean-Claude König[2]

[1] La.M.I., C.N.R.S. EP 738, Université d'Evry Val d'Essonne, Boulevard Mitterrand,
91025 Evry Cedex, France
[2] LIRMM, 161 rue Ada, 34392 Montpellier Cedex 5, France

Abstract. We propose a two-step algorithm that efficiently constructs a schedule of minimum makespan for the precedence multiprocessor constrained scheduling problem in the presence of hierarchical communications and task duplication for UET-UCT graphs.

1   Introduction

Many works deal with the multiprocessor scheduling problem with communication delays ([4],[5],[7]). Usually the parallel program is represented by a directed acyclic graph where each vertex corresponds to a task and each arc to a (potential) communication between a predecessor-task and a successor-task. The objective is to find a feasible schedule minimizing the completion time (makespan).

Last years the notion of hierarchical parallel architecture becomes a reality with the advance of many parallel machines which are based on this concept. Parallel architectures of this type include parallel machines constituted by different multiprocessors, bi-processors connected by myrinet switches, architectures where the processors are connected by hierarchical busses, or point-to-point architectures where each vertex of the topology is a cluster of processors. In this type of architectures, it is clear that there are two levels of communications: the intracluster-communications between two processors of a same machine, and the interprocessor-communications between two processors of different machines, that we call in what follows intercluster-communications. Hence, there is a need for an extension of the classical scheduling model with homogeneous communication delays in order to take into account this important technological evolution.

We study here the simplest case where the tasks have unit execution times, the intercluster-communication delay takes also a unit of time ($c_{ij} = 1$) while the intracluster-communication delay is equal to zero ($\epsilon_{ij} = 0$). We focus on the case where the number of clusters is unrestricted, the number of processors within each cluster is equal to two, and task duplication is allowed. Extending the standard three-field notation of [6], our problem can be denoted as $\bar{P}, 2|prec; (c_{ij}, \epsilon_{ij}) = (1, 0); p_i = 1; dup|C_{max}$. This problem has already been studied in the case where the duplication of tasks is not allowed ([1],[2]). In

P. Amestoy et al. (Eds.): Euro-Par'99, LNCS 1685, pp. 369-372, 1999.

[2], the authors proved that the problem of deciding whether an instance of $\bar{P}, 2|prec; (c_{ij}, \epsilon_{ij}) = (1,0); p_i = 1|C_{max}$ has a schedule of length at most four is NP-complete. As a corollary, no polynomial-time algorithm exists with performance bound smaller than $\frac{5}{4}$ unless $P = NP$. In [1], a heuristic with a relative performance guarantee of $\frac{8}{5}$ has been proposed.

In this paper, we allow the duplication of tasks and we propose an extension of the approach proposed in [4]. Colin and Chrétienne in [4] defined an optimal algorithm for the so-called small (homogeneous) communication times case (SCT), where the minimum execution time of any task is greater than or equal to the biggest communication delay. We focus on the hierarchical communication-case and more precisely on the $\bar{P}, 2|prec; (c_{ij}, \epsilon_{ij}) = (1,0); p_i = 1; dup|C_{max}$ problem, and we propose an optimal scheduling algorithm. Our algorithm is in two steps: we first compute for each task an earliest starting time of any of its copies, we construct a critical graph, and then we build up an earliest optimal schedule. The proofs have been omitted because of space limitations.

2   Problem definition

We consider a directed acyclic graph $G = (V; E)$ which models the precedence constraints. Given a graph G and a task i, we denote by $Pred_G(i)$ (resp. $Succ_G(i)$) the set of immediate predecessors (resp. successors) of the task i in G. More generally, we define $Pred_G^l(i) = Pred_G^{l-1}(Pred_G(i))$, where integer $l > 1$, and $Pred_G^1(i) = Pred_G(i)$. $Pred_G^*(i)$ denotes the transitive closure of $Pred_G(i)$. In the following, if no confusion arises, we will use the notation $Pred(i)$ instead of $Pred_G(i)$.

We may assume w.l.o.g. that all the copies of any task $i \in V$ start their execution at the same time, denoted t_i. Let us denote by Π_i the set of clusters on which we execute the copies of task i. In order to simplify the presentation whenever two tasks i, j (or their copies) such that $(i, j) \in E$, are not executed in consecutive time units, we consider w.l.o.g. that they are executed on different clusters. Any feasible schedule must satisfy the following constraints:

1. at any time, a cluster executes at most two copies;
2. if (i, j) is an arc of G then:

$$t_j \geq t_i + 2 \text{ if } \Pi_i \bigcap \Pi_j = \varnothing$$
$$t_i \geq t_j + 1 \text{ if } \Pi_i \bigcap \Pi_j \neq \varnothing$$

Our objective is to find a feasible schedule with a minimum makespan.

3   An optimal schedule

3.1   Finding the release times of the tasks

The algorithm works in two steps; the first step is the procedure RT (Release Times), which computes a lower bound of the starting time of any copy of task i, denoted by b_i, and a critical graph, denoted in what follows as $G_c = (V, E_c)$. The second step of the algorithm is the procedure ES (Earliest Scheduling), which uses the critical graph to build up a schedule in which any copy of a task starts at its lower bound.

Procedure RT can be stated as follows:

Let $G_c = (V, E_c)$ be the critical graph of $G = (V, E)$. Initially, $E_c = \varnothing$.
For every task $i \in V$ without predecessor, let $b_i = 0$.
While there is a task i which has not been assigned a lower bound and whose all predecessors have been assigned their release times, do the following:
Let $b_{P(i)} = \max\{b_k \backslash k \in Pred(i)\}$, and $S_i = \{k \backslash b_k = b_{P(i)}\}$.

- If $|S_i| = 1$ then $b_i := b_{P(i)} + 1$, and if $S_i = \{k\}$, then $E_c = E_c \bigcup \{(k, i)\}$
- else if $|S_i| \geq 3$ then $b_i := b_{P(i)} + 2$;
- else if $|S_i| = 2$ then w.l.o.g., let $S_i = \{j, k\}$
  - if for every l, $|Pred^l_{G_c}(j) \bigcup Pred^l_{G_c}(k)| \leq 2$, then $b_i := b_{P(i)} + 1$, and $E_c = E_c \bigcup \{(k, i), (j, i)\}$
  - else $b_i := b_{P(i)} + 2$;

We proved in the full version of the paper [3] that RT computes the lower bounds of the starting times of any copy of each task.

Lemma 1. *Let b_i be the starting time computed by procedure RT. For any feasible schedule of G the starting time of any copy of task i is not less than b_i.*

3.2 Construction of the earliest schedule

The second step of the algorithm consists in building up an earliest schedule, that is, a schedule such that each copy is scheduled at its lower bound.

We define an arc (i, j) of E to be a *critical arc*, if $b_i + 1 = b_j$. From this definition, it is clear that if (i, j) is a critical arc, then in any earliest schedule every copy of task j must be preceded by a copy of task i executed on the same cluster. The critical graph $G_c = (V; E_c)$ is defined as the partial graph of G induced by the critical arcs. According to the construction, it is easy to see that the following lemma is true.

Lemma 2. *Every vertex in the critical graph has at most two critical predecessors.*

In the following, for every terminal task i in the critical graph G_c (i.e. a vertex such that $Succ_{G_c}(i) = \varnothing$), we denote by C^i the *critical subgraph* of $G_c = (V; E_c)$ defined as follows:

$$C^i = \{[i], [Pred_{G_c}(i)], [Pred^2_{G_c}(i)], \ldots\} = Pred^*_{G_c}(i).$$

The tasks inside the brackets ([ ]) can be (potentially) executed simultaneously. According to the construction of $G_c = (V; E_c)$, it is clear that for every i and l, we have $|Pred^l_{G_c}(i)| \leq 2$. Notice also that a critical subgraph cannot be a proper subgraph of another critical subgraph. So, we have at most n critical subgraphs where n is the number of tasks, and in addition every task in the initial graph G is an element of at least one critical subgraph.

It is obvious that the critical subgraphs are in one-to-one correspondence with the terminal tasks of G_c, a property that will be used in the second step of the algorithm. This is done by allocating one cluster to each critical subgraph and processing all the corresponding copies at their release times.

The second step of the algorithm is the procedure ES which can be stated simply as follows:
- Assign each critical subgraph of $G_c = (V; E_c)$ to a distinct cluster;
- Start executing each copy of the tasks at its lower bound.

According to the construction, it is not difficult to verify that the obtained schedule is a feasible optimal schedule (see [3] for the proof).

Theorem 1. *The scheduling given above is feasible.*

4   Conclusion and future works

An optimal $O(n^2)$ algorithm for the multiprocessor scheduling in the presence of hierarchical communications and task duplication has been presented. We have considered here the unit-execution-time unit-communication-time problem. We are now working to extend our study in the case of small hierarchical communication delays.

References

1. E. Bampis, R. Giroudeau, and J.C. König. A heuristic for the precedence constrained multiprocessor scheduling problem with hierarchical communications. Technical Report 35, Université d'Evry Val d'Essonne, 1999.
2. E. Bampis, R. Giroudeau, and J.C. König. On the hardness of approximationg the precedence constrained multiprocessor scheduling problem with hierarchical communications. Technical Report 34, Université d'Evry Val d'Essonne, 1999.
3. E. Bampis, R. Giroudeau, and J.C. König. Using duplication for the precedence constrained multiprocessor scheduling problem with hierarchical communications. Technical Report 36, Université d'Evry Val d'Essonne, 1999.
4. P. Chrétienne and J.Y. Colin. C.P.M. scheduling with small interprocessor communication delays. *Operations Research*, 39(3), 1991.
5. P. Chrétienne, E.J. Coffman Jr, J.K. Lenstra, and Z. Liu. *Scheduling Theory and its Applications*. Wiley, 1995.
6. R.L. Graham, E.L. Lawler, J.K. Lenstra, and A.H.G. Rinnooy Kan. Optimization and approximation in determinics sequencing and scheduling theory : a survey. *Ann. Discrete Math.*, 5:287–326, 1979.
7. A. Munier and C. Hanen. Using duplication for scheduling unitary tasks on m processors with communication delays. *Theoretical Computer Science*, 178:119–127, 1997.

Topic 04
Compilers for High Performance Systems

Barbara Chapman

Global Chair

1 Introduction

This workshop is devoted to research on technologies for compiling programs to
execute on high performance systems. Contributions were sought for all related
areas, including parallelization of imperative languages, as well as translation
of object-oriented languages and several other paradigms for parallel systems.
Compiler analysis and transformation increasingly makes use of run-time information; the interaction of compilers with development and execution environments is thus explicitly included in the scope of this workshop.

We received a very healthy crop of submissions to this area, and their topics
reflected the variety of current compiler research in this field. Both shared memory and distributed memory parallelism was targeted. The accepted papers have
been organised into three sessions whose contents are outlined in the following.

2 The Papers

Array privatisation is one of the traditional techniques employed by parallelizing compilers in order to reduce the number of references to shared data. When
applicable, it creates a local copy of the data structure in question for each processor, and if necessary, inserts additional code to resolve control flow. Several
researchers have sought to lower the memory overhead implied by this technique.
The first paper in this workshop, by Cohen and Lefebvre, presents just such an
approach, which generalises the technique of partial array expansion to handle
arbitrary loop nests. Some compilers convert programs into Static Single Assignment form, in order to eliminate all but true dependences. This technique was
recently extended to handle arrays on a per-element basis, enabling its use to
perform constant propagation for array elements, and more. Collard extends this
technology in his contribution to the workshop, in which he develops an array
SSA framework which can be applied to explicitly parallel programs containing
parallel sections, and a range of synchronisation mechanisms. Another existing
analysis framework is extended for use in conjunction with explicitly parallel
programs, in the last paper of this session. Here, Knoop introduces demand-
driven data flow analysis, in which one searches for the weakest preconditions
which must hold if a specific data flow factum is to be guaranteed. He shows how
this may be performed for sequential programs, and then proceeds to develop
an approach to deal with shared memory parallel programs.

P. Amestoy et al. (Eds.): Euro-Par'99, LNCS 1685, pp. 373–374, 1999.
© Springer-Verlag Berlin Heidelberg 1999

The second session begins with two publications related to systolic computing, and to the design of systolic arrays. The first of these, by Fimmel and Merker, considers the problem of minimising the cost of implementing communication channels in silicon, where bandwidth and communication delay may be varied. Their work also applies to the problem of obtaining locality for data dependences on a given interconnection topology. Crop and Wilde also contribute to techniques for the design of systolic arrays in their paper on the scheduling of affine recurrence equations. They use scheduling to add temporal information to the system, which helps determine the placement of pipeline registers, and more. Dependence analysis must accordingly be extended to derive constraints on the timing functions. Laure and colleagues describe the compilation of Opus, a language which permits the expression of coarse grained HPF tasks and their coordination. The role of HPF compilers and the run time system in this approach are discussed. Next, Leair and colleagues describe their implementation of a generalised block distribution in a commercial HPF compiler. The impact of such distributions, where the owner of a specific datum cannot be computed, on compiler optimisation is discussed. Performance comparisons of program versions using standard and general block distributions, respectively, indicate the usefulness of the latter.

There are many applications whose parallelised version requires communication mainly to perform one or more global reduction operations. In the last session of this workshop, Gutierrez et al. present a new method for parallelising irregular reduction operations to run on a shared memory multiprocessor. As with the inspector-executor approach employed for distributed memory systems, their strategy distributes data and assigns the computation in a manner which enforces the locality of writes to the reduction array. Yet it achieves superior performance by reordering accesses to the subscripted arrays. Kandemir and colleagues discuss the impact of I/O on techniques for loop tiling applied to out of core computations. They show that traditional methods which optimise the usage of cache may not be suitable when I/O is performed within the loop, and propose an algorithm which considers both the I/O and the computation within loop nests to obtain an overall tiling strategy. Shafer and Ghose examine the impact of message reordering and task duplication optimisations on a scheduled set of tasks, and propose a strategy to determine whether to merge messages between pairs of tasks. Although it appears that the average payoff is not large, the authors argue that the techniques are worthwhile because they lead to some improvement in a large fraction of cases tested. In the final paper of this workshop, Pfannenstiel proposes an implementation technique for nested parallel programs which combines piecewise execution based on cost-effective multithreading with fusion optimisation. The strategy supports execution of programs with large data sets. Experiments show the importance of the optimisations.

Storage Mapping Optimization for Parallel Programs

Albert Cohen and Vincent Lefebvre

PRiSM, Université de Versailles, 45 avenue des Etats-Unis, 78035 Versailles, France

Abstract. Data dependences are known to hamper efficient parallelization of programs. Memory expansion is a general method to remove dependences in assigning distinct memory locations to dependent writes. Parallelization via memory expansion requires both moderation in the expansion degree and efficiency at run-time. We present a general storage mapping optimization framework for imperative programs, applicable to most loop nest parallelization techniques.

1 Introduction

Data dependences are known to hamper automatic parallelization of imperative programs and their efficient compilation on modern processors or supercomputers. A general method to reduce the number of memory-based dependences is to disambiguate memory accesses in assigning distinct memory locations to non-conflicting writes, i.e. to *expand* data structures. In parallel processing, expanding a datum also allows to place one copy of the datum on each processor, enhancing parallelism. This technique is known as *array privatization* [5, 12, 16] and is extremely important to parallelizing and vectorizing compilers.

In the extreme case, each memory location is written at most once, and the program is said to be in *single-assignment* form (total memory expansion). The high memory cost is a major drawback of this method. Moreover, when the control flow cannot be predicted at compile-time, some run-time computation is needed to preserve the original data flow: Similarly to the *static* single-assignment framework [6], ϕ-functions may be needed to "merge" possible data definitions due to several incoming control paths.

Therefore parallelization via memory expansion requires both *moderation* in the expansion degree, and *efficiency* in the run-time computation of ϕ-functions. A technique limited to affine loop-nests was proposed in [11] to optimize memory management. The systolic community have a similar technique implemented in ALPHA compilers [14]. A different approach [15] is limited to perfect uniform loop-nests, and introduces universal storage mappings. We present a general storage mapping optimization framework for expansion of imperative programs, applicable to most parallelization techniques, for any nest of loops with unrestricted conditionals and array subscripts.

Section 2 studies a motivating example showing what we want to achieve, before pointing out contributions in more detail. Section 3 formally defines storage

P. Amestoy et al. (Eds.): Euro-Par'99, LNCS 1685, pp. 375–382, 1999.

mapping optimization, then we present our algorithm in Section 4. Experimental results are studied in Section 5, before we conclude.

2 Motivating Example

We first study the kernel in Figure 1.1, which appears in several convolution codes[1]; Parts denoted by $\cdots$ have no side-effect on variable x. For any statement in a loop nest, the *iteration vector* is built from surrounding loop counters[2]. Each loop iteration spawns *instances* of statements included in the loop body: Instances of S are denoted by $\langle S, i, j \rangle$, for $1 \leq i \leq n$ and $1 \leq j \leq \infty$.

```
real x                  real D_T[n], D_S[n, m]          real D_TS[n]
for i = 1 to n          for // i = 1 to n               for // i = 1 to n
T  x = ...              T  D_T[i] = ...                 T  D_TS[i] = ...
   for j = 1 to ...        for j = 1 to ...                for j = 1 to ...
S     x = x ...         S     D_S[i] = if(j=1) D_T[i]    S     D_TS[i] = D_TS[i] ...
   end for                    else D_S[i, j-1] ...          end for
R  ... = x ...             end for                      R  ... = D_TS[i] ...
end for                 R  ... = if(j=1) D_T[i]         end for
                              else D_S[i, j-1] ...
1.1. Original program.  end for                         1.3. Partial expansion.

                        1.2. Single assignment.
```

Fig. 1. Convolution example.

2.1 Instance-Wise Reaching Definition Analysis

We believe that an efficient parallelization framework must rely on a precise knowledge of the flow of data, and advocate for *Instance-wise Reaching Definition Analysis* (IRDA): It computes which instance of which write statement defined the value used by a given instance of a statement. This write is the *(reaching) definition* of the read access.

Any IRDA is suitable to our purpose, but Fuzzy Array Data-flow Analysis (FADA) [1] is prefered since it handles any loop nest and achieves today's best precision. Value-based Dependence Analysis [17] is also a good IRDA. In the following, σ is alternatively seen as a function and as a relation. The results for references x in right-hand side of R and S are *nested conditionals*: $\sigma(\langle S, i, j \rangle, \mathbf{x}) =$ if $j = 1$ then $\{T\}$ else $\{\langle S, i, j - 1 \rangle\}$, $\sigma(\langle R, i \rangle, \mathbf{x}) = \{\langle S, i, j \rangle : 1 \leq j\}$.

2.2 Conversion to Single Assignment Form

Here, memory-based dependences hampers direct parallelization via *scheduling* or *tiling*. We need to expand scalar x and remove as many output-, anti- and

---

[1] E.g. Horn and Schunck's 3D Gaussian smoothing by separable convolution.
[2] When dealing with while loops, we introduce artificial integer counters.

true-dependences as possible. In the extreme expansion case, we would like to convert the program into *single-assignment* (SA) form [8], where *all* dependences due to memory reuse are removed.

Reaching definition analysis is at the core of SA algorithms, since it records the location of values in expanded data-structures. However, when the flow of data is unknown at compile-time, ϕ-functions are introduced for run-time restoration of values [4, 6]. Figure 1.2 shows our program converted to SA form, with the outer loop marked parallel (m is the maximum number of iterations that can take the inner loop). A ϕ-function is necessary, but can be computed at low cost: It represents the *last* iteration of the inner loop.

2.3   Reducing Memory Usage

SA programs suffer from high memory requirements: S now assigns a huge $n \times m$ array. Optimizing memory usage is thus a critical point when applying memory expansion techniques to parallelization.

Figure 1.3 shows the parallel program after partial expansion. Since T executes before the inner loop in the parallel version, S and T may assign the same array. Moreover a one-dimensional array is sufficient since the inner loop is not parallel. As a side-effect, no ϕ-function is needed any more. Storage requirement is n, to be compared with $n \times m + n$ in the SA version, and with 1 in the original program (with no legal parallel reordering).

We have built an optimized *schedule-independent* or *universal* storage mapping, in the sense of [15]. On many programs, a more memory-economical technique consists in computing a legal storage mapping *according to a given parallel execution order*, instead of finding a universal storage compatible with any legal execution order. This is done in [11] for *affine* loop nests only.

Our contributions are the following: Formalize the correctness of a storage mapping, *according to a given parallel execution order*, for any nest of loops with *unrestricted conditional expressions and array subscripts*; Show that universal storage mappings defined in [15] correspond to correct storage mappings *according to the data-flow execution order*; Present an algorithm for storage mapping optimization, applicable to any nest of loops and *all parallelization techniques* based on polyhedral dependence graphs.

3   Formalization of the Correctness

Let us start with some vocabulary. A run-time statement instance is called an *operation*. The sequential execution order of the program defines a total order over *operations*, call it $\prec$. Each statement can involve several array or scalar *references*, at most one of these being in left-hand side. A pair (o, r) of an operation and a reference in the statement is called an *access*. The set of all *accesses* is denoted by **A**, built of **R**, the set of all *reads*—i.e. accesses performing some read in memory—and **W**, the set of all *writes*.

Imperative programs are seen as pairs $(\prec, f_e)$, where $\prec$ is the sequential order over all *operations* and f_e maps every *access* to the memory location it either reads or writes. Function f_e is the *storage mapping* of the program (subscript e stands for "exact"). *Parallelization* means construction of a parallel program $(\prec', f'_e)$, where $\prec'$ is a sub-order of $\prec$ preserving the sequential program semantics. Transforming f_e into the new mapping f'_e is called *memory expansion*.

The basis of our parallelization scheme is instance-wise reaching definition analysis: Each *read* access in a memory location is mapped to the last *write* access in the same memory location. To stress the point that we deal with *operations* (i.e. run-time instances of statements), we talk about *sources* instead of *definitions*. In our sense, reaching definition analysis computes a subset of the program *dependences* (associated with Bernstein's conditions). Practically, the source relation σ computed by IRDA is a pessimistic (a.k.a. conservative) approximation: A given access may have several "possible sources".

As a compromise between expressivity and computability, and because our prefered IRDA is FADA [1], we choose affine relations as an abstraction. using tools like Omega [13] and PIP [8].

3.1 Correctness of the Parallelization

What is a *correct parallel execution order* for a program in SA form? Any execution order $\prec'$ (over operations) must preserve the flow of data (the source of an access in the original program executes before this access in the parallel program): $\forall e, \forall (o_1, r_1), (o_2, r_2) \in \mathbf{A} : (o_1, r_1)\sigma_e(o_2, r_2) \Rightarrow o_1 \prec' o_2$ (where o_1, o_2 are operations and r_1, r_2 are references in a statement). Now we want a static description and approximate σ_e for every execution.

Theorem 1 (Correctness of execution orders). *If the following condition holds, then the parallel order is correct—i.e. preserve the program semantics.*

$$\forall (o_1, r_1), (o_2, r_2) \in \mathbf{A} : (o_1, r_1)\sigma(o_2, r_2) \implies o_1 \prec' o_2. \tag{1}$$

Given a parallel execution order $\prec'$, we have to characterize *correct expansions* allowing parallel execution to preserve the program semantics. We need to handle "absence of conflict" equations of the form $f_e(v) \neq f_e(w)$, which are undecidable since subscript function f_e may be very complicated. Therefore, we suppose that pessimistic approximation $\neq$ is made available by a previous stage of program analysis (probably as a side-effect of IRDA): $f_e(v) \neq f_e(w) \Rightarrow v \not\equiv w$.

Theorem 2 (Correctness of storage mappings). *If the following condition holds, then the expansion is correct—i.e. allows parallel execution to preserve the program semantics.*

$$\forall v, w \in \mathbf{W} : \left(\exists u \in \mathbf{R} : v\sigma u \wedge w \not\prec' v \wedge u \not\prec' w \wedge (u \prec w \vee w \prec v \vee v \not\equiv w)\right)$$
$$\implies f'_e(v) \neq f'_e(w). \tag{2}$$

The proof is given in [3]. This result requires the source v of a read u and an other write w to assign different memory locations, when: In the parallel program: w executes *between* v and u; And in the original one: Either w does *not* execute between v and u or w assigns a different memory location from v ($v \neq w$).

Building parallel program $(\prec', f_e')$ resumes to solving (1) and (2) in sequence.

3.2 Computing Parallel Execution Orders

We rely on classical algorithms to compute parallel order $\prec'$ from the dependence graph associated with σ. Scheduling algorithms [7, 9] compute a function θ from operations to integers (or vectors of integers in the case of multidimensional schedules [9]). Building $\prec'$ from θ is straightforward: $u \prec' v \Leftrightarrow \theta(u) < \theta(v)$.

With additional hypotheses on the original program (such as being a perfect loop nest), tiling [2, 10] algorithms improve data locality and reduces communications. Given a tiling function T from operations to tile names, and a tile schedule θ from tile names to integers: $u \prec' v \Leftrightarrow \theta(u) < \theta(v) \vee (T(u) = T(v) \wedge u \prec v)$.

In both cases—and for any polyhedral representation—computing $\prec'$ yields an affine relation, compatible with the expansion correctness criterion.

Eventually, the data-flow order defined by relation σ is supposed (from Theorem 1) to be a sub-order of every other parallel execution order. Plugging it into (2) describes *schedule-independent* storage mappings, compatible with any parallel execution. This generalizes the technique by Strout et al. [15] to *any nest of loops*. *Schedule-independent* storage mappings have the same "portability" as SA with a much more economical memory usage. Of course, tuning expansion to a given parallel execution order generally yields more economical mappings.

4   An Algorithm for Storage Mapping Optimization

Finding the minimal amount of memory to store the values produced by the program is a graph coloring problem where vertices are operations and edges represent interferences between operations: There is an edge between v and w iff they can't share the same memory location, i.e. when the left-hand side of (2) holds. Since classic coloring algorithms only apply to finite graphs, Feautrier and Lefebvre designed a new algorithm [11], which we extend to general loop-nests.

4.1   Partial Expansion Algorithm

Input is the sequential program, the result of an IRDA, and a parallel execution order (not used for simple SA form conversion); It leaves unchanged its control structures but thoroughly reconstitutes its data. Let us define $\mathsf{Stmt}(\langle S, \mathsf{x} \rangle) = S$ and $\mathsf{Index}(\langle S, \mathsf{x} \rangle) = \mathsf{x}$.

1. For each statement S whose iteration vector is x: Build an expansion vector E_S which gives the shape of a new data structure D_S, see Section 4.2 for details. Then, the left-hand side (lhs) of S becomes $\mathsf{D}_S[\mathsf{x} \bmod \mathsf{E}_S]$.

2. Considering σ as a function from accesses to sets of operations (like in Section 2), it can be expressed as a *nested conditionals*. For each statement S and iteration vector x, replace each read reference r in the right-hand side (rhs) with Convert(r), where:

   - If $\sigma(\langle S, x \rangle, r) = \{u\}$, then Convert($r$) = $D_{\text{Stmt}(u)}[\text{Index}(u) \bmod E_{\text{Stmt}(u)}]$.
   - If $\sigma(\langle S, x \rangle, r) = \emptyset$, then Convert($r$) = r (the initial reference expression).
   - If $\sigma(\langle S, x \rangle, r)$ is not a singleton, then Convert(r) = $\phi(\sigma(\langle S, x \rangle, r))$; There is a general method to compute ϕ at run-time, but we prefer pragmatic techniques, such as the one presented in [3] or another algorithm proposed in [4].
   - If $\sigma(\langle S, x \rangle, r) = $ if p then r_1 else r_2, then
     Convert(r) = if p then Convert(r_1) else Convert(r_2).

3. Apply partial renaming to coalesce data structures, using any classical graph coloring heuristic, see [11].

This algorithm outputs an expanded program whose data are adapted to the partial execution order $\prec'$. We are assured that with these new data, the original program semantic will be preserved in the parallel version.

4.2 Building an Expansion Vector

For each statement S, the expansion vector must ensure that expansion is systematically done when the lhs of (2) holds, and introduce memory reuse between instances of S when it does not hold.

The dimension of E_S is equal to the number of loops surrounding S, written N_S. Each element $E_S[p+1]$ is the *expansion degree of* S *at depth* p (the depth of the loop considered), with $p \in [0, N_S - 1]$ and gives the size of the dimension $(p+1)$ of D_S. For a given access v, the set of operations which *may not write* in the same location as v can be deduced from the expansion correctness criterion (2), call it $W_p^S(v)$. It holds all operations w such that:

- w is an instance of S: Stmt(w) = S;
- Index(v)$[1..p]$ = Index(w)$[1..p]$ and Index(v)$[p+1]$ < Index(w)$[p+1]$;
- And lhs of (2) is satisfied for v and w, or w and v.

Let $w_p^S(v)$ be the lexicographic maximum of $W_p^S(v)$. The following definition of E_S has been proven to forbid any output-dependence between instances of S satisfying the lhs of (2) [3, 11].

$$E_S[p] = \max(\text{Index}(w_p^S)[p+1] - \text{Index}(v)[p+1] + 1) \qquad (3)$$

Computing this for each dimension of E_S ensures that D_S has a sufficient size for the expansion to preserve the sequential program semantics.

4.3 Summary of the Expansion Process

Since we consider unrestricted loop nests, some approximations[3] are performed to stick with affine relations (automatically processed by PIP or Omega).

---

[3] Source function σ is a *pessimistic* approximation, as well as $\neq$.

The more general application of our technique starts with IRDA, then apply a parallelization algorithm using σ as dependence graph (thus avoiding constraints due to spurious memory-based dependences), describe the result as a partial order $\prec'$, and eventually apply the partial expansion algorithm. This technique yields the best results, but involves an external parallelization technique, such as scheduling or tiling. It is well suited to parallelizing compilers.

If one looks for a schedule-independent storage mapping, the second technique sets the partial order $\prec'$ according to σ, the data-flow execution order[4]. This is useful whenever no parallel execution scheme is enforced: The "portability" of SA form is preserved, at a much lower cost in memory usage.

5   Experimental Results

Partial expansion has been implemented for Cray-Fortran affine loop nests [11]. Semi-automatic storage mapping optimization has also been performed on general loop-nests, using FADA, Omega, and PIP.

The result for the motivating example is that the storage mapping computed from a scheduled or tiled version is the same as the schedule-independent one (computed from the data-flow execution order). The resulting program is the same as the hand-crafted one in Figure 1.

A few experiments have been made on an SGI Origin 2000, using the mp library (but not PCA, the built-in automatic parallelizer...). As one would expect, results for the convolution program are excellent even for small values of n. The interested reader may find more results on the following web page: http://www.prism.uvsq.fr/~acohen/smo/smo.html.

6   Conclusion and Perspectives

Expanding data structures is a classical optimization to cut memory-based dependences. The first problem is to ensure that all reads refer to the correct memory location, in the generated code. When control and data flow cannot be known at compile-time, run-time computations have to be done to find the identity of the correct memory location. The second problem is that converting programs to single-assignment form is too costly, in terms of memory usage.

We have tackled both problems here, proposing a general method for partial memory expansion based on instance-wise reaching definition information, a robust run-time data-flow restoration scheme, and a versatile storage mapping optimization algorithm. Our techniques are either novel or generalize previous work to unrestricted nests of loops.

Future work is twofold. First, improve optimization of the generated code and study—both theoretically and experimentally—the effect of ϕ-functions on parallel code performance. Second, study how comprehensive parallelization techniques can be plugged into this framework: Reducing memory usage is a good thing, but choosing the right parallel execution order is another one.

---

[4] But $\prec'$ must be described as an *effective order*. One must compute the transitive closure of the symmetric relation: $(\sigma^{-1})^+$.

Acknowledgments: We would like to thank Jean-François Collard, Paul Feautrier and Denis Barthou for their help and support. Access to the SGI Origin 2000 was provided by the *Université Louis Pasteur, Strasbourg*, thanks to G.-R. Perrin.

References

[1] D. Barthou, J.-F. Collard, and P. Feautrier. Fuzzy array dataflow analysis. *Journal of Parallel and Distributed Computing*, 40:210–226, 1997.

[2] L. Carter, J. Ferrante, and S. Flynn Hummel. Efficient multiprocessor parallelism via hierarchical tiling. In *SIAM Conference on Parallel Processing for Scientific Computing*, 1995.

[3] A. Cohen and V. Lefebvre. Optimization of storage mappings for parallel programs. Technical Report 1998/46, PRiSM, U. of Versailles, 1998.

[4] J.-F. Collard. The advantages of reaching definition analyses in Array (S)SA. In *Proc. Workshop on Languages and Compilers for Parallel Computing*, Chapel Hill, NC, August 1998. Springer-Verlag.

[5] B. Creusillet. *Array Region Analyses and Applications.* PhD thesis, Ecole des Mines de Paris, December 1996.

[6] R. Cytron, J. Ferrante, B. K. Rosen, M. N. Wegman, and F. K. Zadeck. Efficiently computing static single assignment form and the control dependence graph. *ACM Transactions on Programming Languages and Systems*, 13(4):451–490, October 1991.

[7] A. Darte and F. Vivien. Optimal fine and medium grain parallelism detection in polyhedral reduced dependence graphs. *Int. Journal of Parallel Programming*, 25(6):447–496, December 1997.

[8] P. Feautrier. Dataflow analysis of scalar and array references. *Int. Journal of Parallel Programming*, 20(1):23–53, February 1991.

[9] P. Feautrier. Some efficient solution to the affine scheduling problem, part II, multidimensional time. *Int. J. of Parallel Programming*, 21(6), December 1992.

[10] F. Irigoin and R. Triolet. Supernode partitioning. In *Proc. 15th POPL*, pages 319–328, San Diego, Cal., January 1988.

[11] V. Lefebvre and P. Feautrier. Automatic storage management for parallel programs. *Journal on Parallel Computing*, 24:649–671, 1998.

[12] D. E. Maydan, S. P. Amarasinghe, and M. S. Lam. Array dataflow analysis and its use in array privatization. In *Proc. of ACM Conf. on Principles of Programming Languages*, pages 2–15, January 1993.

[13] W. Pugh. A practical algorithm for exact array dependence analysis. *Communications of the ACM*, 35(8):27–47, August 1992.

[14] Fabien Quilleré and Sanjay Rajopadhye. Optimizing memory usage in the polyhedral model. Technical Report 1228, IRISA, January 1999.

[15] M. Mills Strout, L. Carter, J. Ferrante, and B. Simon. Schedule-independant storage mapping for loops. In *ACM Int. Conf. on Arch. Support for Prog. Lang. and Oper. Sys. (ASPLOS-VIII)*, 1998.

[16] P. Tu and D. Padua. Automatic array privatization. In *Proc. Sixth Workshop on Languages and Compilers for Parallel Computing*, number 768 in Lecture Notes in Computer Science, pages 500–521, August 1993. Portland, Oregon.

[17] D. G. Wonnacott. *Constraint-Based Array Dependence Analysis.* PhD thesis, University of Maryland, 1995.

Array SSA for Explicitly Parallel Programs

Jean-François Collard*

CNRS & PRISM, University of Versailles St-Quentin
45 Avenue des Etats-Unis. 78035 Versailles, France
jfc@prism.uvsq.fr

Abstract. The SSA framework has been separately extended to parallel programs and to sequential programs with arrays. This paper proposes an Array SSA form for parallel programs with either weak or strong memory consistency, with event-based synchronization or mutual exclusion, with parallel sections or indexed parallel constructs.

1 Introduction and Related Work

Extending reaching definition analyses (RDAs) and SSA to parallel programs has been a very active area of research. Reaching definition analyses (RDAs) for parallel programs were proposed in [4, 5, 2]. SSAs for various kinds of parallel language models were proposed in [8, 7, 9]. Several issues make comparing related papers a tricky task. Questions that arise include: (1) Does the paper apply element-wise to arrays? (Most papers apply to scalars only or to arrays considered as atomic variables, except for [2, 6]). (2) Does the paper consider both the weak and the strong model of memory consistency? (Most papers apply to only one of them) (3) Does the paper allow shared variables? (E.g., [5] doesn't.) (4) Are event-based and mutual-exclusion synchronizations allowed? (5) Are indexed parallel constructs (parallel loops) and parallel sections allowed? Except for [2], related papers apply to parallel sections only.

This paper answers Yes to all of these questions. This paper extends the RDA in [2] to handle strong consistency, event-based synchronization and mutual exclusion. Arrays are considered element-wise, and any array subscript is allowed [1]. We then derive a precise Parallel Array SSA form.

2 Problem Position

In Array SSA [6], each array is assigned by at most one statement. Converting a program to Array SSA form requires, therefore, to rename the original arrays. As in the sequential case, two issues threaten to modify the program's data-flow while converting: Unpredictable control flow, due to e.g. ifs, and array subscripts too intricate to be understood by the static analysis. Classically, restoring the data-flow is then done using "value-merging" ϕ-functions [3].

---

* Partially supported by INRIA A3 and the German-French Procope Program

P. Amestoy et al. (Eds.): Euro-Par'99, LNCS 1685, pp. 383–390, 1999.

However, a third problem occurs in parallel programs only. Consider the parallel sections in Fig. 1.(a). Since the two sections execute in any order, the value read in the last statement is not determined. Now consider the Parallel Array SSA version in Fig. 1.(b). Because arrays are renamed, an auxiliary function is inserted to restore the flow of data. As in [7], this function is called a π-function, to be formally defined in Section 3.3.

```
par
  section
    a(1) = ...
  end section
  section
    a(1) = ...
  end section
end par
.. = a(1)
```

```
par
  section
    a1(1) = ...
  end section
  section
    a2(1) = ...
  end section
end par
a3(1) = π (a1(1),a2(1))
.. = a3(1)
```

```
do i = 1, n
  par
    section
1     x = ...
    end section
    section
2     ... = x
3     x = ...
    end section
  end par
end do
```

(a) Original program

(b) Parallel Array SSA form (This paper)

(c) Illustrating the weak and strong models

Fig. 1. Array SSA with Parallel control flow

Moreover, parallel languages may have a strong or weak model of memory consistency. Consider the program in Fig.1.(c). In a weak model, a consistent view of memory is restored at the exit of the **par** construct only. Therefore, the definition reaching the ith instance of statement 2 is the $(i-1)$th instance of either 1 or 3. (This holds for $i > 1$; otherwise, the reaching definition is undefined, denoted by $\perp$. Remember we reason at the statement instance level.) In a strong model, like in Java threads, a definition in 1 is immediately visible by 2. Therefore, the ith instance of 1 may always reach the ith instance of 2 in this model. In the sequel, parallel sections with weak (resp. strong) consistency will be introduced by **par_weak** (resp. **par_strong**). Keyword **par** means that both models apply.

Indexed Parallel Constructs We consider two indexed parallel constructs: **doall** and **forall**. Several languages have indexed parallel constructs where a statement instance is spawned for each possible value of the index variable. We call this construct **doall**. Conceptually, each instance has its own copy of shared data structures, and all reads and writes are applied to this copy. This corresponds to the **FIRSTPRIVATE** directive in OpenMP. A consistent view of memory is provided on exiting the parallel construct.

Similarly, a **forall** construct spawns as many instances as there are possible values for the index variable. The semantics of **forall** we consider is reminiscent from HPF: Data structures are shared and are updated before any instance of following statements begins. A **forall** construct corresponds, in OpenMP, to a **parallel** do with **shared** arrays and **barriers** between two successive assignments. Indeed, **barrier** implies that shared variables are **flushed** to memory to provide a consistent view of memory.

Parallel sections Parallel sections represent code segments executed by independent tasks. In the **par_strong** construct, we assume strong consistency and interleaving semantics, i.e., the memory updates made by a thread are immediately visible by the other threads. The **par_weak** construct has a weak consistency model. (Both consistency models are discussed to show that our framework captures both views.) Notice that data independence is not required, unlike [5]. Unlike [7], loops may contain parallel sections.

Synchronization Accepted models of synchronization are events (and therefore, barriers) and mutual exclusion (and therefore, critical sections). Busy waiting is not allowed, because busy wait is not syntactically declared and can be hidden in some weird behavior of a thread's code.

Two motivating examples are given in Figure 2. In the first program (a), weak consistency is assumed. Parallel sections and sequential loops are nested. Figure 3.(a) gives its counterpart in Parallel Array SSA. In the second example,

```
S₀ : a(..)= ..                      S₀ :    a(..) = ..
        do i = 1, n                          do i = 1, n
P₁ :    par_weak                    P₁ :       par_strong
            section                              section
                do j = 1, i         I :             if(foo)
P₂ :            par                 S₁ :               a(i) = ...
                    section                          end if
S₁ :                a(i+j) = ...                     Post_e1(i)
                    end section                    end section
                    section                        section
R :                     ... = ..a(i+j-1)..  S₂ :        a(i) = a(i-1)
                    end section                        Post_e2(i)
                end par                             end section
            end do                                  section
        end section                                    Wait_e1(i)
        section                                        Wait_e2(i)
S₂ :        a(i+1) = ...             R :                ... = a(i)
        end section                                 end section
    end par                                      end par
end do                                       end do

(a)                                 (b)
```

Fig. 2. Motivating examples.

we assume strong memory consistency. The example features a conditional with unknown predicate (line *I*) and event synchronization. Its counterpart in Parallel Array SSA appears in Figure 3.(b).

3   Framework

The dimension of a vector x is denoted by $|x|$. The k-th entry of vector x, $k \geq 0$, is denoted by $x[k]$. The sub-vector built from components k to l is written as: $x[k..l]$. If $k > l$, then $x[k..l]$ has dimension 0. $x.y$ denotes the vector built from x and y by concatenation. Furthermore, $\ll$ ($\lll$) denotes the (resp. strict) lexicographical order on such vectors. The *iteration vector* of a statement S is the vector built from the counters of do, forall, doall and while constructs surrounding S. The *depth* M_S of S is dimension of the iteration vector. The iteration domain of S is denoted by $\mathbf{D}(S)$.

Let $C_p(S)$ be the iterative construct (do, while, forall or doall) surrounding S at depth p. (When clear from the context, S will be omitted.) Let $\mathcal{P}(S,R)$ be the innermost par construct surrounding both S and R such that S and R appear in distinct sections of the par construct. Moreover, let M_{SR} be the depth of $\mathcal{P}(S,R)$. (If $\mathcal{P}(S,R)$ does not exist, then M_{SR} is the number of do, forall, doall or while constructs surrounding both S and R.) Finally, $S \lhd R$ holds if S textually precedes R and if S and R do not appear in opposite arms of an if..then..else or where..elsewhere construct, or in distinct sections of a par_weak construct.

3.1   Expressing the Order of Memory Updates

As far as data flows are concerned, we are mainly interested in the order $\prec$ in which computations, and their corresponding writes and reads to memory, occur. The fact that the memory modifications made by $\langle S, x \rangle$ are visible by $\langle R, y \rangle$ in the parallel program is written as: $\langle S, x \rangle \prec \langle R, y \rangle$. By convention, $\bot$ is the name of an arbitrary operation executed before all other operations of the program, i.e. $\forall S, x : \bot \prec \langle S, x \rangle$.

Event-base Synchronizations An *event* is a variable with two states *posted* and *cleared*. Usually, parallel programs manipulate arrays of events, which are therefore labeled with an integer variable. Statement $\mathtt{Post}_e(\mathtt{i})$ sets the ith instance of event e to 'posted', and statement $\mathtt{Wait}_e(\mathtt{i})$ suspends the calling thread until the ith event e is posted. Statements Post and Wait are allowed in par constructs only. Let z be the iteration vector of $\mathcal{P}(\mathtt{Post}, \mathtt{Wait})$ surrounding matching Post and Wait. As before, $M_{\mathtt{PostWait}} = |z|$. If a statement S is in the same section as $\mathtt{Post}_e(\mathtt{i})$ (with $S \lhd \mathtt{Post}_e(\mathtt{i})$), if a statement R is in the same section as $\mathtt{Wait}_e(\mathtt{j})$ (with $\mathtt{Wait}_e(\mathtt{j}) \lhd R$), and if the events are the same ($i = j$) for the same instance of the Post/Wait, then instances of the two statements are ordered: $\langle S, x \rangle \prec_{\mathtt{Wait}} \langle R, y \rangle \equiv S \lhd \mathtt{Post}_e(\mathtt{i})$ $\wedge \mathtt{Wait}_e(\mathtt{j}) \lhd R \wedge x[0..M_{\mathtt{PostWait}} - 1] = y[0..M_{\mathtt{PostWait}} - 1] \wedge i = j$. Similarly, for Barrier statements: $\langle S, x \rangle \prec_{\mathtt{Barrier}} \langle R, y \rangle \equiv S \lhd \mathtt{Barrier} \wedge \mathtt{Barrier} \lhd R \wedge x[0..M_{\mathtt{Barrier}} - 1] = y[0..M_{\mathtt{Barrier}} - 1]$. For example, let us consider S_1 and R in Figure 2.(b). Since $S_1 \lhd \mathtt{Post}_{e1}(\mathtt{i})$ and $\mathtt{Wait}_{e1}(\mathtt{i'}) \lhd R$, we have: $i = i' \implies \langle S_1, i \rangle \prec \langle R, i' \rangle$. The same applies to S_2: $i = i' \implies \langle S_2, i \rangle \prec \langle R, i' \rangle$.

Order $\prec$ is expressed by affine equalities and inequalities.

$$\langle S, x\rangle \prec \langle R, y\rangle \Leftrightarrow \left(\begin{array}{c} \bigvee \\ p = 0..M_{SR} - 1, \quad Pred(p, x, y) \\ C_p = \mathtt{do} \vee C_p = \mathtt{while} \end{array} \right)$$

$$\vee \left(\left(\bigwedge_{p=0..M_{SR}-1} Equ(p, C_p, x, y) \right) \wedge T(S, R) \right)$$

$$\vee \langle S, x\rangle \prec_{\mathtt{Wait}} \langle R, y\rangle \vee \langle S, x\rangle \prec_{\mathtt{Barrier}} \langle R, y\rangle \qquad (1)$$

with $T(S, R) \equiv S \lhd R \vee (\mathcal{P}(S, R) = \mathtt{par\_strong})$ and:

$$Pred(p, x, y) \equiv \left(\bigwedge_{i=0..p-1} Equ(i, C_i, x, y) \right) \wedge x[p] < y[p] \qquad (2)$$

$$Equ(p, \mathtt{forall}, x, y) \equiv \mathbf{true} \qquad (3)$$

$$Equ(p, \mathtt{doall}, x, y) \equiv x[p] = y[p] \qquad (4)$$

and $Equ(p, \mathtt{do}, x, y) \equiv x[p] = y[p]$ and $Equ(p, \mathtt{while}, x, y) \equiv x[p] = y[p]$.

Intuitively, Predicate $Pred$ in (1) formalizes the sequential order of a given do or while loop at depth p. Such a loop enforces an order up to the first par construct encountered at depth M_{SR} while traversing the nest of control structures, from the outermost level to the innermost. The order of sequential loops is given by the strict inequality in (2), under the condition that the two instances are not ordered up to level $p-1$; hence the conjunction on the p outer predicates Equ. Instances of two successive statements inside a forall at depth p are always ordered at depth p due to (3), but are ordered inside a doall only if they belong to the same task (i.e., the values of the doall index are equal, cf (4)). Eq. (1) yields a partial order on statement instances. Notice also that for any predicate P, $\bigvee_{i \in \emptyset} P(i) = \mathbf{false}$ and $\bigwedge_{i \in \emptyset} P(i) = \mathbf{true}$.

For example, let us come back to Figure 2.(b) and consider S_1 and R. Applying (1), we get:

$$\langle S_1, i\rangle \prec \langle R, i'\rangle \Leftrightarrow Pred(0, i, i') \vee (Equ(0, \mathtt{do}, i, i') \wedge S_1 \lhd R) \vee \langle S, i\rangle \prec_{\mathtt{Wait}} \langle R, i'\rangle$$
$$\Leftrightarrow i < i' \vee (i = i' \wedge \mathbf{false}) \vee i = i' \Leftrightarrow i \leq i'$$

For S_2 and R, the result is the same: $\langle S_2, i\rangle \prec \langle R, i'\rangle \Leftrightarrow i \leq i'$. Comparing instances of S_1 and S_2 respectively yields: $\langle S_1, i\rangle \prec \langle S_1, i'\rangle \Leftrightarrow i < i' \vee (i = i' \wedge \mathbf{false}) \Leftrightarrow i < i'$ and $\langle S_2, i\rangle \prec \langle S_2, i'\rangle \Leftrightarrow i < i' \vee (i = i' \wedge \mathbf{false}) \Leftrightarrow i < i'$. However, S_1 and S_2 appear in distinct sections of a common par_strong construct, so: $\langle S_1, i'\rangle \prec \langle S_2, i\rangle \Leftrightarrow i < i' \vee (i = i' \wedge \mathbf{true}) \Leftrightarrow i \leq i'$.

Mutual Exclusion and Critical Sections Statements Lock(l) and Unlock(l) on a lock l are modeled as follows. Let P_1 and P_2 be two distinct sections of code surrounded by matching Lock/Unlock. Then:
$$\forall S \in P_1, \forall T \in P_2, \forall x \in \mathbf{D}(S), \ \forall y \in \mathbf{D}(T) : (\langle S, x\rangle \prec \langle T, y\rangle) \vee (\langle T, y\rangle \prec \langle S, x\rangle).$$

This equation also models a critical section P, with $P = P_1 = P_2$.

3.2   Reaching Definition Analysis

Let us consider two statements S and R, surrounded by two iterations vectors x and y respectively. Suppose that S and R respectively writes into and reads an array a (For the sake of clarity, we assume that a statement has at most one read reference):

$$S : \quad a(f_S(x)) = \ldots$$
$$R : \quad \ldots = a(g_R(y))$$

The aim of an instance-wise RDA on arrays is to find, for a parametric y, the definition of $a(g_R(y))$ that reaches the y-th instance of R. This reaching definition is denoted by $\mathsf{RD}(\langle R, y \rangle)$. We also define: $\mathsf{RD}(\langle R, y \rangle) = \bigcup_{S \in \mathcal{S}(R)} \mathsf{RD}_S(\langle R, y \rangle)$.

3.3   Conversion to Parallel Array SSA

Renaming arrays in the left-hand sides of all statements is easy. In a statement S writing into an array a: $a(f_S(x)) = \ldots$, the l-value is replaced with: $a_S(f_S(x)) = \ldots$. However, read references have to be updated accordingly so as to preserve the initial flow of data.

ϕ and π Functions Restoring the flow of a datum is needed if and only if the set of reaching definitions has more than one element. Indeed, let us consider a read reference $a(g_R(y))$ in some statement R with iteration vector y. If, for all x in $\mathbf{D}(R)$, $\mathsf{RD}(\langle R, y \rangle) = \{\langle S, h(y) \rangle\}$ (mapping h is given by the RDA), then the read is transformed into $a_S(f_S(h(y)))$ after renaming or, equivalently, into $a_S(g_R(y))$.

When $\mathsf{RD}(\langle R, y \rangle)$ contains the undefined symbol $\perp$, the program may read an uninitialized value. When $\mathsf{RD}(\langle R, y \rangle)$ has more than one element, the analysis was unable to decide the reaching definition statically, and some "black-box" needs to be inserted to restore the flow of data. Notice that this point of view does not distinguish between restoring the data flow because of sequential indeterminacies (e.g., branching in ifs) and restoring the data flow because of parallelism. I.e., ϕ and π functions are handled in the same way.

Conceptually, where ϕ or π functions are needed, new arrays are created to store the "merged" values coming from the various reaching definitions. To do this is, we create a new array a_R and insert an assignment Φ_R: $a_R = \phi(a_{S_1}, \ldots, a_{S_n})$ (or: $a_R = \pi(a_{S_1}, \ldots, a_{S_n})$), such that $\forall S_i : \exists y \in \mathbf{D}(R), \exists x \in \mathbf{D}(S_i), \langle S_i, x \rangle \in \mathsf{RD}(\langle R, y \rangle)$.

Placing ϕ and π Functions We place the ϕ and π functions at the earliest point guaranteed to be executed before the read and after all its possible reaching definitions. I.e.:

$$\Phi_R = \min_{\lhd} \{ T \mid \forall y \in \mathbf{D}(R), \forall \langle S, x \rangle \in \mathsf{RD}(\langle R, y \rangle) : \tag{5}$$
$$\exists z \in \mathbf{D}(T), \langle S, x \rangle \prec \langle T, z \rangle \prec \langle R, y \rangle \}$$

Since $\lhd$ is a partial order, (5) may give several locations where the ϕ or π function can be inserted.

4   Back to the Examples

First Example Let us consider the read in R in Program ExP. Possible reaching assignments are S_1 and S_2. For statement S_1: We have $\mathcal{P}(S_1, R) = P_2$, so $M_{S_1R} = 2$. Thus, the execution order between instances of Statements S_1 and R is given by $\langle S_1, i', j' \rangle \prec \langle R, i, j \rangle \Leftrightarrow i' < i \vee (i' = i \wedge j' \leq j)$ In a similar way: $\langle S_1, i', j' \rangle \prec \langle S_1, i, j \rangle \Leftrightarrow i' < i \vee (i' = i \wedge j' < j)$. $\mathrm{RD}_{S_1}(\langle R, i, j \rangle)$ can now be computed by Omega: $\forall i, 2 \leq i \leq n : \mathrm{RD}_{S_1}(\langle R, i, 1 \rangle) = \langle S_1, i - 1, 1 \rangle$ and $\forall i, j, 2 \leq j \leq i \leq n : \mathrm{RD}_{S_1}(\langle R, i, j \rangle) = \langle S_1, i, j - 1 \rangle$.

For statement S_2: We saw that $\mathcal{P}(S_2, R) = P_1$, so $M_{S_2R} = 1$. Thus, the execution order between instances of Statements S_2 and R is given by $\langle S_2, i' \rangle \prec \langle R, i, j \rangle \Leftrightarrow Pred(0, (i'), (i, j)) \vee (Equ(0, (i'), (i, j)) \wedge T_{S_2R})$, which is equivalent to $i' < i$. We compute the reaching definitions $\mathrm{RD}_{S_2}(\langle R, i, j \rangle)$ which are instances $\langle S_2, i' \rangle$ using the Omega Calculator: $\forall i, 2 \leq i \leq n : \mathrm{RD}_{S_2}(\langle R, i, 1 \rangle) = \langle S_2, i - 1 \rangle$. Merging $\mathrm{RD}_{S_1}(\langle R, i, j \rangle)$ and $\mathrm{RD}_{S_2}(\langle R, i, j \rangle)$ gives the assignment Φ_R we insert:

```
if(i >= 2) then if(j >= 2) then a₃(i+j-1) = a₁(i+j-1)
                else a₃(i+j-1) = π (a₁(i+j-1),a₂(i)) endif
else a₃(i+j-1) = a₀(i+j-1) endif
```

Notice that, in Figure 3, (5) does not simply insert Φ_R just before R at the first line of the innermost loop.

Second Example Let us just consider the read in S_2. We get: $\forall i, 1 \leq i \leq n :$ $\mathrm{RD}_{S_1}(\langle S_2, i \rangle) = \{\langle S_1, i - 1 \rangle, \bot\}$ (because $\langle S_1, i - 1 \rangle$ may or may not execute, and other instance of S_1 do not write in a(i-1)) and $\forall i, 1 \leq i \leq n : \mathrm{RD}_{S_2}(\langle S_2, i \rangle) = \langle S_2, i - 1 \rangle$. Merging the two gives:

$$\mathrm{RD}(\langle S_2, i \rangle) = \text{ if } i = 1 \text{ then } S_0 \text{ else } \{\langle S_1, i - 1 \rangle, \langle S_2, i - 1 \rangle\}.$$

This immediately gives the assignment Φ_{S_2} we have to insert to restore the data flow:

```
if(i=1) then a₃(i-1) = a₀(i-1)
        else a₃(i-1) = π (a₁(i-1),a₂(i-1))
```

The resulting code appears in Figure 3.(b). Notice Φ_{S_2} is not placed just before S_2, but is lifted up to the first line in the loop. However, (5) is not allowed to lift Φ_R above the two Waits.

References

[1] D. Barthou, J.-F. Collard, and P. Feautrier. Fuzzy array dataflow analysis. *Journal of Parallel and Distributed Computing*, 40:210–226, 1997.

[2] J.-F. Collard and M. Griebl. Array dataflow analysis for explicitly parallel programs. *Parallel Processing Letters*, 1997.

[3] R. Cytron, J. Ferrante, B. K. Rosen, M. N. Wegman, and F. K. Zadeck. Efficiently computing static single assignment form and the control dependence graph. *ACM Trans. on Prog. Lang. Sys.*, 13(4):451–490, Oct. 1991.

```
        a0(..) = ...
        do i = 1, n
P1 :    par_weak
            section
                do j = 1, i
ΦR :            if(i >= 2)
                then if(j >= 2)
                then a3(i+j-1) =
                    a1(i+j-1)
                else a3(i+j-1) =
                    π (a1(i+j-1),a2(i))
                end if
                else a3(i+j-1)=a0(i+j-1)
                end if
P2 :            par
                    section
S1 :                a1(i+j) = ...
                    end section
                    section
R :                 ... = a3(i+j-1)
                    end section
                end par
            end do
        end section
        section
S2 :        a2(i+1) = ...
        end section
    end par
end do
```

```
        a0(..) = ..
        do i = 1, n
ΦS2 :       if(i=1)
            then a3(i-1) = a0(i-1)
            else a3(i-1) =
                π (a1(i-1),a2(i-1))
            end if
            par_strong
                section
S1 :            if(foo) a1(i)= ... end if
                Poste1(i)
                end section
                section
S2 :            a2(i) = a3(i-1)
                Poste2(i)
                end section
                section
                Waite1(i)
                Waite2(i)
ΦR :            a4(i) = π (a1(i),a2(i))
R :             ... = a4(i)
                end section
            end par
        end do
```

Fig. 3. Array SSA form for the two motivating examples.

[4] J. Ferrante, D. Grunwald, and H. Srinivasan. Computing communication sets for control parallel programs. In *W. on Lang. and Comp. for Par. Comp. (LCPC)*, volume 892 of *LNCS*, pages 316–330, Ithaca, NY, August 1994.

[5] D. Grunwald and H. Srinivasan. Data flow equations for explicitly parallel programs. In *ACM Symp. on Prin. of Prog. Lang. (PoPL)*, pages 159–168, May 1993.

[6] K. Knobe and V. Sarkar. Array SSA form and its use in parallelization. In *ACM Symp. on Prin. of Prog. Lang. (PoPL)*, pages 107–120, San Diego, CA, Jan. 1998.

[7] J. Lee, S. Midkiff, and D. A. Padua. Concurrent static single assignment form and constant propagation for explicitly parallel programs. In *W. on Lang. and Comp. for Par. Comp. (LCPC)*, Aug. 1997.

[8] D. Novillo, R. Unrau, and J. Schaeffer. Concurrent SSA form in the presence of mutual exclusion. In *Int. Conf. on Par. Proc. (ICPP'98)*, pages 356–364, Aug. 1998.

[9] H. Srinivasan, J. Hook, and M. Wolfe. Static single assignment form for explicitly parallel programs. In *ACM Symp. on Prin. of Prog. Lang. (PoPL)*, pages 260–272, Charleston, D.C., Jan. 1993.

Parallel Data-Flow Analysis of Explicitly Parallel Programs

Jens Knoop

Universität Dortmund, D-44221 Dortmund, Germany
phone: ++49-231-755-5803, fax: ++49-231-755-5802
knoop@ls5.cs.uni-dortmund.de
http://sunshine.cs.uni-dortmund.de/~knoop

Abstract. In terms of program verification *data-flow analysis* (*DFA*) is commonly understood as the computation of the *strongest* postcondition for every program point with respect to a precondition which is assured to be valid at the entry of the program. Here, we consider DFA under the dual *weakest* precondition view of program verification. Based on this view we develop an approach for *demand-driven* DFA of explicitly parallel programs, which we exemplify for the large and practically most important class of bitvector problems. This approach can directly be used for the construction of online debugging tools. Moreover, it is tailored for parallelization. For bitvector problems, this allows us to compute the information provided by conventional, strongest postcondition-centered DFA at the costs of answering a data-flow query, which are usually much smaller. In practice, this can lead to a remarkable speed-up of analysing and optimizing explicitly parallel programs.

1 Motivation

Intuitively, the essence of *data-flow analysis* (*DFA*) is to determine run-time properties of a program without actually executing it, i.e., at compile-time. In its conventional peculiarity, DFA means to compute for every program point the most precise, i.e., the "largest" data-flow fact which can be guaranteed under the assumption that a specific data-flow fact is valid at the entry of the program. In terms of program verification (cf. [1]), this corresponds to a strongest postcondition approach: Compute for every program point the *strongest* postcondition with respect to a precondition assured to be valid at the entry of the program.

In this article we consider DFA under the dual, weakest precondition view of program verification, i.e., under the view of asking for the *weakest* precondition, which must be valid at the entry of the program in order to guarantee the validity of a specific "requested" postcondition at a certain program point of interest. In terms of DFA this means to compute the "smallest" data-flow fact which has to be valid at the entry of the program in order to guarantee the validity of the "requested" data-flow fact at the program point of interest.

This dual view of DFA, which has its paragon in program verification, is actually the conceptual and theoretical backbone of approaches for *demand-driven*

P. Amestoy et al. (Eds.): Euro-Par'99, LNCS 1685, pp. 391–400, 1999.

(*DD*) DFA. Originally, such approaches have been proposed for the interprocedural analysis of sequential programs, in particular aiming at the construction of online debugging tools (cf. [3, 7, 16]), but were also successfully applied to other fields (cf. [4, 17]).

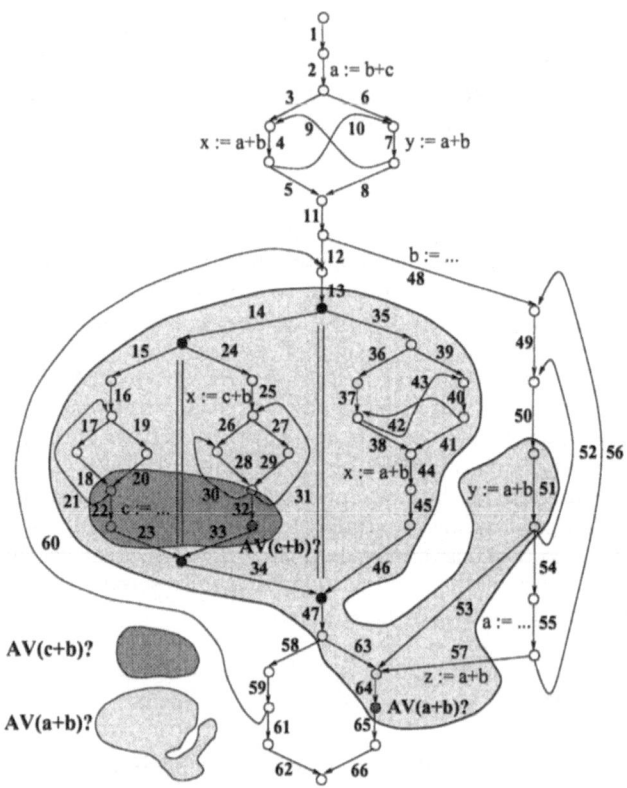

Fig. 1. Demand-drivenly, only small portions of the program need to be investigated.

In this article we apply this approach to the analysis of explicitly parallel programs with shared memory and interleaving semantics. The central benefit of this approach, which leads to the efficiency gain in practice, is illustrated in Figure 1 using the *availability of terms*, a classical DFA-problem (cf. [6]), for illustration. Intuitively, a term is *available* at a program point, if it has been computed on every program path reaching this point without a succeeding modification of any of its operands. Using the conventional approach to DFA, the program of Figure 1 must completely be investigated in order to detect the availability and the inavailability of the terms $a+b$ and $c+b$ at the destination nodes of the edges **64** and **32**, respectively. In contrast, our DD-approach has only to consider small portions of the complete program in order to answer these two data-flow queries. Though generally DD-approaches have the same worst-case time complexity as their conventional counterparts, in practice they are usually much more efficient (cf. [5]). Additionally, unlike its conventional counterpart, the DD-approach we are going to propose here can immediately be parallelized: Different data-flow queries can concurrently be processed. Obviously, this is tailored for a setting offering parallel processes on shared memory because the processes can easily share the answers to data-flow queries computed by other processes, which leads to a further speed-up of the analysis. Additionally, for unidirectional bitvector analyses (cf. [6]), the class of DFAs the framework of [15] is designed for, this approach can directly be used to mimic conventional DFA, for which it is characteristic to decide the DFA-property of interest for every program point. In our approach, this is achieved by answering the corresponding DFA-queries for all program points in parallel. In fact, this yields the information delivered by conventional DFA at the costs of answering a data-flow query, which are usually much smaller. This is important as subsequent program optimizations usually rely on the "full" information.

2   Background: DD-DFA in the Sequential Setting

In DFA programs are commonly represented by flow graphs $G = (N, E, \mathbf{s}, \mathbf{e})$ with node set N, edge set E, and a unique start node $\mathbf{s}$ and end node $\mathbf{e}$, which are assumed to have no incoming and outgoing edges, respectively. Here, we assume that edges represent the statements and the nondeterministic control flow of the underlying program, while nodes represent program points. We denote the sets of immediate predecessors and successors of a node n by $pred(n)$ and $succ(n)$, and the set of all finite paths from a node m to a node n by $\mathbf{P}[m, n]$.

In order to decide some (aspect of a) run-time property by a DFA, one usually replaces the "full" semantics of the program by a simpler, more abstract version (cf. [2]), which is tailored for the problem under consideration. Usually, the abstract semantics is given by a *semantic functional* $[\![\ ]\!] : E \rightarrow (\mathcal{C} \rightarrow \mathcal{C})$, which gives abstract meaning to the elementary statements of a program in terms of a transformation on a complete lattice $\mathcal{C}$ with least element $\bot$ and greatest element $\top$. The elements of $\mathcal{C}$ represent the data-flow facts of interest. Without loss of generality we assume that $\top \in \mathcal{C}$ represents "unsatisfiable (inconsistent)" data-flow information, and is invariant under all semantic functions.

Conventional DFA: Strongest Postcondition View. Conventionally, DFA aims at computing for every program point the strongest postcondition with respect to a precondition $c_\mathbf{s} \in \mathcal{C} \backslash \{\top\}$ assured to be valid at $\mathbf{s}$. Formally, this is expressed by the *meet-over-all-paths* (*MOP*) approach, which induces the *MOP*-solution (cf. [9]). Note that $[\![p]\!]$ denotes the straightforward extension of $[\![\ ]\!]$ to paths, i.e., $[\![p]\!]$, $p = \langle e_1, \ldots, e_q \rangle$, equals the identity $Id_\mathcal{C}$ on $\mathcal{C}$, if $q < 1$, and $[\![\langle e_2, \ldots, e_q \rangle]\!] \circ [\![e_1]\!]$, otherwise.

The *MOP*-Solution: The Strongest Postcondition

$$\forall c_\mathbf{s} \in \mathcal{C} \ \forall n \in N. \ MOP_{([\![\ ]\!], c_\mathbf{s})}(n) =_{df} \sqcap \{ [\![p]\!](c_\mathbf{s}) \mid p \in \mathbf{P}[\mathbf{s}, n] \}$$

Demand-Driven DFA: Weakest Precondition View. Fundamental for the dual, weakest precondition view of DFA is the notion of the *reversed semantic functional* $[\![\ ]\!]_R : E \to (\mathcal{C} \to \mathcal{C})$. It is induced by $[\![\ ]\!]$, and defined as follows:

$$\forall e \in E \ \forall c \in \mathcal{C}. \ [\![\, e\,]\!]_R(c) =_{df} \sqcap \{\, c' \mid [\![\, e\,]\!](c') \sqsupseteq c\,\}.$$

Intuitively, the reversed function $[\![\, e\,]\!]_R$, $e \in E$, maps a data-flow fact $c \in \mathcal{C}$, which plays here the role of a postcondition, to the least data-flow fact, which is sufficient to guarantee the validity of c after executing e. We have (cf. [8]):

Proposition 1. *For all edges* $e \in E$ *we have:* (1) $[\![\, e\,]\!]_R$ *is well-defined and monotonic.* (2) *If* $[\![\, e\,]\!]$ *is distributive, then* $[\![\, e\,]\!]_R$ *is additive.*

Like their counterparts also the reversed functions can easily be extended to capture finite paths. $[\![\, p\,]\!]_R$, $p = \langle e_1, \ldots, e_q \rangle$, equals the identity $Id_{\mathcal{C}}$ on $\mathcal{C}$, if $q < 1$, and $[\![\, \langle e_1, \ldots, e_{q-1} \rangle\,]\!]_R \circ [\![\, e_q\,]\!]_R$, otherwise. Note that this means a "backward" traversal of p as expected when dealing with weakest preconditions.

In order to complete the presentation of the weakest precondition view of DFA we introduce the following simple extension of G, which simplifies the formal development. If $q \in N$ is the program point of interest, i.e., the *query node*, we assume that G is extended by a copy **q** of q having the same predecessors as q, but no successors. Obviously, the *MOP*-solutions for q in G and for **q** in the extended program coincide; a fact, which is important for the correctness of considering the extended program. Denoting the data-flow query by c_q, where the index implicitly fixes the query node, the straightforward reversal of the *MOP*-approach leads from the strongest postcondition view to its dual weakest precondition view. Besides reversing the analysis direction, the duality also requires to consider the dual lattice operation of *joining* data-flow facts. The dual analogue of the meet-over-all-paths approach is thus the *reversed-join-over-all-paths* (*R-JOP*) approach. Intuitively, the solution of the *R-JOP*-approach at a program point is the join of all data-flow facts required by some program continuation reaching **q** in order to assure c_q at **q** (and q).

The *R-JOP*-Solution: The Weakest Precondition

$$\forall c_q \in \mathcal{C} \ \forall n \in N. \ R\text{-}JOP_{([\![\ ]\!], c_q)}(n) =_{df} \bigsqcup \{\, [\![\, p\,]\!]_R(c_q) \mid p \in \mathbf{P}[n, \mathbf{q}]\,\}$$

The Connecting Link. The following theorem establishes the link between the strongest postcondition and the weakest preconditon view of DFA. If the semantic functions $[\![\, e\,]\!]$, $e \in E$, are all distributive (as it will be the case in Section 4!), the *MOP*-solution and the *R-JOP*-solution are equally informative.

Theorem 1 (Link Theorem).
Let all functions $[\![\, e\,]\!]$, $e \in E$, *be distributive. Then for every pair of a node* $q \in N$ *and its copy* **q**, *and for every pair of data-flow facts* $c_s, c_q \in \mathcal{C}$ *we have:*

$$MOP_{([\![\ ]\!], c_s)}(q) \sqsupseteq c_q \iff R\text{-}JOP_{([\![\ ]\!], c_q)}(\mathbf{s}) \sqsubseteq c_s$$

In particular, we have: (1) $R\text{-}JOP_{([\![\ ]\!], MOP_{([\![\ ]\!], c_s)}(q))}(\mathbf{s}) \sqsubseteq c_s$ *and*
(2) $MOP_{([\![\ ]\!], R\text{-}JOP_{([\![\ ]\!], c_q)}(\mathbf{s}))}(q) = MOP_{([\![\ ]\!], R\text{-}JOP_{([\![\ ]\!], c_q)}(\mathbf{s}))}(\mathbf{q}) \sqsupseteq c_q$

Computing Strongest Postconditions and Weakest Preconditions. The *MOP*-solution and the *R-JOP*-solution are both specificational in nature. They generally do not induce an effective computation procedure. In practice, this is taken care of by complementing each approach with a fixed-point approach. These are based on the following two equation systems, which impose consistency constraints on an annotation of the program with data-flow facts, where *ass* and *req-ass* remind to "assertion" and "required assertion," respectively.

Equation System 1 (*MaxFP*-Approach: The *MOP*-Counterpart).

$$\mathbf{ass}(n) = \begin{cases} c_{\mathbf{s}} & \text{if } n = \mathbf{s} \\ \sqcap \{ \, [\![(m,n)]\!] (\mathbf{ass}(m)) \mid m \in pred(n) \, \} & \text{otherwise} \end{cases}$$

Equation System 2 (*R-MinFP*-Approach: The *R-JOP*-Counterpart).

$$\mathbf{req\text{-}ass}(n) = \begin{cases} c_q & \text{if } n = \mathbf{q} \\ \sqcup \{ \, [\![(n,m)]\!]_R (\mathbf{req\text{-}ass}(m)) \mid m \in succ(n) \, \} & \text{otherwise} \end{cases}$$

Denoting the greatest solution of Equation System 1 with respect to the start assertion $c_{\mathbf{s}} \in \mathcal{C}$ by $\mathbf{ass}_{c_{\mathbf{s}}}$, and the least solution of Equation System 2 with respect to the data-flow query c_q at $\mathbf{q}$ by $\mathbf{req\text{-}ass}_{c_q}$, the solutions of the *MaxFP*-approach and the *R-MinFP*-approach are defined as follows:

The *MaxFP*-Solution: $\forall c_{\mathbf{s}} \in \mathcal{C} \ \forall n \in N. \ MaxFP_{([\![\ ]\!], c_{\mathbf{s}})}(n) =_{df} \mathbf{ass}_{c_{\mathbf{s}}}(n)$

The *R-MinFP*-Solution: $\forall c_q \in \mathcal{C} \ \forall n \in N. \ R\text{-}MinFP_{([\![\ ]\!], c_q)}(n) =_{df} \mathbf{req\text{-}ass}_{c_q}(n)$

Both fixed-point solutions can effectively be computed, when the semantic functions are monotonic and the lattice is free of infinite chains. Moreover, they coincide with their pathwise defined counterparts, when the semantic functions are distributive, which implies additivity of their reversed counterparts. We have:

Theorem 2 (Coincidence Theorem).
The MaxFP-solution and the MOP-solution coincide, i.e., $\forall c_{\mathbf{s}} \in \mathcal{C} \ \forall n \in N$. $MaxFP_{([\![\ ]\!], c_{\mathbf{s}})}(n) = MOP_{([\![\ ]\!], c_{\mathbf{s}})}(n)$, if the semantic functions $[\![e]\!]$, $e \in E$, are all distributive.

Theorem 3 (Reversed Coincidence Theorem).
The R-MinFP-solution and the R-JOP-solution coincide, i.e., $\forall c_q \in \mathcal{C} \ \forall n \in N$. $R\text{-}MinFP_{([\![\ ]\!], c_q)}(n) = R\text{-}JOP_{([\![\ ]\!], c_q)}(n)$, if the semantic functions $[\![e]\!]$, $e \in E$, are all distributive.

Hence, for distributive semantic functions the *R-MinFP*-solution and the *R-JOP*-solution coincide. Together with the Link Theorem 1, this means that the *R-MinFP*-solution is equally informative as the *MOP*-solution. Moreover, in the absence of infinite chains it can effectively be computed by an iterative process. This is central for demand-driven DFA, however, its idea goes a step further: It stops the computation *as early as possible*, i.e., as soon as the validity or invalidity of the data-flow query under consideration is detected. Reduced to a slogan, demand-driven (DD) DFA can thus be considered the "sum of computing weakest preconditions and early termination." The following algorithm realizes this concept of DD-DFA for sequential programs. After its termination the variable *answerToQuery* stores the answer to the data-flow query under consideration, i.e., whether c_q holds at node q. In Section 4 we will show how to extend this approach to explicitly parallel programs.

(Prologue: Initialization of the annotation array *reqAss* and the variable *workset*)
FORALL $n \in N \backslash \{q\}$ DO $reqAss[n] := \bot$ OD;
$reqAss[q] := c_q$; $workset := \{q\}$;

(Main process: Iterative fixed-point computation)
WHILE $workset \neq \emptyset$ DO
    LET $m \in workset$
      BEGIN
          $workset := workset \backslash \{m\}$;
          (Update the predecessor-environment of node m)
          FORALL $n \in pred(m)$ DO
             $join := [\![(n,m)]\!]_R(reqAss[m]) \sqcup reqAss[n]$;
             IF $join = \top \ \vee \ (n = \mathbf{s} \wedge join \not\sqsubseteq c_{\mathbf{s}})$
                THEN (Query failed!)
                    $reqAss[\mathbf{s}] := join$;   (Indicate Failure)
                    goto 99   (Stop Computation)
              ELSIF $reqAss[n] \sqsubset join$
                THEN $reqAss[n] := join$;
                    $workset := workset \cup \{n\}$ FI OD END OD;

(Epilogue)
99 : $answerToQuery := (reqAss[\mathbf{s}] \sqsubseteq c_{\mathbf{s}})$.

3   The Parallel Setting

In this section we sketch our parallel setup, which has been presented in detail in [15]. We consider parallel imperative programs with shared memory and interleaving semantics. Parallelism is syntactically expressed by means of a **par** statement. As usual, we assume that there are neither jumps leading into a component of a parallel statement from outside nor vice versa. As shown in Figure 1, we represent a parallel program by an edge-labelled *parallel flow-graph G* with node set N and edge set E having a unique start node **s** and end node **e** without incoming and outgoing edges, respectively.

Additionally, $IntPred(n) \subseteq E$, $n \in N$, denotes the set of so-called *interleaving predecessors* of a node n, i.e., the set of edges whose execution can be interleaved with the execution of the statements of n's incoming edge(s) (cf. [15]). A *parallel program path* is an edge sequence corresponding to an interleaving sequence of G. By $\mathbf{PP}[m,n]$, $m, n \in N$, we denote the set of all finite parallel program paths from m to n.

4   Demand-Driven DFA in the Parallel Setting

As in [15] we concentrate on *unidirectional bitvector (UBV)* problems. The most prominent representatives of this class are the availability and very busyness of terms, reaching definitions, and live variables (cf. [6]).

UBV-problems are characterized by the simplicity of the semantic functional $[\![\ ]\!] : E \rightarrow (\mathcal{B}_X \rightarrow \mathcal{B}_X)$ defining the abstract program semantics. It specifies the effect of an edge e on a particular component of the bitvector, where $\mathcal{B}_X$ is the lattice $(\{\mathit{ff}, \mathit{tt}, \mathit{failure}\}, \sqcap, \sqsubseteq)$ of Boolean truth values with $\mathit{ff} \sqsubseteq \mathit{tt} \sqsubseteq \mathit{failure}$ and the logical "and" as meet operation $\sqcap$ (or its dual counterpart). The element *failure* plays here the role of the top element representing unsatisfiable data-flow information. Hence, for each edge e, the function $[\![\, e\,]\!]$ is an element of the set of functions $\{Cst_{tt}^X, Cst_{\mathit{ff}}^X, Id_{\mathcal{B}_X}\}$, where $Id_{\mathcal{B}_X}$ denotes the identity on $\mathcal{B}_X$, and Cst_{tt}^X and Cst_{ff}^X the extensions of the constant functions Cst_{tt} and Cst_{ff} from $\mathcal{B}$ to $\mathcal{B}_X$ leaving the argument *failure* invariant.

The reversed counterparts $R\text{-}Cst_{tt}^X$, $R\text{-}Cst_{\mathit{ff}}^X$, and $R\text{-}Id_{\mathcal{B}_X}$ of these functions, and hence the reversed semantic functions $[\![\, e\,]\!]_R$, $e \in E$, are defined as follows:

$$\forall b \in \mathcal{B}_X.\ R\text{-}Cst_{tt}^X(b) =_{df} \begin{cases} \mathit{ff} & \text{if } b \in \mathcal{B} \\ \mathit{failure} & \text{otherwise (i.e., if } b = \mathit{failure}) \end{cases}$$

$$\forall b \in \mathcal{B}_X.\ R\text{-}Cst_{\mathit{ff}}^X(b) =_{df} \begin{cases} \mathit{ff} & \text{if } b = \mathit{ff} \\ \mathit{failure} & \text{otherwise} \end{cases}$$

$$R\text{-}Id_{\mathcal{B}_X} =_{df} Id_{\mathcal{B}_X}$$

Note that all the functions Cst_{tt}^X, Cst_{ff}^X, and $Id_{\mathcal{B}_X}$ are distributive. Hence, the reversed functions are additive (cf. Proposition 1(2)).

The *MOP*-approach and the *R-JOP*-approach of the sequential setting can now straightforwardly be transferred to the parallel setting. Instead of sequential program paths we are now considering parallel program paths representing interleaving sequences of the parallel program. We define:

The $PMOP_{UBV}$-Solution: Strongest Postcondition

$$\forall b_0 \in \mathcal{B}_X\ \forall n \in N.\ PMOP_{([\![\ ]\!],b_0)}(n)(b_0) = \bigsqcap \{\, [\![\, p\,]\!](b_0) \mid p \in \mathbf{PP}[\mathbf{s}, n]\,\}$$

The $R\text{-}PJOP_{UBV}$-Solution: Weakest Precondition

$$\forall b_0 \in \mathcal{B}_X\ \forall n \in N.\ R\text{-}JOP_{([\![\ ]\!],b_0)}(n) =_{df} \bigsqcup \{\, [\![\, p\,]\!]_R(b_0) \mid p \in \mathbf{PP}[n, \mathbf{q}]\,\}$$

Note that actually tt is the only interesting, i.e., non-trivial data-flow query for UBV-problems. In fact, ff is always a valid assertion, while *failure* is never.

Like their sequential counterparts, the $PMOP_{UBV}$-solution and the $R\text{-}PJOP_{UBV}$-solution are specificational in nature. They do not induce an effective computation procedure. In [15], however, it has been shown, how to compute for UBV-problems the parallel version of the *MOP*-solution by means of a fixed-point process in a fashion similar to the sequential setting. Basically, this is a two-step approach. First, the effect of parallel statements is computed by means of an hierarchical process considering innermost parallel statements first and propagating information computed about them to their enclosing parallel statements. Subsequently, these results are used for computing the parallel version of the *MFP*-solution. This can be done almost as in the sequential case since in this phase parallel statements can be considered like elementary ones. The complete process resembles the one of *interprocedural* DFA of sequential programs, where first the effect functions of procedures are computed, which subsequently are used for computing the interprocedural version of the *MFP*-solution (cf. [10]).

Important for the success of the complete approach is the restriction to UBV-problems. For this problem class the effects of *interference* and *synchronization* between parallel components can be computed without consideration of any interleaving (cf. [15]). The point is that for UBV-problems the effect of each interleaving sequence is determined by a single statement (cf. [15]). In fact, this is also the key for the successful transfer of the DD-approach from the sequential to the parallel setting. In addition, the two steps of the conventional approach are fused together in the DD-approach. Since in sequential program parts the algorithm works as its sequential counterpart of Section 2, we here consider only the treatment of parallel program parts in more detail focusing on how to capture *interference* and *synchronization*. The complete DD-algorithm is given in [13].

Interference. In a parallel program part, the DD-algorithm checks first the existence of a "bad guy." This is an interleaving predecessor destroying the property under consideration, i.e., an edge e with $[\![e]\!] = Cst_{f\!f}^X$. If a bad guy exists, the data-flow query can definitely be answered with $f\!f$, and the analysis stops. This check is realized by means of the function *Interference*. We remark that every edge is investigated at most once for being a bad guy.

FUNCTION *Interference* ($m \in N$) : $\mathcal{B}$;
BEGIN
   IF *IntPred*(m) $\cap$ *badGuys* $\neq \emptyset$
      THEN (Failure of the data-flow query because of a bad guy!)
        return(*tt*)
      ELSE
        WHILE *edgesToBeChecked* $\cap$ *IntPred*(m) $\neq \emptyset$ DO
          LET $e \in$ *edgesToBeChecked* $\cap$ *IntPred*(m)
            BEGIN
               edgesToBeChecked := *edgesToBeChecked* $\setminus \{e\}$;
               IF $[\![e]\!] = Cst_{f\!f}$
                 THEN
                      badGuys := *badGuys* $\cup \{e\}$; *return*(*tt*) FI END OD;
        return(*f\!f*) FI
END;

Synchronization. In the DD-approach synchronization of the parallel components of a parallel statement takes place when a data-flow query reaches a start node of a parallel component. This actually implies the absence of bad guys, which has been checked earlier. Synchronization reduces thus to checking whether some of the parallel siblings guarantee the data-flow query (i.e., there is a sibling whose start node carries the annotation ff), otherwise, the data-flow query has to be propagated across the parallel statement. This is realized by the procedure *Synchronization*, where pfg, $\mathcal{G}_C$, and $start$ denote functions, which yield the smallest parallel statement containing a node m, the set of all component graphs of a parallel statement, and the start node of a graph, respectively.

PROCEDURE *Synchronization* ($m \in N$);
BEGIN
    IF $\forall\, n \in start(\mathcal{G}_C(pfg(m)))$. $reqAss[n] = tt$
        THEN IF $reqAss[start(pfg(m))] \sqsubset tt$
            THEN $reqAss[start(pfg(m))] := tt$;
                 $workset := workset \cup \{\, start(pfg(m)) \,\}$ FI FI
END;

The following point is worth to be noted. Unless the data-flow query under consideration fails, in the sequential setting the workset must always completely be processed. In the parallel setting, however, even in the opposite case parts of the workset can sometimes be skipped because of synchronization. If a component of a parallel statement is detected to guarantee the data-flow query under consideration (which implies the absence of destructive interference), nodes of its immediate parallel siblings need not to be investigated as synchronization guarantees that the effect of the component carries over to the effect of the enclosing parallel statement. This is discussed in more detail in [13].

5 Typical Applications

The application spectrum of our approach to DD-DFA of explicitly parallel programs is the same as the one of any DD-approach to program analysis. As pointed out in [5] this includes: (1) *Interactive tools* in order to handle user queries, as e.g., online debugging tools. (2) *Selective optimizers* focusing e.g. on specific program parts like loops having selective data-flow requirements. (3) *Incremental tools* selectively updating data-flow information after incremental program modifications. Thus, our approach is quite appropriate for a broad variety of tools supporting the design, analysis, and optimization of explicitly parallel programs.

6 Conclusions

In this article we complemented the conventional, strongest postcondition-centered approach to DFA of explicitly parallel programs of [15] with a demand-driven, weakest precondition-centered approach. Most importantly, this approach is tailored for parallelization, and allows us to mimic conventional DFA at the costs of answering a data-flow query, which are usually much smaller. This can directly be used to improve the performance of bitvector-based optimizations of explicitly parallel programs like *code motion* [14] and *partial dead-code elimination* [11]. However, our approach is not limited to this setting. In a similar fashion it can be extended e.g. to the *constant-propagation* optimization proposed in [12].

References

[1] K. R. Apt and E.-R. Olderog. *Verification of Sequential and Concurrent Programs.* 2nd edition, Springer-V., 1997.

[2] P. Cousot and R. Cousot. Abstract interpretation: A unified lattice model for static analysis of programs by construction or approximation of fixpoints. In *Conf. Rec. 4th Symp. Principles of Prog. Lang. (POPL'77)*, pages 238 – 252. ACM, NY, 1977.

[3] E. Duesterwald, R. Gupta, and M. L. Soffa. Demand-driven computation of interprocedural data flow. In *Conf. Rec. 22nd Symp. Principles of Prog. Lang. (POPL'95)*, pages 37 – 48. ACM, NY, 1995.

[4] E. Duesterwald, R. Gupta, and M. L. Soffa. A demand-driven analyzer for data flow testing at the integration level. In *Proc. IEEE Int. Conf. on Software Engineering (CoSE'96)*, pages 575 – 586, 1996.

[5] E. Duesterwald, R. Gupta, and M. L. Soffa. A practical framework for demand-driven interprocedural data flow analysis. *ACM Trans. Prog. Lang. Syst.*, 19(6):992 – 1030, 1997.

[6] M. S. Hecht. *Flow Analysis of Computer Programs.* Elsevier, North-Holland, 1977.

[7] S. Horwitz, T. Reps, and M. Sagiv. Demand interprocedural dataflow analysis. In *Proc. 3rd ACM SIGSOFT Symp. Foundations of Software Eng. (FSE'95)*, pages 104 – 115, 1995.

[8] J. Hughes and J. Launchbury. Reversing abstract interpretations. *Science of Computer Programming*, 22:307 – 326, 1994.

[9] J. B. Kam and J. D. Ullman. Monotone data flow analysis frameworks. *Acta Informatica*, 7:305 – 317, 1977.

[10] J. Knoop. *Optimal Interprocedural Program Optimization: A new Framework and its Application.* PhD thesis, Univ. of Kiel, Germany, 1993. LNCS Tutorial 1428, Springer-V., 1998.

[11] J. Knoop. Eliminating partially dead code in explicitly parallel programs. *TCS*, 196(1-2):365 – 393, 1998. (Special issue devoted to *Euro-Par'96*).

[12] J. Knoop. Parallel constant propagation. In *Proc. 4th Europ. Conf. on Parallel Processing (Euro-Par'98)*, LNCS 1470, pages 445 – 455. Springer-V., 1998.

[13] J. Knoop. Parallel data-flow analysis of explicitly parallel programs. Technical Report 707/1999, Fachbereich Informatik, Universität Dortmund, Germany, 1999.

[14] J. Knoop and B. Steffen. Code motion for explicitly parallel programs. In Proc. 7th ACM SIGPLAN Symp. on Principles and Practice of Parallel Programming (PPoPP'99), Atlanta, Georgia, pages 13 - 24, 1999.

[15] J. Knoop, B. Steffen, and J. Vollmer. Parallelism for free: Efficient and optimal bitvector analyses for parallel programs. *ACM Trans. Prog. Lang. Syst.*, 18(3):268 – 299, 1996.

[16] T. Reps. Solving demand versions of interprocedural analysis problems. In *Proc. 5th Int. Conf. on Compiler Construction (CC'94)*, LNCS 786, pages 389 – 403. Springer-V., 1994.

[17] X. Yuan, R. Gupta, and R. Melham. Demand-driven data flow analysis for communication optimization. *Parallel Processing Letters*, 7(4):359 – 370, 1997.

Localization of Data Transfer in Processor Arrays

Dirk Fimmel and Renate Merker

Department of Electrical Engineering
Dresden University of Technology

Abstract. In this paper we present an approach to localize the data transfer in processor arrays. Our aim is to select channels between processors of the processor array performing the data transfers. Channels can be varying with respect to the bandwidth and to the communication delay and can be bidirectional. Our objective is to minimize the implementation cost of the channels while satisfying the data dependencies.
The presented approach also applies to the problem of localizing data dependencies for a given interconnection topology. The formulation of our method as an integer linear program allows its use for automatic parallelization.

1 Introduction

Processor arrays (PA) are well suited to implement time-consuming algorithms of signal processing with real-time requirements. Technological progress allows the implementation of even complex processor arrays in silicon as well as in FPGAs. To explore the degrees of freedom in the design of processor arrays automatic tools are required.

Processor arrays are characterized by a significant number of processors which communicate via interconnections in a small neighborhood. Data transfer caused by the original algorithm has to be organized using local interconnections between processors. This paper covers the design of a cost-minimal interconnection network and the organization of the data transfers using this interconnections. A solution of the problem of organizing the data transfers for a given interconnection network is also presented.

The design of processor arrays is well studied (e.g. [2, 7, 8, 10, 13]) and became more realistic by inclusion of resource constraints [3, 5, 12]. But up to now, only some work has been done in the organization of data transfer. Fortes and Moldovan [6] as well as Lee and Kedem [9] discuss the need of a decomposition of global interconnections into a set of local interconnections without consideration of access conflicts to channels. Chou and Kung [1] present an approach to organize the communications in a partitioned processor array, but do not give a solution for the decomposition problem.

In this paper, we present an approach to localize the data transfer in processor arrays. Channels with different bandwidth and latency can be selected to implement the interconnections between processors. The data transfers which are

P. Amestoy et al. (Eds.): Euro-Par'99, LNCS 1685, pp. 401–408, 1999.
© Springer-Verlag Berlin Heidelberg 1999

given as displacement vectors are decomposed into a set of channels. The decomposition of displacement vectors as well as the order of using the channels is determined by an optimization problem. The objective of the optimization problem is to minimize the cost associated with an implementation of the channels in silicon.

The paper is organized as follows. Basics of the design of processor arrays are given in section 2. In section 3 a model of channels between processors is introduced. The communication problem which describes the organization of the data transfers is discussed in section 4. In section 5 a linear programming approach for the solution of the communication problem is presented. An example is given in section 6 to explain our approach followed by concluding remarks in section 7.

2   Design of Parallel Processor Arrays

In our underlying design system of processor arrays we consider the class of *regular iterative algorithms* (RIA) [10]. A RIA is a set of equations S_i of the following form:

$$S_i: \quad y_i[\mathbf{i}] = F_i(\cdots, y_j[f_{ij}^k(\mathbf{i})], \cdots), \quad \mathbf{i} \in \mathcal{I}, \quad 1 \le i,j \le m, 1 \le k \le m_{ij}, \quad (1)$$

where $\mathbf{i} \in \mathbf{Z}^n$ is an index vector, $f_{ij}^k(\mathbf{i}) = \mathbf{i} - \mathbf{d}_{ij}^k$ are index functions, the constant vectors $\mathbf{d}_{ij}^k \in \mathbf{Z}^n$ are called dependence vectors, y_i are indexed variables and F_i are arbitrary operations. The equations are defined in an index space $\mathcal{I}$ being a polytope $\mathcal{I} = \{\mathbf{i} \mid \mathbf{Hi} \ge \mathbf{h}_0\}$, $\mathbf{H} \in \mathbb{Q}^{n_h \times n}$, $\mathbf{h}_0 \in \mathbb{Q}^{n_h}$. We suppose that RIAs have a *single assignment form* (each instance of a variable y_i is defined only once in the algorithm) and that there exists a partial order of the instances of the equations that satisfies the data dependencies.

Next we introduce a graph representation to describe the data dependencies of the RIA. The equations of the RIA build the m nodes $S_i \in \mathcal{S}$ of the reduced dependence graph $\langle \mathcal{S}, \mathcal{E} \rangle$. The directed edges $(S_i, S_j) \in \mathcal{E}$ are the data dependencies weighted by the dependence vectors $\mathbf{d}_{ij}^k$. The weight of an edge $e \in \mathcal{E}$ is called $\mathbf{d}(e)$, the source of this edge $\sigma(e)$ and the sink $\delta(e)$.

The main task of the design of processor arrays is to determine the time and the processor when and where each instance of the equations of the RIA has to be evaluated. In order to keep the regularity of the algorithm in the resulting processor array only uniform affine mappings [10] are applied to the RIA.

A uniform affine *allocation function* assigns an evaluation processor to each instance of the equations and has the following form:

$$\pi_i : \mathbf{Z}^n \to \mathbf{Z}^{n-1} : \quad \pi_i(\mathbf{i}) = \mathbf{Si} + \mathbf{p}_i, \quad 1 \le i \le m, \quad (2)$$

where $\mathbf{p}_i \in \mathbf{Z}^{n-1}$ and $\mathbf{S} \in \mathbf{Z}^{n-1 \times n}$ is of full row rank and assumed to be e-unimodular [11] leading to a dense processor array (see [4] for further details). Since $\mathbf{S}$ is of full row rank, the vector $\mathbf{u} \in \mathbf{Z}^n$ which is coprime and satisfies $\mathbf{Su} = \mathbf{0}$ and $\mathbf{u} \neq \mathbf{0}$ is uniquely defined (except to the sign) and called *projection vector*. The importance of the projection vector is due to the fact that those and only those index points of an index space lying on a line spanned by the projection vector $\mathbf{u}$ are mapped onto the same processor.

Using the allocation function $\pi_i(\mathbf{i})$ dependence vectors $\mathbf{d}(e)$ are mapped onto *displacement vectors* $\mathbf{v}(e)$ by: $\mathbf{v}(e) = \mathbf{Sd}(e) + \mathbf{p}_{\delta(e)} - \mathbf{p}_{\sigma(e)}$.

A uniform affine *scheduling function* assigns an evaluation time to each instance of the equations and has the following form:

$$\tau_i : \mathbf{Z}^n \to \mathbf{Z} : \quad \tau_i(\mathbf{i}) = \boldsymbol{\tau}^T \mathbf{i} + t_i, \quad 1 \leq i \leq m, \tag{3}$$

where $\boldsymbol{\tau} \in \mathbf{Z}^n, t_i \in \mathbf{Z}$.

The following *causality constraint* ensures the satisfaction of data dependencies:

$$\boldsymbol{\tau}^T \mathbf{d}(e) + t_{\delta(e)} - t_{\sigma(e)} \geq d_{\sigma(e)}, \quad \forall e \in \mathcal{E}, \tag{4}$$

where d_i is the time needed to compute operation F_i.

Due to the regularity of the index space and the uniform affine scheduling function, the processor executes the operations associated with that index points consecutively if $\boldsymbol{\tau}^T \mathbf{u} \neq 0$ with a constant time distance $\lambda = |\boldsymbol{\tau}^T \mathbf{u}|$ which is called *iteration interval*.

3   Channels and Decomposition of Displacement Vectors

A set $\mathcal{W} = \{\mathbf{w}_1, \cdots, \mathbf{w}_{|\mathcal{W}|}\}$ of channels $\mathbf{w}_i \in \mathbf{Z}^{n-1}$ between processors is supposed to be given. The subset $\mathcal{W}^* \subseteq \mathcal{W}$ of channels can be used bidirectional, the subset $\mathcal{W} \backslash \mathcal{W}^*$ of onedirectional channels is denoted $\mathcal{W}^+$. Each channel $\mathbf{w}_i \in \mathcal{W}$ is weighted by implementation cost c_i and by a communication time l_i. The communication time l_i can depend on data dependencies, i.e. $l_i = l_i(e)$, which allows to consider both channels and data with different bandwidth. The value $l_i(e)$ is assumed to already contain a multiple use of a channel $\mathbf{w}_i$ if the bandwidth of value $y_{\sigma(e)}$ exceeds the bandwidth of that channel. In our model, the communication on a channel is restricted to one value per time even if the bandwidth of a channel would allow some values of smaller bandwidth per time.

In order to organize the data transfer the displacement vectors $\mathbf{v}(e)$ have to be decomposed into a linear combination of channels $\mathbf{w}_i \in \mathcal{W}$:

$$\mathbf{v}(e) = \sum_{i=1}^{|\mathcal{W}|} b_i(e) \mathbf{w}_i, \quad \forall e \in \mathcal{E}, \tag{5}$$

where $b_i(e) \in \mathbf{Z}$ and $\forall \mathbf{w}_i \in \mathcal{W}^+ . b_i(e) \geq 0$.

Since the summation in (5) is commutative we allow the use of the channels in an arbitrary order.

4   Communication Problem

The communication problem consists in minimizing the implementation cost for the channels subject to a conflict free organization of the data transfer. Due to the regularity of the processor array it is sufficient to consider one iteration interval of one processor to describe the communication problem.

Following the notation of the previous sections the implementation cost for the channels are measured by $C = \sum_{i=1}^{|W|} n_i c_i$, where n_i is the number of instances of channel $\mathbf{w}_i \in W$ leading off from one processor, and c_i is the implementation cost of channel $\mathbf{w}_i$.

The data transfer must be causal which includes the following points for a data dependence $e \in \mathcal{E}$:

- Data transfer can start at time $t^{start}(e) = t_{\sigma(e)} + d_{\sigma(e)}$,
- Data transfer must be finished at time $t^{end}(e) = \tau^T \mathbf{d}(e) + t_{\delta(e)}$,
- When a data transfer e uses several channels successively, the use of that channels must not be overlapping. Suppose that $t_i^j(e)$ and $t_k^l(e)$ are the starting time of the j-th and l-th communication on the channels $\mathbf{w}_i$ and $\mathbf{w}_k$ respectively. Then either $t_i^j(e) \geq t_k^l(e) + l_k(e)$ or $t_k^l(e) \geq t_i^j(e) + l_i(e)$ has to be satisfied depending on which communication is starting first.

The causality of an example data transfer $e \in \mathcal{E}$ is depicted in figure 1. Either $t^{start}(e) \leq t_1^1(e) \leq t_1^1(e) + l_1(e) \leq t_2^1(e) \leq t_2^1(e) + l_2(e) \leq t^{end}(e)$ or $t^{start}(e) \leq t_2^1(e) \leq t_2^1(e) + l_2(e) \leq t_1^1(e) \leq t_1^1(e) + l_1(e) \leq t^{end}(e)$ has to be satisfied.

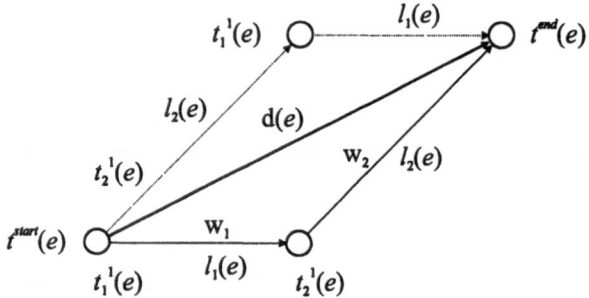

Fig. 1. *Causality of data transfer*

As already mentioned above it is sufficient to consider one iteration interval to ensure a conflict free use of the channels, since the data transfer on each channel is repeated periodically with the period λ. W.l.o.g. we consider the interval $[0, \lambda)$. To this end we decompose the starting time of the j-th communication on channel $\mathbf{w}_i$ used for the data transfer $e \in \mathcal{E}$ into $t_i^j(e) = p_i^j(e)\lambda + o_i^j(e)$, where $0 \leq o_i^j(e) < \lambda$ and $p_i^j(e), o_i^j(e) \in \mathbb{Z}$. A conflict free use of channel $\mathbf{w}_i$ by data transfers $e_1, e_2 \in \mathcal{E}$ is satisfied if:

$$\left. \begin{array}{l} o_i^j(e_1) - o_i^k(e_2) \geq l_i(e_2), \\ \lambda - o_i^j(e_1) + o_i^k(e_2) \geq l_i(e_1), \end{array} \right\} \text{ if } o_i^j(e_1) > o_i^k(e_2),$$
$$\left. \begin{array}{l} o_i^k(e_2) - o_i^j(e_1) \geq l_i(e_1), \\ \lambda - o_i^k(e_2) + o_i^j(e_1) \geq l_i(e_2), \end{array} \right\} \text{ if } o_i^j(e_1) \leq o_i^k(e_2), \tag{6}$$

for all $e_1, e_2 \in \mathcal{E}$, $1 \leq j \leq |b_i(e_1)|$, $1 \leq k \leq |b_i(e_2)|$.

The case $o_i^j(e_1) \leq o_i^k(e_2)$ is depicted in Fig. 2. We have $p_i^j(e_1) = 0$ and $p_i^k(e_2) = 1$ since $t_i^j(e_1) = o_i^j(e_1)$ and $t_i^k(e_2) = \lambda + o_i^k(e_2)$.

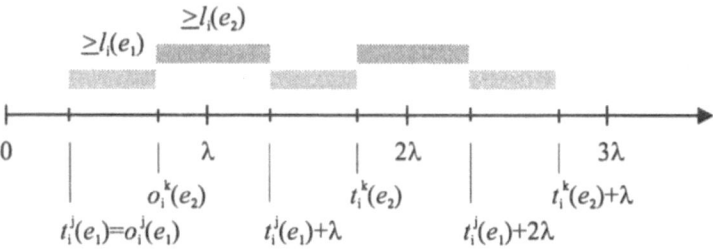

Fig. 2. *Conflict free use of channels*

The grey bars show the delay time that must exceed the communication time l_i of channel $\mathbf{w}_i$. Two remarks have to be added to the conflict free use of channels: 1) The implementation of several instances of channel $\mathbf{w}_i \in \mathcal{W}$ requires the solution of (6) only if e_1 and e_2 use the same instance of $\mathbf{w}_i$. 2) Particularly in the case of bit-serial channels $\mathbf{w}_i$ communication time $l_i(e)$ may exceed the iteration interval λ, i.e. $l_i(e) \geq \lambda$. Then $\lfloor l_i(e)/\lambda \rfloor$ instances of $\mathbf{w}_i$ are permanently used by communication $e \in \mathcal{E}$ and $l_i(e)$ has to be replaced by $(l_i(e) \bmod \lambda)$ in (6). Both remarks are satisfied in the integer linear program formulation presented in the next section.

Next, we give a constraint for the solvability of the communication problem. The minimal communication time $\Delta t(\mathbf{v}(e))$ needed for the data transfer $e \in \mathcal{E}$ under assumption of unlimited instances of channels $\mathbf{w}_i \in \mathcal{W}$ is:

$$\Delta t(\mathbf{v}(e)) = \min_{\mathbf{v} = \sum\limits_{i=1}^{|\mathcal{W}|} b_i(e)\mathbf{w}_i} \sum_{i=1}^{|\mathcal{W}|} |b_i(e)| l_i(e), \quad b_i(e) \in \mathbb{Z}, \quad \forall \mathbf{w}_i \in \mathcal{W}^+ . b_i(e) \geq 0. \quad (7)$$

Note that a fast computation of the value of $\Delta t(\mathbf{v}(e))$ is possible if matrix $\mathbf{W} = (\mathbf{w}_1, \cdots \mathbf{w}_{|\mathcal{W}|})$ is e-unimodular which leads to an integer polyhedron $\{b_i(e) \mid \mathbf{v} = \sum_{i=1}^{|\mathcal{W}|} b_i(e)\mathbf{w}_i\}$ [11].

In order to get a solution of the communication problem the causality constraint (4) has to be replaced by:

$$\tau^T \mathbf{d}(e) + t_{\delta(e)} - t_{\sigma(e)} \geq d_{\sigma(e)} + \Delta t(\mathbf{v}(e)), \quad \forall e \in \mathcal{E}. \quad (8)$$

A value needed for the program formulation in the next section is the maximal number of channels $\mathbf{w}_i = (w_{i1}, \cdots, w_{in-1})^T \in \mathcal{W}$ that can be used for data transfer $e \in \mathcal{E}$. We call this value $b_i^{max}(e)$. If we restrict the data transfer to have no steps backwards in each component of $v_j(e)$, $1 \leq j \leq n-1$, of the displacement vector $\mathbf{v}(e)$ then $b_i^{max}(e)$ can be computed by:

$$b_i^{max}(e) = \max\{ \max_{sign(w_{ij})v_j(e) \geq b_i w_{ij}, 1 \leq j \leq n-1} b_i, 0\}, \quad 1 \leq i \leq |\mathcal{W}|, \forall e \in \mathcal{E}. \quad (9)$$

For $\mathbf{w}_i \in \mathcal{W}^*$ it is easy to prove that $b_i^{max}(e) = 0$ either for $\mathbf{w}_i$ or for $-\mathbf{w}_i$.

minimize: $\sum\limits_{i=1}^{|\mathcal{W}|} n_i c_i,$ $\qquad\qquad n_i \in \mathbf{Z},$ #1

subject to:

$$v(e) = \sum_{i=1}^{|\mathcal{W}|} \sum_{j=1}^{b_i^{max}(e)} \beta_i^j(e)\mathbf{w}_i, \qquad \beta_i^j(e) \in \{0,1\}, \forall e \in \mathcal{E}, \qquad \#2$$

$$\sum_{k=1}^{N_i} \alpha_{ik}^j(e) = 1, \qquad\qquad \alpha_{ik}^j(e) \in \{0,1\}, 1 \le i \le |\mathcal{W}|, 1 \le j \le b_i^{max}(e), \forall e \in \mathcal{E}, \qquad \#3$$

$$n_i \ge \sum_{k=1}^{N_i} k\alpha_{ik}^j(e) - (1 - \beta_i^j(e))N_i + \sum_{e \in \mathcal{E}} \sum_{j=1}^{b_i^{max}} \beta_i^j(e)\lfloor l_i(e)/\lambda \rfloor, \qquad \#4$$

$$1 \le j \le b_i^{max}(e), \quad \forall e \in \mathcal{E}_{(l_i(e) \bmod \lambda) \ne 0}, 1 \le i \le |\mathcal{W}|,$$

$$t^{start}(e) \le p_i^j(e)\lambda + o_i^j(e), \qquad \#5$$
$$t^{end}(e) \ge p_i^j(e)\lambda + o_i^j(e) + l_i(e),$$
$$p_i^j(e) \in \mathbf{Z}, 1 \le j \le b_i^{max}(e), 1 \le i \le |\mathcal{W}|, \forall e \in \mathcal{E},$$

$$p_j^j(e)\lambda + o_i^j(e) - p_k^l(e)\lambda - o_k^l(e) \ge l_k(e) - (\gamma_{ik}^{jl}(e) + (1 - \beta_i^j(e)) + (1 - \beta_k^l(e)))C_k(e), \qquad \#6$$
$$p_k^l(e)\lambda + o_k^l(e) - p_i^j(e)\lambda - o_i^j(e) \ge l_i(e) - ((1 - \gamma_{ik}^{jl}(e)) + (1 - \beta_i^j(e)) + (1 - \beta_k^l(e)))C_i(e),$$
$$\gamma_{ik}^{jl}(e) \in \{0,1\}, 1 \le j \le b_1^{max}(e), 1(+j \text{ if } i = k) \le l \le b_k^{max}(e), 1 \le i \le k \le |\mathcal{W}|, \forall e \in \mathcal{E},$$

$$o_i^j(e_1) - o_i^k(e_2) \ge l_i(e_2) - (\delta_i^{jk}(e_1, e_2) + (2 - \alpha_{il}^j(e_1) - \alpha_{il}^k(e_2)) + (2 - \beta_i^j(e_1) - \beta_i^k(e_2)))(2\lambda + l_i(e_2)), \qquad \#7$$
$$\lambda - o_i^j(e_1) + o_i^k(e_2) \ge l_i(e_1) - (\delta_i^{jk}(e_1, e_2) + (2 - \alpha_{il}^j(e_1) - \alpha_{il}^k(e_2)) + (2 - \beta_i^j(e_1) - \beta_i^k(e_2)))l_i(e_1),$$
$$o_i^k(e_2) - o_i^j(e_1) \ge l_i(e_1) - ((1 - \delta_i^{jk}(e_1, e_2)) + (2 - \alpha_{il}^j(e_1) - \alpha_{il}^k(e_2)) + (2 - \beta_i^j(e_1) - \beta_i^k(e_2)))(2\lambda + l_i(e_1)),$$
$$\lambda - o_i^k(e_2) + o_i^j(e_1) \ge l_i(e_2) - ((1 - \delta_i^{jk}(e_1, e_2)) + (2 - \alpha_{il}^j(e_1) - \alpha_{il}^k(e_2)) + (2 - \beta_i^j(e_1) - \beta_i^k(e_2)))l_i(e_2),$$
$$1 \le l \le N_i, 1 \le j \le b_i^{max}(e_1), 1(+j \text{ if } e_1 = e_2) \le k \le b_i^{max}(e_2), \forall e_1, e_2 \in \mathcal{E}, 1 \le i \le |\mathcal{W}|.$$

Table 1. Integer linear program of the communication problem

5   Solution of the Communication Problem

In this section we present an integer linear program formulation of the communication problem. The entire program is listed in table 1. The constraints in Table 1 are explained in the following:

#2 (Decomposition of displacement vectors): Variable $b_i(e)$ is substituted by $b_i(e) = \sum_{j=1}^{b_i^{max}(e)} \beta_i^j(e)$, where $\beta_i^j(e)$ determines whether channel $\mathbf{w}_i \in W$ is used for data transfer $e \in \mathcal{E}$ or not. Channel $\mathbf{w}_i$ has to be replaced by $-\mathbf{w}_i$ if $\mathbf{w}_i \in W^*$.

#3 (Assignment to instances of channels) : Instance $\sum_{k=1}^{N_i} k\alpha_{ik}^j(e)$ of channel $\mathbf{w}_i$ is used for communication $e \in \mathcal{E}$, N_i is the maximal number of instances of channel $\mathbf{w}_i$ leading away from one processors.

#4 (Number of instances of channels): The first part ensures that instance $\sum_{k=1}^{N_i} k\alpha_{ik}^j(e)$ of channel $\mathbf{w}_i$ has to be considered only if $\beta_i^j(e) = 1$. The second part includes additional instances of channel $\mathbf{w}_i$ if $l_i(e) \ge \lambda$ as discussed in the previous section.

#5, #6 (Causality of data transfer): The binary variables γ_{ik}^{jl} are used to decide whether the j-th communication on channel $\mathbf{w}_i$ starts before or after the l-th communication on channel $\mathbf{w}_k$. Constant $C_i(e)$ can be determined by $C_i(e) = t^{end}(e) - t^{start}(e) + l_i(e)$.

#7 (Conflict free use of channels): The binary variables $\delta_i^{jk}(e_1, e_2)$ decide the if part of constraint (6). As mentioned in the previous section $l_i(e)$ has to be replaced by $(l_i(e) \bmod \lambda)$ if $l_i(e) \ge \lambda$.

The integer program consists of $|\mathcal{W}| + 2\sum_{e\in\mathcal{E}}\sum_{i=1}^{|\mathcal{W}|} b_i^{max}(e)$ integer and $\sum_{e\in\mathcal{E}}\sum_{i=1}^{|\mathcal{W}|}(1 + N_i)b_i^{max}(e) + \sum_{e\in\mathcal{E}}(\sum_{i=1}^{|\mathcal{W}|}b_i^{max}(e)(\sum_{i=1}^{|\mathcal{W}|}b_i^{max}(e) - 1))/2 + \sum_{i=1}^{|\mathcal{W}|}(\sum_{e\in\mathcal{E}}b_i^{max}(e)(\sum_{e\in\mathcal{E}}b_i^{max}(e) - 1))/2$ binary variables. The number of constraints is $3\sum_{e\in\mathcal{E}}\sum_{i=1}^{|\mathcal{W}|}b_i^{max}(e) + \sum_{e\in\mathcal{E}}(\sum_{i=1}^{|\mathcal{W}|}b_i^{max}(e)(\sum_{i=1}^{|\mathcal{W}|}b_i^{max}(e) - 1)) + |\mathcal{E}|(n - 1) + 2\sum_{i=1}^{|\mathcal{W}|}\gamma_i(\sum_{e\in\mathcal{E}}b_i^{max}(e)(\sum_{e\in\mathcal{E}}b_i^{max}(e) - 1))$.

6 Experimental Results

Because of lack of space we present only one example. We assume that we have three data transfers and four available channels. The related data are summarized in Table 2. The iteration interval is given with $\lambda = 8$.

	e_1	e_2	e_3		$\mathbf{w}_1$	$\mathbf{w}_2$	$\mathbf{w}_3$	$\mathbf{w}_4$
$v(e)$	$(1,0)^T$	$(3,2)^T$	$(4,1)^T$		$(1,0)^T$	$(0,1)^T$	$(1,1)^T$	$(2,1)^T$
$t^{start}(e)$	2	4	7		$l_i(e)$	4	4	4	16
$t^{end}(e)$	18	26	28		c_i	8	8	8	2

Table 2. Data transfers and available channels

Application of the integer program leads to the result that minimal cost for the implementation of channels arise using two instances of channel $\mathbf{w}_1$ and one instance of channels $\mathbf{w}_3$ and $\mathbf{w}_4$. The starting time of each use of a channel is shown in Table 3 and depicted as a bar chart in Figure 3.

$t_1^1(e_1)$	$t_3^1(e_2)$	$t_4^1(e_2)$	$t_1^1(e_3)$	$t_1^2(e_3)$	$t_1^3(e_3)$	$t_3^1(e_3)$
11	4	10	7	11	15	24

Table 3. Starting time of communications

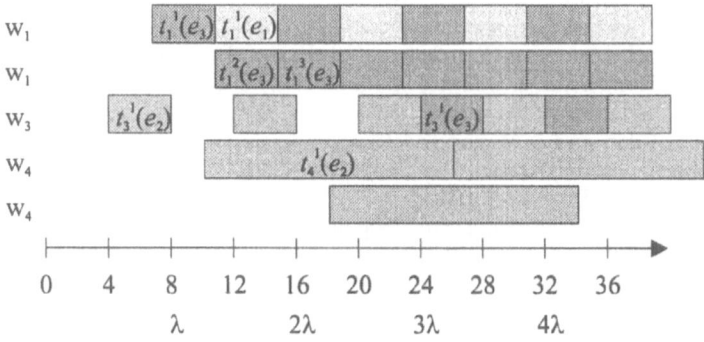

Fig. 3. Bar chart of communications

The linear program has 36 integer and 148 binary variables and consists of 588 constraints if we set $N_i = 3$, $1 \le i \le 4$. The solution takes about 28 seconds on a SUN-SPARC station.

7   Conclusion and Further Research

The presented approach is suitable to implement the data transfer in parallel processor arrays using local communications. The objective of the approach is the minimization of implementation cost for the channels. Only small changes to the integer program are required to match a given interconnection topology. To this end, (#1) in Table 1 can either be replaced by **minimize:** $\sum_{i=1}^{|\mathcal{W}|} n_i$ or left empty, whereas the constraint $n_i \leq N_i$, $1 \leq i \leq |\mathcal{W}|$, has to be added.

The iteration interval λ as well as $t^{start}(e)$ and $t^{end}(e)$ depend on the scheduling function $\tau(\mathbf{i})$. Hence, a more consequent approach consists in solving the communication problem and determining the scheduling function in one problem. This can be done by adopting techniques presented in [5]. In principle, this allows to determine a scheduling function, the functionality of processors and the channels between processors by solving one optimization problem. Unfortunately, the arising integer program tends to long solution time even for small problems.

The presented method also applies to the problem of limited communication support in partitioned processor arrays and extends the approach in [1].

References

[1] W.H. Chou and S.Y. Kung. Scheduling partitioned algorithms on processor arrays with limited communication support. In *Proc. IEEE Int. Conf. on Application Specific Systems, Architectures and Processors'93*, pages 53–64, Venice, 1993.

[2] A. Darte and Y. Robert. Constructive methods for scheduling uniform loop nests. *IEEE Trans. on Parallel and Distributed Systems*, 5(8):814–822, 1994.

[3] M. Dion, T. Risset, and Y. Robert. Resource constraint scheduling of partitioned algorithms on processor arrays. *Integration, the VLSI Journal*, 20:139–159, 1996.

[4] D. Fimmel and R. Merker. Determination of the processor functionality in the design of processor arrays. In *Proc. IEEE Int. Conf. on Application Specific Systems, Architectures and Processors'97*, pages 199–208, Zürich, 1997.

[5] D. Fimmel and R. Merker. Design of processor arrays for real-time applications. In *Proc. Int. Conf. Euro-Par '98*, pages 1018–1028, Southampton, 1998. Lecture Notes in Computer Science, Springer.

[6] J.A.B. Fortes and D.I. Moldovan. Parallelism detection and transformation techniques useful for vlsi algorithms. *Journal of Parallel and Distributed Computing*, 2:277–301, 1985.

[7] R.M. Karp, R.E. Miller, and S. Winograd. The organization of computations for uniform recurrence equations. *Journal of the ACM*, 14:563–590, 1967.

[8] S.Y. Kung. *VLSI Array Processors*. Prentice Hall, Englewood Cliffs, 1987.

[9] P.Z. Lee and Z.M. Kedem. Mapping nested loop algorithms into multidimensional systolic arrays. *IEEE Trans. on Parallel and Distributed Systems*, 1:64–76, 1990.

[10] S.K. Rao. *Regular Iterative Algorithms and their Implementations on Processor Arrays*. PhD thesis, Stanford University, 1985.

[11] A. Schrijver. *Theory of Integer and Linear Programming*. John Wiley & Sons, New York, 1986.

[12] L. Thiele. Resource constraint scheduling of uniform algorithms. *Int. Journal on VLSI and Signal Processing*, 10:295–310, 1995.

[13] Y. Wong and J.M. Delosme. Optimal systolic implementation of n-dimensional recurrences. In *Proc. ICCD*, pages 618–621, 1985.

Scheduling Structured Systems

Jason B. Crop   Doran K. Wilde

Department of Electrical and Computer Engineering
Brigham Young University
Provo, Utah

Abstract. The use of subsystems is fundamental to the modeling of hierarchical hardware using recurrence equations. Scheduling adds temporal information to a system and is thus a key step in the synthesis of parallel hardware from algorithms. It determines such things as the placement of pipeline registers, the latency and throughput of the circuit, and the order and rate that inputs are consumed and outputs produced. This paper will show how to extend usual dependence analysis to derive the additional dependences on the timing function needed when subsystems are used.

Introduction

The only practical way of representing large, complex designs is by using a structural hierarchy of modules. In this paper, we consider the problem of scheduling structured hierarchical systems of affine recurrence equations, a problem encountered during the synthesis of structured parallel hardware from structured recurrence equations.

The interconnection and delays in a system constrain the time when each operation in the system can be computed. When a system is represented hierarchically, constraints between the calling system, and the system being called must also be taken into account when scheduling both systems. These constraints take the form of dependences between variables. The scheduling of systems of recurrence equations is based on constraining the coefficients of affine timing functions and then deriving coefficient values by solving a linear programming problem[4,5]. Scheduling adds temporal information to a system and is thus an important step in the process of synthesizing parallel hardware from algorithms. It determines such things as the placement of pipeline registers, the latency and throughput of the circuit, and the order and rate that inputs are consumed and outputs produced.

Scheduling of hierarchical systems can be done by flattening the structure and then scheduling. However, this leads to huge, difficult to solve linear programming problems, and defeats the advantages of structured systems. After flattening, information about program structure is lost and it is difficult to reconstitute the hierarchy. Structured scheduling is simpler and preserves the hierarchical structure.

We have been using ALPHA[1,2] to model and synthesize parallel hardware. Structured hardware modules map easily onto ALPHA subsystems. Primitive hardware modules have been designed, and corresponding ALPHA models written, to

P. Amestoy et al. (Eds.): Euro-Par'99, LNCS 1685, pp. 409-412, 1999.
© Springer-Verlag Berlin Heidelberg 1999

implement a large set of basic arithmetic operators. When synthesizing hardware using a library of pre-designed hardware modules, the subsystems written to model library modules are necessarily pre-scheduled. We develop the scheduling constraints imposed on a calling system when it instantiates (calls) a subsystem.

As part of the synthesis process, arithmetic operators in recurrence equations are replaced by calls to low-level, pre-scheduled subsystems that model the hardware designed to perform those operations. This leads to the following assumptions:

1. After substituting arithmetic operations with calls to subsystems, all computation is done in the low level subsystems that model and map to hardware. The calling system only instantiates these subsystems and specifies "delay-free" interconnections between them.

2. Before a system is scheduled, all subsystems that it calls have been previously scheduled, that is, timing functions exist for both the outputs (saying when they are computed) and the inputs (saying when they are first used) of every subsystem. All instances of a subsystem use a single common subsystem schedule.

3. The subsystem schedules are available to, and are used by the scheduler when scheduling the calling system.

Defining and Instantiating ALPHA Subsystems

The ALPHA language gives a formal syntax and semantics for writing systems of recurrence equations that maintains the generality and power of the recurrence equation model [1,2]. A recent extension of the language supports the hierarchical and structural representation of algorithms using subsystem calls [3]. To promote generality and reusability, (sub)systems are parameterized by size parameters that are used to specify the size and shape of input and output data arrays. The interface to a subsystem is thus declared as a set of input and output array variables declared over parameterized polyhedral-shaped index domains.

An ALPHA (sub)system [5] is defined as follows:

system xyz : D_P $(I{:}D_I,\cdots)$ returns $(O{:}D_O,\cdots)$; var *locals*; let *body* tel;

where *xyz* is the subsystem name, D_P is the parameter domain which specifies the parameter indices and the restrictions placed on them, I and O are the input and output variable names, and D_I and D_O are their respective index domains, *locals* are local variables and *body* is a system of affine recurrence equations that describe how to compute the outputs as functions of the inputs.

The precomputed schedule for a (sub)system is represented in an ALPHA format as shown to the right. Each input and output has its own schedule (shown as *g* and *h* in the example on the right) that specifies when each input is first used and when each output is computed, relative to each other. This relative time is called *subsystem time*.

```
system xyz : D_P ( ) returns ( );
    var
        TIME : D_TIME ;
        I : D_I of invar;
        O : D_O of outvar;
    let
        I = TIME . g;
        O = TIME . h;
    tel;
```

After having been declared, a subsystem is used by arraying instances of the subsystem, one instance for each index point in a parameterized polyhedral-shaped instantiation domain. A subsystem can be called hierarchically from another system using the syntax:

$$\forall \; z \in \mathcal{D}_E : \text{use } xyz[Tz] \; (A[Px]) \text{ returns } (B[Qy]);$$

This statement instantiates an *array* of instances of xyz using $\mathcal{D}_E$, the *instantiation domain*. One instance of subsystem xyz is created for every integer point z in domain $\mathcal{D}_E$. The matrix T defines an affine map from z and calling system parameters to the subsystem parameters such that $\forall z \in \mathcal{D}_E : Tz \in \mathcal{D}_P$. The matrix P maps index points in the domain of formal input I to the domain of actual input A. Likewise, matrix Q maps index points in the domain of formal output O into the index domain of the actual output B. When instantiated, subsystem variables (formals) are array-extended by crossing (Def: $\mathcal{D} \times \mathcal{E} = \left\{ (x,y) \mid x \in \mathcal{D}, y \in \mathcal{E} \right\}$) the instantiation domain, $\mathcal{D}_E$, with the local domains of the variables declared in the subsystem definition.

Scheduling Subsystems

We extend existing methods for scheduling recurrence equations [4,5] by deriving the additional constraints needed to cover the dependences between calling system and instances of a subsystem.

We assume that the *timing function* of a variable $x[v]$ in a subsystem parameterized by $N \in \mathcal{D}_P$ is $t_x(v,N) = h_1^T v + h_2^T N + h_3$, an affine combination of index vector v and subsystem parameter vector N. If this subsystem is array-instantiated by the statement "$\forall \; z \in \mathcal{D}_E : \text{use } xyz[Tz]$" then variable $x[v]$ must be instance-array extended by index vector z, becoming $x[v,z]$ before it can be used in the calling system. The calling system *extended timing function* $t_x(v,z,M)$ functions maps instance-array extended variable $x[v,z]$ to time in the calling system as a function of calling system parameters M. As a byproduct of instantiation, the subsystem parameter vector N is set to $Tz = T_1 z + T_2 M + T_3$. When scheduling instances of subsystems, the system extended timing function must be *subsystem time invariant*, meaning that the extended schedules cannot change the intrinsic properties of time within the subsystem— a relative delay between two computation events must be the same in both the subsystem and the extended subsystem. A result of this property is that for any given instance (fixed value of z), the timing function for an extended variable is its subsystem timing function plus an instance-specific fixed time displacement. This time displacement function is $t(z,M) = r_1^T z + r_2^T M + r_3$, an affine combination of the instance array index z and the calling system parameter vector M., and evaluates to a constant time for given values of z and M. Subsystem time invariance requires that the coefficient vector $r = \begin{bmatrix} r_1^T & r_2^T & r_3 \end{bmatrix}^T$ be constant for all instances generated by a single use-statement. The derivation of the system extended timing function follows:

$$\forall z \in \mathcal{D}_E: \quad \overbrace{t_x(v,z,M)}^{\text{extended timing function in system}} = \overbrace{t_x(v,N)}^{\text{timing function in subsystem}} + \overbrace{t(z,M)}^{\text{instance specific time displacement}}$$

$$= h_1^T v + h_2^T(T_1\, z + T_2\, M + T_3) + h_3 + r_1^T z + r_2^T M + r_3$$

$$= h_1^T v + (h_2^T T_1 + r_1^T)z + (h_2^T T_2 + r_2^T)M + (h_2^T T_3 + h_3 + r_3)$$

Thus, extended timing functions for formal subsystem input and output variables are obtained by simply adding an affine function of the instance index z to the local subsystem timing function. This function computes the time displacement for each instance of a subsystem. Structured scheduling not only derives a timing function for each variable, but also a time displacement function is also for each use-statement.

The semantics of passing arguments in and out of subsystems give the following equations.

Passing in an input argument: $I[x]$ (arrayed formal) = $A[Px]$ (actual)

Passing out an output argument: $B[Qy]$ (actual) = $O[y]$ (arrayed formal)

A subsystem input must be computed *before* it is first used and an output of a subsystem is computed *at the same time* it is computed in the subsystem. This leads to the following dependences, one for each input argument, and one for each output argument (the symbol $\mapsto$ is read as "depends on"):

$\forall\, x \ni \mathcal{D}_{I \times E}: I[x]$ (arrayed formal input) $\mapsto A[Px]$ (actual input), with delay > 0.

$\forall\, y \ni \mathcal{D}_{O \times E}: B[Qy]$ (actual output) $\mapsto O[y]$ (arrayed formal output), with no delay

and these dependences in turn generate the inequalities that constrain the timing functions for the actual (system) arguments and the extended timing functions (derived above) for the formal (subsystem) arguments. Space does not allow further elaboration here, however additional details and examples can be found in [6].

References

[1] H. Le Verge, C. Mauras, P. Quinton, *The Alpha Language and its Use for the Design of Systolic Arrays*. Journal of VLSI Signal Processing, vol. 3, no. 3, pp. 173-182, September, 1991.

[2] D. Wilde, *The Alpha Language*. Internal publication 827, IRISA, Rennes, France, December 1993. Also published as INRIA research report 2157.

[3] Florent de Dinechin, Patrice Quinton, Tanguy Risset, *Structuration of the Alpha Language*. In Giloi, Jahnichen, Shriver, editors, Massively Parallel Programming Models, pp. 18-24, IEEE Computer Society Press, August 1995.

[4] C. Mauras, P. Quinton, S. Rajopadhye, Y. Saouter, *"Scheduling Affine Parameterized Recurrences by means of Variable Dependent Timing Functions"*, in S. Y. Kung, E. E. Swartzlander, Jr., J. A. B. Fortes, K. W. Przytula, editors, *"Application Specific Array Processors"*, IEEE Computer Society Press, pp. 100-110, Sept. 1990.

[5] P. Feautrier, *Some Efficient Solutions to the Affine Scheduling Problem, Part I, One-Dimensional Time*. International Journal of Parallel Programming, vol. 21, no. 5, Plenum Publishing Co., pp. 313-347, 1992.

[6] J. Crop, D. Wilde, *"Scheduling Structured Systems"*, extended version available at http://www.ee.byu.edu/~wilde/pubs.html

Compiling Data Parallel Tasks for Coordinated Execution*

Erwin Laure[1], Matthew Haines[2], Piyush Mehrotra[3], and Hans Zima[1]

[1] Institute for Software Technology and Parallel Systems, University of Vienna
{erwin,zima}@par.univie.ac.at
[2] Department of Computer Science, University of Wyoming
haines@uwyo.edu
[3] ICASE, NASA Langley Research Center
pm@icase.edu

Abstract. Many advanced scientific applications are heterogeneous and multidisciplinary in nature, consisting of multiple, independent modules. Such applications require efficient means of coordination for their program units. The programming language Opus was designed recently to assist in coordinating the execution of multiple, independent program modules. In this paper we address the problem of how to compile an Opus program such that it can be efficiently executed on a broad class of machines.

1 Introduction

The Opus coordination language [3] was recently designed to support multidisciplinary applications by providing means for coordinating multiple, independent program unites. Opus defines language extensions to HPF that introduce coarse-grained *task parallelism* on top of an existing data parallel model, where each *task* is represented by an independent program module. The central concept of Opus is an abstract data type named *ShareD Abstraction* (or *SDA*) which encapsulates data and the methods acting on this data. Instances of SDAs can be created dynamically and execute in a separate address space. Resource allocation in Opus is controlled by an "on-clause" which can be used to specify the machine and the number of processors on which the SDA should be created. SDAs communicate with each other via *synchronous* or *asynchronous Method Invocations (MIs)*. Means for synchronization are provided in that asynchronous MIs are associated with *event variables* that can either be tested (non-blocking) or waited on (blocking). Furthermore, methods of an SDA may have an associated *condition-clause* that specifies a logical expression which must hold in order

---

* The work described in this paper was partially supported by the Special Research Program SFB F011 "AURORA" of the Austrian Science Fund and by the National Aeronautics and Space Administration under NASA Contract No. NAS1-19840, while the authors were in residence at ICASE, NASA Langley Research Center, Hampton, VA 23681. Matthew Haines was supported in part by NSF CAREER Award ASC-9623904.

P. Amestoy et al. (Eds.): Euro-Par'99, LNCS 1685, pp. 413–417, 1999.

for the method to be executed. Arguments to methods are passed with copy-in/copy-out semantics. See [3] for a detailed discussion of the Opus language design as well as some application examples.

In this paper, we discuss how an Opus program is compiled such that it can be efficiently executed on a broad class of machines. One of the major goals in designing the Opus compiler was its integration with an existing HPF compilation system (the *Vienna Fortran Compiler VFC* [2]) such that we do not have to re-implement any of the HPF compiler work and can thus focus on the coordination features of Opus.

2 The Opus Compilation Process

The transformation of Opus source code into an executable program is a multistage process:

The Opus Compiler transforms the code into blocks of HPF code with calls to the Opus Runtime System (*ORS*). The resulting code is further processed by VFC which produces Fortran 90 code with calls to the VFC runtime system. This code is eventually compiled and linked with both the VFC runtime system and the ORS.

The ORS (cf. [7]) provides runtime representations for SDAs and MIs as well as higher abstractions of low level system components. Moreover, methods for buffering MIs on the callee side, which is due to the conditional execution of methods, are provided. MI-buffers are realized using two FIFO queues called *MI-queues*: the *inqueue* buffers incoming MIs while the *outqueue* keeps track of issued MIs.

The Opus compiler transforms the code of SDAs into separate threads of control to facilitate an efficient execution of multiple SDAs mapped onto overlapping resources. However, within an SDA thread there is the need for multiple threads as well because we need a component that is able to receive MIs independently of ongoing method executions. Hence, one can think of a hierarchical thread structure where an SDA thread is subdivided into two independent threads communicating via a shared memory segment. As shown in Figure 1 of an SDA are:

- A shared memory area, in which the *MI-queues* are allocated.
- The *Server Thread* which receives MIs, constructs execution records (the runtime representation of MIs) and places them into the inqueue.
- The *Execution Thread* which retrieves records from the inqueue, evaluates the condition-clauses and executes the respective methods.

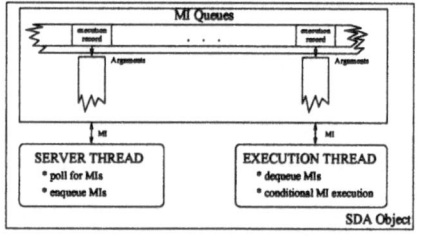

Fig. 1. Structure of an SDA Object

Both threads execute asynchronously parallel; the only synchronization required is a mutually exclusive access to the shared queues. The structure of these threads is fixed in so-called *compilation templates* which are filled in by the compiler.

The Server Thread The body of the server thread consists of a loop that waits on incoming MIs and handles them. The waiting is blocking and the underlying thread system ensures that the server thread is suspended if no messages are available. The server thread fulfills 3 different tasks:

First, it is responsible for *receiving MIs* and their associated input data. The server thread creates an execution record which is inserted into the inqueue. It is guaranteed that all input arguments have been received before the record becomes available to the execution thread. Note, that all communication carried out by the server thread may well be overlapped with useful computation in the execution thread.

The second task is the management of *execution acknowledgement* from MIs. The server thread ensures that all result data has been received and wakes up all SDAs blocked in a *wait* on this method.

Finally, the server thread also responds directly to inquiries about the state of a MI. This allows such queries to be "short-cut" and answered immediately, rather than being enqueued for service by the execution thread. Such queries occur when *waiting* or *testing* an asynchronous MI.

The Execution Thread The execution thread contains the internal SDA data, all of the method code, which is unaltered by the Opus compiler except for the method calls, as well as the code required to evaluate the condition-clauses for conditional SDA method invocation. The execution thread also contains a generic "execution loop" that looks in the inqueue for work. The execution thread is implemented as a Fortran 90 subroutine whose body is the execution loop; the method procedures are compiled to internal subroutines and the condition-clauses are transformed to internal logical functions. Hence, all methods and the condition-clause functions have access to the internal data of an SDA via *host association*. The execution loop repeatedly checks the inqueue for MIs. If none are found, the execution thread is suspended. Otherwise, the procedure associated with the MI is executed, given that the associated condition-clause function evaluates to true. The implementation of the execution thread guarantees that no two method procedures of an SDA can execute in parallel which is in conformance with the monitor-like semantics of SDAs.

Method Invocation SDAs communicate with each other via method invocation requests and synchronize by *waiting* or *testing* an event associated with the interactions are implemented by replacing the respective Opus calls with calls to the ORS: First, a new execution record is created and inserted into the outqueue; if an event was specified, the event variable is bound to this record. The actual invocation is initiated by making a call to the ORS and the needed input data is transferred to the callee. In the case of an asynchronous call, the caller resumes its execution immediately while in the case of an synchronous call it will block until the MI has returned. However, in either case, the caller does not resume its execution before all input data has been transferred in order to ensure that the input data to the invocation is not modified during the process of setting up the MI. A call to the intrinsic routines *wait* and *test* are simply compiled to a respective ORS calls.

MIs between distributed SDAs need some additional synchronization, since we have to ensure that all nodes of both the caller and the callee are ready to exchange the data, and that the callee knows the actual layout of the input data in order to compute the correct communication schedules. This synchronization is achieved by combining all the nodes of an SDA into an SDA-group, where in every group one node acts as a master, and a simple handshake protocol between the masters of two communicating SDAs.

The interested reader is referred to [7] for a more detailed discussion of the Opus compilation process. In that report we discuss also thread-safety issues that arise when integrating pthreads with MPI as well as the coordination aspects in the data exchange of distributed SDAs in more detail.

3   Rational and Related Work

On of the major decision in designing the Opus compilation system was to make use of a restructuring compiler, which implements most of the Opus semantics and is supported by a rather thin runtime system based upon pthreads and MPI. As discussed in [7] the use of higher level runtime packages like Nexus or Chant would have caused integration problems with the existing VFC runtime system. Moreover, Opus concepts like conditional method execution can be more efficiently implemented by using low level services directly.

Related coordination approaches that make use of a compiler include Fx [5], Fortran-M [4], and Orca [1]. Fx does not need any runtime support since it has a quite simple tasking model similar to the tasking facilities in HPF-2 (in fact, HPF-2 took over many ideas introduced in Fx). Fortran-M is based upon Nexus which fits perfectly into the Fortran-M model, however, it is a purely task parallel language thus the problems of Nexus in a data parallel context (cf. [7]) do not arise. Similarly, Orca was initially designed as a task parallel language based upon Panda. Subsequently, Orca was extended towards data parallelism [6] using similar concepts as Opus. However, in contrast to Opus, Orca is not an extension to an existing data parallel language but a new language where both data and task parallel features have been designed in a way that they integrate well. Hence, Panda was designed to support both task and data parallelism whereas ORS has to be integrated with an existing data parallel runtime system.

4   Summary

In this paper, we have presented our approach to compiling Opus programs in order to facilitate an efficient coordination of data parallel tasks. Our approach makes intensive use of a restructuring compiler which generates an HPF-conforming code expressing the Opus semantics. By using a restructuring compiler our implementation is open for future extensions or modifications of the Opus language that might need more advanced compiler techniques (like the exploitation of intra-SDA parallelism). Another benefit of the compiler approach is that we do not need to tackle the interoperability problems of Fortran 90 and C or C++ since all of our resulting code and most of our runtime system is written in Fortran 90. First experiences with a prototype implementation prove the effectiveness of our approach. These results, as well as a more detailed discussion of the Opus system design can be found in [7].

References

[1] H.E. Bal, M.F. Kaashoek, and A.S. Tanenbaum. Orca: A Language For Parallel Programming of Distributed Systems. *IEEE Transactions on Software Engineering*, Vol. 18(No. 3), March 1992.

[2] S. Benkner. VFC: The Vienna Fortran Compiler. *Journal of Scientific Programming*, 7(1):67–81, December 1998.

[3] B. Chapman, M. Haines, P. Mehrotra, J. Van Rosendale, and H. Zima. OPUS: A Coordination Language for Multidisciplinary Applications. *Scientific Programming*, 6/9:345–362, Winter 1997.

[4] I. Foster and K.M. Chandy. Fortran M: A Language for Modular Parallel Programming. *Journal of Parallel and Distributed Computing*, Vol. 26, 1995.

[5] T. Gross, D. O'Hallaron, and J. Subhlok. Task Parallelism in a High Performance Fortran Framework. *IEEE Parallel & Distributed Technology*, 2(3):16–26, Fall 1994.

[6] S.B. Hassen and H.E. Bal. Integrating Task and Data Parallelism Using Shared Objects. In *10th ACM International Conference on Supercomputing*, Philadelphia, PA, May 1996.

[7] E. Laure, M. Haines, P. Mehrotra, and H. Zima. On the Implementation of the Opus Coordination Language. Technical Report TR 99-03, Institute for Software Technology and Parallel Systems, May 1999.

Flexible Data Distribution in PGHPF

Mark Leair, Douglas Miles, Vincent Schuster, and Michael Wolfe

The Portland Group, Inc., USA, http://www.pgroup.com

Abstract. We have implemented the GEN_BLOCK (generalized block) data distribution in PGHPF, our High Performance Fortran implementation. Compared to a BLOCK or CYCLIC distribution, the more flexible GEN_BLOCK distribution allows users to balance the load between processors. Simple benchmark programs demonstrate the benefits of the new distribution format for unbalanced work loads, getting speedup of up to 2X over simple distributions. We also show performance results for a whole application.

High Performance Fortran (HPF) gives programmers a simple mechanism to run parallel applications by distributing arrays across multiple processors. HPF programs are single-threaded, achieving data parallelism through parallel loops and array assignments. The performance of HPF programs depends to a great deal on the data distribution, and on the optimizations in the compiler to recognize and reduce communications.

HPF compilers are responsible for implementing the distributions and generating any necessary synchronization or communication among the processors. In HPF-1, the two main distribution formats were BLOCK and CYCLIC. In a BLOCK distribution, the size of the array was divided as evenly as possible across all processors, with each processor getting a contiguous chunk of the array. In a CYCLIC distribution, the elements of the array are distributed round-robin. The choice of which distribution to use is up to the programmer, and depends on communication patterns and load balancing issues; in both cases, each processor gets roughly equal parts of the array.

HPF-2 introduced a new *generalized BLOCK* or GEN_BLOCK distribution format; a GEN_BLOCK distributed array can have different sized blocks on each processor; the block size for each processor is under user control, allowing flexible load balancing.

At the Portland Group, Inc. (PGI), we have implemented the GEN_BLOCK distribution format in our PGHPF compiler. We report on the changes required to support the new distribution format, and give some performance results.

1 The GEN_BLOCK Distribution Format

Suppose we have a 100×100 matrix A where we distribute the rows across four processors using a BLOCK distribution; each processor then "owns" 25 rows of the matrix. This is specified in HPF with the directive

P. Amestoy et al. (Eds.): Euro-Par'99, LNCS 1685, pp. 418–421, 1999.

```
!hpf$ distribute A(block,*)
```

If the amount of work in the program is the same for each row, this distribution should balance the load between processors. If we are only operating on the lower triangle of the matrix, processor zero operates on 325 elements, while processor three operates on 2200, a very unbalanced workload.

One way to try to balance the workload in a simple example like this is to use a CYCLIC distribution, but this often adds communication overhead as well as additional subscript calculation. Block-cyclic (CYCLIC(N)) distributions add significant communication, index computation, and loop overhead, and its use is discouraged in most HPF compilers. A GEN_BLOCK distribution allows us to retain the advantages of a BLOCK distribution, while distributing the rows in a fashion to balance the workload across processors. If we allocate the rows to processors as shown in the following table, the workload is balanced within 5% across processors:

proc	rows	work
0	1:50	1275
1	51:71	1281
2	72:87	1272
3	88:100	1222

This could be specified in HPF with the statement

```
!hpf$ distribute matrix(gen_block(gb))
```

where **gb** is an integer vector with values $\{50, 21, 16, 13\}$. The GEN_BLOCK vector can be computed at runtime, as long as its values are computed before the matrix is allocated.

Many Fortran applications have irregular data structures implemented using vectors and arrays, where a simple BLOCK distribution causes unnecessary load imbalance and communication. A GEN_BLOCK distribution will improve performance in many of these applications.

2 Implementing GEN_BLOCK in PGHPF

We recently added support for GEN_BLOCK in our PGHPF compiler. Since GEN_BLOCK is so close to BLOCK, the changes in the compiler itself were straightforward. Both are implemented with a local dynamically allocated array on each processor and a local descriptor containing information about the global distribution and the local subarray. Array assignments and parallel loops on the BLOCK or GEN_BLOCK arrays are localized into a single loop over the local subarray. We support *shadow* regions for BLOCK distributions; this was easy to carry forward to GEN_BLOCK distributions as well. When the compiler recognizes nearest-neighbor communication, it issues a single collective communication to exchange shadow regions before entering the computation loop; computation can then proceed without communication.

The compiler must manage the GEN_BLOCK distribution vector (**gb** in our example above); since the distribution vector might be changed after the distributed array is allocated, the compiler generates code to copy the vector to a temporary variable.

One major difference between a GEN_BLOCK distribution and a BLOCK distribution is for random indexing. Given a subscript value i in a BLOCK-distributed dimension the compiler can generate code to determine the processor index that owns i by computing $\lfloor i/b \rfloor$, where b is the block size. In a GEN_BLOCK distributed dimension, there is no simple formula to determine the processor that owns subscript i; the fastest code involves a binary search using the GEN_BLOCK distribution vector, which we implement in a library routine. While this is potentially very costly, most common communication patterns are structured, avoiding the search.

Most of the implementation effort went into augmenting the interprocessor communication library to support communication between GEN_BLOCK and BLOCK or CYCLIC arrays, and between different GEN_BLOCK distributed arrays. There library incurs additional overhead to consult the GEN_BLOCK distribution vector, but in irregular problems, the benefits of load balancing more than covers this cost.

3   Performance Results

We first tested our GEN_BLOCK on a simple triangular matrix operation:

```
integer, dimension(np) :: gb
real, dimension(n,n) :: a,b,c
!hpf$ distribute a(gen_block(gb),*)
!hpf$ align with a(i,*) :: b(i,*), c(i,*)
!... initialize a, b, c
do i = 1,n
  do j = 1,i
    a(i,j) = b(i,j) + c(i,j)
  enddo
enddo
```

We initialized the GEN_BLOCK distribution vector **gb** to balance the load among processors, and compared the resulting performance to that of a simple BLOCK distribution. The following table compares execution time of this loop on 1–32 processors and the two distributions on a Hewlett-Packard Exemplar, an IBM SP-2, and a Cray T3E (we were unable to get time on the Cray for 16 or 32 processors). The matrix sizes are given, and the loop was repeated 100 times.

	processors	1	2	4	8	16	32
HP Exemplar	gen_block	519	126	118	36	44.3	5.4
n=4099	block		524	327	219	146	62.2	46.9
IBM SP-2	gen_block	407	178	81	36.8	17.1	8.2
n=2053	block		407	240	119	49.8	23.2	13.8
Cray T3E	gen_block	173	87	44.7	22.1		
n=2053	block		172	136	77.0	40.9		

Time in Seconds (header for above)

As an example of how GEN_BLOCK distribution can help balance the load, the following table shows the time for each of 8 processors on the Cray T3E with BLOCK and GEN_BLOCK distributions, for 100 iterations the benchmark loop with matrix size 2053 × 2053:

	P0	P1	P2	P3	P4	P5	P6	P7
gen_block	10.8	18.8	20.4	21.1	21.6	21.6	21.9	22.1
block	0.72	4.11	10.2	16.6	22.9	29.0	35.3	40.9

Since GEN_BLOCK is new, few applications have been modified to take advantage of it. We were able to find customer program implementing an adaptive, hierarchical, N-body particle-in-cell code (Hu et al. 1997) that had been modified in anticipation of the availability of GEN_BLOCK. We executed this program with a test input on 32 processors of an IBM SP-2, and compared the performance of GEN_BLOCK distribution to a BLOCK distribution, with time in milliseconds:

	processors	time
block	32	6562
gen_block	32	5000
improvement		1.31

In this application, the GEN_BLOCK distributed version completed 31% faster than the BLOCK version; we note the GEN_BLOCK distribution vector was carefully computed by the author to balance the load.

In conclusion, the compiler modifications to support the GEN_BLOCK distribution were relatively minor; most of the effort was in the communications library. With GEN_BLOCK, most irregular HPF applications can now balance their own workload across processors.

References

[HPF]   High Performance Fortran Language Specification, Version 2.0 (1997)
[Hu]   Hu, Y. C., Johnnson, S. L., Teng, S-H., High Performance Fortran for Highly Irregular Problems, Proc. Sixth Symp. Principles and Practices of Parallel Programming, ACM Press (1997) 13–24

On Automatic Parallelization of Irregular Reductions on Scalable Shared Memory Systems*

E. Gutiérrez, O. Plata, and E.L. Zapata

Department of Computer Architecture, University of Málaga,
P.O. Box 4114, E-29080 Málaga, Spain
{eladio,oscar,ezapata}@ac.uma.es

Abstract. This paper presents a new parallelization method for reductions of arrays with subscripted subscripts on scalable shared-memory multiprocessors. The mapping of computations is based on the conflict-free write distribution of the reduction vector across the processors. The proposed method is general, scalable, and easy to implement on a compiler. A performance evaluation and comparison with other existing techniques is presented. From the experimental results, the proposed method is a clear alternative to the array expansion and privatized buffer methods, usual on state-of-the-art parallelizing compilers, like Polaris or SUIF.

1 Introduction and Background

Irregular reduction operations are frequently found in the core of many large scientific and engineering applications. Figure 1 shows simple examples of reduction loops (*histogram* reduction [10]), with a single reduction vector, A(), updated through single or multiple subscript arrays, f1(), f2(). Due to the loop-variant nature of the subscript array/s, loop-carried dependences may be present. It is usual that this reduction loop is executed many times, in an iterative process. The subscript array/s may be static (unmodified) during all the computation, or may change, usually slowly, through the iterative process.

A general approach to parallelize irregular codes, including reductions, is based on the *inspector-executor model* [9]. However, this strategy is usually highly inefficient as introduces significant overheads due to its generality. A more specific and efficient method may be developed if data affinity is exploited, as in the (LOCALWRITE) technique [5] (although it is not reported as a clear good alternative to parallelize irregular reductions).

In a shared-memory context, academic parallelizers like Polaris [2] and SUIF [4], recognize and parallelize irregular reductions. A number of techniques are available [7, 1]: critical sections, data affinity, privatized buffer (SUIF), array

---

* This work was supported by the Ministry of Education and Science (CICYT) of Spain (TIC96-1125-C03)

P. Amestoy et al. (Eds.): Euro-Par'99, LNCS 1685, pp. 422–429, 1999.

real A(1:ADim)
integer f(1:fDim)
 	
do i = 1, fDim
  r = function(i,f(i))
  A(f(i)) = A(f(i)) + r
end do
 	
(a)
	real A(1:ADim)
	integer f1(1:fDim), f2(1:fDim)
	
	do i = 1, fDim
	  r = function(i,f1(i),f2(i))
	  A(f1(i)) = A(f1(i)) + r
	  A(f2(i)) = A(f2(i)) - r
	end do
	
	(b)

Fig. 1. A single (a) and a multiple (b) irregular histogram reduction

expansion (Polaris) and reduction table. The most efficient techniques, privatized buffer and array expansion, however have scalability problems due to the high memory overhead they exhibit.

Contemporary commercial compilers do not recognize irregular reductions, although, in some cases, they allow to parallelize them, for instance, using data affinity (as in SGI Fortran90 compiler [11]). A similar approach may also be exploited into HPF [6] (for example, using ON HOME) or OpenMP [8]. However, when multiple reduction arrays occur, conditional sentences usually appear inside the reduction loop in order to fulfill the owner compute rule (which may introduce also computation replication). These implementations reduce drastically the performance of the parallel code and compromise its scalability.

We present here a method to parallelize irregular reductions on scalable shared-memory machines (although may be adapted to a message-passing machine), whose efficiency clearly overcomes that of all the previously mentioned techniques. The mapping of computations is based on the conflict-free write distribution of the reduction vector across the processors. The proposed method is general, scalable, and easy to implement on a compiler.

2   Data Write Affinity with Loop-Index Prefetching

2.1   Single Reductions

Array expansion is based on the domain decomposition of the histogram reduction loop (that is, the [1:fDim] domain). This way, and due to the irregular data access pattern to the reduction vector through f() (see Fig. 2 (a)), private copies of such vector are needed (high memory overhead for large domains, and cache locality problem). Such private buffers can be avoided if the domain decomposition of the loop is substituted for a data decomposition of the reduction vector. The reduction vector may be, for instance, block distributed across the processors. Afterwards, the computations of the histogram loop are arranged in such a way that each processor only computes those iterations that update owned reduction vector entries. Data distribution of the reduction vector may be carried out at compile time (using some compiler or language directive), or at runtime, as a consequence of the arrangement of the loop iterations.

A simple form to implement this computation arrangement is called *data affiliated loop* in [7]. Each processor traverses all the iterations in the reduction

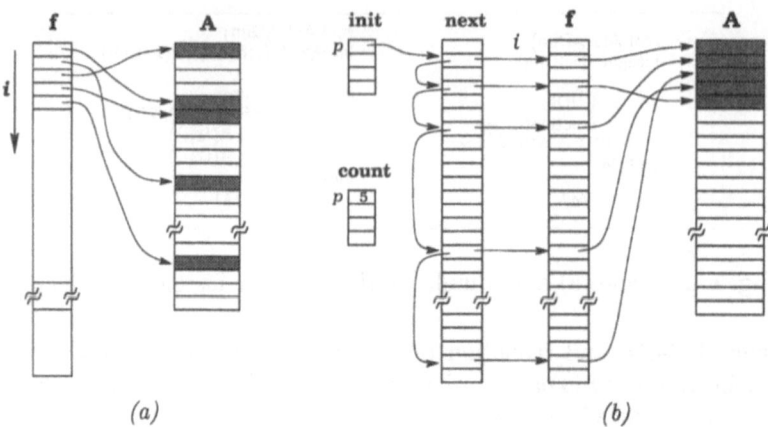

Fig. 2. Graphical depiction for irregular data accesses to the reduction vector (a), and the write data affinity-based access to it by using a loop-index prefetching array (b)

loop and checks whether it owns the reduction vector element referenced in the current iteration. If such case, the iteration is executed; otherwise, the iteration is skipped. The above implementation is not efficient for large iteration domains. A better approach consists in building a loop-index prefetching array, that contains, for each processor, the set of iterations that writes those reduction vector elements assigned to it. In the code in Fig. 3 (a), the loop-index prefetching array is implemented using three arrays, init(), count() and next(), that actually represents a linked list that reorders the subscript array f() (see Fig. 2 (b)). Each processor (thread) has an associated entry in both arrays, init() and count(). The entry in init() points to the first entry in next() owned by that processor (that is, whose index corresponds to a reduction loop iteration that writes in an owned reduction vector element). That entry in next() contains a pointer to the next element that is also owned by the same processor. The process is repeated the number of times stored in count().

In Fig. 3 (b) a simple code that implements the loop-index prefetching array is presented. The second loop in this code contains a histogram reduction on the count() array. As the reduction vector has a size given by the number of threads computing the code, it may be parallelized using array expansion without a significant memory overhead.

In a large class of scientific/engineering problems, the subscript array f() is static, that is, it is not modified during the whole execution of the program. In such codes, the loop-index prefetching array is computed only once and reused without modification in all the reduction loops of the program. Some other problems, on the other hand, have a dynamic nature, which is showed in the periodic updating of f(). The prefetching array has to be recalculated periodically, at least, partially. However, as it is usual that the dynamic nature of the problem changes slowly, the overhead of recalculating that array is partially compensated for its reuse in a number of executions of the reduction loop.

```
real A(1:ADim)                          integer prev(1:NumThreads)
integer f(1:fDim)
integer init(1:NumThreads)              BlockSz = floor(ADim/NumThreads) + 1
integer count(1:NumThreads)             do p = 1, NumThreads
integer next(1:fDim)                       count(p) = 0
                                        end do
doall p = 1, NumThreads                 do i = 1, fDim
   i = init(p)                             block = (f(i)-1)/BlockSz + 1
   cnt = count(p)                          if ( count(block) .eq. 0 ) then
   do k = 1, cnt                              init(block) = i
      r = function(i,f(i))                 else
      A(f(i)) = A(f(i)) + r                   next(prev(block)) = i
      i = next(i)                          end if
   end do                                  prev(block) = i
end doall                                  count(block) = count(block) + 1
                                        end do

          (a)                                        (b)
```

Fig. 3. Parallel single histogram reduction using data write affinity on the reduction vector with loop-index prefetching (a), and the sequential computation
of the prefetching array (b)

2.2   Multiple Reductions

Many real codes include irregular reduction loops containing multiple subscript
arrays indexing the same reduction vector, as shown in Fig. 1 (b). Our approach
can also be applied to this case, but with some modifications. As the reduction
vector A() is block distributed among the processors, there can be some iterations
in the reduction loop that evaluate the subscript arrays f1() and f2() pointing
to two different partitions. As each processor only writes on its partition, then
such iterations have to be replicated on both processors.

To solve this problem, we have split the set of iterations of the reduction
loop into two subsets. The first subset contains all iterations (local iterations)
that reference reduction vector entries belonging to the same partition. The
second subset contains the rest of iterations (boundary iterations). This way, all
local iterations are executed only once in some processor, while the boundary
iterations have to be replicated into two processors.

Figure 4 (a) shows this parallelization strategy applied to the reduction loop
of Fig. 1 (b). Two loop-index prefetching arrays have been built, one referencing
local iterations (init-l(), count-l() and next1()), and the other referencing boundary iterations. The second prefetching array is further split into two subarrays.
The first one, init-b1(), count-b1() and next1(), reorders the subscript array f1()
but restricted to boundary iterations. The other subarray is similar but reordering subscript array f2().

To compute those prefetching arrays, we can use a similar code than for single
reductions, as shown in Fig. 4 (b). This code also can be parallelized using the
expansion array technique (applied to init*() and count*()). Note that, in order
to save memory overhead, the first boundary iteration prefetching subarray uses
the same next1() array than the local iteration prefetching array. This way, we
need only two copies of this large array instead of three (next1() and next2()).

```
real A(1:ADim)                              integer prev(1:NumThreads)
integer f(1:fDim)
integer init-l(1:NumThreads)                BlockSz = floor(ADim/NumThreads) + 1
integer init-b1(1:NumThreads)               do p = 1, NumThreads
integer init-b2(1:NumThreads)                  count-l(p) = 0
integer count-l(1:NumThreads)                  count-b1(p) = 0
integer count-b1(1:NumThreads)                 count-b2(p) = 0
integer count-b2(1:NumThreads)              end do
integer next1(fDim), next2(fDim)            do i = 1, N
                                               block1 = (f1(i)-1)/BlockSz + 1
doall p = 1, NumThreads                        block2 = (f2(i)-1)/BlockSz + 1
   i = init-l(p)                               if ( block1 .eq. block2 ) then
   cnt = count-l(p)                               if ( count-i(block1) .eq. 0 ) then
   do k = 1, cnt                                     init-l(block1) = i
      r = function(i,f1(i),f2(i))                 else
      A(f1(i)) = A(f1(i)) + r                        next1(prev(block1)) = i
      A(f2(i)) = A(f2(i)) - r                     end if
      i = next1(i)                                prev(block1) = i
   end do                                         count-l(block1) = count-l(block1) + 1
   i = init-b1(p)                              else
   cnt = count-b1(p)                              if ( count-b(1,block1) .eq. 0 ) then
   do k = 1, cnt                                     init-b1(block1) = i
      r = function(i,f1(i),f2(i))                 else
      A(f1(i)) = A(f1(i)) + r                        next1(prev(block1)) = i
      i = next1(i)                                end if
   end do                                         prev(block1) = i
   i = init-b2(p)                                 count-b1(block1) = count-b(1,block1) + 1
   cnt = count-b2(p)                              if ( count-b(2,block2) .eq. 0 ) then
   do k = 1, cnt                                     init-b2(block2) = i
      r = function(i,f1(i),f2(i))                 else
      A(f2(i)) = A(f2(i)) - r                        next2(prev(block2)) = i
      i = next2(i)                                end if
   end do                                         prev(block2) = i
end doall                                         count-b2(block2) = count-b(2,block2) + 1
                                               end if
                                            end do
            (a)                                            (b)
```

Fig. 4. Parallel multiple histogram reduction using data write affinity on the reduction vector with loop-index prefetching (a), and the sequential computation of the prefetching arrays (b)

The replication of the boundary iterations introduces an additional computational overhead, which is represented by the third loop inside the doall loop in Fig. 4 (a) and the *else* section of the main *if* in Fig. 4 (b). However, in many realistic applications, there are many more local iterations than boundary ones, and hence, this additional computation overhead is very small.

3 Analysis and Performance Evaluation

The proposed technique has been applied to the code EULER (HPF-2 [3] motivating applications suite), which solves the differential Euler equations on an irregular mesh. In order to avoid other effects, we have selected in our experiments only one of the loops in the code, computing an irregular reduction inside a time-step loop (Fig. 5). The experiments have been conducted on a SGI Origin2000 (32 250MHz R10000 processors with 4 MB L2 cache and 8192 MB main memory), and implemented using the MIPSpro Fortran90 shared-memory di-

```
real vel_delta(1:3,numNodes)
integer edge(1:2,numEdges)
real edgeData(1:3,numEdges)
real velocity(1:3,numNodes)

do i = 1, numEdges
    n1 = edge(1,i)
    n2 = edge(2,i)
    r1 = funct1(edgeData(:,i), velocity(:,n1), velocity(:,n2))
    r2 = funct2(edgeData(:,i), velocity(:,n1), velocity(:,n2))
    r3 = funct3(edgeData(:,i), velocity(:,n1), velocity(:,n2))
    vel_delta(1,n1) = vel_delta(1,n1) + r1
    vel_delta(2,n1) = vel_delta(2,n1) + r2
    vel_delta(3,n1) = vel_delta(3,n1) + r3
    vel_delta(1,n2) = vel_delta(1,n2) - r1
    vel_delta(2,n2) = vel_delta(2,n2) - r2
    vel_delta(3,n2) = vel_delta(3,n2) - r3
end do
```

Fig. 5. An irregular reduction loop from the EULER code

rectives [11]. The array expansion parallel code was obtained using the Polaris compiler. Parallel codes were compiled using the MIPSpro compiler with optimization level 2.

Test irregular meshes have been obtained using the mesh generator included with the EULER code (sizes 783K and 1161K nodes, with connectivity of 8). Two versions of each mesh has been tested: colored and sorted. In the first version, an edge-coloring algorithm has been applied, and the edges of the same color have been placed consecutively in the indirection array. In this case, a low locality in accesses to reduction array would be expected. In the second version, the list of edges has been sorted, improving the access locality.

Fig. 6 depicts the performance for the colored and sorted versions of the mesh of size 1161K nodes. Part (a) shows the execution time (5 iterations of the time-step loop) of both methods, the array expansion and the proposed data write affinity with loop-index prefetching (DWA-LIP). These times excludes the calculation of the prefetching array, as this is done only once, before entering into the iterative loop. Part (b) shows speedups with respect to the sequential code (sequential time is 103.5 sec. and 15.3 sec. for the colored and sorted meshes, respectively). DWA-LIP obtains a significant performance improvement because it exploits efficiently locality when writing in the reduction array.

Fig. 7 shows the efficiencies for the colored (a) and sorted (b) meshes using DWA-LIP. For each class, results from two different mesh sizes show good scalability. The sequential times for the small colored and sorted meshes of sizes 783K nodes are 51.8 sec. and 11.8 sec., respectively. The sequential time costs of computing the loop-index prefetching for both mesh sizes are 2.3 sec. (small mesh) and 3.0 sec. (large mesh). These times are a small fraction of the total reduction time, that can be further reduced parallelizing the code (see Fig. 8).

Array expansion has a significant memory overhead due to the replication of the reduction vector in all the processors ($\mathcal{O}(Q * NumNodes * NumThreads)$), where Q is the number of reduction vectors). DWA-LIP also has memory overhead due to the prefetching array ($\mathcal{O}(K * NumEdges + NumThreads^2)$), where

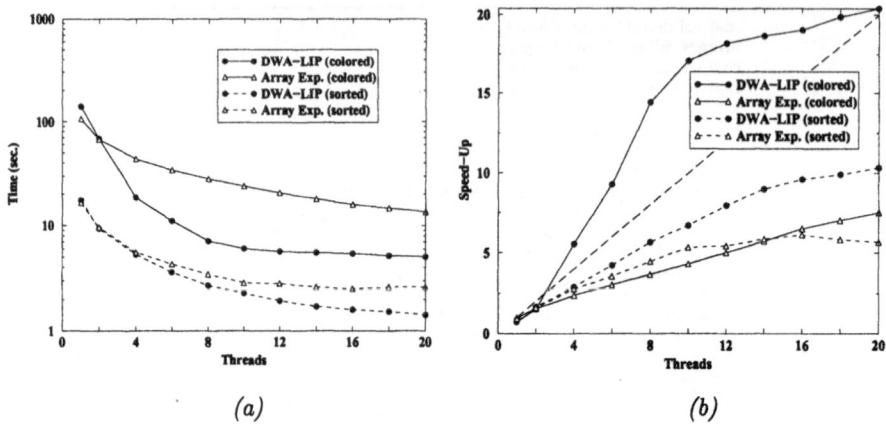

Fig. 6. Parallel execution times (a) and speedups (b) for DWA-LIP and array expansion using colored and sorted meshes with 1161K nodes

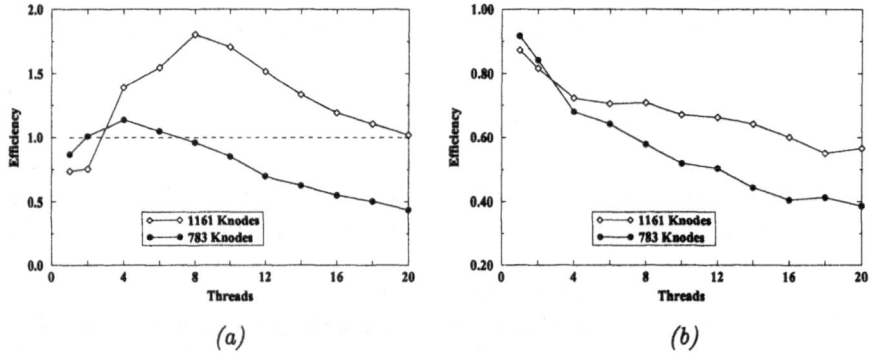

Fig. 7. Parallel efficiencies for colored (a) and sorted (b) meshes using DWA-LIP

K is the number of subscript arrays). In the EULER reduction loop, $Q = 3$ and $K = 2$, being $numEdges \approx 8 * numNodes$. Hence memory overhead in array expansion is greater than in DWA-LIP when more than 5 threads are used. In the EULER code, there are reduction loops with $Q = 5$ and $K = 4$.

4   Conclusions

The method proposed in this work parallelizes an irregular reduction loop exploiting write locality, similar to [5]. In our approach, the accesses to the subscript arrays are reordered using a linked list, instead of a general inspector-executor mechanism. This fact justifies the better performance and scalability of our method.

Compared with array expansion, our method does not need to replicate the reduction vector and exploits better data locality. Thus it has better performance and scalability. Despite the overhead of building the prefetching array, however

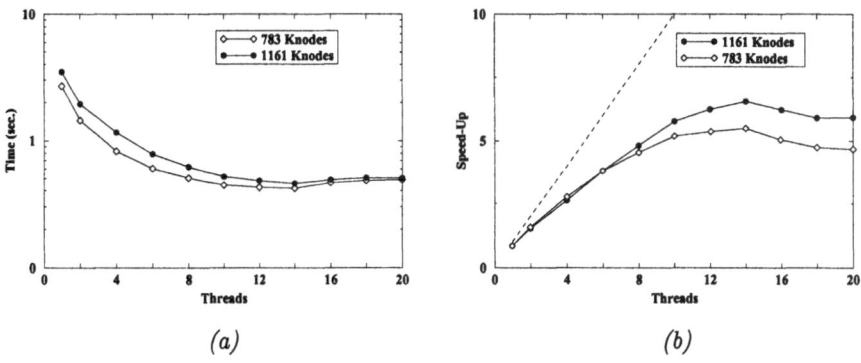

(a) (b)

Fig. 8. Parallel execution times (a) and speedups (b) for the loop-index prefetching

it is computed only once and reused through the whole program. In dynamic systems the prefetching array must be recomputed periodically.

References

[1] R. Asenjo, E. Gutierrez, Y. Lin, D. Padua, B. Pottengerg and E. Zapata, *On the Automatic Parallelization of Sparse and Irregular Fortran Codes*, **TR–1512**, Univ. of Illinois at Urbana-Champaign, Ctr. for Supercomputing R&D., Dec. 1996.

[2] W. Blume, R. Doallo, R. Eigemann, *et. al.*, *Parallel Programming with Polaris*, **IEEE Computer**, 29(12):78-82, Dec. 1996.

[3] I. Foster, R. Schreiber and P. Havlak, *HPF-2, Scope of Activities and Motivating Applications*, *Tech. Rep. CRPC-TR94492*, Rice Univ., Nov. 1994.

[4] M.W. Hall, J.M. Anderson, S.P. Amarasinghe, *et. al.*, *Maximizing Multiprocessor Performance with the SUIF Compiler* IEEE Computer, 29(12), Dec. 1996.

[5] H. Han and C.-W. Tseng, *Improving Compiler and Run-Time Support for Irregular Reductions*, **11th Workshop on Languages and Compilers for Parallel Computing**, Chapel Hill, NC, Aug. 1998.

[6] *High Performance Fortran Language Specification, Version 2.0*. High Performance Fortran Forum, Oct. 1996

[7] Y. Lin and D. Padua, *On the Automatic Parallelization of Sparse and Irregular Fortran Programs*, **4th Workshop on Languages, Compilers and Runtime Systems for Scalable Computers**, Pittsburgh, PA, May 1998.

[8] *OpenMP: A Proposed Industry Standard API for Shared Memory Programming*, OpenMP Architecture Review Board, http://www.openmp.org), 1997.

[9] R. Ponnusamy, J. Saltz, A. Choudhary, S. Hwang and G. Fox, *Runtime Support and Compilation Methods for User-Specified Data Distributions*, **IEEE Trans. on Parallel and Distributed Systems**, 6(8):815–831, Jun. 1995.

[10] B. Pottenger and R. Eigenmann, *Idiom Recognition in the Polaris Parallelizing Compiler*, **9th ACM Int'l Conf. on Supercomputing**, Barcelona, Spain, pp. 444–448, Jul. 1995.

[11] Silicon Graphics, Inc. *MIPSpro Automatic Parallelization.* SGI, Inc. 1998

I/O-Conscious Tiling for Disk-Resident Data Sets

Mahmut Kandemir[1], Alok Choudhary[2], and J. Ramanujam[3]

[1] EECS Dept., Syracuse University, Syracuse, NY 13244, USA
(mtk@top.cis.syr.edu)
[2] ECE Dept., Northwestern University, Evanston, IL 60208, USA
(choudhar@ece.nwu.edu)
[3] ECE Dept., Louisiana State University, Baton Rouge, LA 70803, USA
(jxr@ee.lsu.edu)

Abstract. This paper describes a tiling technique that can be used by application programmers and optimizing compilers to obtain I/O-efficient versions of regular scientific loop nests. Due to the particular characteristics of I/O operations, straightforward extension of the traditional tiling method to I/O-intensive programs may result in poor I/O performance. Therefore, the technique proposed in this paper customizes iteration space tiling for high I/O performance. The generated code results in huge savings in the number of I/O calls as well as the data volume transferred between the disk subsystem and main memory.

1 Introduction

An important problem that scientific programmers face today is one of writing programs that perform I/O in an efficient manner. Unfortunately, a number of factors render this problem very difficult. First, programming I/O is highly architecture-dependent, i.e., low-level optimizations performed with a specific I/O model in mind may not work well in systems with different I/O models and/or architectures. Second, there is very little help from compilers and run-time systems to optimize I/O operations. While optimizing compiler technology [6] has made impressive progress in analyzing and exploiting regular array access patterns and loop structures, the main focus of almost all the work published so far is the so called *in-core* computations, i.e., computations that make frequent use of the cache–main memory hierarchy rather than *out-of-core* computations, where the main memory–disk subsystem hierarchy is heavily utilized. Third, the large quantity of data involved in I/O operations makes it difficult for the programmer to derive suitable data management and transfer strategies.

Nevertheless it is extremely important to perform I/O operations as efficiently as possible, especially on parallel architectures which are natural target platforms for grand-challenge I/O-intensive applications. It is widely acknowledged by now that the per-byte cost (time) of I/O is orders of magnitude higher than those of communication and computation. A direct consequence of this phenomenon is that no matter how well the communication and computation are optimized, poor I/O performance can lead to unacceptably low overall performance on parallel architectures.

A subproblem within this context is one of writing I/O-efficient versions of loop nests. This subproblem is very important as in scientific codes the bulk of the execution times is spent in loop nests. Thus, it is reasonable to assume that in codes that

P. Amestoy et al. (Eds.): Euro-Par'99, LNCS 1685, pp. 430–439, 1999.

perform frequent I/O, a majority of the execution time will be spent in loop nests that perform I/O in accessing disk-resident multidimensional arrays (i.e., I/O-intensive loop nests). Given this, it is important to assess the extent to which existing techniques developed for nested loops can be used for improving the behavior of the loops that perform I/O. Although at a first glance, it appears that current techniques easily extend to I/O-intensive loop nests, it is not necessarily the case as we show in this paper. In particular, tiling [3, 2, 6], a prominent locality enhancing technique, can result in suboptimal I/O performance if applied to I/O-performing nests without modification.

In this paper we propose a new tiling approach, called *I/O-conscious tiling*, that is customized for I/O performing loops. We believe that this approach will be useful for at least two reasons. First, we express it in a simple yet powerful framework so that in many cases it can be applied by application programmers without much difficulty. Second, we show that it is also possible to automate the approach so that it can be implemented within an optimizing compiler framework without increasing the complexity of the compiler and compilation time excessively. Our approach may achieve significant improvements in overall execution times in comparison with traditional tiling. This is largely due to the fact that while performing tiling it takes I/O specific factors (e.g., file layouts) into account, which leads to substantial reductions in the number of I/O calls as well as the number of bytes transferred between main memory and disk.

The remainder of this paper is organized as follows. Section 2 reviews the basic principles of tiling focusing in particular on the state-of-the-art. Section 3 discusses why the traditional tiling approach may not be very efficient in coding I/O-performing versions of loop nests and makes a case for our modified tiling approach. Section 4 summarizes our experience and briefly comments on future work.

2   Tiling for Data Locality

We say that an array element has *temporal reuse* when it gets accessed more than once during the execution of a loop nest. *Spatial reuse*, on the other hand, occurs when nearby items are accessed [6]. Tiling [3, 2, 7] is a well-known optimization technique for enhancing memory performance and parallelism in nested loops. Instead of operating on entire columns or rows of a given array, tiling enables operations on multidimensional sections of arrays at a time. The aim is to keep the active sections of the arrays in faster levels of the memory hierarchy as long as possible so that when an array item (data element) is reused it can be accessed from the faster (higher level) memory instead of the slower (lower level) memory. In the context of I/O-performing loop nests, the faster level is the *main memory* and the slower level is the *disk subsystem* (or *secondary storage*). Therefore, we want to use tiling to enable the reuse of the array sections in memory as much as possible while minimizing disk activity. For an illustration of tiling, consider the matrix-multiply code given in Figure 1(a). Let us assume that the layouts of all arrays are *row-major*. It is easy to see that from the locality perspective, this loop nest does not exhibit good performance; this is because, although array Z has temporal reuse in the innermost loop (the k loop) and the successive iterations of this loop access consecutive elements from array X (i.e., array X has spatial reuse in the innermost loop), the successive accesses to array Y touch different rows of this array. Fortunately,

```
                                                        for kk = 1, N, B
                                                         kkbb = min(N,kk+B-1)
                                                         for jj = 1, N, B
                                                          jjbb = min(N,jj+B-1)
for i = 1, N            for i = 1, N                       for i = 1, N
 for j = 1, N            for k = 1, N                       for k = kk, kkbb
  for k = 1, N            for j = 1, N                       for j = jj, jjbb
   Z(i,j) += X(i,k)*Y(k,j)  Z(i,j) += X(i,k)*Y(k,j)           Z(i,j) += X(i,k)*Y(k,j)
  endfor                   endfor                            endfor
 endfor                   endfor                            endfor
endfor                   endfor                            endfor
  (a)                      (b)                             endfor
                                                          endfor
                                                            (c)
```

```
for kk = 1, N, B                                        for ii = 1, N, B
 kkbb = min(N,kk+B-1)                                    iibb = min(N,ii+B-1)
 ioread X[1:N,kk:kkbb]                                   ioread Z[ii:iibb,1:N]
 for jj = 1, N, B        for ii = 1, N, B                for kk = 1, N, B
  jjbb = min(N,jj+B-1)    iibb = min(N,ii+B-1)            kkbb = min(N,kk+B-1)
  ioread Z[1:N,jj:jjbb]   for kk = 1, N, B                ioread X[ii:iibb,kk:kkbb]
  ioread Y[kk:kkbb,jj:jjbb] kkbb = min(N,kk+B-1)          ioread Y[kk:kkbb,1:N]
  for i = 1, N             for i = ii, iibb               for i = ii, iibb
   for k = kk, kkbb         for k = kk, kkbb               for k = kk, kkbb
    for j = jj, jjbb         for j = 1, N                   for j = 1, N
     Z(i,j) += X(i,k)*Y(k,j)  Z(i,j) += X(i,k)*Y(k,j)        Z(i,j) += X(i,k)*Y(k,j)
    endfor                  endfor                          endfor
   endfor                  endfor                          endfor
  endfor                  endfor                           endfor
  iowrite Z[1:N,jj:jjbb]  endfor                          endfor
 endfor                   (e)                             iowrite Z[ii:iibb,1:N]
endfor                                                   endfor
  (d)                                                      (f)
```

Fig. 1. (a) Matrix-multiply nest (b) Locality-optimized version (c) Tiled version (d) Tiled version with I/O calls (e) I/O-conscious tiled version (f) I/O-conscious tiled version with I/O calls.

state-of-the-art optimizing compiler technology [6] is able to derive the code shown in Figure 1(b) given the one in Figure 1(a). In this so called optimized code array X has temporal reuse in the innermost loop (the j loop now) and arrays Z and Y have spatial reuse meaning that the successive iterations of the innermost loop touch consecutive elements from both the arrays.

However, unless the faster memory in question is large enough to hold the $N \times N$ array Y, many elements of this array will most probably be replaced from the faster memory before they are reused in successive iterations of the outermost i loop. Instead of operating on individual array elements, tiling achieves reuse of array sections by performing the calculations on array sections (in our case sub-matrices). Figure 1(c) shows the tiled version of Figure 1(b). In this tiled code the loops kk and jj are called the *tile loops* whereas the loops i, k, and j are the *element loops*. Note that the tiled version of the matrix-multiply code operates on $N \times B$ sub-matrices of arrays Z and X, and a $B \times B$ sub-matrix of array Y at a time. Here it is important to choose the *blocking factor* B such that all the $B^2 + 2NB$ array items accessed by the element loops i, k, j fit in the faster memory. Assuming that the matrices in this example are in-core (i.e., they fit in main memory) ensuring that the $B^2 + 2NB$ array elements can be kept in cache will be help in obtaining high levels of performance. However, in practice, cache conflicts occur depending on the cache size, cache associativity, and absolute array addresses in memory. Consequently, the choice of the blocking factor B is very important but difficult [3].

One important issue that needs to be clarified is *how* to select the loops that are to be tiled. A straightforward approach of tiling every loop that can be tiled (subject to dependences [6]) may actually degrade the performance. More sophisticated techniques attempt to tile only the loops that carry some form of reuse. For example, the *reuse-driven tiling* approach proposed by Xue and Huang [7] attempts to tile a given loop nest such that the outer untiled loops will not carry any reuse and the inner tiled loops will carry all the reuse and will consist of as few loops as possible. We stress that since after the linear locality optimizations (a step preceding tiling) most of the inherent reuse in the nest will be carried by the *innermost* loop, almost all tiling approaches tile the *innermost* loop, provided it is legal to do so. To sum up, tiling improves locality in scientific loop nests by capturing data reuse in the inner *element* loops. However, a straightforward application of traditional tiling to I/O-intensive codes may not be very effective as shown in the following section.

3   Tiling for Disk-Resident Data

3.1   The Problem

At first glance the traditional tiling method summarized above seems to be readily applicable to computations on disk-resident arrays as well. Returning to our matrix-multiply code, assuming that we have a total in-core memory of size H that can be allocated to this computation and that all the arrays are disk-resident (e.g., out-of-core), we only need to ensure that $B^2 + 2NB \leq H$. This tiling scheme is shown in Figure 1(d). The call ioread is an I/O routine that reads a section of a disk-resident array from file on disk into main memory; iowrite performs a similar transfer in the reverse direction. U[a:b,c:d] denotes a section of $(b-a+1) \times (d-c+1)$ elements in a two dimensional array U; the sections for higher-dimensional arrays are defined similarly. It should also be emphasized that since the sections are brought into main memory by explicit I/O commands, the conflict problem mentioned above for cache memories does not happen here.

While such a straightforward adaptation of the state-of-the-art (in-core) tiling leads to *data reuse* for the array sections brought into memory, it can cause *poor* I/O performance. The reason for this becomes obvious from Figure 2(a) which illustrates the layout of the arrays and representative sections from arrays (bounded by thick lines) that are active in memory at the same time. The lines with arrows within the arrays indicate the storage order—assumed to be *row-major* in our case—and each circle corresponds to an array element. Let us envision now how a section of array Z will be read from file into main memory. Since the array is row-major, in order to read the 16 elements shown in the section, it requires 8 I/O calls to the file system, each for only 2 consecutive array elements. Note that even though a state-of-the-art parallel I/O library allows us to read this rectangular array section using only a single high-level library call, since the elements in the section to be read are not fully consecutive in the file, it will still require 8 internal I/O calls for the said library to read it. It should also be noted that

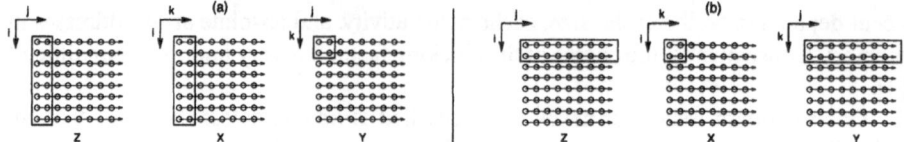

Fig. 2. (a) Unoptimized and (b) Optimized tile access patterns for matrix-multiply nest.

the alternative of reading the whole array and sieving out the unwanted array items is, in most cases, unacceptable due to the huge array sizes in I/O-intensive codes. Similar situations also occur with arrays X and Y. The source of the problem here is that the traditional (in-core) tiling attempts to optimize *what* to read into the faster memory, not *how* to read it. While this does not cause a major problem for the cache–main memory interface, the high disk access times render *"how to read"* a real issue. A simple rule is to always read array sections in a *layout conformant* manner. For example, if the file layout is row-major we should try to read as many rows of the array as possible in a single I/O call. In our matrix-multiply example, we have failed to do so due to the tiling of the innermost j loop, which reduces the number of elements that can be consecutively read in a single I/O call. Since this loop carries spatial reuse for both Z and Y, we should use this loop to read as many consecutive elements as possible from the said arrays. For example, instead of reading an $N \times B$ section from array Z, we should try to read a $B \times N$ section as shown in Figure 2(b) if it is possible to do so.

3.2 The Solution

Our solution to the tiling problem is as follows. As a first step, prior to tiling, the loops in the nest in question are reordered (or transformed) for maximum data reuse in the innermost loop. Such an order can be obtained either by an optimizing compiler or a sophisticated user. In the second step, we tile all the loops by a blocking factor B except the *innermost* loop which is *not tiled*. Since, after ordering the loops for locality, the innermost loop will, hopefully, carry the highest amount of spatial reuse in the nest, by not tiling it our approach ensures that the number of array elements read by individual I/O calls will be maximized.

   As an application of this approach, we consider once more the matrix-multiply code given in Figure 1(a). After the first step, we obtain the code shown in Figure 1(b). After that, tiling the i and k loops and leaving the innermost loop (j) untiled, we reach the code shown in Figure 1(e). Finally, by inserting the I/O read and write calls in appropriate places between the tile loops, we have the final code given in Figure 1(f). The tile access pattern for this code is shown in Figure 2(b). We note that the section-reads for arrays Z and Y, and the section-writes for array Z are very efficient since the file layouts are row-major. It should be stressed that the amount of memory used by the sections shown in Figures 2(a) and (b) is exactly the *same*. It should also be noted that as compared to arrays Z and Y, the I/O access pattern of array X is not as efficient. This is due to the nature of the matrix-multiply nest and may not be as important. Since the sections of array X are read much less frequently compared to those of the other two arrays (because, it has temporal reuse in the innermost loop), the impact of the I/O behavior of array X on the overall I/O performance of this loop nest will be less than that of the other two.

An important issue is the placement of I/O calls in a given tiled nest. There are two subproblems here: (1) *where are the I/O calls placed?* and (2) *what are the parameters to the I/O calls?* All we need to do is to look at the indices used in the subscripts of the reference in question, and insert the I/O call associated with the reference in between appropriate tile loops. For example, in Figure 1(e), the subscript functions of array Z use only loop indices i and j. Since there are only two tile loops, namely ii and kk, and only ii controls the loop index i, we insert the I/O read routine for this reference just after the ii loop as shown in Figure 1(f). But, the other two references use the element loop index k which varies with the tile loop index kk; therefore, we need to place the read routines for these references inside the kk loop (just before the element loops). The write routines are placed using a similar reasoning. For handling the second subproblem, namely, determining the sections to read and write, we employ the method of *extreme values of affine functions*; this method computes the maximum and minimum values of an affine function using Fourier-Motzkin elimination [6]. The details of this are omitted for lack of space.

3.3   Automating the Approach

In the matrix-multiply example it is straightforward to tile the loop nest. The reason for this is the presence of the innermost loop index in *at most* one subscript position of each reference. In cases where the innermost loop index appears in *more than one* subscript positions we have a problem of determining the section shape. Consider a reference such as Y(j+k,k) where k is the innermost loop index. Since, in our approach, we do not tile the innermost loop, when we try to read the bounding box which contains all the elements accessed by the innermost loop, we may end up having to read almost the whole array. Of course, this is not acceptable in general as in I/O-intensive routines we may not have a luxury of reading the whole array into memory at one stroke. This situation occurs in the example shown in Figure 3(a). Assuming *row-major* layouts, if we tile this nest as it is using the I/O-conscious approach, there will be no problem with array Z as we can read a $B \times N$ section of it using the fewest number of I/O calls. However, for array Y it is not trivial to identify the section to be read because the innermost loop index accesses the array in a diagonal fashion. The source of the problem is that this array reference does not fit our criterion which assumes at most a single occurrence of the innermost loop index. In our example reference, the innermost loop index k appears in both the subscript positions. In general we want each reference to an m-dimensional *row-major* array Z to be in either of the following two forms:

$Z(f_1, f_2, ..., f_m)$: In this form f_m is an affine function of all loop indices with a co-efficient of 1 for the innermost loop index whereas f_1 through $f_{(m-1)}$ are affine functions of all loop indices except the innermost one.

$Z(g_1, g_2, ..., g_m)$: In this form all g_1 through g_m are affine functions of all loop indices except the innermost one.

In the first case we have *spatial reuse* for the reference in the innermost loop and in the second case we have *temporal reuse* in the innermost loop. Since, according to our I/O-conscious tiling, all loops except the innermost one will be tiled using a blocking factor B, for the arrays which fit in the first form we can access $B \times ... \times B \times N$ sections (assuming N is the number of iterations of the innermost loop), which minimizes the number of I/O calls per array section. As for the arrays that fit into the second form, we can access sections of $B \times ... \times B \times B$ as all the loops that determine the section are tiled. Our task, then, is to bring each array (reference) to one of the forms given above.

Algorithm In the following we propose a *compiler algorithm* to transform a loop nest such that the resultant nest will have references of the forms mentioned above. We assume that the compiler is to determine the most appropriate file layouts for each array as well as a suitable loop (iteration space) transformation; that is, we assume that the layouts are *not* fixed as a result of a previous computation phase. Our algorithm, however, can be extended to accommodate those cases as well.

In our framework each execution of an n-*deep* loop nest is represented using an *iteration vector* $\bar{I} = (i_1, i_2, ..., i_n)$ where i_j corresponds to j^{th} loop from outermost. Also each reference to an m-*dimensional* array Z is represented by an *access (reference) matrix* L_z and an *offset vector* $\bar{o}_z$ such that $L_z\bar{I} + \bar{o}_z$ addresses the element accessed by a specific $\bar{I}$ [6]. As an example, consider a reference $Z(i+j, j-1)$ to a two-dimensional array in a two-deep loop nest with i is the inner and j is the outer. We have, $L_z = \begin{bmatrix} 1 & 1 \\ 0 & 1 \end{bmatrix}$ and $\bar{o}_z = \begin{bmatrix} 0 \\ -1 \end{bmatrix}$. Also we know from previous research in optimizing compilers [6, 4, 5] that (omitting the offset vector) assuming L is an access matrix, an $n \times n$ *loop* transformation matrix T transforms it to LT^{-1} and an $m \times m$ *data* transformation matrix M transforms it to ML. Given an L matrix, our aim is to find matrices T and M such that the transformed access matrix will fit in one of our two desired forms mentioned above. Notice also that while T is unique to a loop nest, we need to find an M for each disk-resident array.

Since for a given L determining both T and M simultaneously such that MLT^{-1} will be in a desired form involves solving a system of non-linear equations, we solve it using a *two-step* approach. In the first step we find a matrix T such that $L' = LT^{-1}$ will have a zero last column except the element in an rth row which is 1 (for spatial locality in the innermost loop) or we find a T such that the last column of $L' = LT^{-1}$ will be zero column (for temporal locality in the innermost loop). If the reference is optimized for spatial locality in the first step, in the second step we find a matrix M such that this r^{th} row in L' (mentioned above) will be the *last* row in $L'' = ML'$.

The overall algorithm is given in Figure 4. In Step 1, we select a *representative reference* for each array accessed in the nest. Using profile information might be helpful in determining run-time access frequencies of individual references. For each array, we select a reference that will be accessed maximum number of times in a typical execution. Since for each array we have two desired candidate forms (one corresponding to temporal locality in the innermost loop and one corresponding to spatial locality in the innermost loop), in Step 2, we exhaustively try all 2^s possible loop transformations, each corresponding to a specific combination of localities (spatial or temporal) for the

```
for i =            for u =            for u =            for u =            for u =
  for j =            for v =            for v =            for v =            for v =
    for k =            for w =            for w =            for w =            for w =
      Z(i,k+j)=          Z(u,v)=            Z(u,v)=            Z(u+w,v-w)=        Z(u,w)=
        Y(k,i+j+k)         Y(u+v,v-w)         Y(u+v,v+w)         Y(v,u+v)           Y(u,u+v+w)
    endfor             endfor             endfor             endfor             endfor
  endfor             endfor             endfor             endfor             endfor
endfor             endfor             endfor             endfor             endfor
   (a)                (b)                (c)                (d)                (e)
```

Fig. 3. (a) An example loop nest; (b–e) Transformed versions of (a).

arrays. In Step 2.1, we set the desired access matrix L'_i for each array i and in the next step we determine a loop transformation matrix T which obtains as many desired forms as possible. A typical optimization scenario is as follows. Suppose that in an alternative v where $1 \leq v \leq 2^s$ we want to optimize references 1 through b for temporal locality and references $b + 1$ through s for spatial locality. After Step 2.2, we typically have c references that can be optimized for temporal locality and d references that can be optimized for spatial locality where $0 \leq c \leq b$ and $0 \leq d \leq (s - b)$. This means that a total of $s - (c + d)$ references (arrays) will have *no locality* in the *innermost* loop. We do not apply any data transformations for the c arrays that have temporal locality (as they are accessed infrequently). For each array j (of maximum d arrays) that can be optimized for spatial locality, within the loop starting at Step 2.3, we find a data transformation matrix such that the resulting access matrix will be in our desired form. In Step 2.5 we record c, d, and $s - (c + d)$ for this alternative and move to the next alternative. After all the alternatives are processed we select the *most suitable* one (i.e., the one with the largest $c + d$ value). There are three important points to note here. First, in completing the partially-filled loop transformation matrix T, we use the approach proposed by Bik and Wijshoff [1] to ensure that the resulting matrix *preserves* all *data dependences* [6] in the original nest. Second, we also need a mechanism (when necessary) to favor some arrays over others. The reason is that it may not always be possible to find a T such that all L'_i arrays targeted in a specific alternative are realized. In those cases, we need to omit some references from consideration, and attempt to satisfy (optimize) the remaining. Again, profile information can be used for this purpose. Third, even an alternative does not specifically target the optimization of a reference for a specific locality, it may happen that the resultant T matrix generates such a locality for the said reference. In deciding the most suitable alternative we need also take such (unintentionally optimized) references into account.

As an example application of the algorithm, consider the loop nest shown in Figure 3(a) assuming that optimizing array Z is more important than optimizing array Y. The access matrices are $L_z = \begin{bmatrix} 1 & 0 & 0 \\ 0 & 1 & 1 \end{bmatrix}$ and $L_y = \begin{bmatrix} 0 & 0 & 1 \\ 1 & 1 & 1 \end{bmatrix}$. We have four possible optimization alternatives here: (1) temporal locality for both arrays; (2) temporal locality for Z and spatial locality for Y; (3) temporal locality for Y and spatial locality for Z; and (4) spatial locality for both arrays. These alternatives result in the following transformations.

Alternative (i) : $T^{-1} = \begin{bmatrix} 1 & 0 & 0 \\ 0 & 0 & 1 \\ 0 & 1 & -1 \end{bmatrix}$, $M_y = \begin{bmatrix} 0 & 1 \\ 1 & 0 \end{bmatrix} \Rightarrow L''_z = \begin{bmatrix} 1 & 0 & 0 \\ 0 & 1 & 0 \end{bmatrix}$ and $L''_y = \begin{bmatrix} 1 & 1 & 0 \\ 0 & 1 & -1 \end{bmatrix}$.

INPUT: access matrices for the references in the nest
OUTPUT: loop transformation matrix T and data transformation matrix M_i for each disk-
resident array i

(1) using profiling determine a representative access matrix for each array i $(1 \leq i \leq s)$
(2) *for* each of the 2^s alternatives *do*
    (2.1) determine target $L_1', L_2',...,L_s'$
    (2.2) using $L_i T^{-1} = L_i'$ determine a T
    (2.3) *for* each array j with the spatial locality *do*
        (2.3.1) let r_k be the row (if any) containing the only non-zero element in the last
        column for L_i'
        (2.3.2) find an M_j such that $L_j'' = M_j L_j'$ will be in the desired form (i.e., r_k will be
        the last row)
    (2.4) *endfor*
    (2.5) record for the current alternative the number of references with
        temporal locality, spatial locality, and no locality in the innermost loop
(3) *endfor*
(4) select the most suitable alternative (see the explanation in the text)
(5) *I/O-conscious tile* the loop nest and insert the I/O read/write routines

Fig. 4. I/O-conscious tiling algorithm.

$$Alternative\ (ii): T^{-1} = \begin{bmatrix} 1 & 0 & 0 \\ 0 & 0 & -1 \\ 0 & 1 & 1 \end{bmatrix}, M_y = \begin{bmatrix} 0 & 1 \\ 1 & 0 \end{bmatrix} \Rightarrow L_z'' = \begin{bmatrix} 1 & 0 & 0 \\ 0 & 1 & 0 \end{bmatrix} \text{ and } L_y'' = \begin{bmatrix} 1 & 1 & 0 \\ 0 & 1 & 1 \end{bmatrix}.$$

$$Alternative\ (iii): T^{-1} = \begin{bmatrix} 1 & 0 & 1 \\ 0 & 0 & -1 \\ 0 & 1 & 0 \end{bmatrix} \Rightarrow L_z'' = \begin{bmatrix} 1 & 0 & 1 \\ 0 & 1 & -1 \end{bmatrix} \text{ and } L_y'' = \begin{bmatrix} 0 & 1 & 0 \\ 1 & 1 & 0 \end{bmatrix}.$$

$$Alternative\ (iv): T^{-1} = \begin{bmatrix} 0 & 0 & 1 \\ 0 & 1 & 0 \\ 1 & 0 & 0 \end{bmatrix}, M_x = \begin{bmatrix} 0 & 1 \\ 1 & 0 \end{bmatrix} \Rightarrow L_z'' = \begin{bmatrix} 1 & 0 & 0 \\ 0 & 0 & 1 \end{bmatrix} \text{ and } L_y'' = \begin{bmatrix} 1 & 0 & 0 \\ 1 & 1 & 1 \end{bmatrix}.$$

The resulting programs are shown in Figures 3(b)– 3(e) for alternatives (i), (ii), (iii),
and (iv), respectively (the transformed loop bounds are omitted for clarity). It is easy
to see that the alternatives (i) and (ii) are superior to the other two. Alternative (iii)
cannot optimize array Z and alternative (iv) optimizes both arrays for spatial locality.
Our algorithm chooses alternative (i) or (ii) as both ensure temporal locality for the LHS
array and spatial locality for the RHS array in the innermost loop.

4   Conclusions and Future Work

In this paper we have presented an I/O-conscious tiling strategy that can be used by
programmers or can be automated in an optimizing compiler for the I/O-intensive loop
nests. We have shown that a straightforward extension of the traditional tiling strategy
to I/O-performing loop nests may lead to poor performance and demonstrated the necessary modifications to obtain an I/O-conscious tiling strategy. Our current interest is

in implementing this approach fully in a compilation framework and in testing it using large programs.

References

[1] A. Bik and H. Wijshoff. On a completion method for unimodular matrices. Technical Report 94–14, Dept. of Computer Science, Leiden University, 1994.

[2] F. Irigoin and R. Triolet. Super-node partitioning. In *Proc. 15th Annual ACM Symp. Principles of Programming Languages*, pp. 319–329, January 1988.

[3] M. Lam, E. Rothberg, and M. Wolf. The cache performance of blocked algorithms. In *Proc. 4th Int. Conf. Architectural Support for Programming Languages and Operating Systems*, April 1991.

[4] W. Li. *Compiling for NUMA parallel machines*. Ph.D. Dissertation, Cornell University, Ithaca, NY, 1993.

[5] M. O'Boyle and P. Knijnenburg. Non-singular data transformations: Definition, validity, applications. In *Proc. 6th Workshop on Compilers for Parallel Computers*, pp. 287–297, 1996.

[6] M. Wolfe. *High Performance Compilers for Parallel Computing*, Addison-Wesley, 1996.

[7] J. Xue and C.-H. Huang. Reuse-driven tiling for data locality. In *Languages and Compilers for Parallel Computing*, Z. Li et al., Eds., Lecture Notes in Computer Science, Volume 1366, Springer-Verlag, 1998.

Post-Scheduling Optimization of Parallel Programs

Stephen Shafer[1] and Kanad Ghose[2]

[1] Lockheed Martin Postal Systems, 1801 State Route 17 C,
Owego, NY 13827-3998
shafer@cs.binghamton.edu

[2] Department of Computer Science, State University of New York,
Binghamton, NY 13902-6000
ghose@cs.binghamton.edu

Abstract. No current task schedulers for distributed-memory MIMD machines produce optimal schedules in general, so it is possible for optimizations to be performed after scheduling. Message merging, communication reordering, and task duplication are shown to be effective post-scheduling optimizations. The percentage decrease in execution time is dependent on the original schedule, so improvements are not always achievable for every schedule. However, significant decreases in execution time (up to 50%) are possible, which makes the investment in extra processing time worthwhile.

1. Introduction

When preparing a program for execution on a distributed-memory MIMD machine, choosing the right task scheduler can make a big difference in the final execution time of the program. Many static schedulers are available which use varying methods such as list scheduling, clustering, and graph theory for deriving a schedule. Except for a few restricted cases, however, no scheduler can consistently produce an optimal answer. Thus, post-scheduling optimizations are still possible. Assuming that tasks are indivisible, and that the ordering of tasks on each processor does not change, there are two possible sources for improvements in execution time: the tasks themselves, and the inter-processor messages.

There are two different communication-related optimizations that we explore here. First, there is the possibility that pairs of messages from one processor to another may be combined, or merged, so that the data from both are sent as one message. This can result in a reduction of the execution time of the program. The second communication-related optimization deals with those tasks that have multiple messages being sent to tasks on other processors. For those tasks, the order of the messages may be changed so that more time-critical messages are sent first. We show two different orderings and how they affect the final execution time. Finally, we have implemented a task-related optimization that duplicates tasks when the messages that they send are causing delays in the destination tasks.

P. Amestoy et al. (Eds.): Euro-Par'99, LNCS 1685, pp. 440-444, 1999.
© Springer-Verlag Berlin Heidelberg 1999

2. Message Merging

2.1 The Concept of Merging

Wang and Mehrotra [WM 91] proposed a technique for merging messages that correspond to data dependencies. Their approach reduces the number of communications by combining a pair of messages from a source processor to the same destination processor into a single message that satisfies the data dependencies. In [SG 95], we showed that this method can actually increase the execution time, and might even introduce deadlock. We also showed that message merging can reduce execution time without introducing deadlock if the right pairs of messages are merged. In our previous work, however, I/O channel contention was not considered when execution times were calculated, so our results could have been more accurate. The results in this paper include the effects of channel contention. The entire merging process is explained in [SG 95].

2.2 Merging Results

The merging process was performed on sets of randomly created task graphs with varying numbers of tasks, from 100 tasks up to 800 tasks. In addition, several task graphs from actual applications were tested. The results are shown in Figure 1.

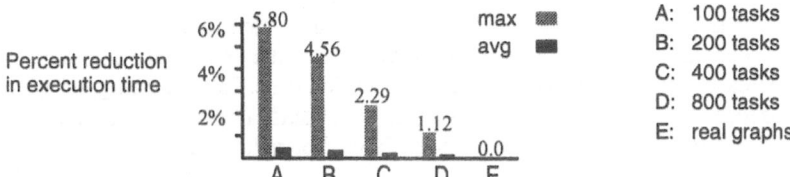

Figure 1. Message merging results on task graphs of varying sizes.

The actual applications fared the worst, with no merges being performed for any of the graphs tested. However, merging does not always guarantee a positive result. Given the small number of real graphs tested (five), it is possible that they just did not have the right circumstances for a profitable merge. The random graphs did much better. The number of graphs tested for each number of tasks was around fifty. That is, there were about 50 graphs with 100 tasks, 50 with 200 tasks, and so on. The results are shown as percent reduction in execution time, with maximums and averages shown. The results of merging depend on the original schedule. We expect that task graphs partitioned and scheduled by hand will show much better results than these.

3. Reordering Outgoing Communications

When multiple messages are being sent by one task, it is possible that more than one needs to be sent on the same link. When that occurs, and one of those messages is on the critical path, it may be delayed waiting for another message before it can get sent. We have implemented two reordering schemes that determine whether the execution time of a schedule can be reduced by changing the order of outgoing communications.

The first reordering consists of sending any messages on the critical path first. This should avoid any extra communication delay caused by the critical path message waiting for the link unnecessarily. The second reordering consists of sorting all outgoing messages from all tasks by their scheduling level, or s-level, which is defined as the length of the longest path from a node to any terminal node. This optimization attempts to send messages first to those tasks that are highest up in the task graph in the belief that the schedule length will be reduced if the higher tasks are executed first.

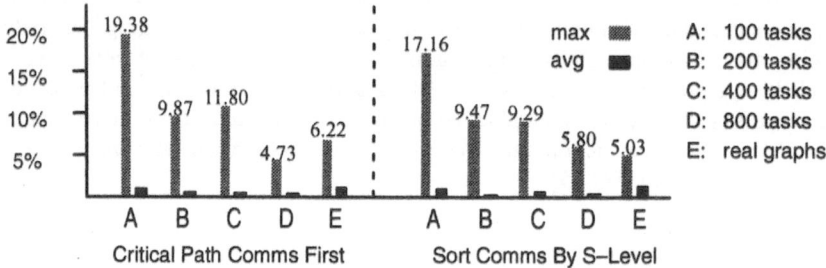

Figure 2. Reorder communications sent by one task.

The results (see Figure 2) show that a delay due to the order of outgoing communications can occur in schedules, and can be avoided. Again, the numbers shown are percent decreases in the execution times of the schedule. Like the results of the merging optimization, the average decreases are small, caused by not being able to improve all schedules. The results here, however, are better than those for merging. The best decrease was a 19.38% reduction in execution time. Like merging, however, the results depend to a great degree on the schedule itself. The average reduction in execution time is rather small, but the chance of a greater reduction is still there.

4. Task Duplication

The study of task duplication has been motivated by the fact that the execution time of a parallel program can be greatly affected by communication delays in the system. Sometimes, these delays cause a task on the critical path to wait for a message from another task. If such a delay could be reduced, or removed, then the execution time of the program can be reduced. Task duplication tries to decrease these delays.

The simplest situation in which to duplicate tasks is when there is only one start node of the task graph, and it has multiple children. Figure 3 shows an example of this where it is assumes there are five processors, and each task and communication take unit time. In this situation, task 1 can be duplicated on every processor, allowing tasks 2 through 6 to start execution without waiting for a message from another processor. The final schedule length has been reduced by the duplication. A more complex example would be where a task with multiple incoming messages and multiple outgoing messages is causing one of its child tasks to be delayed. In this case, when the task is duplicated, all of its incoming messages must also be sent to the copies, and each copy must only send a subset of the outgoing messages, with none of the messages being sent twice.

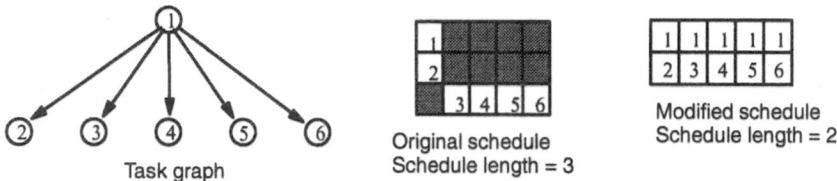

Figure 3. Simple case of task duplication.

4.1 Duplication During Scheduling

There are several schedulers including [CC 91], [CR 92], [SCM 95], and [AK 98] that have used task duplication in the scheduling process. [AK 98] includes a comparison of these and other schedulers. All of them use task duplication during scheduling, which typically searches for an idle time slot on the destination processor in which to put the duplicated task. This ensures that nothing besides the delayed task's start time will be affected by the duplication. We feel that this is too restrictive. By allowing tasks only to be placed in idle processor time, opportunities for duplication are passed up that may have adversely affected some task's start time, but if that task is not on the critical path then it might not matter. The possible benefit of duplication is lost.

If task duplication were to allow a duplicated task to adversely affect some task's start time, however, there is no way to know during scheduling whether or not the duplication has increased the final schedule length. We feel that the answer is post-scheduling task duplication.

4.2 Duplication After Scheduling

Duplicating tasks can reduce the execution time of a program, but it can also increase it. Copies of tasks other than start tasks require duplication of incoming messages also. Before the final schedule is produced, the effects of these extra messages on the schedule length can't be known. If duplication is performed after scheduling, however, all affects of any change can be known before that change is accepted. In addition, unnecessary duplication can be avoided since only the critical path needs to be checked for possible duplications. The major steps that need to be performed are:

- copy the task to the new PE
- duplicate messages sent to the task so that both copies of the task receive them
- have each copy of the task send a disjoint subset of outgoing messages, depending on the destination

The results of testing this approach are shown in Figure 4. There are significant decreases in execution time seen for all categories of task graph tested. One of the actual application graphs, a task graph for an atmospheric science application even had an almost 50% reduction in execution time.

5. Summary

We have shown three different types of post-scheduling optimizations: message merging, communication reordering, and task duplication. None of them promises to be able to reduce the execution time of every schedule. In almost all cases tested, the

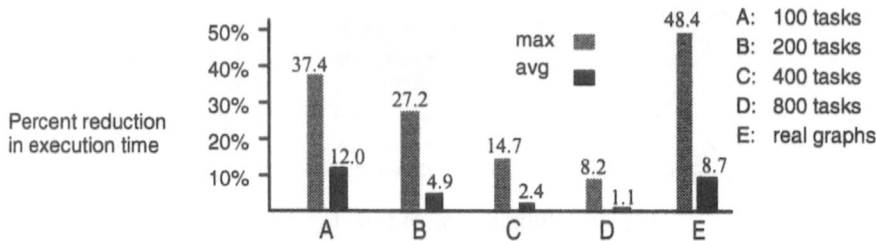

Figure 4. Results from Post–Scheduling Task Duplication

average reduction seen was rather low, around 1 to 2 percent. However, the chances of achieving some reduction ranges from a low of 39.4% (comm reordering) to a high of 82.5% (task duplication). This makes it very likely that the execution time can be reduced to some extent. Given the maximum reductions seen for these optimizations, the possibility of significantly reducing the execution time is worth the extra processing time to try. We believe that these optimizations are best tried after scheduling has already been completed. The actual critical path is known at that point and there is no chance of making some change to the system without knowing what effect it will have on the final execution time.

References

[AK 98]   I. Ahmad, Y.K. Kwok, "On Exploiting Task Duplication in Parallel Program Scheduling", *IEEE Trans. Parallel and Distrib. Systems,* vol. 9, no. 9, September 1998.

[CR 92]   Chung, Y.C., Ranka, S., "Application and Performance Analysis of a Compile-Time Optimization Approach for List Scheduling Algorithms on Distributed-Memory Multiprocessors", *Proc. Supercomputing '92*, pp.512-521, November 1992.

[CC 91]   Colin, J.Y., Chretienne, P., "C.P.M. Scheduling With Small Computation Delays and Task Duplication", *Operations Research*, pp. 680-684, 1991.

[SG 95]   S. Shafer, K. Ghose, "Static Message Combining in Task Graph schedules", *Proc. Int'l. Conf. on Parallel Processing*, August 1995, Vol.-II.

[SCM 95]   Shirazi, B., Chen, H., Marquis, J., "Comparative Study of Task Duplication Static Scheduling versus Clustering and Non-Clustering Techniques", *Concurrence: Practice and Experience*, vol. 7, no. 5, pp. 371-390, August 1995.

[WM 91]   Wang, K.Y., Mehrotra, P., "Optimizing Data Synchronizations On Distributed Memory Architectures, *Proc. 1991 Int'l. Conf. on Parallel Processing*, Vol. II, pp. 76-82.

Piecewise Execution of Nested Parallel Programs - A Thread-Based Approach

W. Pfannenstiel

Technische Universität Berlin
wolfp@cs.tu-berlin.de

Abstract. In this paper, a multithreaded implementation technique for piecewise execution of memory-intense nested data parallel programs is presented. The execution model and some experimental results are described.

1 Introduction

The flattening transformation introduced by Blelloch & Sabot allows us to implement nested data parallelism while exploiting all the parallelism contained in nested programs [2]. The resulting programs usually have a high degree of excess parallelism, which can lead to increased memory requirements. E.g., in many programs, operations known as generators produce large vectors, which are reduced to smaller vectors (or scalars) by accumulators almost immediately afterwards. High memory consumption is one of the most serious problems of nested data parallel languages. Palmer, Prins, Chatterjee & Faith introduced an implementation technique known as *Piecewise Execution* [4]. Here, the vector operations work on vector pieces of constant size only. In this way, low memory bounds for a certain class of programs can be guaranteed. One algorithm in this class is, e.g. the calculation of the maximum force between any two particles in N-body computations [1].

In this paper, we propose a multi-threaded implementation technique of piecewise execution. Constant memory usage can be guaranteed for a certain class of programs. Furthermore, the processor cache is utilized well.

2 Piecewise Execution

The NESL program presented in Figure 1 is a typical example of a matching generator/accumulator pair. The operation [s:e] enumerates all integers from s to e, plus_scan calculates all prefix sums, {x * x : x in b} denotes the elementwise parallel squaring of b, and sum sums up all values of a vector in parallel. Typically, there is a lot of excess parallelism in such expressions that must be partly sequentialized to match the size of the parallel machine. However, if the generator executes in one go, it produces a vector whose size is proportional to the degree of parallelism. In many cases, then, memory consumption is so high that the program cannot execute.

P. Amestoy et al. (Eds.): Euro-Par'99, LNCS 1685, pp. 445–448, 1999.

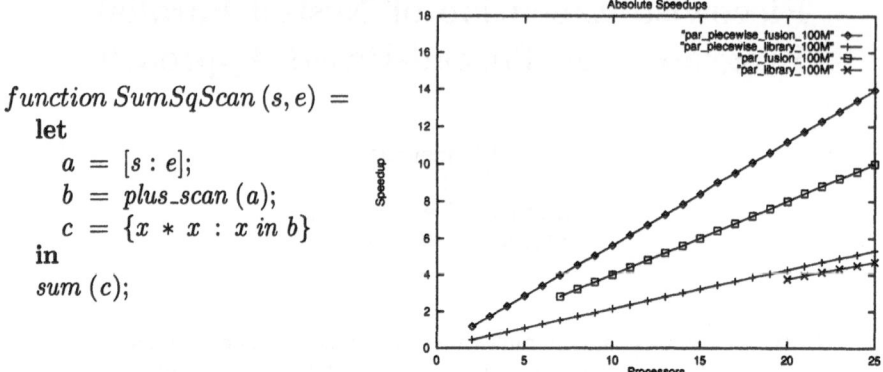

$$function \ SumSqScan \ (s, e) \ =$$
$$\quad let$$
$$\qquad a \ = \ [s : e];$$
$$\qquad b \ = \ plus\_scan \ (a);$$
$$\qquad c \ = \ \{x * x \ : \ x \ in \ b\}$$
$$\quad in$$
$$\qquad sum \ (c);$$

Figure 1: Example program   Figure 2: Speedups

In piecewise execution, each operation receives only a piece, of constant size, of its arguments at a time. The operation consumes the piece to produce (a part of) its result. If the piece of input is completely consumed, the next piece of input is requested. Once a full piece of output is produced, it is passed down as an input piece to the next vector operation. Thus, large intermediate vectors never exist in their full length. However, some means of control are needed to keep track of the computation. In [4], an interpreter is employed to manage control flow. Unfortunately, interpretation involves a significant overhead.

3   Execution Model and Implementation

We propose an execution model for piecewise execution using multiple threads. Threads in our model cannot migrate to remote processors. At the start of a thread, arguments can be supplied as in a function call. A thread can suspend and return control to the thread (or function) from which it was called or restarted using the operation switch_to_parent. Each thread has an identifier, which is supplied to restart_thread to reactivate it from anywhere in the program.

3.1   Consumer/Producer Model

A piecewise operation consumes one input piece and generates one output piece at a time. In general, a generator produces numerous output pieces from one input piece, whereas an accumulator consumes many pieces of input before completing a piece of output. Elementwise vector operations produce one piece of output from one piece of input. In the program DAG, the bottom node (realized as p threads on p processors cooperating in SPMD style) calculates the overall result. It requests a piece of input from the node(s) on which it depends. The demand is propagated to the top thread, which already has its input. Whenever a thread completes a piece of output, it suspends and control switches to the consumer node below. If a thread runs out of input before a full piece of output has been built, it restarts (one of) its producer node(s). In this way, control moves up and down in the DAG until the last node has produced the last piece of the overall result.

Only certain parts of a flattened program should run in a piecewise fashion. Thus, it must be possible to switch seamlessly between ordinary and piecewise execution mode. There are two special thread classes pw_in and pw_out, which provide an entry/exit protocol for piecewise program fragments. A pw_in thread consumes an ordinary vector at once, which is then supplied in pieces to the underlying threads of the piecewise program fragment. Thus, pw_in is always the topmost producer of data. However, it does not really produce data. It copies its argument vector(s) to the output in pieces. It never restarts any thread, because its input is supplied at once at call time. At the other end of the piecewise program fragment, pw_out is the last consumer of data. It collects all data pieces from above until the last piece is finished. Then, the complete result is passed on to the remaining part of the program, which works in an ordinary manner. An instance of pw_out never calls switch_to_parent but uses return after the complete result has been assembled. Often, pw_in and pw_out are not required, if – like in the example – scalars are consumed or produced. They are atomic values requiring constant space, making it impossible to supply them in pieces.

In our execution model the computation is triggered by the last thread, and the demand for data is propagated upwards in the program DAG. This resembles a demand-driven dataflow-like execution. However, to have normal execution order in the whole program, the threads in the piecewise program part are spawned in the same order as the other operations. First the producer is called, then the consumer. When a thread is spawned, it returns a *closure* and suspends. A closure consists of a thread handle and the address of a buffer for its output. The consumer receives the closure of its producer as an argument to be able to restart it if it needs data.

The threads are self-scheduled, i.e control switches explicitly among threads. For every piecewise operation, there is exactly one thread per processor and the structure of calls, suspends and restarts is the same on every processor. The threads constituting a piecewise operation run in sync with one another and use barriers to synchronize.

4 Experiments and Benchmarks

To test the feasibility of our approach, we transformed the example program SumSqScan from Figure 1 manually and implemented it on the Cray T3E. Keller & Chakravarty have proposed a new intermediate language DKL, whose main feature is the separation of local computations from communication of flattened parallel programs [3]. Local computations can be optimized by a set of transformations, the most important optimization being the fusion of consecutive loops. For our experiments, we tried to combine the proposed optimizations with piecewise execution. Thus, we implemented four combinations of features: A plain library version, a fused version in which parallel operations are fused as much as possible and the piecewise variants of both. The programs are written in C in SPMD style using the StackThreads library [5]. Communication and synchronization among processors is realized using the native Cray shmem communication library. We ran the four versions of the program for different parameter ranges and calculated the absolute speedups. The speedups are shown in Figure 2. The piece size was set to the respective optimum of the piecewise version, which corresponds to the size of the processor cache. Surprisingly, the combination of fusion and piecewise techniques yields the fastest results. Fusion leads to a significant improvement, but employing piecewise execution leads to shorter runtimes in this example, too. The optimized use of the processor cache would appear to compensate the overhead incurred by the multithreading implementation. Furthermore, the improvement of fusion and piecewise execution seem to mix well. Both techniques reduce memory requirements in an orthogonal way. The pure library and fusion versions cannot execute in some situations because of memory overflow. The input size of the piecewise versions is limited only by the maximum integer value.

5   Conclusion and Future Work

We have presented a new implementation technique for piecewise execution based on cost-effective multithreading. We have shown that piecewise execution does not necessarily mean increasing runtime. On the contrary, the combination of program fusion and piecewise execution resulted in the best overall performance for a typical example. Piecewise execution allows us to execute a large class of programs with large input sizes that could not normally run owing to an insufficient amount of memory.

We intend to develop transformation rules that automatically find and transform program fragments suitable for piecewise execution for use in a compiler.

References

[1] J. E. Barnes and P. Hut. Error analysis of a tree code. *The Astrophysical Journal Supplement Series*, (70):389–417, July 1989.
[2] G. E. Blelloch and G. Sabot. Compiling collection-oriented languages onto massively parallel computers. *Journal of Parallel and Distributed Computing*, 8(2):119–134, 1990.
[3] G. Keller and M. M. T. Chakravarty. On the distributed implementation of aggregate data structures by program transformation. In *Proceedings of the Fourth International Workshop on High-Level Parallel Programming Models and Supportive Environments (HIPS'99)*. IEEE CS, 1999.
[4] Daniel Palmer, Jan Prins, Siddhartha Chatterjee, and Rik Faith. Piecewise execution of nested data-parallel programs. In *Languages and compilers for parallel computing : 8th International Workshop*. Springer-Verlag, 1996.
[5] K. Taura and A. Yonezawa. Fine-grain multithreading with minimal compiler support - a cost effective approach to implementing efficient multithreading languages. In *Proceedings of the 1997 ACM SIGPLAN Conference on Programming Language Design and Implementation (PLDI)*. ACM, 1997.

Topic 05
Parallel and Distributed Databases

Burkhard Freitag and Kader Hameurlain

Co-chairmen

Although parallel and distributed database have been known for several years and are an established technology, new application domains require a deeper understanding of the underlying principles and the development of new solutions.

Besides such important questions as parallel query processing, optimization techniques, parallel database architectures, load sharing and balancing to name but a few, transaction management, data allocation, and specialized algorithms for advanced applications are topics of high relevance.

The parallel transactional execution of operations is addressed by the following three papers.

Data checkpointing is essential in distributed transaction processing and thus in distributed database systems. Roberto Baldoni et al. address the problem of how to determine consistent checkpoints in distributed databases and present formal solutions.

Armin Fessler and Hans-Jörg Schek investigate the similarities of database transaction management, in particular the scheduling of nested transactions, and parallel programming.

The standard serialization approach to object locking causes a considerable performance loss on shared memory architectures. Christian Jacobi and Cedric Lichtenau describe a serialization free approach to object locking in parallel shared memory architectures.

Data allocation and redistribution is critical for the overall performance of a distibuted or parallel database system.

The article by Holger Märtens explores ways of allocating intermediate results of large database queries across the disks of a parallel system. As a solution, declustering even of self-contained units of temporary data, e.g. hash buckets, is proposed. An analytical model is presented to show that the performance gain by improved parallel I/O justifies the costs of increased fragmentation.

Particular problem classes and/or architectures allow specialized algorithms and thus some extra gain in performance.

Basilis Mamalis et al. describe a parallel text retrieval algorithm designed for the fully connected hypercube. The parallel protocols presented are analyzed both analytically and experimentally.

The papers reflect the vivid discussion and ongoing research and development in the field of distributed and parallel databases which is not only intellectually challenging but also commercially most relevant.

P. Amestoy et al. (Eds.): Euro-Par'99, LNCS 1685, pp. 449–449, 1999.
© Springer-Verlag Berlin Heidelberg 1999

Distributed Database Checkpointing

Roberto Baldoni[1], Francesco Quaglia[1], and Michel Raynal[2]

[1] DIS, Università "La Sapienza", Roma, Italy
[2] IRISA - Campus de Beaulieu, 35042 Rennes Cedex, France

Abstract. Data checkpointing is an important problem of distributed database systems. Actually, transactions establish dependence relations on data checkpoints taken by data object managers. So, given an arbitrary set of data checkpoints (including at least a single data checkpoint from a data manager, and at most a data checkpoint from each data manager), an important question is the following one: "Can these data checkpoints be members of a same consistent global checkpoint?". This paper answers this question and proposes a *non-intrusive* data checkpointing protocol.

1 Introduction

Checkpointing the state of a database is important for audit or recovery purposes. When compared to its counterpart in distributed systems, the database checkpointing problem has additionally to take into account the serialization order of the transactions that manipulate the data objects forming the database. Actually, transactions create dependences among data objects which make the problem of defining *consistent* global checkpoints harder in database systems [5].

When any data object can be checkpointed independently of the others, there is a risk that no consistent global checkpoint can ever be formed. So, some kind of coordination is necessary when local checkpoints are taken in order to ensure they are mutually consistent. In this paper we are interested in characterizing mutual consistency of local checkpoints. More precisely:

(1) We first provide an answer to the following question $Q(S)$: "Given an arbitrary set S of checkpoints, can this set be extended to get a global checkpoint (*i.e.*, a set including exactly one checkpoint from each data object) that is consistent?". The answer to this question is far from being trivial when the set S includes checkpoints only from a subset of data objects.

(2) We then present a non-intrusive protocol that ensures the following property $P(C)$: "Local checkpoint C belongs to a consistent global checkpoint" when C is any local checkpoint.

The remainder of the paper consists of four main sections. Section 2 introduces the database model. Section 3 defines consistency of global checkpoints. Section 4 answers the previous question Q. To provide such an answer, we study the kind of dependences both the transactions and their serialization order create among checkpoints of distinct data objects. Specifically, it is shown that, while some data checkpoint dependences are causal, and consequently can be

P. Amestoy et al. (Eds.): Euro-Par'99, LNCS 1685, pp. 450–458, 1999.

captured on the fly, some others are "hidden", in the sense that, they cannot be revealed by causality. It is the existence of those hidden dependences that actually makes non-trivial the answer to the previous question. Actually, our study extends Netzer-Xu's result [4] to the database context. Then, Section 5 shows how the answer to Q can be used to design a "transaction-induced" data checkpointing protocol ensuring the previous property $\mathcal{P}$. The interested reader will find more details in [1].

2 Database Model

We consider a classical distributed database model. The system consists of a finite set of data objects, a set of transactions and a concurrency control mechanism.

Data Objects and Data Managers. A data object constitutes the smallest unit of data accessible to users. A non-empty set of data objects is managed by a data manager DM. From a logical viewpoint, we assume each data x has a data manager DM_x.

Transactions. A transaction is defined as a partial order on *read* and *write* operations on data objects and terminates with a *commit* or an *abort* operation. $R_i(x)$ (resp. $W_i(x)$) denotes a read (resp. write) operation issued by transaction T_i on data object x. Each transaction is managed by an instance of the transaction manager (TM) that forwards its operations to the scheduler which runs a specific concurrency control protocol. The write set of a transaction is the set of all the data objects it wrote.

Concurrency control. A concurrency control protocol schedules read and write operations issued by transactions in such a way that any execution of transactions is *strict* and *serializable*. This is not a restriction as concurrency control mechanisms used in practice (*e.g.*, two-phase locking 2PL and timestamp ordering) generate schedules ensuring both properties. The *strictness* property states that no data object may be read or written until the transaction that currently writes it either commits or aborts. So, a transaction actually writes a data object at the commit point. Hence, at some abstract level, which is the one considered by our checkpointing mechanisms, transactions execute atomically at their commit points. If a transaction is aborted, strictness ensures no cascading aborts and the possibility to use *before images* for implementing abort operations which restore the value of an object preceding the transaction access. For example, a 2PL mechanism, that requires transactions to keep their write locks until they commit (or abort), generates such a behavior.

Distributed database. We consider a distributed database as a finite set of sites, each site containing one or several data objects. So, each site contains one or more (logical) data managers, and possibly an instance of the TM. TMs and DMs exchange messages on a communication network which is asynchronous (message transmission times are unpredictable but finite) and reliable (each message will eventually be delivered).

Execution. Let $T = \{T_1, T_2, \ldots, T_n\}$ be a set of transactions accessing a set $O = \{o_1, o_2, \ldots, o_m\}$ of data objects (to simplify notations, data object o_i is identified by its index i). An execution E over T is a partial order on all read and write operations of the transactions belonging to T; this partial order respects the order defined in each transaction. Moreover, let $<_x$ be the partial order defined on all operations accessing a data object x, *i.e.*, $<_x$ orders all pairs of conflicting operations (two operations conflict if they access the same object and one of them is a write). Given an execution E defined over T, T is structured as a *partial order* $\widehat{T} = (T, <_T)$ where $T_i <_T T_j$ means:

$$(i \neq j) \wedge (\exists x \Rightarrow (R_i(x) <_x W_j(x)) \vee (W_i(x) <_x W_j(x)) \vee (W_i(x) <_x R_j(x)))$$

3   Consistent Global Checkpoints

3.1   Local States and Their Relations

Each write on a data object x issued by a transaction defines a new version of x. Let σ_x^i denote the i-th version of x; σ_x^i is called a *local state* (σ_x^0 is the initial local state of x). Transactions establish dependences between local states. This can be formalized in the following way. When T_k issues a write operation $W_k(x)$, it changes the state of x from σ_x^i to σ_x^{i+1}. More precisely, σ_x^i and σ_x^{i+1} are the local states of x, just before and just after the execution[1] of T_k, respectively. This can be expressed in the following way by extending the relation $<_T$ to include local states: T_k *changes* x *from* σ_x^i *to* $\sigma_x^{i+1} \iff (\sigma_x^i <_T T_k) \wedge (T_k <_T \sigma_x^{i+1})$.

Let $<_T^+$ be the transitive closure of the extended relation $<_T$. When we consider only local states, we get the following *happened-before* relation denoted $<_{LS}$ (which is similar to Lamport's happened-relation defined on events [3] in the process/message-passing model):

Definition 1. *(Local precedence, denoted $<_{LS}$)* $\sigma_x^i <_{LS} \sigma_y^j \iff \sigma_x^i <_T^+ \sigma_y^j$

As the relation $<_T$ defined on transactions is a partial order, it is easy to see that the relation $<_{LS}$ defined on local states is also a partial order. Figure 1 shows examples of the relation $<_{LS}$. It considers three data objects x, y, and z, and two transactions T_1 and T_2. Transactions are defined in the following way:

$$T_1: \ R_1(x); \ W_1(y); \ W_1(z); \ commit_1$$
$$T_2: \ R_2(y); \ W_2(x); \ commit_2$$

As there is a read-write conflict on x, two serialization orders are possible. Figure 1.a displays the case $T_1 <_T T_2$, while Figure 1.b displays the case $T_2 <_T$

---

[1] Remind that, as we consider strict and serializable executions, "Just before and just after the execution of T_k" means "Just before and just after T_k is committed".

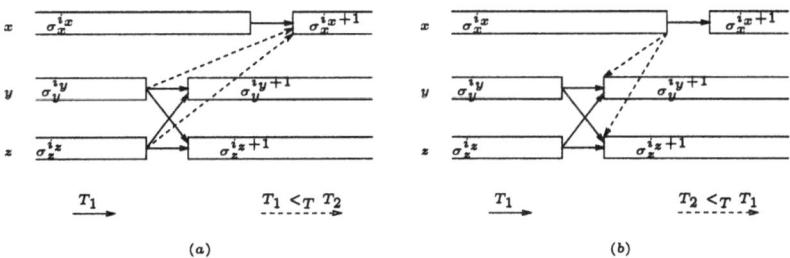

Fig. 1. Partial Order on Local States

T_1. Each horizontal axis depicts the evolution of the state of a data object. For example, the second axis is devoted to the evolution of y: $\sigma_y^{i_y}$ and $\sigma_y^{i_y+1}$ are the states of y before and after T_1, respectively.

Let us consider Figure 1.a. It shows that $W_1(y)$ and $W_1(z)$ add four pairs of local states to the relation $<_{LS}$, namely:

$$\sigma_y^{i_y} <_{LS} \sigma_y^{i_y+1}; \ \sigma_z^{i_z} <_{LS} \sigma_z^{i_z+1}; \ \sigma_y^{i_y} <_{LS} \sigma_z^{i_z+1}; \ \sigma_z^{i_z} <_{LS} \sigma_y^{i_y+1}$$

The relation $<_T$ adds two pairs of local states to $<_{LS}$:

$$\sigma_y^{i_y} <_{LS} \sigma_x^{i_x+1}; \ \sigma_z^{i_z} <_{LS} \sigma_x^{i_x+1}$$

The latter two dependences are due to the serialization order.

Precedence on local states, due to write operations of transactions T_1 and T_2, are indicated with continuous arrows, while the ones due to the serialization order are indicated with dashed arrows. Figure 1.b shows precedences on local states when the serialization order is reversed.

3.2 Consistent Global States

Let T be any set of transactions. Informally, the state reached by a database which includes all the updates made by all transactions in T is a consistent one.

A *global state* of the database is a set of local states, one from each data object. A global state $G = \{\sigma_1^{i_1}, \sigma_2^{i_2}, \ldots, \sigma_m^{i_m}\}$ is *consistent* if it does not contain two dependent local states, *i.e.*, if:

$$\forall x, y \in [1, \ldots, m] \Rightarrow \neg(\sigma_x^{i_x} <_{LS} \sigma_y^{i_y})$$

In Figure 1.a, the two global states $(\sigma_x^{i_x}, \sigma_y^{i_y}, \sigma_z^{i_z})$ and $(\sigma_x^{i_x}, \sigma_y^{i_y+1}, \sigma_z^{i_z+1})$ are consistent. The global state $(\sigma_x^{i_x+1}, \sigma_y^{i_y}, \sigma_z^{i_z+1})$ is not consistent either because $\sigma_y^{i_y} <_{LS} \sigma_x^{i_x+1}$ (due to the fact that $T_1 <_T T_2$) or because $\sigma_y^{i_y} <_{LS} \sigma_z^{i_z+1}$ (due to the fact that T_1 writes both y and z). Intuitively, a non-consistent global state of the database is a global state that could not be seen by any omniscient observer of the database. It is possible to show that, as in the process/message-passing model, the set of all the consistent global states is a lattice.

3.3   Consistent Global Checkpoints

A *local checkpoint* (or equivalently a *data checkpoint*) of a data object x is a local state of x that has been saved in a safe place by the data manager of x. So, all the local checkpoints are local states, but only a subset of local states are defined as local checkpoints. Let C_x^i $(i \geq 0)$ denote the i-th local checkpoint of x; i is called the *index* of C_x^i. Note that C_x^i corresponds to some σ_x^j with $i \leq j$. A *global checkpoint* is a set of local checkpoints one for each data object. It is *consistent* if it is a consistent global state.

We assume that all initial local states are checkpointed. Moreover, we also assume that, when we consider any point of an execution E, each data object will eventually be checkpointed.

4   Dependence on Data Checkpoints

4.1   Introductory Example

As indicated in the previous section, due to write operations of each transaction, or due to the serialization order, transactions create dependences among local states of data objects. Let us consider the following 7 transactions accessing data objects x, y, z and u:

$$T_1 : R_1(u); \ W_1(u); \ commit_1$$
$$T_2 : R_2(z); \ W_2(z); \ commit_2$$
$$T_3 : R_3(z); \ W_3(z); \ W_3(x); \ commit_3$$
$$T_4 : R_4(z); \ R_4(u); \ W_4(z); \ commit_4$$
$$T_5 : R_5(z); \ W_5(y); \ W_5(z); \ commit_5$$
$$T_6 : R_6(y); \ W_6(y); \ commit_6$$
$$T_7 : R_7(x); \ W_7(x); \ commit_7$$

Figure 2.a depicts the serialization imposed by the concurrency control mechanism. Figure 2.b describes dependences between local states generated by this execution. Five local states are defined as data checkpoints (they are indicated by dark rectangles). We study dependences between those data checkpoints. Let us first consider C_u^α and C_y^γ. C_u^α is the (checkpointed) state of u before T_1 wrote it, while C_y^γ is the (checkpointed) state of y after T_6 wrote it (*i.e.*, just after T_6 is committed). The serialization order (see Figure 2.a) shows that $T_1 <_T T_6$, and consequently $C_u^\alpha <_{LS} C_y^\gamma$, *i.e.*, the data checkpoint C_y^γ is causally dependent [3] on the data checkpoint C_u^α (Figure 2.b shows that there is a directed path from C_u^α to C_y^γ). Now let us consider the pair of data checkpoints consisting of C_u^α and C_x^δ. Figure 2.b shows that C_u^α precedes T_1, and that C_x^δ follows T_7. Figure 2.a indicates that T_1 and T_7 are not connected in the serialization graph. So, there is no causal dependence between C_u^α and C_x^δ (Figure 2.b shows that there is no directed path from C_u^α to C_x^δ). But, as the reader can check, there is no consistent global checkpoint including both C_u^α and C_x^δ ([2]). So there is a *hidden*

---

[2] Adding C_y^γ and C_z^β to C_u^α and C_x^δ cannot produce a consistent global state as $C_z^\beta <_{LS} C_x^\delta$. Adding $C_z^{\beta+1}$ instead of C_z^β has the same effect as $C_u^\alpha <_{LS} C_z^{\beta+1}$.

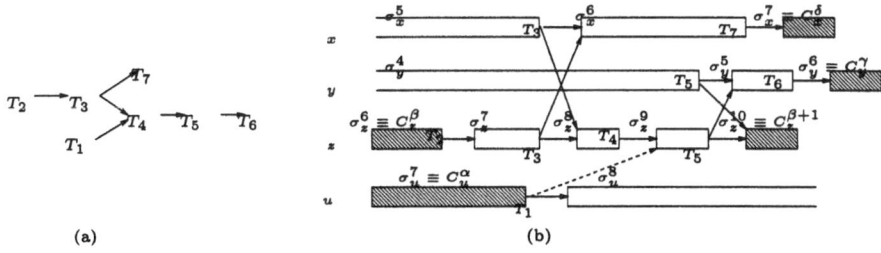

Fig. 2. (a) A Serialization Order and (b) Data Checkpoint Dependences

dependence between C_u^α and C_x^δ which does not allow them to belong to the same consistent global checkpoint.

4.2   Necessary and Sufficient Condition

This section first provides an unified definition of dependences that takes into account both causal and hidden dependences. Then it presents a necessary and sufficient condition for a set of data checkpoints to belong to a consistent global checkpoint.

Definition 2. *(Interval) A checkpoint interval I_x^i is associated with data checkpoint C_x^i. It consists of all the local states σ_x^k such that:*

$$(\sigma_x^k = C_x^i) \vee (C_x^i <_{LS} \sigma_x^k <_{LS} C_x^{i+1})$$

As an example, Figure 2.b shows that I_z^β includes 4 consecutive local states of z. Note that, due to the assumptions on data checkpoints stated in Section 3.3, any local state belongs to exactly one interval. Let us call an edge of the partial order on local states ($<_{LS}$) a *dependence edge*.

Definition 3. *(Dependence Path) There is a dependence path (DP) from a data checkpoint C_x^i to C_y^j (denoted $C_x^i \overset{DP}{\to} C_y^j$) iff:*
i) $x = y$ and $i < j$, or
ii) there is a sequence $(d_1, d_2, \dots, d_r)$ of dependence edges, such that:
ii.1: d_1 starts after C_x^i,
ii.2: $\forall d_q : 1 \le q < r$: let I_z^k be the interval in which d_q arrives; then d_{q+1} starts in the same or in a later interval (i.e., an interval I_z^h such that $k \le h$),
ii.3: d_n arrives before C_y^j.

This definition generalizes the Z-path notion introduced in [4] for asynchronous message-passing systems. In the example depicted in Figure 2.b, the hidden dependence between C_u^α and C_x^δ can be now denoted $C_u^\alpha \overset{DP}{\to} C_x^\delta$ as $C_u^\alpha = \sigma_u^7 <_{LS} \sigma_z^9$ (due to relation $<_T$), $\sigma_z^7 <_{LS} \sigma_x^6$ and $\sigma_x^6 <_{LS} \sigma_x^7 = C_x^\delta$. Note that σ_z^9 and σ_z^7 belong to the same checkpoint interval I_z^β.

Theorem 1. *Let $\mathcal{I} \subseteq \{1, \ldots, m\}$ and $\mathcal{S} = \{C_x^{i_x}\}_{x \in \mathcal{I}}$ be a set of data checkpoints. Then $\mathcal{S}$ is a part of a consistent global checkpoint if and only if:*

$$\forall x, y \in \mathcal{I} \;\Rightarrow\; \neg(C_x^{i_x} \stackrel{DP}{\rightarrow} C_y^{i_y})$$

Proof. Due to space limitation, the proof is omitted. See [1]. ∎

5   A "Transaction-Induced" Checkpointing Protocol

If we suppose that the set S includes only a checkpoint C, the previous Theorem leads to an interesting corollary, namely: C belongs to a consistent global checkpoint if and only if $\neg(C \stackrel{DP}{\rightarrow} C)$. Hence, providing checkpointing protocols ensuring that $\neg(C \stackrel{DP}{\rightarrow} C)$, guarantees the property $\mathcal{P}(C)$ defined in the introduction. These protocols are interesting for two reasons. (1) They avoid wasting time in taking a data checkpoint that will never be used in any consistent global checkpoint. And (2) no domino effect [6] can ever take place as any data checkpoint belongs to a consistent global checkpoint.

Let the *timestamp* of a local checkpoint be a non-negative integer such that:

$$\forall x, i, j : ((i < j) \Rightarrow (timestamp(C_x^i) < timestamp(C_x^j)))$$

A *timestamping function* associates a timestamp with each local checkpoint. (Note that the indexes of local checkpoints define a particular timestamping function).

Moreover, let us consider the following property $\mathcal{TS}$: "Let $\mathcal{S}_n$ be the set formed by data checkpoints timestamped n. If $\mathcal{S}_n$ includes a checkpoint per object, then it constitutes a consistent global checkpoint". In the following we provide a checkpointing protocol that guarantees $\mathcal{P}$ for all local checkpoints, and $\mathcal{TS}$ for any value of n. Actually, this protocol can be seen as an adaptation of a protocol [2] developed for the process/message-passing model.

Local Variables of a Data Manager. The data manager DM_x has a variable ts_x, which stores the timestamp of the last checkpoint of x (it is initialized to zero); i_x denotes the index value of the last checkpoint of x. Moreover, data managers can take checkpoints independently of each other (*basic checkpoints*), for example, by using a periodic algorithm which could be implemented by associating a timer with each data manager (a local timer is set whenever a checkpoint is taken and a basic checkpoint is taken by the data manager when its timer expires). Data managers are directed to take additional data checkpoints (*forced checkpoints*) in order to ensure $\mathcal{P}$ or $\mathcal{TS}$. The decision to take forced checkpoints is based on the control information piggybacked by commit messages of transactions.

Behavior of a Transaction Manager. Let W_{T_i} be the write set of a transaction T_i managed by a transaction manager TM_i. We assume each time an operation of T_i is issued by TM_i to a data manager DM_x, it returns the value of x plus the value of its current timestamp ts_x. TM_i stores in $MAX\_TS_{T_i}$ the maximum value among the timestamps of the data objects read or written by T_i. When transaction T_i is committed, the transaction manager TM_i sends a COMMIT message to each data manager DM_x involved in W_{T_i}. Such COMMIT messages piggyback $MAX\_TS_{T_i}$.

Behavior of a Data Manager. As far as checkpointing is concerned, the behavior of a data manager DM_x is defined by the two following procedures namely take-basic-ckpt and take-forced-ckpt. They define the rules associated with checkpointing.

> take-basic-ckpt:
> > **When** the timer expires:
> > > $i_x \leftarrow i_x + 1;\ ts_x \leftarrow ts_x + 1;$
> > > Take checkpoint $C_x^{i_x};\ timestamp(C_x^{i_x}) \leftarrow ts_x;$
> > > Reset the local timer

> take-forced-ckpt:
> > **When** DM_x **receives** COMMIT$(MAX\_TS_{T_i})$ **from** TM_i:
> > > **if** $ts_x < MAX\_TS_{T_i}$ **then**
> > > > $i_x \leftarrow i_x + 1;\ ts_x \leftarrow MAX\_TS_{T_i};$
> > > > Take a (forced) checkpoint $C_x^{i_x};\ timestamp(C_x^{i_x}) \leftarrow ts_x;$
> > > > Reset the local timer
> > > **endif**;
> > > Process the COMMIT message

From the increase of the timestamp variable ts_x of a data object x, and from the rule associated with the taking of forced checkpoints (take-forced-ckpt), which forces a data checkpoint whenever $ts_x < MAX\_TS_{T_i}$, the condition $\neg(C_x^{i_x} \overset{DP}{\to} C_x^{i_x})$ follows for any data checkpoint $C_x^{i_x}$. Actually, this simple protocol ensures that, if $C_x^{i_x} \overset{DP}{\to} C_y^{i_y}$, then $timestamp(C_x^{i_x}) < timestamp(C_y^{i_y})$.

It follows from the previous observation that if two data checkpoints have the same timestamp value, then they cannot be related by $\overset{DP}{\to}$. So, all the sets S_n that exist are consistent. Note that the take-forced-ckpt rule may produce gaps in the sequence of timestamps assigned to data checkpoints of a data object x. So, from a practical point of view, the following remark is interesting: when no data checkpoint of a data object x is timestamped by a given value n, then the first data checkpoint of x whose timestamp is greater than n can be included in a set containing data checkpoints timestamped by n to form a consistent global checkpoint.

References

[1] Baldoni R., Quaglia F. and Raynal M., Consistent Data Checkpoints in Distributed Database Systems: a Formal Approach. *Tech. Report 22-97*, Dipartimento di Informatica e Sistemistica, Università di Roma "La Sapienza", 1997.

[2] Briatico, D., Ciuffoletti, A. and Simoncini, L., A Distributed Domino-Effect Free Recovery Algorithm, *in Proc. IEEE Int. Symp. on Reliability Distributed Software and Database*, pp. 207-215, 1984.

[3] Lamport, L., Time, Clocks and The Ordering of Events in a Distributed System, *Comm. of the ACM*, 21(7):558-565, 1978.

[4] Netzer, R.H.B. and Xu, J., Necessary and Sufficient Conditions for Consistent Global Snapshots, *IEEE TPDS*, 6(2):165-169,1995.

[5] Pilarski, S. and Kameda, T., Checkpointing for Distributed Databases: Starting from the Basics, *IEEE TPDS*, 3(5):602-610, 1992.

[6] Randell B., System Structure for Software Fault Tolerance, *IEEE TSE*, 1(2):220-232, 1975.

A Generalized Transaction Theory
for Database and Non-database Tasks

Armin Feßler and Hans-Jörg Schek

Institute of Information Systems
ETH Zentrum, CH-8092 Zürich, Switzerland
{fessler,schek}@inf.ethz.ch

Abstract. In both database transaction management and parallel programming, parallel execution of operations is one of the most essential features. Although they look quite different, we will show that many important similarities exist. As a result of a more careful comparison we will be able to point out that recent progress in database transaction management theory in the field of composite stack schedules can improve the degree of parallelism in databases as well as in parallel programming. We will use an example from numerical algorithms and will demonstrate that in principle more parallelism can be achieved.

1 Motivation

Parallel programming (PP) and database transaction management (DBTM) are two separate fields that both provide mechanisms for executing tasks in parallel. At a first glance, these two models look rather different. A more careful inspection, however, reveals much commun ground. Therefore we are interested in a more careful comparison: Does one of these provide more possible parallelism than the other? Can the one model benefit from the other?

These are the questions we want to answer in this paper. In the next section we compare PP with DBTM. It turns out that both models are, in fact, similar in that both control the flow of information in a parallel execution in way that equivalence to a sequential execution is ensured. In both areas there is one level of abstraction where a scheduler (in case of DBTM) or a parallizing compiler (in case of PP) is located and ensures correctness. However, recent progress made in the foundation of database transactions shows that a higher degree of parallelism can be achieved be considering several schedulers at several layers of abstraction. Therefore, we contribute to the comoarison between PP and DBTM and we elaborate the idea of htransforming a single non-DB task into several artificial DB transactions with the aim of increasing the degree of parallelism in PP, too.

Section 2 compares PP and DBTM in more detail. In order to ensure readability, in Section 3 we explain informally the main insights from the theory of schedulers working at several layers of abstraction ("stack schedules"). Finally, in Section 4, we show how this theory could be adopted for the parallel execution of numerical algorithms as an example of a problem that is considered a non-database task.

P. Amestoy et al. (Eds.): Euro-Par'99, LNCS 1685, pp. 459–468, 1999.

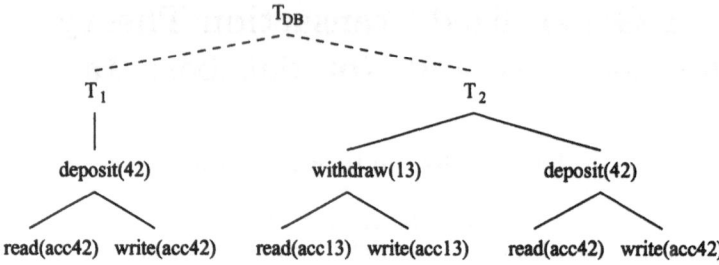

Fig. 1. Example for transaction invocations in a dbs

We are not aware of other work having considered comparisons between PP and DBTM or having applied multi-level transactions to non-DB tasks. In our own previous work [1] we have presented this idea the first time. The paper here gives a more detailed presentation: It contains a comparison between PP dependences and DBTM conflicts and elaborates on the applicability of the DBTM theory using several simple examples from matrix/vector computations. More details are available in [4].

2   Parallelization in Databases and in Parallel Programming

2.1   High-Level Comparison

Let us use simple examples for an overview of database transactions on the one hand, and numerical algorithms on the other hand.

First, we consider database transactions as those shown in figure 1. Traditionally, transactions (like T_1 and T_2) are given to the database system (DBS) – usually by different users. There, each transaction starts operations. Transaction T_1 starts a deposit operation on account number 42, which itself starts a read and a write operation. T_2 – a transfer transaction – starts an operation that withdraws some money from account number 13 and deposits it on account number 42. Each of these operations start a read and a write operation. Now, the scheduler of the DBS has to decide which operations can be executed in parallel, based on the information whether operations of different transactions commute. Since there are two layers of abstraction, there are also two possibilities for placing a scheduler: If it is placed at the upper layer, the scheduler decides on commutativity of the operations of T_1 and T_2. Deposit operations are always commutable. Withdraw commutes with each of the two deposit operations being on another account. If the scheduler is placed at the lower layer (this is the usual layer where DB schedulers work in practice), the scheduler decides on commutativity of the read and write operations as usually. In any case, the scheduler orders pairs of conflicting operations and makes sure that there exists always a sequential execution that orders conflict pairs the same way.

It seems to be a major point that usually transactions are independent and are entered by different clients. However there is nothing wrong if we assume that both transactions (e.g. T_1 and T_2 in our example) are entered by one and the same client, and still the scheduler will parallelize them considering the conflicts between their operations. We therefore can always think of an (artificial) transaction T_{DB} starting all client transactions like T_1 and T_2 in our example. The reason for this "trick" is that DB transactions look more similar to a numerical task that usually is entered by one client.

We now consider the essentials of parallel programs using a simple example. In figure 2 there is an invocation hierarchy of the numerical task T that calculates the sum of two matrices and a matrix vector product as follows:

```
B = A + B;
u = B v;
```

T invokes subprograms (possibly a simple loop) that calculate logical units of work – the sum $B = A + B$ and the matrix-vector product $u = Bv$. The subprograms invoke operations calculating the single elements of B and u. The main point here is that the algorithm T itself is similar to T_{DB} in DBTM. As in DBTM, T invokes subtasks that, in turn, invoke operations. In PP we distinguish two cases: Either the programmer decides whether operations can be executed in parallel using PP languages [3], or a parallelizing compiler (also called super-compiler [7]) transforms a given sequential program into a PP. The programming constructs and outcomes of both methods are the same, so we do not discriminate them in the following. In High Performance Fortran there are two types of parallel loops, FORALL and INDEPENDENT DO, in the example:

```
INDEPENDENT DO i=1, m
    DO j=1, n
        b(i,j) = a(i,j) + b(i,j)
        u(i) += b(i,j) * v(j)
    END DO
END DO
```

The outer loop is a parallel loop which executes its code independently for every i (the row index of B). The inner loop is a sequential loop over the column index j of B. On a high level, we now see that PP and DBTM are very similar. In both areas, a task is divided into subtasks, which possibly start subtasks themselves. But how does the parallel compiler decide which tasks can be parallelized and which not? Does PP have a notion similar to conflict pairs?

2.2   Dependences in Parallel Programs

A parallelizing compiler realizes what (sub)tasks can be executed concurrently by looking at data dependences. It ensures that all dependences of the (sequential) input program are kept in the produced parallel program. Three kinds of dependences are distinguished: flow-, anti-, and output dependence. These are easy to understand by observing that they are essentially the same as the three kinds of conflicts in transaction processing of the read/write model of DBTM.

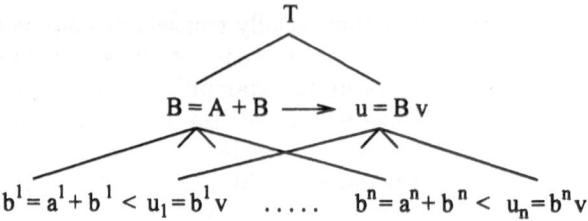

Fig. 2. Interleaved execution of ordered operations within a task

As a thorough analysis proves, the following counterparts between DBTM and PP notions can be found:

Parallel Programming	Database Transactions
variable	data-object
$x = y + z$	r(y) r(z) w(x)
flow dependence	wr-conflict
anti dependence	rw-conflict
output dependence	ww-conflict

In essence, the statement $x = y + z$ in PP is nothing else than reading y and z, adding them, and writing the result into the object x in DBTM terminology. A flow dependence is a type of a data dependence, when a variable is set and then read, in the example $x = 1; y = x;$. Anti dependence is the case, when a variable is read and then set, e.g., $y = x; x = 1;$. Output dependence appears, whenever a variable is set by two operations, e.g., $x = 10; x = 3;$. The order cannot be reversed without changing the result. Obviously, the three cases correspond to a write/read-, read/write-, and write/write-conflict in DBTM, resp.

Traditionally, in both PP and in DBTM, there is only one level, on which tasks are scheduled. As a major departure from this one-layer-scheduling, in DBTM it came out that parallelism can be gained by introducing additional layers of abstraction and allowing a scheduler at every level of abstraction. In our simple introductory example a higher-layer scheduler will not order any pair because there is no conflict between the operations: The two deposits commute since deposits always commute, and deposit(42) commutes with withdraw(13) because it is another account. However, if the intermediate level of the deposit and withdraw operations would be eliminated to achieve a traditional one-layer scheduler, the execution in the figure would not be correct any more. In the following we will informally explain this essential extension to DBTM theory, the formal definitions of which can be found in [1].

3   Stack Schedules

When executing transactions, a scheduler restricts parallelism because it must, first, observe the order constraints between the operations of each transaction

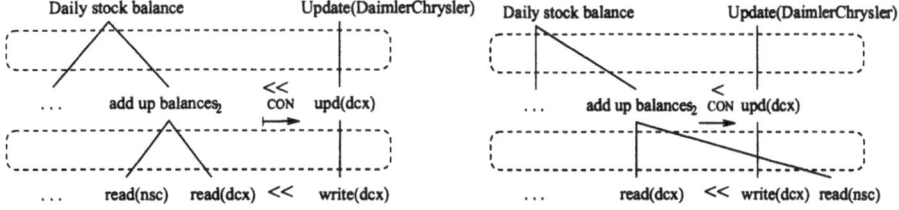

Fig. 3. Restr. parallelism (strong order)   **Fig. 4.** Restr. parallelism (weak order)

and, second, impose order constraints between conflicting operations of different transactions. The restriction in parallelism occurs because, in a conventional scheduler, ordered operations are executed sequentially. As shown in [1] and as we explained just above, this is too restrictive. But parallelizing conflicting operations over several levels requires to relax some of the ordering requirements of traditional schedulers. In addition, a mechanism is needed to specify to a scheduler what is a correct execution from the point of view of the invoking scheduler. For these two purposes we use the notion of *weak* and *strong* orders:

Strong and Weak Order: A and B are any tasks (actions, transactions).

- Sequential (strong) order: $A \ll B$, A has to complete before B starts.
- Unrestricted parallel execution: $A \| B$, A and B can execute concurrently equivalent to any order, i.e., $A \ll B$ or $B \ll A$.
- Restricted parallel (weak) order: $A < B$, A and B can be executed concurrently but equivalent to executing $A \ll B$. $\quad\square$

To have a closer look at this concept, let us consider an example. In figure 3 there are two schedulers, one above the other. On the higher level there are two transactions: One calculates the daily stock balance by invoking subtransactions that calculate certain types of stocks each, possibly on different servers. Here, it suffices to consider only one of these subtransactions, add up balances$_2$, which reads the prices of two stocks, DaimlerChrysler, dcx, and netscape, nsc.

The other transaction updates the price of the DaimlerChrysler stock. The upper schedule knows that these two subtransactions, add up balances$_2$ and upd(dcx), are in conflict. In the case of a conflict, traditional theories sequentialize the regarding operations, which we call a strong order. Because of this strong order, all operations of add up balances$_2$ are executed before the operation of upd(dcx). Here is the main point that we lose parallelism as read(nsc) is strongly ordered before write(dcx).

To prevent a loss of parallelism we make now use of the weak order (figure 4). Conflicting operations are ordered, but only weakly. This weak order is a constraint for the lower-level scheduler: Its serialisation graph has to follow the given weak order. Therefore write(dcx) can be executed in parallel to read(nsc) or in any arder as long as write(dcx) is executed after read(dcx).

3.1   Transactions, Schedules

The distinction between weak and strong order require slight changes in the definitions of traditional notions. A **transaction** is a set of operations with partial strong AND weak order constraints between the operations. A **schedule** receives as input transactions and a weak and a strong input order. The output of a schedule consists of the set of all operations of all input transactions and and of the weak and strong (output) orders. Commutativity of operations is globally given by a conflict predicate. A schedule must weakly order every pair of conflicting operations from different transactions without contradicting the weak input order. Furthermore, all weak and strong transaction orders are contained in the weak and strong output orders, respectively. Naturally, strong input order are propagated from the transactions to their operations, thereby separating the execution tree of strongly ordered transactions.

3.2   Correct Schedule

The distinction between the two orderings requires also to modify the traditional notion of correctness. We will assume, as usual, that a transaction executed in isolation is correct.

A schedule S is correct or conflict consistent (CC), *if there is a serial schedule, whose strong input and weak output order contain the weak input and output order of S, resp.*

The extension to the traditional theory may become clearer when we use the serialisation graph defined as in the classical theory. We have shown (proof see [2]):

A schedule is conflict consistent iff the union of its weak input order and its serialisation graph is acyclic.

3.3   Stack Conflict Consistency

With these preparations we are able to explain **stack schedules**: in a stack of schedules the output of one schedule is directly used as input to the next. Every level has a scheduler S that provides a set $\hat{O}_S$ of operation invocations to be used to build transactions. I.e., an operation of a scheduler can be a transaction of the next lower level scheduler. Every scheduler S has a commutativity specification expressed by the conflict predicate CON_S. Every scheduler works locally ensuring correctness with respect to its (local) CON_S. Fortunately we can prove [2] the following **theorem**:

An n-level stack schedule SS is correct, iff each individual schedule S_i in SS is conflict consistent, for $1 \leq i \leq n$.

We will use SCC as a shorthand for correctness of a stack schedule. Thus, every scheduler, except the lowest one, produces an executable plan for the next lower scheduler. The lowest scheduler, however, has to execute the operations it produces, thereby changing all weak output orders into strong ones. If higher-level schedulers make this change, the produced output is correct, too, but the achievable degree of parallelism is reduced.

In figure 4, the lower level schedule is conflict consistent, since there is a corresponding serial schedule, which is the lower level schedule in figure 3.

4   Parallelization of Numerical Algorithms

4.1   Application to the Introductory Example

SCC provides a larger class of correct executions, and introduces a possibility for parallelism which was not possible neither in previous DBTM nor in PP. This comes from perhaps the most interesting aspect of SCC: the possibility of executing operations in parallel, even if they conflict. This raises the possibility of implementing a *parallel-do* operation. By parallel-do we mean that the operations specified are executed in parallel, while preserving imposed serialisation dependences between them. In the example shown in figure 2, the programmer will specify something like:

$Begin\_Parallel\_Do$
$\quad (B = A + B) \;\rightarrow\; (u = Bv)$
$End\_Parallel\_Do$

thereby indicating that they should be executed in parallel while preserving the given ordering. Now, is the execution shown in the figure correct? Obviously, the upper level schedule is CC. In the lower schedule, every conflict pair $(b^i = a^i + b^i, u_i = b^i v)$ is ordered in the same way. So, there is a serialisation graph edge from $B = A + B$ to $u = Bv$, consistent with the weak input order. Thus, the lower schedule is CC, and hence, the stack is SCC and correct.

4.2   Application to a Complex Example

Traditional Strategies We move to a more complex example and show what we can gain by the approach. Consider the first step of the Gauss-algorithm for $n = 3$. This requires $A = C_1 A,\; A = C_2 A$ with:

$$A = \begin{pmatrix} a^1 \\ a^2 \\ a^3 \end{pmatrix}, \quad C_1 = \begin{pmatrix} 1 & 0 & 0 \\ -\frac{a_{21}}{a_{11}} & 1 & 0 \\ -\frac{a_{31}}{a_{11}} & 0 & 1 \end{pmatrix} = \begin{pmatrix} c_1^1 \\ c_1^2 \\ c_1^3 \end{pmatrix}, \quad C_2 = \begin{pmatrix} 1 & 0 & 0 \\ 0 & 1 & 0 \\ 0 & -\frac{a_{32}}{a_{22}} & 1 \end{pmatrix} = \begin{pmatrix} c_2^1 \\ c_2^2 \\ c_2^3 \end{pmatrix}$$

In parallel programming and parallelizing compilers, this can be parallelized using two parallel and one sequential loops:

```
DO j=1, n-1
    INDEPENDENT DO i=j+1, n
        l(i,j) = a(i,j)/a(j,j)
        INDEPENDENT DO k=j, n
            a(i,k) = a(i,k) - l(i,k) * a(j,k)
        END DO
    END DO
END DO
```

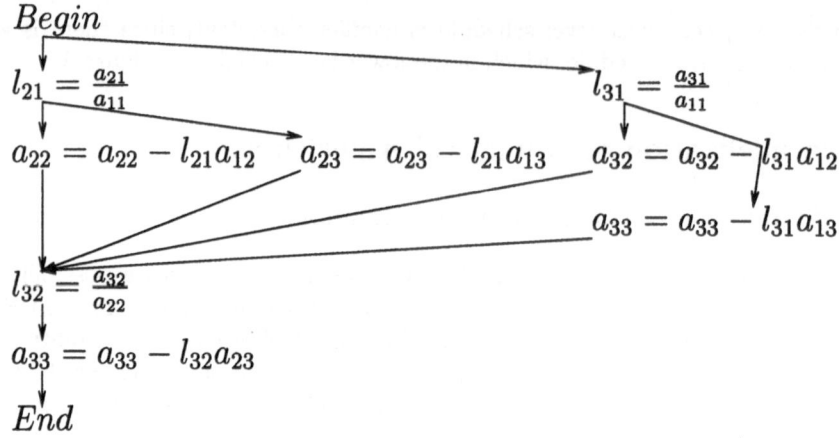

$Begin$

$l_{21} = \frac{a_{21}}{a_{11}}$

$a_{22} = a_{22} - l_{21}a_{12}$ $a_{23} = a_{23} - l_{21}a_{13}$

$l_{31} = \frac{a_{31}}{a_{11}}$

$a_{32} = a_{32} - l_{31}a_{12}$

$a_{33} = a_{33} - l_{31}a_{13}$

$l_{32} = \frac{a_{32}}{a_{22}}$

$a_{33} = a_{33} - l_{32}a_{23}$

End

Fig. 5. Parallelization of a nested loop

However, the dependence graph for this code fragment (figure 5) shows that there are unnecessary operation orders. It does not show the data dependences, but the control dependences of the parallel loop, i.e., the data dependences are coded with the control dependences. As parallelizing compilers can only parallelize loops as a whole, the above outer loop is a sequential one. Therefore, the operation $l_{32} = \frac{a_{32}}{a_{22}}$ is executed after $a_{33} = a_{33} - l_{31}a_{13}$, although there is no data dependence between these two operations. If we make the simplifying assumption that every operation is executed in one time step, and three processors are available, these unnecessary orderings force a total computation of five steps.

Parallelizing Using Stack Conflict Consistency The idea consists in introducing compilers on several layers by decomposing a numerical algorithm into steps and considering them as transactions that run concurrently, observing some given external partial order. Each step, in turn, is decomposed into simpler steps that are considered as operations in the DBTM terminology.

Let us consider what exactly has to be done to be able to apply stack conflict consistency to our example. First, the whole computation is defined as a single transaction T consisting of operations $T_1 : A = C_1 A$ and $T_2 : A = C_2 A$, which is the algorithm in matrix form (figure 6[1]). Since order matters between T_1 and T_2, they are weakly ordered. The interesting part is on the lower scheduler. There, not all operation pairs from different (weakly ordered) transactions are conflicting. Therefore, some of those operations can be executed in parallel: E.g., T_{123} can be executed concurrently with T_{211}, although their parents T_{12} and T_{21} are (weakly) ordered. Since this order is only weak, it does not matter if non-conflicting operations are reversed. This type of parallelism does not exist in conventional PP.

---

[1] To make the presentation clearer, only (weak) input orders are shown

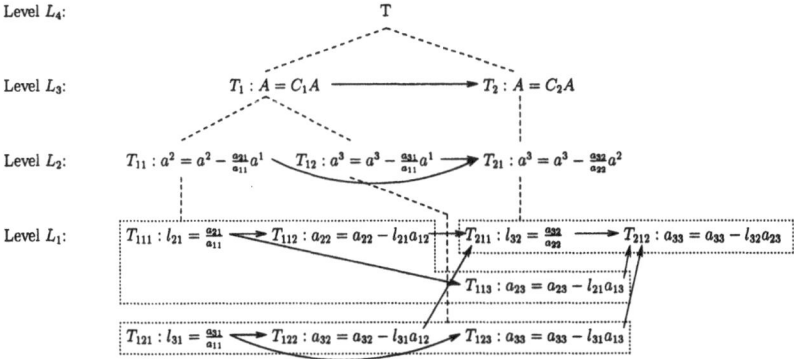

Fig. 6. Interleaved execution of ordered steps of a parallel program.

Now, is the schedule shown in figure 6 stack conflict consistent? We only have to prove whether every schedule of it is CC. The upmost schedule (level $L_4 - L_3$) certainly is conflict consistent, as there is only one transaction. The next schedule (level $L_3 - L_2$) consists of two transactions, T_1 and T_2, connected by a weak input order. Since this input order is consistent with the serialisation graph (T_{11} and T_{12} before T_{21}, resp.), this schedule is also correct. The lowest schedule (level L_2-L_1) is the most complicated. There are two weak input orders (from transactions T_{11} and T_{12} to T_{21}, resp.). As one can prove, all conflicting operation pairs are weakly output ordered (we used here the same symbol as for the input order). All these weak output orders, however, are according to the weak input orders of their parents. For example, T_{112} is ordered before T_{211}, which is consistent with $T_{11} \rightarrow T_{21}$, the weak input order of their parents.

By summing up this example, we find that a stack schedule is similar to conventional parallelizing compilers on several layers of abstraction. Such compilers parallelize loops as a whole by determining if there are any data dependences contradicting the parallelization of this loop. We propose to have several layers of compilers working together like a stack schedule; a higher-level compiler gets the complete (sequential) source code and produces an execution plan, which is the input for next lower schedule that, in turn, produces an execution plan for the next scheduler. Only one scheduler of a stack, the lowest one, is actually executing its operations, by obeying the plan it produces and by only using strong output orders. In order to define parallel loops, every scheduler can make use of a new sort of loop: The *Parallel-Do with a weak order.*

What do we gain by applying stack scheduling to this example? In the last subsection, it came out that a parallelizing compiler would need five time steps to execute this program, given that three processors are available. As we can see in figure 6 only four steps are now necessary to execute all operations of level L_1 (which is the only level on which real computations are done)[2]. It is

---

[2] Note that the time axis is from left to right, as usual in DBTM.

expected that the profit of one time step in this scenario increases in the number of dimensions of the involved matrices, and in the complexity of the algorithm.

4.3 Other Approaches

There is a number of previous approaches to coping with multilevel schedules, the most advanced of which is Weikum's model. Weikum proposes a weaker condition than order preservation for independent scheduling in composite systems. This condition forces conflicting operations at a non-leaf level to have conflicting descendants at all lower levels (axiom 1 in [5]). This restriction, though often natural, restricts the scope of Weikum's model. E.g., it does not hold for multiversion CC algorithms, or in our context the addition of sparse matrices [4].

In level-by-level serialisability [5], when two operations conflict (that is, they are ordered by the weak execution order), they must also be strongly ordered. Consequently, the execution trees of conflicting operations cannot be interleaved. This is not the case for stack conflict consistency.

A rule-based approach to (partly) nested transactions are ULTRA transactions [6]. Similar to parallelizing compilers, operations are collected in an evaluation phase, and performed in a subsequent materialization phase. The ULTRA system executes basic updates simultaneously with the logical evaluation of the transactions. This optimistic method requires the possibility of compensation.

5   Acknowledgement

We would like to thank Gustavo Alonso for many fruitful discussions and the anonymous referees for their valuable comments.

References

[1] G. Alonso, S. Blott, A. Fessler, and H.-J. Schek. Correctness and parallelism in composite systems. In *Proc. of the 16th Symp. on Principles of Database Systems (PODS'97)*, Tucson, Arizona, May 1997.

[2] G. Alonso, A. Feßler, G. Pardon, and H.-J. Schek. Transactions in stack, fork, and join composite systems. In *7th International Conference on Database Theory (ICDT)*, Jerusalem, Israel, Jan. 1999.

[3] S. Brawer. *Introduction to Parallel Programming*. Academic Press, 1989.

[4] A. Feßler. *Eine verallgemeinerte Transaktionstheorie für Datenbank- und Nichtdatenbankaufgaben (to appear)*. Dissertation, Department of Computer Science, ETH Zürich, 1999.

[5] G. Weikum. Principles and Realisation Strategies of Multilevel Transaction Management. *ACM Transactions on Database Systems*, 16(1):132, March 1991.

[6] C.-A. Wichert, A. Fent, and B. Freitag. How to execute ULTRA transactions. Technical Report MIP-9812, Universität Passau (FMI), 1998.

[7] H. Zima and B. Chapman. *Supercompilers for Parallel and Vector Computers*. Addison-Wesley, 1991.

On Disk Allocation of Intermediate Query Results in Parallel Database Systems

Holger Märtens

Universität Leipzig, Institut für Informatik, Postfach 920, D–04009 Leipzig, Germany
maertens@informatik.uni-leipzig.de

Abstract. For complex queries in parallel database systems, substantial amounts of data must be redistributed between operators executed on different processing nodes. Frequently, such intermediate results cannot be held in main memory and must be stored on disk. To limit the ensuing performance penalty, a data allocation must be found that supports parallel I/O to the greatest possible extent.

In this paper, we propose declustering even self-contained units of temporary data processed in a single operation (such as individual buckets of parallel hash joins) across multiple disks. Using a suitable analytical model, we find that the improvement of parallel I/O outweighs the penalty of increased fragmentation.

1 Introduction

In parallel database systems used for advanced applications like data warehousing, complex queries are performed on very large data sets, often in terabyte ranges. Parallel operators executed on different processing nodes exchange substantial amounts of intermediate results, and when the processors' memory capacity is exceeded, temporary data must be stored on disk. The response time problems caused by slow disk access are alleviated by parallel I/O, often using more disks than processors to avoid bottlenecks.

In a shared-disk architecture, intermediate results can be written out by the sender nodes and read directly by the receivers. Such *disk-based data transfer* is convenient and reduces the overhead of communication between processors. But depending on the operators' access patterns, a smart disk allocation is required to limit disk contention.

In most algorithms, data fragments are stored on many disks, but each fragment is kept on a single device. Thus, when a receiver processes its fragments sequentially, it can read from just one disk at a time and parallel I/O is not fully exploited. In this article, we propose declustering individual data fragments across multiple disks to increase the performance of parallel database systems for complex queries on large amounts of data. We develop an appropriate analytical model to show that the benefits of parallel I/O for the receiving operator usually outweigh the additional disk load due to increased fragmentation. Our approach works for several operators and most system architectures.

Our paper is structured as follows: Sect. 2 describes the processing model of a parallel hash join in a shared-disk system, which serves as a case study throughout the text. Sect. 3 is devoted to finding the optimal degree of declustering and includes our analytical model. In Sect. 4, we outline possible extensions of our method to different operators and architectures. Related work is discussed in Sect. 5, and we conclude in Sect. 6.

P. Amestoy et al. (Eds.): Euro-Par '99, LNCS 1685, pp. 469–476, 1999.

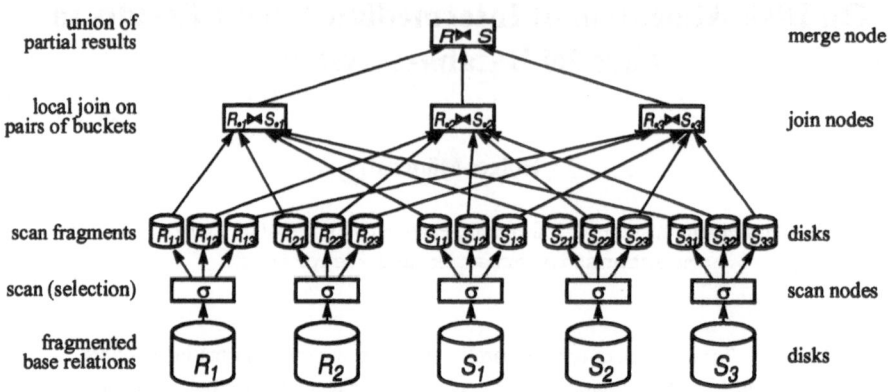

Fig. 1. Processing model of a parallel hash join in a shared-disk architecture

2 Parallel Hash Joins in Shared-Disk Architectures

For a clear presentation, we restrict most of this paper to a concrete case: a parallel two-way hash equi-join in a shared-disk environment. We now describe the basic processing model before we give some heuristics for parameter selection and disk allocation.

2.1 Processing Model

Let R and S be the inner and outer relations of a join query, declustered across r and s disks, respectively. In the *scan phase*, R is read by n scan nodes which apply a selection and partition their output into b buckets. If the scan result is very large, the buckets cannot be held in main memory and are stored on d disks. S is processed similarly, possibly by a different number of scan nodes but with a corresponding partitioning of buckets.

In the *join phase*, m join nodes each process one bucket pair at a time, using hash joins in which a hash table is built from an R-bucket and the matching S-bucket is probed against it. The local results are merged at a specified processor. This model is illustrated in Fig. 1. In the shared-disk environment we assume, the allocation of buckets to processors can be chosen dynamically to balance the workload.

For large data sets, each join node processes several bucket pairs and each disk must hold several buckets. Also, any scan node can contribute to any bucket, creating $b \cdot n$ *scan fragments*. Consequently, disk contention between processors occurs. To limit contention while supporting parallel I/O, buckets must be properly allocated to disks. We introduce the parameter v, denoting the degree of bucket declustering. With each bucket split across multiple disks, parallel reading is enabled. Assuming the same degree of declustering for all buckets, $b \cdot v$ *bucket fragments* are stored on disk.

Thus, the parameters n, m, d, b, and v must be found for a two-way equi-join query.

2.2 Selection of Basic Parameters and Disk Allocation

Our main concern is finding an optimal degree of declustering (v), which we discuss in detail in Sect. 3. Before that, we provide a simple heuristic for the remaining parameters and a disk allocation scheme as a basis for further calculations.

scan nodes

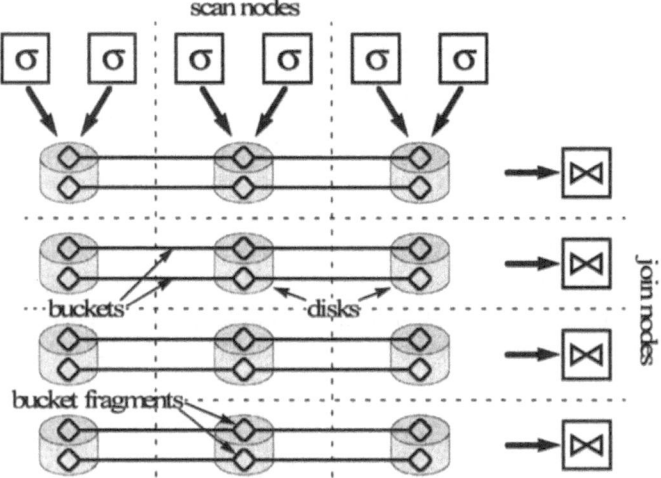

Fig. 2. Example of the processing model and the allocation scheme. Eighty buckets (not all shown) are processed using six scan nodes and four join nodes. The buckets are declustered across twelve disks with a degree of three. To minimize access conflicts, each disk is used by just two scan nodes and one join node. The parameters from Sect. 2.1 are set as follows:

$$n = 6 \qquad m = 4 \qquad d = 12 \qquad b = 80 \qquad v = 3$$

The numbers of processors and disks—n, m, and d—should be set so as to match their processing rates as closely as possible, starting from the degree of declustering r of relation R. These calculations involve some information on system performance (preferably reflecting the current load state) as well as the selectivity of the scan, which is estimated by histograms or sampling. The number of buckets, b, should be large enough to fit each bucket into a single node's available memory (which may be just a fraction of physical memory). When data skew occurs, bucket sizes will vary and b must be selected high enough for the largest bucket to meet the memory restriction.

Example. Let base relation R be scanned by six processors. If each node's output rate is sufficient to keep two disks busy, twelve disks are used to store the buckets. If, in the join phase, one node can process the data delivered by three disks, only four join nodes are required. Assuming a total scan result of 800 MB with 10 MB of memory available on each of the selected processing nodes, 80 buckets are created (no skew). This example is depicted in Fig. 2; the allocation and declustering applied are justified below. □

For a given selection of all five parameters, the following allocation scheme (also exemplified in Fig. 2) can be shown to yield the smallest number of processors accessing the same disks, thus minimizing disk contention. As in most studies, we assume integer proportions between some parameters for simplicity.

1. Arrange the d disks into a matrix of v columns and d/v rows (one disk per cell).
2. Assign $b \cdot v/d$ buckets to each row, declustering every bucket across v disks.
3. Assign n/v scan nodes to each column; make them write to the d/v disks there.
4. Assign $m \cdot v/d$ join nodes to each row; let them read from the v disks in that row.

3 Determining the Degree of Declustering

In this section, which constitutes the core of our study, we trade off parallel I/O against disk contention to determine the optimal degree of bucket declustering, v. We define and analyze basic indicators of processing performance before introducing our analytical model that leads to the final solution. All of these considerations are based on the above allocation scheme. Details omitted due to space restrictions are discussed in [5].

3.1 Performance Indicators

Since disks are normally shared by several nodes, the available degree of I/O parallelism cannot be measured simply by the number of disks a processor can access at a time. Rather, it must be interpreted as the number of disks available divided by the number of nodes accessing them. With each scan node writing to d/v disks and a disk shared by n/v nodes, the available degree of *write parallelism* is $wp = d/n$, which is independent of v. In the join phase, each node reads from v disks to assemble its current bucket. It may have exclusive access to these if $m \cdot v \le d$; otherwise, a disk is shared by $m \cdot v/d$ processors. Thus, the degree of *read parallelism* is $rp = min(v, d/m)$.

Disk contention can be defined by the number of concurrent read and write operations per disk. With k such operations running at a given time, the disk read-write head will have to move between k different positions, and the resulting seek times constitute the allocation-dependent share of I/O cost. In the scan phase, k is the number of bucket fragments per disk. Thus, *write contention* is measured as $wc = b \cdot v/d$. During the join, a disk is accessed by only one node if $m \cdot v \le d$ or shared by $m \cdot v/d$ processors otherwise (as mentioned above). With each node reading from a single position, *read contention* is $rc = max(1, m \cdot v/d)$.

In the example from Sect. 2.2 and Fig. 2, the performance indicators have the following values: $wp = 2$, $rp = 3$, $wc = 20$, $rc = 1$.

Observations. Although these coefficients are not proportional to either performance or response times (their precise effects are analyzed in Sect. 3.2), we can make some general observations: While both write and read contention are best avoided for low values of v, higher degrees of bucket declustering are useful to support read parallelism. Write parallelism, however, is constant; thus, we need not regard wp any further.

Let us examine three common-sense settings of v, viz.: no declustering ($v = 1$), full declustering ($v = d$), and *read-optimal* declustering ($v = d/m$). The latter is so named because it just allows full read parallelism without introducing read contention. As can be inferred from Table 1, the optimal degree of declustering must be between 1 and d/m. In this interval, there is a true trade-off between parallelism and contention. For $v \ge d/m$, however, contention is increased without further gains in parallelism. To find the true optimum within the range of $v \in [1, d/m]$, we have devised an elaborate analytical model which is presented in the next section.

3.2 Analytical Model

We construct a cost function to capture the total disk response time for the I/O of the join buckets. $T = T_w + T_r$ comprises writing and reading in the scan and join phase, respectively. Let p be the total number of pages (or other suitable, uniform I/O *gran-*

Table 1. Development of performance indicators for different degrees of declustering

	contention		parallelism	
declustering	*write*	*read*	*write*	*read*
none ($v = 1$)	low	none		low
read-optimal ($v = d/m$)	medium	none	constant	high
full ($v = d$)	high	high		high

ules) to be written into the buckets. If all d disks are busy all the time (neglecting skew), then $T_w = p/d \cdot t_w$, where t_w is the average time for a single write operation, estimated as $t_w = 1/k_w \cdot t_s + (1 - 1/k_w) \cdot t_l$.

Here, k_w denotes the number of disk access positions used at the time of writing. There is a probability of $1/k_w$ that the disk head need *not* be moved because it is already in the right position from the previous access. This case causes a "short" disk access (t_s); otherwise, a "long" access (t_l), including a track seek operation, occurs. This is the most important distinction to make when modeling disk activity because seek times are known to dominate disk response times [11]. Note that average values of t_s and t_l are quite sufficient for our purposes since we are interested in the overall sum of access times only. Defining $t_\Delta = t_l - t_s$, we can now simplify $t_w = t_l - t_\Delta/k_w$.

The number of access positions on each disk, k_w, corresponds to the number of bucket fragments per disk plus an adequate number of entry points for concurrent queries in multi-user mode, x. The term x can be composed of arbitrary sub-terms; we are only interested in its average total magnitude. While this model of multi-user mode may seem simplistic, we will see later that it is quite sufficient. For now, $k_w = b \cdot v/d + x$.

Note that our formula does not include waiting times caused by write requests not being served immediately. Rather, we assume asynchronous access so that processing can continue while data is (queuing to be) written. We further presume that the disks are kept busy but are not overloaded; this assumption is justified because we specifically selected the ratio of disks and processing nodes so as to match their processing rates (cf. Sect. 2.2). Thus, our model need only capture the actual disk access times.

For read operations in the join phase, we can assume only $m \cdot v$ disks to be used at the same time (each of the m join nodes assembles its current bucket from v disks). Note that $m \cdot v$ cannot exceed d because we have limited v to a maximum of d/m in the previous section. Now, we can define $T_r = p/(m \cdot v) \cdot t_r$ with $t_r = t_l - t_\Delta/k_r$, similar to the scan phase. The number of access positions, however, is lower now because we can exclude contention within the current join: $k_r = 1 + x$.

We assume x to have the same value as in the scan phase to represent the same degree of inter-query contention. After some more transformations, we can write the complete cost formula as a function of v:

$$T(v) = \frac{1}{v} \cdot \left(\frac{p \cdot t_l}{m} - \frac{p \cdot t_\Delta}{m + mx} \right) - \frac{1}{bv + d \cdot x} \cdot p \cdot t_\Delta + \frac{p \cdot t_l}{d} . \qquad (1)$$

To find its minimum within the bounds of $v \in [1, d/m]$, we discern several cases.

Single-User Mode. With $x = 0$ to represent single-user mode, T simplifies to

$$T(v) = \frac{p}{v} \cdot \left(\frac{t_s}{m} - \frac{t_\Delta}{b} \right) + \frac{p \cdot t_l}{d}.$$ (2)

The properties of this function depend on the relationship of t_s/m to t_Δ/b or, rearranging the terms, of b/m to t_Δ/t_s. If both should happen to be equal—in other words: if the number of buckets per join node corresponds to the ratio of disk seek time and short access time—the function is constant and all values of v are equivalent.

If b/m is greater (many buckets), total I/O cost strictly decreases with v because the performance gains from parallel reading outweigh the losses due to write contention. In this case, v should be selected as large as possible, i. e. $v = d/m$. If b/m is less than t_Δ/t_s (few buckets), the opposite applies and disk contention dominates. Now, a small value of v is appropriate, i. e. $v = 1$.

Multi-User Mode. In multi-user mode, the cost function cannot be simplified, and lengthy calculations ensue. However, it can be shown that $T(v)$ is strictly decreasing if

$$\frac{b}{m} \geq \frac{t_\Delta}{t_l - t_\Delta/(1+x)}.$$ (3)

This condition is true for most sensible parameters (i. e. $b/m \geq 2$ and $x \geq 1$). In other words: Unless we are "almost" in single-user mode ($x < 1$, meaning that there is just one competing operation per disk at any given time), or we process just one bucket per join node, we should decluster the buckets with a degree of $v = d/m$.

If no such property can be ensured, more case distinctions are required. We found that for all cases of $x \geq 1$, the degree of declustering should be set to d/m. For some very small values of x, declustering should be avoided. This corresponds to "near"-single-user mode with few buckets per join node as above. There is only a very narrow margin of values of x for which the optimum of v is within the interval $[1, d/m]$.

Analysis. The results for both single- and multi-user mode can be interpreted as follows: For a high number of buckets, disk contention in the scan phase is already severe because there are many fragments on each disk, causing a very low probability of "short" write times. Thus, further increasing k_w through declustering has little effect on the scan phase while the join phase is sped up considerably through parallel reading. This is true even when inter-query contention affects both phases. This result also justifies our choosing a simple coefficient like x: It is unnecessary to use a more complex term that will still exceed the boundary of 1 in any true multi-user system.

With few buckets, there is still a significant share of short write operations that are destroyed by declustering, outweighing the performance gain during the join. Note that the number of data pages, p, does not directly influence the number of seek positions although b usually increases with p. Also, the ratio t_Δ/t_s varies with the size of the read/write granule; the larger the granule, the more useful bucket declustering will be.

Summary. Looking for an optimal degree of bucket declustering, we found that in all practical cases, the read-optimal setting from Sect. 3.1 is favorable. The only notable exception is for small numbers of buckets in single-user mode; in this case, declustering should be avoided. Medium degrees of declustering are not useful to consider.

4 Extensions

To account for the second relation, S, the computations of n, d, and m work as in Sect. 2.2. However, the declustering of base relations they start from, r, must now be replaced with $r + s$, $max(r, s)$, or the like, depending on whether R and S are stored on disjunct disks. If R and S result from sub-queries, their processing rates must be used. For the number of bucket pairs, b, the previous heuristics must be applied to the *inner* relation (usually the smaller one). To find an appropriate value of v, we can also use the previous rules but have to interpret v differently, e. g., $v_R = v_S = v$ (for separate scanning as in standard hash joins) or $v_R + v_S = v$ (for simultaneous scanning as in sort-merge joins).

Our approach is not restricted to two-way equi-joins. In principle, it is applicable to all blocking operators that exchange large amounts of data (exceeding main memory) in a many-to-many relationship. Possible applications include non-equi-joins, distribution sort (cf. Sect. 5), and several types of aggregation, especially when combined with group-by clauses. Some adaptations of the allocation scheme may be required for operators with different access patterns.

Our approach assumes that every processor can access any disk. Thus, our method can be used in shared-disk and shared-everything systems, some hybrid architectures, or certain variants of NUMA. It can even be adapted to shared-nothing architectures by transferring the data through the network and having the receivers write the buckets back to their local disks as above, provided that each node owns at least d/m disks. However, shared-nothing architectures are less flexible in dynamic task allocation, complicating load balancing and/or causing a higher communication overhead [4, 9].

5 Related Work

While parallel I/O in general is naturally applied in parallel database systems, declustering of single data units such as join buckets has received little attention. Of the operators we mentioned, aggregation and grouping have not been associated with this idea.

In the context of joins, most load balancing studies have focused on CPUs and main memory [1, 4, 8, 10, 13]. While the significance of I/O has been asserted, only its overall reduction has actually been addressed [10]; declustering is either not performed or not discussed. For shared-nothing architectures, *bucket spreading* (full declustering) was introduced to equalize skew effects [3], but optimizing I/O was not a primary goal.

Mergesort algorithms naturally provide for parallel reading; in addition, workfiles may be striped across disks. Full striping is indeed found useful [14] especially in multi-user mode if the striping unit corresponds to the read granule; workload is balanced across disks by randomization. These results, however, cannot be easily generalized due to the particular access patterns of the mergesort operator. For distribution sort, which is more similar to joins and aggregation, full striping of single files is used to achieve parallel I/O [7]. Again, optimal declustering is not an aim.

Disk arrays automatically provide parallel I/O. But even though allocation strategies have been developed for various applications [6, 12], disk arrays cannot address the particular allocation requirements of different algorithms. Specifically, independent join buckets are best stored on disjunct devices to allow reading them without contention; automatic (possibly full) striping in disk arrays usually defeats this goal [2].

6 Conclusion

In this paper, we have investigated ways of allocating intermediate results of large database queries across the disks of a parallel system. Based on a well-founded analytical model for the sample case of join queries, we concluded that in most cases, it is useful to decluster even individual join buckets across several disks to enable parallel reading in the subsequent query stage. The benefits of parallelism usually outweigh the penalty of disk contention. The optimal degree of declustering is such that the receiving processors can keep all disks busy without introducing intra-query contention.

Our results are applicable to several different operators and largely independent of the underlying system architecture. To the best of our knowledge, this is the first study that has considered bucket declustering in such a general context. In the future, we plan to validate our results by simulation studies for various architectures and workloads.

Acknowledgment. The author would like to thank Dr. Dieter Sosna for his help in handling the cost function used in Sect. 3.2.

References

1. DeWitt, D.J., Naughton, J.F., Schneider, D.A., Seshadri, S.: Practical Skew Handling in Parallel Joins. Proc. 18th VLDB Conference, Vancouver (1992) 27–40
2. Graefe, G.: Query Evaluation Techniques for Large Databases. ACM Computing Surveys, Vol. 25, No. 2 (1993) 73–170
3. Kitsuregawa, M., Ogawa, Y.: Bucket Spreading Parallel Hash: A New, Robust, Parallel Hash Join Method for Data Skew in the Super Database Computer (SDC). Proc. 16th VLDB Conference, Brisbane (1990) 210–221
4. Märtens, H.: Skew-Insensitive Join Processing in Shared-Disk Database Systems. Proc. IDPT Conference, Vol. 2, Berlin (1998) 17–24
5. Märtens, H.: Disk Scheduling for Intermediate Results of Large Join Queries in Shared-Disk Parallel Database Systems. IfI-Report Nr. 9/98, Universität Leipzig (1998)
6. Merchant, A., Yu, P.S.: Analytic Modeling and Comparisons of Striping Strategies for Replicated Disk Arrays. IEEE Trans. Computers, Vol. 44, No. 3 (1995) 419–433
7. Nodine, M.H., Vitter, J.S.: Deterministic Distribution Sort in Shared and Distributed Memory Multiprocessors. Proc. 5th SPAA, Velen (1993) 120–129
8. Omiecinski, E.: Performance Analysis of a Load Balancing Hash-Join Algorithm for a Shared-Memory Multiprocessor. Proc. 17th VLDB Conference, Barcelona (1991) 375–385
9. Rahm, E.: Dynamic Load Balancing in Parallel Database Systems. Proc. Euro-Par '96 Conference, Lyon (1996) 37–52
10. Rahm, E., Marek, R.: Dynamic Multi-Resource Load Balancing in Parallel Database Systems. Proc. 21st VLDB Conference, Zürich (1995) 395–406
11. Ruemmler, C., Wilkes, J.: An introduction to disk drive modeling. IEEE Computer, Vol. 27, No. 3 (1994) 17–28
12. Scheuermann, P., Weikum, G., Zabback, P.: Data Partitioning and Load Balancing in Parallel Disk Systems. VLDB Journal, Vol. 7, No. 1 (1998) 48–66
13. Wolf, J.L., Dias, D.M., Yu, P.S., Turek, J.: New Algorithms for Parallelizing Relational Database Joins in the Presence of Data Skew. IEEE Trans. Knowl. Data Eng., Vol. 6, No. 6 (1994) 990–997
14. Wu, K.-L., Yu, P.S., Chung, J.-Y., Teng, J.Z.: A Performance Study of Workfile Disk Management for Concurrent Mergesorts in a Multiprocessor Database System. Proc. 21st VLDB Conference, Zürich (1995) 100–109

Highly Concurrent Locking in Shared Memory Database Systems

Christian Jacobi and Cédric Lichtenau

University of Saarland — Computer Science Department
PO Box 151150, 66041 Saarbruecken, Germany
{cj,cls}@cs.uni-sb.de,
http://www-wjp.cs.uni-sb.de/

Abstract. In parallel database systems, conflicts for accesses to objects are solved through object locking. In order to acquire and release locks, in the standard implementation of a lock manager small sections of the code may be executed only by a single thread. On massively parallel shared memory machines (SMM) the serialization of these critical sections leads to serious performance degradation. We eliminate the serialization by decomposing the complex database lock needed for granular locking into basic lock primitives. By doing so, we measured a speedup of a factor 200 on the SB-PRAM. Our method can be ported to any architecture supporting the used lock primitives, as most SMMs do.

1 Introduction

In parallel database systems (DBS), the access to objects (e.g., relation, page, record) must be regulated to ensure transaction isolation. This is done by locking objects before using them. Standard locking implementations rely on a complex database lock control structure. Small sections of the code accessing this structure must be serialized to prevent race conditions between concurrent threads. In massively parallel systems like the TERA MTA or the SB-PRAM running a very large number of threads (in the order of thousand), not only the avoidance of long lock duration becomes crucial. When many short transactions requests a lock on the same object, the serialization in the lock manager leads to heavy performance degradation even if virtually all lock requests are compatible. This situation arises frequently on relation granularity if intentional locks are used. We present a novel approach to the locking problem that prevents this unnecessary serialization. We replace the complex lock control structure by a combination of efficient lock primitives that are typically supported in SMMs. We show that this approach dramatically speeds up the lock manager on massively parallel computers. A speedup factor of 200 was measured on the SB-PRAM, a 64 processors SMM. Our implementation of the lock manager can easily be ported to other SMMs since it requires only standard locks available on most SMMs.

In the next section we introduce the shared memory model and basic lock primitives used later on. In section 3 we shortly describe multi-granular locking. In section 4 we outline the standard implementation of the database lock and

P. Amestoy et al. (Eds.): Euro-Par'99, LNCS 1685, pp. 477–481, 1999.
© Springer-Verlag Berlin Heidelberg 1999

its drawbacks. Section 5 presents our new approach to DBS locking. Finally we compare the performance of both implementations.

2   Hardware Model

In the shared memory model, an arbitrary large number of processors can concurrently access a shared memory. There exist several implementation of this model such as the NYU Ultracomputer [3], the SB-PRAM [2] and the TERA Multithreaded Architecture [1]. These computers have between 8 and 64 processors and are capable of running up to over thousand threads in parallel. These machines also feature hardware support for basic lock primitives in oder to efficiently enable parallel work and fast synchronization of so many threads. We introduce three basic locks that will be used later on:

Binary Lock At any point of time, only a single thread can hold this lock.
Group Lock An n-group lock, $n \in \mathbb{N}$, manages n groups. A thread may request a lock for a group i, $1 \leq i \leq n$. At any time, multiple locks for the same group may be granted, but never for two different groups (cf. Table 1).
Reader/Writer Lock (R/W-lock) The R/W-lock is a variation of the group lock. There are two groups, namely readers and writers. At any time, an R/W-lock can be helt either by multiple readers or by a single writer.

Our locking code was developed and benchmarked on the SB-PRAM, an PRAM emulation with 64 processors running 2048 threads built at University of Saarbruecken. The concurrent locking by an arbitrary number of threads on the SB-PRAM is executed in parallel without any kind of serialization, if the locks are compatible. Other systems like the NYU Ultracomputer or the TERA MTA have little serialization in the memory modules.

3   Locking in DBS

To ensure transaction isolation, virtually all database systems use locking. The basic lock modes are shared (S) and exclusive (X). An S lock on an object r, which is either a relation, a page, or a tupel, allows reading of r. An X lock allows reading and writing on r. To enable granular locking, intentional lock modes IX, IS and SIX are supported in addition to X and S locks. Before requesting an X, IX

requested lock	already granted lock		
	g1	g2	g3
g1	+	−	−
g2	−	+	−
g3	−	−	+

requested lock	already granted lock	
	read	write
read	+	−
write	−	−

Table 1. Three group lock and R/W-lock matrix. + compatible, − incompatible

or SIX lock on fine granuls, there must be IX or SIX locks set on coarser granules. Similarly, before requesting IS or S locks, the coarser granules must be locked IS or IX. Table 2 illustrates the compatibility matrix of the five lock modes X, S, IX, IS and SIX. A survey of isolation concepts is given in [4].

We define a **CC-lock** to be a data structure with operations cc-lock() and cc-unlock() that works as a lock according to this compatibility matrix. In the next sections we will describe two implementations of this CC-lock. The first implementation is well known, but suffers poor performance on massive parallel shared memory architectures. The second implementation eliminates this disadvantage. Both implementations acquire and release locks via operations on the shared memory. In contrast to most distributed database systems, no designated lock manager is used. Molesky and Ramamritham call this *autonomous locking* and showed its superior performance on SM systems [6].

4 Standard CC-Lock Implementation

On the standard implementation of the CC-lock, each CC-lock consists of two lists, one holding granted lock requests, the other one holding waiting lock requests. A new lock for mode φ is granted immediately iff φ is compatible to all granted locks and to all waiting lock request. The new lock request is added to the waiters list, otherwise.

On a lock release, waiting lock requests (if any) are checked one by one for compatibility with the active locks in the same order as they where appended to the waiters list (FIFO). Waiting locks get activated until the first incompatible waiting request is encountered. Activation can for example be done by interrupts or by polling of the waiting threads on a shared memory variable.

To the best of our knowledge, in all implementations following this basic concept, concurrent accesses of multiple threads to the same CC-lock get serialized by a binary lock. The SB-PRAM implementation of this CC-lock has a critical section of 60 instructions [8].

The SB-PRAM supports 2048 hardware threads, about 2000 threads are availabe for transaction processing, the others are used by the operating system. Suppose the 2000 threads concurrently request compatible locks on the same object. The last thread then has to wait $2000 \cdot 60 = 120K$ instructions until it can acquire the requested lock, although all requested locks are compatible. This situation arises frequently with IS/IX locks on coarse granule. For example, each access to a page or tupel needs a previously set IS or IX lock set on relation granularity. This results in a 120K instructions lower bound for transaction length, since the first thread cannot enter the release-lock critical section before the last thread has finished its request-lock critical section. A lower bound of 120K instructions is far too high for short transactions, like debit/credit transactions.

The claimed 120K instructions lower bound only holds, if all threads lock the same object in a round-robin fashion. If the object is not locked round-robin, then 120K instructions is only the expectation.

requested	already granted mode					
lock mode	free	X	S	IS	IX	SIX
X	+	−	−	−	−	−
S	+	−	+	+	−	−
IS	+	−	+	+	+	+
IX	+	−	−	+	+	−
SIX	+	−	−	+	−	−

Table 2. Compatibility matrix of the CC-lock

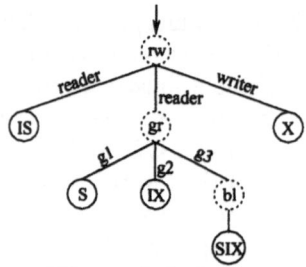

Figure 1. lock tree

5   A New Approach to the CC-Lock

As seen in the previous section, the standard implementation of the CC-lock tends to unnecessarily serialize lock requests. We now describe a new approach for the implementation of CC-locks, which does not serialize compatible lock requests. The main idea is to split the complex CC-lock into basic locks (section 2). A combination of three basic locks will then act as a CC-lock.

According to the compatibility matrix (Table 2), a requested X lock is incompatible to any other granted lock, including granted X locks, and vice versa. Nevertheless, multiple non-X locks may be granted concurrently (e.g., IX and IS locks). If we, in thought, combine the non-X lock modes of the compatibility matrix to one lock mode compatible to itself, we get the compatibility matrix of the R/W-lock. Thus, we can use a R/W-lock (rw) to detach non-X locks from X locks. There remains the problem of testing compatibility of the non-X locks.

According to the compatibility matrix of the four remaining lock modes S, IX, IS and SIX, a requested IS lock is compatible to any other granted lock. Vice versa, any requested lock is compatible to an already granted IS lock. Thus, an IS lock request is granted as soon as the r/w lock rw is succesfully obtained.

Consequently, we just have to look at the three remaining lock modes S, IX, SIX. Suppose that requested SIX locks would be compatible to granted SIX locks, then, the remaining compatibility matrix of these three modes would be a diagonal — the same matrix as of the group lock. We can therefore separate the three remaining modes from each other by using a 3-group lock (gr) and assigning a group number to each mode (e.g., S=g1, IX=g2, SIX=g3). The remaining problem is to serialize concurrent SIX locks. This can be done by using a binary lock (bl).

The described combination of basic locks yields a lock tree (Figure 1). The leaves are labeled with the lock modes of intentional locking. The inner nodes are labeled with basic locks, and the edges are labeled with lock modes of the basic locks. On a lock request, the thread goes from the root to the requested leaf, thereby acquiring the basic locks in the appropriate lock modes. In order to release a lock, the thread goes upward from the leaf to the root releasing the basic locks on the path.

Correctness Since the proof of correctness and fairness of the CC-lock is too long, it is omitted here, but it can be found in [5].

6   Comparisons and Conclusion

We have described a fast and highly concurrent lock manager. The basic idea was to split up the complex database lock structure (CC-lock) into three basic lock primitives. On the SB-PRAM, these lock primitives do not serialize compatible lock requests, and therefore the CC-lock does not serialize, either. Our implementation can easily be ported to any parallel architecture supporting the used lock primitives.

As measurments on the SB-PRAM showed, the new CC-lock has a run time of less than 70 instructions if no waits due to incompatibility to other locks are necessary. The total run time of a lock request thru the lock manager is between 400 and 600 instructions. On relation granularity, where mostly mutually compatible intention locks (IS/IX) are used, nearly all requests can be handled with 600 instructions. In contrast, due to serialization, the standard implementation needs 120K instructions if 2000 threads are running. This results in a speedup of factor 200 for locking on relation granularity for high speed shared memory systems, if our new implementation is used.

We currenctly implement of a rudementary database system on the SB-PRAM. For this, we also developed and implemented a serialization free deadlock detection algorithm [5]. Detailed results are expected for fall 1999.

Acknowledgments: The authors would like to thank Michael Bosch, Arno Formella, Silvia M. Mueller, and Jochen Preiss for their help and valuable discussions.

References

[1] G. ALVERSON, ET AL.: *The Tera Computer System*, Supercomputing, 1990.

[2] F. ABOLHASSAN, J. KELLER AND W.J. PAUL: *On the Cost-Effectiveness and Realization of the theoretical PRAM Model*, SFB-Report 09/91, March 1991.

[3] A. GOTTLIEB, ET AL.: *The NYU Ultracomputer — Designing an MIMD Shared-Memory Parallel Computer*, IEEE Trans. on Computers, pp. 175-189, Feb 1983.

[4] J. GRAY AND A. REUTER: *Transaction Processing: Concepts and Techniques*, Morgan Kaufmann, 1993

[5] C. JACOBI: *A relational Database System for the SB-PRAM. TPC–B and Concurrency-Control*, Master Thesis (german), University of Saarland, 1999.

[6] L.D. MOLESKY AND K. RAMAMRITHAM: *Efficient Locking for Shared Memory Database Systems*, University of Massachusetts Technical Report, March 1994.

[7] J. ROEHRIG: *Implementing P4 on the SB-PRAM*, Master Thesis (german), University of Saarland, 1996.

[8] B. SPENGLER: *A relational Database System for the SB-PRAM. Basic Components*, Master Thesis (german), University of Saarland, 1997.

Parallel Processing of Multiple Text Queries on Hypercube Interconnection Networks[*]

Basilis Mamalis[1,2], Paul Spirakis[1,2], and Basil Tampakas[1,3]

[1] Computer Technology Institute, Kolokotroni 3, 26221, Patras, Greece
[2] Computer Engineering & Informatics Dept., University of Patras,
Rion, 26500, Patras, Greece
[3] Technological Educational Institute of Patras, M. Alexandrou 1,
Koukouli, Patras, Greece
{mamalis, spirakis, tampakas}@cti.gr

Abstract. In this paper, we present a new, efficient, parallel text retrieval algorithm, appropriate for concurrent processing of multiple text queries in very low total execution times. Our approach is based on the widely–known vector space text processing model assuming the existence of a high–capacity interconnection network, such as the fully connected hypercube. We exploit the large number of available communication links of the hypercube and develop efficient parallel protocols under heavy query load, we give their detailed theoretical analysis and prove their optimal performance. We also prove the better performance of our protocols on hypercubes vs. other a) high capacity interconnection topologies like ideal fat–trees and b) single query parallel text retrieval methods. All the above results are also experimentally demonstrated via suitable embeddings on the Parsytec GCel3/512 supercomputer.

1 Introduction

Generally, our work deals with the problem of finding an efficient solution for the parallelization of the text retrieval tasks implied by the Vector Space Model (VSM), in a distributed memory parallel environment and under heavy query load (i.e. imagine a highly accessed search engine over the INTERNET).

More specifically, following this model, any _given_ text collection composed of D documents is finally represented by D corresponding document–vectors consisted of its most significant identifiers. In a distributed memory parallel environment of N working processors, we accept that the whole set of the D document vectors is _appropriately distributed_ and _almost equally balanced_ over the local memories of the N processors. Consequently, supposing that we are _given_ a set X of concurrently arriving text queries which are represented by $|X|$ corresponding VSM–indexed query–term vectors of the same form, _the goal_ is to find an appropriate algorithm and a corresponding well–suited interconnection

[*] This research was partially funded by ESPRIT LTR project No 20244 (ALCOM–IT), and by Basic Research project No. 9072 (GEPPCOM) of the E.U.

P. Amestoy et al. (Eds.): Euro-Par'99, LNCS 1685, pp. 482–486, 1999.
© Springer-Verlag Berlin Heidelberg 1999

network, for the efficient parallelization of the scoring and ranking tasks for all the $|X|$ queries over all the D documents of the given text collection.

A typical approach to the above problem could be the sequential processing (manipulating the queries one by one) with the use of an appropriate single-query parallel algorithm ([1],[4]) for each query. In our work, trying to further minimize the total communication times needed, we have followed a more complex approach, based on the parallel concurrent processing of all the $|X|$ queries in sets of N (equal to the number of processors) queries each time. In this case the communication messages needed for processing N queries together, are gathered in one execution leading to a quite heavy total communication task. Consequently, the appropriate use of high capacity networks that can deliver multiple messages concurrently in very low total communication times would be a proper solution. With respect to the above considerations, we have chosen the use of two widely-known networks, the *hypercube* (outlined as the best case throughout the paper) and the *ideal fat-tree* (a normal case) topologies, as typical paradigms of such well-studied high-capacity network topologies.

As a direct result, the amortized communication time per query is proved to be excessively small and almost constant ($O(1 + \log N/N)$) with regard to the number of processors. Moreover, the total amortized query processing times are proved to be much better than those offered by the usual single-query algorithms. Furthermore, the specific communication protocols used for the multiple message exchanging tasks are proved to be *optimal* for both hypercube and ideal fat-tree topologies. Finally, trying to obtain some experimental evidence for the high performance of our multiple query algorithm on a realistic parallel environment, we have performed a number of well-suited experiments over the GCel Parsytec machine, which have led to very promising results.

Due to lack of space, several important parts of our work (the specific communication protocols used, their optimality, theorems and proofs, experimental evaluation etc.) have not been included in this short version of the paper, however they can be easily found in the full version of the paper held at "http://helios.cti.gr/europar99". Also, with regard to the basic components of our work (Vector Space Model – [5]; Hypercube interconnection networks – [7]; Ideal Fat-trees – [6]), the reader may find complete descriptions in the corresponding references. Finally, a good description of the most significant related work in the field done till now, can be found in our previous work of [2].

2   The Proposed Multiple Queries Processing Method

The use of the proposed general multiple queries algorithm is strongly recommended specially in the case of heavy query load (i.e. a highly accessed search engine over the INTERNET). This is the case when the number of the concurrently arriving queries is much greater than the number of nodes (N) of the underlying interconnection network ($|X| >> N$, i.e. $|X| = a \cdot N \mid a >> 1$). In this case our multiple query algorithm can complete the whole task by processing N queries each time and assigning (virtually) one query to each node. For the above reason we can work with only N queries while presenting the basic functionalities of our algorithm.

Initial Conditions

1. There are D documents and N queries indexed using VSM, thus resulting to D and N corresponding document and query term-vectors respectively.
2. The underlying multiprocessor hardware platform consists of N working processors (each one having its own local memory), connected to each other via a suitable high capacity interconnection network.
3. The D document term–vectors are appropriately distributed over the N processors in such a way that the total amount of data is suitably shared and balanced (leading to almost $\frac{D}{N}$ document vectors for each processor).

The Parallel Algorithm

Phase 1: The host processor sends the N query vectors to all the N processors using appropriate pipelined steps (multiple broadcasting of N different messages). At the end of all steps, each one of the N processors has received the complete set of N queries and he knows that he is *virtually* responsible for a specific query.

Phase 2: Each one of the N processors performs the *scoring* and *ranking* tasks (see [5]) over his own $\frac{D}{N}$ document vectors, and for all the N queries. Moreover, during the ranking task, only the R top–ranked document vectors are kept for immediate further treatment. That results to N ranked relevant document sets ($RD_i \mid i = 1...D$ sets of size R) in each processor (one for each query).

Phase 3: Each processor sends each one of the $N - 1$ relevant document sets (all except the one that refers to the query he is responsible for) to its corresponding, responsible for it, processor. During this phase, all the processors work in parallel (concurrently), thus leading to a heavy and complicated All–to–All personalized communication task.

Phase 4: Each processor merges his own N partial relevant document sets. As a result, during phase 4, a complete relevant document set of size R is finally extracted from each processor, corresponding to the query he is responsible for.

Phase 5: Each processor sends its own merged document set to the host processor. Since all the N processors work in parallel, the corresponding communication task results to the concurrent sending of N different messages from all the processors towards the host processor – one from each processor.

3 Performance Analysis

In this section, we give the performance analysis of the whole multiple–query algorithm. The description of the specific communication protocols developed for the efficient implementation of phases 1,3 and 5 on hypercube topologies, their detailed analysis, as well as their optimality in the sense of exact analysis, can be found in the full version of our work.

The parallelization of the computational tasks of phase 2 is somewhat similar to the one presented in [1] if we consider each query separately. Thus, assuming that T_s and T_{lr} are the required times for scoring and ranking on each processor, then the total computational time for N queries (T_c) is equal to:

$$T_c = N \cdot (T_s + T_{lr}) \tag{1}$$

With regard to the execution of phase 4, assuming that the size of each relevant document set is equal to R and the required time for merging of N ranked lists of size R is $O(N \times R)$, then the total execution time is equal to:

$$T_m = 2 \cdot (N - 1) \cdot R \tag{2}$$

Furthermore, the communication times (in unit–message steps) required for the execution of the proposed protocols, are given by the following equations.

$$T_{hy1} = \lceil \frac{N}{\log N} \rceil + \log N - 1 \tag{3}$$

$$T_{hy5} = \lceil \frac{N - 1}{\log N} \rceil \tag{4}$$

$$T_{hy3} = \frac{N}{2} \tag{5}$$

Concluding, if we add the results given by equations 1 through 5 we obtain that the total execution time of our algorithm for the parallel processing of N concurrently arriving queries in a hypercube $H(N, d)$, is equal to:

$$T_{hy\_total} = \frac{2N - 1}{\log N} + \frac{N}{2} + \log N - 1 \; + \; N \cdot (T_s + T_{lr}) \; + \; 2 \cdot (N - 1) \cdot R \tag{6}$$

Furthermore, if we devide the above total execution time by the number of queries (N), we easily obtain the *amortized* processing time per query:

$$T_{hy\_am} = \frac{1}{2} + \frac{2N - 1}{N \log N} + \frac{\log N - 1}{N} \; + \; (T_s + T_{lr}) \; + \; 2 \cdot R \tag{7}$$

On the other hand, if we consider the existence of an ideal fat–tree topology as the underlying interconnection network, and based on the results given in [2], the corresponding *total execution* and *amortized processing* times, have as follows:

$$T_{fat\_total} = 2N + 4 \log N - 1 \; + \; N \cdot (T_s + T_{lr}) \; + \; 2 \cdot (N - 1) \cdot R \tag{8}$$

$$T_{fat\_am} = 2 + \frac{4 \log N - 1}{N} \; + \; (T_s + T_{lr}) \; + \; 2 \cdot R \tag{9}$$

As it comes out directly from equations 7 and 9, an almost perfect (expected) speed–up is achieved in respect to the amortized processing time per query (especially when the hypercube topology is used), since:

1. The computational tasks (scoring and ranking) are parallelized almost perfectly with the same way as in [1], where $T_s + T_{lr} \simeq (1/P)(t_s + t_{lr}) \mid P$ is the number of processors and t_s, t_{lr} the serial times for scoring and ranking.
2. The computation overhead due to the merging of the results is restricted to $O(R)$ time per query which is extremely small comparing to the other computational tasks.
3. The total communication amortized time, is almost **constant** and equal to 2 and $\frac{1}{2}$ message steps, for the ideal fat–tree and the hypecube topology respectively. Also, the above difference in the amortized communication times (2 vs. $\frac{1}{2}$ message steps), indicates that the use of hypercube topologies instead of the initially used ideal fat–trees, introduces significant improvements in the total performance of our general multiple queries algorithm.

Moreover, it is obvious that the same conclusions hold also in the general case of $a \cdot N$ (multiples of N) queries by simply allowing the multiple query algorithm to operate in 'a' sequential phases by processing (concurrently) one set of N queries at each phase.

Furthermore, the excessively high efficiency of our hypercube–based multiple queries approach can be demonstrated in a by comparing it to the traditional single–query parallel text retrieval approaches developed in the past ([1], [4],[3] etc.). I.e., by comparing to the binary–tree single query algorithm of [1], where the corresponding execution time for processing one query is equal to (relative comparisons should be done between eq. 7 and 10):

$$T_{tb} = (2\log N + 2) + (T_s + T_{lr}) + 4 \cdot R \cdot (\log N - 1), \qquad (10)$$

the reader may notice that, in the single–query algorithm the communication and the merging times are increased **logarithmically** compared to N, while in the hypercube–based multiple query algorithm they remain almost **constant**.

References

[1] P. Efraimidis, C. Glymidakis, B. Mamalis, P. Spirakis and B. Tampakas, *"Parallel Text Retrieval on a High Performance Supercomputer Using the Vector Space Model"*, ACM SIGIR'95 Conference, July 9-13, Seattle, pp. 58-66, 1995.
[2] B. Mamalis, P. Spirakis and B. Tampakas, *"Optimal High Performance Parallel Text Retrieval via Fat Trees"*, (full version), to appear in the Theory of Computing Systems (TOCS) journal, Dec. 1997.
[3] C. Stanfill, *"Partitioned Posting Files: A Parallel Inverted File Structure for Information Retrieval"*, ACM SIGIR'90, pp. 413-428, June 1990.
[4] J. Cringean, M. Lynch, G. Manson and P. Willett, *"Best Match Searching in Document Retrieval Systems Using Transputer Networks"*, London: British Library Research and Developement Department, 1989.
[5] G. Salton and McGill, *"An Introduction to Information Retrieval"*,2nd edition, MacGraw-Hill, c1983.
[6] C. Leiserson, *"Universal Networks for Hardware-Efficient Supercomputing"*, IEEE Transactions on Computers, Vol. C-34, No. 10, October 1985.
[7] F. T. Leighton, *"Introduction to Parallel Algorithms and Architectures: Arrays, Trees and Hypercubes"*, Morgan Kaufmann Publishers, Inc., c1992.

Topics 06 + 20
Fault Avoidance and Fault Removal in Real-Time Systems & Fault-Tolerant Computing

Gilles Motet and David Powell

Local Chairs

Many computerised systems require the implementation of parallel processes to improve their performance (super-computers), to take account of non-deterministic environments (real-time systems), or to cater for geographical separation of resources (distributed systems). However, parallelism increases system complexity and thus makes it more difficult to guarantee correct behaviour. Moreover, these systems often support critical functions whose failure may endanger human life (e.g., embedded systems in aircraft or cars), the economy (e.g., banking systems and automated production systems) or the environment (e.g., control of nuclear or chemical plants). Such systems must be *dependable*, i.e., it must be possible to place *justifiable reliance on the service they deliver* [Laprie 1995]. These systems therefore present two antagonistic requirements: an increased need to ensure the correctness of their behaviour and an increased difficulty in obtaining such assurance.

The potential cause of a system failure is called a *fault*, for example, a physical phenomena affecting a system component, an environmental disturbance, a design defect, etc. The means for ensuring dependability are thus defined in terms of four complementary methods:

- *Fault prevention*, that aims to prevent the introduction of faults, e.g., by constraining the design processes by means of rules.
- *Fault removal*, that aims to detect the presence of faults, and then to locate and remove them.
- *Fault tolerance*, that aims to ensure that the presence of faults does not lead to system failure. Fault tolerance relies primarily on error detection and error correction, with the latter being either backward recovery (e.g., retry), forward recovery (e.g., exception handling) or compensation recovery (e.g., majority voting).
- *Fault forecasting*, that aims to quantify the confidence that can be attributed to a system. Measures of the forecasted dependability can be obtained, for example, through stochastic modelling or through extrapolation of field experience from previously deployed systems.

We must stress that these four viewpoints are indeed complementary. For example, the fact that the tolerance of design faults is envisaged admits that efforts to prevent or remove faults are imperfect. Nevertheless, these activities must have been sufficiently well carried out for it to be possible to assume that residual design faults in, say, different versions of a software function, will manifest themselves independently.

P. Amestoy et al. (Eds.): Euro-Par'99, LNCS 1685, pp. 487–488, 1999.
© Springer-Verlag Berlin Heidelberg 1999

From the very earliest work on fault tolerance, parallelism has been seen as both a means for increasing dependability and as potential source of problems. For example, fault tolerance can be implemented through the parallel execution of multiple replicas (e.g., majority voting for tolerance of environmental perturbations or physical faults) or variants (e.g., N-version programming). The 'domino effect' is an example of a negative impact of parallelism: this is the phenomenon of cascading rollbacks that can occur between parallel processes when backward recovery is implemented inappropriately.

Several issues and solutions associated with the various dependability viewpoints are tackled in the following papers.

The paper by L. Carroll *et al.* describes the dependability expectations of real-time systems designed for robot control applications. It then describes how UML can be used to avoid faults and to facilitate their detection.

Multitasking applications use synchronisation protocols to express precedence relations between tasks. When real-time applications are concerned, deadline constraints are added. A missed deadline constitutes a failure. The paper by C. Mourlas proposes a synchronisation strategy that, combined with a task scheduling analysis accounting for resource allocations, allows an *a priori* analysis of deadline satisfaction. It describes a static analysis tool that signals potential failure occurrences.

The paper by B. Ravindran presents a dynamic analysis of the resource allocation problem, taking into account changes occurring in the external environment or in the availability of internal resources. This analysis is processed at run-time. It detects errors caused by such changes. Moreover, thanks to the proposed resource re-allocation procedure, this paper describes a mechanism to tolerate the impacts of these changes.

If the characteristics of the execution environment are considered early in the design process, it is possible to provide timely detection of the presence of design faults. K. Ghose *et al.* propose a tool to model the execution environment (hardware platform, kernel and interconnection system) used by the real-time applications implemented by distributed multiprocessors. Simulations and analysis can then be carried out to highlight possible failures.

The paper written by R. Meier and P. Nixon also deals with distributed systems. They propose tolerance mechanisms based on CORBA to handle the failures of remote components. To obtain such a result, a client of a failed service is re-connected to a backup providing a similar service.

The paper by D. Avresky and S. Geoghesan tackles fault tolerance at a low level of abstraction. The authors propose a software approach for detecting and reconfiguring failing nodes of a hypercube. Since the proposed solution has a small overhead, it is applicable to real-time parallel system implementations.

[Laprie 1995] J.-C. Laprie, "Dependable Computing: Concepts, Limits, Challenges", in *Special Issue, 25th Int. Symp. on Fault-Tolerance Computing (FTCS-25)*, (Pasadena, CA, USA), pp.42-54, IEEE Computer Society Press, June 1995.

Quality of Service Management in Distributed Asynchronous Real-Time Systems

Binoy Ravindran

The Bradley Department of Electrical and Computer Engineering
Virginia Polytechnic Institute and State University
Blacksburg, VA 24061, USA
binoy@vt.edu

Abstract. This paper presents adaptive resource management techniques that achieve the timeliness quality of service (QoS) requirements of distributed real-time systems that are "asynchronous" – both in the sense that processing and communication latencies do not necessarily have known upper bounds, and in the sense that event arrivals are non-deterministically distributed. Examples of such systems include the emerging generation of computer-based, command and control systems of the U.S. Navy. To enable the engineering of such systems, we present resource management middleware strategies that enforce the timeliness QoS requirements of the system. The middleware performs QoS monitoring and failure detection, QoS diagnosis, and reallocation of resources to adapt the system to achieve acceptable levels of QoS. Experimental characterizations of the middleware using a distributed asynchronous real-time benchmark illustrate its effectiveness for adapting the system for achieving the desired QoS during overloaded situations.

1 Introduction

Real-time, military, computer-based command and control (C2) systems such as the surface combatants of the U.S. Navy are characterized by load patterns that do not have known upper bounds [3]. This is primarily due to the difficulty in estimating upper bounds on the size of data and event streams that they must process. Size of the data stream refers to the number of sensor reports that such systems must periodically process for decision making and reaction. Size of the event stream refers to the arrival rate of events that trigger computations in the system that perform mission critical tasks such as reacting to a hostile threat by detonating weapons. We call such real-time systems "asynchronous" real-time systems, as the processing and communication latencies do not have known upper bounds (due to unknown upper bounds on data stream sizes) and event arrivals are non-deterministically distributed (due to unknown arrival patterns of events).

Classical real-time computing research focuses on "hard" real-time computing for synchronous (in the sense that processing and communication latencies have known upper bounds and event arrivals are deterministically distributed),

P. Amestoy et al. (Eds.): Euro-Par'99, LNCS 1685, pp. 489–496, 1999.
© Springer-Verlag Berlin Heidelberg 1999

device level, sampled data monitoring and regulatory control - usually centralized but occasionally distributed[1, 4, 5, 9]. These techniques cannot be practically employed or adapted for systems that are distributed and asynchronous. Distributed asynchronous real-time computer systems and their applications are inherently dynamic, and thus require end-to-end adaptive real-time resource management [2]. We argue that the conventional "hard"/"soft" dichotomy is too coarse, and timeliness is better treated as a quality of service (QoS) for activities in such systems. The paucity of concepts, theories, methodologies, and techniques that allow engineering of, and reasoning about, asynchronous distributed real-time computing systems–and the consequent absence of appropriate commercial real-time operating system and system software (e.g., middleware) products–forces system engineers to "home-brew" and construct ad hoc special purpose, and consequently expensive, technologies and solutions. A good example of this is the current generation of Navy surface combatants.

In response to the need for an infrastructure for engineering distributed asynchronous real-time systems, we present resource management strategies that can achieve their QoS requirements. The objective of resource management is to adapt the system at run-time to changing resource needs so that acceptable levels of timeliness QoS can be achieved. Adaptation is achieved by identifying and eliminating software bottlenecks and by discovering additional hardware resources. The resource management techniques are implemented as part of a middleware infrastructure for the emerging generation of shipboard computing systems using commercial-off-the-shelf (COTS) workstations. Further, the effectiveness of the techniques are evaluated using a distributed asynchronous real-time benchmark application that functionally approximates a shipboard C2 system.

The rest of the paper is organized as follows: Section 2 presents a generic model of real-time C2 systems. The generic model is derived after a study of Navy's Anti-Air Warfare (AAW) system [6] and is used to reason about asynchronous real-time systems and their load characteristics. Steps involved in the resource management process are described in Section 3. An experimental characterization of the resource management algorithms is presented in Section 4. Finally, the paper concludes with a summary of the current work and on-going efforts in Section 5.

2   A Generic Real-Time C2 System

Figure 1 shows a generic real-time C2 system. The control system consists of functions that perform *assessment* of the environment, *initiation* of actions, and monitoring and *guidance* of the actions to their successful completion. The inter-relationship of the functions with the environment, and the intra-relationship of the functions among themselves are illustrated in Figure 1.

The assessment function repeatedly collects data from the environment through hardware sensing devices. The data is filtered, correlated, classified, and then used to determine the necessity of an action by the system. The assessment function has a *continuous* (or cyclic) behavior, since the data collection and assessment is performed repeatedly throughout the operational life of the system. Typically, there is a timeliness objective associated with the completion of each execution cycle of the function, i.e., a bound on the time taken to review each of the elements of the data stream once. When an action is necessary, the assessment function generates an event that activates the initiation function.

The initiation function determines the action that needs to be taken and causes actuating devices to perform the action. Since the function executes in response to an event that can be triggered at any time, the initiation function has a *transient* behavior. The initiation function has, typically, a maximum latency requirement on its execution time, i.e., a latency requirement on the execution time of the function to process a single event. Upon initiation of the actions by the actuators, the monitoring and guidance function is notified.

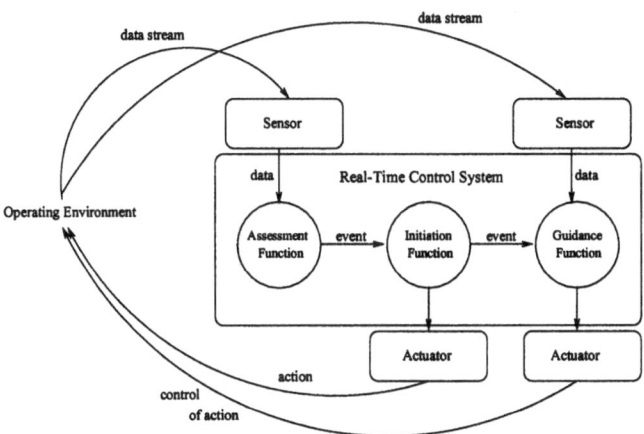

Fig. 1. A Generic Real-Time C2 System

The guidance function repeatedly uses sensor inputs to collect data, to monitor the actions that were initiated, and to guide the actuators to successful completion of the actions. Since the activation of the guidance function can begin and terminate at any time, the function is transient in behavior. Further, the function behaves like an assessment function once it is active, executing in a continuous manner. Hence, the guidance function has a *quasi-continuous* behavior. Observe that each transient activation of the function consists of a set of execution cycles. Guidance functions typically have two timeliness objectives: (1) completion time for one cycle and (2) deactivation time. Typically, it is more critical to perform the required processing before the activation deadline than it is to meet the completion time for each cycle.

After a careful study of Navy's AAW C2 system, we have observed that the load characteristics of control system functions are significantly influenced by: (1) size of the data stream (or the number of data items that the assessment

and guidance functions have to process during a single cycle), and (2) size of the
event stream (or the arrival rate of events that trigger the execution of the initiation and guidance functions). For systems such as the AAW, data stream sizes
(radar tracks) and event (threat) arrivals have neither known upper bounds nor
deterministic distributions, respectively. Thus, we call such real-time systems,
"asynchronous" real-time systems.

3   Adaptive Resource Management Steps

The steps involved in the resource management process are illustrated in Figure 2. The real-time control system components are monitored for conformance
to specified QoS requirements. QoS violations of the system are detected and are
reported to diagnosis functions. Diagnosis functions identify the causes of QoS
violations. Further analysis identifies possible recovery actions to improve the
QoS. Allocation analysis is performed to determine the optimal way to execute
the selected actions. We describe the algorithms that are used in each of these
steps in the subsections that follow.

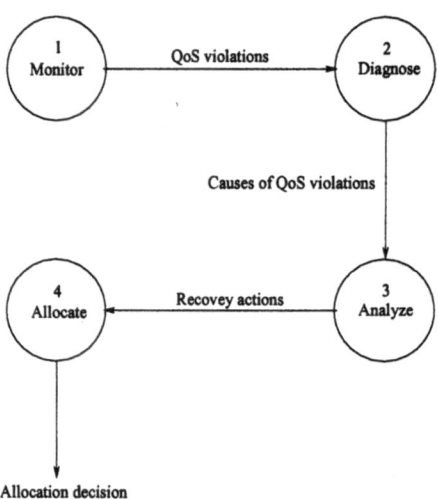

Fig. 2. Adaptive Resource Management Steps

3.1   QoS Monitoring

Monitoring of timeliness QoS involves collection of time-stamped events that are
sent from application programs and synthesis of the events into function-level
end-to-end QoS metrics. Analysis of a time series of the metrics enables detection
of QoS violations. In this paper, we consider timeliness QoS management of the
continuous assessment function.

We define the timeliness QoS of assessment functions in terms of a *minimum slack* value that must always be maintained on the cycle deadline of the function. A timeliness QoS violation of the assessment function is said to occur during an execution cycle when the observed cycle latency of the function exceeds the minimum slack requirement on the cycle deadline of the function. Further, such a violation has to be observed for a certain number of cycles that is larger than the maximum that is allowed in a "window" of recent cycles.

3.2 QoS Diagnosis

When a QoS violation occurs, diagnosis techniques are employed to determine: (1) the cause(s) of the violation, and (2) the possible action(s) that may improve the QoS and recover from the violation.

A timeliness QoS violation of the assessment function takes place due to the increase in execution latencies of some applications of the function, or due to the increase in communication latencies between some pairs of applications (that communicate) of the function, or both. Algorithms that perform timeliness QoS diagnosis identifies applications or communication links that are experiencing significant increase in latencies. In this paper, we consider a "local" QoS diagnosis technique. The technique is local in the sense that it considers only the applications of a control system function that has been reported for QoS violations. The diagnosis technique compares current latency of an application to its least latency in the same host assignment and at the same data stream size. The technique identifies the latency of an application to have increased if the end-to-end cycle latency of the function during the current cycle has exceeded the least function latency that was observed in the past by a significant amount.

3.3 Identifying Recovery Actions

Once the causes of a QoS violation are determined and a set of "unhealthy" applications are identified, further analysis is performed to determine possible recovery *actions* for the applications that will improve the timeliness QoS of the function. We consider an algorithm that analyzes the changes in the data stream size and the host load of an unhealthy application and determines the recovery action as follows:

- If the data stream size of the application has increased significantly and the load of the host of the application has not increased by a large factor, then the recovery action is to *replicate*. The replicas of the application can share the additional data items, process it concurrently, and thereby reduce the end-to-end function cycle latency.
- If the data stream size of the application has not increased significantly and the load of the host of the application has increased by a large factor, then the action is to *migrate*. The application may be residing on a host resource that is heavily "loaded" and may be subjected to increased contention. By migrating the application to a relatively less loaded resource, the resource

contention of the application can be reduced. This can reduce the function latency.

3.4 Allocation Analysis

The QoS analysis algorithm discussed in the previous section produces a set of (*unhealthy-application, action*) pairs. To enact the recovery actions for the unhealthy applications, hardware resources must be allocated to the actions. We consider a resource allocation algorithm that uses a set of candidate (application,action) pairs as its input and selects the hardware resources–host and LAN–for performing the recovery actions.

The algorithm determines the "best" host to replicate or migrate an unhealthy application, and a LAN of the host for the inter-application communication needs of the application. The algorithm only considers the set of hosts where an application is prepared for execution and the set of LANs of each such hosts. The best host and a LAN of the host is determined using a "fitness" function that simultaneously considers host and LAN load information. Load information of hosts and LANs are characterized as trend values over a moving set of load index measurements by determining the slope of a simple linear regression line that plots the load index values as a function of time. The load index value of a host is determined as a function of CPU utilization, CPU ready-queue length, and free available memory, and that of a LAN is determined as a function of the communication activity detected at network interfaces of hosts in the LAN, respectively. The function definitions can be found in [7]. The best host of a candidate application is determined as the host that has the minimum fitness value among all the eligible hosts of the application.

4 Experimental Evaluation

We evaluate the effectiveness of the resource management algorithms through an experimental study. A distributed asynchronous real-time benchmark application is used in the experimental characterizations. The benchmark consists of an assessment function called Sensing that consists of a sensor (a simulator program called Sensor), a filter manager program (called FM), one or more replicas of a filter program (called Filter), an evaluate and decide manager (called EDM), and one or more replicas of an evaluate and decide program (called ED). Details of the benchmark can be found in [8]. The hardware test-bed of the experimental environment consists of a network of SUN Ultra workstations and Pentium-based PCs that are interconnected using 100MBPS Ethernet LANs.

The experiment described here illustrates a scenario where the middleware recovers the Sensing function of the benchmark from timeliness QoS violations during high data stream size situations. For the experiment, we consider an initial data stream size of 1000 data items, a cycle deadline of 140 milliseconds, and a minimum timeliness slack value of 100 milliseconds (i.e., approximately 72% of the deadline). The maximum number of allowable violations and the

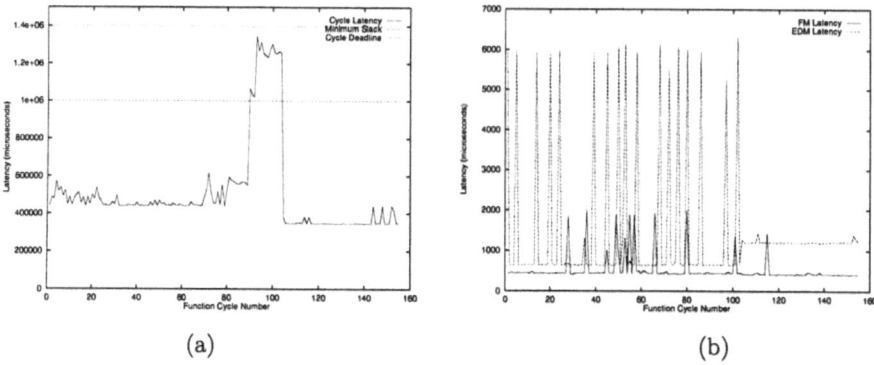

(a)   (b)

Fig. 3. Latencies of (a) Sensing Function and (b) FM and EDM Programs During High Data Stream Sizes

window size is defined as 15 and 20, respectively. The experiment is started by gradually increasing the data stream size from 1000 to 2000 data items. As seen in Figure 3(a), the cycle latency of the function increases and eventually exceeds the minimum slack requirement. Figures 3(b), 4(a), and 4(b) show the corresponding latencies of the FM, EDM, Filter, and ED programs of the function, respectively.

The middleware detects the timeliness QoS violation and performs QoS diagnosis. The result of the diagnosis is to replicate Filter and ED programs, as both the programs are experiencing high data stream sizes. The resource allocation algorithm is used to compute an allocation decision, i.e., the best hosts and LANs for the use of the replica programs. The allocation decision is enacted by the middleware. The newly created replica processes synchronize with other executing programs and start processing the data. Figures 4(a) and 4(b) illustrate the latencies of the replicas of Filter and ED programs, respectively. The replication of the programs and the sharing of the data stream among the replicas cause the function cycle latency to decrease and improves its timeliness QoS. This is observed in Figure 3(a).

5   Conclusions

This paper presents adaptive resource management techniques that achieve the QoS requirements of continuous computations in distributed asynchronous real-time systems. Experimental characterizations of the resource management strategies illustrate its effectiveness in recovering from timeliness QoS violations that occur during high data stream size situations. A benchmark that functionally approximate shipboard C2 systems is used in the experimental study.

The resource management middleware described in this paper has been incorporated into the experimental test-bed of the Navy. On going work includes development of resource management algorithms for transient and quasi-continuous computations that can achieve their desired timeliness QoS.

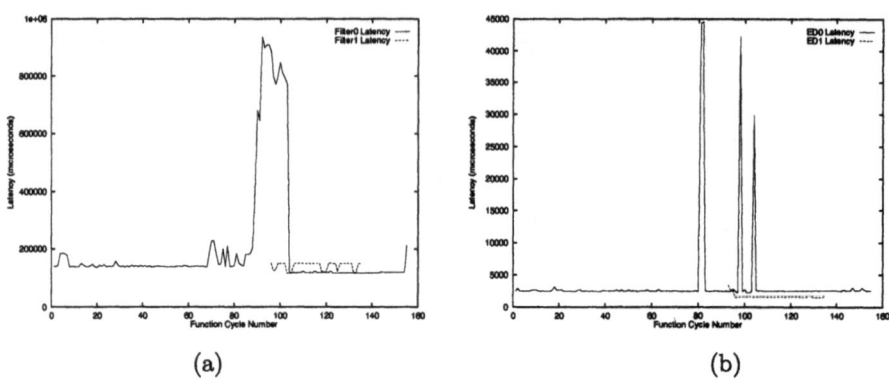

(a)   (b)

Fig. 4. Latencies of (a) Filter and (b) ED Programs During High Data Stream Sizes

References

[1] T. P. Baker. Stack-based scheduling of real-time processes. *Journal of Real-Time Systems*, 3(1):67–99, March 1991.

[2] R. Clark, E. D. Jensen, A. Kanevsky, et al. An adaptive, distributed airborne tracking system. In *Parallel and Distributed Processing*, volume 1586, pages 353–362. Springer-Verlag, April 1999. Lecture Note in Computer Science.

[3] R. D. Harrison Jr. Combat system prerequisites on supercomputer performance analysis. In *Proceedings of The NATO Advanced Study Institute on Real-Time Computing*, volume F127 of *NATO ASI*, pages 512–513, 1994.

[4] K. Ramamritham, J. A. Stankovic, and W. Zhao. Distributed scheduling of tasks with deadlines and resource requirements. *IEEE Transactions on Computers*, 38(8):1110–1123, August 1989.

[5] L. Sha, M. H. Klein, and J. B. Goodenough. Rate montonic analysis for real-time systems. In A. M. van Tilborg and G. M. Koob, editors, *Scheduling and Resource Management*, pages 129–156. 1991.

[6] L. R. Welch. Large-grain, dynamic control system architectures. In *Proceedings of The Workshop on Parallel and Distributed Real-Time Systems*, April 1997.

[7] L. R. Welch et al. Instrumentation, modeling and analysis of dynamic, distributed real-time systems. In *International Journal of Parallel and Distributed Systems and Networks*, 1999. Special Issue on Measurement of Program and System Performance, Accepted for publication. To appear.

[8] L. R. Welch and B. A. Shirazi. A dynamic real-time benchmark for assessment of qos and resource management technology. In *Proceedings of The Fifth IEEE Real-Time Technology and Applications Symposium*, June 1999.

[9] J. Xu and D. L. Parnas. Scheduling processes with release times, deadlines, precedence, and exclusion relations. *IEEE Transactions on Software Engineering*, 16(3):360–369, March 1990.

Multiprocessor Scheduling of Real-Time Tasks with Resource Requirements

Costas Mourlas

Department of Computer Science, University of Cyprus,
75 Kallipoleos str., CY-1678 Nicosia, Cyprus
mourlas@ucy.ac.cy

Abstract. In this paper, we present a new synchronization strategy for real-time tasks executed in a parallel environment. This strategy makes the timing properties of the system predictable since one is able to determine analytically whether the timing requirements of a task set will be met, and if not, which task timing requirements will fail. The proposed synchronization protocol is based on the on-demand paradigm where resources are assigned only when actually required so that the system never wastes unused assignments. Simple formulae for the worst-case determination of task blocking durations as well as for the schedulability analysis of a set of tasks are described. A schedulability analysis example is also presented that illustrates the concepts of scheduling and resource allocation of a set of time critical tasks.

1 Introduction

The term *scheduling of real-time tasks* actually refers to the concept of sequencing the use of any shared resource whose use involves meeting application time constraints [2]. Alternatively, *real-time scheduling* is defined as the ordering of the execution of the tasks such that their timing constraints are met and the consistency of resources is maintained. Thus, the objective of any scheduling algorithm is to find a feasible schedule whenever one exists for a given set of tasks such that each completes its computation before its deadline, and a given set of resource requirements is satisfied. We have to emphasize here that the cpu scheduling is not the main problem of real-time systems but both cpu scheduling and resource allocation.

In this paper, we present a new synchronization strategy for real-time tasks executed in a parallel environment, where each task may have resource requirements, i.e. it can require the use of non-preemptable resources or access shared data. Thus, we focus mainly on the resource allocation part of the scheduling process. This strategy is a continuation of our previous work on synchronization methods for real-time systems [4] where more intelligent techniques and less pessimistic formulae have been investigated and introduced in the current version. The proposed strategy is analyzable and understandable at a high level, meaning that given a set of tasks we know in advance if they will meet their deadlines or not. It is based on the rate monotonic algorithm [6, 1] in the way the priorities

P. Amestoy et al. (Eds.): Euro-Par'99, LNCS 1685, pp. 497–504, 1999.
© Springer-Verlag Berlin Heidelberg 1999

are assigned to tasks, and which also effectively creates an integrated resource allocation model. Thus, we consider our work as a fixed priority architecture based upon the rate-monotonic scheduling, which is the optimal static priority scheduling algorithm with good performance, low complexity and minimal implementation overhead.

The most well known synchronization protocol used by the rate monotonic algorithm is that of priority ceiling protocol described in [7, 5] where the blocking time is deterministic and bounded. It has many advantages in the uniprocessor environment but its multiprocessor extension, i.e. the distributed priority ceiling protocol [5] is complex enough and requires also the remote procedure call to be supported by the underlying software. In our approach, we try to minimize the implementation requirements and the complexity of the scheduling and synchronization strategy in order to have a more efficient scheme that fits well in the parallel environment. Given a blocking duration B of a task τ with period T, then the ratio B/T is a measure of schedulability loss due to blocking. As we'll see in the following sections, we try to minimize this ratio as much as possible.

2   Assumptions and Notation

In this paper, we consider only hard dealines for critical tasks where also these tasks can require the use of non-preemptable resources or access shared data. To guarantee that critical tasks will never miss their deadlines, they are treated as *periodic* processes, where the period of the process is related to the maximum frequency with which the execution of this process is requested.

We assume that critical tasks are assigned priorities inversely to tasks periods. Hence, task t_i with period T_i receives higher priority than t_j with period T_j if $T_i < T_j$. Ties are broken arbitrarily. Tasks are periodic, are ready at the start of each period and have known, deterministic worst-case execution times. We consider in our analysis that all task deadlines are shorter than (or equal to) the corresponding task periods (i.e. $D_i \leq T_i$). We assume that static binding of tasks to processors is used and for the simplicity of our presentation we also assume that every periodic task is allocated on a different processor/node of the parallel system. Each periodic task can require the use of non-preemptable resources or access shared data. Thus, cpu scheduling is not the main problem at this stage, but since tasks are inter-dependent the main problem is task synchronization and resource allocation. In real-time environments, blocking due to synchronization has to be deterministic in order to have nice analysis properties and a high degree of system predictability.

We assume that every critical section is guarded by a binary semaphore but the strategy is also applicable when monitors or Ada rendezvous are used. We also assume that the locks on semaphores will be released before or at the end of a task. The term "critical section" will be used to denote any critical section between a P(S) and the corresponding V(S) statement. P(S) and V(S) denote the indivisible operations *wait* and *signal* respectively on the binary semaphore S. The duration of the critical section is defined to be the time to execute the code between the P and its corresponding V operations when the task executes alone on the processor. A task τ can have multiple non-overlapping critical sections, e.g.

$$\tau = \{\dots \text{P}(\text{S}_1)\dots\text{V}(\text{S}_1)\dots\dots\text{P}(\text{S}_2)\dots\text{V}(\text{S}_2)\dots\}$$

but not any nested critical section.

3   The Proposed Synchronization Protocol

In this section, we present a new synchronization protocol suitable for synchronizing real-time tasks executing on parallel systems that requires only the message-passing scheme to be supported by the underlying software. Consider a fixed set of n periodic tasks $\tau_1, \dots, \tau_n$ each one bound to a different processor. Each task is characterized by four components $(CS^i, C^i_{non-cs}, T_i, D_i)$, $1 \leq i \leq n$, where

CS^i is the set $\{CS^i_{j,k} \mid j, k \geq 1\}$ that includes all the (durations of the) critical sections of the task τ_i. $CS^i_{j,k}$ is the kth critical section in task τ_i guarded by semaphore S_j. This notation is necessary because task τ_i may access the semaphore S_j more than once. $CS^i_{\cdot,k}$ denotes the kth critical section if the corresponding semaphore is not of interest. Similarly, $CS^i_{j,\cdot}$ denotes any critical section of task τ_i guarded by semaphore S_j. We define as C^i_{cs} the total deterministic computation requirement of task τ_i within its critical sections, i.e $C^i_{cs} = \sum_{x \in CS^i} x$,

C^i_{non-cs} is the deterministic computation requirement of task τ_i outside its critical sections,

T_i is the period of task τ_i,

D_i is the deadline of task τ_i.

We also use the notation C_i as the total deterministic computation requirement of task τ_i where $C_i = C^i_{cs} + C^i_{non-cs}$. Moreover, each semaphore S_j can be either *locked* by a task τ_i if τ_i is within its critical section $CS^i_{j,\cdot}$ or *free* otherwise.

Suppose that a task τ_i requires to enter its critical section $CS^i_{j,k}$ issuing the operation $P(S_j)$. Then the following cases can occur:

1. The semaphore S_j is *free*. Then, the semaphore S_j is allocated to the task τ_i, the task τ_i proceeds to its critical section and the state of S_j becomes *locked*.
2. If case 1 does not hold, i.e. semaphore S_j is currently *locked*, then after its release it is allocated to the highest priority task that is asking for its use. The task τ_i will proceed to its critical section if and only if semaphore S_j has been allocated to τ_i.

Note that by the definition of the protocol, a task τ_i can be blocked by a lower priority task τ_j, only if τ_j is executing within its critical section $CS^j_{l,\cdot}$ when τ_i asked for the use of the semaphore S_l.

Theorem 3.1 The proposed synchronization protocol prevents deadlocks.
Proof: By definition, for every task τ_i there is no any nested critical section.
Thus, τ_i will never ask in its critical section for the use of any other semaphore
and so a blocking cycle (deadlock) cannot be formed. □

We can easily conclude that a set of n periodic tasks, each one bound to a different processor $\wp$ can be scheduled using the proposed synchronization protocol if
the following conditions are satisfied:

$$\forall i, 1 \leq i \leq n, \quad C_i + B_i \leq D_i \tag{1}$$

Once B_is have been computed for all i, the conditions (1) can then be used to
determine the schedulability of the set of tasks.

4   Determination of Task Blocking Time

Here, we shall compute the worst-case blocking time that a task has to wait
for its resource requirements. Task τ_i requires in every period T_i to enter ν
different critical sections $\{CS^i_{.,j} \mid 1 < j \leq \nu\}$. For instance, each time that τ_i
attempts to enter its critical section $CS^i_{l,m}$ guarded by S_l, it can find that S_l
is currently held by another task. Thus, τ_i blocks and waits at most $B^i_{l,m}$ time
units until the semaphore S_l has been allocated to τ_i. In other words, $B^i_{l,m}$ is
the worst case blocking time that task τ_i waits to enter its mth critical section
guarded by S_l. The total worst-case blocking duration B_i experienced by task τ_i
is the sum of all these blocking durations, i.e. $B_i = \sum_{1 \leq j \leq \nu} B^i_{.,j}$. If the blocking
time B_i for any task τ_i is non-deterministic then the whole system becomes
unpredictable and difficult to analyze. In real-time environments, the blocking
has to be deterministic and for this reason our approach imposes a specific
structure on blocking to bound the blocking time.

Theorem 4.1 Consider a set of n tasks $\tau_1, \ldots, \tau_n$ arranged in descending order
of priority (i.e. $p_i > p_{i+1}$). Each task is bound to a different processor $\wp_i$ and the
proposed synchronization protocol is used for the allocation of the semaphores.
Let

$$H^i_l = \{CS^j_{l,.} \mid 1 \leq j < i\},$$ - set of critical sections used by tasks having
higher priorities than τ_i and guarded by the same
semaphore S_l

$$L^i_l = \{CS^j_{l,.} \mid i < j \leq n\},$$ - set of critical sections used by tasks having
lower priorities than τ_i and guarded by the same
semaphore S_l

$$\beta^i_l = \max(L^i_l).$$ - blocking time due to lower priority tasks

Then, the worst case blocking time $B^i_{l,.}$ each time task τ_i attempts to enter any
of its critical sections guarded by semaphore S_l is equal to

$$
B_{l,\cdot}^i = \begin{cases} \beta_l^i + \sum_{x \in H_l^i} x + \sum_{y \in \Delta_l^i} y & \text{if } \sum_{x \in H_l^i} x < max(\{T_m \mid CS_{l,\cdot}^m \in H_l^i\}) \\ \\ \infty & \text{otherwise} \end{cases}
\tag{2}
$$

where Δ_l^i is defined as:

$$
\Delta_l^i = \{CS_{l,\cdot}^j \in H_l^i \mid \sum_{x \in H_l^i} x \geq T_j\}
\tag{3}
$$

Proof: The sum in formula 2 above represents the longest blocking time $B_{l,\cdot}^i$ for a task τ_i trying to enter any of its critical sections $CS_{l,\cdot}^i$ at its worst-case task set phasing. At this worst-case phasing of the tasks, when τ_i wants to enter into critical section $CS_{l,\cdot}^i$ critical section, it finds a lower priority task executing within its critical section guarded by semaphore S_l. Then, just before this lower priority task has finished its critical section and has released semaphore S_l, all the higher priority tasks that use this semaphore S_l come one after the other and ask to enter their critical section.

Thus, $B_{l,\cdot}^i$ has two parts. The first part, namely β_l^i is due to L_l^i and it is equal to the maximum duration of critical section among all the critical sections of the lower priority tasks in L_l^i. The second part comes from tasks in H_l^i. If $\sum_{x \in H_l^i} x$ is less than the minimum period T_{min} of the tasks whose critical sections are in H_l^i then the worst case blocking time $B_{l,\cdot}^i$ equals to the sum $\beta_l^i + \sum_{x \in H_l^i} x$ (note that $\Delta_l^i = \emptyset$). Otherwise, task τ_{min} could block again task τ_i in its next period and due to this fact the element $CS_{l,\cdot}^{min}$ is included in Δ_l^i. The same scenario may happen for all the tasks with critical sections in H_l^i whose periods are less than (or equal to) the value $\sum_{x \in H_l^i} x$ (see definition 3). In all cases, the duration of this sum should be less than the maximum period T_m of the tasks with critical sections in H_l^i, otherwise all these higher priority tasks could block repeatedly the task τ_i and in this case $B_{l,\cdot}^i$ can be prohibitively large or even unbounded (condition of formula 2). Hence the Theorem follows. $\square$

One main implicit assumption has been made throughout our previous analysis. We assumed that the points of time at which the successive instances of a periodic task τ_i request their shared resources, i.e. execute the $P()$ operation, are separated by not less than the task period T_i. This assumption has also been made in the distributed priority ceiling protocol where it is noted that a technique called *Period Enforcer* can be used to ensure that the property holds [5].

The total worst-case blocking duration B_i experienced by task τ_i is the sum of all these blocking durations $\{B_{\cdot,j}^i \mid 1 \leq j \leq \nu\}$. Once these blocking terms B_i, $1 \leq i \leq n$, have been determined, conditions (1) give a fairly complete solution for the real-time task synchronization and scheduling in the parallel environment. It is a fairly complete solution in the sense that the proposed protocol provides sufficient conditions to test whether a given task set is schedulable. The general problem of determining whether a given task set where tasks

Parameters of Task Set				
Task	Period	C_{non-cs}	CS		
			$CS^1_{1,1}$	$CS^1_{2,2}$	
τ_1	8	1	1	1	
			$CS^2_{2,1}$	$CS^2_{1,2}$	$CS^2_{3,3}$
τ_2	19	2	1	4	3
			$CS^3_{1,1}$	$CS^3_{3,2}$	
τ_3	24	6	4	2	
			$CS^4_{1,1}$		
τ_4	27	8	1		

Table 1. Parameters of Task Set in Example

share data can meet its timing requirements is NP-hard even in a uniprocessor environment [5].

We have to notice here that the worst-case blocking duration B of a task is a function of critical section durations only and the impact of B is minimal by letting a lower priority task rather than a higher priority task to experience blocking. In order to have a better view of the proposed synchronization protocol and the schedulability analysis for a set of real-time tasks we'll give an illustrative example.

5   A Schedulability Analysis Example

We illustrate the schedulability analysis based on the proposed synchronization protocol with the following example. This walk-through example is also used to illustrate the fact that the evaluated blocking times of the tasks serve as an upper bound and thus the schedulability of these tasks can be determined.

Example 5.1 Consider a parallel environment and one application which is comprised of four tasks and eight critical sections guarded by semaphores S_1,S_2 and S_3. Each task τ_i is bound to processor $\wp_i$, the periods and computation times within and outside critical sections for each task are listed in Table 1 and execute the following sequence of steps in each period:

$\tau_1 = \{ \ldots,P(S_1),\ldots,V(S_1),\ldots,P(S_2),\ldots,V(S_2),\ldots,\}$
$\tau_2 = \{ \ldots,P(S_2),\ldots,V(S_2),\ldots,P(S_1),\ldots,V(S_1),\ldots,P(S_3),\ldots,V(S_3),\ldots,\}$
$\tau_3 = \{ \ldots,P(S_1),\ldots,V(S_1),\ldots,P(S_3),\ldots,V(S_3),\ldots,\}$
$\tau_3 = \{ \ldots,\ldots,P(S_1),\ldots,\ldots,V(S_1),\ldots,\}$

The worst-case blocking durations of each task can be determined using the formula 2 as follows:

Task τ_1:

- $H^1_1 = \emptyset$ (The set of critical sections used by tasks having higher priorities than τ_1 and guarded by the semaphore S_1)

- $L_1^1 = \{CS_{1,2}^2, CS_{1,1}^3, CS_{1,1}^4\}$ (The set of critical sections used by tasks having lower priorities than τ_1 and guarded by the semaphore S_1)
- $\beta_1^1 = \max(\{CS_{1,2}^2, CS_{1,1}^3, CS_{1,1}^4\}) = 4$
- $B_{1,1}^1 = \beta_1^1 + \sum_{x \in H_1^1} x + \sum_{y \in \Delta_1^1} y = \beta_1^1 = 4$ since $H_1^1 = \emptyset$ and $\Delta_1^1 = \emptyset$

- $H_2^1 = \emptyset$
- $L_2^1 = \{CS_{2,1}^2\}$
- $\beta_2^1 = \max(\{CS_{2,1}^2\}) = 1$
- $B_{2,2}^1 = \beta_2^1 + \sum_{x \in H_2^1} x + \sum_{y \in \Delta_2^1} y = \beta_2^1 = 1$

  Thus, the total worst-case blocking duration for task τ_1 is given by the formula $B_1 = B_{1,1}^1 + B_{2,2}^1 = 4 + 1 = 5$.

Task τ_2: Similarly working we take the results $B_{2,1}^2 = 1$, $B_{1,2}^2 = 5$ and $B_{3,3}^2 = 2$.
   Thus, $B_2 = B_{2,1}^2 + B_{1,2}^2 + B_{3,3}^2 = 1 + 5 + 2 = 8$.
Task τ_3: $B_{1,1}^3 = 6$, $B_{3,2}^3 = 3$. Thus, $B_3 = B_{1,1}^3 + B_{3,2}^3 = 6 + 3 = 9$.
Task τ_4:

- $H_1^4 = \{CS_{1,1}^1, CS_{1,2}^2, CS_{1,1}^3\}$
- $L_1^4 = \emptyset$
- $\beta_1^4 = 0$
- $B_{1,1}^4 = \beta_1^4 + \sum_{x \in H_1^4} x + \sum_{y \in \Delta_1^4} y = 0 + (CS_{1,1}^1 + CS_{1,2}^2 + CS_{1,1}^3) + (CS_{1,1}^1) = 0 + (1 + 4 + 4) + 1 = 10$ since $\Delta_1^4 = \{CS_1^1\}$ (notice that $\sum_{x \in H_1^4} x \geq T_1$).

  The fact that Δ_1^4 is not empty indicates that task τ_1 can block in its next continuous execution the task τ_4. More precisely, at this worst-case phasing of tasks, τ_1 will ask to enter its critical section $CS_{1,1}^1$ for a second time after exactly eight time units ($T_1 = 8$).
  Thus, $B_4 = B_{1,1}^4 = 10$.

Now that the blocking times for each task have been determined, the next step is to determine whether each task can meet its deadline during execution. Conditions (1) will be used for this test taking into account the $CS_{\cdot,\cdot}^i$ and C_{non-cs}^i values of tasks listed in Table 1 as well as the equation $C_i = C_{cs}^i + C_{non-cs}^i$. Evaluating the conditions we have:

- i=1 $C_1 + B_1 = 3 + 5 = 8 \leq 8$ ($T_1 = 8$)
- i=2 $C_2 + B_2 = 10 + 8 = 18 \leq 19$ ($T_2 = 19$)
- i=3 $C_3 + B_3 = 12 + 9 = 21 \leq 24$ ($T_3 = 24$)
- i=4 $C_4 + B_4 = 9 + 10 = 19 \leq 27$ ($T_4 = 27$)

We therefore conclude that the above set of tasks $\{\tau_1, \tau_2, \tau_3, \tau_4\}$ is schedulable in a multiprocessor environment where four processors are supported for the execution of the above task set.

6   Conclusions

In this paper, we have presented a fixed priority scheduling mechanism with an integrated resource allocation model that it is analyzable and understandable at a high level, meaning that given a set of tasks we know in advance if they will meet their deadlines or not. This scheduling strategy has been designed for real-time applications that need a parallel computing environment where every real-time task is allocated on a different node or processor and can require the use of global resources. The proposed synchronization protocol places an upper bound on the task blocking duration and once the blocking durations have been computed the conditions (1) can be used to determine the schedulability of a set of tasks.

We have to notice at this point the applicability of the method in other areas except that of real-time systems, like the area of distributed multimedia systems. The proposed scheduling and resource management strategy can be regarded as an important step in meeting the objectives of these systems. A set of distributed multi-media applications running in a network of computers sharing a number of resources such as video and audio units, cameras or file systems can be modelled as a set of periodic tasks that require the use of non-preemptable resources [3]. Hence, with the theory presented in this paper one is able to determine analytically whether the timing and synchronization requirements of a distributed multi-media application will be met, and if not, which requirements will fail.

References

[1] J.P. Lehoczky, L. Sha, J.K. Strosnider, and H. Tokuda. Fixed Priority Scheduling Theory for Hard Real-Time Systems. In A.M. van Tilborg and G.M. Koob, editors, *Foundations of Real-Time Computing: Scheduling and Resource Management*, chapter 1, pages 1–30. Kluwer Academic Publishers, 1991.

[2] C. Douglass Locke. Software Architecture for Hard Real-Time Applications: Cyclic Executives vs. Fixed Priority Executives. *Real-Time Systems*, 4:37–53, 1992.

[3] C. Mourlas, David Duce, and Michael Wilson. On Satisfying Timing and Resource Constraints in Distributed Multimedia Systems. In *Proceedings of the IEEE International Conference on Multimedia Computing and Systems ICMCS'99*. IEEE Computer Society, 1999.

[4] C. Mourlas and C. Halatsis. Task Synchronization for Distributed Real-Time Applications. In *Proceedings of the Ninth Euromicro Workshop on Real-Time Systems*, pages 184–190. IEEE Computer Society, 1997.

[5] Ragunathan Rajkumar, editor. *Synchronization in Real-Time Systems: A Priority Inheritance Approach*. Kluwer Academic Publishers, 1991.

[6] L. Sha, M. Klein, and J. Goodenough. Rate Monotonic Analysis for Real-Time Systems. In A.M. van Tilborg and G.M. Koob, editors, *Foundations of Real-Time Computing: Scheduling and Resource Management*, chapter 5, pages 129–155. Kluwer Academic Publishers, 1991.

[7] L. Sha, R. Rajkumar, and J.P. Lehoczky. Priority Inheritance Protocols: An Approach to Real-Time Synchronization. *IEEE Trans. on Computers*, 39(9):1175–1185, September 1990.

Designing Multiprocessor/Distributed Real-Time Systems Using the ASSERTS Toolkit

Kanad Ghose[1], Sudhir Aggarwal[1], Abhrajit Ghosh[1], David Goldman[1],
Peter Sulatycke[1], Pavel Vasek [1], and David R. Vogel [2]

[1] Department of Computer Science, State University of New York,
Binghamton, NY 13902-6000
ghose@cs.binghamton.edu

[2] Lockhead-Martin Federal Systems, Owego, NY

Abstract. We describe a real-time software engineering toolkit called ASSERTS that provides complementary analysis and simulation capabilities to assist the designer of multiprocessor/distributed real-time systems. ASSERTS allows users to describe the parameters of the hardware platform, interconnection and the kernel, the programming model (shared or distributed memory) and the task system for the purpose of simulation and analysis. The simulation component of ASSERTS is quite detailed and features a number of *built-in* simulation models for practical real-time schedulers, interconnections as well as models of resource access protocols (such as priority inheritance and priority ceiling protocol). Users can describe the behavior of the tasks at various levels of abstraction using a fairly small set of about 20 macroinstructions for the purpose of simulation.

1. Introduction

Real-time system design has emerged as a mature computer science/engineering discipline over the recent years. This has been made possible through the evolution of a science for real-time systems and the availability of formal analysis and simulation tools for the real-time software design process itself [ABRW 94], [LRD+ 93], [StLi 96], [Ti98]. Analytical techniques for testing the validity of timing specifications of real-time typically take the form of schedulability tests. Schedulability tests of a rigorous and sufficiently general nature exists for uniprocessor systems and allow software designers to get a hard guarantee on the timing performance when the test passes. Unfortunately, when the schedulability tests fail for an uniprocessor system, it rarely tells the designer of the sequence of events that lead to the timing violation. With multiprocessor/distributed systems, there is a marked lack of scheduling theories and schedulability tests of a general nature. This is primarily due to the delays introduced by interactions among tasks, which reflect task dependencies and introduce delays that depend on potential contention within the interconnection. While such delays may be easily enumerated for a small number of tasks for the purpose of analysis, the situation quickly gets out of control even with a moderate number of tasks on a system with a few nodes. This is precisely where accurate simulation techniques are useful. We thus believe that analysis and simulation play complementary and useful roles in the design of real-time software, particularly for multiprocessor/distributed real-time systems.

P. Amestoy et al. (Eds.): Euro-Par'99, LNCS 1685, pp. 505-510, 1999.

This paper focuses on facilities within a prototype real-time software design environment called ASSERTS (*A Software Simulation Environment for Real Time Systems*) that supports such capabilities. Additionally, ASSERTS also incorporates an analysis component to do formal schedulability tests where applicable. In particular, the simulation component of ASSERTS can identify the duration of blocking of a task due to dependencies with other tasks (located in possibly different nodes), which can then be used in analytical tests, such as the progressive schedulability test [SRL 88]. ASSERTS is currently available for Solaris, AIX, NT and Windows 95 platforms.

2. Overview of ASSERTS

ASSERTS has been specifically designed to support real-time software design for both uniprocessor and multiprocessor systems and incorporates both analysis and simulation components that can be used by the software designer starting from the early design phase. Figure 1 summarizes the inputs and outputs of ASSERTS. Platform and kernel details include specifications of the CPUs (relative speed, RAM size), real-time kernel details (scheduling discipline, context-switching time etc.) and details of the interconnections used (types, parameters, software layer delay parameters etc.). Details of the real-time tasks include a description of tasks, their placements on the CPUs, inter-task dependencies (through messages or accesses to shared variables). Each task is described using two components - the task attributes (periodicity, period, deadline, static priority - if any, CPU on which task is to be run etc.) and the task body, which is the functional description of the various phases of the task. The task bodies can be progressively refined as the design evolves. The duration of the compute phases of a task are described using the notation {seconds.millisecs.microsecs.nanoseconds}; the time quoted is the duration on a reference CPU. Specified timing durations gets scaled appropriately based on the relative speed of the CPU on which the task is run. The delays of all other phases of a task are obtained through actual simulation.

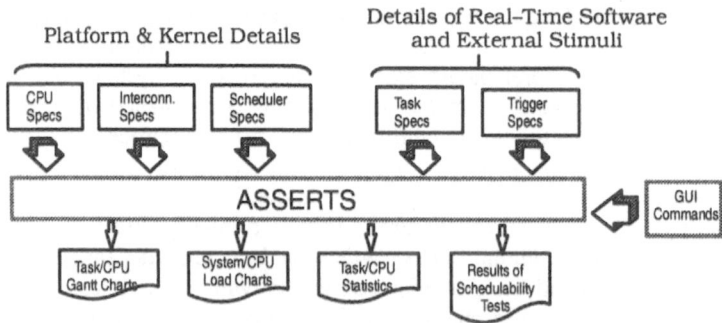

Figure 1. Inputs and outputs for ASSERTS

Most of the descriptions can be entered and altered through a GUI. Users may specify the outputs needed from ASSERTS through the GUI. These outputs could be Gantt charts (for individual tasks, CPUs, interconnection links etc.), load charts (scheduler queue states, blocked task queues, CPU ready task queues etc.), detailed logs or the result of analytical tests (such as the Liu/Layland test, progressive schedulability tests [SRL 88], specific rate-monotonic schedulability tests for PCP etc.) .Other analytical tests are also possible for other types of schedulers (such as EDF, cyclic executives etc.) for uniprocessor systems.

When multiple Ganntt charts are opened, one per window, the cursors in the various windows can be synchronized to allow cause-effect events to be easily identified. These windows also allow users to zoom-in or zoom-out, or scroll along the time scale, again in a synchronized fashion.   Task phases displayed within the task windows are color coded and users can click on any portion of a Ganntt chart to see what macroinstructions they correspond to within the body of a task.   Similar facilities are available for the other types of windows that can be opened (for CPUs, interconnection etc.).

We believe that a particularly useful feature in ASSERTS is the ability to progressively describe the functional behavior of a task at a very high level as its design evolves. This description identifies various phases of a task, such as the duration of computation phases, task interaction (via message passing or shared variable access), conditional waiting etc.   A small set of "macroinstructions" are defined to describe each phase of a task (see [GAV+ 97). Initially, when the details of a task are unclear, the task may be described as a single computation phase; its description may then be progressively refined by adding other phases as the design evolves. Figure 2 shows the progressive refinement of the description of a task. The functional description of a task is used by the simulation component of ASSERTS, along with the task attributes.   The parameters used by the analytical tests are typically gleaned from the task attributes.

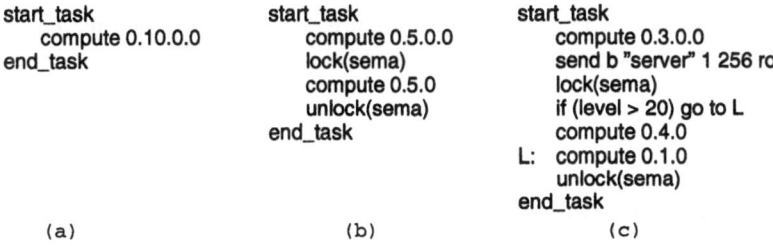

Figure 2. Progressive refinements of the functional behavior of a task from the initial description based on the allocated timing budget shown in (a).

3. Message Passing, Interconnections and Shared Memory Access

At the level of the tasks, user's describe communication phases within tasks using the following send-receive macroinstructions:

```
send <type> <destination> <message_tag>  <message_size> [<waiting_mode>]

receive  <type> <source> <message_tag>  <buffer_mode> [<waiting_mode>]
```

The send and receive macroinstructions specify the type of the primitive used. For the send, the type can be blocking (sending task waits till the message transmission is completed), blocking with an acknowledgement (sending task block till the receiving task sends an acknowledgement back) and non-blocking. For a `receive` the primitive can be either blocking or non-blocking. The recipient task and the sending task are explicitly named in the `<destination>` and `<sender>` fields, respectively. A message tag is also included to match up the sends and receives between a pair of tasks. The `<message_size>` field gives the message size in bytes for a send. For a receive, the `<buffer_mode>` field indicates the mode of reading the buffer (read only the matching message, read the first message in the buffer irrespective of the message tag and read the first message irrespective of its tag). The `<waiting_mode>` field specifies the type of waiting where applicable - busy (i.e., spin) waiting or sleep waiting. The send and receive macroinstructions thus allow a wide variety of communication primitives to be modeled.

The simulation component simulates the basic actions of these primitives to determine the times needed for the task phases described by these macroinstructions, including any contention delays on the interconnection. Another variant of the communication macroinstructions also exist - these are used for describing the actual message contents.

In the interest of accuracy, ASSERTS models both the software, hardware and queueing delays of message passing. The software delays are due to copying delays between the user's and kernel's address space, as is typical in many systems. The number of copying steps may be specified as one (direct DMA between user-space buffers and the network interface) or two (transfers between the user-space buffer and the network interface going through a kernel-level buffer). The delay of each copy step is given by an expression of the form:

$$T_{copy} = \text{flat\_overhead} + \text{constant} * \text{message\_size}$$

Consequently, ASSERTS requires users to input the flat overhead and the constant. The actual software delay(s) is(are) then computed based on the message size. The "hardware" components of the delays in message passing are specific to the type of the interconnection used, as well as on the interconnection topology. For convenience, ASSERTS incorporates standard simulation models of a variety of commonly used interconnections, such as Ethernet, IEEE Futurebus and point-to-point interconnections (such as ATM, Myrinet).

The simulation models for these interconnections capture the impact of contention, arbitration and propagation delays. An extensive C++ class library also exists that allow advanced users to model other types of interconnections. The interconnection topology is specified by the user, as part of the platform description. Relevant delay parameters for each type of interconnection are also specified. These delay parameters can be easily altered through the ASSERTS GUI to undertake what-if studies, such as gauging the impact of moving from 155 Mbits/sec. ATM to 1.2 Mbits/sec ATM etc. or moving from a 10 Mbits/sec. Ethernet to a Futurebus interconnection.

Any queueing delay in message passing is also captured in the simulations. One use of simulation will be to extract accurate blocking times due to dependencies across tasks, which can then be used in analytical schedulability tests. Events occurring on the interconnection links for message passing can be seen on the Ganntt charts for the interconnections. Queue states for the network interface can also be seen through the GUI. These information are extremely useful in identifying communication bottlenecks, errors in programs or in performing post mortems when timing violations are detected.

There are several aspects to modeling shared variables in ASSERTS. First, the location of shared variables within the system are described using a statement of the form:

```
shared <variable_name> <location> <size>
```

The `<location>` field specifies the CPU node where this variable is located. Within the task body (which is the functional description of a task), shared variables are accessed by each of the following macroinstructions:

```
lock(<variable>)  [<wait_type>],  unlock(<variable>)
wait(<condition>)  [<wait_type>]
if (<condition>) go to <label>
```

where the condition is a relational expression involving one or more shared variable

```
<variable_1> = <variable_2>|<literal>  (assignment)
```

Since the shared variable accessed by a task using one of the above macroinstructions can be located on a different node, the appropriate delay has to be accurately modeled. This model depends on the type of primitives that are used in the actual system: the remote access can be implemented on top of message passing primitives or as a direct access to a memory module (as in a bus connected system, with a memory module board and CPU boards). In the first case, the delay in accessing a shared variable may be modeled as a send-receive pair. In the latter case, the delay has to be modeled as a request-response pair on the interconnect to the relevant memory module, with a specification of the access time of the memory module (plus any queueing delay at the module) as service time of the request. In ASSERTS, for this latter case interconnect cycles corresponding to remote reads, writes and a read-modify-write cycle (for locking) are explicitly modeled. Users specify the memory access time and software delays in the access path, if any. Interconnect events of the appropriate type are automatically generated at the time of compiling the macroinstructions that refer to remote shared variables. These events are simulated, as in the case of message passing, for the specified interconnection.

As in the case of message passing, the simulation can provide an accurate estimate of the task blocking times due to shared variable access, which can then be plugged into analytical tests for schedulability. Also, similar to the case for message passing, windows can be opened through the GUI to see the interconnection events due to shared variable access, as well as the state of relevant queues.

4. ASSERTS and Other Real-Time Design Tools

ASSERTS is only one of a large number of tools that have been designed to assist real-time software development and design. PERTS [LRD+ 93], now marketed by Wind River (and its simulation component, DRTSS [StLi 96]), STRESS [ABRW 94] and the TimeWiz tool offered by TimeSys Corp. [Ti 98] offer a comprehensive system-independent analysis and simulation facility, similar to ASSERTS.

The PERTS tool uses a task graph model to describe the task system, where a task is essentially a computation phase preceded and terminated by optional communication phases, making the model somewhat unrealistic. PERTS includes a task allocation component and implements a schedulability test based on assuming idealistic, contention-free communication. Progressive refinement of the task bodies are not possible in PERTS, which also lacks models for interconnections. STRESS and TimeWiz are both quite similar to ASSERTS. STRESS, however, does not include built-in models for interconnections and scheduling algorithms; its GUI also appears to be somewhat limited. TimeWiz is the closest analog of ASSERTS, but like PERTS it fails to support progressive refinements of task bodies. Unlike ASSERTS, TimeWiz incorporates extensive design documentation generation facilities.

Our current efforts include the addition of an allocation component for ASSERTS, a two-level scheduling algorithm and the ability to interface with UML design tools. ASSERTS is also being used within Lockheed-Martin on a limited scale. Examples of the screen shots for ASSERTS, as well as an on-line manual for the ASSERTS 1.0 release can be found using the url: http://cs.binghamton.edu/~ghose/ASSERTS.

References

[ABRW 94] Audsley, N. C., Burns, A., Richardson, M. F., and Wellings, A. J., "STRESS: a Simulator for Hard Real-Time Systems", in Software-Practice and Experience, Vol. 24(6) (June 1994), pp. 543-564.

[GAV+ 97] Ghose, K., Aggarwal., S., Vasek, P., et al, "ASSERTS: A Toolkit for Real-Time Software Design, Development and Evaluation", in Proc. 9-th Euromicro Workshop on Real-Time Systems, 1997, pp. 224-232.

[LRD+ 93] Liu, J. W.-S., Redondo, J.-L., Deng, Z. et al, "PERTS: A Prototyping Environment for Real-Time Systems", in Proc. 14th. IEEE Real-Time Systems Symposium, 1993, pp. 184-188

[SRL 88] Sha, L., Rajkumar, R. and Lehoczky, J. P., "Priority Inheritance Protocols: An Approach to Real-Time Synchronization", Tech. Report, Software Engineering Institute, CMU, May 23, 1988.

[StLi 96] Storch, M. F., and Liu, J. W.-S., "DRTSS: A Simulation Framework for Complex Real-Time Systems", in Proc. IEEE Real-time Technology and Applications Symposium, 1996, pp. 160-169.

[Ti 98] Timesys Inc. web pages at: http://www.timesys.com

UML Framework for the Design of Real-Time Robot Controllers

L. Carroll, B. Tondu, C. Baron, and J.C. Geffroy

LESIA - INSA, Complexe Scientifique de Rangueil
31077 Toulouse Cedex 4, France
Luis.Carroll@insa-tlse.fr

Abstract. New types of robotic applications, where tasks are accomplished in cooperation with humans and under unstructured environments, have led researchers to consider dependability requirements as fundamental design criteria. Particularly, robot controllers must include additional features allowing to cope with system failures in order to improve reliability and safety. In this article, we suggest the use of the UML within a development method including dependability means, for the design of real-time robot controllers.

1 Introduction

Robot controllers integrate the set of high level processes (robot programming languages and task elementary motion planning) and low-level processes (trajectory generation and joint control). Specific properties coming from electronic and mechanical components and specific properties from the application environment lead industry to develop dedicated controllers without generic considerations (flexibility according to upgrade evolutions or environment perturbations, etc.). In addition, the growing complexity of robot functionalities (integration of external data, management of redundant joints, position and force control, etc.) implies the mastering of the design of evolutionary systems. Moreover, in new types of applications, robots have to cooperate with human operators and to share an unstructured workspace. Therefore, safety performances are required.

2 Robotic Systems Requirements

We consider a robotic system as composed of a robot and its controller, and the service it delivers as the completion of a task. The controller provides the manipulator with the intelligence to perform tasks as described by its user. These tasks generally consist of a set of actions intended to position the robot's end-effector for the manipulation of objects. A failure occurs when the robot deviates from the desired trajectory. This failure can be the result of changes in the robot kinematic parameters or problems in the system software or hardware. Failures resulting from changes in robot parameters can be avoided by implementing

P. Amestoy et al. (Eds.): Euro-Par'99, LNCS 1685, pp. 511–514, 1999.

highly robust control laws. Failures due to hardware or software call for a particular analysis on the critical components of the system. Recent research has been done on different aspects of dependability in the field of robotics [1], [2].

The basic robotic systems requirements deal with the final static and dynamic performances and with dependability criteria: *availability*, *reliability*, *maintainability* and *safety*. Particularly, robot safety is concerned with preventing the occurrence of damage to the robot itself and preventing the robot from damaging its environment, e.g., human operators [3]. Rules and guidelines of safety standards and a selective use of engineering technologies may be used in order to guarantee a safe robot behavior according to its environment. Besides, we identify two other groups of criteria associated with the robustness of the software architecture (reusability, modifiability and traceability) and the quality of the development process (verifiability, understandability, development ease and expression capability).

3 Means for Dependability

A fault can cause the system to get into an error state. This error state will lead the system into an undesirable behavior considered as a failure. Faults can be introduced during the development of the system or they can occur during the useful lifetime. In both cases, we must identify the possible techniques to cope with them, and the system components potentially involved in its origins.

Concerning development faults, fault avoidance and fault removal methods must be used during the specification, design and implementation of the controller. These techniques reduce the faults introduced during the system development by humans and the tools they use. Concerning faults occurring during the useful lifetime, we must implement a real-time controller able to cope with system failures. This controller acquires information data of the current system state from the robot sensors or other sensors located in its workspace and from the controller itself. The different software modules inside the controller make the robot react in the desired way. It is then suitable to include on-line testing for fault detection, aimed at detecting faults not only from sensors or actuators, but also from the controller itself.

In order to develop a dependable robot controller, we consider that the use of a real-time system development method must be associated to a dependable methodology including fault prevention, fault removal and fault tolerance processes as proposed by [4]. This methodology deals with the avoidance and removal of faults during the development phase, and also enables the introduction of fault tolerant routines inside the normal life cycle of the controller.

4 A Real-Time Robot Controller

We have chosen the UML [5] for the modeling of a real-time robot controller. The comparative study we have presented in [6] highlights the qualities of the UML for the design of the controller. Besides its description and representation

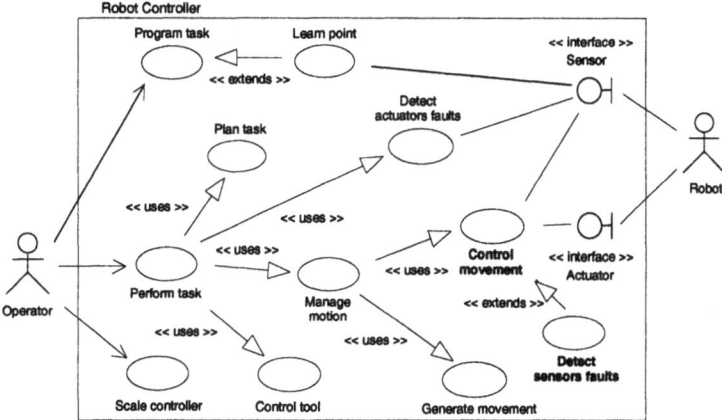

Fig. 1. Robot controller Use Cases Diagram.

capabilities, it offers large possibilities for the improvement of dependability. Particularly, Use Case Diagrams are well dedicated to the capture of requirements [7]. They drive the designer during the development process and assist the functional testing of the software. Figure 1 shows the Use Cases of the robot controller considered (each one represented by an ellipse). Use Cases may stand on their own, they may use capabilities specified in other Use Cases and they may be extended by other Use Cases. These relations ("uses" and "extends") are represented by arrows. Actors (operator and robot) are people or things that interact with the system (robot controller). They may communicate with Use Cases by the way of interfaces (robot sensor and actuators).

A scenario, generally represented by a Sequence Diagram where objects interact for the accomplishment of the requirement, expresses each Use Case. Figure 2 shows the interaction between different objects performing the "Control movement" Use Case. This Use Case is extended by the "Detect sensor faults" Use Case as represented in figure 1. This scenario is particularly interesting because it shows on-line detection of sensor faults within the sampling period of the control loop. The use of functionally redundant sensors allows the Fault detector objet to verify data and communicate faults to the Task Planner object.

These UML visual elements are well adapted to capture dependability requirements. For example, Sequence Diagrams facilitate functional test and help to represent temporal constraints and behavior of objects. This feature is useful during the implementation stage in RTOS.

5 Conclusions and Future Work

The evolution of robot systems leads to new types of applications where human beings can actively interact with robots. Dependability must be considered as an important design criterion, providing people with safety inside the robot workspace. The choice of a development process, from the initial specification to

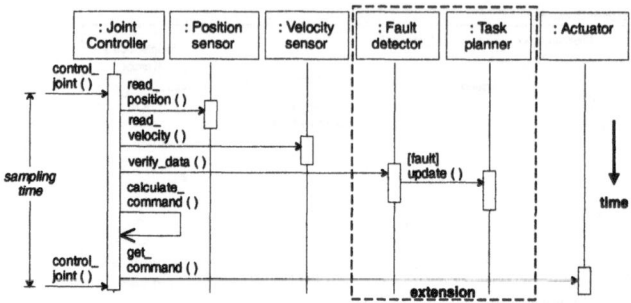

Fig. 2. Sequence diagram for the control of joints including sensors fault detection.

the final robot controller implementation, is important in relation to the functionality and dependability of the whole system. Fault avoidance, fault removal and fault tolerance techniques complement this process during all the system's life cycle.

We have developed a robot controller based on a VxWorks RTOS using the proposed method for the control of a 2 DOF SCARA-type robot. The use of the UML resulted to be appropriated for the controller development and maintainability. It is important to mention that adaptive and perfective maintenance becomes easier for the removal of specification faults and improvement and test of different detection and tolerance routines. This allows the implementation of the different techniques proposed by other authors in the field of fault tolerant robotic systems. Anyhow, it must be completed with formal methods or simulation tools aimed at the verification and validation of the system.

References

[1] Visinsky M.L., Cavallaro J.R., Walker I.D.: Robotic Fault Detection and Fault Tolerance: a Survey, Int. J. Reliability Eng. and System Safety, Vol. 46 (1994) 139-158

[2] Tso K.S., Hecht M., Marzwell N.: Fault-Tolerant Robotic System for Critical Applications, IEEE Int. Conf. Robotics and Automation, Atlanta, GA, USA (1993) 691-696

[3] Dhillon B.S.: Robot Reliability and Safety, Springer-Verlag (1991)

[4] Laprie J.C.: Dependability of Computer Systems: Concepts, limits, improvements, IEEE Int. Workshop CAD, Test and Evaluation for Dependability, Beijing, China (1996) 23-33

[5] Booch G., Jacobson I., Rumbaugh J.: The Unified Modeling Language for Object-Oriented Development, Version 1.0, Rational Software Corporation (1997)

[6] Carroll L., Tondu B., Baron C., Geffroy J.C.: Comparison of Two Significant Development Methods applied to the Design of Real-time Robot Controllers, IEEE Int. Conf. Systems, Man and Cybernetics, LaJolla, USA (1998) 3394-3399

[7] Jacobson, I.: Object-Oriented Software Engineering: an use case driven approach, Addison-Wesley, New York (1992)

Software Implemented Fault Tolerance in Hypercube

D.R. Avresky[1] and S. Geoghegan[2]

[1] Dept. of Electrical and Computer Eng., Boston University
8 Saint Mary's Street, Boston, MA 02215
avresky@bu.edu
[2] Computer Science Dept., Texas A&M University
College Station, TX 77840

Abstract. This paper presents Software Implemented Fault Tolerance[1] (SIFT) for hypercubes which is implemented by means of a software layer. It is written in each node of the nCube parallel computer.The SIFT utilizes an error detection application software and fast reconfiguration algorithm for avoiding faulty nodes. The Balance Spanning Tree (BST) is used for embedding tree-based algorithms into the hypercube topology. Any single faulty node in the hypercube can be tolerated by the software layer. More than 90% of the multiple faults can be tolerated without backtracking. The SIFT approach has been successfully implemented for a quadtree data compression algorithm for 64x64, 128x128 compressible and uncompressible data. The experiments were run on 4 and 16 node nCubes. The time overhead (reconfiguration and recomputation time) incurred by the injected fault was experimentally estimated. The coverage factor, provided by the error-detection software, has been estimated by means of nSOFIT for the quadtree data compression algorithm.

1 Introduction

This paper presents Software Implemented Fault Tolerance (SIFT) for hypercubes which is implemented by means of a software layer, written in each node of the nCube parallel computer, with error-detection application software and fast reconfiguration algorithm for avoiding faulty nodes. The Balance Spanning Tree (BST) is used for embedding tree-based algorithms into the hypercube topology. A spanning tree is a connected graph that spans the nodes of the graph, forming a tree with no cycles [3].

Spanning trees have been shown to be useful in broadcasting and multicasting algorithms [4]. They also have been used in optimal communications and personalized communication in hypercubes [4]. Reconfiguration of faulty hypercubes using GST, BST and CUST are presented in [1]. The Balance Spanning Tree (BST) is used for embedding tree-based algorithms into the hypercube topology.

---

[1] This work was funded by National Science Foundation Grant MIP-963096

P. Amestoy et al. (Eds.): Euro-Par'99, LNCS 1685, pp. 515–518, 1999.
© Springer-Verlag Berlin Heidelberg 1999

Any single-node fault in the hypercube can be tolerated by the software layer. More than 90% of the multiple faults can be tolerated without backtracking. The faults in the application level in each node are injected by means of the parallel version of the Software Fault Injection Tool (nSOFIT) for nCube. The software based fault injection tools are widely used for validating the error detection software and mechanisms [2]. The time overhead (reconfiguration and recomputation time) incurred by the injected fault is experimentally evaluated. The proposed algorithm is scalable, i.e., by increasing the size of the hypercube the reconfiguration time is constant and the recomputation time is decreasing. For reconfiguration and recovery, we make the assumptions, which are typical ones related to fault tolerance of hypercubes [1].

2 Balanced Spanning Tree (BST) and Data Distribution

The reconfiguration algorithm for a BST is based on finding a new path that avoids the faulty node. We have the same assumptions, and if any node f is detected as faulty, the reconfiguration process takes place. The spanning tree is reconstructed by connecting the children of f to the parent of f via one unused link and another link that might have been used before failure.

In [1], the new parent was computed as $v_{new} = v_c \oplus v_f \oplus v_p$ where v_f is the faulty node, v_c is a child of v_f, v_p is the parent of v_f and v_{new} is the replacement for v_f. The path $v_c v_{new} v_p$ is then used to replace the path $v_c v_f v_p$. It's been proven the large portion of the multiple faults and any single faulty link can be tolerated [1].

Construction of a complete reconfiguration algorithm must consider some data distribution and result collection factors. The typical algorithm using the BST involves three phases. First, the data is distributed to all the nodes in the BST. Second, all the nodes process their respective parts of the data. Finally, each node sends its results up to the root of the BST. During any of these steps, a node may fail and the parent and children nodes must invoke the reconfiguration algorithm to tolerate the fault. Thus, the algorithm must ensure that the input data is properly distributed to all the nodes, the data assigned to the faulty node are processed by an alternate node, and the results are collected at the root of the BST. The data must be distributed in such a way that all the nodes receive the data that is to be processed on that node. Following the failure of a node, all the fault-free nodes must still receive the appropriate data. In addition, an alternate node must have access to the data assigned to the faulty node in order to process the data in place of the faulty node. To meet these criteria, the distribution is achieved by first partitioning the data and assigning one partition to each BST node. All partitions are then sent to the root of the BST. Next, the root sends each of its children all the partitions assigned to the nodes in the subtree rooted by that child. This process continues until all the nodes receive the appropriate data. This distribution algorithm ensures that all the ancestors of each node, i, contain the data assigned to node i.

3   Quadtree Data Compression Algorithm

The SIFT approach for hypercube topology, using BST, is suitable for divide-and-conquer, sorting and similar tree based programs. A quadtree can be used to store and/or compress two-dimensional arrays or graphical regions such as those used for image processing [5]. To validate the SIFT approach, the quadtree data compression algorithm was written and executed on a BST embedded in a hypercube topology. Three separate test were run to measure different parameters of the quadtree algorithm. First, the overhead incurred by the reconfiguration step was measured. Next, the overhead incurred by the different fault-tolerance methods was measured. Finally, the percentage of fault detected achieved by the fault tolerance methods was estimated. The time required to execute the reconfiguration algorithm on each node has been measured and is presented in Table 1. Two versions of this experiment were run. In the first test, a 4-node BST was used in which $v_f = 1$, $v_c = 3$, $v_p = 0$, and $v_{new} = 2$. The second experiment involved a 16-node BST in which $v_f = 7$, $v_c = 15$, $v_p = 3$, and $v_{new} = 11$. In each experiment, a timer was placed in the code involved in the reconfiguration process. As the data compression program was executing, v_f was forced to fail by fault injection, using nSOFIT. Three different time overhead values are obtained since each of the three nodes involved in the reconfiguration process performs a different task. As seen by the values in the table, the overhead incurred in each node is very small and does not vary with the size of the BST.

Node	Overhead for 4 nodes (seconds)	Overhead for 16 nodes (seconds)
v_c	0.0005	0.0004
v_p	0.0006	0.0006
v_{new}	0.0004	0.0004

Table 1. Reconfiguration overhead and parameters

The results of different experiments are summarized in Table 2. The speedup in Table 2 is the ratio of the time it takes to run the quadtree algorithm on 16 nodes vs. the execution time on 4 nodes system.

The tool(SOFIT) [2] is designed to evaluate the coverage factor of the implemented error-detection technique by applying a software approach to injecting faults in a computer system while a user application is executing. In this experiment, faults of 10 cycles duration were injected into the address bus, CPU, data bus, and memory. The coverage factor was determined using SOFIT to inject faults into node 7 as the quadtree algorithm was executed on a 16 node hypercube array. The results are summarized in Table 3. These results have been used to estimate the coverage factor, provided by the duplex error- detection scheme.

Nodes (number)	Simplex w/out "I'm Alive"	Simplex with "I'm Alive"	Duplex w/out "I'm Alive"	Duplex with "I'm Alive"	Time for Duplex with "I'm Alive" reconf. and recomp. (s)	Recomp. overhead on vp (0,3)(s)
Uncompressible 64x64 data						
4	0.0759	0.0771	0.2065	0.2095	0.2109	0.0760
16	0.0363	0.0380	0.1252	0.1284	0.1299	0.0191
speedup	2.09	2.02	1.649	1.6316	1.62	3.97
Compressible 64x64 data						
4	0.0503	0.0515	0.1257	0.1283	0.1297	0.0877
16	0.0292	0.0309	0.0681	0.0706	0.0721	0.0221
speeedup	1.72	1.66	1.8458	1.8172	1.798	3.968
Compressible 128x128 data						
4	0.7417	0.7439	2.0905	2.2084	2.2097	0.7786
16	0.0653	0.0671	0.2139	0.2192	0.2207	0.00877
speedup	11	11.09	9.77	10.07	10.01	8.88

Table 2. Duration for quadtree algorithm

	OS Detected		FTAM Detected		Timeout		Incorrect Result		Correct Result	
Address bus	125	83.33%	25	16.67%	0	0.00%	0	0.00%	150	100.00%
Data bus	17	11.33%	108	72.00%	2	1.33%	23	15.33%	127	84.67%
Memory	109	72.67%	26	17.33%	2	1.33%	13	8.67%	137	91.33%
CPU	145	96.67%	5	3.33%	0	0.00%	0	0.00%	150	100.00%

Table 3. Coverage for the quadtree algorithm

References

[1] D. R. Avresky, "Embedding and Reconfiguration of Spanning Trees in Faulty Hypercube," *IEEE Transaction on Parallel and Distributed Systems,* vol 10, no.3, March 1999, USA.

[2] D. R. Avresky, S. J. Geoghegan, and P. K. Tapadiya, "A Software-Based Fault Injecton Tool", *International Journal of Computer Systems Science and Engineering,* vol 13, no. 6. pp. 125-135. November 1998.

[3] T. Cormen, C. Leiserson, and R. Rivest, *Introduction to Algorithms,* McGraw-Hill, pp. 498-513, 1990.

[4] S. Johnsson, and C. Ho, "Optimum Broadcasting and Personalized Communication in Hypercubes," *IEEE Trans. on Comp.,* Vol. C-38, No. 9, pp. 197-202, Sept. 1989.

[5] H. Samet, , "The Quadtree and related Hierarchial Data Structures ," *ACM Comp. Surveys,* 16:2, pp 187-260 1984.

Managing Fault Tolerance Transparently Using CORBA Services

René Meier and Paddy Nixon

Department of Computer Science, Trinity College, Dublin 2, Ireland
{Rene.Meier, Paddy.Nixon}@cs.tcd.ie

Abstract. Fault tolerance problems arise in large-scale distributed systems because application components may eventually fail due to hardware problems, operator mistakes or design faults. Fault tolerance mechanisms must be employed to reduce the susceptibility of a given system to failure. In this paper, we describe the design of an architecture to overcome potential application component failures, using *CORBA*, a distributed object middleware specified by the *OMG*. Of primary importance to this architecture is *OMG's CORBA Object Trading Service* as the mechanism to advertise and manage service offers for fault tolerant application components. This mechanism enables clients transparently to detect a failed connection to a service object, to discover a similar backup service object and to re-connect to it. This improves overall system stability and enables scalability.

1   Introduction

Fault tolerance problems [4], associated with the use of large-scale distributed systems [1][6], arise because application components may eventually fail due to hardware problems, operator mistakes or software faults [2]. Within most environments, and in particular within a banking environment, such failures are not acceptable.

The author [5] identifies four possible causes of application component failure [6] in a large corporate banking system, which is described in Section 2. All of them can be overcome by providing a fault tolerance mechanism that re-connects a client from a failed service to a similar backup service. In Section 3, we introduce such a mechanism, which is described more completely in [5]. It is essential that the suggested mechanism fit into the existing banking system. The prototype implementation of the fault tolerant architecture was successfully tested and evaluated as shown in Section 4.

2   The Target Banking System

In this Section, we present an existing large corporate banking system, into which the suggested fault tolerance architecture must fit. Although the proposed architecture

P. Amestoy et al. (Eds.): Euro-Par'99, LNCS 1685, pp. 519-522, 1999.
© Springer-Verlag Berlin Heidelberg 1999

was designed to fit into this banking system, it can easily be applied to any similar client-server based architecture.

The banking system is based on a three-tier client-server architecture. It is implemented using CORBA [2][3][7], a distributed object middleware specified by the OMG [7]. Clients are grouped together in locations accessing service objects in the middle-tier (service-tier). Service objects are logically associated with locations and may be located on one or more servers. The servers access a distributed database (database-tier) that is kept consistent using a data replication protocol.

The client-tier and the middle-tier are of particular concern. The middle-tier includes the *Service Manager* and the *Notification Service*, two components that will be used in the design of the suggested architecture.

- The *Service Manager* maintains the status of each of the service objects, using *pings*, requests that ask if the service is still running. It will shutdown and / or try to restart any service objects that have failed. The status of the service objects is used by the Notification Service to publish service object status notifications. The Service Manager guarantees the service objects to be fail silent.

- The *Notification Service* publishes several different types of notifications. A particular notification type is published in the event of a service object status change. To receive notifications, clients register with the notification type they are interested in.

2.1   Problem Statement, Requirements, and Assumptions

The banking system lacks a mechanism to re-connect clients from unavailable master services to backup services. Master services may not be available because of failures[1] or for maintenance reasons. Currently, banking system clients connected to an unavailable master service are not able to retrieve data requested by a user, despite the presence of other similar services (i.e. in another location) that could provide the requested data. The basic assumption is that re-connection delays and even a lower access performance are acceptable, but a total loss of a service is unacceptable.

It is essential that the suggested architecture fit into the existing banking system. Fault tolerance issues must be *hidden* from the client application program and therefore from the user and must also be Object Request Broker (ORB) [7] *independent*. The suggested architecture must support *on-line* application component management by configuration adjustment, without rebooting the rest of the system. It should provide a highly available system with a good trade-off between *system scalability* and *service performance*.

It is assumed that the architecture of the given banking system includes the service manager and the notification service components and that it adequately addresses single point of failure.

---

[1] Application component failures can be caused by service, server or network failures. Furthermore, an application component might be unavailable due to maintenance.

3   Managing Fault Tolerance

To solve the fault tolerance problems identified in the previous sections, we propose an architecture that introduces the OMG Object Trading Service [7] to the existing banking system as the mechanism to advertise and manage service offers for fault tolerant application components. The basic architecture is a simple low cost solution that solves the fault tolerance problem without involving the Notification Service. Clients query the Trading Service for master and backup service offers and cache the retrieved service references. If the service in use fails, the service user (client) is re-connected to the backup service. During re-connection, the service user is idle. Then, the client starts pinging the original service and re-connects the service user to it as soon as it is back on-line. Pinging is inefficient and causes unnecessary network traffic. To eliminate this, an *improved architecture* is proposed, that makes use of the Notification Service and a new component: the *Trading Service Manager*. This improvement lets clients use notifications to detect service failures and therefore eliminates the need to ping services[2]. This results in less network traffic and significantly reduced service user idle time due to re-connection in background. The Trading Service Manager also uses notifications to detect service failures and is responsible for marking invalid service offers in the trader.

A more detailed description of the basic and the improved architecture, its components and the configuration of the OMG Object Trading Service is omitted due to the limited space but can be found in [5].

4   Implementation and Evaluation

A prototype of the basic fault tolerance architecture was implemented in Java using Iona's OrbixWeb and OrbixTrader. Client side fault tolerance algorithms were implemented as smart proxy classes, in order to hide them from the client application program. Unfortunately, the smart proxy feature is ORB vendor specific. An ORB-independent way to implement a fault tolerance algorithm is to place it between the IDL interface and the client application program, thus removing transparency.

The prototype implementation was successfully tested on both Solaris and Windows NT platforms. During integration testing, services were killed to force clients to re-connect to backup services and to re-connect to the original service, as soon they were re-started. To evaluate the performance of the implemented prototype, we measured the duration of method invocation on a service object running on a remote server with and without (using bind) the fault tolerance mechanism. Furthermore, we measured the duration of re-connection to the backup and re-connection to the original service object running on a remote server. It was observed that invocation time on a bound service object and on a fault tolerant service object is *essentially identical*. Re-connection to the backup service object (worst case < 2 seconds) and re-connection to the original service object (worst case < 15 seconds) may both cause service user idle time. The measurements show that user idle time is within an acceptable range. The worst case is less than 15 seconds, as opposed to

---

[2] Instead of several clients having to ping the same service, the service manager pings the service and informs client on status change.

minutes or even hours due to a temporary or total loss of a service in absence of a fault tolerance mechanism.

5   Conclusions

This paper has described an architecture to transparently manage fault tolerance in a large-scale distributed system. The presented architecture is designed to fit into an existing banking system and allows the development of fault tolerant application components. Of primary importance to our solution is the inclusion of the OMG CORBA Object Trading Service into the fault tolerance architecture as the mechanism to advertise and manage service offers for fault tolerant application components. The mechanism enables clients transparently to detect a failed connection to a service object, to discover a similar backup service object and to re-connect to it. The architecture allows application component management by configuration adjustments, without re-booting the system, supports the different needs of the banking system's clients and adequately addresses the scalability requirements of the system's infrastructure.

A prototype of the suggested mechanism has been implemented and evaluated. The implementation shows that performance issues are appropriately addressed and that performance can be improved by completely implementing the architecture. The limitation of the implementation is the trade off that had to be made between fault tolerance hiding and ORB independence. In order to hide fault tolerance algorithms from the client application program, an ORB specific feature, the so called smart proxy, was used. This issue is subject of further research. In conclusion, it has been demonstrated that this architecture supports the development of fault tolerant distributed application components, which results in *improved availability*.

Acknowledgements. René Meier completed this project as part of the M.Sc. in Networks and Distributed Systems at Trinity College Dublin and is now a Ph.D. student in the Distributed Systems Group funded by the Higher Education Authority (HEA) of Ireland.

References

[1] J. Bacon, *Concurrent Systems.* Addison-Wesley, 1993.
[2] M. Banâtre and P. A. Lee, *Hardware and Software Architectures for Fault Tolerance.* Springer Verlag, 1994.
[3] S. Landis and S. Maffeis, *Building Reliable Distributed Systems with CORBA.* Theory and Practice of Object Systems, John Wiley, New York, April 1997.
[4] P. A. Lee and T. Anderson, *Fault Tolerance: Principles and Practice (second edition).* Springer Verlag, 1990.
[5] R. Meier and P. Nixon, Managing Fault Tolerance Transparently using CORBA Services. Technical Report TCD-CS-1999-05, University of Dublin, Trinity College, February 1999. http://www.cs.tcd.ie/publications/tech-reports.
[6] S. Mullender, *Distributed Systems.* Addison-Wesley, 1993.
[7] Object Management Group. http://www.omg.org.

Topic 07
Theory and Models for Parallel Computation

Michel Cosnard

Global Chair

This topic intends to discuss the current state of parallel computational and cost models, algorithms and complexity. An emphasis is on research that takes into account properties of real machines like communication cost and asynchrony. Famous examples are BSP, CGM, LogP, and PRAM. Developments, comparisons of, and algorithms for such models as well as studies of complexity issues are studied. Special consideration will be given to scalable, practical algorithms. Algorithmic strategies for abstract parallel models are encouraged, as well as algorithmic tuning for specific architectures, and experimental analysis which provides insight into algorithmic choice.

The topics of interest include cost models, modelling architectures, algorithms and data structures, complexity theory, applications and implementations, experimental analysis.

15 papers have been submitted to this workshop, most of them of good scientific quality. They have been refereed by at least 2 independent reviewers. Among these papers, the programme committee selected 5 papers to be presented during the conference.

Parallel Algorithms for Grounded Range Search and Applications by Michael G. Lamoureux and Andrew Rau-Chaplin : this paper presents a parallel algorithm for solving the grounded range search problem in associative-function mode using a BSP like model referred to as the Coarse Grained Multicomputer (CGM).

Multi-level cooperative search: a new paradigm for combinatorial optimization and an application to graph partitioning by Michel Toulouse

and K. Thulasiraman and Fred Glover : this paper proposes a new design for cooperative algorithms which is also a new parallel problem solving paradigm for combinatorial optimization problems.

A Quantitative Measure of Portability with Application to Bandwidth Latency Models of Parallel Computing by Gianfranco Bilardi, Andrea Pietracaprina and Geppino Pucci : this paper introduces a novel methodology for the quantitative assessment of model properties such as effectiveness and portability.

P. Amestoy et al. (Eds.): Euro-Par'99, LNCS 1685, pp. 523–524, 1999.

A Cost Model For Asynchronous and Structured Message Passing by Emmanuel Melin and Bruno Raffin and Xavier Rebeuf and Bernard Virot : this paper presents a cost model taking into account properties of present time machines, mainly SCLchan, an asynchronous parallel execution model allowing explicit message passing.

A Parallel Simulation of Cellular Automata by Spatial Machines by Bruno Martin : this paper considers the relations between cellular automata and spatial machines.

In name of the programme committee, I would like to thank all the authors and referees for the excellent work done in order to set-up such an interesting programme.

Parallel Algorithms for Grounded Range Search and Applications

Michael G. Lamoureux and Andrew Rau-Chaplin

Faculty of Computer Science
Dalhousie University
P.O. Box 1000, Halifax, Canada B3J 2X4.
Voice: 902-494-2732, Fax 902-492-1517.
mlamour@cs.dal.ca, arc@cs.dal.ca

Abstract. This paper presents a parallel algorithm for solving grounded range search in associative-function mode using the BSP-like Coarse Grained Multicomputer (CGM). Given a set S of n weighted points in the plane, the algorithm requires $O(1)$ communication rounds (h-relations with $h = O(n/p)$), $O((n/p) \log n)$ local computation, and $O(n/p)$ memory per processor ($n/p \geq p$), to solve $m = O(n)$ grounded range search problems. The result implies new or improved solutions to a number of other geometric applications including d-dimensional range search, quadrant search, interval intersection, and chromatic range queries.

1 Introduction

A *grounded range search query* in the plane is a 2-dimensional domain given by the 4-tuple $(x_{min}, x_{max}, -\infty, y_{max})$ which specifies a geometric query range. Let S be a set of n points in the plane with some *weight* $w(v)$ assigned to each $v \in S$, let Q be a set of $m = O(n)$ grounded range search queries, and let $\otimes$ be a binary associative function applied to a subset of weighted points.

The *2D grounded range search* problem consists of determining for each $q \in Q$ either the subset of the points in S contained in the domain of q, or the number of such points, or more generally $result(q, S)$, the result of applying the binary associative function $\otimes$ over such points. The former version of this problem is called the *report mode* while the latter versions are called the *counting* and *associative-function modes*, respectively.

The classical sequential solution to this problem in counting mode combines the results of two dominance queries which can be answered in $O(\log n)$ time and $O(n)$ space using the method of [2]. This reduction from grounded range search to dominance is applicable in this case because addition is not only associative and commutative but also cancellative (i.e. $x \otimes a = y \otimes a \Rightarrow x = y$). However, for many useful non-cancellative operators, like Max, this approach is not applicable.

The classical sequential solution to this problem in associative-function and reporting modes uses the $O(n)$ size priority search tree of McCreight [21] which answers a grounded range search query in $O(\log n)$ and $O(\log n + t)$ time, respectively, where t is the size of the output.

P. Amestoy et al. (Eds.): Euro-Par'99, LNCS 1685, pp. 525–532, 1999.

This paper presents an efficient parallel algorithm for solving the associative-function mode variant of the grounded range search problem for a set of $m = O(n)$ queries and n weighted points in the plane on a p-processor coarse grained multicomputer with arbitrary interconnection network and local memories of size $O(\frac{n}{p})$, $\frac{n}{p} \geq p$, in time $O(\frac{n \log n}{p} + T_h(n, p))$, where $T_h(n, p)$ is the time required to route a single h-relation with $h = O(n/p)$ [22, 23]. Note that the local computation time of this algorithm is optimal and that it requires only a constant number of communication rounds.

This algorithm permits efficient parallel solutions to d-dimensional range search, quadrant search, interval intersection search, and chromatic range search.

This paper reduces the query time required for $m = O(n)$ d-dimensional range queries from $O(\frac{n \log^d n}{p} + T_h(s, p))$ to $O(\frac{n \log^{d-1} n}{p} + T_h(s, p))$ in associative-function mode without increasing the storage requirement, thus improving upon the solution of [14] by a $\log n$ factor in search time.

This paper also presents algorithms to solve $m = O(n)$ associate-function mode quadrant search, interval intersection search, or chromatic range search queries on a set of n weighted points in the plane using a p-processor coarse grained multicomputer with arbitrary interconnection network and local memories of size $O(\frac{n}{p})$, $\frac{n}{p} \geq p$, in time $O(\frac{n \log n}{p} + T_h(n, p))$. These are, to the best of our knowledge, the first coarse grained parallel solutions to these problems.

2   The Coarse Grained Multicomputer Model

We are using a variation of the BSP model, referred to as the *Coarse Grained Multicomputer*, CGM [4, 5, 6, 7, 8, 9, 10, 11, 13, 14, 15]. It is comprised of a set of p processors $P_1, \ldots, P_p$ with $O(n/p)$ local memory per processor and an arbitrary communication network (or shared memory). The term "coarse grained" refers to the fact that we assume that the size $O(n/p)$ of each local memory is "considerably larger" than $O(1)$. Our definition of "considerably larger" will be that $n/p \geq p$.

All algorithms consist of alternating local computation and global communication rounds. Each communication round consists of routing a single h-relation with $h = O(n/p)$, i.e. each processor sends $O(n/p)$ data and receives $O(n/p)$ data. We require that all information sent from a given processor to another processor in one communication round is packed into one message. In the BSP model, a computation/communication round is equivalent to a superstep with $L = (n/p)g$ (plus the above "packing" and "coarse grained" requirement).

Finding an optimal algorithm in the coarse grained multicomputer model is equivalent to minimizing the number of communication rounds as well as the total local computation time. This considers all parameters discussed above that are affecting the final observed speedup, and it requires no assumption on g. Furthermore, it has been shown that minimizing the number of supersteps also leads to improved portability across different parallel architectures [9, 22, 23]. The above model has been used in parallel algorithm design for various problems ([3, 5, 6, 7, 8, 10, 13, 20]) and has demonstrated good practical timing results.

We now list some operations required by our algorithms. Each of these operations reduces to $O(1)$ communication rounds for $n/p \geq p$.

All-to-all broadcast: Every processor sends one message to all other processors [6] $(O((n/p))$ local computation).

Personalized all-to-all broadcast: Every processor (in parallel) sends a different message to every other processor [6] $(O((n/p))$ local computation).

Partial sum (Scan): Every processor stores one value, and all processors compute the partial sums of these values with respect to some associative operator [6] $(O((n/p))$ local computation).

Global sort: Sort $O(n)$ data items stored on a CGM, n/p data items per processor, with respect to the CGM's processor numbering. As shown in [17], for $n/p \geq p$ it is possible to sort in $O(1)$ communication rounds with $O(n)$ memory per processor and $O((n/p)\log n)$ local computation.

Global integer sort: Sort $O(n)$ integers in the range $1, \ldots, n^c$ for fixed constant c stored on a CGM, n/p data items per processor, with respect to the CGM's processor numbering. The sort algorithm in [17] is based on Cole's merge sort [16]. The $O((n/p)\log n)$ local computation in [17] is due to a constant number of local sorts. Hence, by applying radix sort for the integer case, we obtain $O(n/p)$ local computation without increasing the number of communication rounds. A CGM integer sorting algorithm with 9 communication rounds, $O(n/p)$ memory per processor and $O(n/p)$ local computation can be found in [4].

Load Balance: Given a set $\bar{Q}$ of $m = O(n)$ queries where associated with each query is a value $next(q)$ which is the name of the substructure it next requires and a distributed data structure $\bar{S}$ which consists of p substructures $\bar{S}_i$ $(1 \leq i \leq p)$ of size $O(n/p)$ stored with $\bar{S}_i$ on processor p_i of a $CGM(n, p)$. This operation balances queries and structures such that each query $q \in \bar{Q}$ is stored on a processor which also stores a copy of the substructure $next(q)$.

Algorithm 1 "Load Balance$(\bar{S}, \bar{Q})$".

Architecture: A p-processor coarse grained multicomputer, $CGM(n, p)$, with arbitrary interconnection network and local memories of size $O(\frac{n}{p})$, $\frac{n}{p} \geq p$.

Input: Each processor p_i stores a set $\bar{Q}_i$ of n/p queries from $\bar{Q}$ and a substructure $\bar{S}_i$ of size $O(n/p)$ from $\bar{S}$. Associated with each query is the value $next(q)$ which is the name of the substructure it next requires.

Output: Each query $q \in \bar{Q}$ is stored on a processor which also stores a copy of the substructure $next(q)$.

1 Globally compute $c(\bar{S}_i) = \lceil \frac{|\{q \in \bar{Q} | next(q) = i\}|}{\frac{n}{p}} \rceil$

2 Make $c(\bar{S}_i)$ copies of S_i and distribute them evenly such that each processor stores at most two substructures.

3 Redistribute $\bar{Q}$ evenly so that every query $q \in \bar{Q}$ is stored on a processor that also stores a copy of the element of $\bar{S}$ which q is visiting.

— End of Algorithm —

Note that this algorithm evenly distributes queries and substructures $\bar{S}_i$, such that each processor has $O(1)$ copies of each. This approach to load balancing is based on a CGM technique described and analyzed in [6] and parallel integer sorting, it requires $O(1)$ communication rounds and $O(n/p)$ local computation.

3   A Parallel Algorithm for Grounded Range Search

Consider a set of p horizontal lines h_i which partition S into p subsets H_i of $\frac{n}{p}$ points each (with h_i below H_i, and h_{i+1} above H_i). Analogously, consider p vertical lines l_j which partition S into p subsets V_j of $\frac{n}{p}$ points each (with l_j to the left of V_j, and l_{j+1} to the right of V_j). For a subset $A \subset S$ let $w(A) = \sum_{a \in A} w(a)$. Let V_{ij} be the set of points of V_j that are below h_i. Let H and V be the set of points in H_i $(1 \leq i \leq p)$ and V_j $(1 \leq j \leq p)$, respectively which are assumed to be in general position without loss of generality. (See Figure 1.)

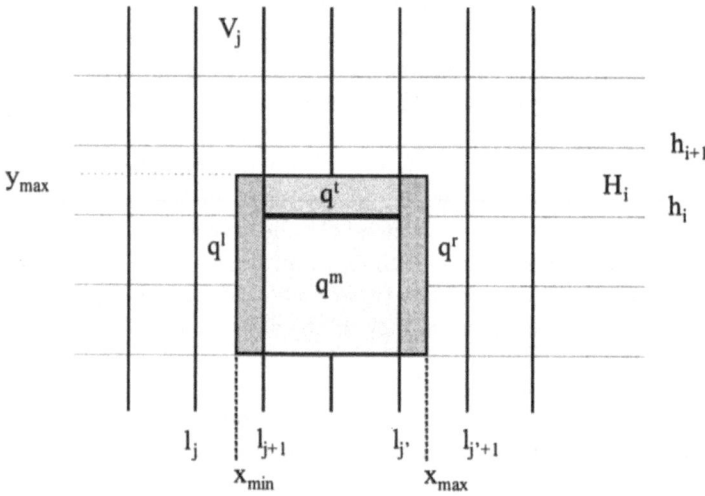

Fig. 1. A grounded range search query with respect to a set of horizontal and vertical partitions of the the plane.

Observation 1 *For each query $q \in Q$ with $x_{min} \in [l_j, l_{j+1}]$, $x_{max} \in [l_{j'}, l_{j'+1}]$, and $y_{max} \in [h_i, h_{i+1}]$ let $q^l = (x_{min}, l_{j+1}, -\infty, y_{max})$, $q^r = (l_{j'}, x_{max}, -\infty, y_{max})$, $q^t = (l_{j+1}, l_{j'}, h_i, y_{max})$, and $q^m = (l_{j+1}, l_{j'}, -\infty, h_i)$. Note that $result(q, S) = result(q^l, V_j) \otimes result(q^r, V_{j'}) \otimes result(q^t, H_i) \otimes result(q^m, \cup_{k=j+1}^{j'-1} V_{ik})$.*

Algorithm 2 "Grounded Range Search".
Architecture: A p-processor coarse grained multicomputer, $CGM(n, p)$, with arbitrary interconnection network and local memories of size $O(\frac{n}{p})$, $\frac{n}{p} \geq p$.

Input: Each processor stores $\frac{n}{p}$ points of S.
Output: Each processor stores $result(q, S)$ for each of its $\frac{n}{p}$ queries $q \in Q$.

1 Globally sort S by x-coordinates such that processor p_j stores V_j and l_j. Perform an all-to-all broadcast, where processor p_j sends l_j to all other processors. Every processor now stores all vertical lines $l_1, \ldots, l_p$.

2 Each processor p_j uses $l_1, \ldots, l_p$ to constructs subqueries q^l and q^r and computes $next(q^l)$ and $next(q^r)$ for each query q it stores. Let Q^{lr} denote the set of all q^l and q^r queries.

3 Call Load-Balance(V, Q^{lr}) and then solve all queries in Q^{lr} sequentially associating the result with the query.

4 Globally sort a copy of S by y-coordinates such that processor p_i stores H_i and h_i. Perform an all-to-all broadcast, where processor p_i sends h_i to all other processors. Every processor stores now all horizontal lines $h_1, \ldots, h_p$.

5 Each processor p_j which stores V_j uses horizontal lines $h_1, \ldots, h_p$ to construct q^t and q^m and compute $next(q^t)$ and $next(q^m)$ for each query q it stores. Let Q^{tm} denote the set of all q^t and q^m queries. Each processor also computes $w(V_{ij})$ for $i \in h_1, \ldots, h_p$. Perform an all-to-all broadcast, where processor p_j sends $w(V_{ij})$ to processor p_{i+1}, $1 \le i < p$.

6 Each processor p_i which stores H_i and a part of Q^{tm} now also stores $V_{i-1,j}(1 \le j \le p)$ associated with H_i. Load-Balance(H, Q^{tm}) and solve the q^m queries by performing a partial sum in $V_{i-1,j}(1 \le j \le p)$ and the q^t queries by sequential grounded range search. Associate the results with the query.

7 Globally sort $Q^{lr} \cup Q^{tm}$ by query index such that for each original query q the subqueries q^l, q^r, q^t, q^m are contiguous. Use a scan operation with $\otimes$ to compute the final result for each original query.

— End of Algorithm —

Theorem 1. *The grounded range search problem in associative-function mode for a set of $m = O(n)$ queries and n weighted points in the plane can be solved on a p-processor coarse grained multicomputer with arbitrary interconnection network and local memories of size $O(\frac{n}{p})$, $\frac{n}{p} \ge p$, in time $O(\frac{n \log n}{p} + T_h(n, p))$, where $T_h(n, p)$ is the time required to route a single h-relation with $h = O(n/p)$.*

Proof. The correctness of Algorithm 2 follows from Observation 1. Steps 1-3 solve the q^l and q^r queries in $O(\frac{n \log n}{p})$ local computation (from sorting and sequential grounded range search) and $O(1)$ communications rounds (from sorting, load-balancing, the distribution of vertical cutting lines). Since only $O(1)$ copies of points and queries are made and there are at most p cutting lines the space requirement is $O(\frac{n}{p} + p) = O(\frac{n}{p})$ per processor. Steps 4-6 solve the q^t and

q^m queries in $O(\frac{n \log n}{p})$ local computation (again from sorting and sequential grounded range search) and $O(1)$ communications rounds (from sorting, load-balancing, and the distribution of $w(V_{ij})$ and the horizontal cutting lines). Again, since only $O(1)$ copies of points and queries are made, and since there are at most p cutting lines, the space requirement is $O(\frac{n}{p} + p) = O(\frac{n}{p})$ per processor.

The local computation time for each step is $O(\frac{n}{p} \log n)$. The global communication in each step reduces to a constant number of global sorts and communication operations listed in Section 2 and the time complexity follows.   □

4   Applications

This section describes efficient parallel solutions to d-dimensional range queries, quadrant queries, interval intersection queries, and chromatic range queries using grounded range search.

The authors of [14] demonstrate how to construct a distributed range tree T on a d-dimensional set S of n points on a coarse grained multicomputer in $O(\frac{n \log^{d-1} n}{p} + T_h(s,p))$ time to answer a set Q of $m = O(n)$ range queries in time $O(\frac{n \log^d n}{p} + T_h(s,p))$ in associative-function mode. If we use grounded range search queries in the last two structural dimensions, we can reduce the query time by a $\Theta(\log n)$ factor to $O(\frac{n \log^{d-1} n}{p} + T_h(s,p))$ in associative-function mode.

Theorem 2. *The d-dimensional range search problem in associative-function mode for a set of $m = O(n)$ queries and n weighted points in d-space can be solved on a p-processor coarse grained multicomputer with arbitrary interconnection network and local memories of size $O(\frac{n \log^{d-1} n}{p})$, $\frac{n \log^{d-1} n}{p} \geq p$, in time $O(\frac{n \log^{d-1} n}{p} + T_h(n,p))$, where $T_h(n,p)$ is the time required to route a single h-relation with $h = O(\frac{n \log^{d-1} n}{p})$.*

Proof. If we use the algorithm of [14] to build a modified d-dimensional range tree where the structures in the dth dimension are simply stored as point sets, we can solve a d-dimensional range query by solving a (d-2)-dimensional range query on the first (d-2) dimensions of the d-dimensional structure and solving two grounded range search queries on the point sets associated with the left and right children of each node in range in the dimension (d-1) substructures using the method of Edelsbrunner [12].   □

Given a point v=(x,y) in the plane, a quadrant query asks for all points that lie in one of the four quadrants defined by the point. Since the quadrants are defined by semi-infinite ranges in the x and y directions, a quadrant query may be viewed as a grounded range search query with one side open.

Theorem 3. *The quadrant search problem in associative-function mode for a set of $m = O(n)$ queries and n weighted points in the plane can be solved on a p-processor coarse grained multicomputer with arbitrary interconnection network and local memories of size $O(\frac{n}{p})$, $\frac{n}{p} \geq p$, in time $O(\frac{n \log n}{p} + T_h(n,p))$, where $T_h(n,p)$ is the time required to route a single h-relation with $h = O(n/p)$.*

Proof. Omitted.

Given a set S of n weighted line segments, the associative interval intersection problem asks for the result of a binary associative operator applied to the weights of each pair of intervals in the set S that intersect. Grounded range search may be used to construct a very elegant solution to this problem. If we follow the precedent of McCreight [21] and map the intervals $[a, b]$ in S to the points (a, b) in S' and the query intervals $[u, v]$ in Q to the points (u, v) in Q', we can solve our interval intersection queries by performing a quadrant search on $[u, \infty) * (-\infty, v]$ for each query point $q' = (u, v)$ in Q.

Theorem 4. *The interval intersection search problem in associative-function mode for a set of $m = O(n)$ queries and n weighted points in the plane can be solved on a p-processor coarse grained multicomputer with arbitrary interconnection network and local memories of size $O(\frac{n}{p})$, $\frac{n}{p} \geq p$, in time $O(\frac{n \log n}{p} + T_h(n, p))$, where $T_h(n, p)$ is the time required to route a single h-relation with $h = O(n/p)$.*

Proof. Omitted.

Janardan and Lopez [19] define a chromatic query as a query on a set S of n geometric objects which belong to g disjoint groups, each labeled with a color, and it is the groups, and not the objects, which are of interest. In associative mode, the groups are weighted and a chromatic range query is a range query which asks for the result of the repeated application of a binary associative operator to each group that contains a datapoint located in the given range.

Theorem 5. *The chromatic search problem in associative-function mode for a set of $m = O(n)$ queries and n weighted points in the plane can be solved on a p-processor coarse grained multicomputer with arbitrary interconnection network and local memories of size $O(\frac{n}{p})$, $\frac{n}{p} \geq p$, in time $O(\frac{n \log n}{p} + T_h(n, p))$, where $T_h(n, p)$ is the time required to route a single h-relation with $h = O(n/p)$.*

Proof. If we use the technique of Gupta, Janardan, and Smid [18] and transform each point p in S to the point $p' = (p, pred(p))$ in S' where $pred(p)$ is its immediate predecessor of the same color (or $-\infty$ if the point p has no predecessor). Note that this transformation is such that there is a point p of color c in the query interval $q = [l, r]$ if and only if there is a point p' of color c in the grounded query rectangle $q' = [l, r] * (-\infty, l)$. Thus, we can use a grounded range search on the transformed point set S'.   □

References

[1] S.G. Akl and K.A. Lyons, "Parallel Computational Geometry", Prentice Hall, 1993.
[2] J.L. Bentley, "Multidimensional Divide and Conquer", Comm. ACM 23(4):214-229, 1980.

[3] G.E. Blelloch, C.E. Leiserson, B.M. Maggs and C.G. Plaxton, "A Comparison of Sorting Algorithms for the Connection Machine CM-2", in Proc. ACM Symp. on Parallel Algorithms and Architectures, 1991, pp. 3-16.

[4] A. Chan, F. Dehne, "A Note on Coarse Grained Parallel Integer Sorting," Technical Report TR-98-06, School of Computer Science, Carleton University.

[5] F. Dehne, A. Fabri and C. Kenyon, "Scalable and Architecture Independent Parallel Geometric Algorithms with High Probability Optimal Time", in Proc. 6th IEEE Symp. Parallel and Distributed Processing, 1994, pp. 586-593.

[6] F. Dehne, A. Fabri and A. Rau-Chaplin, "Scalable Parallel Computational Geometry for Coarse Grained Multicomputers", in Proc. ACM Symp. Computational Geometry, 1993, pp. 298-307.

[7] F. Dehne, X. Deng, P. Dymond, A. Fabri and A.A. Kokhar, "A randomized parallel 3D convex hull algorithm for coarse grained parallel multicomputers", in Proc. ACM Symp. on Parallel Algorithms and Architectures, 1995.

[8] X. Deng, "A Convex Hull Algorithm for Coarse Grained Multiprocessors", in Proc. 5th Int. Symp. on Algorithms and Computation, 1994.

[9] X. Deng and P. Dymond, "Efficient Routing and Message Bounds for Optimal Parallel Algorithms", in Proc. Int. Parallel Proc. Symp., 1995.

[10] X. Deng and N. Gu, "Good Programming Style on Multiprocessors", in Proc. IEEE Symp. on Parallel and Distributed Processing, 1994, pp. 538-543.

[11] O. Develliers and A. Fabri, "Scalable Algorithms for Bichromatic Line Segment Intersection Problems on Coarse Grained Multicomputers", (WADS'93), Springer Lecture Notes in Computer Science, 1993, pp. 277-288.

[12] H. Edelsbrunner, "A Note on Dynamic Range Searching", Bulletin of the EATCS 15:34-40, 1981.

[13] A. Ferreira, A. Rau-Chaplin, S. Ubeda, "Scalable 2D Convex Hull and Triangulation for Coarse Grained Multicomputers", in Proc. 6th IEEE Symp. on Parallel and Distributed Processing, San Antonio, 1996.

[14] A. Ferreira, C. Kenyon, A. Rau-Chaplin, S. Ubeda, "d-Dimensional Range Search on Multicomputers", Proc. 11th Int. Parallel Processing Symposium (IPPS'97), April 1997, pp. 616-620.

[15] A.V. Gerbessiotis and L.G. Valiant, "Direct Bulk-Synchronous Parallel Algorithms", in Proc. 3rd Scandinavian Workshop on Algorithm Theory, Lecture Notes in Computer Science, Vol. 621, 1992, pp. 1-18.

[16] R. Cole, "Parallel Merge Sort", SIAM J. Comp. 17(4):770-785, 1998.

[17] M.T. Goodrich, "Communication Efficient Parallel Sorting", in Proc. 28th ACM Symp. on Theory of Computing (STOC'96), 1996.

[18] P. Gupta, R. Janardan, M. Smid, "Further results on generalized intsersection searching problems: counting, reporting, dynamization", J. Algs 19:282-317, 1995.

[19] Janardan, Ravi and Mario Lopez, "Generalized Intersection Searching Problems", Int. J. Comp. Geom. and Applications 3(1):39-69, 1993.

[20] H. Li and K.C. Sevcik, "Parallel Sorting by Overpartitioning", in Proc. ACM Symp. on Parallel Algorithms and Architectures, 1994, pp. 46-56.

[21] McCreight, E.M., "Priority Search Trees" SIAM J. Comput. 14(2):257-276, 1985.

[22] L.G. Valiant, "A Bridging Model for Parallel Computation", Comm. ACM 33:103-111, 1990.

[23] L.G. Valiant et. al., "General Purpose Parallel Architectures", Handbook of Theoretical Computer Science, Edited by J. van Leeuwen, MIT Press/Elsevier, 1990, pp. 943-972.

Multi-level Cooperative Search: A New Paradigm for Combinatorial Optimization and an Application to Graph Partitioning

Michel Toulouse[1], Krishnaiyan Thulasiraman[2], and Fred Glover[3]

[1] Department of Computer Science
University of Manitoba
[2] School of Computer Science
University of Oklahoma
[3] Graduate School of Business
University of Colorado

Abstract. Cooperative search is a parallelization strategy for search algorithms where parallelism is obtained by concurrently executing several search programs. The solution space is implicitly decomposed according to the search strategy of each program. The programs cooperate by exchanging information on previously explored regions of the solution space. In this paper we propose a new design for cooperative search algorithms which is also a new parallel problem solving paradigm for combinatorial optimization problems. Our new design is based on an innovative approach to decompose the solution space which is inspired from the modeling of cooperative algorithms based on dynamical systems theory. Our design also gives a new purpose to the sharing of information among cooperating tasks based on principles borrowed from scatter search evolutionary algorithms. We have applied this paradigm to the graph partitioning problem. We describe the parallel implementation of this algorithm on a cluster of workstations and compare our results with other well known graph partitioning methods.

1 Introduction

Given that a finite amount of computational time is allocated to find a near optimal solution, local search methods often stand to benefit from parallelization to find better solutions. It is common to base the parallelization on *restarts* of the local search from different initial points in the solution space and/or to use different search criteria to improve the quality of the solutions returned by the search method. Restarting a search method yields an *implicit decomposition* of the solution space, and if these restarts do not require knowledge of the outcome of other restarts, we can exploit them by creating independent tasks for a concurrent exploration of the solution space using parallel computers. Following the principle underscored in tabu search that the history of the solution process is important for determining decisions made at various stages, it has been observed that the search methods can benefit from information gathered by the

P. Amestoy et al. (Eds.): Euro-Par'99, LNCS 1685, pp. 533–542, 1999.
© Springer-Verlag Berlin Heidelberg 1999

other concurrent search threads in a parallel search. This has given rise to a parallelization paradigm called *cooperative search* (Toulouse, Crainic and Sansó 1997).

In this paper we propose an innovative design for cooperative search algorithms which is also a new parallel problem solving paradigm for combinatorial optimization problems. Current design of cooperative search algorithms are mostly based on search programs working with the same data set and exploring the solution space according to an implicit partition obtained from different search strategies. In our proposed design, each search program works with a different data set, each object of each data set is a partial solution of the optimization problem and each partial solution of a data set is obtained from objects of the previous data set in the hierarchy and also serves as a building block of objects in the next data sets in the hierarchy. The second aspect in which our design departs from current cooperative algorithms is relative to the sharing of information. In our new design, shared information is used to reorganize the partial solutions in the data sets. Given a particular instance of a hierarchy of data sets, the exploration of the problem solution space is limited to regions that can be reached by the neighborhood functions defined by the set of partial solutions at each level. By reorganizing the partial solutions in the data sets, i.e. destroying some partial solutions and creating others, new data sets (and consequently new neighborhood functions) are created at each level where such a reorganization occurs, which provides for new regions of the solution space to be explored. This strategy uses principles borrowed from *scatter search* evolutionary algorithms (Glover 1977) and *vocabulary building* procedures (Glover 1992, Glover and Laguna 1993, Glover and Laguna 1997) to guide the "combination" of shared information originating from different sources.

We have applied this problem solving method and cooperative search design to the k-way graph partitioning problem (GPP). In terms of the quality of the solutions, our approach turns out to be by far the most powerful heuristic for this optimization problem. The performance of this approach is also consistent, always outperforming the other algorithms for all the problem instances tested. This outcome demonstrates that the present design is highly successful in adapting the dynamical behavior of the system to the optimization objective of the problem.

The rest of paper is organized as follows. In Section 2 we introduce a formalization of the GPP and classify the existing graph partitioning methods. In Section 3 we introduce our multi-level cooperative search framework. In Section 4 we describe an application of our multi-level cooperative search to the GPP. In Section 5 we give the resulting edge cuts from a parallel implementation of the algorithm to several graphs. Concluding remarks are provided in Section 6.

2   The Multi-level Cooperative Search Framework

Cooperative algorithms are relatively new, and most of the attention thus far has focused on finding useful mechanisms for sharing information to improve the search behavior of cooperating programs. However, cooperation issues are more subtle than simply considering the impact of reused information on the performance of cooperating programs. Shared information among the search programs creates an inextricable network of dependencies which are most likely to be at the source of the divergent behavior of cooperating algorithms when compared to the individual search strategies that compose them. Our concern is to organize the dependencies among cooperating programs in a more manageable control structure in terms of how it affects the search of cooperative algorithms.

Similar issues have been addressed previously by scatter search evolutionary methods and their derivatives, notably the vocabulary building procedures. Scatter search approaches are population based search heuristics that seek to exploit information contained in "elite solutions" by combining them to derive new solutions. The role of the elite solutions in scatter search as a means to control the exploration of the solution space is based on strategies whose roots can be traced in part to OR and AI proposals at the end of the 50s and early 60s. (For a review see Glover 1997 and Glover and Laguna 1997.) Vocabulary building constitutes a variation of scatter search which focuses on components of elite solution vectors (partial solutions) rather than complete solutions. Procedures using this design have been implemented in a sequential programming setting by Rochat and Taillard (1995), Kelly and Xu (1995), Taillard et al. (1995) and Lopez, Carter and Gendreau (1996). Elite solutions are identified using local search methods. Partial solutions are extracted from the elite solutions to form a pool of *solution fragments*, which in turn are combined to form larger fragments until complete solutions are generated.

By adapting the data sets of the search programs to embody a hierarchical structure of the form proposed above, we have been able to convert current cooperative algorithms into multi-level structures. Our design is based on assuming that solutions of the combinatorial optimization problem can be represented as sequences or partitions. Decision variables that define the objective function serve as the logical choice to constitute the lowest level components of the multi-level structure, and we progressively combine these variables to provide larger aggregates of variables. Thus the original decision variables are building blocks for the second level of the structure, whose elements are larger sequences or partitions of solutions. These larger aggregates again function as building blocks, to be combined into higher order assemblies, forming a set of hierarchically structured complexes.

A large number of multi-level structures can be obtained by combining the decision variables in different ways – a choice that is available at each level of aggregation. Each combination strategy defines an *instance* of a multi-level structure that organizes the original data set of the problem. Each set of aggregates of a given instance of this structure, which provides decision criteria for all levels, is used as an input (a "data set") for one of the search programs composing a cooperative algorithm.

If we use only one multi-level structure instance (as in an effort to design multi-level graph partitioning algorithms), higher level controls may not be able to give very favorable guidelines to lower control levels, and thus may yield a poor exploration of the solution space. A better approach is to explore the solution space according to different instances of the multi-level structure. Instances can be generated statically (using different combination strategies) with or without the knowledge of the results of the exploration of previous instances, or they can be generated dynamically. We have favored the latter option in our design. We generated instances dynamically by using interactions occurring among the levels being explored by the search programs. In our current design we have defined three different kinds of local interactions which are implemented in terms of inter-task operators: an *interpolation operator*, a *destroy operator* and a *create operator*. As in other cooperative algorithms, these local interactions have a cumulative impact and give rise to diffusion processes among the programs of the multi-level structure. However, unlike previous designs, the local interactions are part of a hierarchical control structure which evolves according to the optimization logic of the search methods. This helps to create dynamics in the system favoring a better exploration of the solution space of the optimization problem.

3   Application to the Graph Partitioning Problem

The graph partitioning problem is a combinatorial optimization problem of practical interest in many engineering fields. Given a graph $G = \{V, E\}$ where $|V| = n$ and $|E| = m$, the graph partitioning problem seeks to partition the set of vertices V into k subsets $V_1, V_2, \ldots, V_k$ in order to minimize

$$\sum_{i=1}^{i=m} w(e_i) \mid e_i \text{ connects vertices belonging to different subsets}$$

subject to $V_i \bigcap V_j = \emptyset$ $(i \neq j)$; $|V_i| - |V_j| \leq 1$, $\forall\, i, j \in \{1, \ldots, k\}$; and $\bigcup_{i=1}^{i=k} V_i = V$. In this section we introduce a parallel implementation of a cooperative algorithm for the graph partitioning problem which incorporates the design principles described in the previous section. First we describe the initialization phase of the procedure.

3.1   Initialization

The initial set of components is based on the decision variables of the optimization problem. For the graph partitioning problem, this initial set corresponds to the vertices of the graph. To obtain the components (aggregates) at the second level of the multi-level data set to exploit the building block effect, we aggregate the initial set of components into partial solutions. For the graph partitioning problem, this consists of creating aggregates of vertices which become the components to be manipulated at the second level of the structure.

In our current implementation of the multi-level cooperative algorithm for the GPP, we have a single hierarchy of data sets. These data sets are obtained using a maximal matching algorithm (Hendrickson and Leland 1993 and Karypis and Kumar 1998) which consists of finding a maximal set of edges in the graph such that no two edges are incident to the same vertex. (The term "maximal", in contrast to "maximum", refers to a matching that is locally optimal in a constructive sense, i.e., it is contained in no larger matching.) The two vertices i and j of an edge (i, j) in the matching are merged into a single vertex. The new vertex weight equals the sum of the weights of its constituent vertices and the vertex also retains the edges adjacent to its constituent vertices. Edge weights are left unchanged, except where vertices i and j both are joined by edges to a common vertex k before being merged. Then the two edges (i, k) and (j, k) are replaced by a single edge whose weight equals the sum of the weights of (i, k) and (j, k).

Denote graph $G = \{V, E\}$ by $G_0 = \{V_0, E_0\}$. The aggregates built from the initial set of vertices V_0 using the maximal matching algorithm defines a new graph $G_1 = \{V_1, E_1\}$ where $|V_1| \approx |V_0|/2$. The maximal matching is then applied to graph G_1, and iteratively to the graph obtained from G_1 until the coarsened graphs are small enough to be easy to explore. Assuming that p graphs have been generated by iteratively applying the maximal matching algorithm to the last generated graph, we then have after $p - 1$ aggregation operations, p graphs $G_0, G_1, \ldots, G_{p-1}$.

3.2   The Control Structure

The control structure represents the problem solving method. In this multi-level algorithm there are two types of control structure (two types of neighborhood functions). The first one consists of the graph partitioning algorithms applied to the data set at each level (the traditional neighborhood function defined by the search move used, swap, 2-opt, 3-opt, etc.). Essentially we use constructive or local improvement partitioning algorithms to search the graphs at each level of the structure, therefore performing the search of all the levels in parallel.

The interactions among the data sets at different levels constitute the second type of control structure. While the search space of the levels $i > 0$ is increasingly smaller and easier to search, the search processes at these levels can only explore partially the solution space of the problem and may not yield very good solutions. We need to create new multi-level structure instances. This is achieved through interactions among consecutive levels of the structure, changing the sets of aggregates. At this point we elaborate the three operators which are used to implement the interactions among the cooperating programs of the multi-level structure.

Destroy operator: It determines whether the vertices of an aggregate in G_i lie in the same set of the best partition in graph G_{i-1}. If not, destroy the aggregate in graph G_i so that each of the vertices of the aggregate constitute a separate set.

Each process P_i has a data structure which records which vertices of the graph G_i have been merged together in a single vertex (aggregate) of graph G_{i+1}. To implement the destroy operator, processes $P_i, i = 0, \ldots, p - 2$ compare a good partition of graph G_i with the set of aggregates in the graph G_{i+1}. If an aggregate in graph G_{i+1} has components (vertices) from graph G_i which are not in the same set in the good partition, this aggregate is marked to be destroyed. This step is followed by a communication phase where process P_i sends the set of marked aggregates to process P_{i+1}. Once a process $P_i, i = 1, \ldots p - 1$ has received the information about which aggregates have to be destroyed, it proceeds to remove these aggregates from the graph G_i and to create a new set of vertices in graph G_i from the components of the destroyed aggregates. The outcome of executing the destroy procedure is a new multi-level structure instance consisting of modified data sets potentially at each level of the structure. These data sets yield new graphs in the sequence of graphs (except for graph G_0). The new graphs are based on different sets of partial solutions, therefore providing a new set of neighborhood functions to search the graph G_0. Processes need then to resume the exploration of the solution space of G_0.

Create operator: Given vertices which are often in the same set relative to a collection of good partitions in graph G_i, create an aggregate with these vertices and make it a new aggregate (vertex) of graph G_{i+1}.

The sequence of operations of the procedure to implement the "create" operator is similar to the sequence for the "destroy" operator. New aggregates for graph G_i are identified based on elite solutions found when process P_i explores graph G_i. A variety of strategies and guidelines have been proposed in the literature on scatter search methods (and their path relinking generalizations) to combine such elite solutions. In our current implementation, when a good solution is visited, we record in which set of the partition each vertex is found. Then we use a simple frequency memory to identify "consistent variables" following a scatter search theme that is shared with tabu search. Specifically, before applying the create operator, those vertices which have often been together in the same set are considered candidates to be merged in a single vertex of graph G_{i+1}, i.e. to become the building blocks or components of new aggregates in graph G_{i+1}. Following this initial phase of the "create" procedure, the sequence of operations of the "create" operator is identical to the sequence for the "destroy" operator.

Interpolation operator: The current best partition in G_{i+1} is transformed by this operator to become an initial solution for the exploration of graph G_i. In particular, the operator places nodes of G_i in the same sets of the partition as their aggregate in the best partition in G_{i+1}.

The procedure that implements the "interpolation" operator for process P_i, receives a good partition of graph G_{i+1} from process P_{i+1}. Assume aggregate a of graph G_{i+1} is in the set A_{i+1} of the good partition received from process P_{i+1}. The aggregate a is then uncoarsened and all components of the aggregate a are placed in the set A_i of the partition for graph G_i. This yields an initial solution for process P_i from which a local search is launched to explore the solution space of graph G_i.

3.3   The Parallel Implementation

In our current implementation, these three operators are in sequence in the main loop of the program. The main loop iterates in a synchronized manner, applying first the destroy operator at all levels in parallel, followed by the interpolation and create operators again at all levels, before entering in the next iteration. The implementation and experiments described in this paper have been run on a cluster of Sparc workstations. The multi-level structure is a sequence of graphs. If the multi-level structure has p levels, we reserve a set of p processors and define an interconnection topology among them corresponding to an array. Let processors P_{i-1} and P_{i+1} be the neighbors of processor P_i in the array topology. This means that our communication software need to send messages only to processors P_{i-1} and P_{i+1} from processor P_i. In this array topology, P_{i-1} is not defined for P_0 and P_{i+1} is not defined for P_{p-1}. Our mapping strategy is to associate the process which searches graph G_i to processor P_i in the array such that levels G_{i-1}, G_i and G_{i+1} are neighbor tasks in the logical array configuration of processors. All data exchanges among the processes use our own communication software, which is stream based using the TLI protocol interface. (We have also realized an implementation of this multi-level cooperative search under PVM, and a second one that runs on the Origin 2000 shared memory parallel computer.)

This parallel algorithm is an implementation of the framework described in section 2. We believe that similar implementations can be realized for several optimization problems where solutions can be expressed as a combination of partial solutions.

4   Experiments

The experimental results show that with only a summary implementation, our new method is actually quite effective. The tests have been performed using the set of graphs in Table 1 as benchmark set (which represents the complete set of graphs that we have tested).

Graphs	# of vertices	# of edges	Graphs	# of vertices	# of edges
3elt	4720	13722	tooth	78136	452591
whitaker3	9800	28989	rotor	99617	662431
4elt2	11143	32818	ocean	143437	409593
4elt	15606	45878	auto(ef_589)	110971	741934
sphere	16386	49152	m144	144649	1074393
bcsstk31	35588	572914	m14	214765	1679018
brack2	62631	366559			

Table 1. Characteristics of the graphs in the benchmark set

These graphs have been widely used as benchmarks for graph partitioning algorithms. Most of these graphs are triangular and quadrangular unstructured meshes related to fluid dynamics, structural mechanics, or combinatorial optimization problems, obtained from the web site:

$ftp : // ftp.u - bordeaux.fr/pub/Local/Info/Software/Scotch/Graphs$

Graphs	Methods	Number of sets in the partitions / Edge cut								
		2	4	8	16	32	64	128	256	512
3elt	C2.0	91	225	386	629	1102	1736			
	M3.0	90	221	386	631	1098	1723			
	MO	90	205	347	578	818	1649			
whit	C2.0	132	392	700	1218	1879	2814			
	M3.0	131	427	716	1219	1886	2811-1			
	MO	128	382	667	1131	1754-1	2661-1			
4elt2	C2.0	130	351	656	1154	1809	2814			
	M3.0	130	359	658	1122	1784	2830-2			
	MO	130	351	617	1048	1710-1	2632-1			
4elt	C2.0	147	384	647	1111	1880	2994			
	M3.0	142	380	645	1043	1738	2938			
	MO	139	337	557	1038	1665	2726			
sphere	C2.0	430	824	1273	2016	2893	4144			
	M3.0	424	864	1333	2059	2943	4183			
	MO	386	776	1202	1847	2695-1	3924-2			
btk31	C2.0	3396	9423	17262	28656	45291	67134			
	M3.0	2815	7879	14426	26366	43665	67533-1			
	MO	2795	7736	14356	25175	40264-1	61563-1			
brack2	C2.0	734	3365	7851	13126	19975	29725	42472	60215	83498
	M3.0	742	3426	7753	13289	19940	29409	42560-30	59036-14	81001-8
	MO	731	3143	7669	12109	18412-1	27790	41061-1	57472-1	79429-1
tooth	C2.0	4307	7940	13332	20061	29282	40417	54186	72593	96930
	M3.0	4618	7854	13329	21249	29061-1	40260-1	53224-38	71256-19	94355-11
	MO	3877	7241	12580	19238	27243	37427	50583-1	67929-1	91472-1
ocean	C2.0	468	2089	5126	9496	15207	23138	32110	43904	59122
	M3.0	509	2054	5114	9079	15614	23745-1	34630-33	46848-34	61096-28
	MO	472	1928	4468	9057	14439	22212-1	31185	42715-1	58169-1
rotor	C2.0	2157	8300	15108	24171	37546	54101	75373	104949	142131
	M3.0	2647	8354	15359	24841-1	36860-1	52626	74283-50	100063-24	134213-14
	MO	2113	7762	13686-1	21720	33585-1	49243-1	70084-1	96960-1	131490-1
auto	C2.0	2486	8855	17925	30885	45261	64576	89532	120717	159149
	M3.0	2444	8631	18028	29268	45716	64094	86200-52	115127-27	151134-14
	MO	2421	8309	16988-1	28137	42480	61102	83031	112113-1	147322-1
m144	C2.0	7494	18278	30012	45085	63532	88457	121865	163048	217250
	M3.0	6919	17647	29791	43658	63856	89576-1	117613-66	157126-35	205550-16
	MO	6761	16494	27261	40931	60368	84484-1	113908-1	150929-1	201402-1
m14	C2.0	3966	13838	28231	52499	77621	111343	157092	215337	290045
	M3.0	4109	14322	28128	50410	74169	110465-1	152160-1	207093-52	277622-25
	MO	3893	13600	27027	45579	71586	106225	147802	203111	272514-1

Table 2. Experimental results

For comparison purposes, we report tests with version 2.0 of Chaco (C2.0) Hendrickson and Leland (1995) and version 3.0 of Metis (M3.0) Karypis and Kumar (1997), which are both well known graph partitioning software packages. Tests performed with Chaco use the multi-level algorithm implemented in its software. Tests performed with Metis use the program "pmetis" when the number of sets in the partition is smaller than 65 and the program "kmetis" when the number of sets is larger than 64. We have performed tests with several other partitioners, but their results were generally dominated by these two partitioning packages, therefore for lack of space we don't report the results obtained with the other partitioners.

In Table 2, the first column refers to the graphs tested ("whit" stands for whitaker3 and "btk31" stands for bcsstk31) and the second column refers to the graph partitioning methods used. The other columns refer to the values of the edge cuts according to the number of sets in the partition. A dash followed by a number indicates an imbalance in the sets of the best partition. For example, "-1" indicates an imbalance of one node between the set(s) having the larger number of nodes and the one(s) having the smaller number of nodes. Chaco always reports balanced partition while Metis and our algorithm sometime report a best solution corresponding to an imbalance partition. In a single case (Ocean, 2-way), our results are not better or as good as the two other partitioners. On small graphs and small number of sets, we obtain same solutions or small improvements. For larger graphs or when the number of sets in the partitions increase, we distance ourself from the two other approaches with sometime substantial improvements both in relative and absolute values.

Our algorithm requests a substantial amount of computational time. For the results in Table 2, Chaco and Metis ended their computation 5 to 20 times faster than our algorithm. Generally, our algorithm need about twice more time than Chaco or Metis to get a solution which is better than these partitioners, the rest of the time is spent to improve the quality of the solution. On the other hand, running any other partitioner for as much time as our algorithm (using several different coarsening factors or any other adjustable parameters) didn't get solutions competitive with those shown in Table 2 for our algorithm.

5   Conclusion

The main goal in the design of our current approach was to enhance the control structure of the cooperative procedure (the set of search programs as a whole, their global behavior). Our strategy to achieve this goal has been to simplify the cooperation scheme in terms of the neighborhood structure allowed among cooperating programs (a linear array), to have a more explicit definition of the search space that a process is allowed to explore (the neighborhood functions), and to decrease the probability that shared information will mislead the search programs, by establishing better control over the cost function for selecting reused information (the three inter-task operators). But most importantly, this new design redefines the purpose of the search programs $P_i > 0$ in the multi-level structure. Reused information helps to modify the data sets defining the search space available to the search methods rather than directly influencing the search parameters of the cooperating programs. The main purpose of the search programs is then to find information useful to neighbor levels in the control structure, rather than to directly generate solutions for the optimization problem. The search programs $P_i > 0$ become part of a complex multi-level control structure which spans several computers. Their goal is to assist the search performed by program P_0, which is the only program that can find explicit solutions to the optimization problem. In this regard, the current design is an example of how we can transform the complex interactions among cooperating programs into a useful explicit control structure of cooperative procedures.

References

[1] F. Glover. Heuristics for Integer Programming Using Surrogate Constraints. *Decision Sciences*, 8(1):156–166, 1977.

[2] F. Glover. Ejection Chains, Reference Structures and Alternating Path Methods for the Traveling Salesman Problem. Report, University of Colorado, Boulder, 1992.

[3] F. Glover. A Template for Scatter Search and Path Relinking. In J.K. Hao, E. Lutton, E. Ronald, M. Schoenauer and D. Snyers, editors, *Lecture Notes in computer Science*, 1997.

[4] F. Glover and M. Laguna. Tabu Search. In C. Reeves, editor, *Modern Heuristic Techniques for Combinatorial Problems*, pages 70–141. Blackwell Scientific Publishing, 1993.

[5] F. Glover and M. Laguna. *Tabu Search*. Kluwer Academic Publishers, 1997.

[6] B. Hendrickson and R. Leland. A Multilevel Algorithm for Partitioning Graphs. Report SAND93-1301, Sandia National Laboratories, 1993.

[7] B. Hendrickson and R. Leland. The Chaco User's Guide: Version 2.0. Report SAND95-2344, Sandia National Laboratories, 1995.

[8] G. Karypis and V. Kumar. A Fast and High Quality Multilevel Scheme for Partitioning Irregular Graphs. *SIAM Journal on Scientific Computing*, to appear.

[9] G. Karypis and V. Kumar. A Software Package for Partitioning Unstructured Graphs, Partitioning Meshes, and Computing Fill-Reducing Orderings of Sparse Matrices: Version 3.0. Report, University of Minnesota, 1997.

[10] J.P. Kelly and J. Xu. Tabu Search and Vocabulary Building for Routing Problems. Technical report, Graduate School of Business Administration, University of Colorado at Boulder, 1995.

[11] L. Lopez, M.W. Carter, and M. Gendreau. The Hot Strip Mill Production Scheduling Problem: A Tabu Search Approach. Report, Center for Research on Transportation, Université de Montréal, 1996.

[12] Y. Rochat and E. Taillard. Probabilistic Diversification and Intensification in Local Search for Vehicle Routing. *Journal of Heuristics*, 1(1):147–167, 1995.

[13] E. Taillard, P. Badeau, M. Gendreau, F. Guertin, and J.-Y. Potvin. A New Neighborhood Structure for Vehicule Routing with Time Window. Report CRT-95-66, Center for Research on Transportation, Université de Montréal, 1995.

[14] M. Toulouse, T.G. Crainic, and B. Sansó. An Experimental Study of Systemic Behavior of Cooperative Search Algorithms. In I.H. Osman S. Voss, S. Martello and C. Roucairol, editors, *Meta-Heuristics: Theory and Applications*, pages 373–392. Kluwer Academic Publishers, 1997.

A Quantitative Measure of Portability with Application to Bandwidth-Latency Models for Parallel Computing*

(Extended Abstract)

Gianfranco Bilardi, Andrea Pietracaprina, and Geppino Pucci

Dipartimento di Elettronica e Informatica, Università di Padova, Padova, Italy.
{bilardi,andrea,geppo}@artemide.dei.unipd.it

Abstract. We introduce a novel methodology for the quantitative assessment of the effectiveness and portability of models of parallel computation. Specifically, we relate the effectiveness of a model M, adopted for algorithm design, with respect to a platform M', where algorithms developed for M are ultimately executed, to the product of cross-simulation slowdowns between M and M'. The portability of M with respect to a class of platforms can be estimated by its minimum effectiveness over the platforms in the class. We apply our methodology to assess the portability of enhanced variants of the BSP model with respect to processor networks, with particular emphasis on multidimensional arrays.

1 Introduction

It is widely recognized that the choice of a model for parallel computation is made particularly hard by the desire of simultaneously achieving usability, effectiveness, and portability. *Usability* refers to the ease of algorithm design and analysis. *Effectiveness* means that efficiency of algorithms in the model translates into efficiency of execution on some given platform. *Portability* denotes the ability of achieving effectiveness with respect to a wide class of target platforms. These properties appear, to some extent, incompatible. For instance, effectiveness requires modeling a number of platform-specific aspects that affect performance (e.g., data layout, task assignment, scheduling), at the expense of portability and usability.

The quest for a suitable model of parallel computation has been ongoing for over two decades, as witnessed by the proliferation of models in the literature (see [MMT95] for a survey). One elusive aspect of this quest is the lack of a systematic framework to compare and evaluate candidate models at a quantitative level. Typically, the usability of a model is illustrated by exhibiting efficient algorithms for key computational problems, while effectiveness and portability are argued on the basis of qualitative considerations on the essential features to be expected of present and future machines. Simulations are sometimes advocated as a tool to compare models. For example, there are many results of the form: "model M_1 is more powerful than model M_2", meaning that M_1 can simulate M_2 with constant slowdown, but not vice versa. However, such characterization of

---

* This research was supported, in part, by MURST of Italy under Project MOSAICO, and by CNR of Italy under grant CB97.05085.CT12.

P. Amestoy et al. (Eds.): Euro-Par'99, LNCS 1685, pp. 543–551, 1999.

power does not directly relate to effectiveness: in fact, if M_1 derives its greater power by capabilities that are available on the target platforms, then M_1 is likely to be more effective than M_2, but when M_1 derives its greater power by assuming lower cost for primitives that are hard to implement on the target machines, M_2 might well be more effective than M_1.

In this paper, we propose and begin to explore a methodology for studying effectiveness and portability in a quantitative way. Namely, in Section 2 we argue that, if a model M and a platform M' can simulate each other with respective slowdowns $S(M, M')$ and $S(M', M)$, then the quantity $\delta(M, M') = S(M, M')S(M', M)$ provides an upper measure of effectiveness for M with respect to M'. Namely, effectiveness decreases with increasing $\delta(M, M')$ and is highest for $\delta(M, M') = 1$. When maximized over all platforms M' in a given class, this quantity provides an upper measure of the portability of M with respect to the class. Next, we apply our methodology to compare the effectiveness and portability of the *Bulk-Synchronous Parallel (BSP)* model [Val90] and one of its variants, the *Decomposable BSP (D-BSP)* [DK96], with respect to the class of *processor networks*.

A BSP(p, g, ℓ) machine consists of a set of p processor/memory pairs communicating through a router, whose computation is organized in supersteps. Each superstep embodies a phase where processors compute on locally available data, a phase where processors exchange messages, and a phase of global synchronization. The superstep is charged a cost of $w + gh + \ell$, where w (resp., h) is the maximum number of operations performed (resp., messages sent/received) by any processor in that superstep, and g and ℓ are parameters capturing bandwidth-latency characteristics of the router. D-BSP is defined similarly, with an additional provision that allows the computation to proceed independently within subsets of processors, each subset being characterized by its on bandwidth and latency parameters. This mechanism aims at capturing the locality of communication and synchronization. (See Section 3 for a formal definition of D-BSP.) Both BSP and D-BSP attempt to strike a balance between usability, effectiveness and portability by allowing programs to be written in terms of the bandwidth and latency parameters. Specific values for these parameters are selected at run-time to adapt to the characteristics of the target platform. Conceivably, parameterization affords higher portability, while it introduces only a moderate burden on the algorithm developer.

It is straightforward to show that suitable choices of the BSP or D-BSP parameters allow these models to be simulated on most networks of interest with only constant slowdown (with the above notation, $S(\text{BSP}, M')$, $S(\text{D-BSP}, M') = O(1)$, with M' a processor network). Hence, to study the effectiveness of such models, in Section 4 we concentrate on the simulation of an arbitrary processor network M' on BSP and D-BSP, determining a lower bound to $S(M', \text{BSP})$ and an upper bound on $S(M', \text{D-BSP})$. When M' is a multidimensional array, $S(M', \text{D-BSP})$ turns out to be considerably smaller than the optimal achievable slowdown $S(M', \text{BSP})$. Based on these results, we conclude that, under the δ metric, D-BSP is more effective than BSP for such networks.

Finally, in Section 5, we show that although no specific provisions are made in D-BSP to deal with unbalanced communication patterns, the model, unlike BSP, can efficiently cope with such patterns, which is an indication that further complications to the model, as introduced in other BSP variants (e.g., Y-BSP and E-BSP [DK96, JW96]), may not be needed to achieve higher effectiveness.

2   A Quantitative Approach to Effectiveness and Portability

Let us consider a model M where designers develop and analyze programs, which we call M-*programs*, and a platform M' onto which M-programs are translated and executed. We call M'-*programs* the programs that are ultimately executed on M'. During the design process, choices between different programs (possibly coding alternative algorithms for the same problem) will be clearly guided by model M, by comparing their running times as predicted by the model's cost function. Intuitively, we consider M to be effective with respect to M' if the choices based on M turn out to be the right ones in relation to program performance on M'. In other words, we hope that the relative performance of any two M-programs reflects the relative performance of their "counterparts" on M'. The difficulty with this approach is that of clearly defining the association between M-programs and M'-programs, which we do as follows.

We postulate the existence of an equivalence relation ρ among both M-programs and M'-programs and restrict our attention to ρ-*optimal* programs, where a ρ-optimal M-program (resp., M'-program) is the fastest among all M-programs (resp., M'-programs) ρ-equivalent to it. Let us define Π and Π' to be the sets of ρ-optimal M-programs and M'-programs, respectively. We also define a *translation function* σ that associates a program $\pi \in \Pi$ with its ρ-equivalent counterpart $\sigma(\pi) \in \Pi'$. In other words, for every M-program, σ identifies a translation into an M'-program which is consistent with the choice of ρ. Several choices for ρ are possible. For instance ρ-equivalent programs could be those that have the same input-output map (which, however, makes σ rather difficult to obtain in practice), or those that implement the same high-level algorithm (which makes σ more realistic). For simplicity, we will not formally specify ρ but, when applying our methodology, we will only give an informal description of its properties.

For $\pi \in \Pi$ and $\pi' \in \Pi'$ we let $T(\pi)$ and $T'(\pi')$ denote their running times on M and M', respectively. We propose the following *inverse* measure of effectiveness

$$\eta(M, M') = \max_{\pi_1, \pi_2 \in \Pi} \frac{T(\pi_1)}{T(\pi_2)} \cdot \frac{T'(\sigma(\pi_2))}{T'(\sigma(\pi_1))} \ . \tag{1}$$

Note that $\eta(M, M') \geq 1$, where a value close to 1 implies a high effectiveness of M with respect to M', while a large value indicates that the relative performance of programs on M may not be preserved on M'. If, rather than a single platform M', we consider a class $\mathcal{C}$ of target platforms, then the quantity $\eta_{\mathcal{C}}(M) = \max_{M' \in \mathcal{C}} \eta(M, M')$ provides an *inverse* measure of portability of model M.

Next, we show that an upper estimate of $\eta(M, M')$ can be obtained based on the ability of M and M' to simulate each other. Consider an algorithm that takes any M-program π and simulates it on M' as an M'-program π' ρ-equivalent to π. (Note that in neither π nor π' needs be ρ-optimal.) We define the *slowdown* $S(M, M')$ of the simulation as the ratio $T'(\pi')/T(\pi)$ maximized over all possible M-programs π. In practice, $S(M, M')$ represents the (worst-case) cost for supporting model M on M'.

We define $S(M', M)$ in a similar fashion, by interchanging the roles of M and M'. Then, we define the *distance* between M and M' to be the quantity

$$\delta(M, M') = S(M, M')S(M', M) \ , \tag{2}$$

and show that the following key inequality is satisfied:

$$\eta(M, M') \leq \delta(M, M') \ . \tag{3}$$

Indeed, since the simulation algorithms considered in the definition of $S(M, M')$ and $S(M', M)$ preserve ρ-equivalence, it is easy to see that for any $\pi_1, \pi_2 \in \Pi$, $T'(\sigma(\pi_2)) \leq S(M, M')T(\pi_2)$ and $T(\pi_1) \leq S(M', M)T'(\sigma(\pi_1))$. Thus, we have that $(T(\pi_1)/T(\pi_2)) \cdot (T'(\sigma(\pi_2))/T'(\sigma(\pi_1))) \leq \delta(M, M')$, which implies that $\delta(M, M')$ is an upper bound to $\eta(M, M')$.

3   D-BSP: Capturing Network Proximity in BSP

The *Decomposable Bulk Synchronous Parallel* model (D-BSP) was originally intro-duced in [DK96] as an extension of BSP. In this paper we refer to a weaker vari-ant[1] of the model which can be defined as follows. Let $g = (g_0, g_1, \ldots, g_{\log p})$ and $\ell = (\ell_0, \ell_1, \ldots, \ell_{\log p})$. A D-BSP (p, g, ℓ) is a collection of p processor-memory pairs communicating through a router. For $0 \leq i \leq \log p$, the p processors are partitioned into 2^i fixed, disjoint *i-clusters* $C_0^{(i)}, C_1^{(i)}, \cdots, C_{2^i-1}^{(i)}$ of $p/2^i$ processors each, where the processors of a cluster are able to communicate and synchronize among them-selves independently of the other clusters. The clusters form a binary decomposition tree for the D-BSP: specifically, $C_j^{(\log p)}$ contains only Processor j, for $0 \leq j < p$, and $C_j^{(i)} = C_{2j}^{(i+1)} \cup C_{2j+1}^{(i+1)}$, for every $0 \leq i < \log p$ and $0 \leq j < 2^i$. Parameters g_i and ℓ_i are related to the bandwidth and latency guaranteed by the router when communication occurs within *i*-clusters, for $0 \leq i \leq \log p$.

Analogously to BSP, a p-processor D-BSP computation consists of a sequence of *labeled supersteps*, where labels range in the set $\{0, 1, \ldots \log p\}$. In a superstep of la-bel i, communication and synchronization occurs exclusively within *i*-clusters. If, in an *i*-superstep, each processor performs at most w local operations, and the messages exchanged form an h-relation, then the cost of the superstep is $w + hg_i + \ell_i$.

Note that the standard BSP(p, g, ℓ) can be regarded as a D-BSP (p, g, ℓ) with $g_i = g$ and $\ell_i = \ell$ for every i, $0 \leq i \leq \log p$, hence a valid D-BSP algorithm is also a valid BSP algorithm. Vice versa, any valid BSP(p, g, ℓ) algorithm is also valid for any D-BSP (p, g, ℓ) with $g_0 = g$ and $\ell_0 = \ell$. This ensures full compatibility between the two models, which differ only with respect to their cost functions. In particular, D-BSP introduces the notion of proximity in BSP through clustering, and groups h-relations into specialized classes associated with different costs.

For suitable choices of the parameters, both BSP and D-BSP can be supported (i.e., simulated) with no loss of efficiency on processor networks. For example, by setting

---

[1] Our variant is weaker in the sense that the algorithms for the variant immediately translate into algorithms for the original model of [DK96] with no loss in performance.

$g = \ell = p^{1/d}$ for BSP and $g_i = \ell_i = (p/2^i)^{1/d}, 0 \leq i \leq \log p$, for D-BSP, both models can be simulated with constant slowdown on a d-dimensional array. (See [DK96] for other examples.)

4 Effectiveness of Bulk Synchronous Models for Processor Networks

In this section, we consider the issue of simulating processor networks on BSP and D-BSP. In the light of our previous discussion, our goal is to provide quantitative evidence of how the richer structure exposed in D-BSP improves its effectiveness with respect to the "flat" BSP.

BSP Let G be a connected N-processor network, where in one *step* each processor executes a constant number of local operations and may send/receive one point-to-point message to/from each neighboring processor (multi-port regimen). We can prove the following general lower bound for the simulation of G by BSP.

Theorem 1. *For any connected N-node network G of bounded degree and every integer $T \geq 1$ there exists a T-step computation of G that requires time*

$$\Omega \left(T \left(\frac{N}{p} + \min\{g, T, N\} \right) + \ell \right)$$

to be simulated by $BSP(p, g, \ell)$. The lower bound holds even if at the beginning of the simulation each BSP processor knows the initial state of every node of G, and if the operations performed by a node of G may be simulated by more than one processor.

Proof (Sketch). Since G embeds the N-node linear array A_N with constant dilation, it suffices to derive a lower bound for BSP simulations of A_N. The terms $T(N/p)$ and ℓ are obtained immediately. Next, observe that an arbitrary T-step computation of A_N can be modeled as a a a dag $\Delta_T(A_N)$ whose nodes (i.e., operations) are all pairs (v, t), with $v \in V$ and $0 \leq t \leq T$, and whose arcs (i.e., state dependencies) connect pairs $(v_1, t), (v_2, t+1)$, where $t < T$ and $|v_1 - v_2| \leq 1$. Simulating a T-step computation of A_N on BSP amounts to *execute* $\Delta_T(A_N)$. During an execution of $\Delta_T(A_N)$, an operation associated to a dag node can be executed by a BSP processor q if and only if q knows the state of all the node's predecessors.
 Consider the case $T \leq N$. Then, $\Delta_T(A_N)$ contains the $\lceil T/2 \rceil \times \lceil T/2 \rceil$ *diamond dag* $D_{T/2}$ as a subgraph. The result in [ACS90, Th. 5.1] implies that any BSP execution of $D_{T/2}$ either requires an $\Omega(T)$-relation or an $\Omega(T^2)$-time sequential computation performed by some processor. Hence, any BSP execution of $D_{T/2}$ requires $\Omega(T \min\{g, T\})$ time. For the case $T \geq N$, it suffices to observe that $\Delta_T(A_N)$ contains a vertical stack of $\Theta(T/N) \lceil N/2 \rceil \times \lceil N/2 \rceil$ disjoint copies of $D_{T/2}$, whose first executions do not overlap in time, thus requiring $\Omega(T \min\{g, N\})$ BSP time.

The next theorem, whose proof is omitted for brevity, provides a lower bound on the reverse simulation.

Theorem 2. *Let G be an N-node network of bisection width B. Then, any simulation of $BSP(p, g, \ell)$ on G requires slowdown*

$$S\left(BSP(p, g, \ell), G\right) = \Omega\left(\frac{\min\{N, p\}}{gB}\right) \ .$$

D-BSP Let us now turn to network simulations on D-BSP. In what follows, we assume w.l.o.g. that G has a decomposition tree $\{G_0^{(i)}, G_1^{(i)}, \cdots, G_{2^i-1}^{(i)} : \forall i, 0 \le i \le \log p\}$, where each $G_j^{(i)}$ (i-subnet) has $N/2^i$ nodes and is connected to the rest of the network by at most b_i boundary links; moreover, $G_j^{(i)} = G_{2j}^{(i+1)} \cup G_{2j+1}^{(i+1)}$. In order to simulate G on a p-processor D-BSP, we adopt the following strategy. Partition the nodes of G among the D-BSP processors so that the nodes of $G_j^{(i)}$ are assigned to the processors of i-cluster $C_j^{(i)}$, for every i and j. Let $M_{i,j}^{\text{out}}$ (resp., $M_{i,j}^{\text{in}}$) denote the messages that are sent (resp., received) by nodes of $G_j^{(i)}$ to (resp., from) nodes outside the subnet. Since the number of boundary links of an i-subnet is at most b_i, we have that $|M_{i,j}^{\text{out}}|, |M_{i,j}^{\text{in}}| \le b_i$. Let also $\bar{M}_{i,j}^{\text{out}} \subseteq M_{i,j}^{\text{out}}$ denote those messages that from $G_j^{(i)}$ go to nodes in its sibling $G_{j'}^{(i)}$, with $j' = j \pm 1$ depending on whether j is even or odd. The idea behind the simulation is to guarantee, for each cluster, that the outgoing messages be balanced among the processors of the cluster before they are sent out, and, similarly, that the incoming messages destined to any pair of sibling clusters be balanced among the processors of the cluster's father before they are acquired.

More precisely, after a first superstep where each D-BSP processor simulates the local computation and communications internal to the subnet assigned to it, the following two cycles are executed.

1. For $i = \log p - 1$ down to 0 do in parallel within each $C_j^{(i)}$, for $0 \le j < 2^i$:
   (a) Send the messages in $\bar{M}_{i+1,2j}^{\text{out}}$ (resp., $\bar{M}_{i+1,2j+1}^{\text{out}}$) from $C_{2j}^{(i+1)}$ (resp., $C_{2j+1}^{(i+1)}$) to $C_{2j+1}^{(i+1)}$ (resp., $C_{2j}^{(i+1)}$), so that each processor receives (roughly) the same number of messages.
   (b) Balance the messages in $M_{i,j}^{\text{out}}$ among the processors of $C_j^{(i)}$. (This step is vacuous for $i = 0$.)
2. For $i = 1$ to $\log p - 1$ do in parallel within each $C_j^{(i)}$, for $0 \le j < 2^i$:
   (a) Send the messages in $M_{i,j}^{\text{in}} \cap M_{i+1,2j}^{\text{in}}$ (resp., $M_{i,j}^{\text{in}} \cap M_{i+1,2j+1}^{\text{in}}$) to the processors of $C_{2j}^{(i+1)}$ (resp., $C_{2j+1}^{(i+1)}$), so that each processor receives (roughly) the same number of messages.

A formal proof of correctness of the above procedure will be provided in the full version of this abstract. As for its running time, let $h_i = \lceil b_i/(p/2^i) \rceil$, for $0 \le i \le \log p$, denote the average number of incoming/outgoing messages for an i-cluster. The balancing operations performed by the algorithm guarantee that iteration i of either cycle entails a $\max\{h_i, h_{i+1}\}$-relation within i-clusters. Finally, we can optimize the above simulation by running it entirely within within a cluster of $p' \le p$ processors of the p-processor D-BSP, where p' is the value that minimizes the overall slowdown. The following theorem summarizes the above discussion.

Theorem 3. *For any N-node network G with a decomposition tree of parameters $(b_0, b_1, \ldots, b_{\log p})$, one step of G can be simulated on a D-BSP (p, g, ℓ) in time*

$$
O\left(\min_{p' \le p} \left\{ \frac{N}{p'} + \sum_{i=\log(p/p')}^{\log p - 1} (g_i \max\{h_i, h_{i+1}\} + \ell_i) \right\} \right) ,
$$

where $h_i = \lceil b_{i-\log(p/p')}/(p/2^i) \rceil$, for $\log(p/p') \le i < \log p$.

Portability of BSP vs D-BSP on multidimensional arrays First, we remark that all the simulation results described above deal with simulation algorithms which preserve ρ-equivalence for most realistic definitions of ρ (e.g., same input-output map, or same high-level algorithm). Let now M be a BSP(p, g, ℓ), and G be a p-node d-dimensional array. Then, from Theorems 1 and 2, it follows that $\delta(M, G) = \Omega\left(p^{1/d}\right)$. Consider now a D-BSP (p, g, ℓ) with $g_i = \ell_i = (p/2^i)^{1/d}$, for $0 \le i \le \log p$. Such D-BSP (p, g, ℓ) can be simulated on G with constant slowdown. Since G has a decomposition tree with subnets $G_j^{(i)}$ that have $b_i = O\left((p/2^i)^{(d-1)/d}\right)$, the D-BSP simulation presented above yields a slowdown of $O\left(p^{1/(d+1)}\right)$ per step, when entirely run on a cluster of $p' = p^{d/(d+1)}$ processors. In conclusion, letting M' be the D-BSP with the above choice of parameters, we have that $\delta(M', G) = O\left(p^{1/(d+1)}\right)$. Therefore, under the δ metric, D-BSP is asymptotically less distant than BSP from a multidimensional array.

One drawback of the D-BSP simulation is that it requires that the memory size of each D-BSP processor be a $\Theta\left(p^{1/(d+1)}\right)$ factor larger than the memory size of an array node, which might be unreasonable in practice. In the full version of this paper, we will exhibit a subtler simulation strategy of multidimensional arrays on D-BSP which reduces the memory blowup to a mere $O(\log \log p)$ factor, at the expense of an extra $O(\log p)$ factor in the simulation slowdown.

Finally, if we allow the simulation of a step of the network to start before the simulation of previous steps is completed, we can set up a more complex recursive strategy for simulating arrays with constant-size memory at each node on D-BSP, whose slowdown is exponentially smaller than the one obtained by applying the step-by-step simulation of Theorem 3. We have:

Theorem 4. *For $T = \Omega\left(p^{1/d}\right)$, any T-step computation of a p-node d-dimensional array G (d constant) with constant-size memory at each node can be simulated by a D-BSP (p, g, ℓ) with $g_i = \ell_i = (p/2^i)^{1/d}$, for $0 \le i \le \log p$, in time $O\left(T 2^{a\sqrt{\log p}}\right)$, with $a = 2d + 6$.*

(The simulation algorithm and the proof of Theorem 4 are omitted.) Recall that the lower bound on network simulations by BSP (Theorem 1) does not depend on the size of the memory modules local to each node of G and, in particular, it holds even if such modules have constant size. Therefore, Theorem 4 implies that the δ metric for D-BSP and multidimensional arrays with constant-size memory at each node is exponentially smaller than the respective metric for BSP.

5   Exploiting D-BSP Locality for Unbalanced Communication

An (h, m)-relation is a routing instance where each processor sends/receives at most h-messages, and a total of m messages are exchanged [JW96]. On several networks, the time to route an (h, m) relation decreases with m, for fixed h. This does not happen in BSP, where *any* (h, m)-relation requires $gh + \ell$ steps, regardless of m. Motivated by these observation, some authors have proposed the introduction of m as an explicit parameter of the model (see E-BSP [JW96] and Y-BSP [DK96]).

In this section, we argue that D-BSP can route (h, m) relations in ways that take advantage of small values of m. In particular, we show that, when a d-dimensional array simulates the routing of an (h, m)-relation on D-BSP, the resulting network performance is within a constant factor of optimal. Therefore, for such networks, there is no increase of effectiveness by augmenting D-BSP with m-sensitive capabilities. This is an indication that the complication of making a model which already captures network proximity also sensitive to unbalance may not be warranted by the accruing advantages.

The key insight that leads to the exploitation of cluster locality to deal with unbalance is similar to the one employed in Section 4 to simulate one step of a network, and is inspired by the efficient routing protocols known for specific architectures (e.g., see [PPS94]). Specifically, messages are progressively balanced within clusters of increasing size so that the degree of the relation decreases as the cost of communication within the clusters increases. More specifically, let k be such that $2^k \leq hp/m < 2^{k+1}$ (note that $0 \leq k \leq \log p$). The (h, m)-relation is routed in two phases. In the first phase (*ascend*), messages are iteratively balanced within larger and larger clusters (starting from k-clusters). In the second phase (*descend*) messages are kept balanced and iteratively dispatched to arbitrary processors within their destination k-cluster and from there to their final destinations. We have:

Theorem 5. *An (h, m)-relation can be routed on a D-BSP (p, g, ℓ) in time*

$$
T(h, m) = O \left(\sum_{i=0}^{\lfloor \log(hp/m) \rfloor} \left(T_{\mathrm{pr}}(i) + g_i \left\lceil \frac{m}{(p/2^i)} \right\rceil + \ell_i \right) \right),
$$

where $T_{\mathrm{pr}}(i)$ denotes the time of a prefix-sum operation within an i-cluster.

To appreciate the potential of the above strategy, consider again the case of a p-node d-dimensional array, which is characterized by parameters $g_i = \ell_i = p^{1/d}$, for $0 \leq i \leq \log p$. For such values of the parameters, $T_{\mathrm{pr}}(i) = O\left((p/2^i)^{1/d}\right)$ [DK96]. Hence, by Theorem 5 we get

$$
T(h, m) = O \left(p^{1/d} + m^{1/d} h^{1-1/d} \right),
$$

which is smaller than the BSP time by a factor $(ph/m)^{1/d}$ when $p/h^{d-1} \leq m \leq ph$. More surprisingly, a simple bandwidth argument shows that the above time is optimal for routing an arbitrary (h, m)-relation directly on a d-dimensional array. Hence, for such communication patterns, the effectiveness of D-BSP is maximum, in the sense that there is no loss of efficiency in running the D-BSP algorithm on the network with respect to running the best network-specific algorithm. A corollary of this result is that for multidimensional arrays D-BSP is as effective in treating unbalanced communication as E-BSP [JW96], where the treatment of unbalanced communication is a primitive of the model.

References

[ACS90]   A. Aggarwal, A.K. Chandra, and M. Snir. Communication complexity of PRAMs. *Theoretical Computer Science*, 71:3–28, 1990. See also *Proc. of ICALP '88*, 1–17.

[DK96]   P. De la Torre and C.P. Kruskal. Submachine locality in the bulk synchronous setting. In *Proc. of EUROPAR 96*, LNCS 1124, pages 352–358, August 1996.

[JW96]   B.H.H. Juurlink and H.A.G. Wijshoff. The E-BSP model: Incorporating general locality and unbalanced communication into the BSP model. In *Proc. of EUROPAR 96*, LNCS 1124, pages 339–347, August 1996.

[MMT95]   B. Maggs, L.R. Matheson, and R.E. Tarjan. Models of parallel computation: A survey and synthesis. In *Proc. of the 28th Hawaii International Conference on System Sciences (HICSS)*, volume 2, pages 61–70, January 1995.

[PPS94]   A. Pietracaprina, G. Pucci, and J. Sibeyn. Constructive deterministic PRAM simulation on a mesh-connected computer. In *Proc. 6th ACM Symp. on Parallel Algorithms and Architectures*, pages 248–256, Cape May, NJ, June 1994.

[Val90]   L.G. Valiant. A bridging model for parallel computation. *Communications of the ACM*, 33(8):103–111, August 1990.

A Cost Model For Asynchronous and Structured Message Passing*

Emmanuel Melin, Bruno Raffin, Xavier Rebeuf, and Bernard Virot

LIFO - Université d'Orléans, F-45067 Orléans, Cedex 02 - FRANCE
Phone: (+33) 2 38 41 73 18 Fax: (+33) 2 38 41 71 37
virot@lifo.univ-orleans.fr

Abstract. We present a cost model that relies on an asynchronous and structured parallel execution model for message passing. We show that it is possible to define a complexity function for programs that yields a symbolic date for each communication event. By ordering these dates, an upper bound for the network load can be computed. In contrast to classical approaches this cost computation can handle asynchronism for message passing and communication/computation overlap.
Keywords: Cost Model; Symbolic Date; Parallel Programming Languages; Asynchronous Execution; Structural Clock.

1 Introduction

An important problem of parallel programming is to define a model of parallel complexity to a priori quantify the efficiency of an algorithm. Several complexity models have been proposed for synchronous computations, but the prevailing one is the PRAM model [3]. The cost associated to an algorithm is the number of synchronous elementary steps performed. The PRAM model allows straightforward cost evaluation for complex algorithms. Nevertheless, it is not well suited for actual distributed memory machines since it does not take into account communication overhead.

We defined in [5] a structured communication model called $\mathcal{SCL}$-chan. The communication semantics, driven by the syntactic structure of the program, yields a simple synchronous programming model and an efficient loosely synchronized execution model. In this paper, we show that it is possible to define a cost model for $\mathcal{SCL}$-chan programs. We propose a complexity function that yields a symbolic date for each communication event. These dates can be ordered, allowing to compute, for each date, an upper bound for the network load. In contrast to classical approaches, such as BSP [4], [8], the cost computation can handle asynchronism for message passing and communication/computation overlap.

---

* This work has been partly supported by the "Institut du Développement et des Ressources en Informatique Scientifique", Orsay, France. Project 990829.

P. Amestoy et al. (Eds.): Euro-Par'99, LNCS 1685, pp. 552–556, 1999.

2   A Structured Message Passing Language

The abstract machine is a distributed memory asynchronous parallel machine with P processors. Processors are also called *indices* and denoted by u, v,... The value of a variable X stored in the memory of an index u is denoted by $X|_u$. Data exchanges between indices take place through explicit send/receive. Data sent and not yet received are stored in *channels* handled as LIFO (Last-In First-Out) lists. There exists exactly one channel by variable and by pair of indices. We denote by (u, v, X) the channel associated to the sender u, the receiver v and the variable X. $\mathcal{SCL}$-chan programs rely on SPMD paradigm. Only the receive instruction can block, other instructions are executed asynchronously.

$X := E$. An index u updates its value $X|_u$ in its memory with the value $E|_u$.

where B do S end. If an index u evaluates the pure condition $B|_u$ to *true*, then it executes the instruction block S.

$S; T$. When an index finishes the execution of S, it starts the execution of T.

forwhere $I := A$ to B do S end. An index u repeatedly executes S until its counter $I|_u$ reaches the value $B|_u$.

send X to A. This instruction is non blocking. A sending index u sends the value $X|_u$ to the index $v = A|_u$, i.e. it adds this value into the channel (u, v, X).

receive X from A into Y. A receiving index u waits that the sending index $v = A|_u$ has executed all the send instructions coming before (syntactically, unfolding the loops) this receive. Then, index u extracts the last of the corresponding values from (v, u, X).

3   A Complexity Function for $\mathcal{SCL}$-chan Programs

We now define a complexity function for $\mathcal{SCL}$-chan programs. We first present the computation of the communication time of a datum. We model the network by an input and an output buffer for each index and a global router. First the sender copies the data from its memory into the output buffer. Then the router transmits the message from the output buffer of the sender to the input buffer of the receiver. Finally the receiver copies the message from its input buffer into its memory. The time required to execute an elementary instruction defines the time unit we use for all indices. To quantify the communications, we introduce a few architecture parameters. We denote by β_o (resp. β_i) the start-up time and by τ_o (resp. τ_i) the copy rate to access the output (resp. input) buffer. We denote by τ_r the transmission rate through the router. The transmission time of a message depends on the other messages currently in transit through the router. The communication time for a message M, assuming that n messages $M_1, ..., M_n$ are currently in transit, is given by $\beta_o + size(M)\tau_o + l\tau_r + \beta_i + size(M)\tau_i$ with $l = size(M) + \sum_{k=1}^{n} size(M_k)$.

To model the execution of a $\mathcal{SCL}$-chan program, we consider a weighted causality graph where each node $(inst, u)$ corresponds to an instance of the instruction $inst$ executed by the index u. The weight of each arrow corresponds to the minimum number of time units between the executions of the two corresponding instances. Note that these weights are symbolic expressions depending on the architecture parameters. We associate a symbolic execution date to each node by computing the longest path ending at this node. The execution cost for an algorithm S denoted by $Cost(S)$ is the maximum, for all indices, of the final execution dates.

To evaluate the load l, we first deduce from the causality graph a set Net gathering all sent messages labelled by their dates of emission. To each node corresponding to a sending instruction, we associate an element of Net defined as a pair composed by the date where the message enters the network and the size of the message. The entering date corresponds to the date of the node augmented by the time of copy to the output buffer. We denote by $l(d)$ the network load associated to the execution date d. The notation l will be used instead of $l(d)$ in case the date d is not ambiguous. The value of $l(d)$ corresponds to messages sent before d and not yet received. To compute an upper bound for $l(d)$, we have to define an order on dates which subsumes all possibles execution orders. Let d and d' be two symbolic dates. We say that d is lower than d', denoted by $d < d'$, if the expression d' is an upperbound for the expression d. Since two expressions may be not comparable, the order on dates is only a partial order. The load function $l(d)$ considers all messages with an entering date lower than d or not comparable to d and then, it selects all messages arrived at a date later than d or not comparable to d.

$$l(d) = \Sigma\{s/(s,d') \in Net \ \wedge \ not(d < d') \ \wedge \ not(d' + l(d')\tau_r < d)\}$$

Example1. Let us consider the following program where each index sends a value to its neighbor.

> where $(This \ mod \ 2) = 0$ do; $X := 1$; end;
> send X to $(This + 1)$; receive X from $(This - 1)$ into Y

The left hand side of the following figure displays the corresponding causality graph. We suppose that the size of the message is 10. We denote by $\delta(u)$ the final execution date for the index u. The max expression encodes the longest path search when a communication occurs.

$$\delta(u) = \begin{cases} \beta_o + 10\tau_o + max\{3, 2 + l(2 + \beta_o + 10\tau_o)\tau_r\} + \beta_i + 10\tau_i & \text{if } u \text{ is even} \\ \beta_o + 10\tau_o + max\{2, 3 + l(3 + \beta_o + 10\tau_o)\tau_r\} + \beta_i + 10\tau_i & \text{otherwise} \end{cases}$$

The right hand side of the figure displays the ordering of the dates. We now construct the set Net. As each odd (resp. even) index sends X at the date $2 + \beta_o + 10\tau_o$ (resp. $3 + \beta_o + 10\tau_o$). We obtain: $Net = \{(2 + \beta_o + 10\tau_o, 10), ..., (3 + \beta_o + 10\tau_o, 10), ...\}$.

The load for the date $2 + \beta_o + 10\tau_o$ have only to take into account the messages sent by the odd indices, we have $l(2 + \beta_o + 10\tau_o) = 5P$. Since we cannot ensure that these messages exit the network before the date $3 + \beta_o + 10\tau_o$, the load for the date $3 + \beta_o + \tau_o$ has to take into account all the messages, then $l(3 + \beta_o + 10\tau_o) = 10P$. Thus $Cost(S) = max\{\delta(u)\} = 3 + \beta_o + 10\tau_o + 10P\tau_r + \beta_i + 10\tau_i$.

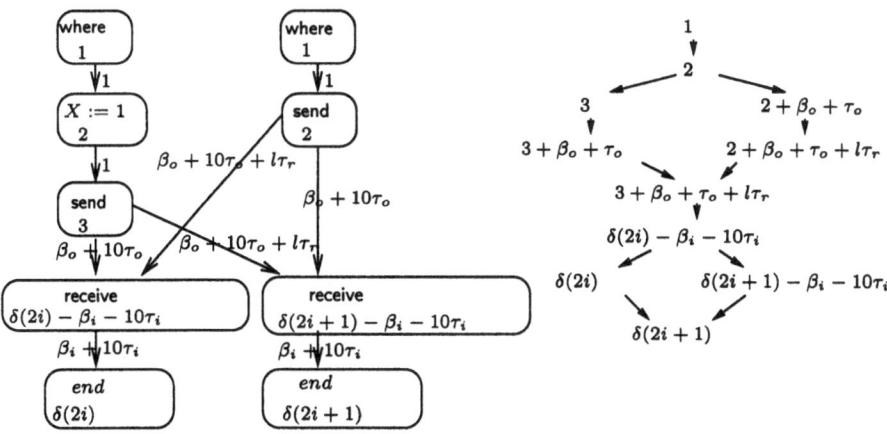

Example2. We now study a scan algorithm. The N data are distributed by block. We demote by $M = N/P$ the size of each block. The algorithm can be divided into three parts as shown by the following array. As no communication occurs between two different parts, the corresponding instances of instructions can be labeled independently. We denote by $\delta_{inst}(u)$ the resulting date of the execution of the instance $inst$ by the index u. The array displays the construction of the execution dates.

Inst.	Program	$\delta_{Inst}(u)$
$2I - 3$	forwhere $I := 2$ to M do	$2I - 3$
$2I - 2$	$\quad X[I] := X[I] + X[I - 1];$	$2I - 2$
$2(M - 1) + 1$	end;	$2(M - 1)$
1	$Y := X[M];$	$2M - 1$
2	send Y to $This + 1;$	$2M + \beta_o + size(Y)\tau_o$
$4I - 1$	forwhere $I := 1$ to $\lfloor log_2(This) \rfloor$ do	$\delta_{4I-2}(u) + 1$
$4I$	$\quad$ receive Y	$max\{\delta_{4I-1}(u); \delta_{4I-3}(u - 2^{(I-1)}) +$
	$\quad\quad$ from $This - 2^{(I-1)}$	$l(\delta_{4I-3}(u - 2^{(I-1)}))\tau_r\} +$
	$\quad\quad$ into $Z;$	$\beta_i + size(Y)\tau_i$
$4I + 1$	$\quad Y := Y + Z;$	$\delta_{4I}(u) + 1$
$4I + 2$	$\quad$ send Y to $This + 2^I;$	$\delta_{4I+1}(u) + \beta_o + size(Y)\tau_o$
$4\lfloor log_2(u) \rfloor + 3$	end;	$\delta_{4\lfloor log_2(u) \rfloor + 2}(u)$
$2I - 1$	forwhere $I := 1$ to M do	$\delta_{4\lfloor log_2(u) \rfloor + 2}(u) + 2I - 2$
$2I$	$\quad X[I] := X[I] + Y - X[M]$	$\delta_{4\lfloor log_2(u) \rfloor + 2}(u) + 2I - 1$
$2M + 1$	end	$\delta_{4\lfloor log_2(u) \rfloor + 2}(u) + 2M$

In the second part of the algorithm, communications occur. We first construct the set Net: $Net = \{(\delta_{4I+2}(u), size(Y))/u \in [1; P] \wedge I \in [0; log_2(u)]\}$. Since we have $\delta_{4I-2}(u - 2^{(I-1)}) < \delta_{4I+3}(u)$, the network load of a given iteration only has to take into account the messages sent during the previous iteration. We obtain $l(2M + 1) = Psize(Y)$ and $l(\delta_{4I+2}(u)) = (P/2^I)size(Y)$. By induction on the iterations of the forwhere loop we obtain the final date $\delta(u) = 4M + \beta_o + \tau_o + log_2(u)[2 + \beta_i + \beta_o + size(Y)(\tau_i + \tau_o)] + (2 - 2^{-\lfloor log_2(u) \rfloor - 1})Psize(Y)\tau_r$, and hence $Cost(S) = max\{\delta(u)\} = \delta(P)$.

By using an implementation of $\mathcal{SCL}$-chan on a Cray T3E, we measured the values of parameters via a ping pong test repeated one hundred times. We have estimated the cost of the scan on a Cray T3E for 32 processors. Then we measured the execution time. The following array displays results in μs.

N (word)	64	320	3200	32000
$Cost(S)$ ($10^{-6}sec.$)	756	756	757	759
$Measured(S)$ ($10^{-6}sec.$)	694	701	718	792

4   Conclusion

We presented a complexity model well suited for asynchronous structured message passing. Thanks to our structured communication semantics, we were able to build a complexity time function relying on the program structure and on the sequential complexity on each index. It yields a symbolic expression encoding the communication scheme. This result is used to evaluate an upper bound for the network load taking into account the asynchronism of communication events. By using this evaluation, we can deduce a symbolic cost associated to a given execution of the algorithm. Note that our model is general enough to be easily adapted to another modelization of the network, for example by introducing several routers or an adaptative routing.

In contrast to models like BSP, CGM, MINIBSP [2], [4], [7], [8], asynchronous communication-computation overlapping can be taken into account. In contrast to LogP [1],structured communication schemes can allow to compute worst case costs [6].

Experimental results on Cray T3E highlight accurate cost predictions. The complexity model appears to be a good trade-off between usability and effectiveness.

References

[1] D. Culler, R. Karp, D. Patterson, A. Sahay, K. E. Schauser, E. Santos, R. Subramonian, and V. E. Thorsten. LogP: Towards a realistic model of parallel computation. In *Proceedings of the Fourth ACM SIGPLAN Symposium on Principles & Practice of Parallel Programming*, pages 1–12, May 1993.

[2] F.Dehne, A.Fabri, and A. Rau-Chaplin. Scalable Parallel Geometric Algorithms for Coarse Grained Multicomputer. *ACM 9th Symposium on Computational Geometry*, 1993.

[3] J. Gabarro J. Balcazar, J. Diaz. *Structural Complexity II*. Springer-Verlag, 1989.

[4] W F McColl. General purpose parallel computing. In A M Gibbons and P Spirakis, editors, *Lectures on Parallel Computation. Proc. 1991 ALCOM Spring School on Parallel Computation*, pages 337–391. Cambridge University Press, 93.

[5] Emmanuel Melin, Bruno Raffin, Xavier Rebeuf, and Bernard Virot. A Structured Synchronization and Communication Model Fitting Irregular Data Accesses. *Journal of Parallel and Distributed Computing*, 50(1):3–27, 1998.

[6] Emmanuel Melin, Bruno Raffin, Xavier Rebeuf, and Bernard Virot. A Cost Model For Asynchronous and Structured Message Passing (Extended Version). Technical Report RR99-07, LIFO, Orléans, France, April 1999.

[7] D. B. Skillicorn. miniBSP: A BSP Language and Transformation System. http://www.qucis.queensu.ca/home/skill, 1996.

[8] L. G. Valiant. A Bridging Model for Parallel Computation. *Communications of the ACM*, 33(8):103–111, August 1990.

A Parallel Simulation of Cellular Automata by Spatial Machines

Bruno Martin

I3S-UNSA, UPRES-A 6070 CNRS, 930 Route des Colles,
B.P. 145, F-06903 Sophia-Antipolis Cedex

Abstract. We consider the relations between cellular automata and spatial machines. Spatial machines have been introduced as a bridging model for parallel computation in order to replace PRAM. We propose here a parallel simulation of cellular automata by a family of spatial machines.

Introduction

The success of the Turing model of sequential computation can be attributed to the existence of an efficient bridge between model and machines architecture. A common belief growing in importance is the lack of an analogue bridge for parallel machines. There are a few models of parallel computation but the unifying one seems to be the PRAM (Parallel Random-Access Machines), introduced in [4], which captures the notion of circuits [5] (another model of parallel computation).

Unfortunately, PRAM have two major drawbacks. First, they do not take into account neither the topology of a network between the processors nor the communication time between any two processors. Second, PRAM are unreasonably strong: some sublinear algorithms can be run exponentially faster on PRAM [3]. An alternative model is cellular automata, whose notion is due to S. Ulam and J. von Neumann [7], but it requires a huge number of simple processors.

The model of spatial machines has been introduced in [3] in order to model more properly realistic parallel machines where the question of the power of spatial machines was also answered.

1 Definitions and Notation

1.1 Cellular Automata

Definition 1. *A* cellular automaton *is a doubly-infinite array of identical cells indexed by* $\mathbb{Z}$*, the set of integers. Each cell is a finite state machine* $C = (Q, \delta)$ *where* Q *is a finite set of states and* δ *is a mapping* $\delta : Q \times Q \times Q \to Q$.

δ is called the *local transition function* and has the following meaning: the next state of cell i at time t is function of the state of cells $i - 1$, i and $i + 1$ at time $t - 1$. If we denote by $c(i, t)$ the state of cell i at time t, we can write $c(i, t) = \delta(c(i - 1, t - 1), c(i, t - 1), c(i + 1, t - 1))$. For a fixed t, the sequence of

P. Amestoy et al. (Eds.): Euro-Par'99, LNCS 1685, pp. 557–560, 1999.

all the values $c(i)$ for $i \in \mathbb{Z}$ of the cells is called the *configuration* of the cellular automaton at time t. It is a mapping $c : \mathbb{Z} \to Q$ which assigns a state of Q to each cell of the cellular automaton. Usually, cellular automata use only a finite part of the array to make computations and the rest of the cells are in a *quiescent state* $q \in Q$ with the property $\delta(q, q, q) = q$.

If we denote by $\mathbb{C} = Q^{\mathbb{Z}}$ the set of all possible configurations of $C = (Q, \delta)$, from the local transition function δ, one can build $\Delta : \mathbb{C} \to \mathbb{C}$ a *global transition function* as: $\forall c \in \mathbb{C}, \forall i \in \mathbb{Z}, \Delta(c)_i = \delta(c_{i-1}, c_i, c_{i+1})$.

We will also use *totalistic cellular automata* which have a structured set of states.

Definition 2 ([1]). *A cellular automaton with set of states Q and transition function $\delta : Q \times Q \times Q \to Q$ is called* totalistic *if $Q \subset \mathbb{N}$ and if there is a partial function f over $\mathbb{N}$ such that $\forall a, b, c \in Q, \delta(a, b, c) = f(a + b + c)$*

Let us remark that totalistic cellular automata can be made to differentiate their own states from the states of their left and right neighbors. This comes from:

Lemma 1 ([1]). *For every cellular automaton, there exists a totalistic one which simulates it without loss of time and has at most four time as many states.*

1.2   Spatial Machines

Informally, a spatial machine is a finite set of processors moving and exchanging messages in a 3 dimensional Euclidean space. Each processor is like a Turing machine's unit of control where the tape has been suppressed. They communicate through sequences of bits moving in the space. Several distinctions about the model have been made according to the number of processors and their mobility. Let us denote by *stationary* a spatial machine whose processors are all required to be immobile and by *sequential* a two-processor spatial machine, one mobile and one immobile. It has been proved in [3] that sequential stationary spatial machines are not more powerful than finite automata and that sequential spatial machines are as powerful as Turing machines.

Here, we present the model informally. A formal definition can be found in [3]. The spatial machine model is usually defined on an infinite 3D Cartesian grid of unit cells. It can be defined as well on a 1D or 2D Cartesian grid. The model computes at discrete steps with all components performing each computation step synchronously. A finite number of the unit cells are filled with a *processor*. The communications are achieved in the following way :

- every unit cell (empty or filled) has six *flying bits*, one for each direction (positive and negative). Each bit takes its values into $\{0, 1, B\}$, where B stands for the blank symbol ;
- in each computation step of the spatial machine, each flying bit is copied onto the adjacent cell according to its flying bit.

Flying bits are not stable memory. They represent the ability of empty space to temporarily store and transmit data bits.

Each processor is a (possibly different) finite control with a finite set of states and a transition function. In each computation step of the spatial machine, according to the current state of the finite control and the value of the six flying bits of the cell, each processor may perform one or all of the following operations :

- change the state of the finite control ;
- write new values in any or all of the six flying bits of the cell (prior to their copying to other cells) ;
- move the processor to an adjacent cell in one of the directions.

2 Simulation Results

2.1 Sequential Simulations

Theorem 1 ([3]). *Any given Turing machine can be simulated by a sequential spatial machine. Furthermore, if the Turing machine time-complexity is $T(n)$, the corresponding spatial machine time-complexity will be $O(T^2(n))$.*

Observe that in this simulation, the Turing machine's transition function has been completely encoded into the spatial machine's transition function.

Theorem 2 ([6]). *There exists a spatial machine capable of simulating a -non-totalistic- k-state cellular automaton. Moreover, if the cellular automaton time-complexity is $T(n)$, the time-complexity of the corresponding spatial machine is $(T^2(n) + (n-1)T(n))(\lceil \log_2 k \rceil + O(1))$.*

Here also, the cellular automaton's transition function is completely encoded into the simulating spatial machine.

2.2 Parallel Simulation of Cellular Automata

Theorem 3. *There exists a spatial machine capable to simulate any k-state totalistic cellular automaton whose running time is a factor $O(k\lceil \log_2 k \rceil)$ slower.*

Proof. Let A be the k-state totalistic cellular automaton to be simulated by S, the parallel spatial machine and Q its set of states. Let $\text{CLARGE}_A(x)$ be the complexity measure associated to the maximum number of cells necessary for the computation. Let $\mathbb{C} = Q^{\mathbb{Z}}$ denote the set of all the configurations of a cellular automaton and $F : \mathbb{C} \to \mathbb{C}$ the global transition function of A.

$$\text{CLARGE}_A(x) = \max\Big\{|x|, \max_{i,j}\{|i-j| : \exists t, t' : F^t(c_0)_i \neq q \wedge F^{t'}(c_0)_j \neq q\}\Big\}.$$

This complexity measure gives the number of processors required by S to simulate A. It has been introduced in [2].

Each processor of S has the same behavior and let P_i denote the processor simulating the behavior of cell i. To each processor P_i, we also associate two reflectors (that is two processors whose behavior is just to reflect the flying bits). The top reflector serves as a memory for the state of "cell" i and the bottom one as a memory for the transition table.

We next describe the encodings of the states and of the transition table. The states of A will be encoded as a binary number with a fixed number of bits equal to η which is equal to $\beta = \lceil \log_2 k \rceil$ if β is even and to $\beta + 1$ if β is odd. If we assume that the local transition function is totalistic, it can be encoded in binary as $f(0), f(1), f(2), \ldots, f(l)$ where the argument corresponds to the sum of the cell's value plus the values of the neighbors. The total length of the description of A will be $l\eta$. For the synchronization of the processors, we add two more blanks in the coding of the description.

The distance between the processor simulating cell i and its state reflector will be $\eta/2$, the distance between the processor simulating cell i and its description reflector $(l + 2)\eta$ and the distance between two processors η.

The behavior of P_i is then the following:

- receive the bits from P_{i-1}, P_{i+1} and from the state reflector and compute the sum s;
- wait for the sth entry from the description loop;
- replace on line the contents of the state loop by the new value;
- at the first blank of the description loop, forward the contents of the state loop to the neighbor processors.

This behavior implies the initial configuration to be stored in S. The total time required by S to simulate one transition of A is thus $(l+2)\eta$ where $l = O(k)$. □

Observe that in the previous proof the spatial machine can simulate any k-state l-transition and n-cell cellular automaton. Thus, we have a family of spatial machines which can simulate any arbitrary cellular automaton which can easily be described by the number of processors and the three distances described above thus easily in logspace on a Turing machine. Thus, it forms a *uniform* family of spatial machines.

References

[1] J. Albert and K. Čulik II. A simple universal cellular automaton and its one-way and totalistic version. *Complex Systems*, 1:1–16, 1987.

[2] H. Ben-Azza. *Automates cellulaires et pavages vus comme des réseaux booléens.* PhD thesis, Université Claude Bernard (Lyon I), 1995.

[3] Y. Feldman and E. Shapiro. Spatial machines, a more realistic approach to parallel computation. *Communications of the ACM*, 35(10):61–73, 1992.

[4] S. Fortune and J. Willie. Parallelism in random access machines. In *10th Ann. ACM Symp. on Theory of Computing*, pages 114–118, 1978.

[5] R. M. Karp and V. Ramachandran. Parallel algorithms for shared-memory machines. In *Handbook of Theoretical Computer Science*, volume A, chapter 17. Elsevier, 1990.

[6] B. Martin. Machines spatiales universelles. *Technique et Science Informatiques*, 14(5):551–566, 1995.

[7] J. von Neumann. *Theory of Self-Reproducing Automata.* University of Illinois Press, 1966.

Topic 08
High-Performance Computing and Applications

Wolfgang Gentzsch

Vice Chair

This topic aims at the development, implementation and experience of real industrial, engineering and scientific applications on parallel and distributed systems.

In many HPC projects in the past couple of years, one of the most important task in industrial high-performance computing has been porting and parallelization of existing real application codes to mostly moderate parallel SMP and DMP systems. This was clearly driven by the need to save the huge investments spent in the past in software development, and also to further reduce development cycles and thus time-to-market for industrial products.

On the other hand, today, we see more and more new compute- and data-intensive applications for which software and algorithms still have to be developed, and the unique chance exists to take the new target parallel computers into account, already in the early design phase of industrial applications software.

Thus, topics of interest for this Workshop include:

- existing industrial applications in CFD, FE, chemistry, and others,
- new applications like multimedia support, data compression, geographical information systems, cognitive recognition
- parallel design and implementation for real applications
- dynamic distribution, load balancing, and domain decomposition
- applications on networks of workstations and PCs

For this Workshop, we selected 18 papers, with the main focus on applications on networks of workstations and PCs, load balancing and domain decomposition, and a group of interesting applications in the fields of CFD, FEM, rendering, geographical information systems, biological populations, and neural networks. Accordingly, the Workshop has been structured into the following subdomains:

- Applications on Networks of Workstations and PCs
- Domain Decomposition and Loadbalancing
- Parallelization of Several Applications

P. Amestoy et al. (Eds.): Euro-Par'99, LNCS 1685, pp. 561–561, 1999.
© Springer-Verlag Berlin Heidelberg 1999

Null Messages Cancellation Through Load Balancing in Distributed Simulations

Azzedine Boukerche and Sajal K. Das

Department of Computer Sciences, University of North Texas, Denton, TX. USA

Abstract. This paper presents the results of an emperical study of the effects of null messages cancellation through load balancing in distributed simulations. Our null-message-cancellation scheme is a modification of the Chandy-Misra protocol, wherein old null messages are discarded. We propose two load balancing strategies based upon a process migration and study its scalability on an Intel paragon machine. The experimental results show the impact of our load balancing schemes on the null message concellation protocol, where a significant reduction of the null-message overhead was observed.

1 Introduction

Due to its importance, load balancing is a well studied problem in parallel and distributed systems in general, and a significant body of literature exist. Most of the research focused on designing effective partitioning and load balancing algorithms with as low overhead as possible. The methods proposed range from analytical study plus simulation to implementation plus testing. However, most of the load balancing algorithms studied in distributed systems are not readily suitable for distributed simulations. This is due the causality and the synchronization constraints[1] that exacerbate dependencies between the logical processes.

In this paper, we make use of Chandy-Misra protocol [4] which employs null messages in order to avoid deadlocks and to increase the parallelism of the simulation. When an event is sent on an output link a null message bearing the same time stamp as the event message is sent on all other output links. As is well known, it is possible to generate an inordinate number of null messages under this scheme, nullifying any performance gain [1,2]. To increase the efficiency of this basic scheme, we employ the following cancellation approach. In the event that a null message is queued at an LP and a subsequent message arrives on the same channel, we overwrite the old null message with the new message. We associate one buffer with each input channel at an LP to store null messages, thereby saving space and the time required to perform the queuing and dequeuing operations associated with null messages.

---

[1] Recall that in the conservative approach, a logical (LP) process is not allowed to process an event unless it is certain that it will not receive an earlier event message.

P. Amestoy et al. (Eds.): Euro-Par'99, LNCS 1685, pp. 562–569, 1999.

2   Null-Message Cancellation Through Load Balancing

We propose to implement the load balancing facility as two kinds of processes: *load balancer* and *process migration* processes. In our implementation, both processes belong to the same processor[2]. A load balancer makes decision on *when* to move *which* process to *where*, while migration process carries out the decision made by the load balancer to move processes between processors. In other words, the load balancer plays the policy role while the migration process supplies the mechanism in the load balancing facilities. This separation between policy and migration provides us a flexibility of various load balancing schemes such as preemptive and non-preemptive strategies [1]. Several approaches to both information gathering and decision making were investigated. We propose to use a centralized and a multi-level load balancing strategy due mainly to reduce the message traffic over a fully distributed approach.

• **Centralized Load Balancing (CL)**: The load balancer sends periodically a message $< Request\_Load >$ to each processor (Pr_k) requesting the load of each logical process, LP_i, which is assigned to Pr_k. When all processors have responded, the load balancer computes the new load balance. later, we will define the load, and show how to compute the load. Once the load balancer makes the decision on *when to move which LP to where*, it sends $< Migrate_{Request}, Which :$ $LP, Where\_To : Processor >$ to the migration process which in turns initiates the migration mechanism. Note that the performance of this approach can be degraded significantly by the fact that all request/reply messages are routed through the LBF. Hence, this scheme may lead to a major bottleneck.

• **Multi-Level Load Balancing (ML)**: In this approach, the goal is to prevent the bottleneck and reduce the message traffic over a fully distributed approach. Several hierarchical strategies were investigated, in which processors are grouped and work loads among processors are balanced hierarchically through multiple levels. For simplification, we settle with a Two-Level scheme. In level 1, we use a centralized approach that makes use of a *(Global) Load Balancing Facility, (GLBF)*, and the global state consisting of process/processors mapping table. The load balancer $(GLBF)$ sends periodically a message $< Request\_Load >$ to a specific processor, called *First_Processor*, within each cluster requesting the average load of all processors within that cluster.

In level 2, the processors are partitioned into clusters. The processors within each cluster are structured as a virtual ring and operate in parallel to collect the work loads of processors. We choose to use a distributed scheme at level 2 to improve the efficiency and the performance of our load balancing algorithm. A virtual ring is designed to be traversed by a token, which originates at a particular processor, called *First_Processor*, passes through intermediate processors, and ends its traversal at a preidentified processor called *Last_Processor*. Each of the rings will have its own circulating token, so that information (i.e., work loads) gathering within the rings is concurrent. As the token travels through the processors of the ring, it accumulates such information as the load of each

---

[2] One could as well use two processors instead.

processors and *id* of the processors that contain the highest/lowest loads, so that when it arrives at the *last_processor*, information have been gathered from all processors of the ring. The token is generated at *First_Processor* whenever the *first_processor* receives a *Request_Load* message from the *GLBF*. It is destroyed at the *Last_Processor*. However, since the *Last_Processor* possess the information requested by the *GLBF*, *Last_Processor* sends a reply message with the updated information about the ring to *GLBF*. When the *Last_Processor* (s) of each cluster have responded, the load balancer (*GLBF*) computes the new load balance. Once the load balancer makes the decision on *when* to move *which* LP to *where*, it sends $< Migrate_{Request}, Which : LP, Where\_To : Processor >$ to the migration process which in turns initiates the migration mechanism.

Process Migration: A process migration facility in a distributed system dynamically relocates running processes among the component machines. Such relocation can help cope up with dynamic fluctuations in load and service needs. The use of process migration for optimistic distributed simulation may also be found in [6]. We choose the following implementation for the process migration. Let us denote by $MAP(LP_i)$ the identity of the processor handling the logical process LP_i where each processor maintains the mapping table. We denote by S_{LPi} and D_{LP_i} the source and the destination processors, which respectively indicate from *where* to move LP_i to *where*. In our mechanism, when LP_i migrates from processor S_{LP_i} to D_{LP_i}, the process transfer occurs in three phases:

i) Establishment: Processors S_{LP_i} and D_{LP_i}, after being told by the LBF (i.e., the process migration) to migrate the logical process LP_i, agree to the transfer. D_{LP_i} will keep all messages that should be sent to LP_i (hence to S_{LP_i}). That is possible only if LP_i has neighbors that are assigned to D_{LP_i}. If D_{LP_i} does not keep these messages, they would have been sent to S_{LP_i} which will forward them back to D_{LP_i} since LP_i does not reside anymore in S_{LP_i}. In order to avoid this, we decided to keep these messages and then forward them to LP_i which resides now in Processor D_{LP_i}. When the migration of LP_i from S_{LP_i} to D_{LP_i} is terminated, D_{LP_i} will forward these messages to LP_i.

ii) Transfer: S_{LP_i} sends a message of type *migrate*, i.e., $< Migrate_{LP}, Which : LP_i, Where\_To : D_{LP_i} >$ to D_{LP_i}. This message will include all information regarding LP_i, i.e., data structure pertaining to the migrant process including its local simulation time (LST), and the event messages waiting in its input queues. Note that the transfer is made only when a process finishes processing its event message (and not while it is processing the event).

iii) Notification: S_{LP_i} must inform its neighboring processors about LP_i-migration. Upon receipt of such message, each processor will update its mapping table, MAP. In our implementation, S_{LP_i} will notify it's neighbors only after the migration procedure is finished. This will avoid any confusion, such as having a node being notified about the transfer before actually performing the migration.

Once S_{LP_i} has sent the message of type *migrate*, any messages received later by S_{LP_i} and scheduled for LP_i will be forwarded to D_{LP_i}. In order to decrease the communication overhead, we store the messages sent to LP_i and check periodically if S_{LP_i} receives messages for LP_i, which will be forwarded to

D_{LP_i}. Our experiments show that this simple mechanism decreases significantly the network traffic and the communication overhead.

Load Calculation: The computational load in our distributed simulation model consists of executing the null and the real messages. We choose to measure the current load at each processor and then balance the load at run time so that all available processors will be assigned the same load.

Let R_{avg}^k and N_{avg}^k the average CPU-queue lengths for real messages and null messages at each processor Pr_k. In order to distribute the null messages and the real messages among all all the processors, let us define the (normalized) load at each processor Pr_k as a function, $\mathcal{F}$, of R_{avg}^k and N_{avg}^k. It is defined as follows :
$Load_k = \mathcal{F}(R_{avg}^k, N_{avg}^k) = \alpha R_{avg}^k / R_{avg} + (1 - \alpha) N_{avg}^k / N_{avg}$; where α and $(1 - \alpha)$ are respectively the relative weights of the corresponding parameters.

The value of α was determined empirically. In our experiments, $\alpha = 0.25$ yielded good results. An intuitive explanation lies in the nature of the Chandy-Misra protocol and the cancellation scheme described at the beginning of this section. Since the Chandy-Misra protocol generates an inordinate number of null messages, the success and the efficiency of our scheme depends on the number of null messages present during the simulation. The more the null messages, the more likely that the scheme will overwrite the old null messages, thereby reducing the null messages that would have been generated.

The load is well balanced if $Load_k$ is equal to 1, i.e., $\alpha + (1 - \alpha)$. If the load balancing algorithm requires that each processor strives to adjust its load to be exactly equal to the average, processor *thrashing* may result. The migration of a process to an underloaded processor may increase its load to the point of making it overloaded, necessitating the migration of that LP to yet another processor. We avoid this by defining a *tolerance* factor, δ, within which the load of all the processors should fall. The value of δ is to be determined empirically. (In our experiments, $\delta = 0.2$ yielded good results.) Thus, if we suppose that the maximum deviation δ from an equal distribution is allowed in the simulation, then $1 - \delta \leq Load_k \leq 1 + \delta$,   where $k = 1, ..., K$, and K is the number of processors. Consequently, the processor is overloaded if $Load_k > 1 + \delta$. The tolerance factor determines the responsiveness as well as the communication cost of the load balancing algorithm. However, a wider tolerance allows more variation in load among the processors. Choosing the appropriate tolerance allows the load balancing algorithm to adapt in the distributed simulation environment. Our approach is then to balance the workload among the processors while minimizing the inter-processor communication. One way to decrease the communication overhead is to combine logical processes together. So we choose the following method: The load balancer selects a (heavily overloaded) process LP_s from the heavily overloaded processor, such that LP_s has at least r neighbor processes $\{LP_1, LP_2, ..., LP_r\}$, and has at least one process LP_i, where $i = 1...r$, assigned to a different processor than the one containing LP_s. Here r is can be determined either empirically or from the network topology under consideration Then, the migration process sends this selected process to the lightly underloaded neighboring processor.

3   Simulation Experiments

The experiments were conducted on an Intel Paragon at CalTech. The Paragon is a distributed memory multicomputer, consisting of 72 nodes, arranged in a two-dimensional mesh. In this paper, we consider a distributed communication network, and a traffic flow network. The communication and the traffic flow networks are realistic simulation models that mimic many real world problems.

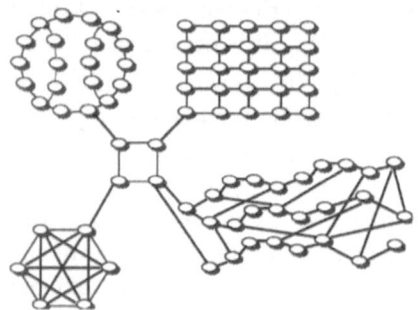

Fig. 1. Distributed Communication Network

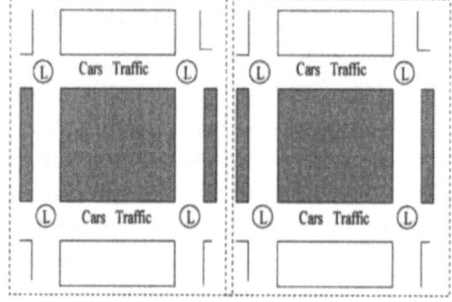

Fig. 2. Traffic Flow Network

An initial set of experiments was performed in which the routing probability was characterized by a uniform distribution, In this case, the obtained results indicated that our dynamic load balancing did not provide any significant improvement over a static partitioning algorithm. These results led us to ask if these results are applicable to *asymmetric* workload as well. Consequently, we selected an unbalanced network with the following routing probability: at the beginning of each simulation run, each process randomly selects one link as a *favorite* link which will receive twice as many messages as the other(s).

The Distributed communication model [6] used in our experiments, models a national network consisting of four regions which are connected through four centrally located delays. Final message destination is not known when a message is created. Hence, no routing algorithm is simulated. Instead one third of messages arrival are forwarded to neighboring nodes. A uniform distribution is employed to select which neighbor receives a message. Messages may flow between any two nodes, possibly through several paths. Nodes receive messages at varying rates that are dependent on traffic patterns. There are numerous deadlocks in this model, and hence it provides a stress test for any conservative synchronization mechanism. Various simulation conditions were created by mimicking the daily load fluctuation found in large communication network operating across time zones. Therefore, in the pipeline region, for instance, we arranged the sub-region into stages, and processes in the same stages perform the same normal distribution with a standard deviation 20%. The time required to execute a message is significantly greater than the time to raise an event message in the next stage (message passing delay). We use a pipeline sub-model with 26 processes, a toroid sub-model with 25 processes, a fully connected sub-model with 5 processes, and a circular sub-model with 20 processes. We choose a shifted exponential service time distribution for the 3 sub-regions (circular, toroid, and fully connected sub-regions).

In our next model, we represent the traffic network as a square mesh composed of streets running in the horizontal and vertical directions with traffic lights at the intersections of the street. This model is partitioned into sub-systems; which we refer to as grids. Each grid is assigned to a logical process (LP). The cars enter the simulation and travel within and between the grids using message passing. To reflect traffic flow network model, we define a light interval to be a triple $< r, g, y >$; where r is the number of clock ticks that the light is red, g is the number of clock ticks that the light is green, and y is the number of clock ticks that the light is yellow. Fig. 2 illustrates an example of network traffic with two grids, where each grid contains four lights. A car enters the simulation at a construction site labeled a *Source_Sink*. All of the boundary streets, as shown in this figure, are sources and sinks where cars are generated according to a probability distribution. A car might travel either North, South, West, or East with a fixed probability distribution of changing directions at any given intersection. We also consider the contention problem at the intersection, i.e., if a car is turning left into a path of a car going straight, then a contention mechanism will inhibit one of the cars until the other car clears the intersection. In our experiments, we choose an average traffic flow of 100,000 cars. The number of lights is the traffic network is held constant at 576 lights for controlling the workload of the system. To make the simulation more interesting, we choose 20 hot spots uniformly distributed among all the grids, where the probability of car is generated at a source is 0.25.

• Performance Results

The experimental results were obtained by averaging several trial runs. First, we present in the form of graphs the results for the execution time (in seconds). Next, we study the synchronization overhead as *the null message ratio* (NMR) which is defined as the number of null messages processed by the simulation using Chandy-Misra null-message approach divided by the number of real messages processed. It is important to understand that there are no existing (dynamic) load balancing algorithms for the conservative simulation paradigm against which we can compare the performance our algorithm. Hence comparisons were made to a static partitioning algorithm. The static algorithm used for comparisons is based upon [5]. The algorithm basically attempts to minimize the communication overhead by uniformly distributing the execution load among the processors. We assume that there are as many clusters as there are processors, and each cluster is assigned to one processor. The algorithm starts with an initial random partition, and then iteratively moves processes between clusters until no improvements can be found (a local optimum). The processes are moved among the clusters so that the total cut-size is reduced and the cluster sizes remain balanced. All possible moves for each process are considered, and the process which contributes to the maximum gain is chosen. A process is moved only if it does not violate the cluster size constraints.

Let us now turn to our results. Fig. 3 show respectively the run times for distributed communication network and flow traffic models. The trends of the curves and the relative performances are relatively similar to the other simulation models. We see a reduction of 20-25% in the execution time when we increase the number of processors from 2 to 8, and 40-45% when we increase the number of processors from 16 to 64. We observe a significant decrease in the running time when the CL and ML load balancing schemes are used, for both methods and for both models. We observe 50% reduction in the execution time when ML strategy is used compared to the static algorithm. The results also indicate that the CL scheduling performs worse than the ML load balancing method for all four simulation workload models. This is due to the nature of the centralized scheme that leads to a major bottleneck, thereby slowing down the simulation. On the other hand, the ML scheduling tries to avoid the bottleneck and reduces the message traffic by undertaking a fully distributed approach.

We now examine the synchronization overhead,, namely the null-message ratio (NMR). Fig. 4 displays the NMR as a function of the number of processors employed in the network models. We observe a significant reduction of null-messages for all load balancing schemes in all four models. Also NMR increases as the number of processors increases for all four simulation models. For instance, in Fig. 8, if we confine ourselves to less than 8 processors, an approximately 20-25% reduction of the synchronization overhead using the CL dynamic load balancing algorithm is observed over the static one. Increasing the number of processors from 8 to 32, we observe about 30-40% reduction of NMR. The hierarchical scheme strategy always outperforms the centralized one. We observe a 45-50% reduction using ML strategy over the static one, when we increase the number of of processor from 32 to 64. Similar results were obtained with the fully connected communication model, the distributed communication network, and the traffic flow model. The trends of the curves and the relative performances are relatively similar. These results conclude that the multi-level load balancing strategy significantly reduces the synchronization overhead when compared to a centralized strategy. In other words, a careful dynamic load balancing improves the performance of a conservative distributed simulation.

4   Conclusions

In this paper, we have presented a null-messages cancellation scheme through load balancing in distributed simulation. Our results show that a significant reduction of null message using our scheme. A significant decrease in run-time was also obtained with the use of our proposed cancellation scheme as compared to the use of a static partitioning algorithm. We note that a hierarchical approach seems to be a promising solution in reducing further the execution time of the simulation.

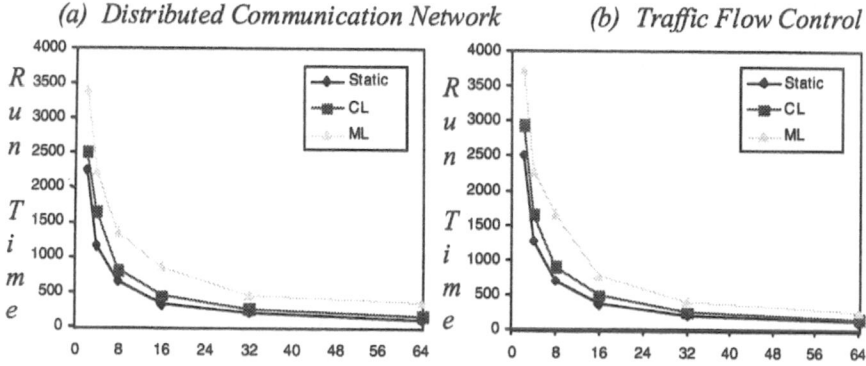

Fig. 3. *Run-Time Vs. Number of Processors*

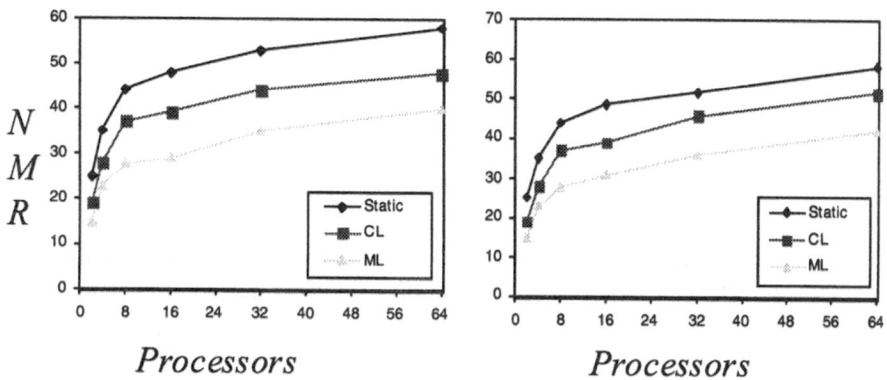

Fig. 4. *Synchronization Overhead Vs. Number of Processors*

References

[1] Boukerche, A.: "Time Management in Parallel Simulation", in High Performance Cluster Computing, Vol. 2, Prentice Hall, 1999. (Eds. R. Buyya, and M. Baker).

[2] Boukerche, A., and Tropper C.: "A Static Partitioning and Mapping Algorithm for Conservative Parallel Simulations", IEEE/ACM PADS'94, 1994, 164–172.

[3] Fujimoto, R. M.: "Parallel Discrete Event Simulation", CACM, 33(10) 1990.

[4] Misra, J., "Distributed Discrete-Event Simulation", ACM *Comp. Surveys*, 1986.

[5] Nandy, B., and Loucks, W. M., "An Algorithm for Partitioning and Mapping Conservative Parallel Simulation onto Multicomputer", PADS'92, 139–146.

[6] Glazer, D., and Tropper, C.,"On Process Migration and Load Balancing in Time Warp", *IEEE Trans. on Parallel and Distributed Systems*, 4(3), 1993, 318–327.

Efficient Load-Balancing and Communication Overlap in Parallel Shear-Warp Algorithm on a Cluster of PCs

Frédérique Chaussumier[1], Frédéric Desprez[2], and Michel Loi[1]

[1] LHPC, ENS-Lyon, F-69364 Lyon cedex 07, France
[2] LIP, ENS-Lyon, F-69364 Lyon cedex 07, France
[fchaussu,desprez,mloi]@ens-lyon.fr

Abstract. In the medical field, volume rendering provides good quality 3D visualizations but is still not interactive enough for a day-to-day practice. The most efficient sequential algorithm is the Shear-Warp algorithm. It renders up to 10 images per second for a small dataset. The goal of this paper is to present an efficient parallel implementation of the Shear-Warp algorithm for a distributed memory architecture, a cluster of PCs connected with a high speed network.

1 Introduction and Motivations

Real-time rendering is an important goal in visualization applications. Most of these applications require the generation of a sequence of images for different orientations of the volume. Consequently, real-time rendering should enable a continuous visualization of the volume as its orientation changes. Moreover, higher and higher resolution datasets combined with the high computational cost of direct volume rendering makes it difficult, if not impossible, for sequential implementations to deliver the required level of performance. Therefore, such applications have been parallelized not to trade off image quality for speed. Lacroute and Levoy [LL94] developed the Shear-Warp algorithm that exploits coherence in the volume and image space. This algorithm is currently acknowledged to be the fastest sequential volume rendering algorithm.

The goal of this paper is to present an efficient parallel implementation of the Shear-Warp algorithm for a distributed memory architecture, a cluster of PCs connected with a high-speed network and using a light weight and fast communication layer. This new parallel implementation is load balanced and overlaps communication with computation using MPI asynchronous communications.

This paper is organized as follows: in the next section, we describe and analyze the Shear-Warp algorithm. The third section exhibits the main problems associated with the parallel formulation. It focuses on architecture, task partition and communications patterns. In the fourth section, we propose a new dynamic load-balancing algorithm for the Shear-Warp algorithm tuned to interactivity. In order to improve the scalability of the algorithm, we discuss, in the fifth section, the possibility of implementing communication overlap in this algorithm.

P. Amestoy et al. (Eds.): Euro-Par'99, LNCS 1685, pp. 570–577, 1999.

The last section, gives our experimental results in terms of load-balancing and scalability.

2   The Shear-Warp Algorithm

Volume rendering [Kau96] is the process of creating a 2D image directly from 3D volumetric data so that no information contained within the data is lost during the rendering process. For example, in computed tomography scanned data, useful informations are not only contained on the surfaces but also within the data. Therefore, it must have a volumetric representation, and must be displayed using volume rendering techniques.

Lacroute and Levoy [LL94] described a fast volume rendering algorithm called the Shear-Warp factorization. It is based on an algorithm that factors the viewing transformation into a 3D shear (parallel to the data slices), a projection to form an intermediate but distorted image, and finally a 2D warp to form an undistorted final image. The Shear-Warp factorization has the property that rows of voxels in the volume are aligned with rows of pixels in the intermediate image. Consequently, a scanline-based algorithm has been constructed that traverses the volume and the intermediate image synchronously, taking advantage of the spatial coherence present in both volume and image. Lacroute and Levoy optimized the original algorithm by using spatial data structures based on run-length encoding for both the volume and the image and also taking advantage of early ray termination. This data structure is a sparse data structure that contains only non-transparent voxels for the object and non-saturated pixels for the image. Using a RLE, we skip empty voxels and saturated pixels. This implies that scanlines may have widely different amount of data associated with them.

An implementation running on an SGI Indigo workstation renders a 256^3 voxel data set in one second. This algorithm is based on three main steps: the computation of the shading lookup table, the projection of the volume data into the intermediate image, and finally the warping of the intermediate image. The projection of the volume data into the intermediate image dominates the cost of the sequential algorithm. It takes over 80% of the total amount of time for a whole execution [AGS95]. Therefore, we focus in this paper on the compositing step which is the projection of the volume data into the intermediate image.

Data repartitions in both object and image are highly irregular. They depend on the scene and the viewpoint. For instance, Figures 1, 2, and 3 illustrate how the data repartition in the intermediate image depends on the viewpoint. The default viewpoint is the zero-degree rotation angle. We compare the data distribution of respectively 5, 10 and 20-degree rotation angle to the default viewpoint. We can notice that the bigger the rotation is, the more different the data repartition is.

Finally, we outline that the Shear-Warp algorithm leads to a highly irregular application because of the RLE data structure. Accordingly, in the parallel algorithm, computation and communication are very irregular as well.

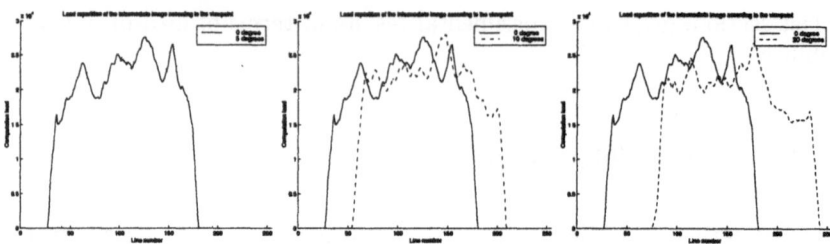

Fig. 1. Data reparti-tion in the intermedi-ate image (5-deg. rota-tion angle). **Fig. 2.** Data reparti-tion in the intermediate image (10-deg. rotation angle). **Fig. 3.** Data repartition in the intermediate image (20-deg. rotation angle).

3 Parallel Algorithm

In this section we exhibit the main problems associated with the parallel implementation of the Shear-Warp algorithm. The Shear-Warp algorithm augmented with early ray termination and run-length encoding yields to an excellent per-frame sequential rendering times. This algorithm forms the basis of our parallel formulation. The critical issues in any parallel algorithm are concurrency, minimization of communication overhead, and a good load-balancing among processors.

3.1 Related Work

Some authors have already proposed parallel formulations of this algorithm for both shared and distributed memory architectures.

To get real-time performance, both Lacroute [Lac95] and then Jiang and Singh [JS97] parallelized the Shear-Warp algorithm on a 16-processor SMP SGI Challenge. Unlike distributed memory architectures, this architecture supports fine-grain and low-latency communications adapted to the irregular communication and computation patterns of the Shear-Warp algorithm. They render a 256^3 voxel data set at over 10 frames per second. Amin et al. [AGS95] have implemented the Shear-Warp algorithm for a distributed memory architecture. With a 128-processor TMC CM5, they could render a 256^3 voxel data set at 12 frames per second. It is comparable to the results obtained on the 16-processor shared memory architecture. However, they restricted the utilization of the Shear-Warp algorithm by allowing only one-degree rotations to change the viewpoint. Despite of this restriction, their algorithm is not scalable: the speedup is 30 for 128 processors.

The two general types of task partitions for parallel volume rendering algorithms are object partitions and image partitions. In an object partition, each processor gets a specific subset of the volume data to resample and to composite. The partial results from each processor must then be composited together to form the image. In contrast, using an image partition, each processor has to compute a specific portion of the image. Each image pixel is computed only by one processor, but the volume data must be moved to different processors as the viewing transformations changes. It is very important not to limit the size of the data volume. In the medical field, standard volumes are composed of 512^3 voxels, which leads to at least 135 Mbytes without compression. Thus, we chose to distribute the data on every processor because the replication for such volumes is impossible on standard machines. All existing implementations have designed their parallelization using an image partition that takes full advantage of the optimizations of the rendering algorithm. Moreover, for a shared memory architecture, data movement is less significant. The partitioned image is the intermediate image created during the shear step. The unit of work can be individual pixels, scanlines of pixels, or rectangular pixels. In [Lac95], it is shown that the best shape is scanlines of pixels because it minimizes the overhead due to decoding the run-length data structure. It also maximizes the spatial locality both in the intermediate image space and object space.

Given that the fundamental unit of work is a group of contiguous scanlines of the intermediate image, minimizing load imbalances leaves three options for the partition: static, static interleaved, and dynamic partitions. For a shared memory architecture, Lacroute chose to use a distributed task queue and a dynamic work stealing. This solution is too expensive for a distributed memory architecture because of its prohibitive communication overhead. Therefore, for such an architecture, Amin et al. determined heuristics based on adaptative load-balancing scheme. But because their utilization restriction that considers only one degree rotations, they finally conclude that they only needed a static load-balancing.

Our approach is to implement the parallel version of Shear-Warp algorithm on a distributed memory architecture (a cluster of PCs interconnected with a high-speed Myrinet) because of its good scalability. On one hand, one of our major goals is to achieve real-time performances with higher resolution data sets (particularly 512^3). On the other hand, we believe that it is important not to restrict the user utilization and to allow him to change arbitrarily the viewpoint. Thus, our new implementation proposes a dynamic load-balancing that do not depend on the previous rendering.

3.2 Parallelization

Because of the distributed memory architecture, we had to determine an explicit data distribution (and redistribution) that minimizes communication but keeps a good load-balancing.

Explicit data distribution is a difficult problem when an image partition is used because the portion of the volume required by a particular processor depends on the viewpoint. To distribute data volume in an intermediate image partition the volume is first sheared and then distributed by slices orthogonal to the rays. Each processor can compute its portion of the intermediate image through its assigned volume segment. The resulting intermediate images

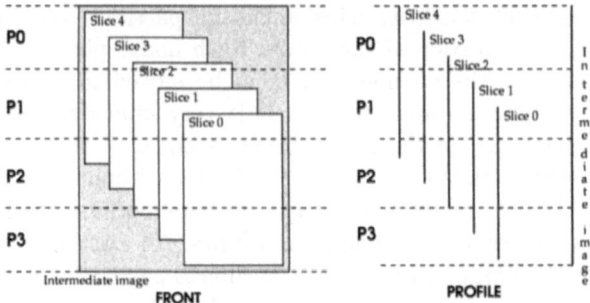

Fig. 4. Volume distribution in an image partition.

on different processors are disjoint and can be independently warped. Figure 5 shows a simple intermediate image partition with 4 processors. The corresponding sheared volume, made up of 5 slices, is partitioned as illustrated on the figure: processor 3 owns a few scanlines in the first slice, processor 2 owns scanlines in every slice, ...

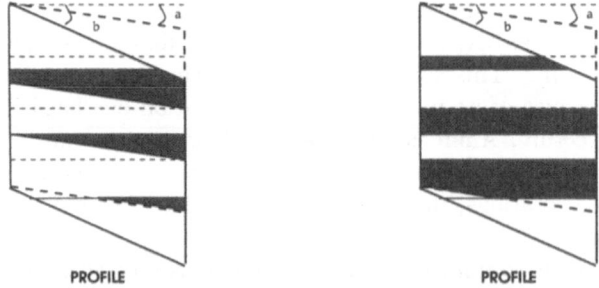

Fig. 5. Data received respectively from previous and next processors.

The main overhead of this algorithm results from communications of volume data when the volume is sheared. The generated communications are shown in black in Figure 5. Every processor receives data corresponding to the shaded scanlines from its neighbor processors except the first and the last ones.

Communication Patterns In our parallel Shear-Warp algorithm, we need two types of communications: a gather of partial images into the final image, and a personalized all-to-all communication for the data redistribution when the viewpoint has changed. We implemented the personalized all-to-all communication in $p-1$ steps, where p is the number of processors, as follows: at each step each processor sends data to a step-far processor in the increasing processor number and receives data from a step-far processor in the decreasing processor number.

Overall Algorithm We implemented the overall algorithm but we only focused on the data distribution and redistribution and the composition that are written in italic (that takes most of the computation time).

```
Procedure Render()
    InitialDistribution()
    Foreach viewpoint do
        Computation of a part of the shading lookup table (LUT)
        Multidistribution of the shading LUT
        Foreach voxels' slices from front to back do
            If I own the data
                Composite(data, part_image)
        EndFor
        image = Warp(part_image)
        Gather(image, root)
        If p == root
            Display(image)
        Personalized-all2all(volume)
    EndFor
End
```

4 Dynamic Load-Balancing

The requirement of an arbitrary rotation of the voxels' cube implies that we should implement a dynamic load-balancing mechanism. As we shown in Figures 1, 2, and 3, load-balancing strictly depends of the viewpoint.

In an image partition, every processor has to compute a specific portion of the image. This portion of image results from the projection of the volume data into this portion of image. Therefore, a naive partitioning of the image that assigns an equal portion to each processor yields to a bad load-balancing. Furthermore, it is impossible to determine in a static way the accurate amount of voxels needed to generate this portion. Consequently, we used the elastic load-balancing algorithm given in [MR91] to determine the load and to get a good load-balancing accordingly. This algorithm consists in computing a local partial load for each processor. Then, each processor broadcasts its partial value and adds its value with the ones received. At this moment, every processor knows the global load. By dividing this global load by the number of processors, each processor finds its elementary load. Then every processor has to get the data necessary to its computation.

In the Shear-Warp algorithm, every processor has to compute an array containing its local contribution for each line of the intermediate image. This array is then broadcast. Each processor adds the arrays received with its own. The resulting array contains the computational load for each line of the intermediate image. They can then obtain the global load. By linearly distributing the intermediate image, processors can balance the load over the processors.

5 Overlapping Communication with Computation

In order to reduce the overhead due to data redistribution, we studied the possibility of introducing communication overlap in the compositing phase. Because of the irregularity of the application, this presents a considerable challenge. In addition to the irregular communication and computation patterns of the Shear-Warp algorithm, we had to deal with communication layers. As a matter of fact, none of our implementations of the MPI standard provide a real asynchronous communication routine.

So far, every communication generated by the data redistribution is done before the volume composition. Consequently it is possible to overlap the communication of the slice $k+1$ with the composition of the slice k. The first overlap optimisation consists in starting the computation of a slice as soon as every processor sent its corresponding data. The second overlap optimisation waits for a processor to send its data and begins immediately its computation corresponding to the received part of slice.

6   Results

The cluster of PCs we used has 8 PowerPC 604e clocked at 200 Mhz interconnected with a high speed network Myrinet. The communication layer BIP (Basic Interface for Parallelism) [Pry98] implemented on the Myrinet network delivers the maximal performance achievable by the hardware to the application. We did experiments on the overlap of communications in both communication layers. The results (described in [CDL99]) shown that a real overlap is only available using the native BIP interface. Moreover, the BIP interface only allows one communication at a time to take a full advantage of the Myrinet network. Because of this restrictions, we only could implement the second possibility of overlapping presented before.

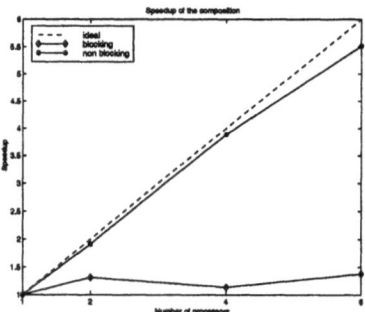

Fig. 6. Speedup of the compositing phase.

Using a static allocation, we distribute the slices with a block-cyclic distribution the lines. Central processors have the whole load. On the contrary, with the dynamic load-balancing algorithm, the data is well balanced. We have some slight variations due to the granularity of the data.

We first implemented the compositing step using blocking MPI primitives. The very bad scalability of this implementation, as shown in the Figure 6, is due to data redistribution overhead. Then we added communication overlap. The figure shows that the implementation using asynchronous communications is almost perfectly scalable. We have a very good overlap of the communications because of the BIP layer. The curves were obtained on a colored high resolution dataset with 512^3 voxels. The total execution time for such a dataset is 1.5 s on 4 processors.

7   Conclusion

The first goal of this application is to provide to physicians a good quality and resolution visualization from medical datasets in real-time with a low-cost distributed memory machine.

In this paper we have presented an high performance and scalable version of the Shear-Warp algorithm implemented on a cluster of PC. Our parallel approach of the Shear-Warp algorithm improves the interactivity of the application by using an adapted load balancing algorithm and by overlapping communication. It then allows the user to get a 3D representation with any viewpoint in real-time. Even with a sparse data-structure and irregular communication patterns, we are able to get performances that are comparable to implementations on "classical" parallel machines.

The optimizations presented in this paper can also be used in other irregular applications, and thus we would like to create a library to overlap communications and computations for this kind of applications.

References

[AGS95]   Minesh B. Amin, Ananth Grama, and Vineet Singh. Fast Volume Rendering Using an Efficient Parallel Formulation of the Shear-Warp Algorithm. In *Proceedings 1995 Parallel Rendering Symp.*, 1995.

[CDL99]   Frédérique Chaussumier, Frédéric Desprez, and Michel Loi. Efficient load-balancing and communication overlap in parallel shear-warp algorithm on a cluster of pcs. Technical Report 99-28, LIP ENS Lyon, May 1999.

[JS97]   Dongming Jiang and Jaswinder Pal Singh. Improving Parallel Shear-Warp Volume Rendering on Shared Adress Space Multiprocessors. In *Proc. of the 1997 ACM SIGPLAN Symp. on Princ. and Prac. of Par. Progr.*, June 1997.

[Kau96]   Arie E. Kaufman. Volume Visualization. *ACM Computing Surveys*, 28(1):165–167, March 1996.

[Lac95]   Philippe Lacroute. Real-Time Volume Rendering on Shared Memory Multi-processors Using the Shear-Warp Factorization. In *Proceedings of the 1995 Parallel Rendering Symposium*, 1995.

[LL94]   Philippe Lacroute and Marc Levoy. Fast Volume Rendering Using a Shear-Warp Factorization of the Viewing Transformation. In *Computer Graphics*, volume 28, pages 451–458. Stanford University, July 1994.

[MR91]   Serge Miguet and Yves Robert. Elastic Load Balancing for Image Processing Algorithms. In H.P. Zima, editor, *Par. Comput.* 1st Int. ACPC Conf., 1991.

[Pry98]   Loïc Prylli. *BIP Messages User Manual for BIP 0.94*, June 1998. http://www-bip.univ-lyon1.fr/bip.html.

A Hierarchical Approach for Parallelization of a Global Optimization Method for Protein Structure Prediction

S. Crivelli[1], T. Head-Gordon[1], R. Byrd[2], E. Eskow[2], and R. Schnabel[2]

[1] Lawrence Berkeley National Laboratory, Berkeley CA 94720, USA
crivelli@global.lbl.gov,TLHead-Gordon@lbl.gov
[2] Computer Science Department, University of Colorado, Boulder CO 80309, USA
{richard,eskow,bobby}@cs.colorado.edu

Abstract. We discuss the parallelization of our protein structure prediction algorithm on distributed-memory computers. Because the computation can be represented as a search through a vast tree of possible solutions, a hierarchical approach that assigns subtrees to different groups of processors allows us to partition the work efficiently and maintain information updated without incurring significant communication overhead. Our results show that a dynamic strategy for load balancing outperforms the static one.

1 Introduction

We present a new dynamic strategy for the implementation of our protein structure prediction algorithm on distributed-memory computers [1, 5]. The protein structure prediction problem is to determine the tertiary structure of a protein given its primary sequence of amino acids. Assuming that tertiary structure occurs at the global minimum of the free energy surface that corresponds to the amino acid sequence, the problem can be formulated as a global minimization problem [8, 11]. The problem is difficult because the function presents many low-lying local minima whose number is assumed to grow exponentially with the number of amino acids in the sequence [8, 11].

Our stochastic/perturbation with soft constraints (SPSC) method makes predictions of certain aspects of the protein structure such as α-helices, β-sheets, and coil regions by neural network techniques and then manifests them as soft constraints to use within two phases. The first phase locates some local minimizers through local minimizations with soft constraints. The second phase improves those structures through expensive global minimizations in the subspace of the dihedral angles predicted to be coil. The SPSC algorithm is based on the stochastic/perturbation approach developed by Byrd et. al [2] which has been successfully applied to Lennard-Jones and water clusters and homopolypeptides [3] using alternative heuristics for defining the subspace than those described here and elsewhere [1, 5].

P. Amestoy et al. (Eds.): Euro-Par'99, LNCS 1685, pp. 578–585, 1999.

The SPSC algorithm has been parallelized hierarchically [1, 2, 3, 5], with a central scheduler that keeps the current list of minimizers updated and distributes it to the supervisors according to some heuristic. Each supervisor and its assigned workers perform a global minimization on a subspace chosen by the central scheduler and the supervisor returns a number of the resulting minimizers to the central scheduler. The process repeats for a number of iterations until no significant changes in the best structures and/or energy occur. Because the work associated with the search cannot be predicted a priori it is impossible to make an efficient partition of the work that keeps the load balanced. Thus, a hierarchical load balancing strategy has been implemented in which idle supervisors reassign their workers to busy supervisors upon determining their current state. Such dynamical load balancing has been effective on model problems [6], while the SPSC algorithm provides for a coarser level of granularity to investigate the benefits of a dynamic parallel implementation.

2   The Algorithm

The potential energy function used in this research is the AMBER molecular mechanical force field [4] defined as

$$
E_A = \sum_i^{bonds} k_b(b_i - b_0)^2 + \sum_i^{angles} k_\theta(\theta_i - \theta_0)^2 +
$$

$$
\sum_i^{dihedrals} k_\chi[1 + \cos(n\chi + \delta)] + \sum_{i<j}^{atoms} \left\{ \frac{q_i q_j}{r_{ij}} + \epsilon \left[\left(\frac{\sigma_{ij}}{r_{ij}} \right)^{12} - \left(\frac{\sigma_{ij}}{r_{ij}} \right)^6 \right] \right\}.
$$

A solvation term, E_S, has been added to E_A that accounts for hydrophobic effects, and consists of a small number of terms of the form

$$
E_S = \sum_{i,j \in \mathcal{A}} K_s \exp\left(\frac{-(r_{ij} - \rho)^2}{v} \right), \tag{1}
$$

where the sum is taken over all aliphatic carbon pairs of the protein. This potential introduces a stabilizing force for forming hydrophobic cores and it is computationally tractable and differentiable [5].

2.1   Phase I: The Local Minimization Algorithm

Phase I generates local minimizers to be used as starting points by the global optimization algorithm. It starts with an extended conformer with no secondary or tertiary structure, and performs local minimizations using a constrained energy function first and then the unconstrained potential energy function.

The soft constraints are derived from protein structure predictions based on neural networks trained on a large data bank of known proteins [12]. Given the primary sequence, the neural network predicts whether the secondary structure of each amino acid should be α-helix, β-sheet, or coil, and provides an indicator of the strength of each prediction. These predictions are utilized within two biasing functions. The first function is a bias of the backbone torsional angles of a residue according to

$$E_{\phi\psi} = \sum_{dihedrals} k_\phi[1 - cos(\phi - \phi_0)] + k_\psi[1 - cos(\psi - \psi_0)], \qquad (2)$$

where ϕ_0 and ψ_0 are dihedral angles of a perfect α-helix or β-sheet, and k_ϕ and k_ψ are force constants related to the strength of the predictions. Because we are initially considering only α-helical proteins, a second soft constraint encourages hydrogen bonds to form between the oxygen of residue i and the hydrogen of residue $i + 4$, which are predicted to be helical. It has the form

$$E_{HB} = -q_i^O q_{i+4}^H / r_{i,i+4}, \qquad (3)$$

where q_i^O and q_{i+4}^H are also related to the strength of the predictions.

The local minimizations on the constrained function encourage the formation of α-helices in the regions where predictions of α-helix are strong. Because the predictions are not exact, the local minimizations on the unconstrained function allow the entire configuration to change, in an attempt to correct at least partially, those areas in which predictions are wrong.

2.2   Phase II: The Global Optimization Algorithm

Our algorithm starts with the list of α-helical minimizers of the objective function and attempts to improve those minimizers by using a small-dimensional global optimization method developed by Rinnooy-Kan [10]. Because this algorithm works well on small spaces (4-10 variables), we use secondary structure predictions to limit the search to the space of dihedral angles predicted to be coil. From this set, the algorithm randomly selects a subset and performs a small scale global optimization using the selected dihedral angles as variables while keeping the rest temporarily fixed at their current values. This optimization produces a number of local minimizers to the unbiased energy function on the subspace of dihedral angles chosen. A fixed number of those minima with the lowest energy values is selected for local minimizations on the full variable space. The new minimizers obtained from the local minimizations are merged into the current list and the second phase starts again. The process repeats for a number of iterations until no progress can be made.

To guarantee that the search through the space of possible solutions is well balanced in breadth and depth, a heuristic is used to determine which local minimizer to pick from the list of local minimizers. Thus, Phase II is performed in two stages. In stage 1 or balancing stage, the work is balanced over a fixed number of trees. A tree consists of an initial minimizer and all the minimizers generated from it by applying a global optimization on a small subspace of dihedral angles followed by a local minimization over the entire space. At each iteration of the balancing stage, the tree with the least amount of work performed on its members so far is selected, and the best configuration in this tree that has not already been used is chosen. Configurations are rated in terms of their energy value, and best refers to the lowest in energy value. In stage 2 or nonbalancing stage, the best configuration is selected, regardless of which tree it comes from. The combination of breadth and depth in the search of the configuration space contributes to the success of this method [3].

3   A Hierarchical Parallel Implementation

Phase II of our algorithm, which accounts for most of the computational effort, can be parallelized at two different levels. The first level lets one central processor handle the list of local minimizers and assign them to a group of supervisors one at a time. The second level lets the supervisors and their workers work on improving their assigned minimizers by applying small-scale global optimizations to them followed by full-scale local minimizations. We first describe a static approach that keeps a fixed number of workers per supervisor throughout the computation and then we present a dynamic approach that reassigns workers dynamically to improve the performance.

3.1   The Static Approach

Our initial strategy consists of a central scheduler assigning different configurations and randomly selected subspaces of the dihedral angles to N supervisors. Each supervisor divides up the given subspace into M regions and assigns one region to each of its M workers for the global search. The supervisor then schedules the workers to perform the tasks associated with a small-scale global minimization over the selected dihedral angles. For each iteration of the global optimization, the workers generate sample points within their assigned regions, select start points for small-scale local minimizations, and perform the minimizations themselves. The supervisor gathers the partial results from all the scheduled tasks and keeps a list of the local minimizers found so far in the particular subspace.

Although the number of sample points to generate per region is fixed, the number of small-scale minimizations to perform and the number of iterations involved in each minimization depend on the region and cannot be determined a priori. In order to keep the load balanced at the supervisor-workers level, each supervisor assigns one task at a time to its workers upon request. Because the computation involves some number of iterations, priority is given to the tasks according to their iteration number.

Following a fixed number of iterations, the supervisor assigns its workers and itself a number of the best small-scale minimizers for local minimization on the entire space. Finally, the supervisor sends the list of full-scale local minimizers to the central scheduler and waits for a new configuration to proceed. This process repeats until the best configuration and/or the best energy value do not show significant change. In practice, because the computation may take days, it is divided into smaller jobs by reducing the maximum number of iterations performed by the central scheduler.

3.2 A Dynamic Approach

Although the static approach keeps the load balanced at the supervisor-workers level, it does not attempt the same at the central scheduler-supervisors level. At this level, the central scheduler assigns minimizers to the supervisors upon request until it reaches a maximum number of iterations. Because the amount of computational work assigned to each supervisor and its workers is considerable but unknown, it is hard to make an efficient static assignment of the work. Thus, some of the supervisors and their workers may be working for quite some time while other supervisors and their workers may be sitting idle waiting for them to finish. This issue becomes particularly significant as the computational load and the number of workers working for each supervisor grow.

We have implemented a load balancing strategy that reassigns workers to new supervisors as they become idle. The idea is that with a larger number of workers assigned to them, the busy supervisors will complete their tasks in less time. The hierarchical approach of our algorithm makes it easy and cost efficient to reassign processors to new supervisors. In fact, rather than letting the idle workers find a busy supervisor, the load balancing strategy lets the idle supervisors reassign their workers. Thus, when the central scheduler reaches the maximum number of iterations, it sends an idle message to the supervisors that contact it in search for new work. In the message, the central scheduler sends the supervisors a list of the supervisors that are still busy. Upon receiving the idle message, a supervisor splits its workers among the busy supervisors and deletes them from its list of workers. To avoid assigning a worker to a supervisor that becomes idle before accepting it, the former supervisor waits for an acknowledgement from its workers before deleting them from its list.

4 Numerical Results

We have tested the static and dynamic algorithms on the prediction of the A-chain of uteroglobin, a helical protein composed of 70 amino acids (1125 atoms.) Phase I generates a configuration with secondary structure using the procedure described in Section 2. A local minimization is performed on the biased function $E_A + E_S + E_{\phi\psi} + E_{HB}$ using the information obtained from the neural network predictions to determine the values of the force constants and the probabilities that the residues are alpha helical. The biased minimization is followed by an unbiased minimization on $E_A + E_S$ using a limited-memory BFGS [9] local minimization algorithm.

We use the structure that results from Phase I as the starting configuration for Phase II. Each iteration of this phase performs global optimizations in subspaces of dihedral angles that are chosen from all residue ϕ, ψ, χ torsion triplets predicted to be coil. The global optimization method [10] explores a given subspace to optimally determine the region most likely to contain the global minimum. The algorithm is general in the sense that subspaces of arbitrary dimension can be explored. However, the probability of finding the global minimum becomes much weaker and the cost associated with it becomes more expensive as the size of the subspace grows. Currently, we use a subspace of six dihedral angles determined by defining a pool of triplets among the set of 28 predicted coil residues and chosing two triplets randomly from that pool. The chosen subspace is divided into M regions, where M is the number of workers. Each worker randomly generates a given number of sample points over a uniform distribution in its assigned region of the domain. Small-scale minimizations are performed by the workers on selected start points. From those small-scale minimizers, $M + 1$ are selected (one for each worker plus the corresponding supervisor) for a full-scale local minimization.

We have run the static and dynamic algorithms on the T3E at NERSC using 19, 34, 68, and 127 processors. We have kept the starting number of workers per supervisor fixed to 8 and varied the number of supervisors. Thus, we have used 2, 4, 7, and 14 supervisors respectively. Notice that in the dynamic algorithm, the number of workers varies dynamically as they are reassigned during the computation. Figure 1 compares the times to complete 14 iterations of Phase II using the static and dynamic algorithms. Although the times for 10 processors, i.e., 8 workers, 1 supervisor, and 1 central scheduler, are the same (dynamic reallocation of workers is not possible with a single supervisor), the dynamic approach shows significant gains for larger number of processors.

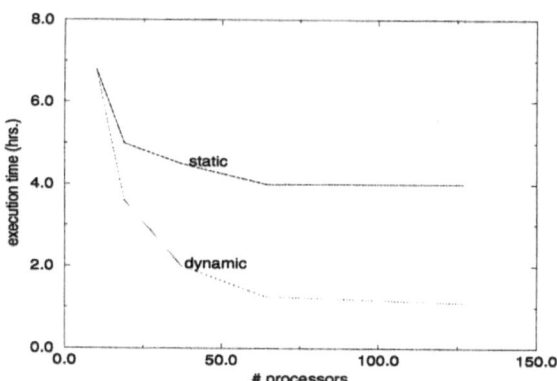

Fig. 1. Execution times for the static and dynamic approaches using different number of processors

The converged structure of Phase II is shown in Figure 2, along with the crystal structure of uteroglobin. The root mean squared (r.m.s.) deviation of our folded structure from the crystal structure, computed by measuring distances between corresponding carbon-alpha atoms, is 7.4Å. This result shows that the

SPSC algorithm is able to find configurations with tertiary structures close to that of the crystal structure. The result was obtained after 5 runs of Phase II using 64 processors. The time required to compute each Phase II iteration varied between 10 minutes and 50 minutes. This gap explains the advantage of the dynamic over the static approach.

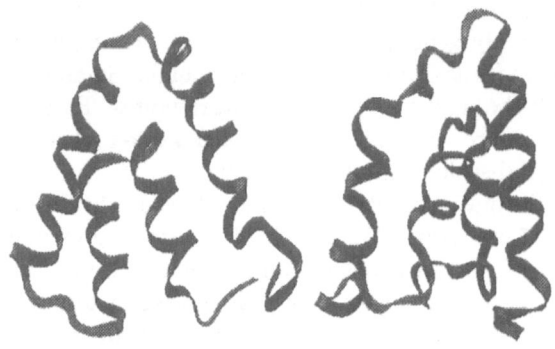

Fig. 2. Crystal structure of uteroglobin and converged structure from Phase II

5   Conclusion and Future Directions

The SPSC algorithm combines neural network predictions of secondary structure with a sophisticated global optimization strategy. This combination narrows the conformational search space by finding good starting configurations and by focusing the work in the subspace of the dihedral angles predicted to be coil. The parallel implementation uses a hierarchical approach that defines two levels of work. This approach allows us to distribute the load and share information without significant overhead. A hierarchical dynamic load balancing strategy keeps the processors busy at the two levels. Our results show that the dynamic approach outperforms the static one using different number of processors.

The SPSC algorithm provides a realistic example of a task-parallel computation. Efficient programming of this type of problem, where the number and execution times of the computational tasks can vary unpredictably demands an asynchronous and adaptive approach. In this type of approach, however, such fundamental programming issues as load balancing, data sharing, and termination detection can present difficult programming problems [7]. To deal with those issues we have developed the PMESC programming library [6]. We plan to implement our SPSC algorithm using PMESC and compare the PMESC results against those obtained with the hand-coded version. In addition, because the library provides different load balancing strategies that can be easily changed, we plan to use this feature to test a variety of load balancing strategies that have only be tested using model problems.

Acknowledgements. Crivelli and Head-Gordon thank the Laboratory Directors Research Development fund and the Office of Mathematical and Information Computing Services, U.S. Department of Energy contract number DE-AC-03-76SF00098, and the National Energy Research Supercomputer Center. Byrd, Eskow, and Schnabel were supported by Air Force Office of Scientific Research Grant F49620-94-1-0101, Army Research Office Contract DAAH04-94-C-0228, and National Science Foundation Grants ASC-9307315 and CDA-9502956.

References

[1] Azmi A., Byrd R., Eskow E., Schnabel R., Crivelli S., Phillips T., Head-Gordon T.: Predicting protein tertiary structure using a global optimization algorithm with smoothing. International Conference on Optimization in Computational Chemistry and Molecular Biology: Local and Global Approaches. (1999)

[2] Byrd R., Eskow E., van der Hoek A., Schnabel R., Shao C.-S., Zou Z.: Global optimization methods for protein folding problems. Proceedings of the DIMACS Workshop Global Minimization of Nonconvex Energy Functions: Molecular Conformation and Protein Folding. Pardalos P., Shalloway D., Xue G. eds., American Mathematical Society, **23** (1996) 29–39

[3] Byrd R., Eskow E., Schnabel R.: A new large-scale global optimization method and its application to Lennard-Jones problems. Technical Report CU-CS-630-92, Dept. of Computer Science, U. of Colorado, Boulder, (1995)

[4] Cornell W., Cieplak P., Bayly C., Gould I., Merz K., Ferguson D., Spellmeyer D., Fox T., Caldwell J., Kollman P.: A second generation force field for the simulation of proteins, nucleic acids, and organic molecules. J. Am. Chem. Soc. **117** (1995) 5179–5197

[5] Crivelli S., Byrd R., Eskow E., Schnabel R., Yu R., Phillips T., Head-Gordon T.: A global optimization strategy for predicting tertiary structure: α-helical proteins. Submitted to Proteins: Structure, Function, and Genetics, (1998)

[6] Crivelli S., Jessup E.: The PMESC programming library for distributed-memory MIMD computers. To appear in Journal of Parallel and Distributed Computing.

[7] Crivelli S., Jessup E.: Task parallelism: What a tool can provide and what should be left to the user. Proceedings of Euro-Par '96. Lecture Notes in Computer Science Series. Springer Verlag (1996)

[8] Eisenhaber F., Persson B., Argos P.: Protein structure prediction: recognition of primary, secondary, and tertiary structural features from amino acid sequence. Crit. Rev. Biochem. Mol. Biol. **30** (1995) 1–94

[9] Liu D., Nocedal J.: On the limited memory BFGS method for large scale optimization methods. Mathematical Programming **45** (1989) 503–528

[10] Rinnooy-Kan A., Timmer G.: A stochastic approach to global optimization. Numerical Optimization, Boggs P., Byrd R., Schnabel R. eds.,

[11] Vasquez M., Nemethy G., Scheraga H.: Conformational energy calculations on polypeptides and proteins. Chem. Rev. **94** (1994) 2183–2239

[12] Yu R., Head-Gordon T.: Neural network design applied to protein secondary structure prediction. Phys. Rev. E**51** (1995) 3619–3627

Parallelization of a Compositional Simulator with a Galerkin Coarse/Fine Method*

Geir Åge Øye and Hilde Reme

Department of Mathematics, University of Bergen,
Johannes Bruns gt.12
5008 Bergen, NORWAY
geiro@mi.uib.no

Abstract. We describe a simulator-parallel method for parallelization of a large industrial code. Our approach is based on domain-decomposition, with a coarse grid operator, generated by a Galerkin technique. The technique gives an easy and flexible method to include local grid refinement in already existing simulators, it also gives a high level of parallelization. The code is run on a Origin 2000 computer for testing parallel performance.

1 Introduction

In this paper we will study a parallel version of a simulator for secondary oil migration (SOM) [4]. SOM is a general three dimensional, multi-component and three phase simulator, which has three sets of primary variables: the temperature T, the water pressure p_w, and all the molar masses N_ν, $\nu = 1, 2, ..., n_c$. Here n_c denotes the number of components in the water-hydrocarbon mixture. For readers interested in a more detailed description of the mathematical background for this simulator, we refer to [7], [4] and references therein.

A control volume finite difference (CVFD) discretization has been used to create numerical approximation to the equation representing the primary variables. For time-discretization, a Forward-Euler method has been used. A Newton-Raphson method are used for linearization. To decouple these equations, we are using a sequential implicit pressure and explicit composition (IMPEC) procedure. An iteration between the pressure and all the molar mass equations are obtained until we have a convergence in the residual volume R. See [7] for a more detailed outline of these subjects.

2 Domain Decomposition and Local Grid Refinement

A multi-block formulation together with use of local refined grid, would give us a very flexible and efficient methodology for describing our physical problem.

---

* This work has received support from the Research Council of Norway (Programme for Supercomputing) through a grant of computing time.

P. Amestoy et al. (Eds.): Euro-Par'99, LNCS 1685, pp. 586–594, 1999.

The refinement strategy in SOM starts with the construction of a coarse, possibly non-regular grid, denoted by Ω_0. Then we define refined sub-domains from a selected set of coarse cells, denoted by Ω_{f_i}; $i = 1, 2, ..., p$. Non-regular and non-matching refinements, which create several different boundary conditions are accepted, see Figure (1). Using regular grid both for the coarse domain and each of the refined sub-domains, a substantial part of the resulting algorithm only requires operations on regular grids, even though the resulting composite grid is non-regular. This means that only minor changes has to be made for solving the resulting discrete composite system. This avoids a complete redesign of the code and saves considerable development costs.

There has been a large work for (CVFD) discretization techniques for locally refined grids, and special care has to be done in the interface between two non-matching interfaces. An ongoing project for implementing a MPFA [1] discretization over these interfaces is under development.

Fig. 1. This figure shows the LGR strategy. (a) shows the underlying coarse grid, in (b) some of the coarse grid blocks are refined and in (c) all the coarse grid blocks are refined.

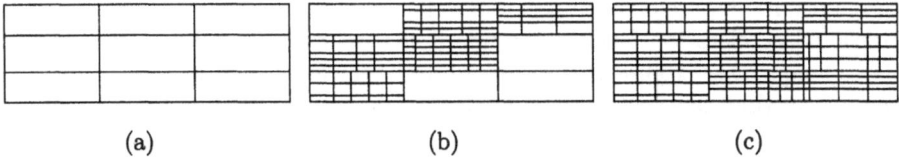

(a)   (b)   (c)

3   A Composite Grid Solution Strategy

Assume that the equations for the primary variable, have been discretized on a composite grid shown in Figure (1 b,c). This would give us following linear system to solve, her given for the temperature $\mathbf{T}$, since the method for the pressure would be analogous.

$$J\Delta\mathbf{T}^n = \mathbf{y}^n, \quad \Delta\mathbf{T}^n \in \mathcal{R}^N. \tag{1}$$

Here J is the composite Jacobian grid matrix, $\Delta\mathbf{T}^n$ and $\mathbf{y}^n$ is the composite grid solution and right hand side vector respectively for time t^n. N is the total number of grid cells in the composite grid.

A global numbering of the nodes for this composite problem would destroy the nice banded structure, which a regular grid would give. This section describe an iteration technique that allows us in some sense to decouple the refined patches and the underlying coarse grid; the coupling is through the internal boundary between coarse and refined regions. This means that we can use fast regular solvers for each refined sub-domain. To attain this, we therefor partition the composite grid Ω into different disjoint sets; Ω_c consists of grid points in the non-refined region and $\Omega_f = \bigcup_{i=1}^{p} \Omega_{f_i}$ is the union of all grid points in the refined sub-domains. After this partitioning the equation (1) can be written

$$\begin{bmatrix} J_{cc}^n & J_{cf}^n \\ J_{fc}^n & J_{ff}^n \end{bmatrix} \begin{bmatrix} \Delta \mathbf{T}_c^n \\ \Delta \mathbf{T}_f^n \end{bmatrix} = \begin{bmatrix} \mathbf{y}_c^n \\ \mathbf{y}_f^n \end{bmatrix} , \tag{2}$$

where

$$J_{ff}^n = \begin{bmatrix} J_{f_1 f_1}^n & \cdots & J_{f_1 f_p}^n \\ \vdots & \ddots & \vdots \\ J_{f_p f_1}^n & \cdots & J_{f_p f_p}^n \end{bmatrix} , \quad J_{cf}^n = \begin{bmatrix} J_{cf_1}^n \\ \vdots \\ J_{cf_p}^n \end{bmatrix}^T , \quad J_{fc}^n = \begin{bmatrix} J_{f_1 c}^n \\ \vdots \\ J_{f_p c}^n \end{bmatrix} .$$

In equation (2), $\Delta \mathbf{T}_c^n$ and $\mathbf{y}_c^n$ are the solution vector, and right hand side respectively, for the coarse region Ω_c, and $\Delta \mathbf{T}_f^n = [\Delta \mathbf{T}_{f_1}^n, ..., \Delta \mathbf{T}_{f_p}^n]^T$ and $\mathbf{y}_f^n = [\mathbf{y}_{f_1}^n, ..., \mathbf{y}_{f_p}^n]^T$ are the solution vector, and the right hand side, for the refined region $\Omega_f = \bigcup_{i=1}^{p} \Omega_{f_i}$. While J_{cf}^n and J_{fc}^n denote the coupling between the coarse and fine cells and the coupling between fine and coarse cells, respectively.

There are several ways for solving the composite linear system above, and several precondition techniques has been presented i.e. the BEPS: Bramble, Ewing, Pasciak and Schatz preconditioner [2] and the FAC: Fast Adaptive Composite preconditioner [6], are efficient techniques for solving this system. The Galerkin variational acceleration formulation was first used by Wachspress in [9]. Later this technique was extended by Fladmark in [5] for local grid refinement. It has been shown that this technique is analogous to the FAC method presented in [6].

Let $\hat{\Omega}$ denote the grid without any refinement, i.e. the regular coarse grid Ω_0. This means that we only have one degree of freedom on each refined sub-domain. Let $\hat{N}$ be the number of grid blocks in $\hat{\Omega}$ and let $M_{\hat{i}} = \{$ fine cells within coarse cell number $\hat{i}$ $\}$, $\hat{i} = 1, 2, ..., \hat{N}$. Define $\mathbf{\Psi}_{\hat{i}}$ as the basis vector for coarse cell no. $\hat{i}$. This basis vector denotes a vector of order N and it consists of only zero elements, except for unit elements at all $i \in M_{\hat{i}}$.

Let P be the prolongation operator (constant interpolation with our choice of $\mathbf{\Psi}_{\hat{i}}$), that maps the coarse grid function in $\hat{\Omega}$, to the composite grid functions in Ω. In our case each column of P consists of the basis vectors $\mathbf{\Psi}_{\hat{i}}$, so $P \in \mathcal{R}^{N \times \hat{N}}$. R is a restriction operator that maps composite grid functions on Ω to coarse grid function in $\hat{\Omega}$. In our case the restriction operator would be $R = P^T$. Note that these linear operators are not build explicitly, only the action of them (i.e. summing and restriction).

Let (s) be the iteration index between coarse and refined solutions and let $\Delta T^{(s)n}$ be the composite solution vector after (s) iterations. The update of the composite solution vector, between two iterate, can be defined as:

$$\Delta \mathbf{T}^{(s+1)n} = \Delta \mathbf{T}^{(s)n} + P \hat{\mathbf{d}}^{(s)n}. \tag{3}$$

Here $\hat{\mathbf{d}}^{(s)n}$ is the updated defect from the coarse grid $\hat{\Omega}$. Note that $\mathbf{T}_c^{(s+1)n} = \mathbf{T}_c^{(s)n} + \mathbf{d}^{(s)n}$. Substitute the expression for the update (3), into the composite linear system and restrict the composite grid functions to the coarse grid $\hat{\Omega}$ gives us:

$$RJ^n P\hat{\mathbf{d}}^{(s)n} = R\mathbf{y}^n - RJ^n \Delta\mathbf{T}^{(s)n}. \tag{4}$$

This would give us the coarse linear system to solve

$$\hat{J}^n \hat{\mathbf{d}}^{(s)n} = \hat{\mathbf{y}}^n, \tag{5}$$

where

$$\hat{J}^n = RJ^n P, \quad \hat{\mathbf{y}}^n = R(\mathbf{y}^n - J^n \Delta\mathbf{T}^{(s)n}).$$

The right hand side is the restriction of the composite grid residual to the coarse grid $\hat{\Omega}$. After solving for the coarse grid we update the composite solution vector by the expression (3). For solving the temperature $\mathbf{T}_{f_q}$ on each of the refined sub-domains Ω_{f_q}; $q = 1, 2, ..., p$, we use the last updated temperatures $\mathbf{T}_c^{(s+1)n}$ and $\mathbf{T}_{f_q}^{(s)n}$ as boundary conditions. This would be a kind of a block Jacobi formulation, and all the sub-domains Ω_{f_q} can be solved in parallel. So for each sub-domain Ω_{f_q}; $q = 1, 2, ..., p$, the sub-problems, which can be solved in parallel are given by:

$$J_{f_q f_q}^n \Delta\mathbf{T}_{f_q}^n = \mathbf{y}_{f_q}^n; \quad \Delta\mathbf{T}_{f_q}^n \in \mathcal{R}^{N_q}, \tag{6}$$

where

$$\mathbf{y}_{f_q}^n = \mathbf{y}_{f_q}^n - \sum_{l \neq q} J_{f_q f_l}^n \mathbf{T}_{f_l}^{(s)} - J_{f_q c}^n \mathbf{T}_c^{(s+1)}.$$

After solving equation (6) for $\mathbf{T}_{f_q}^n$ we get a new temperature iterate for each of the refined sub-domain Ω_{f_q}: $\mathbf{T}_{f_q}^{(s+1)n} = \mathbf{T}_{f_q}^n + \Delta\mathbf{T}_{f_q}^n$; $q = 1, 2, ..., p$. The updated composite solution vector is given by: $\mathbf{T}^{(s+1)n} = [\mathbf{T}_c^{(s+1)n}, \mathbf{T}_f^{(s+1)n}]^T$. The iteration goes back to the formulation of the coarse problem, and the iteration proceeds until the error on the composite solution vector has converged.

Finally we can present the iterative solution algorithm for solving the composite grid problem (1):

Algorithm:

- for($q = 0$ to p) in parallel
    - build $J_{f_q f_q}^n$ and $\mathbf{y}_{f_q}^n$
    - build J_{cc}^n and $\mathbf{y}_c^n$
- end for
- while $\left(\left| \frac{\mathbf{T}^{(s+1)n} - \mathbf{T}^{(s)n}}{\mathbf{T}^{(s)n}} \right| > \epsilon \right)$

    - build $\hat{J}^n$ and $\hat{\mathbf{y}}^n$ i.e.
        $\hat{J}^n = RJ^n P$
        $\hat{\mathbf{y}}^n = R(\mathbf{y}^n - J^n \Delta\mathbf{T}^{(s)n})$
    - solve $\hat{\mathbf{d}}^{(s)n} = \hat{J}^{n^{-1}} \hat{\mathbf{y}}^n$
    - interpolate the correction to the composite grid such that:
        $\Delta\mathbf{T}^{(s+1)n} = \Delta\mathbf{T}^{(s)n} + P\hat{\mathbf{d}}^{(s)n}$
    for($q = 1$ to p) in parallel

        • update $\mathbf{y}_{f_q}^n$ for boundary conditions
        • solve $J_{f_q f_q} \Delta \mathbf{T}_{f_q}^n = \mathbf{y}_{f_q}^n$
   end for
      • update $\Delta \mathbf{T}_{f_q}^{(s+1)n}$
end while loop
• set $\mathbf{T}^{n+1} = \mathbf{T}^{(s+1)n}$

This algorithm is completely analogous to the FAC algorithm given in [6]. Note that this method can be viewed as a two-level block Jacobi domain decomposition method, where the information is passed through the internal boundary's between the coarse and refined regions. This method for solving a composite grid problem is relatively easy to implement into a already existing simulator, since we only solves on rectangular grids. Again this means that we can use already existing fast linear solvers for each sub-domain, in this case a GMRES solver. Also this algorithm is favourable for use together with parallel computers. Each processor has control for each sub-domain and the corresponding linear system.

4 Parallel Implementation

Traditionally DD parallelization of PDE is done at the level of local matrix/ vector operations. However we are interested in a parallelization strategy at the level of sub-domain simulator. In [3] they have proposed a simulator-parallel model for parallelizing existing sequential PDE simulators, which is a kind of a Single Program Multiple Data (SPMD) model.

Basically the simulator-parallel model based on domain-decomposition are: *Assign one processor with one or several sequential simulators each responsible for one sub-domain.* The coordination of the computation among the processors is left to a global administrator implemented at a high abstraction level close to the mathematical formulation of the Galerkin method. In this programming model we only work with local problems. The data distribution are implied by the non-overlapping partitions, so there is no explicit need for global representation of data. The simulator-parallel models strongly promote code reuse, which minimises the development cost, and we have a good control of performance parameters as work balance, communication topology and domain decomposition. Our generic framework consists of three main parts. The sequential sub-domain simulator, a communication part and a global administrator. See Figure (2). The **sub-domain simulator** is the most important building block in the model. It consists mainly of code from the sequential simulator, with some ad-on functions to handle the parallel computing and to be able to use communicated boundary values. The basic functionalities for each sub-domain are the numerical discretization scheme and the assembling of the local linear system. Each sub-domain also have local functions for updating all the physical variables. During a parallel simulation the concrete communication between processors is in form of exchanging messages and it is all handled by the class **Communicator**. To handle the message passing, we are using the C++ library Object

Oriented MPI (OOMPI) [8]. A manager class, named **SOM_Manager**, are administrating each of the sub-domain solvers. The class is controlled from the main program, and it has the global control on the parallel solution process.

Fig. 2. The figure shows the object-oriented framework in the simulator-parallel model. The three main parts are: the sub-domain simulator with local data, the global administrator and the communicator class.

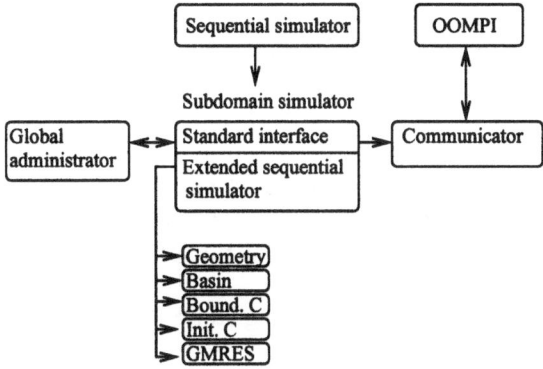

5   Some Implementation Issues

There are at least three sources of overheads that affects the parallel efficiency. Namely: overhead of communication, overhead of synchronisation and overhead of coarse grid correction.

The communication overhead are determined by both the amount of information each sub-domain needs to exchange with all it's neighbours and the number of neighbours it has. This is because the cost of sending and receiving a message between two processors roughly consists of two parts: a constant start up cost and a message send cost that is proportional to the message size. The user must therefor keep the number of neighbours of each processor at a minimum and also the number of data which have to be sent.

The synchronisation overhead is primarily due to the difference in sub-grid size and/or problem. Even a perfect partition of the grid which may gives a good work balance, may be destroyed by an unbalanced convergence.

The Galerkin coarse operator described section **3**, requires that the coarse system are built from each of the sub-problems (summation of the refined elements). This means that the size of the coarse grid requires some attention. A finer coarse grid gives normally better overall convergence, see [7], that may come at the cost of lower parallel efficiency, see Table (1). Since this is a two level method, the processors controlling the refined sub-domains has to be idle while they are waiting for the coarse update. So concerning the overheads of the coarse grid correction, the coarse grid must be relatively small, compared to each of the sub-problems, to reduce the up-dating and the communication costs. Under this assumption it would be sufficient to use a sequential solution procedure for the coarse grid correction.

6   One Computer Experiment

In this section we report parallel simulation results of one numerical experiment. The purpose is as follows: Examine the parallel efficiency of the implemented parallel simulator, due to scalability and speedup. We have carried out the experiments on a SGI/Cray Origin 2000, with R10000 processors, to explore the behaviour of the method.

We study the migration process on a rectangular domain and use uniform spatial discretization. The partition of the domain Ω is as follows. We divide the grid blocks in equally sized sub-domains M and connect one processor to one sub-domain. The decomposition of the grid Ω is done in 3D. In addition we use one processor for the coarse grid, such that the total number of processors used are $P = M + 1$.

For this computer experiment we are using static boundary conditions and uniform lithology. The coarse/fine iteration are carried out until we have an error less than 10^{-7} on the composite solution vector. When solving each linear system we use a GMRES solver, with an Incomplete LU preconditioner.

For each grid cell we have 8 unknowns: 6 components, the temperature and the water pressure. From Table (1) we can see that the parallel efficiency of our parallel version of the compositional simulator is good. When increasing the number of processors by a factor of two, we also approximately decrease the CPU time by a factor of two. This means that the parallel code are very efficient due to the computing/communication relation. The scalability, given by the expression $T(P, N) = T(kP, kN)$, where T is the CPU time, P is the number of processors, N is the total number of unknowns and k is the scaling-factor, are from the table almost given. The load on the Origin 2000 computer vary from time to time, which can explain some of the jumps in the given CPU time in the Table 1.

The table has also a number for speedup, which is the result by comparing the parallel version against the sequential version of the code. The table also shows that we have a superlinear speedup in this experiment. This is because each CPU are now working on a smaller set of memory and the memory are located closer to the CPU. The problem data handled by any one CPU fits better in cache, so each CPU executes faster than the single CPU could do.

7   Conclusion and Further Work

We have discussed the parallelization of a large industrial sequential compositional simulator for Secondary Oil Migration. The work is based on the object-oriented simulator-parallel model, which gives a high level of parallelization. By using this model and with relatively small development costs, we are able to

Table 1. The used CPU time(in seconds) and the parallel performance for different number of processors P and different problem-sizes N. NX $\times$ NY $\times$ NZ are the number of grid cells in each direction. In each grid cell we have 8 unknowns: 6 components, the water pressure and the temperature. Here P=M+1, where M is the number of sub-domains.

#P	Speedup	#N	NX $\times$ NY $\times$ NZ	CPU(s)
9	9.7	36×10^3	$40\times30\times30$	214
9	9.5	72×10^3	$80\times30\times30$	534
9	10.6	144×10^3	$80\times60\times30$	1085
17	16.5	36×10^3	$40\times30\times30$	125
17	18.7	72×10^3	$80\times30\times30$	262
17	17.8	144×10^3	$80\times60\times30$	598
33	32.6	36×10^3	$60\times60\times10$	60
33	33.4	72×10^3	$60\times60\times20$	132
33	33.7	144×10^3	$60\times60\times40$	306
49	48.8	144×10^3	$80\times60\times30$	220
49	48.3	288×10^3	$80\times60\times60$	433

create a simulator for parallel computing. Numerical experiments, done on an Origin 2000 computer, shows very nice parallel performance.

We feel that more numerical experiments need to be carried out in the future and simulations on different parallel platforms are also necessary to verify the parallel performance. The parallel code will also be used in larger industrial applications.

References

[1] Aavatsmark I., Barkve T., Bøe Ø., Mannseth T.: *Discretization on Non-Orthogonal, Quadrilateral Grids for Inhomogeneous, Anisotropic Media.* J. Comput. Phys., 127:2-14, 1996.

[2] Bramble J. H., Ewing R. E., Pasciak J. E., Schatz A. H.: *A preconditioning technique for the efficient solution of problems with local grid refinement.* Comput. Meth. Appl. Mech. Eng. 67, 149-159 (1988).

[3] Bruaset Are Magnus , Cai Xing, Langtangen Hans Petter and Tveito Aslak: *Numerical Solution of PDEs on Parallel Computers Utilising Sequential Simulators.* In: Dr. Scient. Thesis 13, July 1998. University of Oslo, Department of Informatics.

[4] Elewaut, E.F.M.: *Secondary Oil Migration Project, An Integrated Geomechanical and Quantative Modelling Approach for Understanding and Predicting Secondary Oil Migration.* Netherlands Institute of Applied Geoscience - TNO, National Geological Survey, Dec 1996.

[5] Fladmark G. E.: *A numerical method for local refinement of grid blocks applied to reservoir simulation.* Paper presented at the Gordon Research Conference for Modelling of Flow in Permeable Media. Andover, USA, August 9-13, 1982.

[6] McCormick S.F., Thomas J.W.: *The fast adaptive composite grid (FAC) method for eliptic equations.* Math. Comput. 46, 439-456, (1986).

[7] Reme H., Øye G.Å.: *Use of local grid refinement and a Galerkin technique to study Secondary migration in fractured and faulted regions.* Computing and Visualisation in Science. Springer-Verlag, 1999.

[8] Squyres Jeffrey M, McCandless Brain C, Lumsdaine Andrew: *Object Oriented MPI: A Class Library for the Message Passing Interface* http://www.cse.nd.edu/ lsc/reasearch/oompi/

[9] Wachspress E.L.: *Iterative Solutions of Elliptic Systems.* Prentice-Hall International, 1966.

Some Investigations of Domain Decomposition Techniques in Parallel CFD

F. Chalot[1], G. Chevalier[2], Q.V. Dinh[1], and L. Giraud[2]

[1] Dassault Aviation - 78 Quai M. Dassault - 92 214 St. Cloud
[2] CERFACS - 42 Av. Coriolis - 31057 Toulouse

Abstract. Domain decomposition methods in finite element applied aerodynamics provides a real speed-up of the convergence and good parallel scalability, even with the minimum overlap approach used here. Furthermore, a new variant of Restricted Additive Schwarz procedure is tested and shows a very attractive scalability property.

1 Introduction

With the emergence of the parallel distributed memory computers, certain techniques of convergence acceleration, expensive in memory, present a renewed interest for the industrial applications. Exploitation of these techniques on this type of architecture led to the use of domain decomposition decomposition techniques. This article presents the results obtained on a code of aerodynamics where we successively consider:

1. Convergence acceleration on uni-processor with a basic preconditioner (i.e. ILU(0)).
2. The analyze of the behavior of the iteration process in parallel with a domain decomposition method while trying to minimize the communication between processor with a minimum overlap approach.
3. The study of the scalability properties of the resulting code, and the test of a new approach of the additive Schwarz procedure.

We will conclude with a summary of the results obtained and a presentation of the different research path highlighted during that study.

2 Numerical Methods

2.1 Navier Stokes Equations, Discretisation

The presentation of the equations will be brief here, because no modification of the physics was made compared to the former publications related to this code [4].

The compressible Navier-Stokes equations are expressed in conservative form:

$$U_{,t} + F_{i,i} = F_{i,i}^d + \mathcal{F} \quad , \tag{1}$$

P. Amestoy et al. (Eds.): Euro-Par'99, LNCS 1685, pp. 595–602, 1999.
© Springer-Verlag Berlin Heidelberg 1999

where U is the conservative variable array: $U^T = (\rho, \rho.u, \rho v, \rho w, e)$, F_i the Eulerian fluxes , F_i^d the viscous or diffusive fluxes and $\mathcal{F}$ a source term.

One can rewrite the previous system in its quasi-linear form:

$$U_{,t} + A_i U_{,i} = (K_{ij} U_{,j})_{,i} + \mathcal{F} \quad , \tag{2}$$

where $A_i = F_{i,U}$ the ith Jacobian matrix of the Euler equations and K_{ij} the diffusivity matrix.

The space-time discontinuous Galerkin approach is used and allows simultaneous space and time integration formulations. The stabilization of the method is ensured by using a Galerkin Least-Square operator. Moreover a shock-capturing term is included.

Currently we use a pseudo-time iterative scheme (predictor/corrector) because we only seek for a stationary solution. Thus the time stepping procedure, which will be also called "nonlinear iterations", is established, and leads to solve at each step a linear system $\mathbf{A}.\mathbf{x} = \mathbf{b}$. For a more complete description of the formulation and scheme, refer to [7] and the former publications [4].

2.2   Linear System

The linear system to solve at each time step is large, sparse and non-symmetric. Within the framework of this study we consider a preconditioned Krylov method for its solution. Taking into account the memory architecture on the targeted machines and the coarse grain parallelism based on the 3D unstructured grid partitioning, the domain decomposition techniques appeared as the most natural way to develop effective preconditioners. For our application, we cannot afford direct methods for the subdomain problem solutions. In that context, we considered the family of methods, generally referred under the generic name of Additive Schwarz methods, for which we can easily set up variants requiring only an approximate solution of the local Dirichlet problem. Theses techniques can be considered algebraically like local/block preconditioners with an overlap between the blocks. In order to minimize the complexity of the calculation, only one element overlap at most is permitted.

2.3   Formulation of the Additive Schwarz Methods

In this section we briefly present the formulation of the additive Schwarz method. Readers should refer to [8] and the references therein, for a more detailed presentation of the domain decomposition techniques.

With a partition of the unstructured finite element mesh, $\mathcal{M}$ can be decomposed in N disjoints sets of elements (sub-domains), or equivalently into N overlapping sets of nodes. Let's call $\{W_i\}_{i=1,N}$ the subsets of nodes and W the complete set of nodes over the entire grid. This drives to: $W = \bigcup_{i=1}^{N} W_i$, which can be called a "partition with a minimum overlap" of Of that element partition, one can deduce a strict node partition by allocating each interface node to a single subset of elements $W_i^0 \subset W_i$. This leads to: $W = \bigcup_{i=1}^{N} W_i^0$ with $W_i^0 \cap W_j^0 = \emptyset$ for $i \neq j$.

We can then associate with each of the W_i^0 a canonic restriction operator R_i^0. We can define in the same way the restriction operators R_i for the sets W_i. Using the above notations the local problems is defined by:

$$A_i = R_i A R_i^T , \tag{3}$$

where R_i^T is a prolongation operator.

These notations allows the formulation of the basic Schwarz additive preconditioner like: $M_{AS} = \sum_{i=1}^{N} R_i^T A_i^{-1} R_i$.

These notations allows to describe the preconditioner recently introduced by [2, 1] whom called it Restricted Additive Schwarz: $M_{RAS} = \sum_{i=1}^{N} (R_i^0)^T A_i^{-1} R_i$. In the case of the partition of the elements in a Finite Element formulation, the construction of the local problems as defined by the Equation (3) requires the assembly of the stiffness matrix for the interface nodes.

The exact solution of the local problems which appear with the M_{AS} et M_{RAS} being too costly for an industrial problem, we substituted them by approximate resolutions calculated via an ILU(0). In order to limit the communications during the assembly of the local stiffness matrices A_i on the interface nodes, a new variant have been implemented, which consists in assembling only the diagonal block associated with each of the interface nodes. Thus we carried out experiments with three preconditioners:

1. $M_{ILU(0)-AS}$: Additive Schwarz with an inexact local solution using ILU(0).
2. $M_{ILU(0)-dAS}$: Additive Schwarz with inexact local solution using ILU(0) and only the assembly of the interface nodes diagonal block.
3. $M_{ILU(0)-dRAS}$: Restricted Additive Schwarz with inexact local solution using ILU(0) and only the assembly of the interface nodes diagonal block.

These preconditioners have been implemented in order to accelerate the GMRES (Generalized Minimum RESidual) method introduced by Y. Saad and M. Schultz [5]. The size of the Krylov space is limited to 20 in 2D and at 10 in 3D. Moreover the tolerance ε controlling the decrease of the 2-norm of the GMRES' residuals is fixed at 10^{-1}.

3   Implementation & Results

3.1   Sequential Linear System : ILU(0) as Preconditioner

We present here the results obtained in sequential (mono-processor) on a 2D turbulent test case representative of real flow conditions. It is RAE2822 test case number 9 (cf. [3]). The test conditions are those of a turbulent transonic flow: $R_{e_c} = 6.5 \, 10^6$, $\alpha = 2.79°$, $M = 0.73$.

The grid is the cut in triangles of a structured fine grid with 362×64 or 22900 points and 45000 elements. In the results presented here, we used a two-layer $k - \varepsilon$ turbulence model.

One can report the following observation on the sequential tests:

1. logically, the number of Krylov spaces (*i.e.* the number of matrix-vector products) necessary to the solution of the linear systems decreases dramatically when the ILU(0) preconditioner is applied, with a fixed tolerance ε and CFL condition (*i.e.* with the same "speed" speed in the time marching procedure),

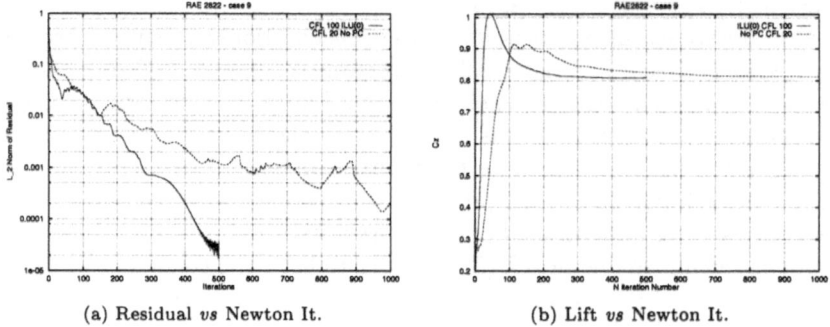

(a) Residual *vs* Newton It.   (b) Lift *vs* Newton It.

Fig. 1. RAE 2822 Case 9 - dashes: CFL=20 No Prec. - solid: CFL=100 ILU(0)

2. the preconditioning by itself brings little to the nonlinear process (the "outer" pseudo-time-stepping iterations) if we do not modify the CFL condition when compared to the unpreconditioned case. Indeed one can then have a gain or a lost of CPU time according to the ill-conditioning of the linear systems, but overall the evolution of the nonlinear residues is similar.
3. If, on the opposite, we take advantage of the improvement of the robustness of the preconditioned system to increase the CFL, then the nonlinear residuals converge much more quickly ($\div 2$ on the CPU time here). Figure 1(a) compare the evolution of the residuals for a unpreconditioned calculation with a maximum CFL condition of 20, and a ILU(0) left-preconditioned computation with a maximum CFL of 100.
   The same acceleration of convergence can be noted on Figure 1(b) which present the evolution of the lift coefficient during the iterations.
   Acceleration comes primarily from the increase in the CFL condition authorized by the use of the preconditioner. Eventually and unfortunately we encounter a limit of stability of the scheme and in practice the CFL lie between 50 (coarse grid) and 300. Beyond that, oscillations appear and calculation does not converge any more. Nevertheless the multiplication by 5 or 15 of the CFL is sufficient to obtain a 2 to 3 fold acceleration over the computational time.
4. For *steady calculations* the value of the tolerance ε with constant CFL does not seem to have a radical effect on the global time-marching process of the nonlinear iterations: this considering the transient phase or on the level of final residue. Nevertheless the use of the preconditioner ILU(0) facilitates

the resolution of the linear system. Let us note that that should present an interest for the unsteady simulations where the value of ε must be relatively low ($\leq 10^{-2}$).

3.2   Study of the M6 Wing Test Case on 4 to 32 Processors

Parallel : Efficiency of the ILU(0) with $M_{ILU(0)-dAS}$ The approach known as domain decomposition is used to reduce the degradation induced in the preconditioner by the localization of the data on each processor; that localization is mandatory for effective parallelism. The procedure of additive Schwarz with minimal overlap used here constitutes an ab-initio approach of domain decomposition and can be developed without the choice and use of dedicated libraries.

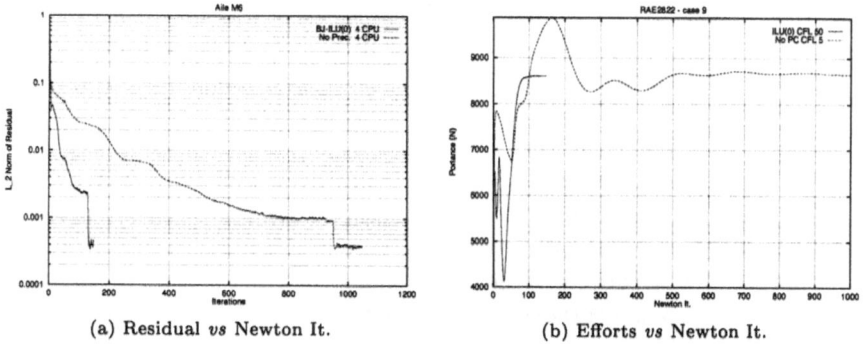

(a) Residual *vs* Newton It.   (b) Efforts *vs* Newton It.

Fig. 2. M6 wing - 4CPU - dashed: CFL=5 Non Prec. - solid: CFL=100 - $M_{ILU(0)-dAS}$

We will use thereafter the case of the ONERA M6 wing to test the parallel applications for the 3D turbulent flows. The simulation parameters are: $R_e = 11.7\,10^6$, $M = 0.844$, $\alpha = 4.56°$ [6]. The grid consists of 77000 points, divided into 4, 8, 16 and 32 subdomains. The size of problem prevented the realization of a sequential test, so the comparison was made between the calculation on 4 processors and the later decompositions. The decomposition itself have been carried out by Dassault-Aviation, taking care to balance the distribution of the points *and* elements over the domains. The stopping criterion of the GMRES is $\varepsilon = 10^{-1}$, the size of the Krylov space is 10 with and we allow only 2 more restarts of the GMRES.

Figure 2(a) compares the residuals of a non-preconditioned calculation with a CFL of 5 with those of a $M_{ILU(0)-dAS}$ calculation with a CFL of 50 (these two CFL are roughly the acceptable maximum for each calculations). We can note that the lift history (Figure 2(b)), shows clearly the acceleration in the establishment of the flow, as do the convergence history.

Finally, we can note that the CPU time has been decreased by a factor of 3 with the use of $M_{ILU(0)-dAS}$: from a total of 7500 units of time for the unpreconditioned case we ended to 2000 units of time with the preconditioner, both of the calculations have been performed on 4 CPU.

Parallel : Scalability of the $M_{ILU(0)-dAS}$ Globally, there is an independency of the residual evolution during nonlinear iterations with respect to the number of processors. Furthermore the CPU time gathered in the Table 1 reveal a good efficiency, in spite of a minimum overlap and a basic preconditioner. The efficiency is calculated by taking as reference time the calculation on 4 processors.

Table 1. M6 wing - CFL=50 - $M_{ILU(0)-dAS}$

	4 Domains	8 Domains	16 Domains
CPU Equiv. Time	1922 u	1031 u	558 u
Acceleration	– 1 –	1.86	3.44
Efficiency	– 1 –	0.93	0.86
# of Krylovs	1259	1423	1493

This is not a sufficient indication to characterize the scalability of the preconditioner; Table 1 also presents the total number of internal GMRES iterations needed to get the appropriate convergence.

We can see that, according to the total number of Krylov spaces, the quality of the local preconditioner degrades with the number of subdomains. The relatively good scalability in CPU times is partly due to a better memory access resulting of the reduction of the problem's size (mainly cache effects).

These results are nevertheless encouraging; they prove that an domain decomposition approach with a minimal overlap strategy presents an sufficient parallel efficiency for applications up to 16 and probably 32 processors.

Effects of Periodic Preconditioner Estimation We have constated that it is no necessary to recompute the preconditioner for each nonlinear iteration, we use the an "old" evaluation of the preconditionner for the current iteration.

For example on the 4 domains M6 wing test case, the quality of the preconditioner suffers when we compute it only every five Newton iterations (1409 Krylovs vs. 1259). But that degradation is largely compensated by the CPU savings of the non-evaluation of the ILDU(0): In our case the gain is clear: on 4 CPU with $M_{ILU(0)-dAS}$ it drives the CPU time from 1922 units to 1626.

That strategy was verified for all the domain decomposition techniques considered here, and on 4, 8, 16 or 32 processors.

Comparison of $M_{ILU(0)-dAS}$ **and** $M_{ILU(0)-AS}$ The $M_{ILU(0)-dAS}$ preconditioner quality can be improved if we not only assemble the diagonal block, but all the blocks of the line in the matrix for each interface node. These additional assemblies require the enrichment of the information that describes the interface (additional handling of the grid) as well as a notable increase in the communications to build the local systems.

The use of $M_{ILU(0)-AS}$ allows a gain of approximately 15% of the number of matrix-vector products for each nonlinear iteration. Moreover, the additional information to exchange increases the size of the message but not the number of them. Thus on hight-bandwidth machines, the overcost of communication related to a higher volume of data exchanged can be neglectible.

The saving of time CPU is thus 10 to 15% compared to $M_{ILU(0)-dAS}$. Nevertheless, the complexity of the connectivities sorting procedure, the non-control of the user on the interface region, makes its use cumbersome without the use of a domain decomposition library dealing with these aspects at the mesh partitioning stage.

3.3 Restricted Additive Schwarz

We use here the Restricted form of the additive Schwarz procedure as introduced in the section 2.3, wich differs only at the preconditioning step in the GMRES: we no more assemble the result of the preconditioning, but rather affect the value of the node's domain to the others.

$M_{ILU(0)-dRAS}$ **Scalability**   We consider a similar study as the one described in Section 3.2 but for the $M_{ILU(0)-dAS}$ with the restricted additive Schwarz procedure and a periodic evaluation of the ILU(0) every other five Newtons iterations. As previously reported, the residual evolution against the nonlinear iteration is not influenced by the successive division into 4, 8, 16 or 32 subdomains. On the other hand, and in a surprising way, the internal iteration count of the GMRES *decreases* when one goes from 4 to 8 then 16 fields (cf Table 2). This is not observed any more on the switch from 16 to 32 processors.

Table 2. M6 wing - CFL=50 - $M_{ILU(0)-dRAS}$ - Periodicity : 5

	4 Domains	8 Domains	16 Domains	32 Domains
CPU equiv. Time	1583 u	795 u	398 u	225
Acceleration	– 1 –	1.99	3.97	7.02
Efficiency	– 1 –	0.995	0.993	0.878
# of Krylovs	1244	1202	975	1222

The CPU times gathered in Table 2 show a remarkable scalability of the method.

4   Conclusion & Future Work

During this study, it was possible to adapt a semi-industrial aerodynamic code in order to test quickly some domain decomposition ideas and preconditioners for practical simulation using coarse grained parallelization on distributed memory architectures.

The choices made during this study were directed towards a minimization of the number of exchanges between processors. Communications are only need by the matrix-vector products, preconditioning, scalar products and for the diagonal block assembly of the Jacobian matrix on the interface nodes. Moreover the ILU(0) was choosed with the same simplicity/robustness requirements. Despite all its intrinsic limitations compared to more traditional approaches of domain decomposition, that technique showed its attracting potential on various aerodynamics simulation test cases. In this study, a 2 to 3 fold acceleration has been observed when compared to the best unpreconditioned implicit calculations, and the very good scalability of the $M_{ILU(0)-dRAS}$ has been demonstrated.

Most of the acceleration gain is a consequence of the increase in the robustness which allowed the increase in the CFL. Depending on the flows/grids tested, the CFL limit lies between 50 and 300. One can question whether further increase the CFL limits would still be so beneficial? And, how to achieve these CFL conditions (integration schemes, linearization schemes, boundary conditions).

References

[1] X.-C. Cai, C. Farhat, and M. Sarkis. A minimum overlap restricted additive Schwarz preconditioner and applications in 3D flow simulations. In *Proceedings of the 10th Intl. Conf. on Domain Decomposition Methods*, pages 479–485, Providence, 1998. AMS.

[2] X.-C. Cai and M. Sarkis. A restricted additive schwarz preconditioner for general sparse linear systems. Technical Report Tech Report CU-CS-843-97, Dept. of Comp. Sci., Univ. of Colorado at Boulder, 1997. to appear in SIAM SIAM J. Sci. Comput.

[3] P.H. Cook, MA McDonald, and MCP Firmin. Aerofoil RAE 2822 - pressure distributions and boundary layer and wake measurements. In *AGARD AR 138*, pages A6, 1–77, 1979.

[4] Thomas J.R. Hughes, Leopoldo P. Franca, and Michel Mallet. A new finite element formulation for computational fluid dynamics: VI. convergence analysis of the generalized SUPG formulation for linear time-dependant multidimentional advective-diffusive systems. *Comp. Meth. in Applied Mechanics and Engineering*, (63):97–112, 1987.

[5] Y. Saad and M. Schultz. GMRES: A generalized minimal residual algorithm for solving nonsymmetric linear systems. *SIAM J. Sci. Stat. Comput.*, 7:856–869, 1986.

[6] V. Schmitt and F. Charpin. Pressure distribution on the ONERA M6 wing at transonic mach numbers. In *AGARD R-702*, pages B1, 1–44, 1982.

[7] Farzin Shakib, Thomas J.R. Hughes, and Zdeněk Johan. A new finite element formulation for computational fluid dynamics: X. the compressible euler and navier-stokes equations. *Comp. Meth. in Applied Mechanics and Engineering*, (89):141–219, 1991.

[8] B.F. Smith, P. Bjørstad, and W. Gropp. *Domain Decomposition, Parallel Multilevel Methods for Elliptic Partial Differential Equations*. Cambridge University Press, New York, 1st edition, 1996.

A Parallel Ocean Model for High Resolution Studies

Marc Guyon[1], Gurvan Madec[2], François-Xavier Roux[3], Maurice Imbard[2]

[1] IDRIS, bât. 506, BP 167, 91403 Orsay Cedex, France.
guyon@idris.fr
[2] LODYC, UMR 7617, CNRS-UPMC-ORSTOM, Université Pierre et Marie Curie, Case 100,
4, place Jussieu, 75252 Paris Cedex 05, France.
{madec, mi}@lodyc.jussieu.fr
[3] ONERA, BP 72, 9322 Châtillon Cedex, France.
roux@onera.fr

Abstract. Domain Decomposition Methods is used for the parallelization of the LODYC ocean general circulation model. The local dependencies problem is solved by using a pencil splitting and an overlapping strategy. Two different parallel solvers of the surface pressure gradient, a preconditioned conjugate gradient method and a Dual Schur Complement method, have been implemented. The code is now used for the high resolution study of the Atlantic circulation by the CLIPPER research project and is one of the components of the operational oceanography MERCATOR project.

1. Motivations

The study of the ocean and its influence on the global climate system require to know more precisely physical processes characterized by a broad range of spatial and temporal scales which encompass the relevant dynamics and involve longer integrations at finer resolution. One of the key to investigate a new physics is to design numerical models that can use state of art high performance computers. As computer technology advances into the age of massively parallel processors, the new generation of OGCM has to offer an efficient parallel tool to exploit memory and computing resources of distributed architectures. Developers need to adapt the software structure to benefit from these new generation of computers.

2. The OPA Model and Its Main Numerical Characteristics

The ocean is a fluid which can be described by the Navier-Stokes equations plus the following additional hypothesis: spherical Earth approximation; thin-shell approximation; turbulent closure hypothesis; Boussinesq hypothesis; hydrostatic hypothesis; incompressibility hypothesis. In oceanography, we refer to this set of

P. Amestoy et al. (Eds.): Euro-Par'99, LNCS 1685, pp. 603-607, 1999.
© Springer-Verlag Berlin Heidelberg 1999

equations as Primitive Equations (hereafter PE). A precise description of the different terms of PE and their parameterization in the OPA model are detailed in [1].

The ocean mesh is defined by the transformation that gives the geographical coordinates as a function of (i, j, k). The arrangement of variables is based on the Arakawa-C grid. The model equations have been discretized using a centered second-order finite difference scheme. For the non-diffusive processes, the time stepping is achieved with a leapfrog scheme whose the computational noise is controlled with an Asselin time filter. For diffusive processes, an Euler forward scheme is used, but an Euler backward scheme can be used for vertical diffusive terms to overcome the strength of the vertical eddy coefficients.

With the second order finite difference approximation chosen, the surface pressure gradient contribution (SPG) has to satisfy a matrix equation of the form:

$$\mathbf{E\,x} = \mathbf{b} \,, \tag{1}$$

where $\mathbf{E}$ is a positive-definite symmetric sparse matrix, and $\mathbf{x}$ and $\mathbf{b}$ are the vector representation of the time derivative of a volume transport streamfunction associated to SPG and of the vertical curl of the collected contribution of the Coriolis, hydrostatic pressure gradient, non linear and viscous terms of PE, respectively. The equation (1) is solved using a Preconditioned Conjugate Gradient (PCG) algorithm.

3. Parallelization Choices

The space dependencies associated with the resolution of the PE allows to identify that: computing SPG requires a knowledge over the whole horizontal domain, the vertical physics (thereafter VP that includes parameterization of convective processes, 1.5 vertical turbulent closure, time stepping on the vertical diffusion terms) involves each ocean water column as a whole, while for the remaining part of the model (hereafter PE') the computation remains local. The pencil splitting is an elegant solution to solve the dependencies problem of VP, to fit with the vertical boundary conditions of the ocean and preserve the parallel efficiency. The OPA finite-difference algorithms are solved using an Arakawa-C grid, so that the complication to specify the interface between neighbouring subdomains for the PE' leads to the easiest programming solution: a data substructuring with overlapping boundaries, the interface matching conditions are then related to usual Dirichlet boundary condition.

4. The Surface Pressure Gradient

In the PE model, the calculation of the SPG leads to a two dimensional horizontal problem that has to be solved with an iterative algorithm, either in time for subcycling time scheme [2], or in space for the other methods that all lead to an elliptic equation. One of the two algorithms used in OPA to solve the equation (1) is an adaptation to massively parallel computer of the diagonal PCG-algorithm. The parallelization of the PCG algorithm is rather straightforward. Indeed, since the preconditioner is a diagonal matrix, communication phases are only required in the calculation of a matrix-vector product and two dot products. The matrix vector product is computed using the same overlapping strategy as those used on PE' in order to solve the dependencies problem as an open Dirichlet boundary one. The local dot products are computed in parallel on each interior subdomain and sum over the whole domain through a reduction-diffusion operation to generate the coefficients associated with the two dot product. The ratio between computation and communication will therefore be a crucial point for SGP calculation and appear to stay strongly machine dependent. One looks for an algorithm [3] that minimizes the communication phase and/or the number of iterations in order to obtain good granularity tasks to benefit from massively parallel computers. The global domain Ω which boundary is $\partial\Omega$ is divided into a set of N_p subdomains $(\Omega_p)_{p=1,N_p}$ which boundaries are $(\partial\Omega_p)_{p=1,N_p}$, and a non-overlapping interface between the subdomains:

$$\Gamma = \bigcup_{p=1,N_p} \partial\Omega_p - \partial\Omega .$$

For purpose of simplification the elliptic equation satisfied by SPG contribution [1, 4] is rewritten as (2a) with Dirichlet boundary conditions (2b):

$$E(x) = \nabla(\kappa \nabla(x)) = b \quad \text{on} \quad \Omega , \tag{2a}$$

$$x = x^{\partial\Omega} \quad \text{on} \quad \partial\Omega . \tag{2b}$$

Let us consider the set of the local second order elliptic operators $(E_p)_{p=1,N_p}$ associated to the SPG operator E on $(\Omega_p)_{p=1,N_p}$, the Dirichlet-boundary problem satisfied by the global field x associated to the source term b (2) is equivalent, in its variational form, to the set of N_p local problems with the continuity conditions, (3c) and (3d), on the interface Γ :

$$E_p(x_p) = b_p \quad \text{on} \quad \Omega_p , \tag{3a}$$

$$x_p = x_p^{\partial\Omega} \quad \text{on} \quad \partial\Omega \cap \partial\Omega_p , \tag{3b}$$

$$x_p = x_{p_i} \quad \text{on} \quad \partial\Omega_p \cap \partial\Omega_{p_i} , \tag{3c}$$

$$\kappa_p \nabla(x_p) n_p + \kappa_{p_i} \nabla(x_{p_i}) n_{p_i} = 0 \quad \text{on} \quad \partial\Omega_p \cap \partial\Omega_{p_i} , \tag{3d}$$

where x_p, b_p, κ_p and $x_p^{\partial\Omega}$ are the values of x, b, κ and $x^{\partial\Omega}$ in the subdomain Ω_p, n_p is the outer normal vector of $\partial\Omega_p \cap \Gamma$. As $(\Omega_{p_i})_{i=1,N_a^p}$ is the set of the N_a^p adjacent subdomains of Ω_p, x_{p_i} and κ_{p_i} are the values of x and κ in the subdomain Ω_{p_i}, n_{p_i} is the outer normal vector of $\partial\Omega_{p_i} \cap \Gamma$. Let us introduce the Lagrange multiplier λ associated to the continuity condition (3c):

$$\kappa_p \nabla(x_p) n_p = s\,\lambda \quad \text{on} \quad \Gamma \cap \partial\Omega_p \,, \tag{4}$$

where the factor $s = \pm 1$. The change of sign indicates that the outer normal derivatives of x_p and x_{p_i} are opposite each other.

The FETI methods consists in finding λ the value of the normal fluxes along the interfaces Γ for which the solution of the local boundary-value problems, (3a) and (3b), satisfies the matching condition (3c). Then, the global field x, whose restriction x_p to each subdomain Ω_p is defined as the solution of the local Neumann problem, is continuous and also the normal fluxes. This is therefore the solution of the global problem. Mathematical proprieties, technical difficulties and a precise version of the FETI algorithm are detailed in [3].

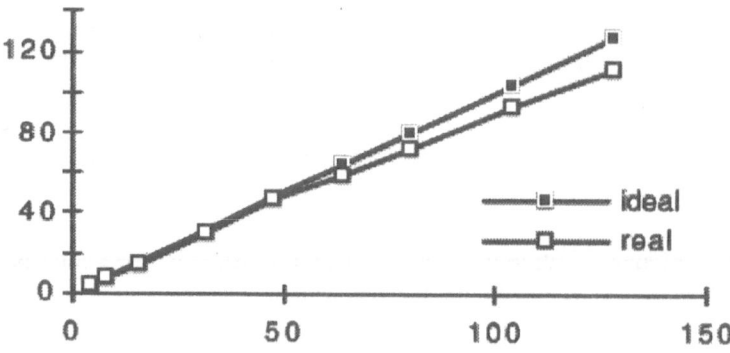

Fig.1. Speed up of the OPA code versus the number of subdomains for one hundred time steps integration on a Cray T3E system.

5. Conclusions

The ideas underlying domain decomposition methods have been integrated in the OPA model release 8. The portability of the code on multi-processors and mono-processor platform is preserved: the code has run on a large set of different platforms (Intel Paragon, Cray T3D, Cray T3E, IBM SP2, Fujitsu VPP) with different message passing communication libraries (NX, SHMEM, PVM, MPI). The straightforward parallelization of the conjugated gradient algorithm offers the advantage of an easy implementation. The FETI method exploits largest granularity computations and its numerical behaviour is more robust [3]. The speed up is nearly linear due to the halo effect when the number of subdomains increases and proves the total scalability [5]. It has been noticed that the initial code was quite well vectorized: a performance of 500 Mflops was obtained on a C90 processor whose peak performance is 1 Gflop. This code is interesting on massively parallel and vector machines: when tasks granularity is preserved it exploits the parallel efficiency of the domain decomposition implementation and/or the potential of the vector processors. A fully optimized version of the FETI algorithm is going to be evaluated in the context of large size computations [6]. Then the parallel model with the PCG algorithm [5] is confronted to the hardship of two major ocean experiments.

The CLIPPER project aims to obtain a High Resolution modelling of the ocean circulation in the Atlantic (1/6° at equator) forced by or coupled with the atmosphere. The program is aiming to model the oceanic circulation in the whole Atlantic Basin (from Iceland to Antartica) with the parallel OPA [7]. The basin extends to an area situated from 98.5° West to 30° East in longitude and from 75° North to 70° South in latitude. The grid size is 1/6° with 43 vertical levels and the resulting model mesh has 773*1296*43=43,077,744 grid points. The MERCATOR project aims to develop an eddy resolving global data assimilation system which has to become fully operational and has to contribute to the development of a climatic prediction system relying on a coupled ocean-atmosphere model [8]. The first objective is to represent the North Atlantic and the Mediterranean circulations with a 1/12° and a 1/16° grid size respectively and 42 vertical levels, the resulting model mesh has 1288*1022*42=55,286,112 grid points.

References

1. MADEC, G., DELECLUSE, P., IMBARD, M. and LEVY, C., "OPA version 8.1, Ocean General Circulation Model, reference manual", Notes du Pôle de Modélisation n°11, IPSL, 91p, France, 1999.
2. KILLWORTH, P. D., STAINFORTH, D., WEBB, D. J. and PATERSON, M., The development of a free-surface Bryan-Cox-Semtner ocean model, Journ. of Phys. Oceanogr., 21, 1333, 1991.
3. FARHAT, C. and ROUX, F. -X., Implicit parallel processing in structural mechanics, Computational mechanics advances, Vol 2, N° 1, June 1994.
4. ROULET, G. and MADEC, G.,"A variable volume formulation conserving salt content for a level OGCM: a fully nonlinear free surface", in preparation, 1999.
5. GUYON, M., MADEC, G., ROUX F. X., HERBAUT, C., IMBARD, M. and FRAUNIE, P., Domain decomposition as a nutshell for a massively parallel ocean modelling with the OPA model, Notes du Pôle de Modélisation, N°12, 34 pp, IPSL, Paris, France, 1999.
6. GUYON, M., MADEC, G., ROUX F. X., IMBARD, M. and FRAUNIE, P., A fully optimized FETI solver to perform the surface gradient in an ocean model, in preparation.
7. BARNIER, B., LE PROVOST, C., MADEC, G., TREGUIER, A.-M., MOLINES, J.-M. and IMBARD M., The CLIPPER project: high resolution modelling of the oceancirculation in the Atlantic forced by or coupled with the atmosphere, in the Proc. of the International Symposium "Monitoring the Oceans in the 2000s: an integrated approach", TOPEX/POSEIDON Science Team Meeting, Biarritz, France, 1997.
8. COURTIER, P., The MERCATOR project and the future of operational oceanography, in the Proc. of the International Symposium "Monitoring the Oceans in the 2000s: an integrated approach", TOPEX/POSEIDON Science Team Meeting, Biarritz, France, 1997.

Nonoverlapping Domain Decomposition Applied to a Computational Fluid Mechanics Code

Paulo B. Vasconcelos and Filomena D. d'Almeida

Faculdade de Engenharia da Universidade do Porto
4099 Porto codex, Portugal
pjv@fe.up.pt, falmeida@fe.up.pt

Abstract. The purpose of this paper is the description of the development and implementation of the linear part of a numerical algorithm for the simulation of a Newtonian fluid flow and the parallelization of that code on several computer architectures. The test problem treated is the steady state, laminar, incompressible, isothermic, 2D fluid flow (extendible to 3D case), the Navier-Stokes equations being discretized by a fully coupled finite volume method. For this problem, sparse data structures, nonstationary iterative methods and several preconditioners are applied. The numerical results allow the conclusion that the fully coupled version can compete with the decoupled classic SIMPLE method (Semi-Implicit Method for Pressure-Linked Equations), by using the Krylov subspace methods. Parallel versions of the coupled method based on nonoverlapping domain decomposition are discussed.

1 Problem Specification and Brief Description of the Algorithms

The test problem used is the steady state two-dimensional incompressible laminar flow in a square lid-driven cavity. This academic problem describes the nonlinear and elliptic behavior of many flows studied in engineering and it is used to test the efficiency of many algorithms [6].

The Navier-Stokes equations governing the flow in this case are

$$\frac{\partial}{\partial x_i}(U_i) = 0 \tag{1}$$

$$\rho \frac{\partial}{\partial x_j}(U_i U_j) = -\frac{\partial P}{\partial x_i} + \frac{\partial}{\partial x_j}(\tau_{ij}), \quad \tau_{ij} = \mu \left(\frac{\partial U_i}{\partial x_j} + \frac{\partial U_j}{\partial x_i} \right) \tag{2}$$

for i=1,2 (in 2-D case) and i=1,2,3 (in 3-D case), respectively, the continuity equation (mass conservation) and the momentum equations, where ρ stands for the density, μ for the dynamic viscosity, Re for the Reynolds number, U for the velocity field where U_i for the velocity along x_i direction ($i = 1, 2$ for the 2-D case), and P for the static pressure.

P. Amestoy et al. (Eds.): Euro-Par'99, LNCS 1685, pp. 608–612, 1999.

The algorithms reported here were also applied to other related mechanical problems, mainly the laminar backward-facing step flow, with similar behaviour.

The classical SIMPLE [3] algorithm solves in a decoupled way the governing differential equations. The resulting linear systems, one for each velocity and pressure, are solved by TDMA which is derived from Gaussian elimination adapted to the particular case of tridiagonal systems.

The DIRECTO [1], solves the fluid flow equations as a completely coupled system. The resulting square coefficient matrix is large, sparse, block tridiagonal and unsymmetric. The matrix size is $n \times n$, $n = 3 \times ni \times nj$ for a $ni \times nj$ grid. So at each stage of the DIRECTO procedure such a linear system is to be solved and this is the most time consuming part of the algorithm. Another important key of the method is that there are no null entries on the diagonal of the coefficient matrix due to the existence of pressure terms on the continuity equation. For more details about relations to other similar methods see [6] [7] and references in them and for the accuracy of DIRECTO see [2].

Domain decomposition techniques were used for the parallelization of DIRECTO. The procedure is distributed among all processors, each capable of generating its part of the grid, to build the corresponding matrix blocks including the computation of the coefficients belonging to its subdomain and the calculation of the nodes relating its domain to the neibourghing ones. The basic idea of this technique is the partition of the domain into subdomains, Ω_i, interconnected by interfaces, Γ_j, solving the original problem corresponding to each domain and then, concatenate the parcial solutions. This concatenation is done via the Schur complement. Given the original system, $A\phi = b$, reordered accordingly to the domain partition

$$\begin{bmatrix} A_{\Omega\Omega} & A_{\Omega\Gamma} \\ A_{\Gamma\Omega} & A_{\Gamma\Gamma} \end{bmatrix} \begin{bmatrix} \phi_\Omega \\ \phi_\Gamma \end{bmatrix} = \begin{bmatrix} b_\Omega \\ b_\Gamma \end{bmatrix} \tag{3}$$

we obtain the block corresponding to the Schur complement, S, and the new right hand side, $\widehat{b}$, by block Gaussian elimination on (3), $S = A_{\Gamma\Gamma} - \sum A_{\Gamma\Omega} A_{\Omega\Omega}^{-1} A_{\Omega\Gamma}$ and $\widehat{b}_\Gamma = b_\Gamma - A_{\Gamma\Omega} A_{\Omega\Omega}^{-1} b_\Omega$. The reduced system solution $S\phi_\Gamma = \widehat{b}_\Gamma$ is used to update the solution on the subdomains by $A_{\Omega\Omega}\phi_\Omega = b_\Omega - A_{\Omega\Gamma}\phi_\Gamma$.

As the explicit computation of S is expensive we have to use it implicitly by computing its action on a given vector. Iterative methods are used for the Schur complement, as it is not known explicitly. Here we used preconditioned nonstationary iterative methods such as (GMRES+ILUT) [5] [4] everywhere requiring a good enough precision in the subdomains.

As DIRECTO is a nonlinear procedure it is not necessary to solve the linear systems very accurately. The use of direct methods is also possible but it is an expensive approach for large systems.

Different partitions of the domain were tested: (H) for horizontal stripes, (V) for vertical stripes and (P) for rectangular boxes [7]. Fig. 1 shows the communication and matrix patterns used for the (P) partition.

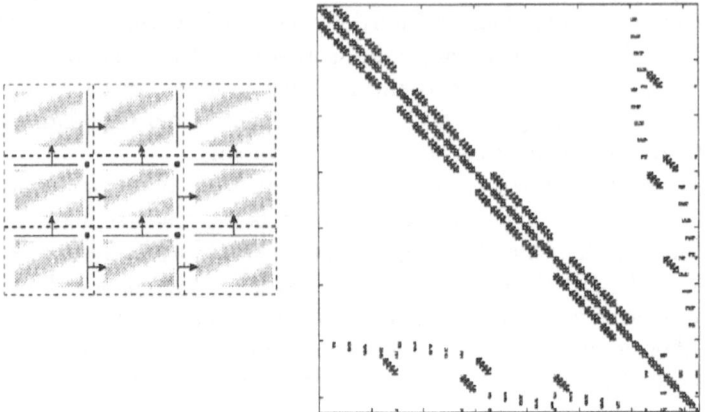

Fig. 1. Interface distribution, communication and matrix pattern for nested dissection of the domain

2   Numerical Results

The following libraries were used: LAPACK, SPARSKIT and PVM (version 3.3.7). Special versions of BLAS routines adapted to the sparse case were built based on SPARSKIT.

The tests were run on a heterogeneous cluster of 5 RISC workstations, 2 DEC AXP 3000/500, 1 DEC AXP 3000/600 and 2 DEC AS 1000 4/233 and 4/266.

The following test parameters were used: $epsout = 1.0e - 5$ for the outer iteration tolerance on the 1-norm of the residuals, $eps = 5.0e - 6$ for the inner iteration tolerance on the 2-norm of the residuals of the non-restarted GMRES method with Krylov subspace dimension 30, $tol = 5.0e - 4$ and $lfil = 30$ for the ILUT preconditioner.

Due to the characteristics of the linear systems the best of the nonstationary methods used was GMRES as shown in Fig. 2. The effect of preconditioning with ILUT is illustrated by its action on the eigenvalues, Fig. 3.

The difficulty with this code is to find an equilibrium between the number of outer iterations (to be minimized) and the precision required for the solution of the linear systems. The contradictory aims to keep in balance are (i) use a small number of GMRES iterations and require small precision, as suggested by the small number of sweeps of TDMA (used in the linearization procedure), to reduce the cost of each nonlinear iteration and (ii) solve the linear systems accurately with GMRES, as suggested by the use of direct methods, to have few nonlinear iterations. This balance was achieved by the use of GMRES with incomplete LU factorization as a preconditioner. Comparing SIMPLE+TDMA, DIRECTO+GMRES with several preconditioners (ILU(0),ILU(1),ILU(2),ILUT), we can see in [6], that ILUT is the best for fine grids, because the number of outer iterations is greatly reduced using this approach. For SIMPLE, the con-

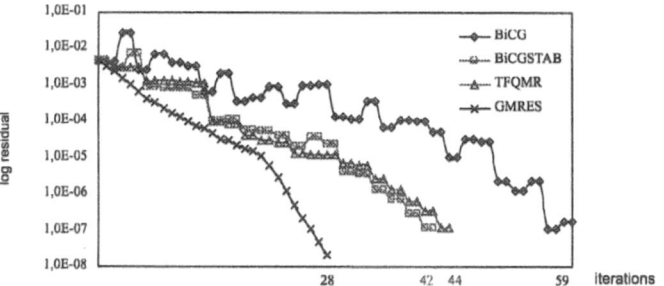

Fig. 2. Evolution of the log of the residuals vs. the number of iterations in a typical linear stage, for several nonstationary methods ($eps = 5.0e - 6$; $tol = 5.0e - 4$; $lfil = 30$)

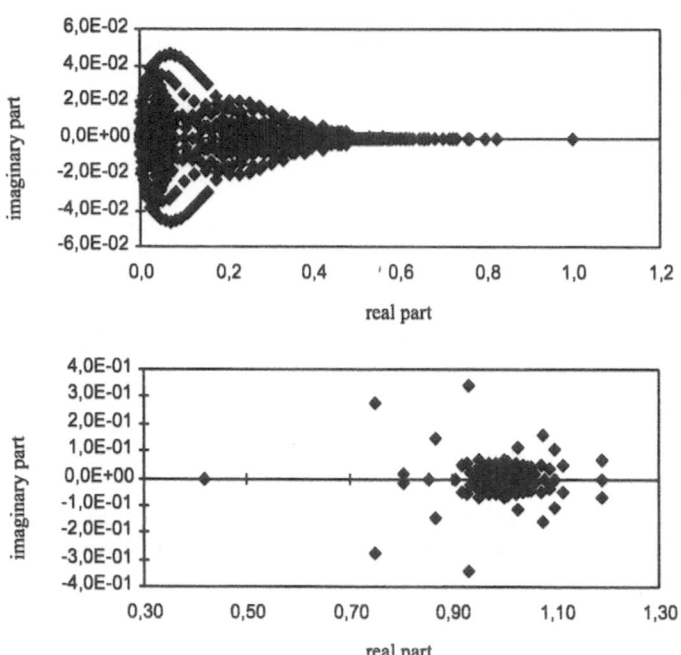

Fig. 3. Eigenvalue distribution before and after ILUT ($5.0e - 4$, $lfil = 30$) preconditioner ($Re = 400$)

Table 1. Elapsed times with 4 subdomains for several grids and partitions, $Re = 1000$.

	SPMD model			
	elapsed time with partitions:			
grid	H M/S	H	V	P
32×32	92.0	70.2	54.0	48.8
64×64	606.8	531.2	425.3	376.2
96×96	1770.1	1498.0	1279.9	1249.3
128×128	4694.3	3841.7	3105.7	2813.6

vergence behaviour was slow because of decoupled treatment of the equations results on a poor convergence for the pressure, and so the velocity field keeps iterating until convergence of the pressure.

Table 1 reports the elapsed times for several partitions of the domain and several grids using 4 subdomains. The parallelization of the code and the use of preconditioned iterative methods working on sparse structures allowed the execution for fine grids (that yield large linear systems) and significant gains in elapsed time comparing to the initial version of the code DIRECTO.

Tests were also run on other computer architectures, [7], to show that the algorithm is well ballanced. To extend the computation to a larger number of processors, the use of multigrid techniques and the migration to the MPI standard will be considered.

References

[1] Castro, F.A.: *Método de cálculo acoplado para a resolução das equações de Navier-Stokes e continuidade*. PhD thesis, Fac. Engenharia da Universidade do Porto, 1997.

[2] D'Almeida, F.D., Castro, F.A., Palma, J.M.L.M., Vasconcelos, P.B.: Development of a parallel implicit algorithm for CFD calculations. In *AGARD Progress and Challenges in CFD Methods and Algorithms*, 1996.

[3] Patankar, S.: *Numerical Heat Transfer and Fluid Flow*. Hemisphere, Washington, DC, 1980.

[4] Saad, Y.: ILUT: A dual threshold incomplete LU factorization. *Num. Lin. Alg. Appl.*1, pages 387–402, 1994.

[5] Saad, Y., Schultz, M.: GMRES: A Generalized Minimum Residual Algorithm for solving nonsymmetric linear systems. *SIAM, J. Sci. Statist. Comput.* **7**, pages 856–869, 1986.

[6] Vasconcelos, P.B., D'Almeida, F.D.: Preconditioning iterative methods in coupled discretization of fluid flow problems. *IMA Journal of Numerical Analysis*, 18:385–397, 1998. CERFACS Technical Report TR/PA/96/04.

[7] Vasconcelos, P.B.: *Paralelização de Algoritmos de Álgebra Linear Numérica com Aplicação a Mecânica de Fluidos Computacional*. PhD thesis, Fac. Engenharia da Universidade do Porto, 1998.

A PC Cluster with Application-Quality MPI

M. Gołębiewski, A. Basermann, M. Baum, R. Hempel, H. Ritzdorf, J. L. Träff

C & C Research Laboratories, NEC Europe Ltd.,
Rathausallee 10, D-53757 Sankt Augustin, Germany
`golebiewski@ccrl-nece.technopark.gmd.de`

Abstract. This paper presents an implementation of MPI on a cluster of Linux-based, dual-processor PCs interconnected by a Myricom high speed network. The implementation uses MPICH for the high level protocol and FM/HPVM for the basic communications layer. It allows multiple processes and multiple users on the same PC, and passes an extensive test suite. Execution times for several application codes, ranging from simple communication kernels to large Fortran codes, show good performance. The result is a high-performance MPI interface with multi-user service for this PC cluster.

Keywords: Workstation cluster; Myrinet; MPI; FM; Numerical applications.

1 Introduction

Continuous price reductions for commodity PC equipment and increasing processor speeds have made PC clusters an attractive low-price alternative to MPP systems for parallel applications. We therefore decided to set up such a system to provide our applications programmers with the required compute power for a large variety of projects, most of them in the area of numerical simulation. Our machine is based on the Local Area MultiProcessor (LAMP) developed at the NEC Research Institute (NECI) [6]. With its symmetric multiprocessor (SMP) architecture, the LAMP was well-suited for our intended experiments with clustered shared memory machines, and the Myrinet [15] provided sufficient communication speed.

The cluster installed at our laboratory consists of 16 dual-processor SMP nodes interconnected by a Myrinet network. Dual 8-Port Myrinet-SAN switches, each serving 4 PCs, provide a tetrahedral network topology. Each SMP node has two Pentium Pro 200MHz CPUs installed on a Tyan S1668 motherboard with Intel's 440FX chip-set and 512 KB of L2 cache per CPU. The PCs have 128 MB of EDO RAM each, so that the cluster has a total of 32 CPUs and 2 GB of RAM. The nodes are equipped with two network interface cards: an Ethernet card and a LANai board for accessing the Myrinet network. The LANai cards are basic 32 bit PCI models equipped with only 256 KB of SRAM. The cluster runs under Linux (Red Hat 5.1 distribution) with kernel version 2.0.34 (SMP). By choosing Unix, we have full support for services such as remote debugging and

P. Amestoy et al. (Eds.): Euro-Par'99, LNCS 1685, pp. 613–623, 1999.

administration, and multiple users on each node. The cluster is complemented by a Pentium II PC (300 MHz, 64 MB RAM) with SCSI disk drives, which is used as a file and compilation server.

Most of our application codes are written in Fortran (some in C) and parallelized using the Message Passing Interface (MPI) [20] which thus had to be installed as the application programming interface on the LAMP. Since similar PC clusters were already in use we attempted to use existing software. The search was guided by the following requirements: reliable, error-free data transmission, multi-user support, support for multiple processes per node, a complete and correct implementation of MPI, including the Fortran77 binding, and good communication performance at the MPI level. As can be seen our goal was to use the LAMP as a reliable compute server for real applications, and not as an experimental system.

In this paper we briefly discuss our survey of existing software (Section 2), followed by a description of our own MPI development (Section 3). Finally, in Sections 4 and 5 we present application results, ranging from kernel benchmarks to real-world codes, and compare the performance of the LAMP cluster with that of MPP systems.

2   Survey of Existing Software

Most PC clusters so far use Fast-Ethernet networks and high-overhead protocols for communication, such as TCP or UDP. This results in a large performance gap between the processing speed of a single PC and the communication between them. So, till now only very coarse-grained parallel applications could use such clusters efficiently. However, recent developments in high-speed networking hardware as, for instance, the Scalable Coherent Interface (SCI) [11] or Myrinet [15] extend the application domain towards problems with medium-, or even fine-grained, parallelism.

We tested all available communication interfaces for Myrinet under Linux:

1. Myricom - Myricom's native API. Only TCP/IP interface provided as basis for third-party MPI [15].
2. Basic Interface for Parallelism (BIP) from Ecole Normale Supérieure de Lyon, France [16].
3. Virtual Memory Mapped Communication (VMMC) developed at NECI. A new, much improved, version requires LANai boards with 1 MB [5].
4. BullDog Myrinet Control Program (BDM) from Mississippi State Univ. [12].
5. High Performance Virtual Machine (HPVM) developed at Univ. of Illinois in Urbana-Champaign [4].

None of them fulfilled all our requirements. Some were optimized for performance, but didn't conform to the MPI standard and didn't pass our test programs. Others did not allow multiple users or check for transmission errors. Table 1 is a summary of our survey, details can be found in [7]. To meet our requirements we concluded that we had to develop our own MPI implementation.

Communication interface	Myricom	BIP	VMMC	BDM	HPVM
max. number of nodes	16	8	16	16	> 16
multiple users/cluster	yes	no	no	no	yes
multiple processes/node	yes	no	no	no	yes
MPI (availability)	third-party	yes	no	yes	yes
MPI (standard compliance)	n/a	low	n/a	low	low
MPI (performance)	slow	fast	n/a	fast	fast

Table 1. Comparison of existing communication interfaces for Myrinet.

3   Implementation

We decided to base our new MPI library on the HPVM low-level interface FM 2.1 which met our requirements at that level. We could thus save the substantial effort needed to develop a new low-level library. For the high-level part, we used the MPICH software [9] and connected it with FM by our own device driver at the generic channel interface level. This device driver should have as little overhead as possible. We developed two versions:

1. The single-threaded (ST) device driver is optimized for minimum latency. It implements only the blocking channel interface primitives.
2. In the multi-threaded (MT) version both blocking and non-blocking primitives are implemented by having a communication thread for each application thread. This also allows to overlap computation with communication.

Both drivers use different protocols for three message length regimes, called *short*, *eager* and *rendezvous*, and switch between protocols at the same message lengths. Details of these three MPICH standard protocols are described in [10].

Communication in a program using the FM library may block when the receiver does not extract incoming messages from the network. Since in the ST version all FM calls are issued directly from the application processes, this may block the sender when the receiver is busy in a long computation. In the MT version the application thread can continue to execute while the communication thread is blocked. We thus expected this version to be more efficient in applications with poor load balancing, in which corresponding send and receive operations are not well synchronized. For details of the driver implementations, see [7].

4   First Test Results

In this section we present the results of experiments performed to assess the maturity of our MPI implementation.

4.1   Completeness and Correctness

We tested the MPI library, using both our ST and MT device drivers. Both versions passed the whole MPICH test suite. These programs not only check the correct semantics of the MPI functions, but also test the behavior of the implementation under intense network traffic. ST additionaly passed a much more demanding test suite from the GMD Communications Library (CLIC) for block-structured grids [17].

4.2   Performance Measured with Kernel Benchmarks

The basic performance indicators, latency and bandwidth, as measured with the ping-pong benchmark *mpptest* from the MPICH distribution, are shown in Fig. 1. We compare our implementations with the MPI interface of HPVM, because, despite of the many failures detected by our test suite, it was the existing software package that came closest to the user requirements.

The ST driver has very low latency, for zero byte messages exceeding that of the HPVM/MPI library by only about $2\,\mu s$. With more than $100\,\mu s$, the latency is much larger for the MT driver. The reasons for this high latency are the use of mutual exclusion and synchronization primitives from the POSIX threads library [3], as well as the necessary request queue manipulations.

Both our versions and HPVM/MPI exhibit similar bandwidth. The maximum bandwidth reached by our implementations is higher than that obtained with HPVM/MPI. ST performance decreases for very long messages using the rendezvous protocol. MT is slow for shorter messages, but in the rendezvous protocol is the fastest. The decrease in bandwidth seen for all three implementations for very long messages is caused by cache effects, and our implementations seem to suffer more from them than HPVM/MPI. This suggests that there is still room for optimization in our device drivers for the rendezvous protocol.

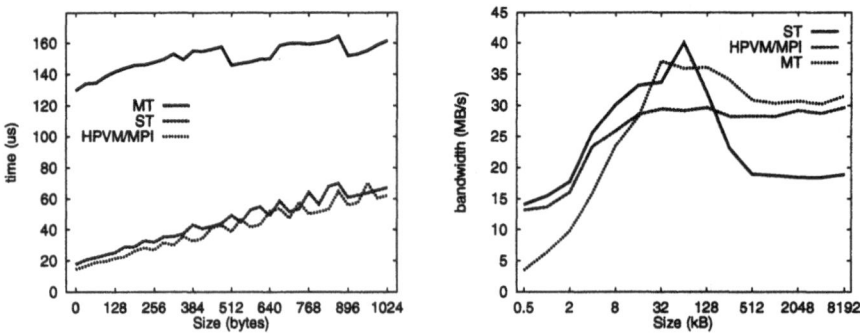

Fig. 1. Latency and bandwidth obtained with the ping-pong benchmark.

4.3   Effect of Load Imbalance on Performance

While basic test programs can be used to measure some basic performance characteristics of an MPI implementation, they cannot replace tests with full application codes. We have tested our implementations with the L_iSS package [18], a Fortran environment for solving partial differential equations, in our case the incompressible Navier-Stokes equations, on general two-dimensional block-structured domains. The CLIC library handles the inter-process communication for L_iSS. Together the packages contain some 85,000 lines of Fortran code.

By choosing the geometric domain appropriately, complicated communication patterns and various degrees of load imbalance can be generated. This latter feature was used to test our expectation that the MT device driver would improve the performance in the presence of load imbalance.

Table 2 gives parallel efficiencies and wall clock times for the ST and MT version of the MPI library for a problem with 66,240 grid points and good load-balance. Since HPVM/MPI does not have a usable Fortran binding, only our MPI libraries could be used for the L_iSS tests. For a comparison with an MPP computer, we included results achieved on a NEC Cenju-3 [14] for the same problem. In all test runs, 9 processes were used in a master-slave fashion. The tests were repeated with two slave processes per node to evaluate the performance loss caused by sharing the LANai boards, and the connection to a Myrinet switch, by two processes. Table 3 reports timings and efficiencies for a problem with 14,212 grid points and a high degree of load-imbalance. 30 processes were used for this set of tests, so either one or two application processes were assigned to each SMP node.

Our two implementations perform very well for the load-balanced test case, with ST always being the better choice. The efficiencies are well above 90%, except for MT with two processes per node. This is caused by the increase in CPU load by the additional threads. For ST we do not observe any significant performance drop by sharing the network interface between two processes. Apparently, there is not enough data traffic to saturate either the PCI bus or the network interface. The slower Cenju-3 performs worst in this test because its 75 MHz VR4400SC RISC CPUs are significantly slower than the 200 MHz PentiumPros of the PC cluster, and the interconnection network of the Cenju-3 is slower than the Myrinet. The unbalanced test case results in much lower performance than the well balanced problem. The parallel efficiencies for both ST and MT are only slightly above 22%. This shows that even for the situations for which the MT device driver was designed, it has no advantage over the ST version. We therefore selected the ST device driver for all further tests, and dropped further MT development.

5   Results for Complete Applications

This section summarizes some experiences with full application codes. The examples presented differ considerably in memory access patterns, communication

driver/ platform	procs per node	accumulated comm. time [s]	total wall clock time [s]	Efficiency [%]
ST	1	35.31	62.90	92.98
ST	2	34.43	62.97	96.16
MT	1	42.70	63.71	91.62
MT	2	145.65	77.04	76.37
Cenju-3	1	399.75	164.58	69.64

Table 2. Load-balanced problem with 66,240 grid points. Communication times are accumulated over all processes.

driver/ platform	accumulated comm. time [s]	total wall clock time [s]	Efficiency [%]
ST	919.45	40.89	22.45
MT	979.72	43.50	22.33
Cenju-3	1690.99	74.82	22.06

Table 3. Load-unbalanced problem, 14,212 grid points.

behavior, and computation to communication ratio, and illustrate well the performance of the LAMP for a wide range of potential applications and compared with other parallel computers. The SP-2 used for all three applications is configured with the High Performance Switch for MPI communication.

5.1 Large Sparse Eigenvalue Problems

The simulation of quantum chemistry and structural mechanics problems requires the solution of computation intensive, large sparse real symmetric or complex Hermitian eigenvalue problems. The JADA package [1] uses the JAcobi-DAvidson method [19] and iterative preconditioners for convergence acceleration to solve them.

JADA's parallelization strategy is matrix and vector partitioning with a data distribution and a communication scheme exploiting the sparsity of the matrix. Grouping of inner products and norm computations within the preconditioners and the basic Jacobi-Davidson iteration reduces the synchronization overhead. In a preprocessing phase, data distribution and communication scheme are automatically derived from the sparsity pattern of the matrix. This provides load-balance and makes it possible to overlap computations with data transfers. The JADA code is written in Fortran77 and C with MPI for message passing. It uses non-blocking communications in the multiplication of sparse matrices with dense vectors and MPI reduce operations for vector reductions.

Figure 2 illustrates the timing and speedup behavior of JADA in determining the four smallest eigenvalues and -vectors for an electron-phonon coupling simulation problem [22] on three different platforms: NEC Cenju-3, LAMP, and

IBM 9076 SP2. The matrix of order 98,800 with 966,254 nonzeros is real symmetric; the sparsity pattern is highly irregular. The result is a marked variation in message lengths and numbers of communication partners per processor. This relatively small problem was chosen because a sizable fraction of the execution time is spent on communication, and thus the effect of the MPI performance is more pronounced.

In Figure 2 (left) JADA has the best performance on the SP2 and the worst on the Cenju-3, with the LAMP results somewhere in between. For 1 processor, the SP2 is 1.6 times faster than the LAMP, which in turn is 1.9 times faster than the Cenju-3. The scaling behavior, as shown in Figure 2 (right), is best for the Cenju-3, which has the best communication to computation ratio. The reason for the marked efficiency loss between 16 and 32 processors for the LAMP is the use of both processors of the SMP nodes. Currently, internode communication is not handled optimally. Due to the good scaling, the LAMP times approach those of the SP2 for up to 16 processors, whereas on 32 processors the execution time on the LAMP is only slightly slower than on the Cenju-3.

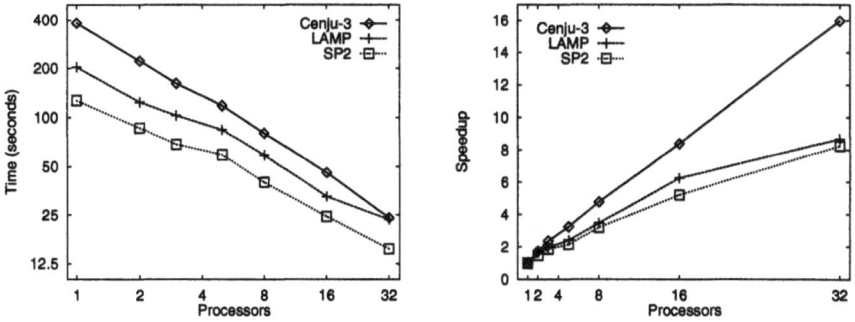

Fig. 2. Wall clock time (note the log-log scale) and parallel speedup for JADA.

5.2   Numerical Simulations of Complex Flows

Cellular automata methods represent an alternative approach to finite volume or finite element techniques for the numerical prediction of fluid flows. The basic idea of these methods is the numerical simulation of simplified molecular dynamics derived from the microscopic description of the fluid, instead of solving macroscopic governing equations. This is done by evaluating a time and space discrete Boltzmann equation, the so called *lattice Boltzmann equation* [2] describing the dynamics of the particle density function by two basic mechanisms, (1) *particle propagation* and (2) *particle collision*. The collision step of this procedure enables the specification of boundary conditions at fluid/solid interfaces in a way that overcomes the usual difficulties of classical CFD-methods in generating suitable grids. A simple bounce-back procedure fulfills the velocity boundary conditions for this type of interface and makes it possible to deal with arbitrarily complex geometries of these interfaces, while ensuring unconditional stability for the overall procedure. This capability to overcome the complex and time-consuming grid-specification procedure of classical approaches make the *Lattice Boltzmann Automata* (LBA) techniques especially appealing for many industrial applications. LBA techniques have been implemented in the Fortran program BEST [2] with MPI for data communication during parallel execution of the code. While particle collision is performed locally on each lattice, particle propagation implies transfer of data between neighboring points. The combination of an explicit time marching method and the restriction to next-neighbor dependencies simplifies the development of an efficient parallel code. However, the very small number of integer and floating point operations per lattice (about 150 flops per lattice per iteration) often leads to situations where the execution time is dominated by communication. The code is parallelized using the domain partitioning technique, with neighboring partitions being stored with an overlap of one lattice layer. The values in the overlap areas are updated once per iteration with all data for each destination process combined into a single message.

Since BEST is communication rather than computation bound for mid-size problems, it is an excellent benchmark for the communication network of parallel platforms under production conditions. Figure 3 shows the performance of BEST for a problem with 32 lattices in each space direction on the LAMP, an IBM SP2, and a NEC Cenju-3. Additionally for the LAMP, results are given for the largest problem that fits into a single node memory, with $256 \times 32 \times 32$ lattices.

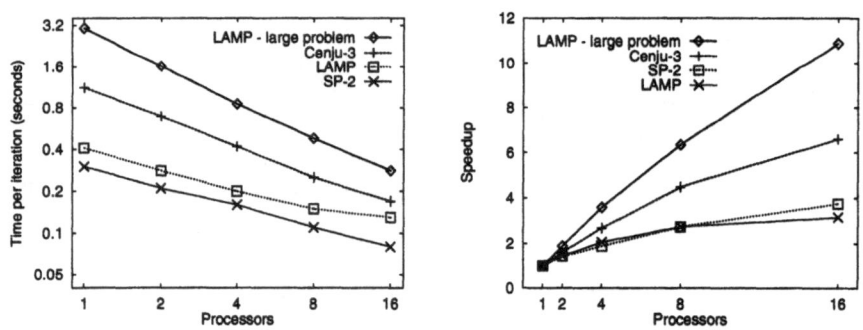

Fig. 3. BEST performance.

5.3 Distributed List Ranking

The last example is a kernel problem in parallel graph algorithms. Given a singly linked list, the *list ranking problem* consists in computing for each list element the number of elements that follow it by a traversal of the links until the last element is reached. It is assumed that the list elements are stored in an array but not necessarily in the order in which they are encountered during traversal of the list. The problem is important in a parallel setting because it makes it possible to perform reduction operations on lists [13].

The sequential problem is trivial, and a careful implementation requires only two scans of the list. The small amount of computation makes it difficult to achieve any speed-up in a parallel program, and thus provides us with another hard test case for evaluation of the LAMP system. We consider the two algorithms discussed in [21]. The list is assumed to be evenly distributed among the processors in a random fashion, with N being the total number of list elements, p the number of processors, and n the number of elements per processor. The first algorithm is the standard pointer jumping algorithm [13]. Depending on the implementation of the all-to-all communication operation, the execution time per processor can be decomposed into $3(p-1)\lceil\log N\rceil$ start-ups, $4n\lceil\log N\rceil$ words communicated, and $O(n\log N)$ local computations (with a very small constant factor). The other, fold-unfold, algorithm works without all-to-all communication. Here the execution time can be decomposed into $2(p-1)$ start-ups per processor, $4n\log p$ words communicated, and $O(n\log p)$ local computations (again with a very small constant factor).

In Figure 4 we give results obtained on an IBM SP2 and the LAMP for lists of length $N = pn$ with n fixed to $1\,000\,000$ elements. The total execution time is compared with the sequential time for the corresponding problem size, optimistically estimated as p times the time to rank the sublists residing at one processor. The two systems show comparable performance. The fold-unfold algorithm performs significantly better than the pointer jumping algorithm, but in terms of achieved speed-up the results are unsatisfactory.

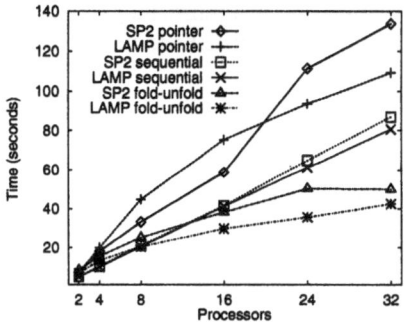

Fig. 4. List ranking: lists with fixed local length $n = 1\,000\,000$.

6   Outlook

The work presented in the paper is an ongoing project. We recently completed a multi-device version of the MPI library using shared memory communications between processes running on the same node. Another important enhancement for the nearest future is to implement an ADI2 device that will replace the current driver for the generic channel device. We also plan to migrate to the new MPICH 1.1.1 version, which integrates the MPI-2 [8] parallel I/O functions. The main objective of the project, however, has been reached already: providing our researches with an efficient and reliable platform for development of complex MPI applications on our PC cluster in a multi-user-environment.

References

[1] A. Basermann, Parallel Preconditioned Solvers for Large Sparse Hermitian Eigenvalue Problems, *Proceedings of the Third International Meeting on Vector and Parallel Processing (VECPAR'98)*, (1998) 31–44.

[2] J. Bernsdorf, F. Durst and M. Schäfer, *Comparison of Cellular Automata and Finite Volume Techniques*, International Journal for Numerical Methods in Fluids **29**, (1999) 251–264.

[3] D.R. Butenhof, *Programming with POSIX Threads*, Addison-Wesley (1997).

[4] A. Chien, S. Pakin, M. Lauria, M. Buchanan, K. Hane, L. Giannini and J. Prusakova, High Performance Virtual Machines (HPVM): Clusters with Supercomputing APIs and Performance, *Proceedings of the 8th SIAM Conference on Parallel Processing for Scientific Computing (PP97)* (1997).

[5] C. Dubnicki, A. Bilas, K. Li and J. Philbin, *Design and Implementation of Virtual Memory-Mapped Communication on Myrinet*, NECI Technical Report (1996).

[6] J. Edler, A. Gottlieb and J. Philbin, The NECI LAMP: What, Why, and How, *Heterogeneous Computing and Multi-Disciplinary Applications, Proceedings of the 8th NEC Research Symposium*.

[7] M. Gołębiewski, M. Baum and R. Hempel, High Performance Implementation of MPI for Myrinet, *ACPC'99*, LNCS 1557, (1999) 510–521.

[8] W. Gropp, S. Huss-Lederman, A. Lumsdaine, E. Lusk, B. Nitzberg, W. Saphir, and M. Snir. *MPI –The Complete Reference*, volume 2, The MPI Extensions. MIT Press, 1998.

[9] W. Gropp and E. Lusk, *A High-Performance, Portable Implementation of the MPI Message Passing Interface Standard*, Parallel Computing, **22(6)**, (1996) 789–828.

[10] W. Gropp and E. Lusk, *MPICH Working Note: The implementation of the second generation MPICH ADI*, Argonne National Laboratory internal report.

[11] D. Gustavson, The Scalable Coherent Interface and related standards projects, *IEEE Micro*, **12(1)** (1992) 10–22.

[12] G. Henley, N. Doss, T. McMahon and A. Skjellum, *BDM: A Multiprotocol Myrinet Control Program and Host Application Programmer Interface*, Technical Report (Mississippi State University, 1997).

[13] Joseph JáJá, *An Introduction to Parallel Algorithms*, Addison-Wesley (1992).

[14] N. Koike, NEC Cenju-3: A Microprocessor-Based Parallel Computer, *Proceedings of the 8th IPPS* (1994) 396–403.

[15] Myricom, *Myrinet Documentation*, http://www.myri.com/scs/documentation

[16] L. Prylli and B. Tourancheau, BIP: new protocol designed for high performance networking on Myrinet, *IPPS/SPDP'98 Workshops*, Lecture Notes in Computer Science **1388** (1998), 472–485.

[17] H. Ritzdorf and R. Hempel, *CLIC - The Communications Library for Industrial Codes*, http://www.gmd.de/SCAI/num/clic/clic.html

[18] H. Ritzdorf, A. Schüller, B. Steckel and K. Stüben, L_iSS - An environment for the parallel multigrid solution of partial differential equations on general 2D domains, *Parallel Computing* **20** (1994) 1559–1570.

[19] G. L. G. Sleijpen and H. A. van der Vorst, A Jacobi-Davidson Iteration Method for Linear Eigenvalue Problems, *SIAM J. Matrix Anal. Appl.* **17** (1996) 401–425.

[20] M. Snir, S. Otto, S. Huss-Lederman, D. Walker, and J. Dongarra. *MPI –The Complete Reference*, volume 1, The MPI Core. MIT Press, second edition, 1998.

[21] J. L. Träff, Portable randomized list ranking on multiprocessors using MPI, *5th European PVM/MPI Users' Group Meeting*, Lecture Notes in Computer Science **1497** (1998) 395–402.

[22] G. Wellein, H. Röder and H. Fehske, Polarons and Bipolarons in Strongly Interacting Electron-Phonon Systems, *Phys. Rev.* **B 53** (1996) 9666–9675.

Using Network of Workstations to Support a Web-Based Visualization Service

Wilfrid Lefer and Jean-Marc Pierson

Laboratoire d'Informatique du Littoral
LIL, B.P. 719
62228 Calais, France
{lefer,pierson}@lil.univ-littoral.fr

Abstract. Nowadays, huge amount of data can be retrieved thanks to the World Wide Web, raising to the need for new services for online data exploration and analysis. In the other hand, scientific visualization offers varying techniques to represent any kind of data visually, taking advantage of the natural skills of the human brain to analyze complex phenomena through visual representations. These techniques are computationally high demanding and can not be handled by a single machine. A Web-based visualization service would rapidly overload the machine and slow down the Web service dramatically. This paper describes *viscWeb*, a distributed visualization architecture, which allows us to use a pool of workstations connected through Internet as a computational resource for a Web-based visualization service. This general architecture could be used to support any kind of Web-based computation service.
Key Words: Distributed Architecture, Web Based Visualization Services, NOW.

Introduction

Nowadays, the World Wide Web is the most powerful facility to access to the largest information space and raises the need for effective visualization systems to speed up the process of searching and analyzing information. The most commonly used way to visualize information gathered on the Web consists in downloading the data onto a local workstation and then applying visualization algorithms to compute images of the data. But this process raises a number of problems. The memory necessary to store all the data could easily exceed the local disk capacity and it could take a long time to download the whole data set. Indeed, the user more likely would like to be able to browse the data for identification of a particular area of interest before deciding to download part of the data for further exploration and analysis.

At LIL we developed *viscWeb* [5], a distributed visualization architecture which allows us to consider a pool of machines connected through Internet as a networked resource to support a Web-based visualization service. This article presents the main changes in the architecture, our scheme of automatic parallelization as well as new results. The outline of the article follows : the first

P. Amestoy et al. (Eds.): Euro-Par'99, LNCS 1685, pp. 624–632, 1999.

part presents some related works, the second gives an overview of the service, the third describes the system architecture while part four details some parts of the implementation. Finally, we purpose response times for the service and we analyze the benefit of the parallelism from the user point of view.

1   Previous Work

Previously published works include distributed architectures for computational steering and Web-based visualization systems. The European project PAGEIN [3] was initiated to set up and evaluate environments for distributed Computer Supported Collaborative Work in a high performance computing and networking context. The project led to the COVISE [14] environment for supporting distributed visualization. Recent extensions to COVISE include a Web-based interface and a distributed VR environment. Shastra [1] is a distributed visualization environment supporting collaborative work. There are other systems especially tuned for computational steering [12] [4].

The concept of a Web-based visualization service has been studied by several authors. The objective is to put visualization on every desk. Typically the user has access to a visualization service using a Web browser, and sends visualization requests by means of HTML (Hyper Text Markup Language) forms. Several scenarios have been investigated, including *thin client*, where the server computes an image of the data and sends it back to the user, and *intelligent client*, where a Java-based visualization program is downloaded on the client site to make the visualization [15][8][13].

2   Overview of viscWeb

With *viscWeb* the user can submit parameterizable requests through HTML forms and receives images or VRML descriptions (Virtual Reality Modeling Language), which are visualized by the browser itself or an external application. *viscWeb* is not handled by a single machine but rather takes advantage of a pool of Unix machines connected by Internet. The features currently supported by *viscWeb* are:

- user sessions: each user is assigned his own user session and at least one visualization server is entirely dedicated to that user,
- dynamic code placement: when a new visualization server has to be started, a machine is elected among the pool of machines available for the service as a function of various parameters, including machine charge and number of currently logged users.
- dynamic code migration: servers can be moved from one machine to another at any time, for instance if a charge imbalance occurs or a regular user logs in the machine.

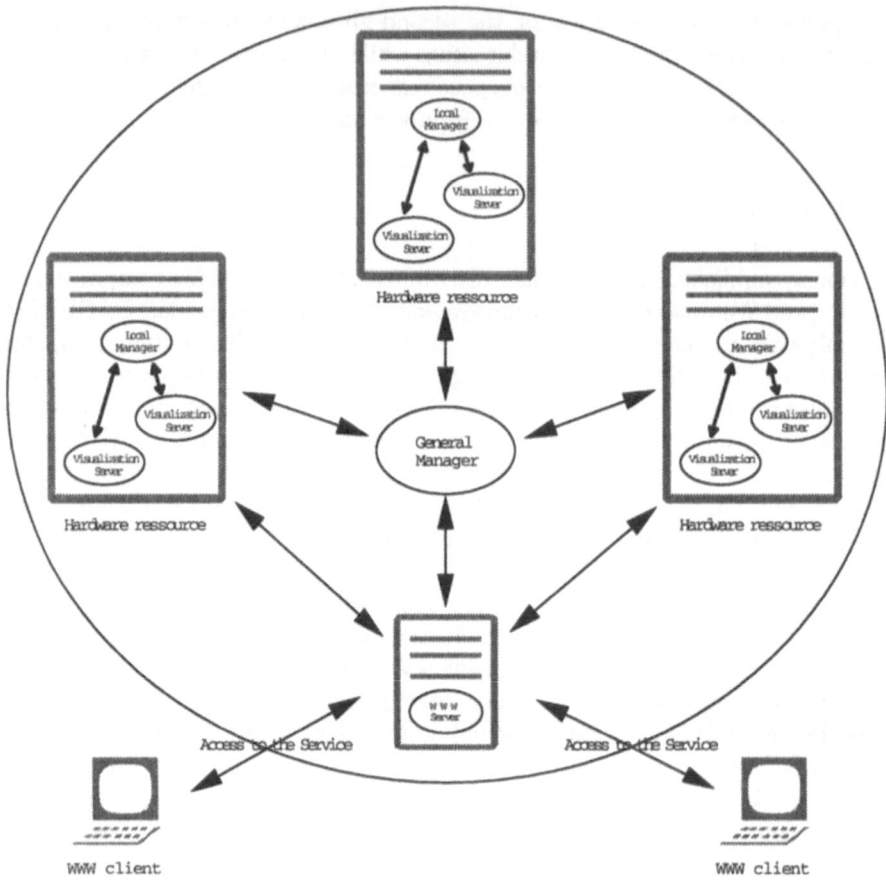

Fig. 1. *viscWeb* architecture.

– automatic parallelization: depending on the number of machines available, several of them can be used to process a visualization request in a parallel scheme.

3   System Architecture

The architecture of the system is shown on figure 1. A number of visualization servers are running to process requests from currently connected users. The distribution of the servers among the machines is managed by a special program called *General Manager*. In addition, on every machine a *Local Manager* is running. Each Local Manager provides the service with the same basic functionalities, including launch a new server, get information about all servers currently running on the machine, stop a server, ... Communications occur directly be-

tween CGI (Common Gateway Interface) programs executed on the machine
supporting the Web service and visualization servers executed on remote machines. The Local Manager communicates only with local visualization servers
and the Global Manager. Local Manager communications are only service messages, for instance to launch a new visualization server or to inform the Global
Manager about the situation on the local machine. All communications in *viscWeb* are based on sockets and TCP (Transport Control Protocol) to ensure
reliable transmissions. The visualization part of *viscWeb* has been written with
VTK [10] (The Visualization ToolKit) because it is freely available on all major
systems as source code.

4   System Features

4.1   User Sessions and Server Placement

When a user connects to the service, a machine is elected and a visualization
server dedicated to that user is launched on this machine. This is achieved by a
RPC call (Remote Procedure Call). This implies that the visualization service
should have been declared on each machine or a least a registered daemon should
run on each machine. Data transmissions in *viscWeb* rely on the XDR format
(eXternal Data Representation), which allows transparent data exchange between binary incompatible systems. Thus *viscWeb* can handle any machine that
runs a Unix system.

The machine chosen to launch a new visualization server is elected as a function of a number of parameters. In the current version we check whether a regular
user session is currently running on the machine and we evaluate the average
workload of the machine. In the future, additional parameters will be taken into
account, such as the bandwidth between the machine and the Web server, the
amount of available memory, ... This visualization server is attached to the user.
Because parameter values are subject to changes over time, we have developed a
mechanism to dynamically move a server when necessary (see section 4.2). The
Global Manager maintains a table which contains, for each visualization server,
all necessary information concerning the link between this server (identity, IP
address, socket connection) and the related user.

In *viscWeb* a new visualization server is not started each time a visualization
request is received by the Web server but rather a new user session is initiated
when a user connects to the service and this session keeps running until the
user disconnects. The advantage of this solution is that it avoids recomputing
all the visualization pipeline each time a parameter is changed, which is the way
visualization pipelines are maintained by traditional softwares such as VTK or
IRIS Explorer. The drawback is that if the connected user does not disconnect,
its environment on the machine running its visualization server is not properly
cleared, and a mechanism to detect zombies servers must be installed.

4.2   Dynamic Code Migration

The parameters used to choose a machine when a new user session is started change over time. For instance a regular user of the machine may log in. In this case it is not possible to complete the task on the same machine. The Local Manager detects the user login and asks the Global Manager for a new location where to move its visualization servers. The algorithm used by the Global Manager is the same as for a new user connection. When new machines have been found to handle the work, the Local Manager stops visualization servers (by means of Unix signals) and gives them their identities. When a new server is launched, it receives from the current server all data and information necessary to complete the task, then the old server terminates.

4.3   Automatic Parallelization

When a visualization request has to be processed, several machines could be available for the task. The final goal being to ensure response times as low as possible, *viscWeb* includes automatic parallelization. The Local Manager chooses a subset of machines that can be used to process the data among the whole pool of available machines, using the same parameters as for a new user connection. In the current version, a data parallel scheme is used to distribute computation and data over available machines. This mechanism is used for surface extraction with the Marching Cubes and volumetric ray-tracing (both available as sequential codes in VTK). These algorithms are easily parallelizable but additional investigations are needed for a general extension to automatic parallelization. Nevertheless, for a good parallelization, the algorithm might take into account such parameters as the network bandwidth, the individual power of the machines, ... in order to deal with heterogeneous load balancing. In the current version, the Local Manager chooses all machines that are not used at the instant of the request. If a second visualization request occurs in the service, a Local Manager is likely to select the same machines since the parameters won't have change much between two consecutive requests.

This problem also will be addressed in future version of viscWeb, where a limited number of machines will be chosen, sparing some of them for a following request. Moreover, some algorithms become inefficient when being too much parallelized, if the data is too partitioned and too many machines used. Many known works point out this issue in purely parallel scopes. Indeed a balance has to be found between increasing the number of machines and the computational power, and increasing the control, hence the number of messages in the network.

4.4   Portability to Other Systems

Our software is fully written in C++, it is based on sockets, TCP, XDR and RPC calls for communications, and VTK for the graphic stuff. Because all these technologies are Unix standards, the portability of viscWeb to any Unix system is straightforward. Currently we have machines running under SGI IRIX and

Linux. The portability toward non-Unix worlds, such as Windows, more likely involves using Java. In fact, since C++ and VTK are available on Windows, only the communication part should be rewritten in Java, for which we plan to use Java RMI (Remote Method Invocation), and the native code in C++ be reused thanks to the JNI (Java Native Interface). This is part of the on-going work concerning this project.

The client part of the software has been tested successfully on IE4.x and Netscape4.x, what is an expected result since it is web pages, forms, and CGI communication based on standard HTTP (Hyper Text Transfer Protocol). We plan to rewrite the client part with Java to make te interface more interactive, which would allow us for instance to validate the parameters before the form be send to the server.

5   Results - Timings

One of the main issues of Web-based visualization concerns the time necessary for the user to get the results. Important factors which affect significantly the response time are the network bandwidth, the server charge and the computational requirements of the visualization. In order to decrease the response time, it is necessary to consider all these three aspects. It is not in the scope of this article to deal with algorithmic issues of scientific visualization and indeed in *viscWeb* we did not develop any specific visualization method. We fully rely on the VTK toolkit, using well known algorithmic approach to parallelize our code.

The network problem is mainly a hardware issue. Today high bandwidth networks such as the 622Mbps MBone offer a suitable solution for Web-based visualization services, assuming no real-time interaction is required. VRML viewers provides us with some kind of real-time interaction but interaction is limited in the navigation through pre-computed scenes, that is only the rendering phase is interactive, the visualization phase remaining static. A suitable approach to enhance interactivity is progressive downloading of hierarchical structures [2] (either images or 3D descriptions), although VRML 2.0 does not support progressive downloading and visualization. One of the goals of this work is to propose a distributed architecture which allows us to take advantage of a pool of machines to decrease the response time of the system.

In order to evaluate the effectiveness of our system, we had made two series of measures for two different requests. A SPECT (Single Positron Emission Computed Tomography) volume of the brain of resolution $64 \times 64 \times 10$ is visualized. This technique is used to detect disfunctionalities in the brain by imaging the blood flow.

The Marching Cubes [7] algorithm is used to extract an iso-surface from the volume which produces 14,000 triangles. Two kind of results had been produced : an image of the volume of resolution 500×500 (image of size 10 kbytes), and a VRML description of size 1 mega-bytes. Timings had shown, for a sequential scheme, two main properties : the time to compute the image is highly prominent as compared to the times to send the HTML form to the Web server, to send

	T1[a]	T2[b]	T3[c]	T4[d]	T5[e]	T6[f]
Req 1 (1 host)	1,213.8	432.9	102,341.4	139.98	36.26	104,164.47
Std deviation	7.21%	5.86%	6.45%	61.52%	16.79%	6.41%
% of total	1.16%	0.41%	98.24%	1.34%	0.03%	100%
Req 2 (2 hosts)	1,322.08	574.79	81,087.19	157.32	40.30	83,181.71
Std deviation	22.76%	43.21%	13.58%	66.65%	16.83%	13.18%
% of total	1.58%	0.69%	97.48%	0.18%	0.04%	100%
Req 3 (4 hosts)	1,120.08	480.79	73,491.76	76.80	28.19	75,197.64
Std deviation	7.82%	9.88%	3.11%	67.28%	83.12%	2.92%
% of total	1.48%	0.63%	97.73%	0.1%	0.03%	100%

[a] Time spent by the CGI program to analyze the request.
[b] Event transmission from the CGI program to the visualization server.
[c] Visualization time.
[d] Result transmission from the visualization server to the CGI program.
[e] Result transmission from the CGI program to the Web server.
[f] Total time necessary to process the visualization request.

Table 1. Impact of parallelism in *viscWeb*. Times are in milliseconds.

the data to the visualization server or to send back the result to the client (from 86% for the VRML description up to 99 % for the GIF image. Please refer to [6] for further details.

We decided obviously to figure out the benefit obtained using a parallel scheme compared to the previous sequential one. We used a bigger data set of resolution $64 \times 64 \times 64$, which produces up to 70,000 triangles. Three requests have been executed with 1, 2 and 4 machines to compute these triangles. In this scheme, only the Marching Cube part of the algorithm is parallelized, where each machine handles the same number of voxels (this will have to be improved since it might lead to an unbalanced system where each machine compute different number of resulting triangles).

Table 1 details some of the results obtained for each request. We used the workstations of our lab to test the system: SGI R4600PC with 32Mbytes for the Web server and SGI R5000SC with 256Mbytes for the visualization server with the sequential scheme, and up to four SGI R4600PC with 32Mbytes for the parallel one. Since the charge of our LAN and the workstations change continuously each request has been executed five times and the given values are the average values. Requests 1, 2 and 3 outline timings using respectively 1, 2 and 4 machines, associated to compute the final result.

The values in rows 2, 5, 8, 11 and 14 give the standard deviation and rows 3, 6, 9, 12, 15 give the percentage of the total time spent in each stage. A first remark concerns the high standard deviation which is due to the fact that machine and network charges change over time. The network also is shared between the members of the laboratory : we did the tests during the day, when others were using the network, leading to high unbalance in the network related times of the result table. We did not want to have especially artificial good results, which could have been achieved using the visualization server during the night, since we wanted to test its behavior over a realistically loaded network.

We can easily see that most of the time is spent in the visualization time, with 1, 2 or 4 machines. The speedup we obtain for 2 processors is about 1.25 and reach 1.4 with 4 processors. The total time to compute the result is lowered by about 30 seconds, which is a great improvement for the user connected to the service. The results could be even better if the final stage of the pipeline visualization (rendering of the produced triangles) was also parallelized.

6   Conclusion and Future Work

viscWeb allows us to enhance the quality of a Web-based visualization service by increasing the computational power used to support the service. Statistical studies have been made about the usage of workstations in a LAN, which has shown that the CPU usage is generally far below average [9][11]. The goal of this version of *viscWeb* is to use the available power in a LAN to support a Web-based visualization service. The next step is to propose a flexible Internet-wise environment to provide visualization services to end-users with a basic equipment. Depending on the hardware and software requirements of a user request, the visualization pipeline will have to be distributed over the global Internet. The placement of the visualization servers will depend on the location of the data, on the availability of the software components necessary to process the request, on the location of the computational resources, on the network bandwidth, ... and we should include the possibility for the user to handle part of the pipeline on its own machine. Moreover such a flexible environment should be perfectly interfaced with DBMS (Data Bases Management System). This involves taking into account all recent standards and technologies for data and code exchange over heterogeneous platforms at the Internet level, such as Java and Corba.

References

[1] V. Anupam and C. Bajaj. Shastra: an architecture for development of collaborative applications. *International Journal of Intelligent and Cooperative Information Systems*, pages 155–172, jul 1994.

[2] Andrew Certain, Jovan Popoviè, Tony DeRose, Tom Duchamp, David Salesin, and Werner Stuetzle. Interactive multiresolution surface viewing. *Computer Graphics*, pages 91–98, aug 1996. Proc. of SIGGRAPH '96.

[3] Michel Grave. PAGEIN: Pilot application in a gigabit european integrated network. Technical Report R2031/ONE/DI/034/b1, ONERA, Châtillon, France, feb 1996.

[4] J.J. Hare, J.A. Clarke, and C.E. Schmitt. The distributed interactive computing environment. Proc. of the Workshop on Distributed Visualization Systems, Research Triangle Park, NC, oct 1998.

[5] Wilfrid Lefer. A distributed architecture for a web-based visualization service, apr 1998. Proc. of Nineth Eurographics Workshop on Visualization in Scientific Computing, Blaubeuren, Germany.

[6] Wilfrid Lefer and Jean-Marc Pierson. A thin client architecture for data visualization on the world wide web. In *International Conference on Visual Computing (ICVC99), Goa, India*, feb 1999.

[7] William E. Lorensen and H. E. Cline. Marching cubes: A high resolution 3D surface construction algorithm. *Computer Graphics*, 21(3):163–169, jul 1987. Proc. of SIGGRAPH '87.

[8] Cherilyn Michaels and Michael Bailey. Vizwiz: A java applet for interactive 3D scientific visualization on the web. In Roni Yagel and Hans Hagen, editors, *Proc. of IEEE Visualization '97*, pages 261–267. IEEE Press, Los Alamitos, CA, oct 1997.

[9] D. A. Nichols. Using idle workstations in a shared computing environment. *ACM Operating System Review*, 21(5):5–12, nov 1987.

[10] William Schroeder, Kenneth Martin, and Bill Lorenzen. *The Visualization Toolkit - An Object-Oriented Approach to 3D Graphics*. Prentice Hall, 1996.

[11] M. M. Theimer and K. A. Lantz. Finding idle machines in a workstation-based distributed environment system. *IEEE Transactions on Software Engineering*, 15(11):1444–1458, nov 1989.

[12] Robert van Liere, J.A. Harkes, and W. C. de Leeuv. A distributed blackboard architecture for interactive data visualization. In *Proc. of IEEE Visualization '98, Research Triangle Park, North Carolina, Oct 19-24, 1998*. IEEE Press, Los Alamitos, CA, oct 1998.

[13] Kiril Vidimce, Viktor Miladinov, and David C. Banks. Simulation and visualization in a browser. Proc. of the Workshop on Distributed Visualization Systems, Research Triangle Park, NC, oct 1998.

[14] A. Wierse, U. Lang, and R. Rühle. A system architeture for data-oriented visualization. In John P. Lee and Georges G. Grinstein, editors, *Database Issues for Data Visualization*, number 871 in Lecture Notes in Computer Science, pages 148–159. Springer-Wien-New-York, 1993. Proc. of IEEE Visualization '93 Workshop, San Jose, California.

[15] Jason Wood, Ken Brodlie, and Helen Wright. Visualization over the world wide web and its application to environmental data. In Roni Yagel and Gregory M. Nielson, editors, *Proc. of Visualization '96*, pages 81–86. IEEE Press, Los Alamitos, CA, oct 1996.

High-Speed LANs: New Environments for Parallel and Distributed Applications

Patrick Geoffray   Laurent Lefèvre   CongDuc Pham   Loïc Prylli[†]
Olivier Reymann[†]   Bernard Tourancheau   Roland Westrelin

RHDAC, Université Lyon 1, Claude Bernard
[†]LIP, Ecole Normale Supérieure de Lyon
e-mail: `bip-team@lhpca.univ-lyon1.fr`

Abstract. As the technology for high-speed networks has incredibly evolved this last decade, the interconnection of workstations at gigabits rates and low prices has become a reality. These clusters, based on regulars workstations (e.g. PCs), can now be used in place of traditional parallel computers with no possible comparison on the prices! In this article, 3 applications (high performance computing, distributed shared memory system and parallel simulation) that were traditionally executed on expensive parallel machines are ported on a Myrinet-based cluster of PCs. The results show that the performances of these new architectures can be very close to those obtained on state-of-the art parallel computers.

1 New technologies for parallel applications

For a long time parallel computers were the solution for people with high computation needs. If the processing units of such massively parallel processors can now be taken from the commodity market, the interconnection networks and the software are still highly customized. While sequential computers have always seen a dramatic cut down in their prices every year, parallel computers took the opposite direction because of the decreasing demand.

However, there are not so many solutions for having more computation power: one has to use several processors. The choice resides on how to make the parallel system: from custom or standard products. If the first choice was before justified by the low-quality of products from the commodity market, this is not the case any more. There is the possibility to go a complete step farther by building parallel systems entirely from standard components. Processing units are just regular workstations or PCs that can be bought in any supermarket for a few thousands of dollars, interconnection networks are also taken from the high-speed regular market of LANs (local area network) such as Gigabits Ethernet, ATM, and Myrinet. These architectures are often referred to as Network Of Workstations (NOW). Several research teams have launched projects dealing with NOWs used as parallel machines. Early experiments with IP-based implementations have shown disappointing performances so the goal of most of these teams is to design the software needed to make clusters built with commodity components and high-speed networks really efficient. The NOW project of UC

P. Amestoy et al. (Eds.): Euro-Par'99, LNCS 1685, pp. 633-642, 1999.
© Springer-Verlag Berlin Heidelberg 1999

Berkeley [1] was one of the first one. Its main contribution is the Active Messages [12] layer that provides high performance access to the network. In this sense, it is very similar to the BIP software we are developing on our cluster.

The objective of the paper is two-folds: (*i*) to show experiments on different types of LANs with several communication softwares and (*ii*) to compare our Myrinet-based test-bed with the BIP software to traditional parallel computers. To do so, 3 applications covering very different areas of distributed and parallel computing are ported on NOWs with a focus on Myrinet-based networks. The paper is organized as follows: Section 2 presents the high-speed environment based on a Myrinet network and the BIP communication software. Section 3 compares state-of-the-art parallel computers and our Myrinet test-bed with the NAS benchmarks. Section 4, 5 and 6 presents the applications (high performance computing, distributed shared memory system and parallel simulations) that were ported on Myrinet. Conclusions are given in Section 7.

2   The high-speed LAN environment based on Myrinet

The Myricom LAN[3] target was chosen for its performance over the Gbits/s, its affordable price and its software openness (all the software and specifications are freely available for customers). There are several features that make this kind of technology much more suitable than a traditional commodity network:

- the hardware provides an end-to-end control flow that guarantees a reliable delivery and alleviates the problem of implementing in software the reliability on top on an lossy channel. As message losses are exceptional, it is possible to use algorithms that focus on very low overheads in the normal case,
- the interface card has a general purpose processor which is powerful enough to handle most of the communication activity without interrupting the main processor. In particular, it provides an efficient overlapping of communication and computation.
- the interface card has up to one megabyte memory for buffers. As the network being as fast as the computer bus, this memory isolates transfers on the network from transfer on the bus.

2.1   BIP

BIP stands for Basic Interface for Parallelism. The idea was to build it with a library interface accessible from applications that will implement a high speed protocol on the Myrinet network. BIP provides only a protocol with low level functionalities. Although specialized parallel applications could interface directly with it, it is intended as the base of other protocol layers like IP, and higher level APIs like the well established MPI and PVM. All the applications described later on in this paper actually use our MPI implementation based on top of BIP. Highly optimized, the raw communication performance for BIP is about $5\mu s$ latency one-way. With large messages, the bandwidth goes up to 125MByte/s. The

MPI layer adds about $5\mu s$ for small messages. As will show the other examples in the paper, this allows applications to scale more by allowing the use of a finer grain of computation. BIP is described in more details in [11].

2.2 Experimentation platform

Our test-bed for the experiments presented in the next sections consists of 8 nodes, each with an Intel Pentium Pro 200 MHz, 64MBytes of RAM, a 440FX chipsets and a myrinet board with LANai 4.1 and 256Kbytes of memory. The test-bed is provided by the LHPC (Laboratoire pour les Hautes Performances en Calcul), a cooperation between ENS-Lyon and Matra Système Information.

2.3 Running the NAS benchmarks

The NAS parallel benchmarks are a set of parallel programs designed to compare the performance of supercomputers. Each program try to test a given aspect of parallel computation that could be found in real applications. These benchmarks are provided both as single processor versions and parallel codes using MPI. We selected 3 different benchmarks and compiled them with MPI-BIP. They are then run on 1, 4 and 8 nodes of our Myrinet cluster described previously. Table 1 gives our measurements and comparisons with several parallel computers. Data for parallel computers come from http://science.nas.nasa.gov/Software/NPB.

IS (Integer Sort) sorts 8388608 keys distributed on the processors. The test uses a lot of small message communications and needs few processing power.

LU solves a finite difference discretization of the 3D compressible Navier-Stokes equations. A 2-D partitioning of the $64 \times 64 \times 64$ data grid is done. Communication of partition boundary data occurs after completion of computation on all diagonals that contact an adjacent partition. We have a relatively large number of small communications of 5 words each.

SP solves three sets of uncoupled systems of equations of size $64 \times 64 \times 64$, first in the x, then in the y, and finally in the z direction. This algorithm uses a multi-partition scheme. The granularity of communications is kept large and a few messages are sent.

The speedup values summarized in table 1 show the very good performances of MPI-BIP and Myrinet. Super-linear speedups in our case can be explained by the large cache size and the super-scalar architecture of the PPro. Reported super-linear speedups for the Sun Enterprise show that a super-linear speedup is no exception. The favorable super-linear speedup for the LU test may be related to the fact that this test has been compiled with f2c that apparently produce a slower code than a commercial Fortran software. The behavior of MPI-BIP is excellent both for small messages and large messages. The speedup for the IS test with MPI-BIP is always better than those obtained on parallel machines. However, when a lot of computational power is needed (SP), a gap appears between traditional parallel machines and our cluster: the weak floating-point unit of the PPro 200 shows its limits!

	MPI-BIP on PPro 200		IBM SP (66/WN)		Cray T3E-900		SGI Origin 2000-195		Sun Enterprise 4000	
	Mop/s	Speedup	Mop/s	Speedup	Mop/s	Speedup	Mop/s	Speedup	Mop/s	Speedup
IS @ 4 proc	2.44	4.5	2.2	3.1	3.2	N/A	2.1	3.8	1.6	3.8
IS @ 8 proc	2.11	8.8	2.0	5.6	3.8	N/A	2.4	8.6	1.5	6.9
LU @ 4 proc	23.68	6	59.0	3.7	67.6	N/A	96.8	N/A	36.9	4.1
LU @ 8 proc	22.73	11.6	57.2	7.1	66.4	N/A	103.3	N/A	37.3	8.4
SP @ 4 proc	10.10	3.4	42.1	3.6	43.0	N/A	60.3	N/A	25.0	3.7

Table 1. Performance for NAS Benchmarks on various platforms.

3   High performance computing: the Thesee application

Thesee is a 3D panel method code, which calculates the characteristic of a wing in an inviscid, incompressible, irrotational, and steady airflow, in order to design new paragliders and sails. Starting from a sequential version [6], *Thesee* has been parallelized using the ScaLAPACK[2] library routines to be run on NOWs. The parallelization process has been done in a systematic manner to keep the development cost low. 3 parts have been identified:

- **Part P1** is the fill-in of the element influence matrix from the 3D mesh. Its complexity is $O(n^2)$ with n the mesh size (number of nodes). Each matrix element gives the contribution of a double layer (source + vortex) singularity distribution on facet i at the center of facet j.
- **Part P2** is the LU decomposition of the element influence matrix and the resolution of the associated linear system ($O(n^3)$), in order to calculate the strength of each element singularity distribution.
- **Part P3** is the speed field computation. Its complexity is $O(n^2)$ because the contribution of every node has to be taken into account for the speed calculation at each node. Pressure is obtained using Bernoulli equations.

Each of these parts are parallelized independently and are linked together by the redistribution of the matrix data. For each part, the data distribution is chosen so as to insure the best possible efficiency of the parallel computation. The ScaLAPACK library uses a block cyclic data distribution on a virtual grid of processors. This solution provides a good load-balance because the processors receive matrix elements from different locations of the original matrix (as opposed to a classical full block decomposition). Communication overheads are reduced to a minimum by preserving the row and the column shape of the matrix (most of communication of 1D arrays can then happen without a complex index computation). Tests are run beforehand to choose the best grid shape and the best block size of the data distribution for our problem on each of the platforms. In the parallel version of *Thesee*, we used the following parameters for the data distribution of the LU factorization: the block size is 32 × 32 and the processor grid shape is a 1D-grid that gives the best overall computation timings.

3.1   Performance results

We ran tests with 3 different interconnection networks on 2 different platforms: (i) Ethernet and ATM network of SUN Sparc5 85MHz with Solaris and (ii) Ethernet and Myrinet network of Pentium-Pro 200MHz under Linux. On both systems, PVM and MPI were used. For IP over Myrinet, the LAM implementation of MPI is used. Otherwise it is the MPI-BIP user level implementation based on MPICH. The overhead for MPI or PVM is similar regarding the small set of primitives involved.

The efficiency of the fill-in of the influence matrix (part 1) and the speed field computation (part 3) is roughly the same on each configuration. The code for these parts is "embarrassingly" well suited for parallel execution and thus the speed-up is almost linear. On the other hand, the LU factorization that involves a lot more communications is highly dependent on the network software and hardware performances.

For space consideration, only results for Ethernet vs. Myrinet are shown in the paper. For Ethernet and ATM, the test shows that high-speed networks designed for long-distance communication are not well-suited to system-oriented communication. Although ATM provides more throughput, the gain obtained is small because the initialization time for each communication on this network is similar to the one on Ethernet. Unfortunately, this initialization time represents the larger part of the communication delay.

| | | Timings (seconds) [speedups] | | |
	System size	PVM/IP/Ethernet (on SUN Sparc)	PVM/IP/Myrinet (on Pentium Pro)	MPI/BIP/Myrinet (on Pentium Pro)
Sequential	902 × 902	10.0		
2 Proc	902 × 902	9.6 [1.03]	6.7 [1.49]	5.0 [1.97]
4 Proc	902 × 902	10.2 [0.97]	4.7 [2.10]	3.1 [3.24]
Sequential	1722 × 1722	87.4		
2 Proc	1722 × 1722	56.4 [1.54]	44.3 [1.97]	38.1 **[2.29]**
4 Proc	1722 × 1722	45.9 [1.90]	28.1 [3.10]	21.3 **[4.09]**

Table 2. Timings and speedups with Myrinet (using the best block size).

Comparison between Ethernet and Myrinet Table 2 presents the timings results obtained on the Myrinet test-bed along with the BIP software. Since the communication/computation ratio is rather high, we expect better performances than those obtained with ATM. The best results are achieved with MPI-BIP and we can see that IP-based implementation can not fully exploit the low latencies of the Myrinet hardware. With MPI-BIP, the low latency for a basic send communication ($9\mu s$ for this case) has an incredible impact on the speedup when compared to the Ethernet run. Super-linear speedup with 4 processors can be explained by a better cache hit ratio in the parallel version of the code. As the matrix is distributed cyclically on the processors, the computation occurs on blocked data that fits better in the cache during the LU decomposition, leading to a better use of the processor's pipeline units. One more reason is an

increase in the overlapping of computation over communications in the parallel
LU decomposition since the communication cost is greatly reduced with the
BIP/Myrinet platform.

4 Distributed Shared Memory System (DSM)

The purpose of DSM is to implement, on top of a distributed memory architecture, a programming model allowing a transparent manipulation of virtually
shared data. Thus, in practice, a DSM system has to handle all the communications and to maintain the shared data coherence. We have developed an object-
based DSM system called DOSMOS (Distributed Objects Shared MemOry
System) that allows processes to share in a transparent way a set of objects
distributed and replicated over distant processors. DOSMOS integrates novel
features:

- **DOSMOS Processes**: a DOSMOS application is composed of two types of
  processes: **Application processes (A.P.)** contain and execute application
  code; **Memory processes (M.P.)** manage the whole DSM system, i.e. they
  provide A.P. with the objects they request and maintain data coherence.
- **Array allocation**: DOSMOS allows to manipulate both basic type variables
  (integer, float, char...) and distributed arrays which can be split into several
  "system objects", replicated among the processors. Various splittings are
  provided: by row, by column, by block and by cyclic block.
- **Weak consistency protocols**: for efficiency and scalability purposes, DOS-
  MOS gives the opportunity to duplicate shared objects. These replicas have
  to be kept coherent. DOSMOS implements a weak protocol: the release consistency which provides two synchronization operators: *acquire* and *release*.
- **Hierarchical structuring of the application processes**: processes can
  be grouped into groups and sub-groups in order to optimize the management
  of the data coherence.

Previously developed on top of PVM and experimented on Ethernet networks [8], DOSMOS is now available on top of MPI and experimented on MPI-
BIP on top of Myrinet networks.

4.1 Gram-Schmidt Application

We based our experiments on the Gram-Schmidt application which has been
completely studied on Ethernet in [4]. We propose four parallel implementations
of the Gram-Schmidt application (Fig. 1) :

- Trivial version: with no splitting of matrix and use of strong consistency
  protocols. Accesses to shared object are sequentialized.
- Release Consistency and Object Splitting: this version takes the benefit of
  splitting the large matrix in small adapted shared objects.

- Synchronization 1: also based on Release Consistency protocols and matrix splitting but the matrix object is also used for synchronization of processes.
- Synchronization 2: this optimized version requires an independent object to synchronize processes.

4.2 Experiments

Figure 1 demonstrates the benefit of using a high-speed LAN for a DSM system like DOSMOS. Improvements added to the original algorithm are not linked with the application performances. Due to low latencies provided by the Myrinet network combined with MPI-BIP, applications with small improvements provide better speedups than highly optimized versions (synchro2). However, fast communications are not enough for trivial applications (*trivial version* in Fig. 1). Figure 2 shows the scalability provided by DOSMOS on top of MPI-BIP. By increasing the problem size, we also increase the efficiency of applications which benefit from large bandwidth provided by the Myrinet network.

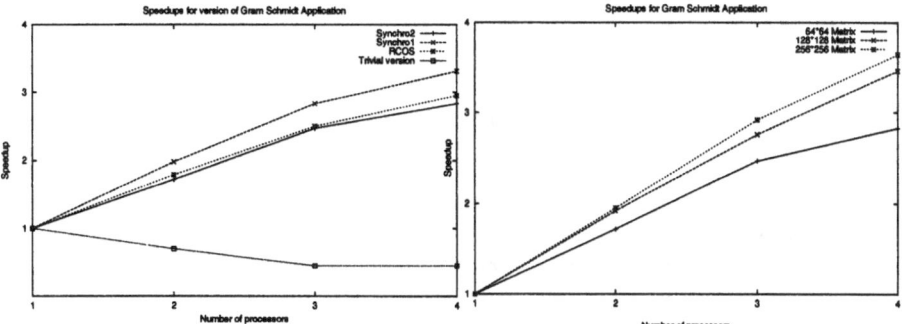

Fig. 1. Various improvements of Gram-Schmidt.

Fig. 2. Speedup of Gram-Schmidt by increasing the problem size.

DSM systems require low-latency networks to provide high performance to applications. Figure 3 shows that DOSMOS can provide high performance distributed objects when combined to low-latency communication protocols like MPI-BIP (the sequential time is computed by using one DOSMOS AP and one MP). However, management of distributed objects (consistency, synchronization...) adds a latency to each distant access combined with the latency of the network (Figure 4). Distant operations concern A.P. which access a shared object managed by their M.P. while Long Distant operations involve one A.P. and two M.P. to access distant objects. The access to a distant object requires a ping-pong communication between two nodes (which takes around $25\mu s$ with MPI-BIP). An additional latency is added by DOSMOS to manage complex objects (like arrays) to provide splitting and a transparent access to distributed arrays. As can be seen, the latency added by DOSMOS is kept reasonable and still compatible with the need of high performances by the applications.

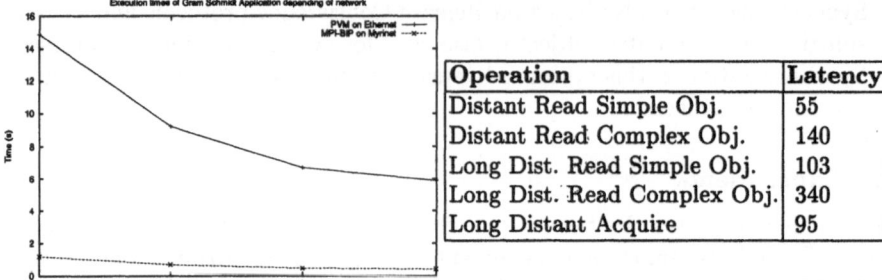

Operation	Latency
Distant Read Simple Obj.	55
Distant Read Complex Obj.	140
Long Dist. Read Simple Obj.	103
Long Dist. Read Complex Obj.	340
Long Distant Acquire	95

Fig. 4. Latency in μs added by DOSMOS to shared object access.

Fig. 3. Gram-Schmidt with Ethernet and Myrinet Networks.

By using a DSM model, programming a parallel application is made easier but the added overheads usually penalizes the users. These experiments on a high-speed LAN show that designing a parallel application with a DSM system like DOSMOS can merge together ease of programming with high performances.

5   Parallel simulation of communication networks

With the emerging of Asynchronous Transfer Mode (ATM) technology, the performance evaluation of future large scale B-ISDN networks based on ATM is a great challenge. Parallel simulation techniques have been proposed during the last 2 decades to reduce the simulation time. These methods fall in two categories: the conservative approach [5] and the optimistic approach [7].

The parallelization process of the network model consists in spatially partitioning the entire network in n regions and to assign each region to a processor. Cells transfer from one region to another are represented by timestamped messages exchanged between the processors. The messages synchronization is performed conservatively. The system under study includes a routing algorithm with the introduction of link cost and dynamic routing functions that provide load-balancing routing features. The simulation of a routing algorithm implies to simulate (i) the mechanism that consists in constructing and updating the routing tables and (ii) the flow of cells that is handled by the network [10].

From the parallel simulation perspective, this application presents a very small granularity and requires a lot of small messages to be exchanged between processors (as opposed to more traditional high-performance computing applications where the computation part is larger). These messages are either model messages, e.g. cell exchanges between ATM switches, or synchronization messages, e.g. null-messages from the conservative kernel. Therefore the application typically requires very low latencies from the communication system. Previous tests on LANs such as Ethernet were quite disappointing and so far only parallel computers were capable of showing interesting speedups.

5.1   Experimental results

In order to compare parallel computers and NOWs, experimental results on a Cray T3E and a Myrinet cluster are presented. The Cray is provided by the *Institut du Développement et des Ressources en Informatique Scientifique* (IDRIS) and consists of 256 processing nodes interconnected by a very low latency 3D torus. Each node on the Cray is a DEC Alpha EV5 with 128 Mo memory running at 300Mhz. The communication libraries are the native SHMEM (SHared MEMory) and MPI. The first one provides a latency of approximately $7\mu s$ while MPI shows a latency of $13\mu s$. The experimental NOW consists of Pentium Pro 200MHz interconnected by the Myrinet network described previously. The communication stack we used is native BIP and MPI-BIP. The latency is about $10\mu s$ with BIP and $17\mu s$ with MPI-BIP (for the message size of our application, 70 bytes). Table 3 summarizes the results for a 78-switch network model. The simulation time has been set to 500,000 time slots that represent 0.31s of the real system. More than 50 millions of events are simulated.

np.	time SHMEM (s)	speedup SHMEM	time MPI (s)	speedup MPI
1	488	-	488	-
4	131	3.72	253	2.07
8	72	6.77	133	3.66
np.	time BIP (s)	speedup BIP	time MPI (s)	speedup MPI
1	321	-	321	-
4	139	2.30	210	1.52
8	59	5.44	88	3.64

Table 3. Comparison between Cray T3E and Myrinet cluster.

As can be seen, the sequential version run faster on a Pentium Pro 200MHz than on a DEC Alpha EV5 300MHz. The main advantage of using standard products is the availability of the most recent processors. As we stated before, parallel computers were interesting because they typically use customized high performance interconnection networks. The low latency on the Cray has a direct impact on the performances of the simulator. Using BIP on Myrinet exploits the small latency provided by the network and therefore the parallel simulation shows very interesting speedups compared to the cost of the hardware used. We have no doubt that this kind of architecture is very promising because of its excellent performance/price ratio.

6   Conclusions

Networks of workstations represent a serious alternative to expensive parallel computers. In this paper, we mainly focus on a Myrinet-based cluster with the BIP software for optimized low-level communications. Although the technology has been ready for some time, it is still used only in a confidential manner outside of the high-speed network research community. Also, even if a lot of people use clusters, a vast majority of them stay with the traditional IP-based implementations of PVM or MPI. Experiments have shown that if the hardware technology has the potential for high performance communications, the communication software layer must be well designed to fully deliver the maximum of performances to the application. One of the principal aims of this paper is to show through a number of applications the maturity of the more recent technologies, and to provide data to help them widespread to the end-users. Generalizing the use of the faster systems available will provide to the end-user community a way to reach a higher level of scalability, and to make parallel solutions usable for a wider range of applications, especially those that require a fine grain decomposition.

References

1. Anderson, T. E., Culler, D. E., Patterson, D. A., the NOW Team: The Case for Networks of Workstations. IEEE Micro Magazine, February 1995.
2. Blackford, Choi, Cleary, d'Azevedo, Demmel, Dhillon, Dongarra, Hammarling, Henry, Petitet, Stanley, Walker, and Whaley: ScaLAPACK Users' Guide. SIAM, 1997. http://www.netlib.org/scalapack/.
3. Boden, Cohen, Feldermann, Kulwik, Seitz, Seizovic, and Su. MYRINET: A Gigabit per second Local Area Network. *IEEE-Micro*, 15:29–36, February 1995.
4. Brunie, L., Restivo, N., Reymann, O.: Programmation d'applications parallèles sur systèmes à mémoire distribuée partagée et expérimentations sur réseaux hautes performances. In *Calculateurs Parallèles*, 9(4), pages 417–433.
5. Chandy, K., Misra, J.: Distribution Simulation: A Case Study in Design and Verification of Distributed Programs. Trans. on Soft. Eng., 5(5) 440–452.
6. Giraudeau L., Perrot G., Petit S., Tourancheau B.: 3-d air flow simulation software for paragliders. TR 96-35, LIP-ENS Lyon, 69364 Lyon, France, 1996.
7. Jefferson, D. R. : Virtual Time. ACM Trans. on Prog. Lang. and Sys., 7(3) (July 1985) 405–425.
8. Lefèvre, L.: Parallel programming on top of DSM Systems : An Experimental Study. In *Parallel Computing - Environments and Tools for Parallel Scientific Computing III*, 23(1-2), pages 235–249. April 1997.
9. Mainwaring,   A.   M.,   Culler, D. E.: Active messages: Organization and applications programming interface. http://now.cs.berkeley.edu/Papers/Papers/am-spec.ps, 1995.
10. Pham, C. D., Brunst, H., Fdida, S.: Conservative Simulation of Load-Balanced Routing in a Large ATM Network Model. In Proceedings of PADS'98, May 26-29 1998, Banff, Canada, pp142-149.
11. Prylli, L., Tourancheau B,: BIP: a new protocol designed for high performance networking on myrinet. In *Parallel and Distributed Processing, IPPS/SPDP'98*. bf 1388 of *LNCS*, pages 472–485. Springer-Verlag, April 1998.
12. von Eicken, T., Culler, D. E., Goldstein S. C., Schauser, K. E.: Active Messages: a Mechanism for Integrated Communication and Computation. Proc. of the 19th Int'l Symp. on Comp. Architecture, May 1992.

Consequences of Modern Hardware Design for Numerical Simulations and Their Realization in FEAST

Ch. Becker, S. Kilian, S.Turek, and the FEAST Group

Institut für Angewandte Mathematik, Universität Heidelberg, Germany
telephone: +49 6221 545446 fax: +49 6221 545634
featflow@gaia.iwr.uni-heidelberg.de
http://www.iwr.uni-heidelberg.de/featflow

Abstract. This paper deals with the influence of hardware aspects of modern computer architectures to the design of software for numerical simulations. We present performance tests for various tasks arising in FEM computations and show under which circumstances performance loss can occur. Further we give key ideas for our new software package FEAST, which is especially designed for high performance computations.

1   Introduction

The situation in Scientific Computing is mainly influenced by following topics. Performing realistic calculations for instance in the field of Computational Fluid Dynamics requires still lots of computing resources like CPU–time and storage memory. On the other hand in the last years huge improvements in the hardware sector have taken place. The processors have become faster and faster and the memories bigger and bigger. This development will be continued in the future.

In about ten years one today so called "PC" will be so powerful as a nowadays complete Cray T3E. An important fact is that the memory access time will not shrink as fast as the processing speed grows. The gap between these main performance indicators will grow in the future! Therefore the performance potential is available, but are the nowadays scientific codes able to benefit from this potential?

On the other hand only to look at the implementation side is not enough. The best hardware adapted code cannot produce satisfying results, if the underlying numerical and algorithmic parts are not optimal. This has been shown by several benchmark computations like the DFG–Benchmark "Flow around a cylinder" ([6]). Improvements of the discretization and error control, which lead to less unknowns, are also necessary as efficient and robust solver strategies to achieve better convergence rates.

In the following we will examine the performance behavior of several numerical linear algebra routines for the solution of stiffness matrices arising in FEM computations. These routines are often the most time consuming part in the simulations. We will show that there are many traps to loose performance.

P. Amestoy et al. (Eds.): Euro-Par'99, LNCS 1685, pp. 643–650, 1999.

Special strategies are necessary to avoid these. To illustrate the strategies we will give some key notes about our software project FEAST (Finite Element Analysis & Solution Tools).

2 Typical Examples for Performance Losses

One of the main components in iterative solution schemes, as for instance Krylov space methods or multigrid solvers, are Matrix–Vector (MV) applications. They are needed for defect calculations, smoothing, step–length control, etc., and often they consume 60 - 90% of CPU time and even more. Hereby *sparse MV* concepts are the standard techniques in Finite Element codes (and others) also well known as *compact storage* technique. Depending on the programming language the matrix entries plus index arrays or pointers are stored as long arrays containing the nonzero elements only. While this general *sparse* approach can be applied to general meshes and arbitrary numberings of the unknowns, no explicit advantage of (possible) highly structured parts can be exploited. Consequently a massive loss of performance with respect to the possible peak rates may be expected since — at least for large *sparse* problems with more than 100,000 unknowns — no "caching" and "pipelining" can be exploited such that the higher cost of memory access will dominate.

To demonstrate this failure we start with examples from our FEATFLOW code [11] which seems to be one of the most efficient simulation tools for the incompressible Navier–Stokes equations on general domains (see the results in [6]). We apply FEATFLOW to the following configuration of "2D flow around a car" and we measure the resulting MFLOP/s rates for the matrix–vector multiplication inside the multigrid solver for the momentum equation (see [8] for a more precise mathematical and algorithmic description). Here we apply the typical (for FEM approaches) "two level" (TL) numbering, a version of the bandwidth–minimizing Cuthill–McKee (CM) algorithm and an arbitrary "stochastic" numbering of the unknowns, which simulates adaptive refinement.

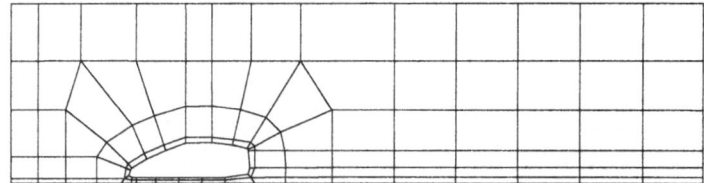

Fig. 1. Coarse mesh for "flow around a car"

All numbering strategies have common that based on the standard sparse *Compact Storage Rowwise (CSR)* technique the cost for arithmetic operations, for storage and for the number of memory accesses are identical. However the resulting timings in FORTRAN77 on a Sun Enterprise E450 (about 250 MFLOP/s peak performance!) can be very different as table 1 shows.

computer	#unknowns	TL	CM	stochastic
	13,688	20	22	19
SUN E450	54,256	15	17	13
($\sim$ 250 MFLOP/s)	216,032	14	16	6
(*CSR*)	862,144	15	16	4

Table 1. MFLOP/s rates of sparse MV multiplication for different numberings

These and numerous similar results can be concluded by the following statements which are quite representative for many other numerical simulation tools:

- **different** numbering strategies can lead to **identical** numerical results and work (w.r.t. arithmetic operations and memory access), but at the same time to **huge** differences in elapsed CPU time,
- sparse MV techniques are **slow** (with respect to possible peak rates) and depend massively on the problem size and the kind of memory access,
- in contrast to the mathematical theory most multigrid implementations will **not** show a realistic run–time behavior which is directly proportional to the mesh level.

Figure 2 which in contrast is based on the highly structured "blocked banded" MV techniques in FEAST shows that the same application can be performed in principle much faster: we additionally exploit **vectorization** facilities and **data locality**. Additionally we can even further differ between the case of *variable (var)* matrix entries and *constant bands (const)* as typical for Poisson–like PDE's. In comparison the same task is performed for a matrix in CSR and adaptive technique. Figure 2 is an excerpt from the FEAST INDICES ([9]) which is a general framework of performance measurements for many processors. It contains measurements for several vector, matrix-vector, smoothing and complete multigrid operations in 2D/3D.

Finally to show that high performance is achievable in more sophisticated algorithms, table 2 shows the performance rates for a complete multigrid cycle. As it can be seen a multigrid with about 450 MFLOP/s is possible. For more information about sparse banded BLAS, see [10].

3D case	N	MV-V	MV-C	TRIGS-V	TRIGS-C	MG-V	MG-C
DEC 21264	17^3	446	765	211	300	342	500
(500 MHz)	33^3	240	768	136	310	233	474
'DS20'	65^3	249	713	109	319	196	447
IBM RS6000	17^3	288	730	130	185	214	367
(200 MHz)	33^3	216	737	104	193	186	380
'PWR3'	65^3	174	710	91	198	154	363

Table 2. MFLOP/s rates of different sparse banded BLAS components (matrix–vector operation, TriGS smoother and multigrid cycle for variable/constant matrix entries)

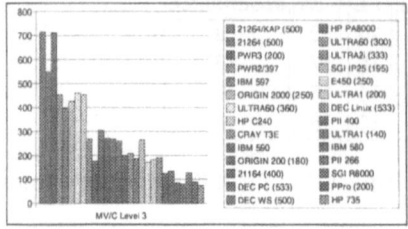
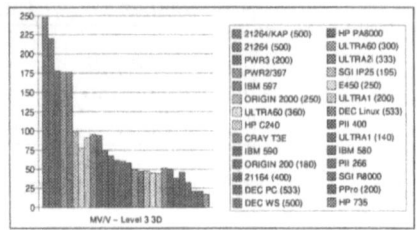

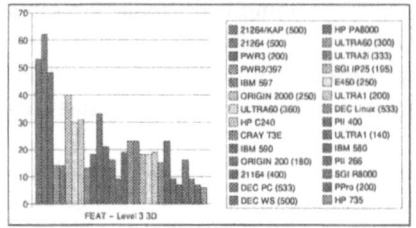
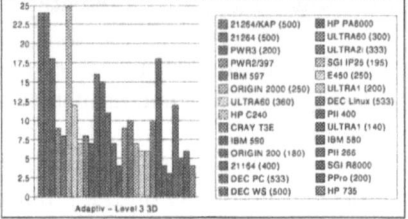

Fig. 2. The figures show the MFLOP/s for the discussed MV multiplications (MV-C/MV-V (blocked banded MV), FEAT (CSR MV), ADAP ("stochastic" MV)) which lead for certain computers as the IBM PWR2's to performance differences of almost a factor of 100! While modern workstation processors show a huge potential of supercomputing power for structured data, they loose for unstructured data in combination with sparse MV techniques.

3   FEAST: Main Principles

3.1   Hierarchical Data, Solver and Matrix Structures

One of the most important principles in FEAST is to apply consequently a *(Recursive) Divide and Conquer* strategy. The solution of the complete global problem is recursively split into smaller "independent" subproblems on "patches" as part of the complete set of unknowns. Thus the two major aims in this splitting procedure which can be performed by hand or via self–adaptive strategies are:

- *Find locally structured parts.*
- *Find locally anisotropic parts.*

Based on "small" structured subdomains on the lowest level (in fact, even one single or a small number of elements is allowed), the "higher–level" substructures are generated via clustering of "lower–level" parts such that algebraic or geometric irregularities are hidden inside the new "higher-level" patch.

Figure 3 illustrates exemplarily the employed data structure for a (coarse) triangulation of a given domain and its recursive partitioning into several kinds of substructures.

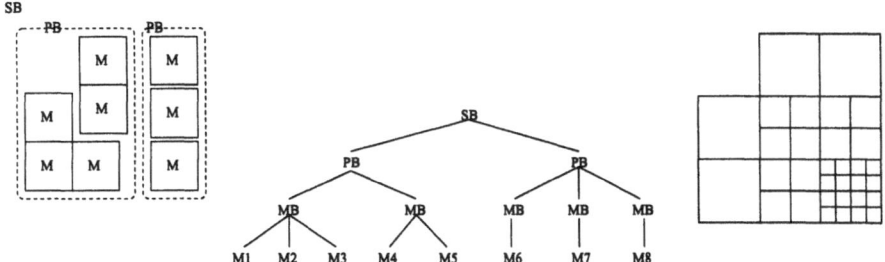

Fig. 3. FEAST domain structure

According to this decomposition, a corresponding data tree – the skeleton of the partitioning strategy – describes the hierarchical decomposition process. It consists of a specific collection of elements, macros (Mxxx), matrix blocks (MB), parallel blocks (PB), subdomain blocks (SB), etc.

The *atomic units* in our decomposition are the "macros" which may be of type **structured** (as $n \times n$ collection of quadrilaterals (in 2D) with sparse banded BLAS data structures) or **unstructured** (any sparse collection of elements). These "macros" (one or several) can be clustered to build a "matrix block" which contains the "local matrix parts": only here is the complete matrix information stored. Higher level constructs are "parallel blocks" (for the parallel distribution and the realization of the load balancing) and "subdomain blocks" (with special conformity rules with respect to grid refinement and applied discretization spaces). They all together build the complete domain, resp. the complete set of unknowns. It is important to realize that each stage in this hierarchical tree can act as independent "father" in relation to its "child" substructures while it is a "child" at the same time in another phase of the solution process (inside of the SCARC solver).

3.2 Generalized Solver Strategy SCARC

In view of their typically "excellent" convergence rates, multigrid methods seem to be most suited for the solution of PDE's. However as the examples of the previous chapter have shown, multigrid on general domains has often poor computational efficiency, at least if the implementation is based on the standard sparse techniques. As a result from our performance measurements (see [9]), the realistic MFLOP/s rates are often in the range of 1 MFLOP/s only, even on very modern high performance workstations. Moreover the linear relationship between problem size and CPU time is hardly realizable, due to the problem-size dependent performance rates of the sparse components. Additionally the robust treatment of complex mesh structures with locally varying details is hard to satisfy by typical "black box" components, even for ILU smoothing (see [1]).

Concerning the parallelization, some further serious problems occur: The parallel efficiency is often bad and far beyond peak, since the solution of the coarse grid problem leads to a large communication overhead. Furthermore the relation between "local" arithmetic operations and "global" data transfer is poor. Besides these computational aspects, the parallelization of the global smoothers (SOR, ILU) cannot be done efficiently because of their recursive character. So smoothing can only be performed block by block, which may lead to a deterioration of the convergence rates. Further the behavior of such block by block smoothing is not clear for complicated geometries with local/global anisotropies.

Motivated by these facts, we started to develop a new strategy for solving discretized PDE's which should satisfy several conditions:

The parallel efficiency shall be high due to a non–overlapping decomposition and a low communication overhead. The convergence rates are supposed to be independent of the mesh size h, the complexity of the domain and the number of subdomains N, and they should be in the range of typical multigrid convergence rates (as $\rho_{MG} \sim 0.1$). Further the method should be easily implementable and use only existing standard methods. The approach should guarantee the treatment of complicated geometries with local anisotropies (large aspect ratios!), without impairment of the overall convergence rates.

The underlying idea is *hiding recursively all anisotropies in single subdomains* combined with a corresponding *block Jacobi/Gauß-Seidel smoothing* within a standard multigrid approach. This approach is based on the numerical experience that the "simple" block Jacobi/Gauß-Seidel schemes perform well as soon as all occurring anisotropies are "locally hidden", that means if the local problems on each block are solved (more or less) exactly. This procedure ensures the global robustness! On the other hand this means that the "local" solution quality in each block can significantly improve the "global" convergence behavior!

These ideas are combined with the previously explained hierarchical data and matrix structures, which prefer tensorproduct–like meshes on each *macro* and which allow to achieve very high performance rates for the necessary numerical linear algebra components in the local (multigrid) solvers. Consequently all solution processes are recursively performed via sequences of more "local" steps until the lowest level, for instance a single *macro* with the described generalized tensorproduct mesh is reached.

The complete SCARC approach (see [2], [3], [4]) can be characterized via:

- **Scalable** (w. r. t. quality and number of local solution steps at each stage)
- **Recursive** ("independently" for each stage in the hierarchy of partitioning)
- **Clustering** (for building blocks via fixed or adaptive blocking strategies)

Numerous examples for this solver strategy can be found in the above mentioned papers. In this paper we shortly present a test calculation for SCARC on a "flow around the cylinder" topology (figure 4).

In table 3 the quantity N_{local} denotes the local number of elements in one space direction per *macro*. We perform $l = 1$ and $l = 2$ (global) smoothing steps and solve the local problems on each *macro* "exactly" via local multigrid. The *macros* around the circle are anisotropically refined in direction towards the circle, such that a finest mesh size $h_{min} \approx 2 \cdot 10^{-7}$ is obtained. The resulting

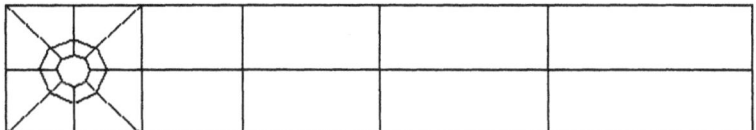

Fig. 4. 'Flow around the cylinder' topology (diameter of cylinder = 0.1)

(parallel) convergence rates are the same for all kinds of local refinement, independent of the amount of local anisotropy ("AR" means "aspect ratio"): The rates are all in the range of $\rho \sim 0.1$, for $l = 1$, and still much better for $l > 1$.

N_{local}	l	isotropic		moderately anisotropic		highly anisotropic	
		#ITE	ρ_{ScaRC}	#ITE	ρ_{ScaRC}	#ITE	ρ_{ScaRC}
16	1	7	0.121	7	0.112	6	0.070
	2	4	0.018	4	0.017	3	0.007
32	1	7	0.124	6	0.091	6	0.066
	2	4	0.023	4	0.018	3	0.005
64	1	7	0.121	6	0.081	6	0.065
	2	4	0.026	4	0.017	3	0.003
		$AR = 3$		$AR = 50$		$AR = 3,000$	
		$h_{min} \approx 6.1\text{-}4$		$h_{min} \approx 1.2\text{-}5$		$h_{min} \approx 1.9\text{-}7$	

Table 3. SCARC results for different degrees of local anisotropy

It is obvious that this parallel approach requires a modified concept of load balancing. Since in general there will be more *macros* than processors - realistic coarse meshes in 3D for complex applications require at least 1,000 up to 10,000 elements - such that each processor will provide the data for several *macros*. First of all the amount of storage on each *macro* is not identical if for instance the case of variable or constant matrix coefficients can be exploited on the tensorproduct–like meshes, or if auxiliary storage for the sparse techniques is needed. Moreover the application of ILU requires different storage cost than Jacobi– or Gauß-Seidel smoothing. However much more decisive is the actual CPU time on each *macro* for the solution of the local problems: The elapsed time depends significantly on the corresponding computational efficiency and the numerical efficiency, that means the actual convergence rates which determine the number of multigrid sweeps, and hence the total number of arithmetic operations. Additionally the problem size on each *macro* may vary.

Hence parallel load balancing is much more complex for this SCARC approach and <u>cannot</u> be determined via a priori strategies, for instance by equilibrating the number of unknowns on each processor. Instead we have to perform a posteriori load balancing techniques, analogously to the adaptive approaches, which are based on the numerical and computational run–time behavior of the last iterate to equilibrate better the total required CPU time on each processor.

4 Conclusions and Outlook

Our numerical examples show the advantages of the proposed method SCARC with respect to efficiency and robustness. This method allows parallelization and the use of standard numerical methods as basic modules for smoothing and solving. Further it makes it possible to use very regular data structures which enables high performance facilities.

Our computational examples have shown that there is a large gap between the LINPACK rates of several hundreds of MFLOP/s as typical representatives for direct solvers and the more realistic results for iterative schemes. Moreover certain properties of cache management and architecture influence the run time behavior massively, which leads to further problems for the developer. On the other hand the examples show that appropriate algorithmic and implementation techniques lead to much higher performance rates. However massive changes in the design of numerical, algorithmic and implementation ingredients are necessary.

References

[1] Altieri, M.: *Robuste und effiziente Mehrgitter-Verfahren auf verallgemeinerten Tensorprodukt-Gittern*, Diploma Thesis, to be published
[2] Becker, Ch.: *The realization of Finite Element software for high-performance applications*, PhD Thesis, to appear
[3] Kilian, S.: *Efficient parallel iterative solvers of* SCARC-*type and their application to the incompressible Navier-Stokes equations*, PhD Thesis, to appear
[4] Kilian, S., Turek, S.: *An example for parallel* SCARC *and its application to the incompressible Navier-Stokes equations*, Proc. ENUMATH-97, Heidelberg, October 1997.
[5] Rannacher, R., Becker, R.: *A Feed-Back Approach to Error Control in Finite Element Methods: Basic Analysis and Examples*, Preprint 96–52, University of Heidelberg, SFB 359, 1996.
[6] Schäfer, M., Rannacher, R., Turek, S.: *Evaluation of a CFD Benchmark for Laminar Flows*, Proc. ENUMATH-97, Heidelberg, October 1997.
[7] *The national technology roadmap for semiconductors*, 1997 edition, http://www.sematech.org/public/roadmap/index.htm
[8] Turek, S.: *Efficient solvers for incompressible flow problems: An algorithmic approach in view of computational aspects*, LNCSE, Springer-Verlag, 1998.
[9] Turek, S. et al.: *The FEAST INDICES - Realistic evaluation of modern software components and processor technologies*, to appear
[10] Turek, S. et al.: *Proposal for Sparse Banded BLAS techniques*, to appear
[11] Turek, S.: *FEATFLOW : Finite element software for the incompressible Navier-Stokes equations: User Manual*, Release 1.1, 1998

A Structured SADT Approach to the Support of a Parallel Adaptive 3D CFD Code

Jonathan Nash, Martin Berzins, and Paul Selwood

School of Computer Studies, The University of Leeds
Leeds LS2 9JT, West Yorkshire, UK

Abstract. The parallel implementation of unstructured adaptive tetra-hedral meshes for the solution of transient flows requires many complex stages of communication. This is due to the irregular data sets and their dynamically changing distribution. This paper describes the use of Shared Abstract Data Types (SADTs) in the restructuring of such a code, called PTETRAD. SADTs are an extension of an ADT with the notion of concurrent access. The potential for increased performance and simplicity of code is demonstrated, while maintaining software portability. It is shown how SADTs can raise the programmer's level of abstraction away from the details of how data sharing is supported. Performance results are provided for the SGI Origin2000 and the Cray T3E machines.

1 Introduction

Parallel computing still suffers from a lack of structured support for the design and analysis of code for distributed memory applications. For example, the MPI library supports a portable set of routines, such that applications can be more readily moved between platforms. However, MPI requires the programmer to become involved in the detailed communication and synchronisation patterns which the application will generate. The resulting code is hard to maintain, and it is often difficult to determine which code segments might require further attention in order to improve performance or to obtain good performance on new platforms.

Abstract Data Types (ADTs) have been used in serial computing to support modular and re-usable code. An example is a Queue, supporting a well-defined interface (Enqueue and Dequeue methods) which separates the functionality of the Queue from its internal implementation. Whereas an ADT supports information sharing between the different components of an application, a Shared ADT (SADT) e.g [1, 2] can support sharing between applications executing across multiple processors. High performance in a parallel environment is supported by allowing the concurrent invocation of the SADT methods, where multiple Enqueue and Dequeue operations can be active across the processors.

The clear distinction between functionality and implementation leads to portable application code, and portable performance, since alternative SADT implementations can be examined without altering the application. The potential to generate re-usable SADTs means that greater degrees of investment, care

P. Amestoy et al. (Eds.): Euro-Par'99, LNCS 1685, pp. 651–658, 1999.

and optimisation can be made in the implementation of an SADT on a given platform. In addition, an SADT can be parameterised by one or more user serial functions, in order to tailor the functionality of the SADT to that required by the application. This user parameterised form of the SADT is particularly useful in dealing with different parts of complex data structures in different ways.

This paper describes work [1] investigating the use of SADTs in a parallel computational fluid dynamics code, called PTETRAD [3, 4]. The unstructured 3D tetrahedral mesh, which forms the basis for a finite volume analysis, is partitioned among the processors by PTETRAD. Mesh adaptivity is performed by recursively refining and de-refining mesh elements, resulting in a local tree data structure rooted at each of the original base elements. The initial mesh partitioning is carried out at this base element level, as is any repartitioning and redistribution of the mesh when load imbalance is detected.

A highly interconnected mesh data structure is used by PTETRAD, in order to support a wide variety of solvers and to reduce the complexity of using unstructured meshes. Nodes hold a one-way linked list of element pointers. Nodes and faces are stored as a two-way link list. Edges are held as a series of two-way linked lists (one per refinement level) with child and parent pointers. In addition, frequently used remote mesh objects are stored locally as halo copies, in order to reduce the communications overhead. The solver, adaptation and redistribution phases each require many different forms of communication within a parallel machine in order to support mesh consistency (of the solution values and the data structures), both of the local partition of the mesh and the halos. PTETRAD currently uses MPI to support this.

In this paper it will be shown how parts of PTETRAD may be used in SADTs based on top of MPI and SHMEM, instead of MPI directly, thus leading to software at a higher level of abstraction with a clear distinction between the serial and parallel parts of the code. Section 2 describes an SADT which has been designed to support the different mesh consistency protocols within an unstructured tetrahedral mesh. A case study in Section 3 will describe the use of the SADT in supporting the mesh redistribution phase. A brief overview of the implementation techniques for the SADT will be given in Section 4, together with performance results for the SADT and for PTETRAD. The paper concludes by pointing to some current and future work.

2   An SADT for Maintaining Data Partition Consistency

The SADT described in this paper has focused on the problem in which a data set has been partitioned among p processors, with each processor holding internal data (the shared area), and overlapping data areas which must be maintained in a consistent state after being updated. PTETRAD maintains an array of pointers to base and leaf elements, which can be used to determine the appropriate information to be sent between partitions. For example, after mesh adaptation,

---

[1] Funded by the EPSRC ROPA programme - Grant number GR/L73104

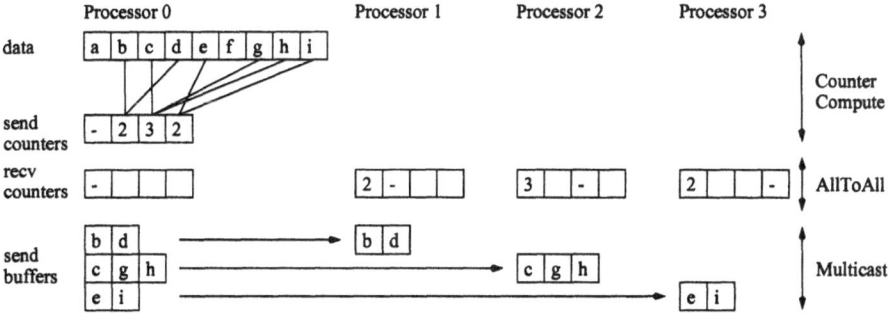

Fig. 1. The SADT communications stages

any refined elements will require that their halo copies also be refined. Also, in the redistribution phase, the base element list can be used to determine which elements need to be moved between partitions. In constructing an SADT for this pattern of sharing, four basic phases of execution can be identified. The SADT contains a consistency protocol which specifies these phases of execution, with the generic form:

```
void Protocol (in, out)    /* interface with in and out data lists*/
{ int send[p], recv[p];    /* Counters used in communications.  */
  pre-processing;          /* Initialisation of internal data.  */
  communications preamble; /* Identifying data exchanges    ... */
  data communications;     /*  Exchanging data between partitions.*/
  post-processing;}        /*  Format the results.              */
```

The protocol is called with a set of user-supplied input and output data lists (*in* and *out*), for example the lists of element pointers described above, and each phase requires the user to supply a number of application-specific serial functions. The SADT is thus parameterised by these user functions which allow the communications phases to be tuned by evaluating various condition functions, for example to determine whether a data item is to be communicated to a given partition (if it has been refined or needs to be redistributed). The SADT also contains basic functions to pack/unpack selected fields of data items to/from message buffers, and to process the new data items which are received.

Figure 1 shows an example of the operation of the protocol for a typical processor. For reasons of clarity the pre-processing and post-processing phases are removed. The protocol can make use of a communications library for global communications operations (denoted by Comms_Function), and may require one or more user-defined serial functions (denoted by User_Function).

(i) The communications preamble is given by:
   (a) Comms_CounterCompute: For each data item, User_CounterCondition decides if it is to be communicated, and User_CounterIndexing will update the associated values in the counters send[] and recv[].

654   Jonathan Nash, Martin Berzins, and Paul Selwood

    (b) `Comms_AllToAll`: An all-to-all communication is executed, in which the other processors note the expected number of items to be received from processor i in `recv[i]` (if `Comms_CounterCompute` is able to determine the counter values in `recv[]` then this communication can be avoided).

(ii) The actual data communications phase is given by `Comms_Multicast`:

    (a) For each input item and each processor in turn, `User_SendCondition` decides if the item should be sent to the processor. `User_PackDatum` will choose the selected fields of the data item to send, and place them in a contiguous memory block, so that it can be copied into the message buffer for that processor (`User_DatumSize` allows the system to allocate the required total send and receive buffer space).

    (b) Once the buffers have been communicated between the processors, each item is removed in turn, using `User_UnpackDatum`, and the local data partition is updated, based on this item, with `User_ProcessDatum`.

3 Case Study: Mesh Redistribution

At the application level of PTETRAD [4], local mesh access is supported by a library of mesh routines. The mesh repartitioning strategy is handled by linking in parallel versions of either the Metis or Jostle packages. The global mesh consistency is handled by making calls to the SADT, which also performs local mesh updates through the mesh library. The coordination between processors is supported by a small communications library which supports common traffic patterns, from a simple all-to-all exchange of integer values, up to the more complex packing, unpacking and processing of data buffers which are sent to all neighbouring mesh partitions.

The new SADT-based approach makes use of MPI, so that it may be run on both massively parallel machines and on networks of workstations, and also uses the Cray/SGI SHMEM library, to exploit the high performance direct memory access routines present on the SGI Origin 2000 and the Cray T3D/E. The use of an alternative communications mechanism is simply a matter of writing a new communications library (typically around 200 lines of code), and linking the compiled library into the main code.

The operation of the redistribution phase can be divided into four stages. The new mesh partitions at the base element level, are computed in the **repartition** stage using the parallel versions of Metis or Jostle. The local and halo owner fields for elements, edges, nodes and faces are updated in the **assign owners** stage. The data to be moved and the new halo data is communicated in the **redistribute** stage. Finally the **establish links** stage destroys any old communications links between local and halo mesh objects, and create the new links. The following examples focus on the second stage of assigning the new owners for edges in the partitioned mesh, in which the halo edges must be updated with the new owner identifiers. PTETRAD maintains an array of edge lists, with each list holding the edges at a given level of refinement in the mesh. An edge stores pointers to the halo copies which reside on other processors. This array is used

Pack the datum into a buffer

```
void User_PackDatum (PTETRAD_Edge *edge, char *buf, int *pos, int pe)
/* mesh edge list, storage space at buf[*pos], processor pe */
{ PTETRAD_EdLnk *halo;                /* edge halo pointer */
  int size = User_DatumSize();        /* the amount of storage required */
  while (edge) {                      /* inspect each edge */
    halo = edge → halo;               /* inspect the halos of the edge */
    while (halo) {                    /* for each edge halo */
      if (Edge_HaloHome (halo, pe)) { /* is the halo on processor pe ? */
        /* PACK THE HALO */
        Comm.ed = halo → edge;        /* note the halo's local address... */
        Comm.own = edge → owner;      /* on processor pe, and the owner */
        Pack (&Comm, size, buf, pos); /* pack this into the buffer area */
      } halo = halo → next;           /* go on to the next halo */
    } edge = edge →next;              /* go on to the next edge */
  } }
```

Fig. 2. SADT data communication functions: sending side

as the input to the SADT protocol, with the user functions processing each edge list in turn.

The data transmission SADT functions Comms_Multicast makes use of the user function shown in Figure 2 . This is part of the SADT consistency protocol relating to the packing, communication, unpacking and processing of the actual mesh edges. Figure 2 shows the initial packing stage. User_DatumSize returns the size of the "flattened" data structure, which forms the contiguous memory area to be communicated. In this case, it is an address of a halo edge on a remote processor, and the new owner identifier to be assigned to it. These details are held in the variable *Comm*. For a particular processor, User_PackDatum is used to search a list of edges at a given mesh refinement level, and determine if any edge halos are located on that processor. Those halos are packed into a contiguous buffer area, ready for communication. The "Pack" function is a standard call within the SADT library, which will copy the data into the communication buffer. At this stage, the data buffers have been filled, and Comms_Multicast will carry out the communication between the processors. Once the data has been exchanged the SADTs User_UnpackDatum will transfer the next data block from the communications buffer into the *EdgeOwner* variable. User_ProcessDatum will use the *ed* field to access the halo edge, and set its owner identifier to the new value.

4   Implementation Details and Performance Results

The amount of source code in the original and new PTETRAD versions, for the mesh redistribution phase, was reduced from 9,080 to 5,220 lines, by using the SADT approach. A significant reduction in the amount of application code has also been achieved by supporting the stages of global mesh consistency as

SADT calls, and implementing the mesh access operations within a separate library. The mesh access library is also being re-used during the restructuring of the solver and adaptation phases. As a typical example, the code for the communication of mesh nodes during redistribution is reduced from 340 lines to 100 lines, with only around 20 of these lines performing actual computation.

Within the SADT library, the main *Update* SADT, for maintaining mesh consistency, contains 200 lines of code, and an *Exchange* SADT (for performing gather/scatter operations) contains 25 lines. The MPI and SHMEM communications interfaces each have their own library which support the Comms_Function operations (see Section 2 and below). New libraries can easily be written to exploit new high performance communications mechanisms, without any changes to the application code.

4.1   The SADT Communications Library

The communications operations employed by an SADT are supported by a small communications library, as outlined in Section 2 and above. This contains operations such as an all-to-all exchange (Comms_AllToAll), and the point-to-point exchange and processing of data buffers representing new or updated mesh data (Comms_Multicast).

Figure 3(a) shows the performance of Comms_AllToAll for the platforms being studied. On the Origin, the difference between using MPI send-receive pairs and the collective routine *MPI_Alltoall* is quite small. Pairwise communication performs better up to around 8 processors. Collective communications take $2,300$ $\mu secs$ on 32 processors, as opposed to $2,770$ $\mu secs$ for the pairwise implementation. On the T3E, pairwise communication performs very similarly, but the collective communications version performs quite substantially better in all cases (eg 258 $\mu secs$ down from 1377 $\mu secs$ on 32 processors), outperforming the Origin version. For both platforms, it can be seen that the pairwise implementation begins to increase greater than linearly, due to the p^2 traffic requirement, whereas the collective communications version stays approximately linear. A third implementation, using the SHMEM library, outperforms pairwise communication by at least an order of magnitude, due to its very low overheads at the sending and receiving sides, taking 30 $\mu secs$ on 32 processors for the T3E, and 110 $\mu secs$ on the Origin. In PTETRAD, this stage of communication represents a small fraction of the overall communications phase, so it is not envisaged that alternative implementation approaches will have any real impact on performance.

The performance of Comms_Multicast, in which each processor i exchanges N words with its four neighbours $i-2$, $i-1$, $i+1$ and $i+2$ (in the case of $p \leq 4$, exchange occurs between the $p-1$ neighbours) (the user (un)packing and processing routines are null operations). On the Origin, performance reaches a ceiling of around 20 MBytes/sec for $N = 1000$ or larger using MPI, and 33 MBytes/sec using SHMEM, across the range of processors. For very small messages, the overheads of MPI begin to have an impact as the number of processors increase. On the T3E, the achievable performance using MPI was significantly higher, supporting 88 MBytes/sec on 32 processors, for large messages. Using

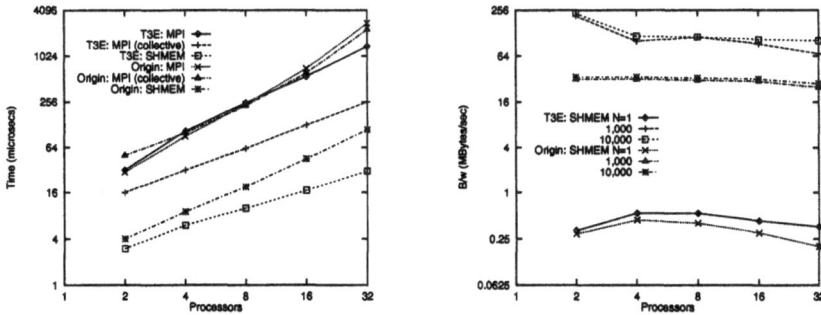

Fig. 3. (a) Comms_AllToAll; (b) Comms_Multicast: SHMEM

	Adaption	Imbalance	Repartition	**Redistribute**	Imbalance	Solve
PTETRAD (4)	5.87	19 %	0.99	**3.26**	11 %	1.23
SADT-MPI (4)	5.88	19 %	0.98	**3.04**	11 %	1.23
SADT-COLL (4)	5.89	19 %	0.98	**3.10**	11 %	1.23
SADT-SHMEM (4)	5.86	19 %	0.97	**2.78**	11 %	1.23
PTETRAD (32)	4.40	26 %	0.38	**2.12**	11 %	0.20
SADT-MPI (32)	4.38	26 %	0.38	**1.92**	11 %	0.20
SADT-COLL (32)	4.45	26 %	0.37	**1.92**	11 %	0.20
SADT-SHMEM (32)	4.44	26 %	0.37	**1.96**	11 %	0.20

Table 1. PTETRAD performance results (SGI/CRAY T3E:times in seconds)

SHMEM, this increases to 103 MBytes/sec, as well as improving the performance for smaller messages. Since this benchmark is measuring the time for all processors to both send and receive data blocks, the bandwidth results can be approximately doubled in order to derive the available bandwidth per processor. Figure 3(b) shows the SHMEM results for both machines. PTETRAD typically communicates messages of size $10K - 100K$ words, by using data blocking, and the above results show that this should make effective use of the available communications bandwidth.

4.2   PTETRAD Performance Results

A number of small test runs were performed using the original version of PTE-TRAD, and the SADT version of the redistribution phase, using pairwise MPI, collective MPI communication and SHMEM. A more comprehensive description of the performance of PTETRAD can be found in [3, 4]. Table 1 shows some typical results on the T3E, for a gas dynamics problem described in [4], using 4 and 32 processors.

Results for the Origin (not given here) show an 8% reduction in redistribution times on 4 processors. The use of the SHMEM library doesn't improve performance any further in this case, since the high level of mesh imbalance means that

local computation is the dominant factor. Thus, the performance improvements
when using the SADT approach originate from the tuning of the serial code. The
other timings are approximately equal, pointing to the fact that the improved redistribution times are real, rather than due to any variation in machine loading.
The T3E results in Table 1 show a reduction in times of between 7% and 10%
using MPI, and a reduction of 15% on 4 processors by linking in the SHMEM
communications library. The lower initial mesh imbalance, coupled with the very
high bandwidths available using SHMEM, result in this significant performance
increase. The slight increase in time on 32 processors using SHMEM seems to
be due to a conflict between SHMEM and MPI on the T3E.

5   Conclusions and Future Work

This paper shows how performance can be improved using three complementary
approaches. involving the use of shared abstract data types (SADTs) to structure parallel applications. The use of SADTs has made it easier to examine an
existing communications library, MPI, to determine if alternative operations can
be used, such as collective communications. Different communications libraries,
such as SHMEM, have been linked in to provide high performance SADT implementations on the Cray T3E and SGI Origin 2000 platforms. Finally, due to the
clear distinction between the parallel communications and local computation,
the serial code executing on each processor can also be more readily tuned. The
amount of code has also been significantly reduced, since the SADT used to
support mesh consistency can be re-used in many parts of the code. In the case
of PTETRAD, the routines to determine the mesh data to redistribute were updated, to reduce the amount of searching of the local mesh partition. and some of
the local data access methods were optimised. This shows up in the performance
results by an immediate increase in performance when moving to the SADT
version which still uses the MPI pairwise communications. Currently, the mesh
redistribution phase has been completed, with the solver and adaptation stages
due for completion in the near future. The intention is to support the proposed
SOPHIA applications interface [3], which provides an abstract view of a mesh
and its halo data.

References

[1] J. M. Nash, P. M. Dew and M. E. Dyer, *A Scalable Concurrent Queue on a Message Passing Machine*, The Computer Journal 39(6), 483-495, 1996.
[2] Jonathan Nash, *Scalable and predictable performance for irregular problems using the WPRAM computational model*, Information Proc. Letters 66, 237-246, 1998.
[3] P.M. Selwood, M. Berzins, J. Nash and P.M. Dew, *Portable Parallel Adaptation of Unstructured Tetrahedral Meshes*, Proceedings of Irregular'98: The 5th Inter-national Symposium on Solving Irregularly Structured Problems in Parallel (Ed. A.Ferreira et al.), Springer Lecture Notes in Comp. Sci., 1457, 56-67, 1998.
[4] P.Selwood and M.Berzins, *Portable Parallel Adaptation of Unstructured Tetrahe-dral Meshes*. Submitted to Concurrency 1998.

A Parallel Algorithm for 3D Geometry Transformations in OpenGL

J. Sébot Julien, A. Vartanian, J-L. Béchennec, and N. Drach-Temam

Université de Paris-Sud,
Laboratoire de Recherche en Informatique, Bâtiment 490,
F-91405 Orsay Cedex, France

Abstract. We have designed an algorithm which allows the OpenGL geometry transformations to be processed on a multiprocessor system. We have integrated it in Mesa, a 3D graphics library with an API which is very similar to that of OpenGL. We get speedup up to 1.8 for biprocessor without any modification of the application using the library. In this paper we study the issues of our algorithm and how we solve them.

1 Introduction

Most 3D polygonal rendering engine have a similar order of operation, a series of processing stages called the 3D pipeline. It is made of two main stages: geometry and reasterization. High end 3D accelerators provide hardware to accelerate the geometry transformations, low end cards delegate them to the software. Unfortunately, current CPU are not able to keep pace with latest low end 3D accelerators. Geometry transformations can be accelerated using fine grain parallelism with SIMD floating point instruction sets. Coarse grain parallelism is massively used in hardware geometry accelerators like SGI's InfiniteReality [3]. Our objective is the use of SMP computers to accelerate the geometry stage. The first solution is to write a parallel library which uses a non standard API adapted to 3D parallel computing. This is the main idea of the Argus project [1]. The major drawback is the need to rewrite 3D applications to use the new API and benefit from the parallelization. The second is to modify a library that uses a standard API to parallelize the geometry transormation computations. We have chosen to study this approach which let each application using this API benefit from the parallelism. We have modified Mesa [5], a 3D graphics library with an API which is very similar to that of OpenGL [2]. It's distributed with the sources under the GNU licence. We have written a parallel version of Mesa called PMesa, which use a scalable parallel algorithm. PMesa can theorically deal with p processors and a theoriticcal speedup close to p. In the rest of the paper, we present PMesa and its algorithm, theorical performance and performance measured on benchmarks.

2 PMesa Overview

Describing a scene with OpenGL means modifying the state of a few parameters for each frame, and submiting the OpenGL objects. An OpenGL object is a

P. Amestoy et al. (Eds.): Euro-Par'99, LNCS 1685, pp. 659–662, 1999.
© Springer-Verlag Berlin Heidelberg 1999

list af vertice, colors, normals and texture coordinate between a call to *glBegin*
and *glEnd*. Each object is stored into a vertex buffer (VB). Mesa computes the
geometry transformations after each *glEnd*. We have distinguished three main
stages in the geometry transformations, the *filling stage* where the vertices are
accumulated, the *computations stage* where they are tranformed and the *empty
stage* where they are sent to the video card for rasterization. The *filling stage*
and the *empty stage* must be done in order but the *computations stage* can be
done out of order.

We have choosen to increase the length of the *computations stage* by accumu-
lating a lot of V.B. This way we will improve the efficiency of our parallelization.
We will use threads on shared memory SMP systems. The optimal placement
for this problem is to have one *filling-computation-empty* stage per thread, and
one thread per processor. We can not work with this placement because we are
working into a shared library. As OpenGL is a state machine, a command behavior can be affected by previous commands. Some of the informations required by
commands execution may reside within the state maintained by the interface,
and some commands modify the state. To get good performances with a parallel
OpenGL library means to find a balance between the duplication of these states
and synchronizing the threads. This has been observed by Igehy in [1] and has
been particulary optimized in the SGI RealityEngine.

We have designed the following algorithm for p processors: Each time, a
master thread processes p *filling* and *empty* stages into p buffers containing G
vertices, and p slaves process the *computations* stage of these p buffers. Our
code is executed after each glEnd, so the thread synchronization overhead grows
lineary with the number of V.B. submitted and not with the number of vertices.
We use a statical load balancing algorithm.

3   Performances Evaluations

3.1   Theorical Performances

r is the *filling* and *empty stage* time for G vertices. c is the computation time for
G vertices. We suppose $r_{seq} = r_{par}$ for one vertex, and $c_{seq} = c_{par}$ for one vertex,
that means we suppose the filling and the computation stage takes the same time
in PMesa as in Mesa when working on one vertex. In Mesa and MITS c_s is the
time elapsed to compute one V.B. N is the number of buffers filled between two
synchronizations and T_N the time needed to transform $N \times G$ vertices (total
vertices between two synchronizations). SPU is the speedup of PMesa against
Mesa, $SPU = \frac{TN_{seq}}{TN_{par}}$.

1. Mesa (sequential):

rs	Cs	rs	Cs	master thread

$$T_{Nseq} = N(c_s + r_s)$$

2. Our algorithm (PMesa):
   - if $r > (p-1) \times c$, the parallel time is longer than the sequential time:

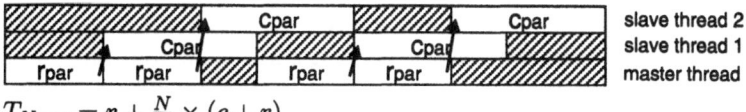

$$T_{Npar} = r + \frac{N}{p} \times (c + r)$$
$$lim_{N \to \infty} SPU_{par} = p$$
We get a speedup close to p for N great enough.

   - if $r < (p-1) \times c$, the sequential time is greater than the parallel time:

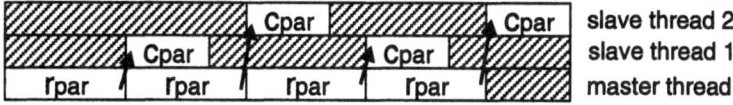

$$T_{Npar} = N \times r + c$$
$$SPU_{par} = 1 + \frac{c \times (N-1)}{N \times r + c}$$
Theorical speedup is close to $1 + \frac{c}{r}$ for N great enough.

The following factors limit the performance:

1. $(p-1) \times \frac{c}{r}$: if $(p-1) \times \frac{c}{r} > 1$ the sequential code is shorter than the parallel time and our placement is optimal.
2. N: When it is is too low, the pipeline is flushed frequently and sequential times grows. The solution is to decrease the grain enough or duplicate more states variables to limit the number of synchronizations.
3. $\frac{r_{par}}{r_{seq}}$: This is the slowdown of sequential code (*filling-empty*) due to PMesa . We observe it is the major factor limiting the performances of PMesa. When the grain grows, the data set we are working on grows and memory hierarchy tends to raise this number. This factor grows when the number of vertices per OpenGL object becomes too low.
4. $\frac{c_{par}}{c_{seq}}$: This is the growth of the parallel code (*computations*) introduced by PMesa. We are using data structures bigger than those of Mesa because we accumulate a lot of vertices, so the memory hierarchy is a performance lowering factor compared to Mesa.
5. The synchronizations between the threads limit the performance. The number of switching depends on the grain. Bigger is the grain fewer are the switching.
6. Bigger the size of OpenGL objects is, shorter is the time passed in our code. We can not control this size because it is application dependant.

3.2 Performance on Benchmarks

All the tests on PMesa has been done on a bi-PII400 512Mo, the rasterization which depends on the 3D accelerator used was off. To verify that our algorithm is efficient for any VB size we have made a microbenchmark called Glutspeed where the number of vertices per OpenGL object is adjustable. It draws a scene with 9,000,000 vertices and measure the time elapsed. 1. The grains between 500 and 1100 vertices give the best performances for all the VB sizes1. When grain grows performance increases because there are less synchronizations, but when the grain becomes higher than 1000 the performances decrease because of the memory hierarchy.

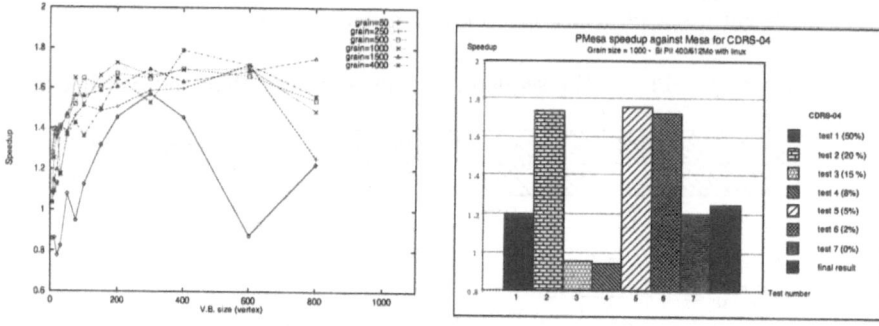

Fig. 1. PMesa performance on Glutspeed and Viewperf CDRS

Viewperf is a set of OpenGL benchmarks written in C. It is the actual standard benchmark for the 3D industry. The CDRS-04 program [4] is the most commonly used. On this benchmark we get speedups between 0.95 to 1.8 when we are using a grain of 1000. The slowdowns are due to the important number of state changes into these benchmarks.

4   Conclusion

On a biprocessor, PMesa gives a speedup of up to 1.8 against Mesa when rasterization is off. This speedup is near the theorical speedup of the algorithm. We have evaluated the PMesa's limiting factors. PMesa is currently working with software rasterization or whith hardware accelerated rasterization (3dfx voodoo[2]). PMesa uses the whole OpenGL API and can also be used by any OpenGL program working on Unix or Windows NT.

References

[1] Gordon Stoll Homan Igehy and Pat Hanrahan. The design of a parallel graphics interface. In *Computer Graphics (Proc. SIGGRAPH '98)*, 1998.
[2] Kurt Akeley Mark Segal. *The OpenGL Graphics System: A Specification.* Silicon Graphics inc., 1996.
[3] John S. Montrym, Daniel R. Baum, David L. Dignam, and Christopher J. Migdal. Infinitereality : A real-time graphics system,. In *Computer Graphics (1997 Symposium on Interactive 3D Graphics)*, pages 293–302, 1997.
[4] OPC. Cdrs-04. http://www.specbench.org/gpc/opc.static/cdrs.html, 1999.
[5] Brian Paul. The mesa 3d-graphic library. http://www.mesa3d.org, Jan 1999.

Parallel Implementation in a Industrial Framework of Statistical Tolerancing Analysis in Microelectronics

Salvatore Rinaudo[1], Francesco Moschella[1], and Marcello A. Anile[2]

[1] ST Microelectronis, Stradale Primosole 50, 95121 Catania (Italy)
[2] Dipartimento di Matematica, Viale Andrea Doria, 95125 Catania (Italy)

Abstract. The aim of this work is to report on a parallel implementation of methods for tolerance analysis in the framework of a microelectronics design center. The methods were designed to run parallelly on different platforms which could have different computational performances. In order to distribute the computations over a network of workstations, the algorithm was designed not by using a parallel compiler, but by using a RPC multi-server network. We have used essentially two methods. The first is the Monte Carlo approach, the second is based on an approximation by numerical integration or quadrature technique [1, 2, 3, 4], which requires far less function evaluations than the Monte Carlo method. These two approaches have been implemented in a parallel algorithm to be used on a cluster of multivendor workstations.

1 Introduction

Starting from a known distribution of some input parameters the statistical distribution of output parameters is evaluated using the statistical approaches for tolerance analysis. Each value of the input component is generated according to the chosen method for estimating moments, and the response function is computed by the numerical simulators. Two applications have been considered:

- The determination of the effects of some process parameters, as implant dose and energy, diffusion time and temperature of termal processes etc. made in HSB technology [5].
- The determination of the effects of resistors and capacitors fluctuations on the center frequency of a Chebyshev filter.

2 The Methods

Several methods are available for tolerance analysis. Some of them (named *conventional*) as **Worst-case** and **root-sum squares**, are inadequate. Worst-case methods give results that are pessimistic, while root-sum squares give optimistic results. Many others are available in the literature [4, 6], but many require the

P. Amestoy et al. (Eds.): Euro-Par'99, LNCS 1685, pp. 663–667, 1999.

knowledge of an explicit relationship between the input parameters and the output variables in analytical form. If we express the relationship between the input and generic output parameters in the following form:

$$y = f(x_1, x_2, ..., x_n) \tag{1}$$

where the x_i are random variables (assumed here independently distributed for the sake of simplicity) and y is the generic output parameter; our problem is to estimate the moments of distribution of the assembly response function $\mathbf{y}$, given the moments of the distribution of the component parameters. which (the first four) are defined as

$$
\begin{array}{ll}
\mu_i = \int x_i w_i(x_i) dx_i & \gamma_i \sigma_i^3 = \int (x_i - \mu_i)^3 w_i(x_i) dx_i \\
\sigma_i^2 = \int (x_i - \mu_i)^2 w_i(x_i) dx_i & \Gamma_i \sigma_i^4 = \int (x_i - \mu_i)^4 w_i(x_i) dx_i
\end{array}
\tag{2}
$$

with $w(x_i)$ the probability density for the ith variable, where μ is the mean value, σ^2 the variance, γ the skewness and Γ the kurtosis.

2.1 Monte Carlo Method

The most popular and simplest method for non linear tolerance analysis is the **Monte Carlo** one. Random values for each random variable are generated according to its distribution and the value of the response function is computed for each set of input values. At the end we have a population of $\mathbf{y}$ and all the moments can be computed using standard statistical formulae. The greater disadvantage of this method is that it requires a great number of computations to have an acceptable precision. A generally accepted rule is: *for an expected probability level of* 10^{-N}, *the size of the samples must be between* 10^{N+2} *and* 10^{N+3}.

2.2 Approximation by Numerical Integration

This technique was studied by Evans [1, 2, 3, 4]. The basis of this technique is that for any function $h(x_1, x_2, ..., x_n)$ the expected value of $\mathbf{h}$ is given by the integral

$$I\{h\} = \int_{-\infty}^{+\infty} ... \int_{-\infty}^{+\infty} h(x_1, x_2, ..., x_n) \prod_{i=1}^{n} w_i(x_i) dx_i \tag{3}$$

where x_i are independent random variables with known density functions $w_i(x_i)$, i=1,2,...,n. The integral in equation (3) can be approximated with the quadrature expression $Q\{h\}$ given in [4] The approximation of the moments of $y = f(x_1, x_2, ..., x_n)$ can be computed if we set

$$h_\nu(x_1, x_2, ..., x_n) = [f(x_1, x_2, ..., x_n) - K]^\nu \tag{4}$$

then $Q\{h_\nu\}$ approximates the νth moment of y with respect to K. This method has several advantages. The first is that for not too many parameters it requires fewer function evaluations than the Monte Carlo simulation. The second (and a paramount one) is that once the function has been evaluated at $2n^2 + 1$ points it is possible to reuse the function values in order to compute the new results corresponding to a change in the moments of the input parameters. In fact only $2n$ new function evaluations are required for this purpose [3]. Also, the results due to a further change of the moments of the input parameters can be obtained without any new evaluations of the function.

When the random variables $(x_1, x_2, ..., x_n)$ are not independently distributed we can generalize the quadrature formula (3) by decomposing the random vector $X = (x_1, x_2, ..., x_n)^T$ into its principal components [7].

3   Implementation on Multivendor CPUs

The independence of the input parameters allows the computation to be distributed on a parallel architecture. In order to distribute the computations over a network of workstations, the methods described above were designed not by using a parallel compiler, but by using a RPC multi-server network [8]. When there is a request to evaluate a function, the algorithm (the client) asynchronously asks a server to run the simulation. The client process does not wait. The server requests are sent out with a high-level *callrpc()*. So, the client needs to register an **RPC** deamon, running a service itself to catch the replies returned by the servers.

3.1   Efficiency of the Algorithm on Distributed CPU

To verify the efficiency of the algorithm on distributed CPUs, the same computation was done using up to 10 CPUs. The CPUs were very different and with different loads. The final results of the algorithm are summarized in table 1. We remark that the definition we used of speed-up is not the theoretical one. In fact our workstations have different performances (see table 1) and the speed-up we have used is the ratio to the fastest single WS (in an heterogeneus environment this amounts to a lower bound on the theoretical speed-up).

Nodes	OS	CPU speed R.	N. CPU	SpeedUp	Eff.	SpeedUp×Eff.
Sun Ultra-E	SunOS 5.5.1	4.06	1	1.00	1.000	1.00
Sun Ultra-1	SunOS 5.5.1	3.98	2	2.00	1.000	2.00
IBM Power2	AIX 3.2.5	3.03	3	2.82	0.940	2.65
Sun Sparc20	SunOS 4.1.4	2.50	4	3.41	0.853	2.91
Sun Sparc20 (2)	SunOS 5.5.1	2.49	5	3.74	0.748	2.80
Sun Sparc20 (3)	SunOS 4.1.4	2.38	6	4.11	0.685	2.81
HP 9000/712	HP-UX B.10.10	2.18	7	4.85	0.690	3.36
Sun Sparc10	SunOS 4.1.3	1.64	8	5.12	0.640	3.28
IBM PowerPC	AIX 3.2.5	1.54	9	5.58	0.620	3.46
Sun Sparc10 (2)	SunOS 4.1.3	1.00	10	6.00	0.600	3.60

table 1 - CPU nodes description and algorithm preformance

4   An Application Example

In this section we perform a tolerance analysis on a submicrometer bipolar transistor. We studied the statistical distribution of process parameters which affect the doping profiles of the device. For each set of input process parameters (eg. dose and energy implant, time and temperature of thermal processes, thicknes of the polysislicon, ect.), the doping profile is reproduced using the 2D process simulation Suprem4, and from each set of doping profiles the microwave performance is determined, after a device simulation with Pisces, using tools and models for microwave characterization has been done [9]. In the end we have 14 process parameters which we assume influence the behaviour of the microwave device. We assume that all the parameters are normally distributed and uncorrelated. We have also performed a Monte Carlo analysisas discussed in 2.1. We find that the Numerical Integration approach fails compared to the Monte Carlo one in this case. We ascribe the failure of the Numerical Integration approach to the fact that in this case the function evaluation is not sufficiently accurate. In fact in this case each function evaluation corresponds to executing the process and device simulators (which give as output numerical results which are not very accurate and in some instances may even not converge) and afterwords an optimizer for extracting the AC parameters. Therefore, since the Numerical Integration technique requires an accurate determination of the function values, it is not surprising that in this case it fails. The quadrature formula will not fail when the function can be evaluated accurately. We remark that the Monte Carlo method, being based on statistical averaging, is more robust and does not need a very accurate evaluation of the function.

5   A Second Example

This example deals with a fourth order Chebyshev active filter which is designed to have a 10KHz center frequency and a 1.5KHz bandwidth. The components have a statistical fluctuation of 1% for resistors and 5% for capacitors. We focused our study on the analysis of the statistic tolerance of the center frequency. The response function was computed using the circuit simulator PSPICE which incorporates a statistical analysis based on the Monte Carlo method (limited, however, to a maximum of 400 function evalutations, clearly insufficient for obtaining reliable results). The study was done using both Numerical Integration technique and the Monte Carlo approach. Both methods give very similar results for the averages and variances (see table 2 [1]), the difference was the computational cost, 201 runs (10 input parameters) for the Numerical Integration

---

[1] Since the moments in the Numerical Integration technique were computed with respect to the response function calculated at the mean values and those in the MC approach with respect to the mean, to compare the two calculations fro the variances it is necessary to make the translation (where K has been previously defined in 2.2) $\sigma_*^2 = \sigma^2 - (\mu - K)^2$.

techniquecompared to more than 2000 Monte Carlo evaluations. The small differences between the results of the two methods could be explained by considering that the techniques used to extract the output values are affected by numerical errors. These errors in circuit simulation are generally low. This limits the precision of the Numerical Integration technique, while in the Monte Carlo approach it has a smaller effect.

Quad. Mean	Quad. σ^2	MC Mean	MC σ^2
10.42	0.053 (0.045)	10.33	0.043

table 2 - Results on II example

6 Conclusion

In this work two methods for tolerance analysis, the Monte Carlo one and one based on a Numerical Integration technique were investigated. These two methods where implemented in a parallel algorithm to run on a cluster of multivendor workstations and used to estimate the first four moments of the most important parameters affecting the behaviour of a microwave transistor(due to the spread of process parameters) and of a Chebyshev active filter (due to the fluctuations in the resistors and capacitors). The higher accuracy of circuit simulators (compared to process and device ones) has made the use of the Numerical Integration technique practical for the statistical analysis of a Chebyshev active filter, leading to a much reduced computing time. By using a parallel algorithm on a cluster of multivendor WorkStations has made it possible to perform tolerance analysis in an effective way in the framework of an industrial design center.

References

[1] David H. Evans *An Application of Numerical Integration Techniques to Statistical Tolerancing* Tecnometrics Vol. 9, No. 3 August 1967.
[2] David H. Evans *An Application of Numerical Integration Techniques to Statistical Tolerancing, II - A Note on the Error* Tecnometrics Vol. 13, No. 2 May 1971
[3] David H. Evans *An Application of Numerical Integration Techniques to Statistical Tolerancing, III - General Distribution* Tecnometrics Vol. 13, No. 2 May 1971
[4] David H. Evans *Statistical Tolerancing: The State of the Art. Part II. Methods for Estimating Moments* Journal of Quality Technology, Vol. 7, No. 1, January 1975
[5] Lombardo, S.; Pinto, A.; Raineri, V.; Ward, P.; La Rosa, G. Privitera, G.; Campisano, S.U. *Si/Ge/SUB X/Si/SUB 1-X/ Heterojunction Bipolar Transistors with the Ge/SUB X/Si/SUB 1-X/ Base Formed* IEEE Electron Device Letters, Vol. 17, Issue: 10, Oct. 1996
[6] Swami D. Nigam and Joshua U. Turner *Review of Statistical approaches to tolerance analysis* Computer-Aided Design. Vol. 27, No 1, pp. 6-15, 1995
[7] B. Flury, *A First Course in Multivariate Statistic*, Springer, 1997
[8] John Bloomer. *Power Programming with RPC* O'Reilly & Associates, Inc., (1992)
[9] Rinaudo, S.; Privitera, G.; Ferla, G.; Galluzzo, A. *Small-Signal and Noise Modeling of Submicrometer Self Aligned Bipolar Transistor* Radio and Wireless Conference, 1998. RAWCON 98. 1998 IEEE (1998)

Interaction Between Data Parallel Compilation and Data Transfer and Storage Cost Minimization for Multimedia Applications

Chidamber Kulkarni[1], Koen Danckaert[1], Francky Catthoor[1,2], and
Manish Gupta[3]

[1] IMEC, Kapeldreef 75, B-3001 Leuven, Belgium
[2] Professor at the Katholieke Universiteit Leuven
[3] IBM T.J. Watson Research Center, Yorktown Heights, NY

Abstract. Real-time multi-media applications need large processing power and yet require a low-power implementation in an embedded programmable parallel processor context. Our main contribution in this context is the proposal of a formalized DTSE (data transfer and storage exploration) methodology, which allows to significantly reduce system bus load and hence overall system performance and also power consumption. We demonstrate the complementarity of this methodology by coupling the DTSE with a state-of-the-art performance optimizing and parallelizing compiler. Experiments on two real-life video and image processing applications show that this combined approach heavily reduces the memory accesses and bus-loading and hence power and also significantly reduces the total execution time. Decomposing the detailed parallelization and DTSE issues into two different stages is important to obtain the benefits of both the stages without exploding the complexity of solving all the issues simultaneously.

1 Introduction and Related Work

Parallel machines were mainly, if not exclusively, being used in scientific communities until recently. Lately, the rapid growth of real-time multi-media applications have brought new challenges in terms of the required processing (computing) power and power requirements. For this type of applications, especially video, graphics and image processing, the processing power of traditional uniprocessors is no longer sufficient. This has lead to the introduction of small- and medium-scale parallelism in this field too, but then mostly oriented towards single chip systems for cost reasons. Today, many weakly parallel video and multi-media processors are emerging (see [14] and its references), increasing the importance of parallelization techniques. Applications on these processors are parallelized manually even now, which can be tedious and error-prone. This paper presents evidence that parallelizing compilers can be used effectively to deal with this problem, if they are combined with other techniques.

Indeed, the cost functions to be used in these new emerging application fields are no longer purely performance based. Power is also a crucial factor, and has to

P. Amestoy et al. (Eds.): Euro-Par'99, LNCS 1685, pp. 668–676, 1999.

be optimized for a given throughput. Real time multi-media processing (RMP) applications are usually memory intensive and a significant part of the power consumption is due to the data transfers i.e. in the memory hierarchy [16].

In a parallel processor context most of the research effort in the community so far addresses the problem of parallelization and processor partitioning [2]. Existing approaches do not sufficiently take into account the background storage and transfer related cost. A first approach for more global memory optimization in a parallel processor context was described by us in [6]. Although many software compilers try to come up with the best array layout in memory for optimal cache performance (see e.g. [5] and its references for a few formal approaches based on compile-time analysis) they do not try to directly reduce the storage requirements as memory is allocated based on the available variable declarations. However, in general, this can lead to a large over-allocation, compared to the maximal amount of memory which is really needed over time. We have shown in [4, 9] that aggressive in-place mapping of array signals, based on a detailed life-time analysis, can heavily reduce data storage requirements and improve the cache performance significantly.

The above issues have been a motivation for us to carry out this study. In this paper, we apply our global DTSE (data transfer and storage exploration) methodology on two real-life RMP applications and then couple this to a state-of-the-art optimizing and parallelizing compiler using directives. In the process, we show that the results are very promising (see section 4).

The remaining paper is organized as follows. In section 2, we identify the problem and present the cost functions used in this paper. Section 3 presents the design methodology used in this work. This is followed by discussion and experimental results on two real-life demonstrators in section 4. Conclusions from this work are provided in section 5.

2   Problem Exploration and Cost Functions

In this section we will identify the problems for which we will provide promising solutions. Also the cost functions that we have used in this paper are presented.

As stated in section 1, the emerging real-time multi-media applications demand large processing power and a low-power implementation. Thus in the context of programmable multimedia solutions on parallel processors, the following problems become evident :

1. Do current multi-media applications lend themselves easily to automatic (or directives-based) parallelization?
2. What are the effects of power-oriented program transformations on the performance of (parallel) multi-media applications?

In the sequel, we will provide answers for the above issues. Moreover, we will also show that optimizations during the DTSE stage help in reducing the communication cost (memory accesses) between processor and the (shared) memory.

Two cost functions are used in this paper for evaluating the effectiveness of our methodology. The first one is the execution time of the concerned application. This time is the total execution time of the program measured using UNIX and C time functions. The second cost function is the power consumption. The power consumption is estimated by evaluating the total number of accesses to the off-chip memories. We will calculate both the total number of accesses and size of memories for the demonstrators in section 4.

3   Design Methodology

In this section we present a brief discussion of the main steps in the DTSE stage and the optimizing and parallelizing compiler. Note that the emphasis here is not on the detailed issues which have been presented elsewhere but just to highlight the main benefits of each individual stage.

3.1   Data Transfer and Storage Exploration

The five steps comprising our methodology for system-level power optimization for real-time multi-media applications are illustrated in figure 1. All the steps of our methodology are currently performed manually in this paper but each individual step has been applied systematically in the proposed sequence. Moreover, we are building a prototype compiler supporting these steps. Below is a brief discussion of the different steps in our methodology :

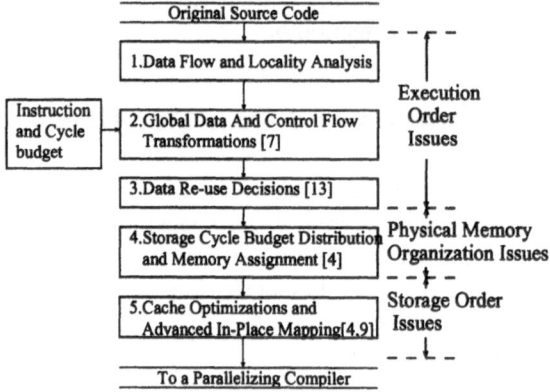

Fig. 1. Data Transfer and Storage Exploration (DTSE) methodology

The program transformation methodology comprises two phases : in the first phase all the transformations, steps 2, 3, 4 and 5 in figure 1, are chosen based on the amount of reduction in the number of memory accesses to the larger off-chip memories. In the second phase we obtain feedback on the effects of these transformations on performance (delay), static instruction count and code size. Note that these phases are local to each step i.e. the real-time feedback is local to each step. Parts of code exhibiting larger delay are transformed again for improved performance. In addition, in order to reduce the adverse impact of complex conditional expressions and addressing after power oriented transformations, we perform advanced code motion and modulo reductions prior to conventional instruction-level compiler(s) using techniques as described in [12]. Also note that steps 2, 3 and 4 modify the execution order of the application, whereas step 5 only modifies the storage order. This approach has significant benefits as illustrated in [10].

3.2   Performance Optimization and Parallelization

We have used a prototype version of the IBM C compiler (xlc) for AIX 4.2 in our experiments to demonstrate the complementarity to the state-of-the-art parallel compiler. This compiler uses a language-independent optimizer, the Toronto Portable Optimizer (TPO), for global program optimizations and parallelization, and finally, an optimizing back-end for target-specific optimizations. TPO performs (inter-procedurally) classical data flow optimizations like constant propagation, copy propagation, dead code elimination, and loop-invariant code motion [11]. It also performs loop transformations like fusion, unimodular transformations, tiling, and distribution to improve data locality and parallelism detection [18]. Finally, TPO supports parallelization of loops based on both data dependence analysis and user directives. It supports various static and dynamic scheduling policies for nested parallelization of loops [8].

Our experiments have shown that the key analysis needed to detect parallelism automatically in the critical loops of the application was a sophisticated form of array privatization analysis [17]. While TPO does perform interprocedural array privatization analysis, it failed to detect the privatizability of arrays indexed by expressions involving modulo operations. The analysis of these kinds of array references has not received much attention in the literature, as they do not appear very frequently in scientific programs. The DTSE steps, on the contrary, often introduce modulo arithmetic in the computation of array subscripts in the transformed loops. Even though the compiler was unable to detect the parallelism in many loops automatically, it was quite easy to make it parallelize any loop by simply using a *parallel loop* directive. The compiler automatically privatizes all variables declared inside the loop being parallelized. Finally, since the computations in these applications tend to be quite regular, we used the default, static scheduling policy for scheduling parallel loops.

4   Experimental Results

In this section we address the questions identified in section 2 in the context of implementation of RMP applications on parallel processors.

4.1   Experimental Set-Up

The experimental set-up comprises two main stages (described in section 3): First the concerned application is optimized for data transfer (power) and storage using the DTSE approach. This application is now analyzed for the power and performance cost using the criteria described in section 2. Next the DTSE-optimized application is now fed to the optimizing and parallelizing compiler (OPC). The resulting code from OPC is executed on a 4-way SMP machine (each node is a Power PC 604 with clock frequency of 112 MHz) to observe the performance characteristics of the parallelized code.

Thus we obtain a power estimation of the original and power-optimized code and performance values for DTSE-optimized and OPC-DTSE-optimized code. This allows us to do a fair comparison of the effects of the DTSE stage on performance and parallelization related optimizations.

4.2   Cavity Detection

The cavity (or edge) detection technique is an important step in many (especially medical) image processing applications [3]. This algorithm mainly consists of a sequence of four distinct steps, each of which computes new matrix information from the output of the previous step as shown in figure 2. In this paper, we assume that the image enters and leaves the system in the conventional row-wise fashion.

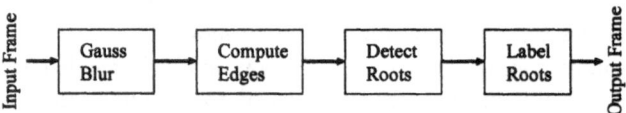

Fig. 2. The Cavity Detection algorithm

Table 1 gives an overview of the reduction of data transfers and the number of memories, accomplished by the different system-level DTSE transformations. The initial algorithm requires 36 accesses per pixel to the frame memory. Whereas the locally optimized algorithm needs 10 accesses to the frame memory and the globally optimized algorithm needs at most two accesses to the input/output buffers. Thus with DTSE transformations, a significant reduction in power and area is obtained for this example. Table 3 gives the execution times for different optimization levels for the parallelized code on different number of processors. Here we observe that the total execution time decreases consistently after the DTSE optimizations, and the optimized version performs better for larger number of processors. Thus apart from reducing the power cost, our DTSE approach heavily enhances the performance on parallel machines for data dominated applications by reducing the potential inter-processor communication.

Version	Parallelism	# Main memory	# Frame transfers
Initial	data	2 (+1)	36
	task	8 (+1)	36
Locally Optimized	data	1 (+1)	10
	task	4 (+1)	10
DTSE optimized	data	0	0(2)
	task	0	0(2)

Table 1. Data transfer and main memory storage requirements of different cavity detection mapping alternatives on a parallel processor. If the input frame buffers are included, the numbers between () are obtained. Frame transfers are counted per pixel.

4.3 Quad-Tree Structured Differential Pulse Code Modulation (QSDPCM)

The QSDPCM [15] technique is an interframe adaptive coding technique for video signals. A global view of the QSDPCM coding algorithm is given in figure 3. Most of the submodules of the algorithm operate in cycles with an iteration period of one frame (see table 3 for frame sizes). It is worth noting that the algorithm spans 20 pages of complex C code. Hence it is practically impossible to present all the detailed dependencies and profile of the algorithm. A more detailed explanation of the different parts of the algorithm is available in [15]. Table 2, gives the number (and size) of memories required and the correspond-

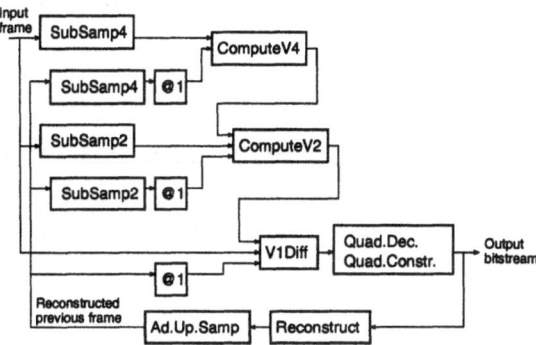

Fig. 3. The QSDPCM algorithm

ing number of accesses for the QSDPCM algorithm. We observe that there is a significant reduction in the total number of off-chip accesses for both the memories. Hence the total power consumption, for the global DTSE-transformed case is heavily reduced compared to the original. Table 3 shows that the DTSE-transformed parallelized code has better execution times compared to that of the original parallelized code. Note that the execution time for the original case

reduces by a factor three with four processors, whereas for the DTSE transformed case there is even a reduction in execution time by an almost ideal factor
3.9. This once again shows that DTSE is complementary and is even augmenting the effect of the more conventional performance and parallelization related
optimizations.

Version	Memory Size	# Accesses
Initial	206K	5020K
	282K	8610K
DTSE	206K	4900K
transformed	1334	8548K

Table 2. Amount of memory and the corresponding total number of accesses
to each of these memories for the QSDPCM algorithm for different optimization
levels. The memory size is in bytes.

Version	Frame Size	P=1	P=2	P=3	P=4
Cavity Detection					
Original	640 × 400	0.87	0.65	0.63	0.63
	1280 × 1000	4.47	3.19	3.01	3.06
	12800 × 10	4.07	3.00	2.94	3.03
DTSE	640 × 400	0.32	0.27	0.21	0.18
transformed	1280 × 1000	1.71	1.30	0.98	0.83
	12800 × 10	3.11	1.92	1.79	1.61
QSDPCM					
Original	288 × 528	5.89	3.11	2.26	2.00
	576 × 1056	22.83	12.32	9.09	7.62
DTSE	288 × 528	3.48	1.75	1.18	0.98
transformed	576 × 1056	13.45	7.28	4.79	3.47

Table 3. Performance of parallelized Cavity detection and QSDPCM algorithm
for various cases using a Power PC 604 based 4-way SMP machine. Here P
represents the number of processors and the execution time is in seconds.

4.4 Summary

Below is a summary of the main results from the experiments with the two
real-life applications :

1. We are able to parallelize both the codes using manual directives to guide
   the parallelizing compiler.
2. The performance results for both the demonstrators show that DTSE transformations are complimentary to and even augmenting the effect of the performance related optimizations (a gain between factor 2.5 and 3.9 for four

processors).This comes on top of the power and bus load reduction effects of the DTSE step.

3. Automatic parallelization of these multimedia demonstrators is possible but requires a sophisticated array privatization scheme (see section 3.2). Modulo operations introduced by DTSE transformations do not effect the inherent parallelism in the code but requires a more complex analysis to detect the parallel loops. Modulo reduction as described in [12] has large benefits. Moreover, using manual directives it is very well possible to obtain large speed-ups in execution times as illustrated in the results.

5 Conclusions

The main conclusions of this paper are : (1) System bus-load and power optimizing DTSE program transformations are complementary to the pure performance related program transformations in data parallel compilers and (2) DTSE transformations do not effect the inherent parallelism present in an application and even enhance the performance itself by reducing the inter-processor communication. So our DTSE approach is fully complementary to the conventional performance related design approaches, and combining the two approaches has major benefits. This is demonstrated on two real-life parallel multimedia demonstrators which have been mapped on a IBM 4-way SMP machine.

References

[1] A.Agarwal, D.Krantz, V.Nataranjan, "Automatic partitioning of parallel loops and data arrays for distributed shared-memory multiprocessors", *IEEE Trans. on Parallel and Distributed Systems*, Vol.6, No.9, pp.943-962, Sep. 1995.

[2] U.Banerjee, R.Eigenmann, A.Nicolau, D.Padua, "Automatic program parallelization", *Proc. of the IEEE*, invited paper, Vol.81, No.2, Feb. 1993.

[3] M.Bister, Y.Taeymans, J.Cornelis, "Automatic Segmentation of Cardiac MR Images", *Computers in Cardiology*, IEEE Computer Society Press, pp.215-218, 1989.

[4] F.Catthoor, S.Wuytack, E.De Greef, F.Balasa, L.Nachtergaele, A.Vandecappelle, "Custom Memory Management Methodology – Exploration of Memory Organization for Embedded Multimedia System Design", ISBN 0-7923-8288-9, Kluwer Acad. Publ., Boston, 1998.

[5] M.Cierniak, W.Li, "Unifying Data and Control Transformations for Distributed Shared-Memory Machines", *Proc. of the SIGPLAN'95 Conf. on Programming Language Design and Implementation*, La Jolla, pp.205-217, Feb. 1995.

[6] K.Danckaert, F.Catthoor and H.De Man, "System-level memory management for weakly parallel image processing", *In proc. EUROPAR-96*, Lecture notes in computer science series, vol. 1124, Lyon, Aug 1996.

[7] E.De Greef, F.Catthoor, H.De Man, "Program transformation strategies for reduced power and memory size in pseudo-regular multimedia applications", accepted for publication in *IEEE Trans. on Circuits and Systems for Video Technology*, 1998.

[8] S. Hummel and E. Schoenberg, "Low-overhead scheduling of nested parallelism",*IBM Journal of Research and Development*, 1991.

[9] C.Kulkarni, F.Catthoor, H.De Man, "Hardware cache optimization for parallel multimedia applications", *In Proc. of EuroPar'98*, Southampton, pp. 923-931, Sept 1998.

[10] C.Kulkarni, D.Moolenaar, L.Nachtergaele, F.Catthoor, H.De Man, "System-level energy-delay exploration for multi-media applications on embedded cores with hardware caches", Accepted for *Journal of VLSI Signal Processing*, special issue on SIPS'97, No.19, Kluwer, Boston, pp., 1999.

[11] S. Muchnick, "Advanced compiler design and implementation", *Morgan Kaufmann Publishers Inc.* , ISBN 1-55860-320-4, 1997.

[12] M.Miranda, F.Catthoor, M.Janssen, H.De Man, "High-level Address Optimization and Synthesis Techniques for Data-Transfer Intensive Applications", *IEEE Trans. on VLSI Systems*, Vol.7, No.1, March 1999.

[13] J.Ph. Diguet, S. Wuytack, F.Catthoor, H.De Man, "Formalized methodology for data reuse exploration in hierarchical memory mappings", *In Proc. Int'l symposium on low power electronics and design*, pp.30-36, Monterey, Ca., Aug 1997.

[14] P.Pirsch, H-J.Stolberg, Y-K.Chen, S.Y.Kung, "Implementation of Media Processors", *IEEE Signal Processing Magazine*, No.4, pp.48-51, July 1997.

[15] P.Strobach, "QSDPCM – A New Technique in Scene Adaptive Coding," *Proc. 4th Eur. Signal Processing Conf.*, EUSIPCO-88, Grenoble, France, Elsevier Publ., Amsterdam, pp.1141–1144, Sep. 1988.

[16] V.Tiwari, S.Malik and A.Wolfe, "Instruction level power analysis and optimization of software", *Journal of VLSI signal processing systems*, vol. 13, pp.223-238, 1996.

[17] P. Tu and D. Padua, "Automatic array privatization", *Proc. 6th Workshop on Languages and Compilers for Parallel Computing*, Portland, OR, August 1993.

[18] M. J. Wolfe, "Optimizing Supercompilers for Supercomputers", *The MIT Press*, 1989.

Parallel Numerical Simulation of a Marine Host-Parasite System

Michel Langlais[1], Guillaume Latu[2], Jean Roman[2], and Patrick Silan[3]

[1] UFR MI 2S, ESA 5466, Université Bordeaux II,
146 Léo Saignat, 33076 Bordeaux Cedex, France
Michel.Langlais@mi2s.u-bordeaux2.fr
[2] LaBRI, UMR CNRS 5800, Université Bordeaux I & ENSERB
351, cours de la Libération, F-33405 Talence, France
{latu|roman}@labri.u-bordeaux.fr
[3] UPR CNRS 9060, Université Montpellier II, Station Méditerranéenne de
l'Environnement Littoral, 1 Quai de la Daurade, 34200 Sète, France
silan@univ-montp2.fr

Abstract. *Diplectanum aequans*, an ectoparasite of the sea bass, can cause pathological problems especially in fish farms. A discrete mathematical model describes the demographic strategy of such fish and parasite populations. Numerical simulations based on this model mimic some of the observed dynamics, and supply hints about the global dynamics of this host-parasite system. Parallelization is required because execution times of the simulator are too long. In this paper, we introduce the biological problem and the associated numerical simulator. Then, a parallel solution is presented, with experimental results on an IBM SP2. The increase in speed has allowed us to improve the accuracy of computation, and to observe new dynamics.

1 Introduction

Host-parasite systems present such a complex behavior that general deterministic models do not provide tools powerful enough to give detailed analysis of any specific system; this is the case of models based on infinite series of ordinary differential equations [8]. Because such a study concerns a pathological problem in fish farming, a validation of the underlying model is expected. In a recent article, Langlais and Silan [6] described a model which is both deterministic and stochastic. In this model, detailed phenomena are reproduced, like spatial and temporal heterogeneities. Numerical simulations and subsequent quantitative analysis of the results can be done. Our first goal in this research and of this simulation is to discover the hierarchy of various mechanisms involved in the host-parasite systems. The second is to understand the sensitivity and stability of the model with respect to the initial conditions. The third is to provide results that can be applied to devise prophylactic methods for fish farming, and more generally in epidemiology. A simulator has already been developed [2, 3], but computation

P. Amestoy et al. (Eds.): Euro-Par'99, LNCS 1685, pp. 677–685, 1999.

time for one simulation can take longer than one month. It is very important to reduce this time to have a usable tool. In this work, we present the main issues concerning the parallel version of the preceding simulator. We point out the complexity of computations and a sequential optimization. Then we develop the parallel algorithmic solution [4, 5] and investigate its performance. Finally we present an overview of the biological results found.

This project is a collaborative effort in an interdisciplinary approach (population dynamics with C.N.R.S, mathematics with Université Bordeaux II, computer science with Université Bordeaux I). This work was supported by C.N.R.S through the Program "Environnement, Vie et Sociétés" and the Program "Architecture, Réseaux et Système, et Parallélisme".

2   Description of the Problem

2.1   Numerical Simulation of Population Dynamics

The numerical simulation is mainly intended to describe the evolution of two populations, fish and parasite, over one year (the time step is $\Delta t = 2$ days). Environmental conditions, like water temperature, and parasite data, are used in the simulations. The final goal is to obtain values of the different variables at each time step. The global algorithm is sketched in Fig. 1.

```
1.  read input parameters;
2.  initialize, compute initial values of data;
3.  for t := 0 to 366 with a time step of 2
    3.1  updating environmental data;
    3.2  lay of eggs by adult parasites;
    3.3  updating the egg population (aging);
    3.4  hatching of eggs (giving swimming larva);
    3.5  updating the larvae population (aging);
    3.6  recruitment of larva by hosts (juvenile parasites);
    3.7  updating the parasite population on hosts (aging);
    3.8  death of overparasitized hosts;
    3.9  output of relevant data;
    endfor
```

Fig. 1. Global algorithm.

We consider that only a surfeit of parasites can cause host death. A detailed age structure of the parasites on a given host is required because only adult parasites lay eggs, while both juvenile and adult parasites have a negative impact on the survival rate of hosts.

2.2   Biomathematical Model

To deal with heterogeneities in both populations, the model has a three dimensional data structure (ignoring the time dimension). Let N be that structure, and let $N_k(l, i, t)$ be the distribution of hosts having l parasites, i being older than $k\Delta t$. Thus, there is an age-structure for parasites (variable k); the fish population is structured into subpopulations having the same total number of parasites (variable l); the parasite distribution on a fish is given by the variable i. The complexity of that dataset comes from the detailed level of modelling we want to reach. The elementary events in the algorithm steps $3.6, 3.7, 3.8$ are quantified into probabilistic fonctions (p, φ, ϕ, ψ) which drive the update of N.

The data N is then modified only once per time step with the following formula :

$$\forall l \in [0, S], \forall i \in [0, l], \forall k \in [2, K]$$

$$N_k(l, i, t + \Delta t) = \sum_{m=i}^{l} \sum_{c=m}^{S} \sum_{d=i}^{c-m+i} N_{k-1}(c, d, t)\, f(l, i, m, c, d) \qquad (1)$$

in which

$$f(l, i, m, c, d) = p(c)\, \varphi(l - m, c)\, \phi(c - m, c)\, \psi(d - i, d, c - m, c) \ .$$

S is limited to the minimum number of parasites which is lethal for a fish (currently $S \leq 700$); K is the number of age classes used ($K = 10$). The accumulated time of this update can take up to 99% of the global simulation time. Our effort of speeding up the application has been mostly concentrated here.

3   Sequential Analysis

Firstly we reorganized computations: some precomputations allow direct memory access to probabilistic functions, while basic factorizations and inversions of sums have been done to reduce computation cost. Evaluating (1) leads to a polynomial complexity $W(S, K)$ with a highest degree term given by $\frac{3K}{40} S^5$. This means, that for one time step, the cost of one update of N can reach $W(700, 10) = 115$ TFLOP. This cost is very high, especially since a complete simulation needs 183 updates. In order to improve performance, our approach has been to vectorize the update of N in order to use BLAS subroutines [1]. A matrix formulation of the algorithm allows us to use BLAS 3 intensively and leads to large speedups. Moreover, it permits us to avoid some computation redundancies in the partial sums. Then, we gain one order of complexity; $W(S, K)$ has a higher degree term of $\frac{K}{4} S^4$ and $W(700, 10)) = 560$ GFLOP. It leads to a new algorithm where intermediate matrices A_1, A_2, A_3 must be filled from the precalculated arrays $\varphi, p, \psi_1, \psi_2, \phi$. The update of N can be now re-written[1] as illustrated in Fig. 2.

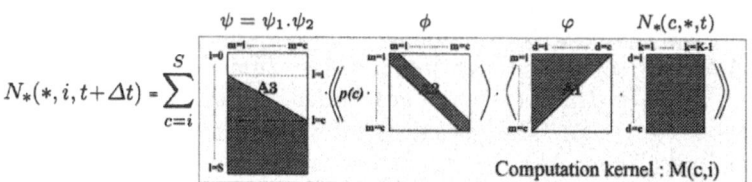

Fig. 2. Matrix form of update of N.

---

[1] The $*$ notation means to consider all elements of one given dimension.

Intermediate matrices contain by construction many zero elements which are represented by light grey sectors in Fig. 2. Since the initial formulation of update of N, we have strongly reduced its cost and we can now study a parallel version.

4   Parallel Solution

4.1   Introduction

For the sake of BLAS performance, it is interesting to parallelize without cutting the computation kernel $M(c, i)$ (defined as the set of matrix multiplications inside the sum, see Fig. 2). Hence, we are led to the following algorithm:

```
Procedure N_Update(S) {
    For i:=0 to S do {
    .   For c:=i to S do {
    .   .   Initialization of A₁, A₂, A₃ matrices;
    .   .   Aux ← A₃.((p(c).A₂).(A₁.N∗(c,∗,t)));   /* task M(c,i) */
    .   .   N∗(∗,i,t+Δt) ← N∗(∗,i,t+Δt) + Aux;
    .   }
    .  }
}
```

Fig. 3. Algorithm for update of N.

If we consider the elementary tasks $M(c, i)$, the algorithm has a medium grain with $(S+1)(S+2)/2$ tasks to distribute over P processors. In parallezing only one loop of the algorithm, we have $S+1$ tasks to map and a coarse grain algorithm. These different possibilities have been investigated; we present in this paper a solution with the parallelization of the outer loop i.

4.2   Data and Computation Distribution

Hereafter, we will refer to the data $N_*(c, *, t)$ as a "row plate", and to $N_*(*, i, t)$ as a "column plate". At iteration $(i = i_0, c = c_0)$ of the algorithm (Fig. 3), the row plate (c_0, t) is used to compute a contribution to column plate $(i_0, t + \Delta t)$. The data are row plates and the results are column plates. The column plate $(i_0, t + \Delta t)$ depends on all row plates (c, t) for which $c \geq i_0$ (Fig. 5). This characteristic implies that two data structures for N are needed at each update (one for time t, the other one for $t + \Delta t$). That means 38 Mb (using simple-precision reals) to dispatch on a parallel machine.

Concerning computations, individual tasks M are independant. As $M(*, i_0)$ compute contributions to column plate $(i_0, t + \Delta t)$, it is interesting to map together on the same processor the data $(i_0, t + \Delta t)$ and all tasks $M(*, i_0)$ in order to execute the matrix addition of the algorithm locally. Indices i_0 are mapped onto processors with a *snake distribution*. Hence, we get a well-balanced work load. The algorithm is re-formulated to take care explicitly of this mapping (Fig. 4); the mapC(c) function returns the processor number that owns the row plate (c, t) (idem for mapI(i) and column plate $(i, t+\Delta t)$). Because row plates are used and column plates are generated, a transposition step is needed to perform several successive updates. The cost in communication of this transposition is the same as the storage amount of $N_*(*, *, t)$ ($\frac{KS^2}{2}$) and is achieved using the MPI subroutine [7] MPI_AllToAll. The broadcast (line 4 of algorithm of Fig. 4) used in the algorithm is pipelined in practice and uses asynchronous communication to take advantage of communication overlap by computation.

For this model, we have proved that because of the snake distribution the computation cost for each row plate is a decreasing function of processor number; the proof uses the theoretical complexity of $M(c, i)$. So for the entire update of N, processor 0 has the heaviest computation task and processor $P-1$ has the lightest one. On the hypothesis that the implementation manages to send the row plates (with the broadcast) to other processors efficiently, the execution time of processors must be a decreasing function of processor number. Our implementation has these properties as shown in run-time experiments of Fig. 6.

```
Procedure N_Update(S) {
  For c:=0 to S do {
  . If (mapC(c)=my_id) {
  .   Broadcast N_*(c, *, t)
  . }
  . For i:=0 to c do {
  . . If (mapI(i)=my_id) {
  . . Initialization of A_1, A_2, A_3 matrices;
  . . Aux ← A_3.((p(c).A_2).(A_1.N_*(c, *, t)));
  . . N_*(*, i, t+Δt) ← N_*(*, i, t+Δt) + Aux;
  . .}
  . }
  }
  Transpose N from column
  plates to row plates;
}
```

Fig. 4. Algorithm for update of N.

$N_*(*, *, t + \Delta t)$ $N_*(*, *, t)$

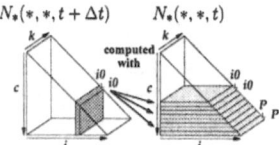

Fig. 5. Data dependencies.

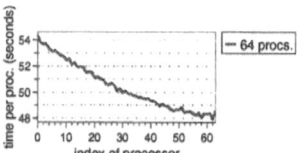

Fig. 6. Time of computation per node for one update of N (S=700)

4.3 Algorithmic Study

The asynchronous pipelined broadcasts of row plates on the P processors mean a communication cost of $\frac{KPS^2}{2}$. That quantity is less than the computation cost $W(S, K)$ and a full computation-communication overlap is obtainable. The cost in communication for transposition of N (*i.e.* $\frac{KS^2}{2}$) is negligible compared to $W(S, K)$. The implementation has confirmed these two elements. The only source of possible load imbalance is work distribution. In the next following part of this section, we will study the efficiency and the scalability.

We now consider the complexity in time $T_{in}(S, K, c, i)$ of one inner iteration (algorithm of Fig. 3), $T_{out}(S, K, i)$ the complexity of one outer iteration of the algorithm, and the linked total sequential complexity in time of one update of N: $T_{seq}(S, K)$. For three reasons, we cannot evaluate $T_{in}(S, K, c, i)$ by simply counting the number of operations performed: the use of BLAS modifies the cost, initialization of intermediate matrices is not negligible, and finally caching effects have an impact. We deduce that complexity in time $T_{seq}(S, K)$ differs from the number of operations $W(S, K)$. Because of the order of matrix multiplications in $M(c, i)$ (see Fig. 2), and because K is a shared dimension of these multiplications and K is constant, the cost in time $T_{in}(S, K, c, i)$ cannot go further than a homogenous polynomial of degree 2 in c, i and S. We have

$$T_{out}(S, K, i) = \sum_{c=i}^{S} T_{in}(S, K, c, i) \ , \qquad T_{seq}(S, K) = \sum_{i=0}^{S} T_{out}(S, K, i) \ .$$

Let $T_{//}(S, K, P)$ be the parallel execution time on P processors. For simplicity, let assume that $S+1$ is a multiple of $2P$. The parallel execution time corresponds to the execution time on the processor 0, because of the snake distribution:

$$T_{//}(S, K, P) = P. \sum_{j=0}^{\frac{S+1}{2P}-1} T_{out}(S, K, 2Pj) + T_{out}(S, K, 2Pj + 2P - 1) \ .$$

$T_0(S, K, P) = P.T_{//}(S, K, P) - T_{seq}(S, K)$ is the overhead induced by load imbalance. Using the polynomial $T_{in}(S, K, c, i)$, this expression yields[2] $T_0(S, K, P) \approx \Theta(P^2 S^2)$. This shows that the overhead cost for each update increases quadratically in the number of processors P and in the size S. $W(S, K)$ is of the same order as $T_{seq}(S, K)$. Let $\varepsilon(K)$ such as $T_{seq}(S, K) \approx \varepsilon(K) S^4$. Let $\gamma(K)$ be a factor which remains to be found, and $E(S, K, P)$ be the efficiency (as defined in [5])

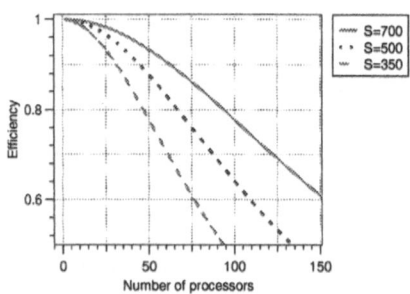

$$\frac{T_0(S, K, P)}{T_{seq}(S, K)} \approx \gamma(K) \frac{P^2}{S^2} \ , \qquad (2)$$

$$E(S, K, P) = \frac{1}{1 + \frac{T_0(S, K, P)}{T_{seq}(S, K)}} \ ,$$

Fig. 7. Approximated efficiency with $\gamma(K = 10) = 14$.

$$E(S, K, P) \approx \frac{1}{1 + \gamma(K) \frac{P^2}{S^2}} \ . \qquad (3)$$

---

[2] This is a tedious derivation performed using *Maple*.

For $K = 10$, experimental executions give for a lower bound of the observed efficiency, $\gamma(10) = 14$. The resulting approximate efficiency is plotted in Fig. 7. If we fix the minimal acceptable efficiency at 80%, we deduce from (3) the inequality $\frac{S}{P} > \sqrt{56} \approx 7.5$. This relation between S and P highlights the conditions for an efficient execution. In the case of 32 processors, the parallel update of N is efficient for $S \in [240, 700]$ (*i.e.* problems between 8 GFLOP and 560 GFLOP); for 64 processors, S must be in the range $[479, 700]$ (123 to 560 GFLOP). For up to 64 processors, we notice that really costly updates can be run with parallel efficiency; for this reason we say the algorithm is scalable. As $T_{seq}(S, K) \approx \varepsilon(K)S^4$, and $T_0(S, K, P) \approx \Theta(P^2 S^2)$, then $T_0(S, K, P) \ll T_{seq}(S, K)$ when P is negligible compared to S. Under this assumption, we conclude the algorithm is cost-optimal, that is, $P.T_{//}(S, K, P) = \Theta(T_{seq}(S, K))$. In deriving the formula of isoefficiency from (3) ($E(K, S, P)$ and K fixed), we get $S = \Theta(P)$ which implies a scalable algorithm. When the number of processors is doubled, S must also be doubled to have the same efficiency (this result is illustrated in Fig. 7 for $S = 350$, $S = 700$).

5   Numerical Experiments

5.1   Performance

Simulations have been performed on two IBM SP2. At LaBRI[3], the machine has 16 Power 2 thin nodes (66 MHz), with 64 or 128 MBytes. At CNUSC[4], there are 160 Power 2 Super Chip thin nodes (120 MHz), 256 MBytes of memory; an SP switch TB3 manages the interconnection of nodes. The results presented in this paper comes from the use of the machine of CNUSC. The code has been developed in FORTRAN 90 with the XL Fortan compiler and using the MPI message-passing library (IBM proprietary version).

	Number of processors				
	4	8	16	32	64
Execution time	782.4 s	392.0 s	196.8 s	100.4 s	54.3 s
Relative efficiency	1.00	1.00	0.99	0.97	0.90
Performance (MFLOPS)	179	178	178	174	161

Table 1. Execution time, relative efficiency and average number of operations per second and per processor for the update of N ($S = 700$, $K = 10$ *i.e.* 560 GFLOP).

	Number of processors			
	8	16	32	64
Execution time	38633.8 s	19511.4 s	10014.8 s	5534.8 s

Table 2. Execution time for one complete and costly simulation (88 time steps with $S = 700$).

---

[3] Laboratoire Bordelais de Recherche en Informatique.
[4] Centre National Universitaire Sud de Calcul.

Beyond the main update of N, other parts of the code have been parallelized, and few sequential operations remain. For realistic simulations the value of S remains 700 during much of the time because of the requirements of the model and the upper limit of S: $S \leq 700$. It is important to have good performance in these circumstances, which is seen in table 1. One update of N with $S = 700$ reaches a relative efficiency of 0.90 on 64 nodes. For a complete simulation we remark that the time is roughly divided by two when the number of processors is doubled. However, the performance per processor reaches roughly 33% of peak performance. There are two reasons for that: firstly the filling of the intermediate matrices (A_1, A_2, A_3) takes much time; secondly, BLAS multiplications uses matrices with one dimension equal to $K-1=9$ which does not lead to large speedups.

5.2   Biological Results

When changing the number of processors, parallel computations are ordered slightly differently, which causes approximation errors. When using double-precision reals for computation, all variables observed during the entire simulation remain identical up to at least the first 7 significant digits (and 4 significant digits for simple-precision reals). Hence, double-precision improves numerical stability but the cost in computation is nearly 40% higher. Numerical stability is important from the biologists and the mathematicians points of view: the observed phenomena in numeric simulations must come from the model and not from computation artifacts.

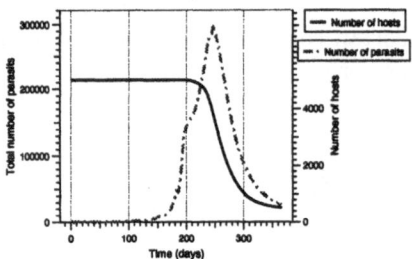

Fig. 8. Temporary endemic state at the end of simulation.

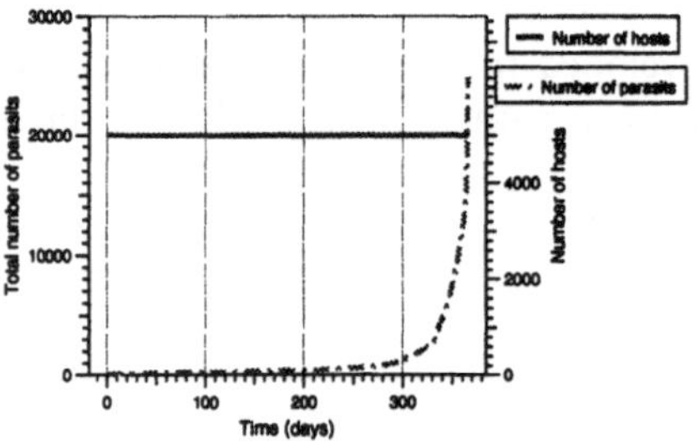

Fig. 9. Growth of parasite population without epidemic.

With the sequential simulator only two dynamic behaviors were observed: first, host extinction because of a parasite epidemic, second, parasite extinction without any host death. With the parallel version, the computations are more accurate (approximations are removed), and fast computations allow to simulate many realistic dynamics. Thus, we have obtained rapidly two other dynamic behaviors (Fig. 8 and 9) corresponding to observed biological cases:

- the fish population does not go extinct after introduction of parasites. A host population regulation is observed and a temporary endemic state is achieved;

- the parasite population develops, but it has no impact on the host population dynamics.

The execution times for different number of processors of the simulation of Fig. 8 are presented in table 2.

6 Conclusion

A complete and costly simulation lasts only 1h30 on 64 nodes (simple-precision reals) versus a month for the sequential simulation. The performance analysis has established the efficiency and scalability (checked until 64 nodes) of the parallel algorithm. These results allow us to improve to a more refined biological model, the number of age classes of parasites increasing from $K = 10$ to $K = 50$.
In forthcoming developments, other distributions of data and computations will be investigated.

References

[1] E. Anderson, Z. Bai, J. Bischof, J. Demmel, J.J. Dongarra, J. DuCroz, A. Green-baum, S. Hammarling, A. McKenney, S. Ostrouchov, and D. Sorensen. *LAPACK User's Guide*, SIAM Publications, 2nd edition, 1995.

[2] C. Bouloux. *Modélisation, simulations et analyse mathématique de systèmes hôtes-parasites*. PhD thesis, Université Bordeaux I, December 1997.

[3] C. Bouloux, M. Langlais, and P. Silan. A marine host-parasite model with direct biological cycle and age structure. *Ecological Modelling*, 107:73–86, 1998.

[4] I. Foster. *Designing and Building Parallel Programs*. Reading, MA: Addison-Wesley, 1995. Available over the World Wide Web at http://www.mcs.anl.gov/dbpp.

[5] A. Grama, A. Gupta, V. Kumar, and G. Karypis. *Introduction to Parallel Computing: Design and Analysis of Algorithms*. Benjamin/Cummings Publishing Company, Redwood City, 1994.

[6] M. Langlais and P. Silan. Theoretical and mathematical approach of some regulation mechanisms in a marine host-parasite system. *Journal of Biological Systems*, 3(2):559–568, 1995.

[7] Message Passing Interface Forum. MPI: A Message-Passing Interface Standard. version 1.1 June 12, 1995. Available by anonymous ftp from ftp.mcs.anl.gov.

[8] P. Silan, M. Langlais, and C. Bouloux. Dynamique des populations et modélisation : Application aux systèmes hôtes-macroparasites et à l'épidémiologie en environnement marin. In C.N.R.S eds, editor, *Tendances nouvelles en modélisation pour l'environnement*. Elsevier, 1997.

Parallel Methods of Training for Multilayer Neural Network

El Mostafa Daoudi* and El Miloud Jaâra

Université Mohammed 1er, Faculté des Sciences
Département de Mathématiques et d'Informatique
60 000 Oujda, Maroc (MOROCCO)
{mdaoudi,jaara}@sciences.univ-oujda.ac.ma

Abstract. In this paper, we present two parallel techniques of training for multilayer neural network. One technique is based on the duplication of the network but the other one is based on the distribution of the multilayer network onto processors. We only have implemented the first parallel technique under PVM, but the parallel implementations for the second one are in progress .

1 Introduction

Various parallel methods of training for multilayer neural networks on distributed memory architectures have been proposed in the literature. Among them:

a. The partitioning method of the network [4]: it consists in partitioning the cells (neural) of each layer to different processors. The distribution of weights is determined by the distribution of the cells. For example, the input weights of a cell are affected to the same processor which contains this cell.
b. The duplication method of the network [2]: it consists in duplicating the same network in different processors. The update of the weights, in each processor, is realized after collection of weights from the other processors.

2 Duplication Method of the Network

This method consists in assigning the same network to each processor of a distributed memory multiprocessor. The n examples, $n = pk$, taken in the basis of training, are uniformly distributed on the p processors in such a way that, every network learns, simultaneously on each processor, a set of k examples. After the locally training phase, the global modification of weights, on each processor, is done by the average sum of the locally weights collected from the other processors which require a total exchange (communication phase) between processors.

---

* Supported by the European project INCO-DC "DAPPI"

P. Amestoy et al. (Eds.): Euro-Par'99, LNCS 1685, pp. 686–690, 1999.

Therefore, after modification, all processors will have the same weights. So, we deduce that the total communication time of training is :

$$m(p-1)\left(\beta + \left(\sum_{j=1}^{N+1} n[j].n[j-1]\right)\tau\right)$$

where m is the number of the necessary communication phases, N is the number of layers in the network, n[j] is the number of cells of the layer j, β is the start up time and τ is the transfer time of one data.

The remaining of this section is devoted to present a technique which improves the previous method. It is based on the same principle as the duplication method but with a different distribution of examples. Instead of training a packet of k examples on each processor P_i with $0 \le i \le p-1$, we propose to affect at each processor P_i a set of $(k + k')$ examples with $0 \le k' \le k$, in such a way that the intersection of the affected examples between two neighbouring processors is not empty. We note that the duplication method of the network is obtained for $k' = 0$. The principle of this distribution consists, at first, in assigning to each processor a packet of k examples, according to the previous distribution, then we increase the number of examples, in each processors P_i, by k' examples chosen from processor $P_{(i+1)\bmod p}$. Using this technique, the obtained weight matrices in each processor, after the locally training phase, can be very close. Therefore, the number of communication phases is reduced to $m' \le m$ (see table 1).

The experimental results are obtained on the distributed memory machine T-node (TN310) using PVM. For the set of tests, we have used a basis of 24 examples, each one represents a character coded on a grid of 7x7 cells[1](see figure 1). The input layer is composed of 49 cells, the intermediate layer is composed of 8 cells and the output layer is composed of 24 cells. Following

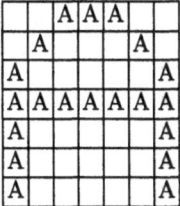
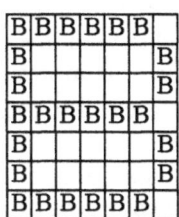

Fig. 1. Matrices representation of characters A and B

the values of k', we have compared the global execution time and the number of communication phases of training according to the number of processors. The obtained results (Table 1) show that the proposed technique is better than the duplication method and also show a clear improvement of the number of communication phases, therefore, a reduction of communication time with regard to the first technique. We note that the values of communication parameters on our system under PVM are $\beta = 0.02$ seconds and $\tau = 22\mu$ seconds.

Table 1. The obtained results following values of p and k'

k'	Execution time (seconds)					Number of communication phases				
	0	$\lceil \frac{k}{4} \rceil$	$\lceil \frac{k}{3} \rceil$	$\lceil \frac{k}{2} \rceil$	k	0	$\lceil \frac{k}{4} \rceil$	$\lceil \frac{k}{3} \rceil$	$\lceil \frac{k}{2} \rceil$	k
$p = 2$	50	70	56	60	43	113	175	72	83	3
$p = 4$	76	61	58	70	48	267	226	183	227	87
$p = 8$	265	206	206	136	74	865	710	710	433	186

3   Distribution Method of the Network

During each communication phase of the duplication method, a total exchange of matrices, each one of size $n \times n$, is needed, which leads to a very high communication overhead when n is very large. Another drawback is the use of the memory space since each network is affected to one processor. In order to avoid the duplication of the network and decrease the exchanged data volume, we propose an improved approach of the partitioning method of the network. Without loss of generality, we assume that p divide $n[k]$.

- **Propagation phase:** it consists, for each layer k ($k \geq 1$), in assigning at each processor $\frac{n[k]}{p}$ cells of the layer k and the corresponding input connection weights with the layer $(k - 1)$ and their corresponding states. The computation of the states of the cells in layer k can be done simultaneously in each processor, since the computation of the states needs only the input connection weights from layer $(k - 1)$ and their states. Then, for computing the states of the cells in layer $(k + 1)$ a total exchange between processors is needed since each processor only contains $\frac{n[k]}{p}$ cells of layer k and their corresponding states.
- **Back-propagation phase:** for the last layer (layer N), during the propagation phase, each processor also computes the gradients relatively to each local cells. During this phase, for every layer k ($k = N, \cdots, 2$), each processor computes the gradients of the local cells and updates the corresponding weights. Since each processor contains $\frac{n[k]}{p}$ gradients and the corresponding $\frac{n[k]}{p}$ weights, a total exchange of ($\frac{n[k]}{p}$ gradients and $\frac{n[k]}{p}$ weights) between processors is needed for every layer k ($1 < k \leq N$).

This technique is very simple to analyze and to implement on real distributed machine. Figure 2 illustrate the distribution of a network composed of 3 layers with $N = 3$ and for $p = 3$.

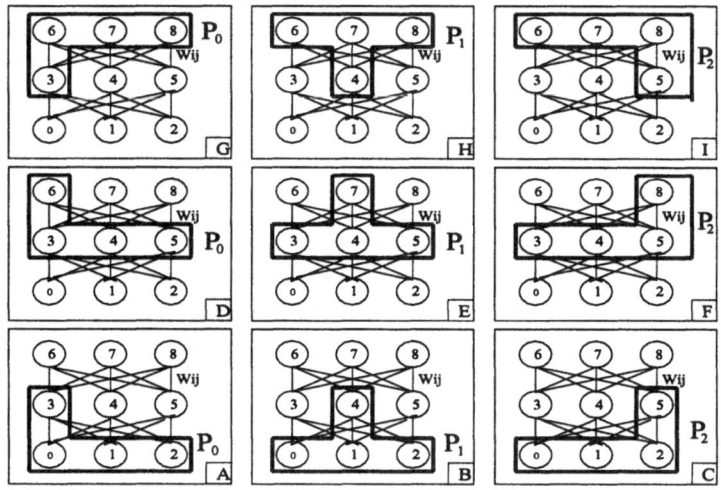

Fig. 2. Distribution of weights and cells on processors

Propagation Phase

1st **stage** (In intermediate layer)

- Processors P_0, P_1 and P_2 compute simultaneously the states of the cells 3, 4 and 5.

- Total exchange of states of the cells 3, 4 and 5 between processors P_0, P_1 and P_2.

2nd **stage** (In output layer)

- Processors P_0, P_1 and P_2 compute simultaneously the states of the cells 6, 7 and 8.

Back-Propagation Phase

- Processors P_0, P_1 and P_2 compute simultaneously the gradient of the cells 6, 7 and 8 and their corresponding weights.

- Total exchange of the gradients of the cells 6, 7 and 8 and their corresponding weights between processors P_0, P_1 and P_2.

3rd **stage** (In intermediate layer)

- Processors P_0, P_1 and P_2 compute simultaneously the states and the gradients of cells 3, 4 and 5 and update the corresponding weights.

The process is repeated for another example until convergence of network.

3.1 Communication Time

In this section we estimate the quantity of the data which will be exchanged during the training of one example.

- **Propagation phase:** in order to compute the states of the cells in layer $(k + 1)$ $(k=1,2,...N-1)$, each processor needs the states of all cells from layer k. Since each processor contains only $\frac{n[k]}{p}$ states, this stage needs a total exchange between processors in order to collect the remaining states from other processors. We deduce that the volume of exchanged data is : $\sum_{k=1}^{N-1} \frac{n[k]}{p}$

- **Back-propagation Phase:** for every layer k $(k = N, \cdots, 2)$, each processor computes the gradient of the local cells and updates the corresponding weights. Since each processor contains $\frac{n[k]}{p}$ gradients and the corresponding $\frac{n[k]}{p}$ weights, a total exchange of ($\frac{n[k]}{p}$ gradients and $\frac{n[k]}{p}$ weights) between processors is needed. Therefore the volume of exchanged data is : $2\sum_{k=2}^{N} \frac{n[k]}{p}$

In the following, we give the communication times, for one iteration of the training on one example, estimated on three different architectures :

Ring: $(p - 1) \left(2(N - 1)\beta + \frac{1}{p} \left(n[1] + 3 \sum_{k=2}^{N-1} n[k] + 2n[N] \right) \tau \right)$

Hypercube: $\left(2log_2(p)(N - 1)\beta + \frac{p-1}{p} \left(n[1] + 3 \sum_{k=2}^{N-1} n[k] + 2n[N] \right) \tau \right)$

Complete network: $\left(2(N - 1)\beta + \frac{1}{p} \left(n[1] + 3 \sum_{k=2}^{N-1} n[k] + 2n[N] \right) \tau \right)$

4   Conclusion

We have presented, at first, a parallel implementation based on the duplication of the layer network which, experimentally, reduces the communication cost compared to the one proposed in the literature. Then we have proposed another technique based on the distribution of the network which theoretically reduces the volume of exchanged data. The implementations of the second technique on distributed memory, more difficult than the first one, are in progress.

References

[1] H. NAIT CHARIF, " A Fault Tolerant Learning Algorithm for Feedforward Neural Networks " *Conférence FTPD 1996, Hawaï.*
[2] H. PAUGAM-MOISY, "Réseaux de Neurones Artificiels : Parallélisme, Apprentissage et Modélisation " *Habilitation à Diriger des Recherches, Ecole Normale Supérieure de Lyon, 6 Janvier 1997.*
[3] H. PAUGAM-MOISY et A. PÉTROWSKI, " Parallel Neural Computation Based on Algebraic Partitioning " *In I. Pitas, editor, Parallel Algorithms for Digital Image Processing, Computer Vision and Neural Networks p. 259-304. John Wiley, 1993.*
[4] S. WANG, " Réseaux Multicouches de Neurones Artificiels ", *Thèse de Doctorat, Institut National Polytechnique de Grenoble, 26 Septembre 1989.*

Partitioning of Vector-Topological Data for Parallel GIS Operations: Assessment and Performance Analysis

Terence M. Sloan[1], Michael J. Mineter[2], Steve Dowers[2], Connor Mulholland[1], Gordon Darling[1], and Bruce M. Gittings[2]

[1] EPCC, University of Edinburgh, EH9 3JZ, UK,
tms@epcc.ed.ac.uk,
http://www.epcc.ed.ac.uk/
[2] Department of Geography, University of Edinburgh EH8 9XP, UK

Abstract. Geographical Information Systems (GIS) are able to manipulate spatial data. Such spatial data can be available in a variety of formats, one of the most important of which is the vector-topological. This format retains the topological relationships between geographical features and is commonly used in a range of geographical data analyses. This paper describes the implementation and performance of a parallel data partitioning algorithm for the input of vector-topological data to parallel processes.

1 Introduction

This paper describes a parallel algorithm for the partitioning of vector-topological geographical data and provides some initial results on its performance. This work has been undertaken as part of a project to develop parallel algorithms for analyses of geographical data. This project is a collaboration between EPCC and GIS-PAL both of whom are located in the University of Edinburgh, UK. It is funded by the UK funding body EPSRC.

2 Vector-Topological Data

The vector-topology model is at the core of most commercial GIS, ranging from traditional products such as ESRI's ARC/INFO through to the more recent object-oriented systems such as Smallworld and LaserScan Gothic. The vector representation of real-world objects by combinations of points, lines and areas allows the description of objects' occupancy of space. The addition of topology allows the expression of the relationships between these components and provides a formal description of the spatial relationships between the real-world objects [1]. Vector-topological datasets are wide-spread and increasingly available from a variety of sources. They include administrative, political, social and environmental data such as census enumeration districts, wards, counties, parishes, national parks, land use, land cover, geology, soil etc. Vector-topological datasets are thus also at the core of a wide range of applications.

P. Amestoy et al. (Eds.): Euro-Par'99, LNCS 1685, pp. 691–694, 1999.

3   The Parallel Data Partitioning Algorithm

The data partitioning algorithm utilised in this paper is based on the design described by Sloan and Dowers in [2]. The algorithm entails partitioning data into sub-areas and assigning one or more to each processor. This approach is preferred to a "pipeline" approach as it maximises the exploitation of existing sequential algorithms. It also allows large datasets to be analysed since the dataset size can be greater than that which can reside in one processor.

The partitioning of vector-topological data is not straightforward since the data comprises multiple types of records, interrelated by means of record identifiers. Typically the vertices are held in one record type, the relationship between polygons and vertices may be held in another and the attribute data are related to polygons by a third record type. Moreover, geographical operations on vector-topological data generally require : the spatial coordinates of the boundaries between geographical features where these boundaries must be sorted by their uppermost y coordinate and the attributes of the geographical features on either side of these boundaries.

The algorithm proceeds through three phases to extract and efficiently merge-sort the data from one or more datasets, according to these requirements. These phases are the Sort, Join and Geom-Attribute Distribution (GAD) phases that partition the data into sub-areas from the multiple input datasets. Parallel processes are used to extract and sort information efficiently. For the Sort and Join phases a minimum of two processes is required. A minimum of three processes is necessary to perform the different roles in the GAD phase. During the Sort and Join phases the participating processes are split into two different groups, the Source and Pool Groups. There is only one process in the Source group which coordinates the actions of the processes in the Pool groups during the Sort and Join phases and gathers the information on data distribution. The Pool processes perform the extraction and merge-sorting of data from the input datasets. During the GAD phase the Pool processes are further divided into the GeomAttributeServer group and the Worker group. Here the processes of the GeomAttributeServer transfer to a Worker process all the data necessary for a GIS operation to be applied on a part of the geographical area of interest.

4   The Implementation and Its Performance

The Sort and Join Phases of this algorithm have been implemented in ANSI C with the MPICH 1.1.1 version [3] of the MPI message passing standard. The original design makes extensive use of parallel I/O facilities where a file is partitioned across a number of disks and accessed by concurrent processes. This implementation does not make use of such facilities. In the Sort phase, the data input file resides on a single disk. The Source process handles all read accesses to the input file from all the other processes. The outputs from the Sort and Join phases are a number of files each containing a sorted list of records of the same type. For each record type there can be one or more files. The number of files is dependent upon the memory available to each Pool process. The memory available is a tunable parameter.

The implementation was executed during a period of exclusive access to a 4 processor Sun Enterprise 3000 with 1 GB of shared memory. Each processor is a 250 MHz UltraSparc. All executions were repeated to ensure timings were consistent. Three datasets of 2.3 MB, 4.1 MB and 11.6 MB in size of geographical data residing on a local disk were used in the timings.

The graph in figure 1 displays the overall Sort/Join elapsed times for 2, 3 and 4 processors with 1 process per processor. (Note the Sort and Join phases require a minimum of 2 processes.) It is clear from the graph that the algorithm is not scalable and in fact degrades at 4 processes. Figures 2 and 3 display the elapsed times for the Sort and Join phases respectively. From figure 2 it can be seen that there is some speed-up in the Sort elapsed time as the number of processors increase. However this speed-up is insignificant compared to the overall time. Software profiling tools reveal the lack of speed-up is due to the the high number of small messages (less than 30 bytes) passing between the Source process (acting as the file server) and the other processes.

In figure 3 it is clear that Join performance degrades as the number of processors increase. This is due to the small number of input files to the phase. The Join phase allocates the files to the Pool processes for merge-sorting. When there are few files, most of the Pool processes stand idle. This is a feature of the design since it was originally intended for use on systems with a parallel file I/O facility and where the input dataset was much larger than available memory [2]. This situation is simulated by reducing the previously mentioned *memory available parameter* within the Pool processes. When this is reduced more input files for the Join phase are created. Since there is no parallel file I/O facility in the implementation these files must be opened directly by the processes. However, the number of file pointers a process can have open is fixed by the operating system. When this number is exceeded, the code fails. This means that for large datasets, when a small number of processes are used the Join phase fails. It will, however, succeed for larger numbers of processes. This can be demonstrated by the fact that the implementation was not able to operate on a 45 MB input dataset when there were 2 and 3 processes but succeeded when then there were 4 processes.

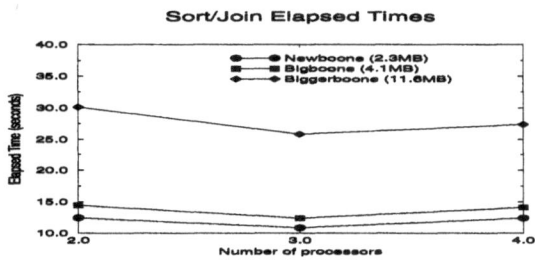

Fig. 1. Elapsed Times for Sort/Join

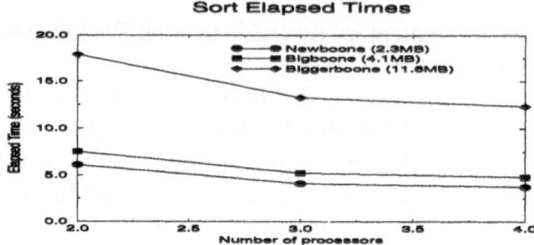

Fig. 2. Elapsed Times for Sort

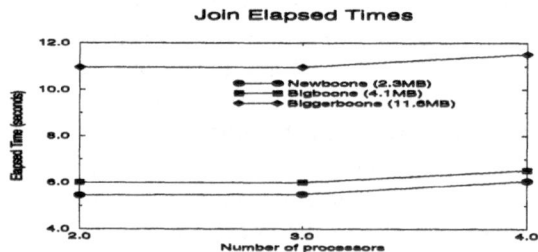

Fig. 3. Elapsed Times for Join

5   Conclusions and Future Work

The performance of the implementation could be improved further by altering the Source process in the Sort phase to buffer records into larger messages to be sent to the Pool processes. Furthermore the provision of a parallel file I/O facility where the input dataset was spread across a number of disks - as intended in the original design - would also improve performance. In the Join phase, such a parallel file I/O facility would also prove beneficial since it would allow valid comparisons of performance between small and larger numbers of processes on large datasets.

References

[1] Worboys, M.: *Representation and Algorithms*. GIS: A Computing Perspective. Taylor and Francis (1995)
[2] Sloan, T.M., Dowers,S.: *Parallel Vector Data Input*. Parallel Processing Algorithms for GIS. Taylow and Francis (1998) 151–178
[3] Gropp, W., Lusk, E.: User's guide for mpich, a Portable Implementation of MPI. University of Chicago ANL/MCS-TM-ANL-96/6 (1998)

Topic 09
Parallel Computer Architecture - What Is Its Future?

Chris Jesshope

Global Chair

The Past

It gives me great pleasure to write this introduction to the Workshop 9 on Parallel Computer Architecture. This subject has a long and interesting history and is still of great interest to me even though I have been involved in it since the mid 1970s, having been one of the first users of the pioneering Illiac V Computer, which was installed at NASA Ames. Much has happened since those early pioneering times. I am sure that you are as aware as I am of the driver of the rapid changes in this field over its first two decades. This has primarily been the exponential growth of the underlying silicon technology. Ironically the braking factor, in innovation at least, over the last decade has been the market forces in the same industry and its economies of scale. It is now no longer viable to set up a silicon foundry and design a processor and compete in the market, the investment is too large. What has happened therefore is that we have seen the diversity in this field dissipate, due to economic factors. But not only this, I must add, because there is a maturity of the knowledge in this area. Indeed the impression gained of the subject over the last decade has been one of consolidation and the application of sound engineering. Parallel Computer Architecture has come of age but the outcome is not as the early pioneers may have imagined it. The major commercial thrust is in the business sector where the development of server architectures has provided much of the innovation, using commodity microprocessors and bus-based symmetric multi-processor architecture. There is also development of massively parallel "supercomputers" but the number of manufacturers now in this sector of the business has been severely curtailed and again the commodity processor rules the roost.

The Future

So, where are we going from here? For what it may be worth, I will give you my opinion of the longer term future, so that you can put the papers to be presented in this workshops in perspective. I believe that the continued growth and abundance of silicon real estate has allowed inefficiency to creep into the design of microprocessors. The current thrust has been to extract as much instruction-level parallelism as possible from existing code and to use out-of-order execution

P. Amestoy et al. (Eds.): Euro-Par'99, LNCS 1685, pp. 695–697, 1999.
© Springer-Verlag Berlin Heidelberg 1999

and speculation in order to overcome the many dependencies that are to be found in compiled code. Just looking at the silicon real estate used in support of these techniques, one is led to wonder why a more explicitly parallel solution has not evolved. The silicon area used in these current, complex microprocessors would support many simpler pipelined datapaths (and their register sets). Why then has such an explicitly parallel approach not been followed, when at best, even with 6-way, superscalar issue it is seemly impossible to get an IPC of more than 2. The problems in this approach are soluable; multi-threading in microprocessor design allows efficient use of high latency shared memories and caching can limit the parallel slackness required to exploit a number of parallel datapaths on a single chip. The answer is that such a scheme would require recompilation, or at least very good analysis tools that could extract explicit parallelism from compiled code. If we look at the history of microprocessor design however, it has made one major paradigm shift, from CISC to mainly RISC based design. A shift from implicit to explicit parallelism, which would allow the development of on-chip, parallel architectures, would not therefore be so far fetched and the knowledge required is already being published in today's literature, including these proceedings. Another area that the abundance of silicon has aided is the massive advance in custom computing. This field exploits the FPGAs that were initially developed for integrating custom logic onto a single die. Because these chips can be programmed from an array of uncustomised logic gates, at system design time, there are the same economies of scale that have made microprocessors so inexpensive. What started as a means to compress glue logic on motherboards was quickly taken up by a community of researchers looking for custom solutions to (usually) highly parallel problems. Today there is much research being undertaken in this area and it is also represented in this workshop.

The Workshop Papers

Let us now look at the papers in this workshop. They can be divided into three groups. In the first we have a collection of papers concerning conventional parallel computer architecture. These include instruction set architectures for vector computing, cache-coherence protocols, multi-threading and graphics engines:

- Vector ISA Extension for Sparse Matrix-Vector Multiplication - Stamatis Vassiliadis, Sorin Cotofana and Pyrrhos Stathis
- Implementing snoop coherence protocol for future SMP architectures - Wissam Hlayhel, Jacques Collet and Laurent Fesquet
- An Adaptive Limited Pointers Directory Scheme for Cache Coherence of Scalable Multiprocessors - Cheol Ho Park, Jong Hyuk Choi, Kyu Ho Park and Daeyeon Park
- Two Schemes to Improve the Performance of a Sort-last 3D Parallel Rendering Machine with Texture Caches - Alexis Vartanian, Jean-Luc Bechennec and Nathalie Drach-Temam

- A Study of a Simultaneous Multithreaded Processor Implementation - Dominik Madon, Eduardo Sanchez and Stefan Monnier

In the second group we have a number of papers looking at the general area of custom computing:

- The MorphoSys Parallel Reconfigurable System - Guangming Lu and Hartej Singh, Eliseu M. C. Filho and Nader Bagherzadeh
- Design and Analysis of Fixed-size Systolic Arrays for Modular Multiplication - Hyun-Sung Kim, Sung-Woo Lee, Jung-Joon Kim, Tae-Geun Kim and Kee-Young Yoo
- ManArray Processor Interconnection Network: An Introduction - G. G. Pechanek and S. Vassiliadis and N. P. Pitsianis

And finally we have two theoretical papers:

- A graph-oriented task manager for small multiprocessor systems - Xavier Verians, Jean-Didier Legat, Jean-Jacques s, Quisquater and Benoit Macq s
- The Algebraic Path Problem Revisited - Sanjay Rajopadhye, Claude Tadonki and Tanguy Risset

I commend these papers to you and welcome you to this workshop on Parallel Computer Architecture.

The Algebraic Path Problem Revisited

Sanjay Rajopadhye[1], Claude Tadonki[2], and Tanguy Risset[1]

[1] IRISA, Rennes, France, http://www.irisa.fr/cosi/Rajopadhye
[2] University of Yaounde, Cameroon, cmtado@uycdc.uninet.cm

Abstract. We derive an efficient linear SIMD architecture for the algebraic path problem (APP). For a graph with n nodes, our array has n processors, each with $3n$ memory cells, and computes the result in $3n^2 - 2n$ steps. Our array is ideally suited for VLSI, since the controls is simple and the memory can be implemented as fifos. I/O is straightforward, since the array is linear. It can be trivially adapted to run in multiple passes, and moreover, this version *improves* the work efficiency. For any constant α, the running time on $\frac{n}{\alpha}$ processors is no more than $(\alpha+2)n^2$. The work is no more than $(1+\frac{2}{\alpha})n^3$ and can be made as close to n^3 as desired by increasing α.

Keywords: transitive closure, shortest path, matrix inversion, Warshall-Floyd & Gauss-Jordan elimination, systolic synthesis, scheduling, space-time mapping, recurrence equations.

1 Introduction

The algebraic path problem (APP) unifies a number of well-known problems into a single algorithm schema. It may be stated as follows. We are given a weighted graph $G = \langle V, E, w \rangle$ with vertices $V = \{1, 2, \ldots n\}$, edges $E \subseteq V \times V$ and a weight function $w : E \to S$, where S is a closed semiring $\langle S, \oplus, \otimes, *, \mathbf{0}, \mathbf{1} \rangle$ (closed in the sense that $*$ is a unary "closure" operator, defined as the infinite sum $x* = x \oplus (x \otimes x) \oplus (x \otimes x \otimes x) \oplus \ldots$. A path in G is a (possibly infinite) sequence of nodes $p = v_1 \ldots v_k$ and the weight of a path is defined as the product $w(p) = w(v_1, v_2) \otimes w(v_2, v_3) \otimes \ldots \otimes w(v_k, v_{k-2})$. Let $P(i, j)$ denote the (possibly infinite) set of all paths from i to j. The APP is the problem of computing, for all pairs i, j, such that $0 < i, j \leq n$, the value $d(i, j)$ defined as follows ($\bigoplus$ is a "summation" operator for S)

$$d(i, j) = \bigoplus_{p \in P(i,j)} w(p)$$

Warshall's transitive closure (TC) algorithm, Floyd's shortest path (SP) algorithm and the Gauss-Jordan matrix inversion (MI) algorithm are but instances of a single, generic algorithm for the APP (henceforth called the WFGJ algorithm), the only difference being the semiring involved. It's sequential complexity (and hence the work) is n^3 semiring operations.

There has been considerable research on implementing the APP (or particular instances thereof) on systolic arrays (see Table 1). Most of the early work was ad

P. Amestoy et al. (Eds.): Euro-Par'99, LNCS 1685, pp. 698–707, 1999.

Authors	Application	Area	Time
Guibas et al. [5]	TC	n^2	$6n$
Nash-Hansen [9]	MI	$3n^2/2$	$5n$
Robert-Tchuent [13]	MI	n^2	$5n$
Kung-Lo [6]	TC	n^2	$7n$
Rote [15]	APP	n^2	$7n$
Robert-Trystam [14]	APP	n^2	$5n$
Kung-Lo-Lewis [7]	TC & SP	n^2	$5n$
Benaini et al. [1]	APP	$n^2/2$	$5n$
Benaini-Robert [2]	APP	$n^2/3$	$5n$
Scheiman-Cappello [16]	APP	$n^2/3$	$5n$
Clauss-Mongenet-Perrin [4]	APP	$n^2/3$	$5n$
Takaoka-Umehara [17]	SP	n^2	$4n$
Rajopadhye [11]	APP	n^2	$4n$
Risset-Robert [12]	APP	n^2	$4n$
Chang-Tsay [3]	APP	n^2	$4n$
Djamegni et al. [18]	APP	$n^2/3$	$4n$
Linear arrays (systolic and otherwise)			
Kumar-Tsay [10]	APP	n^2	$7n^2$
Kumar-Tsay [10]	APP	n^2	$7n^2$
Myoupo-Fabret [8]	APP	n^2/α	$(3+\frac{4}{\alpha})n^2$

Table 1. Systolic implementations for the WFGJ algorithm and its instances (SP, TC and MI respectively denote shortest path, transitive closure and matrix inversion). The table is not exhaustive but in rough chronological order.

hoc, while later work used systematic design methods. All the implementations take $\Theta(n)$ time on $\Theta(n^2)$ processors. In the early 90's a new localization that reduced the running time from $5n$ to $4n$ was proposed independently by a number of authors [3, 11, 12, 17]. The techniques of Scheiman-Cappello and Benaini-Robert [2, 16] can be used to reduce the number of processors in these fast arrays to $n^2/3$, as shown by Djamegni et al. [18]. However, all the arrays to date sacrifice work efficiency by a constant factor.

In practice this is an important limitation. A direct implementation of any of these architectures by partitioning to run on a dedicated VLSI array will not improve the work complexity.

In this paper, we seek an SIMD VLSI architecture that does not suffer from this problem. We formally derive an architecture with only $\Theta(n)$ processors and $\Theta(n^2)$ time and also improve the work efficiency. Specifically, the array has n processors, and computes the result in $3n^2 - 2n$ time steps. The array can be easily adapted to run in multiple passes on $p < n$ processors, without sacrificing but rather, with a *gain* in work efficiency. With α passes (on $\frac{n}{\alpha}$ processors) the running time is no more than $n^2(\alpha + 2)$, and hence, the work is at most $(1+\frac{2}{\alpha})n^3$. Thus by increasing the number of passes we can approach as close to a work optimal implementation as desired. Finally, on a *fixed* number of processors (i.e., where $\alpha = n/p$ is not *constant* but proportional to n) our implementation is work optimal.

Our architecture is not "purely" systolic because there are three shift registers of length n in each processor. Note however, that since a shift register is simply a linear array of registers, one can argue that our architecture is really a two dimensional $n \times n$ systolic array where only the boundary processors do the actual computation—internal "processors" are just registers. Pragmatically, the array is very modular and ideally suited for VLSI implementation.

The remainder of this paper is organized as follows. The following section describes the notation and necessary background. Section 3 presents and intuitive description of our linear SIMD array. Next, Sect. 4 presents its formal derivation and analysis of its performance. We conclude in Sect. 5.

2   Notation and Background

We now recall the Warshall-Floyd-Gauss-Jordan algorithm for the APP and describe some of its important properties. The algorithm is based on the following SRE, where $D_0 = \{i, j, k \mid 0 < i, j \leq n; 0 \leq k \leq n\}$ is the domain of F.

$$d(i,j) = \{i, j \mid 0 < i, j \leq n\} : F(i, j, n) \tag{1}$$

$$F(i,j,k) = \begin{cases} D_0 \cap \{i, j, k \mid k = 0\} & : a_{i,j} \\ D_0 \cap \{i, j, k \mid i = j = k\} & : F(i, j, k-1)* \\ D_0 \cap \{i, j, k \mid i = k \neq j\} & : F(k, k, k) \otimes F(i, j, k-1) \\ D_0 \cap \{i, j, k \mid j = k \neq i\} & : F(i, j, k-1) \otimes F(k.k.k) \\ D_0 \cap \{i, j, k \mid i \neq k; j \neq k\} & : F(i, j, k-1) \oplus \\ & \quad (F(i, k, k) \otimes F(k, j, k-1)) \end{cases} \tag{2}$$

Apart from the initialization plane, $k = 0$ the domain of F is an $n \times n \times n$ cube. For a given value of k, we will call the point $[k, k, k]$ on the line $i = j = k$ as the *pivot*, and points on the respective planes $i = k \neq j$ and $j = k \neq i$ (i.e., points of the form $[k, j, k]$ and $[i, k, k]$, respectively) as the *pivot row* and *pivot column*. The remainder of the points are called the *interior* points. Observe that the four main (apart from the initialization) clauses of (2) correspond to the pivot, the pivot row, the pivot column and the interior points, respectively. We also call the point $[k+1, k+1, k]$ for $j \neq k$, as the *first interior point* of the k-th plane, and the points $[k+1, j, k]$ as the *first interior row*.

Observe that for any plane $k = $ const, we have the following computation order (assuming an unbounded number of processors). First, the pivot $[k, k, k]$ is computed, since it depends on a value from the "preceding" plane. Next, the rest of the pivot column, namely the points $[i, k, k]$ for $i \neq k$ (respectively, the pivot row—points $[k, j, k]$ for $j \neq k$) are computed, since they depend directly on the pivot. Finally, the interior points are computed since they all depend directly on the pivot column. Also note that the pivot of plane $k+1$ depends directly on the first interior point, and hence no computation in the next plane can start before it. In other words, there is a critical path of length three between two successive pivots, namely $[k, k, k] \rightarrow [k + 1, k, k] \rightarrow [k + 1, k + 1, k] \rightarrow [k + 1, k + 1, k + 1]$. Hence the fastest running time of the algorithm is $3n$. Indeed, it is well known [11] that the optimal parallel schedule for the SRE is as follows.

$$t_f(i, j, k) = \begin{cases} D_0 \cap \{i, j, k \mid k = 0\} & : 0 \\ D_0 \cap \{i, j, k \mid i = j = k\} & : 3k - 2 \\ D_0 \cap \{i, j, k \mid i = k \neq j\} & : 3k - 1 \\ D_0 \cap \{i, j, k \mid j = k \neq i\} & : 3k - 1 \\ D_0 \cap \{i, j, k \mid i \neq k; j \neq k\} & : 3k \end{cases} \tag{3}$$

This schedule assumes an unbounded number of processors, $\Theta(n^2)$, to be precise, with global communication: in particular, broadcasts of the pivot (and the pivot column) are necessary. Table 1 illustrates that the locality of communication required for a systolic implementation imposes a slowdown from $3n$ to $4n$ or more.

Before we proceed further, let us derive a modified version of the SRE (1-2), which we shall use in the remainder of this paper. For notational convenience, we define the operator $\dotplus$ by $i \dotplus j = (i + j) \bmod n$. Now we apply the transformation $(i, j, k \rightarrow (i - k) \bmod n, (j - k) \bmod n, k)$ to all points in the (sub) domain $\{i, j, k \mid 0 < i, j, k \leq n\}$ of D_0. By applying the standard transformation rules for recurrence equations, we obtain the following SRE, where the new domain of F is $D_1 = \{i, j, k \mid 0 \leq i, j < n; 0 < k \leq n\} \cup \{i, j, k \mid k = 0 < i, j \leq n\}$.

$$d(i, j) = \begin{cases} \{i, j \mid 0 < i, j < n\} & : F(i, j, n) \\ \{i, j \mid 0 < i < n; j = n\} & : F(i, 0, n) \\ \{i, j \mid i = n; 0 < j < n\} & : F(0, j, n) \\ \{i, j \mid i = j = n\} & : F(0, 0, n) \end{cases} \tag{4}$$

$$F(i, j, k) = \begin{cases} D_1 \cap \{i, j, k \mid k = 0\} & : a_{i,j} \\ D_1 \cap \{i, j, k \mid i = j = 0\} & : F(i \dotplus 1, j \dotplus 1, k - 1) * \\ D_1 \cap \{i, j, k \mid i = 0 < j\} & : F(i \dotplus 1, j \dotplus 1, k - 1) \otimes F(0, 0, k) \\ D_1 \cap \{i, j, k \mid i > 0 = j\} & : F(0, 0, k) \otimes F(i \dotplus 1, j \dotplus 1, k - 1) \\ D_1 \cap \{i, j, k \mid i, j > 0\} & : F(i \dotplus 1, j \dotplus 1, k - 1) \oplus \\ & \quad (F(i, 0, k) \otimes F(1, j \dotplus 1, k - 1)) \end{cases} \tag{5}$$

What we have just applied is the well known reindexation transformation of Kung Lo and Lewis [7]. It effectively brings the pivot to the vertical line $i = j = 0$ and the pivot rows and columns to the sides ($i = 0$ and $j = 0$) of the domain D_1. It may be visualized as a toroidal shift of each plane relative to the previous (this also explains why the old $F(i, j, k - 1)$ argument is systematically replaced by $F(i \dotplus 1, j \dotplus 1, k - 1)$).

3 A Linear Systolic Array: Intuitive Description

We now present the main intuition behind our array. Specifically, we describe the order in which computations are performed by each processor (the local schedule) without entering into details of whether and why it is valid.

In our architecture, the k-th plane is executed by processor k. Thus each processor computes a $n \times n$ matrix as defined by the SRE (4-5). However, the order in which the processor computes its values is rather special. The elements are computed "principal submatrix" by "principal submatrix" as follows.

– The pivot element $F(0, 0, k)$ is computed first. This is the base case.
– After the $(i-1)$-th principal submatrix has been computed, the i-th row and column of the i-th principal submatrix are calculated in an alternating manner: $F(i, 0, k)$, $F(0, i, k)$, $F(i, 1, k)$, $F(1, i, k)$, $\ldots F(i, i-1, k)$, $F(i-1, i, k)$. Finally the i-th diagonal element, $F(i, i, k)$ is computed. This completes the computation of the i-th principal submatrix.

We emphasize that this specifies the *relative* order of computations performed on a given processor, and that there may be intervening idle cycles. Figure 1-a illustrates this schedule for the first processor (for which there are no idle cycles).

Also note that this is merely the order in which the elements of the k-th slice are computed. They are not **sent** to the next processor in this order (this is what causes idle cycles). Rather, the processor first performs a toroidal shift and then sends the elements (of this shifted matrix) in the same "principal submatrix" order described above. The reason for this is the $(i\dot{+}1, j\dot{+}1, k-1)$ dependency—every computation depends (at least) on the element that would be "just below it" if toroidally shifted. Thus the first interior point is the first value sent, and the pivot is the *last*. In the penultimate $2n-2$ steps, the pivot row and column are sent.

Let us now look at the implications of this on the schedule of processor 2 (Fig. 1.b). Since the first element that it can compute is its own pivot, and this depends on $F(1, 1, 1)$ which is itself computed at $t = 4$, processor 2 starts only at $t = 5$. However, for the next two steps, it is idle, because (i) the values computed by processor 1 are not sent, and moreover (ii) the next computation that processor 2 can perform is $F(1, 0, 2)$, which depends on $F(2, 1, 1)$, and that value is computed only at $t = 7$. From here on, the 2×2 principal submatrix is finished as per the above schedule. At the start of the third principal submatrix, there are again two idle cycles. This repeats at the start of each row (as indicated by the 2 ∎ in Fig. 1.b). However, there are no idle cycles at the start of the *last* row because this corresponds to the previous processor's pivot row and column, which were computed well in advance.

A close look at the schedule of processor 3 (Fig. 1.c) will now make the pattern clear. We see that there are now two additional idle cycles at the start of each row (processor 2 imposes and additional latency of two cycles on each row). However, we also see that in the penultimate row, the processor can now catch up (there are only 2 idle cycles instead of 4). This is because there were no idle cycles in the last row of the *previous* processor. As always, the schedule in the order of the "principal submatrix" and every point $F(i, j, k)$ is computed immediately after the point $F(i\dot{+}1, j\dot{+}1, k-1)$ is sent by the previous processor.

We leave blank the table for processor 4 (Fig. 1.d), as an exercise for the reader. Please fill it up before proceeding further, study it carefully, and in particular, try to determine a closed form formula that gives the value of the

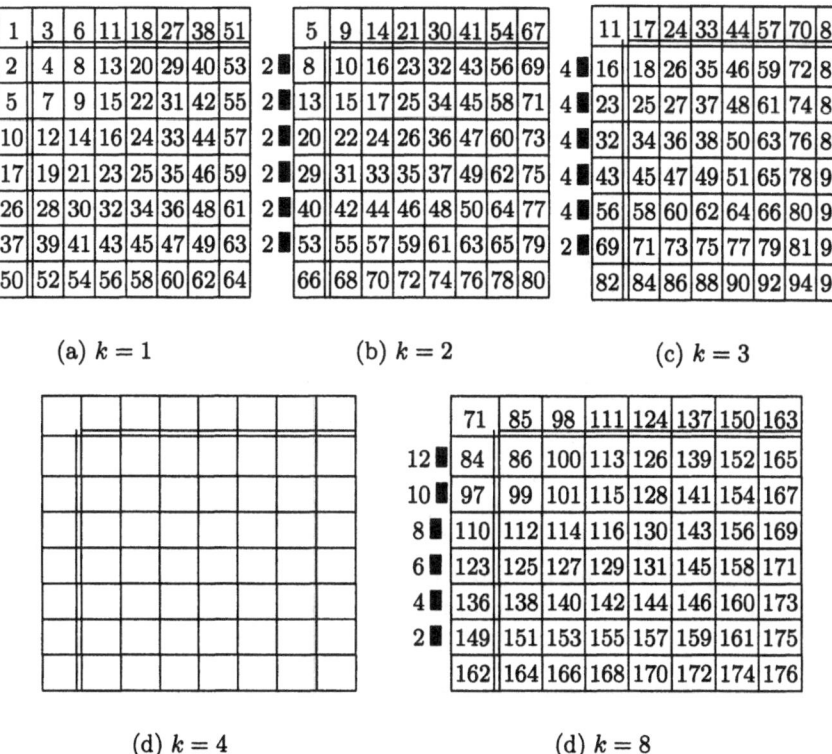

(a) $k = 1$ (b) $k = 2$ (c) $k = 3$

(d) $k = 4$ (d) $k = 8$

Fig. 1. Illustration of the schedule for $n = 8$. Each table shows the time instants at which the elements of $F(i, j, k)$ are computed by processor k (for $k = 1 \ldots 4$, and $k = 8$ the final step). The x∎'s denote x idle ticks during which the processor does nothing. The table for $k = 4$ is left blank as an exercise for the reader.

time instant as a function of i, j and k (hint: first concentrate only on points on the diagonals of each plane, and remember that i and j range from 0 to $n - 1$).

The following remark can now be easily (after the little exercise) verified.

Remark 1. The closed form of the function described in Fig. 1 is given as follows.

$$t(i, j, k) = \begin{cases} \text{if } k = 0 & \text{then} \quad 0 \\ \text{if } i \geq j & \text{then} \quad t_d(i, k) - 2(i - j) \\ \text{if } i < j & \text{then} \quad t_d(j, k) - 2(j - i) + 1 \end{cases} \tag{6}$$

where

$$t_d(i, k) = \begin{cases} \text{if } i + k \geq n & \text{then} \quad n^2 + 2n(i + k - n) + n - i - 1 \\ \text{if } i + k \leq n & \text{then} \quad (i + k)^2 + k - 1 \end{cases} \tag{7}$$

Observe that $t_d(i, k)$ is the value of $t(i, j, k)$ on the diagonal (i.e., where $i = j$).

4   Formal Derivation of the Array

We have so far seen the processor allocation and an intuitive explanation of the schedule of our architecture. However, we do not have a proof of correctness of its validity. We now resolve this question.

Theorem 1. *The function $t(i, j, k)$ defined by (6-7) is a valid schedule for the variable F in the SRE (4-5).*

Outline of Proof: We need to show that whenever a point $(i, j, k) \in D_1$ depends on another $(i', j', k') \in D_1$, then $t(i, j, k) > t(i', j', k')$. Since the dependency function (and even the number of dependencies) is different in different subdomains of D_1, the proof must follow the structure of SRE (4-5). In fact, the schedule is valid iff we show that the (boolean) recurrence (8) can be reduced to a tautology. By substituting from (6) this reduces to showing a number of implications, each of the form that a set of affine constraints (bounds of a subdomain) imply that a polynomial is strictly positive. This is tedious but simple. Indeed, it can even be done automatically with a mechanical theorem prover.   ∎

$$X(i, j, k) = \begin{cases} D_1 \cap \{i, j, k \mid i = j = 0\} : t(i, j, k) > t(i\dot+1, j\dot+1, k - 1) \\ D_1 \cap \{i, j, k \mid i = 0 < j\} : t(i, j, k) > \max(\\ \qquad\qquad\qquad t(i\dot+1, j\dot+1, k - 1), t(0, 0, k)) \\ D_1 \cap \{i, j, k \mid i > 0 = j\} : t(i, j, k) > \max(\\ \qquad\qquad\qquad t(0, 0, k), t(i\dot+1, j\dot+1, k - 1)) \\ D_1 \cap \{i, j, k \mid i, j > 0\} \quad : t(i, j, k) > \max(t(i\dot+1, j\dot+1, k - 1), \\ \qquad\qquad\qquad t(i, 0, k), t(1, j, k - 1)) \end{cases}$$

$$(8)$$

Recall that the processor allocation function is defined as follows.

$$\forall (i, j, k) \in D_1, \quad p(i, j, k) = k \qquad (9)$$

Now, we must show that the schedule and allocation function are not in conflict, i.e., no two distinct points in D_1 are mapped the the same processor at the same time. The proof follows directly from Remark 2 below, which itself can be easily verified.

Remark 2. $t(i, j, k) = t(i', j', k) \Rightarrow i = i'$ and $j = j'$.

Remark 3. The running time of the array is given by $t(n - 1, n - 1, n) = 3n^2 - 2n$ and hence its work efficiency is $\frac{1}{3}$.

The algorithm can be implemented on only $p < n$ processors by using multiple passes. However, the analysis of the running time is a little involved. In the single pass array, the k-th processor starts the first pass at $t_s(k) = t(0, 0, k) = k^2 + k - 1$ and is active until $t_f(k) = t(n - 1, n - 1, k) = n^2 + 2n(k - 1)$. It is thus active for precisely $t_a(k) = t_f(k) - t_s(k) + 1 = n^2 + 2n(k - 1) - (k^2 + k) + 2$, and since $n \geq k$ this *increases* with k. Indeed, this should be obvious since each processor has exactly n^2 computations to perform, but as k increases, the processor has more and more idle cycles. Hence if we start the next pass as soon as the *first* processor is free (i.e., at $t = n^2 + 1$), there will be conflicts.

Ideally, we desire that the last processor, p, start its next pass exactly at $t_f(p) + 1$ so that there will be no wasted time on this processor. Since there are exactly $t_s(p)$ cycles from the start of a pass (on the first processor) to the time at which the last processor starts the pass, we can achieve our goal if the *first* processor starts its second pass at $t_f(p) + 1 - t_s(p)$, i.e., at $t_a(p) + 1$. Since the first processor is active on the first pass for the first n^2 of these cycles, the number of idle cycles for the first processor between the first two passes is $2n(p-1) + 2$, which is of the same order as the total memory (registers) in the array.

The argument holds between any two passes, and since there are $\frac{n}{p}$ passes, we have the following.

Theorem 2. *The total running time on p processors is given by*

$$T_p = \frac{n^3}{p} + 2n^2\frac{p-1}{p} - (n-p)(p+1) + \frac{2n}{p} - 2 \qquad (10)$$

Corollary 1. *For any (positive integer) constant α, the total running time on $p = \frac{n}{\alpha}$ processors is as given below, assuming that $n \approx n + \alpha \approx n - \alpha$, and ignoring constant terms.*

$$T_p \approx n^2\left(\alpha + 2 - \frac{\alpha - 1}{\alpha^2}\right) \leq n^2(\alpha + 2) \qquad (11)$$

The work of the multiple pass array on $\frac{n}{\alpha}$ processors is $n^3(1 + \frac{2}{\alpha} - \frac{\alpha - 1}{\alpha^3})$. Hence, we can improve the work efficiency simply by increasing α, i.e., by sacrificing running time by a constant factor with a corresponding decrease in the number of processors. It is important to note that the savings arise because the first processor is never idle *during* any pass (see Fig. 2), though it emulates a processor which would have idle cycles in the single pass array. A formal proof of correcness of the multiple pass implementation is omited due to space constraints.

The above analysis assumes that α is a constant, i.e., we always use $\Theta(n)$ processors, scaling the architecture with the problem size. On the other hand, it is also clear from (10) that with only a fixed number of processors, we have a work optimal implementation, with a running time dominated by the $\frac{n^3}{p}$ term.

5 Conclusion

We have presented an SIMD architecture for the APP, which unifies unifies many well-known problems into a single algorithm schema. Unlike most previous work which proposed arrays with $\Theta(n^2)$ processors and $\Theta(n)$ time, all entailing a loss of work optimality, our architecture is a linear array of n processors and a running

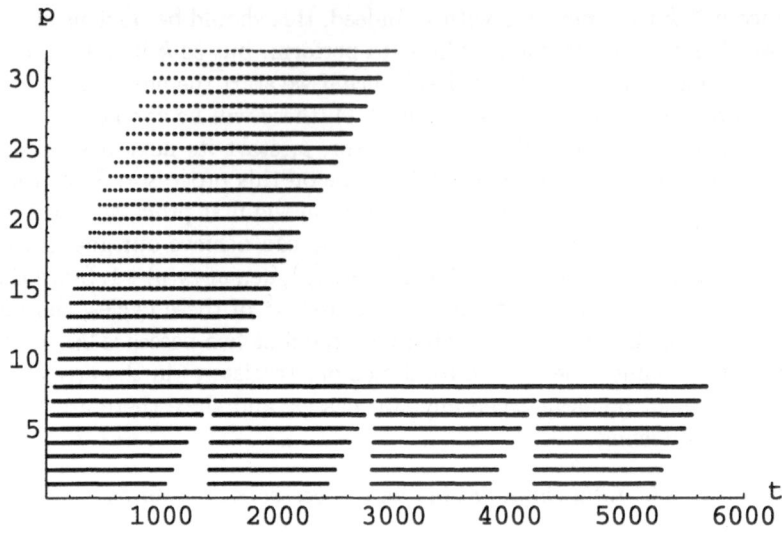

Fig. 2. Illustration of the "space-time" behavior of two arrays (for $n = 32$), one for a single pass which finishes at $t = 3008$, and one with 8 processors performing 4 identical passes finishing at $T = 5678$. A processor is active during the solid line and idle during the gaps. Observe how the first processor has no idle time *during* any pass.

time of $3n^2$ (thus performing three times the work of the sequential algorithm). However, a simple multipass version of the array on $\frac{n}{p}$ processors achieves a running time that is as close to a work optimal as desired. The architecture is derived using the well known systolic synthesis methods but with extensions involving polynomial schedules. Each processor in the array has 3 fifos of n-registers, and the architecture is well suited for direct VLSI implementation. We also note that a simple extension is a "block" version of the architecture, which would be more suitable for implementation on a general purpose parallel machine. For such an implementation, and interesting problem is the choice of the block size that optimizes the total running time, a problem we treat elsewhere.

Acknowledgments C. Tadonki was partially supported by the *Microprocessors and Informatics* Program of the United Nations University, Tokyo, Japan, and the French agency *Aire Dveloppement* through the project *Calcul Parallle*.

References

[1] A. Benaini, P. Quinton, Y. Robert, Y. Saouter, and B. Tourancheau. Synthesis of a new systolic architecture for the algebraic path problem. *Science of Computer Programming*, 15:135–158, 1990.

[2] A. Benaini and Y. Robert. Space-time minimal systolic arrays for gaussian elimination and the algebraic path problem. In *ASAP 90: International Conference on Application Specific Array Processors*, pages 746–757, Princeton, NJ, September 1990. IEEE Press.

[3] P. Y. Chang and J. C. Tsay. A family of efficient regular arrays for the algebraic path problem. *IEEE Transactions on Computers*, 43(7):769–777, July 1994.

[4] P. Clauss, C. Mongenet, and G-R. Perrin. Synthesis of size-optimal toroidal arrays for the algebraic path problem: A new contribution. *Parallel Computing*, 18:185–194, 1992.

[5] L. Guibas, H. T. Kung, and Clark D. Thompson. Direct VLSI implementation of combinatorial algorithms. In *Proc. Conference on Very Large Scale Integration: Architecture, Design and Fabrication*, pages 509–525, January 1979.

[6] S. Y. Kung and S. C. Lo. A spiral systolic algorithm/architecture for transitive closure problems. In *ICCD 85: International Conference on Circuit Design*, pages 622–626, Rye Town, NY, 1985. IEEE.

[7] S. Y. Kung, S. C. Lo, and P. S. Lewis. An optimal systolic design for the transitive closure and the shortest path problems. *IEEE Transactions on Computers*, C-36(5):603–614, May 1987.

[8] J. F. Myoupo and C. Fabret. A modular systolic linearization of the warshall-floyd algorithm. *IEEE Transactions on Parallel and Distributed Systems*, 7(5):449–455, may 1996.

[9] J. G. Nash and S. Hansen. Modified faddeew algorithm for matrix multiplication. In *Proc., SPIE Workshop on Real-Time Signal Processing*, pages 39–46. SPIE, 1984.

[10] V. K. Prasanna Kumar and Y-C Tsai. Designing linear systolic arrays. *Journal of Parrallel and Distributed Computing*, (7):441–463, may 1989.

[11] S. V. Rajopadhye. An improved systolic algorithm for the algebraic path problem. *INTEGRATION: The VLSI Journal*, 14(3):279–296, Feb 1993.

[12] T. Risset and Y. Robert. Synthesis of processors arrays for the algebraic path problem: Unifying old results and deriving new architectures. *Parallel Processing Letters*, 1:19–28, 1991.

[13] Y. Robert and M. Tchuent. Rsolution systolique de systmes linaires denses. *RAIRO Modlisation et Analyse Numrique, Technique et Sciences Informatiques*, 19(2):315–326, 1985.

[14] Y. Robert and D. Trystam. Systolic solution of the algebraic path problem. In W. Moore, A. McCabe, and R. Urquhart, editors, *Systolic Arrays*, 1, pages 171–180, Oxford, UK, 1987. Adam Hilger.

[15] Günter Rote. A systolic array algorithm for the algebraic path problem (shortest paths; matrix inversion). *Computing*, 34(3):191–219, 1985.

[16] Chris J. Schieman and Peter R. Cappello. A processor-time minimal systolic array for transitive closure. In *International Conference on Application Specific Array Processors*, pages 19–30, Princeton, NJ, September 1990. IEEE Computer Society, IEEE Computer Society Press.

[17] T. Takaoka and K. Umehara. An efficient VLSI circuit for the all pairs shortest path problem. *Journal of Parallel and Distributed Computing*, 16:265–270, 1992.

[18] C. Tayou Djamegni, P. Quinton, S. Rajopadhye, and T. Risset. Derivation of systolic algorithms for the algebraic path problem by recurrence transformations. *Parallel Computing*, To appear, 1999.

Vector ISA Extension for Sparse Matrix-Vector Multiplication

Stamatis Vassiliadis, Sorin Cotofana, and Pyrrhos Stathis

Delft University of Technology, Electrical Engineering Dept.,
P.O. Box 5031, 2600 GA Delft, The Netherlands
{Stamatis,Sorin,Pyrrhos}@Plato.ET.TUDelft.NL

Abstract. In this paper we introduce a vector ISA extension to facilitate sparse matrix manipulation on vector processors (VPs). First we introduce a new Block Based Compressed Storage (BBCS) format for sparse matrix representation and a Block-wise Sparse Matrix-Vector Multiplication approach. Additionally, we propose two vector instructions, Multiple Inner Product and Accumulate ($MIPA$) and LoaD Section (LDS), specially tuned to increase the VP performance when executing sparse matrix-vector multiplications.

1 Introduction

In many areas of scientific computing the manipulation of sparse matrices constitutes the kernel of the solvers. Improving the efficiency of sparse operations such as Sparse Matrix-Vector Multiplication (SMVM) has been and continues to be an important research topic. Several compressed formats for sparse matrix representation [5] and algorithms to improve sparse matrix multiplication performance on parallel [3] and vector machines [4] have been developed. Moreover sparse matrix solving machines, e.g., $(SM)^2$ [1], or Finite Element Method (FEM) [7] computations dedicated parallel architectures, e.g., the White Dwarf [6], SPAR [11], have been proposed.

For reasons related to memory bandwidth and to avoid trivial multiplications with zero values sparse matrices are operated upon on compressed formats. Such compressed matrix formats consist of two main data structures: first, the matrix element values, consisting mainly of the non-zero matrix elements and second, the positional information for the first data structure. A compact representation however implies a loss of data regularity and this deteriorates the performance of vector and parallel processors when operating on sparse formats. For example, as suggested in [11], when executing FEM computations a CRAY Y-MP vector supercomputer operates at less than 33% of its peak floating-point unit throughput.

Up to our best knowledge, with the exception of [8], not much has been done to improve the architectural support that vector processors may provide to sparse matrix multiplication. In this paper we propose a vector ISA extension and an associated sparse matrix organization to alleviate the previously mentioned problem. The main contributions of our proposal can be summarized as follows:

P. Amestoy et al. (Eds.): Euro-Par'99, LNCS 1685, pp. 708–715, 1999.
© Springer-Verlag Berlin Heidelberg 1999

- A Block Based Compressed Storage (BBCS) sparse matrix representation format which requires a memory overhead in the order of $\log s$ bits per nonzero matrix element, where s is the vector processor section size, is introduced.
- A Block-wise SMVM scheme to compute the product of an $n \times n$ sparse matrix with a dense vector in $\frac{n}{s}$ vectorizable loops is described. All the trivial with zero multiplications are eliminated and the amount of processed short vectors is substantially reduced.
- Two vector instructions to support the vectorization and execution of the Block-wise SMVM scheme, *Multiple Inner Product and Accumulate (MIPA)* and *LoaD Section (LDS)* are proposed.

The presentation is organized as follows: First, in Section 2 we introduce assumptions and preliminaries about sparse matrices and SMVM. Section 3 describes the sparse matrix compact storage format to be used in conjunction with the proposed ISA extension. Section 4 discusses a Block-wise SMVM method and the vector ISA extension. Finally, in Section 5 we draw some conclusions.

2   Problem Statement, Assumptions, & Preliminaries

The definition of the multiplication of a matrix $A = [a_{i,j}]_{i,j=0,1,\ldots,n-1}$ by a vector $b = [b_i]_{i=0,1,\ldots,n-1}$ producing a vector $c = [c_i]_{i=0,1,\ldots,n-1}$ is as follows:

$$c = Ab, \qquad c_i = \sum_{k=0}^{n-1} a_{i,k}b_k, \qquad i = 0,1,\ldots,n-1 \tag{1}$$

Let now consider the execution of the multiplication in Equation (1) on a vector architecture [9]. More in particular we assume a register type of organization, e.g., $IMB/370$ vector facility [2, 10], with the section size of s elements per register. When executed on such a VP the inner loop, i.e, the computation of $c_i = \sum_{k=0}^{n-1} a_{i,k}b_k$, can be vectorized. Ideally, if the section size s is large enough, i.e., $s \geq n$, one loop could be replaced with just one inner product instruction which multiplies the A_i vector, i.e., the ith row of the matrix A, with the b vector and produces as result the ith element of the c vector. In practice the section size is usually smaller than n and the A_i and b vectors have to be split into segments of at most s element length to fit in vector registers. Consequently, the computation of c_i will be achieved with a number of vector instructions in the order of $\lceil \frac{n}{s} \rceil$. Although the speedup due to vectorization obviously depends on the section size value, due to issues like lower instruction fetch bandwidth, fast execution of loops, good use of the functional units, VPs perform quite well when A is dense. When operating on sparse matrices however VPs are not as effective because the lack of data regularity in sparse formats leads to performance degradation. Consider for instance that A is sparse and stored in the Compressed Row Storage (CRS)[1]. To process this format each row of non-

---

[1] The CRS format consists of the following sets: *Value* (all the $a_{i,j} \neq 0$ elements row-wise stored), *Column_Index* (the $a_{i,j} \neq 0$ column positions), and *Row_Pointer* (the

zero values forms a vector and it is accessed by using the index vector provided by the same format. As the probability that the number of non-zero elements in a matrix row is smaller than the VP's section size is rather high many of the vector instructions manipulate short vectors. Consequently, the processor resources are inefficiently used and the pipeline start-up times dominate the execution time. This makes the vectorization poor and constitutes the main reason for VPs performance degradation. An other source for performance degradation might be related to the use of the indexed load/store instructions which, in principle, can not be completed as efficient as a the standard load vector instruction.

The occurrence of short length vectors relates to the fact that vectorization is performed either on row or column direction but not on both. The number of A matrix elements within a vector register can be increased if more than one row (column) is loaded in a vector register at a time. Such vectorization schemes however are not possible assuming the VPs state of the art as such an approach introduces operation irregularity on the elements of the same vector register. Whereas the data irregularity can be easily dealt with by using index based operations and/or masked execution, the operation irregularity requires specialized architectural support and this is exactly what our vector ISA extension does: it enables the operation on multiple matrix rows at once.

3   Sparse Matrix Storage Format

Before describing the proposed vector ISA instruction extension we first introduce a new Block Based Compressed Storage (BBCS) format to be used for the representation of the A sparse matrix.

The $n \times n$ A matrix is partitioned in $\lceil \frac{n}{s} \rceil$ Vertical Blocks (VBs) A^m, $m = 0, 1, \ldots, \lceil \frac{n}{s} \rceil - 1$, of at most s columns, where s is the VP section size. For each vertical block A^m, all the $a_{i,j} \neq 0$, $sm \leq j < s(m+1)$, $i = 0, 1, \ldots, n-1$, are stored row-wise in increasing row number order. At most one block, the last block, can have less than s columns in the case that n is not a multiple of s. An example of such a partitioning is graphically depicted in Figure 1. In the discussion to follow we will assume, for the simplicity of notations, that n is divisible by s and all the vertical blocks span over s columns.

The rationale behind this partitioning is related to the fact that when computing the Ab product the matrix elements a_{ij} are used only once in the computation whereas the elements of b are used several times depending on the amount of non-zero entries in the A matrix in the corresponding column, as it can be observed in Equation (1). This implies that to increase performance it is advisable to maintain the b values within the execution unit which computes the sparse matrix-vector multiplication for reuse and only stream-in the a_{ij} values. As to each vertical block A^m corresponds an s element section of the b-vector, $b^m = [b_{ms}, b_{ms+1}, \ldots, b_{ms+s-1}]$ we can multiply each A^m block with its cor-

---

pointers in the first two sets to the first non-zero element of each row that contains at least one non-zero element) [5].

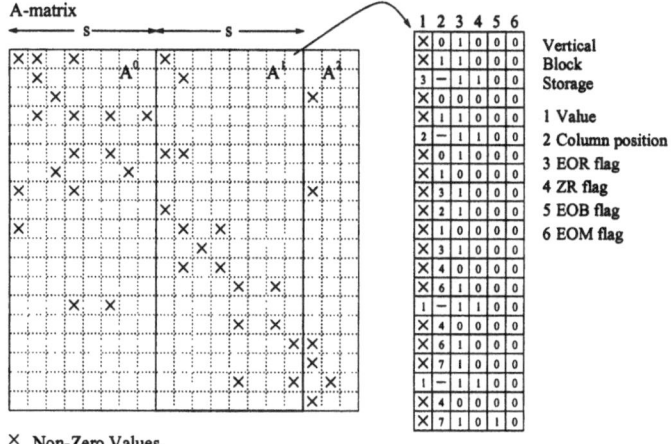

Fig. 1. Block Based Compressed Storage Format

responding b^m section of the b-vector without needing to reload any b-vector element.

Each A^m, $m = 0, 1, \ldots, \frac{n}{s} - 1$, is stored in the main memory as a sequence of 6-tuple entries. The fields of such a data entry are as follows:

1. **Value**: specifies the value of a non-zero $a_{i,j}$ matrix element if $ZR = 0$. Otherwise it denotes the number of subsequent block rows[2] with no non-zero matrix elements.
2. **Column-Position (CP)**: specifies the matrix element column number within the block. Thus for a matrix element $a_{i,j}$ within the vertical block A^m it is computed as $j \mod m$.
3. **End-of-Row Flag (EOR)**: is 1 when the current data entry describes the last non-zero element of the current block row and 0 otherwise.
4. **Zero-Row Flag (ZR)**: is 1 when the current block row contains no non-zero value and 0 otherwise. When this flag is set the *Value* field denotes the number of subsequent block rows that have no non-zero values.
5. **End-of-Block flag (EOB)**: when 1 it indicates that the current matrix element is the last non-zero one within the VB.
6. **End-of-Matrix flag (EOM)**: is 1 only at the last entry of the last VB of the matrix.

The entire A matrix is stored as a sequence of VBs and there is no need for an explicit numbering of the VBs.

When compared with other sparse matrix representation formats, our proposal requires a lower memory overhead and bandwidth since the index values associated with each $a_{i,j} \neq 0$ are restricted within the VB boundaries and can be

---

[2] By block row we mean all the elements of a matrix row that fall within the boundaries of the current VB.

represented only with $\log s$ bits instead of $\log n$ bits. The flags can be explicitly 4-bit represented or 3-bit encoded and, depending on the value of s, they may be packed with the CP field on the same byte/word.

The other multiplication operand, the b-vector, is assumed to be dense than no special data types or flags are required. The b_k, $k = 0, 1, \ldots, n - 1$, values are sequentially stored and their position is implicit. The same applies for the result, i.e., the c-vector.

4   Block-Wise SMVM and ISA Extension

Assume a register vector architecture with section size s and the data organization described in Section 3. The Ab product can be computed as $c = \sum_{m=0}^{\frac{n}{s}-1} A^m \times b^m$.

To vectorize each loop computing the $A^m \times b^m$ product and because generally speaking A^m can not fit in one vector register, we have to split each A^m into a number of subsequent VB-sections A_i^m each of them containing at most s elements. Under the assumption that each vertical block A^m is split into $\#s_m$ VB-sections A_i^m the c-vector can be expressed as $c = \sum_{m=0}^{\frac{n}{s}-1} \sum_{i=0}^{\#s_m} A_i^m \times b^m$. Consequently, c can be iteratively computed within $\frac{n}{s}$ loops as $c_m = c_{m-1} + \sum_{i=0}^{\#s_m} A_i^m \times b^m$, $m = 0, 1, \ldots, \frac{n}{s} - 1$, where c_m specifies the intermediate value of the result vector c after iteration m is completed and $c_{-1} = 0$.

Assuming that $A_i^m = [A_{i,0}^m, A_{i,1}^m, \ldots, A_{i,s-1}^m]$ and $b^m = [b_0^m, b_1^m, \ldots, b_{s-1}^m]$ a standard vector multiplication computes the $A_i^m \times b^m$ inner product as being $c_i^m = \sum_{j=0}^{s-1} A_{i,j}^m b_j^m$ which is the correct result only if A_i^m contains just one row. As one VB-section A_i^m may span over $r_i \geq 1$ rows of A^m the $A_i^m \times b^m$ product should be an r_i element vector. In particular, if $A_i^m = [A_i^{m,0}, A_i^{m,1}, \ldots, A_i^{m,r_i-1}]$, with $A_i^{m,j} = [A_{i,0}^{m,j}, A_{i,1}^{m,j}, \ldots, A_{i,\#r_j-1}^{m,j}]$, $j = 0, 1, \ldots, r_i - 1$, and $\#r_j$ being the number of elements within the row j, $c_i^m = A_i^m \times b^m$ has to be computed as follows:

$$c_i^m = [c_{i,0}^m, c_{i,1}^m, \ldots, c_{i,r_i-1}^m], \qquad c_{i,j}^m = A_i^{m,j} \times b^m, \quad j = 0, 1, \ldots, r_i - 1 \quad (2)$$

Consequently, to compute c_i^m, r_i inner products, each of them involving $\#r_j$ elements, have to be evaluated. Moreover when executing the evaluation of $A_i^{m,j} \times b^m$ inner product the "right" elements of b^m have to be selected. Therefore, $c_{i,j}^m$ is evaluated as follows:

$$
\begin{aligned}
c_{i,j}^m = A_{i,0}^{m,j} \cdot b_{CP(A_{i,0}^{m,j})}^m &+ A_{i,1}^{m,j} \cdot b_{CP(A_{i,1}^{m,j})}^m + \cdots \\
&+ A_{i,\#r_j-1}^{m,j} \cdot b_{CP(A_{i,\#r_j-1}^{m,j})}^m, \quad j = 0, 1, \ldots, r_i - 1
\end{aligned}
\quad (3)
$$

As each $c_{i,j}^m$ contributes to the c vector element in the same row position as $A_i^{m,j}$ and row position information is not explicitly memorized in the BBCS format bookkeeping related to index computation has to be performed. Moreover as A_i^m does not contain information about the row position of its first entry, hence not

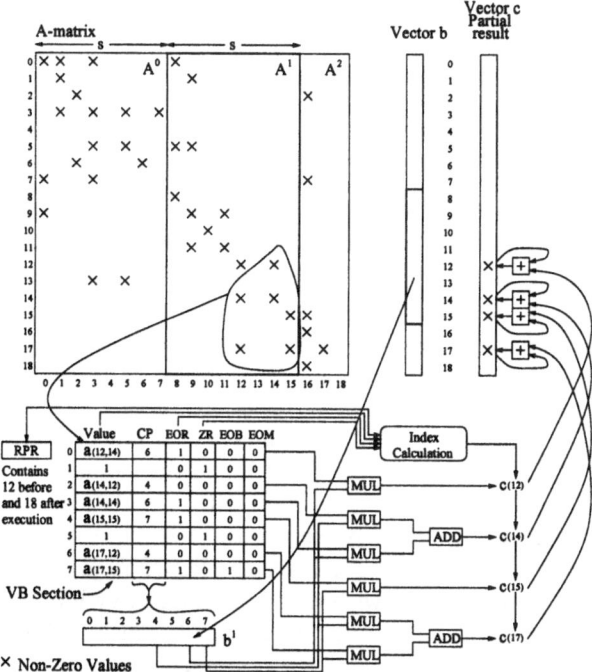

Fig. 2. Sparse Matrix-Vector Multiplication Mechanism

enough information for the calculation of the correct inner product positions is available, a Row Pointer Register (RPR) to memorize the starting row position for A_i^m is needed. The RPR is reset every time the processing of new A^m is initiated and updated by the index computation process.

To clarify the mechanism we present in Figure 2 an example. We assume that the section size is 8, the VB-section contains the last 8 entries of A^1, b^1 contains the second 8-element section of b, and that an intermediate c has been already computed though we depict only the values of c that will be affected by the current step. First the inner product calculation and the index calculation are performed. For the inner product calculation only the a(i, j) elements with $ZR = 0$ are considered. Initially they are multiplied with the corresponding b^1 elements, the CP field is used to select them, and some partial products are obtained. After that, all the partial products within the same row have to be accumulated to form the inner products. As the EOR flags delimit the accumulation boundaries they are used to configure the adders interconnection. The index calculation proceeds with the RPR value and whenever a entry with $ZR = 1$ is encountered RPR is increased with the number in the $Value$ field. When an EOR flag is encountered the RPR is assigned as index to the current inner product and then increased by one. Second the computed indexes are used to select the c elements to whom the computed inner products should be accumulated and the accumulation is performed.

To be able to execute the previously discussed block-wise SMVM we propose the extension of the VP instruction set with 2 new instructions: *Multiple Inner Product and Accumulate* (*MIPA*) and *LoaD Section* (*LDS*).

MIPA is meant to calculate the inner product $A_i^m \times b^m$. Furthermore, it also performs the accumulation of the c_i^m elements to the c-vector values in the corresponding locations. The instruction format is *MIPA VR1,VR2,VR3*. The vector register $VR1$ contains A_i^m, $VR2$ contains b^m, and $VR3$ contains initially those elements of the c-vector that correspond to the non-empty rows of A_i^m and after the instruction execution is completed the updated values of them.

LDS is meant to load an A_i^m VB-section from the main memory. The instruction format is *LDS @A,VR1,VR2*. @A is the address of the first element of the VB-section to be loaded. After an *LDS* instruction is completed $VR1$ contains all the non-zero elements of the VB-section starting at @A, $VR2$ contains the indexes of the VB-section rows with non-zero elements (to be used later on as an index vector to load and store a c section), and the Column Position Register (CPR), a special vector register, is updated with the CP and flag fields of the corresponding elements in $VR1$. To execute the *LDS* instruction the VP Load Unit has to include a mechanism to analyze the BBCS flags in order to compute the indexes and filter the entries with $ZR = 1$. In this way $VR1$ always contains only nonzero matrix elements and no trivial calculations will be performed.

Even though the previously discussed approach guarantees an efficient filling of the vector registers, it may suffer a performance degradation due to the use of indexed vector load/store instructions as such operations, depending on the implementation of the main memory and/or load unit, may create extra overhead. However, the use of indexed load/stores can be avoided if instead of partitioning the VBs in s-element VB-sections, they are divided in $\lceil \frac{n}{ks} \rceil$ ks-element hight *segments*[3] where $k \geq 1$. To support this new division the BBCS has to include an extra flag, the End of Segment (*EOS*) flag. This flag is 1 for the entries which describe the last element of a segment. Under this new assumption when a VB-section is loaded with the *LDS* instruction the loading stops after s nonzero elements or when encountering the *EOS*, *EOB*, or *EOM* flag. By restricting the VB-section's row span within a VB segment we guarantee that all the $A_i^m \times b^m$s will contribute only to elements within a specific ks-element wide section of the vector c. Consequently, the c-vector can be manipulated with standard load/store instructions.

5   Conclusions

In this paper we proposed a vector ISA extension and an associated sparse matrix organization to alleviate some of the problems related to sparse matrix computation on vector processors, e.g., inefficient functional unit utilization due to short vector occurrence and increased memory overhead and bandwidth.

---

[3] All of them are $s \times ks$ segments except the ones on the right and/or bottom edge of the matrix which might be truncated if the dimension of the matrix n is not divisible by s.

First we introduced a Block Based Compressed Storage (BBCS) sparse matrix representation format which requires a memory overhead in the order of $\log s$ bits per nonzero matrix element, where s is the vector processor section size. Additionally, we described a Block-wise SMVM scheme to compute the product of an $n \times n$ sparse matrix with a dense vector in $\frac{n}{s}$ vectorizable loops. To support the vectorization of the Block-wise SMVM scheme two new vector instructions, *Multiple Inner Product and Accumulate* ($MIPA$) and *LoaD Section* (LDS) were proposed. They eliminate all the trivial multiplications with zero and substantially reduce the amount of processed short vectors. Implementation and quantitative evaluations of the performance of the proposed schemes via simulations on sparse matrices benchmarks constitutes the subject of future research.

References

[1] H. Amano, T. Boku, T. Kudoh, and H. Aiso. (SM)^2-II: A new version of the sparse matrix solving machine. In *Proceedings of the 12th Annual International Symposium on Computer Architecture*, pages 100–107, Boston, Massachusetts, June 17–19, 1985. IEEE Computer Society TCA and ACM SIGARCH.

[2] W. Buchholz. The IBM System/370 vector architecture. *IBM Systems Journal*, 25(1):51–62, 1986.

[3] R. Doallo, J. T. no, and F. Hermo. Sparse matrix operations in vector and parallel processorse. *High Performance Computing*, 3:43–52, 1997.

[4] I. S. Duff. The use of vector and parallel computers in the solution of large sparse linear equations. In *Large scale scientific computing. Progress in Scientific Computing Volume 7*, pages 331–348, Boston, MA, USA, 1986. Birkhäuser.

[5] I. S. Duff, A. M. Erisman, and J. K. Reid. *Direct Methods for Sparse Matrices*. Clarendon Press, Oxford, UK, 1986.

[6] A. W. et al. The white dwarf: A high-performance application-specific processor. In *Proceedings of the 15th Annual International Symposium on Computer Architecture*, pages 212–222, Honolulu, Hawaii, May–June 1988. IEEE Computer Society Press.

[7] T. J. R. Hughes. *The Finite Element Method*. Prentice-Hall, Englewood Cliffs, NJ, 1987.

[8] R. N. Ibbett, T. M. Hopkins, and K. I. M. McKinnon. Architectural mechanisms to support sparse vector processing. In *Proceedings of the 16th ASCI*, pages 64–71, Jerusalem, Israel, June 1989. IEEE Computer Society Press.

[9] P. M. Kogge. *The Architecture of Pipelined Computers*. McGraw-Hill, New York, 1981.

[10] A. Padegs, B. B. Moore, R. M. Smith, and W. Buchholz. The IBM System/370 vector architecture: Design considerations. *IEEE Transactions on Computers*, 37:509–520, 1988.

[11] V. E. Taylor, A. Ranade, and D. G. Messerschitt. SPAR: A New Architecture for Large Finite Element Computations. *IEEE Transactions on Computers*, 44(4):531–545, April 1995.

A Study of a Simultaneous Multithreaded Processor Implementation

Dominik Madoń[1], Eduardo Sánchez[1], and Stefan Monnier[2]

[1] Logic Systems Laboratory, Swiss Federal Institute of Technology of Lausanne
INN-Ecublens 1015 Lausanne, Switzerland
[2] Computer Science Dept., Yale University
51 Prospect Street, New-Heaven CT 06520, USA

Abstract. This paper describes an approach to the implementation and the operation of a Simultaneous Multithreaded processor. We propose an architecture which integrates a software mechanism to handle contexts, a rapid communication system, as well as a locking system to ensure mutual exclusion. We explain how the architecture manages the running threads as well as the software interface visible to the programmer. Finally, we provide a few indications on the efficiency of such an architecture.

1 Introduction

Over the past few years numerous publications have shown the advantages [2] [4] and sometimes the disadvantages [3][5] of Simultaneous Multithreaded (SMT) architectures. These architectures are based on a superscalar architecture to which several physical execution contexts are added. Such processors can thus simultaneously execute (and even issue) instructions originating from different processes.

The advantage of this type of processor lies in its better exploitation of functional units and above all in its ability to more efficiently handle the available memory bandwidth. While one process is busy fetching instructions from the main memory (prefetch), others can still execute instructions from the cache or can launch their own prefetches, using a non-blocking memory bus. That's also where the improved use of memory bandwidth comes from. The latency between a fetch from the main memory or from a second-level cache can be hidden by the execution of instructions coming from other processes.

To date, a considerable amount of research has analyzed the advantages and disadvantages of this type of architecture. Yet, to the best of our knowledge, while several papers [6] have been published on possible instructions sets or on accessing registers of different contexts, nobody has studied the problems related to the design of a complete SMT processor, including all aspects of the implementation and in particular the creation of a process, its movement from memory to cache and vice versa, the destruction of a process, their synchronization, and the use of locking and other blocking mechanisms.

P. Amestoy et al. (Eds.): Euro-Par'99, LNCS 1685, pp. 716–726, 1999.

Also while the operation of SMT architectures in heterogeneous multitask-ing mode where multiple independent programs are executed has been studied quite extensively, little information is available on its operation in homogeneous multitasking mode [3]. In the latter mode, all the processes are created within a given application, requiring all the basic concurrent programming commands (for example, means to communicate and to synchronize efficiently and cheaply, in addition to a locking system for the protection of sensitive data).

This paper would like to shed some light on the above issues by proposing an original approach to the use of SMT processors. We describe an architec-ture which includes a management system which uses a ports unit to implement explicit multithreading. Next, we will discuss the implementation of the commu-nication between contexts by presenting a solution for the rapid transmission of data which we will then extend to the creation of locks. Then we will explain how to physically implement these two features, starting with a description of a functional simulator which we use to validate our hypothesis and obtain some preliminary results. In this section we also discuss the physical limitations of the architecture which, up to a certain point, determine the possible choices for the implementation. This article ends with a conclusion on the future direction of our research.

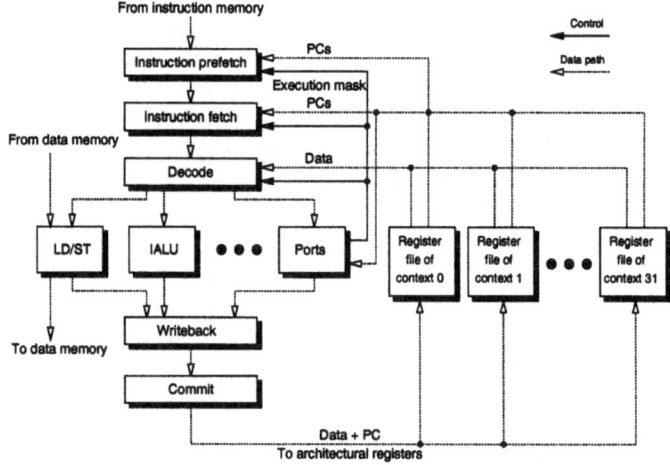

Fig. 1. Our *Simultaneous Multithreaded* Architecture

2 Processor Architecture

2.1 Superscalar Foundation

For this study, we chose as a starting point a well-known 5-stage (fetch, de-code/dispatch, execute, writeback, commit) superscalar architecture. The in-structions are executed out-of-order, using speculative branching and loading, with a one-level prediction table for the branches. The processor contains 4 integer ALU, 1 integer MULT/DIV unit, 2 lad/store units, 4 fp ALU and 1 fp MULT/DIV unit.

The processor holds 32 32-bit integer registers, 32 32-bit floating-point registers (usable by pairs as 64-bit registers), a floating-point state register, and a PC.

The processor holds two first level cache memories, one for instructions and the other for data, and a unified second level cache memory.

2.2 Proposed Extensions

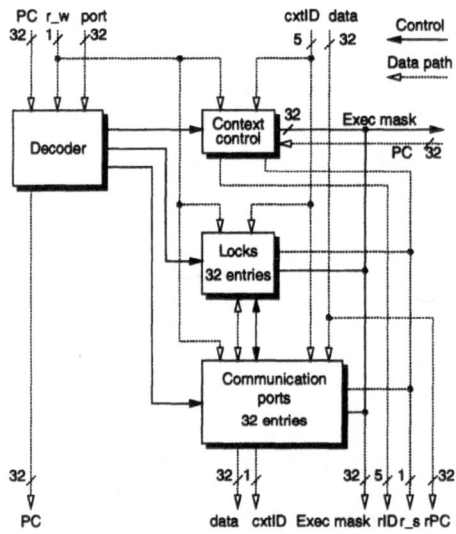

Fig. 2. Functional unit managing contexts/ports/locks

The first modification to migrate to a SMT architecture was to duplicate 32 times the physical context of the original processor, thus allowing 32 execution contexts to execute simultaneously.

We then added a *prefetch* stage to the pipeline to determine which tasks' instructions will be loaded from the memory (the first level cache) into the queue of instructions to be decoded the next clock cycle(fetch stage). This stage, using a mechanism that detects the presence of the instructions the in cache, can foresee cache misses and thus give precedence to tasks which will not block the pipeline. This is the stage that best exploits the data independence between contexts executing different program segments [2](figure 1).

We also added a functional unit in the execution stage which allows for three types of instruction: (1) Instructions which affect physical contexts (e.g. to stop, start, read, or write the PC of a given context, or to access the port on which a context is waiting). (2) Instructions to exchange data between and synchronize two physical contexts. (3) Instructions to exploit physical locks.

This functional unit takes care of informing the various stages of the processor of the activation stage of the contexts execution mask. The instructions belonging to active contexts can go through the pipeline to be executed. The suspended contexts do not have access to the pipeline. The instructions already executed (i.e., in the writeback or commit stages) before the arrival of blocking instruction have the right to complete, while the others are flushed out of the pipeline, exactly as for a branch prediction error.

2.3   Operation of the Extensions

The placement of these three functions in the same unit has been motivated by the similarity of their actions on the ordering of the instructions (see figure 2). Locks and synchronous data exchanges both require the ability to start and stop the execution of physical contexts (i.e. access to the context's PC). This control should be centralized as it implies the suppression of instructions in the decode/dispatch, writeback, and commit stages (invalid instructions).

This unit should be used in the same fashion as the load/store unit when it has to write a word back into the main memory: the instructions sent to this unit have to be executed solely in non-speculative mode to avoid inconsistent results. For instance, if a context sends a message immediately after the execution of a branch, the message can be delivered only if that branch is taken. In fact, the simultaneous execution of instructions coming from different contexts does not guarantee that the message will not be read between the branch and the speculative delivery of the message. If the branch was wrongly predicted, it would then be necessary to correct the context which read the message and invalidate all the instructions since the reading of the message.

2.4   Instruction Set Extensions

To access the context control and communication/lock unit, we added two instructions to the instruction set allowing a context to read from (rport Rd, Rp) and write to a port (wport Rs, Rp). The register Rp holds the number of the port to access. The register Rd holds the value read from the port. The Register Rs holds the value to send to the port. Any of the 32 general purpose registers may be used for Rp, Rd et Rs.

The port numbers are coded on 32 bits. However, some codes are reserved for special use: data exchanges through ports, for instance, require numbers greater than 256 (0xff). The port numbers belong to three categories described in the following sections.

These operations must be idempotent to avoid problems in context switching. We will see later how this characteristic is implemented and how to best exploit it.

It is also useful to note that the three categories of functions could be identified through the coding of the instruction. This solution is of course preferable provided that the instruction set allows it. For the user, this would simplify the access to the communication ports and to the locks, the numbers no longer identifying the function but only the particular port or lock.

2.5   Execution Control of Another Context

The execution or the suspension of a process may be controlled by the process itself or by a different process. When the request is external, it is very often the case that it was motivated by a context switch. Similarly when a context is started by another, it is very likely that a context switch just happened (unless it is as a result of an exchange of messages between two processes in two different physical contexts or of the release of a lock).

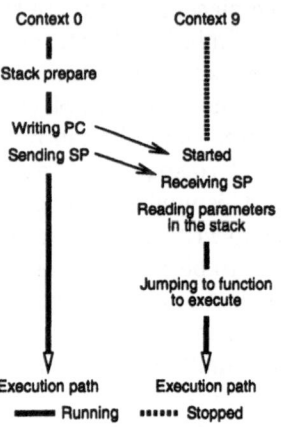

Context 0   Context 9

Stack prepare

Writing PC
Sending SP   Started
                    Receiving SP
                    Reading parameters
                    in the stack

                    Jumping to function
                    to execute

Execution path   Execution path
——— Running   ······ Stopped

Fig. 3. Process starting

The context switches, that is to say the transfer of registers from a physical context to the memory or the contrary, may be controlled by the context that is leaving. The access (both writing and reading) to another context's PC and the existence of a shared memory area are sufficient to perform this task. More precisely, the two operations needed are: (1) Reading a context's PC gives back its value and suspends it on the instruction pointed by the PC. The context will then resume from that same instruction address. (2) Writing into a context's PC causes it to resume.

Starting a new process in a context (figure 3) is done by preparing the associated stack and inserting in it the execution context. After that, the address of a minimal loading procedure is written into the PC of the context which is going to execute the process. That process starts and waits on a specified message port. The initiating process then writes to this particular port the stack pointer. The starting context reads this value, loads the execution context into the registers and jumps to the portion of code it should execute.

A context switch occurs in the same way. A process (which we will call *master*) which decides to suspend another process (the *slave*) writes into the PC of the *slave*'s context the address of the procedure which moves the context back to the memory. This causes the *slave*'s registers to be written to the stack and its stack pointer to be sent as a message back to the *master*. At this point the physical context is ready for use by another process. It is

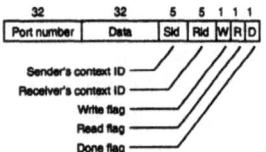

32   32   5   5 1 1 1

| Port number | Data | Sid | Rid | W | R | D |

Sender's context ID
Receiver's context ID
Write flag
Read flag
Done flag

Fig. 4. Entry in the table of communication ports

then sufficient to write into the PC the address of the loading subroutine for the incoming process. This subroutine first reads a message on the port where the *master* process writes the value of the appropriate stack pointer. With it, it is possible to retrieve the correct execution context. Once the value of all the registers has been restored, the *master* can write into the PC the address of the instructions executed by the resuming process at the time it was suspended.

Thus, to stop, start, or switch a context, the ability to access to another context's PC and the existence of a data transmission mechanism are sufficient.

2.6   Synchronous Communication by Port

The transmission of data between contexts can be done, as we mentioned, by sharing a portion of memory. This mechanism, however, is not efficient when transferring only a few words: nothing guarantees that the information sent will remain in the cache until it is requested by the destination context.

A solution consists of using distributed caches, which require a snoopy mechanism to detect when a context is trying to access data which is stored in the cache of another context (and whose value is not necessarily the same same as in the main memory). The use of the cache in writethrough mode solves this cache coherency problem but increases the transfer time of the data (which will have to go from the processor to the memory and back again).

The exchange mechanism that we propose relies on the notion of communication ports. A communication port is a one-word-wide data register, associated to a port number. Reading or writing to a port is blocking (until execution of the corresponding write resp. read) allowing synchronous exchanges of data.

The context management unit thus contains an associative table of 32 entries (it is useless to have more entries than contexts) that allows the implementation of these communication ports. The format of the table's entries is described in figure 4.

Each line contains a field for the number of the port used, a field for the value, two fields for the physical numbers of the contexts which are using this entry (the sender and the recipient), and three status flags.

This entry format allows several contexts to use the same port number, in which case the requests will be satisfied in an arbitrary order.

Sending and receiving are almost symmetrical from the point of view of their operation. What follows is a description of what occurs when context 0 writes in port 3055 and context 9 later reads from the same port (figure 5).

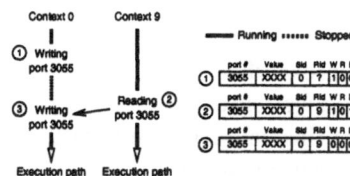

Fig. 5. Communication by synchronous port

1. The first instruction executed by context 0 writes to an available entry (all flags are 0) the port value 3055, the datum XXXX to be transferred, and its physical number (0) in the Sid field. It then sets the W flag to 1, indicating that a context is waiting for this port to be read. The execution of the sender (context 0) is then suspended.
2. When the recipient is ready to read the datum, it writes its number in the appropriate field (Rid), sets the D flag (done) to 1, reads the value, and prompts the sender to resume its execution. Flag D prevents all other processes from reading this datum and will tell the sender that the value has been read and that it may clear the entry.
3. Following its resumption, the sender executes again the instruction, which, seeing that the write flag F and the done Flag D of port 3055 are both 1 and that the Sid field contains its own physical number, will clear all flags and thus free the port.

It will be noted that this structure allows any context to be suspended if it is waiting on a port. The process, when it resumes, will execute the same instruction (send or receive), which will again result in either the exchange of data or another suspension. It will be noted that a context which owns an entry where two flags (W and D or, R and D) are 1 will not be stopped or will not be able to change PCs until the re-execution of the communication instruction which erases all the flags. This mechanism ensures that a process cannot stop a context that has completed a message exchange but has not yet re-executed the rport or wport instruction in order to increment its PC.

2.7 Hardware Locks

Synchronous ports are meant for message exchange but can of course be used to implement locks. This solution is inconvenient in that it requires auxiliary contexts (since a communication port links two contexts, a sender and a recipient, while locks require only a single process). Moreover, locks should also operate as fast as possible. For these two reasons we have created a *lock table*, whose entries have the format described in figure 6.b. These locks allow critical sections to be protected using the access instructions wport and rport of the context control unit. The suspension of a context when a lock prevents access, uses communication ports as a chained list (queue) of waiting contexts. This method allows us both to reuse a structure already available and to transfer to memory contexts waiting on a lock.

Each entry from the lock table (32 in all) contains a communication port number for next context wishing to access the critical section, a communication port number for the last context on standby in the queue, the number of the context which has set the lock, and a flag S (set) indicating if the lock is available or busy.

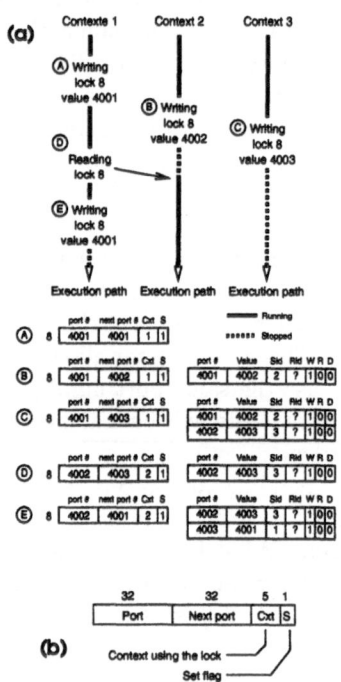

Fig. 6. (a) Example of operation of locks. (b) Locks table's entry

To explain the operation of the lock, let us imagine the following situation: three contexts (1, 2, and 3) ask in turn to set lock 8. Context 1 then releases the lock, and immediately request to again(figure 6.a):

A. Context 1 sets the lock, formerly available, with the following instructions:

```
wport 4001, 104      # grabbing port 8 (104 - 96)
    :                # critical section
rport    4001, 104
```

This instruction sets lock number 8 (corresponding to port 104) by setting the Set flag to 1, writing its context number (1) in the Cxt field, and setting the next port field to 4001, indicating that the next context wishing to access the locked section should use port 4001. It then proceeds and starts executing the critical section.

B. The second context wishing to access the critical section (context 2) will try to write to port 104, find the Set flag is 1, indicating that the critical section is not immediately accessible. It will then read the next port number (4001) to find where the queue ends. It will create a new port entry, using the port number it has retrieved (4001), its own context number (2) for the Sid field, and setting the write flag to 1. To indicate that the end of the queue has changed It will also write the new port number (4002) both in the next port field of the lock and in the value field of the port. Context 2 will then be suspended until port 4001 is released.

C. When the third context asks to access the critical section, step 2 above is repeated (a port entry for port number 4002 is created and the end of the queue moves to 4003).

D. To exit the critical section, context 1 will issue the following instruction: rport 4001,104. This instruction will access port 104, find that the port (4001) and the next port field (4003) are different (i.e., that at least one other context is waiting to access the critical section). It will then access the port entry corresponding to port number 4001, where it will find the number of the context on standby (2) and the port number for the next context in the queue (4002). it will modify the lock entry using these two numbers, and then release the port by clearing the write flag, which will cause context 2 to resume execution. Upon resumption, context 2 will access port 104 and find that it has control of the critical section.

E. Context 1's request to enter anew into the critical section will be handled like any other request, and the modifications to the two tables will occur as above.

3   Simulator Description

To rapidly obtain experimental results without physically implementing a processor and its corresponding software, we have decided to use a set of tools developed for the construction of processor architectures: the SimpleScalar tool set [1]. This package contains many C libraries allowing the construction of a functional processor simulator. It also contains a C compiler (gcc) allowing the generation of machine code with a predefined (but extensible and modifiable) set of instructions.

The architecture described above has been implemented using SimpleScalar. The simulator, called SSMT (SimpleScalar Multithreaded), has been tested with the help of a few benchmarks. We have thus verified that results obtained were similar, in terms of instructions per cycle (IPC) performance, to those of other teams [2]. Although the simulator allows an infinite number of different configurations, certain architectural parameters remained unchanged. We concentrated particularly on adapting the simulator to our extensions and on their test. For the latter we have developed several routines in assembler to exploit the processor, as well as the corresponding C interface.

The results which interest us are derived, for the most part, from the traces obtained from the execution of programs. The analysis of these traces allows us to measure the performance of different configurations and average them. In fact, since the SMT architecture allows the execution of multiple processes, the execution of a process in a context changes depending on the number of contexts already running and on what they are executing.

4 Preliminary Evaluation

The architecture we have described was first tested to determine its efficiency in the execution of independent processes (heterogeneous multitasking), which has allowed us to verify the important results obtained by Tullsen et al. We then began to evaluate the homogeneous multitasking performance.

The central idea of this architecture is based on the execution of several concurrent processes. Performance superior to that of superscalar processors cannot be reached without a sufficiently important degree of parallelism. Nevertheless, we know that the use of parallelism involves a certain increase in the number of instructions to be executed (creation and destruction of processes, message passing, etc). The efficiency, therefore, depends also on the tools at the programmer's disposal. If the creation of a process is not too time-consuming (compared to the calling of a function, for example), function calls might be replaced by parallel execution, where possible.

To compare their performance we have generated several traces on programs executing either function calls or spawning processes. It is difficult to produce accurate numbers since the execution depends on the surrounding instructions, and on other tasks in the case of multitasking. Here we have measured the number of cycles necessary for a function call (from the jump instruction to the first instruction of the function) for a context executing by itself. For the creation of a process, we used two active contexts: the first creates the second which executes the function. In both cases the instructions were in the cache.

The table of the figure ?? shows the ratio of the number of cycles between the two methods (process creation/function call). As an example, the creation of a process required a minimum of 17 cycles.

We have also measured the performance with a greater number of contexts. In this case, with explicit context priorities not implemented, the results show a greater disparity in the number of cycles necessary. A simple function call may require 24 cycles while the creation of a context require 111. In this configuration, the worst ratio is 4.63. Evidently this type of measurements do not allow precise conclusions to be drawn regarding the performance of the creation of processes. The measurements of the speed of context switches in our architecture, compared with a simply superscalar version, also exhibits this type of variance which depends exclusively on the execution of other contexts.

Nevertheless, if the synchronization by port, the message passing, and the use of hardware locks require few execution cycles (the ports management unit executes an instruction in at most two clock cycles), the measurements taken at this level are afflicted by the same type of problems. For all these reasons we are orienting ourselves to other types of measures based on multi task kernel written in C. These will allow us to obtain higher-level information such as, for example, the efficiency of our processor for a classical paradigm such as the producer-consumer problem.

5   Conclusions

We have studied most of the issues related to the implementation of an SMT processor. Our solutions have been validated on a simulator and the first results are very encouraging, pushing us to complete a physical implementation.

The evaluation of the performance of our solution remains incomplete, as we have mentioned. We are currently comparing our architecture with single-processor and SMP (Symmetric Multiprocessor) solutions as well as with other types of multitasking architectures. This comparison is based on high-level concepts (related to the paradigms of concurrent programming).

Nevertheless, the scarcity of SMP or multitasking systems makes it difficult to obtain a good evaluation of the performances of SMT hardware as opposed to those of classical superscalar solutions, particularly since a good compiler generating parallel code remains elusive. While other contributions have shown that the effective output of an SMT processor is greater than that of a standard superscalar processor, it cannot be stated with certainty that the result will be as good in homogeneous multitasking mode with code generated by a parallelizing-compiler. In this case, in fact, the quality of the produced code (the decomposition in different processes) and the efficiency of process creation/destruction tools play an important role.

The SMT simulator and the software tools we have developed allow us, at the present time, to continue our work in two parallel directions, the first being the completion of the processor through the addition of interruptions, and the second a more exhaustive evaluation of the processor.

Acknowledgments

We would like to thank the many peoples who have contributed to this project, and more particularly Professor Eduard Auguadé, of the Polytechnic University of Catalonia (UPC) at Barcelona, for his advice.

References

[1] Douglas C. Burger and Todd M. Austin. The simplescalar tool set, version 2.0. Technical Report CS-TR-97-1342, University of Wisconsin, Madison, June 1997.

[2] S. J. Eggers, J. Emer, H. M. Levy, J. L. Lo, R. Stamm, and D. M. Tullsen. Simultaneous multithreading: A platform for next-generation processors. Technical Report TR-97-04-02, University of Washington, Department of Computer Science and Engineering, April 1997.

[3] A. Farcy and O. Temam. Improving single-process performance with multithreaded processors. In *Proceedings of the 1996 International Conference on Computing*, pages 350–357, New York, May25–28 1996. ACM.

[4] Manu Gulati and Nader Bagherzadeh. Performance study of a multithreaded superscalar microprocessor. In *Proceedings of the Second International Symposium on High-Performance Computer Architecture*, pages 291–301, San Jose, California, February 3–7, 1996. IEEE Computer Society TCCA.

[5] Sébastien Hily and André Seznec. Contention on 2nd level cache may limit the effectiveness of simultaneous multithreading. Technical Report PI-1086, IRISA, University of Rennes 1, 35042 Rennes, France, February 1997.

[6] Mat Loikkanen and Nader Bagherzadeh. A fine-grain multithreading superscalar architecture. In *Proceedings of the 1996 Conference on Parallel Architectures and Compilation Techniques (PACT '96)*, pages 163–168, Boston, Massachusetts, October 20–23, 1996. IEEE Computer Society Press.

The MorphoSys Parallel Reconfigurable System

Guangming Lu[1], Hartej Singh[1], Ming-hau Lee[1], Nader Bagherzadeh[1],
Fadi Kurdahi[1], and Eliseu M.C. Filho[2]

[1] Department of Electrical and Computer Engineering
University of California, Irvine
Irvine, CA 92715 USA
{glu, hsingh, mlee, nader, kurdahi}@ece.uci.edu
[2] Department of Systems and Computer Engineering
Federal University of Rio de Janeiro/COPPE
P.O. Box 68511 21945-970 Rio de Janeiro, RJ Brazil
eliseu@lam.ufrj.br

Abstract. This paper introduces MorphoSys, a parallel system-on-chip which combines a RISC processor with an array of coarse-grain reconfigurable cells. MorphoSys integrates the flexibility of general-purpose systems and high performance levels typical of application-specific systems. Simulation results presented here show significant performance enhancements for different classes of applications, as compared to conventional architectures.

1 Introduction

General-purpose computing systems provide a single computational substrate for applications with diverse characteristics. These systems are flexible but, due to their generality, they may not match the computational needs of many applications. On the other hand, systems built around Application-Specific Integrated Circuits (ASICs) exploit intrinsic characteristics of an algorithm that lead to a high performance. However, the direct architecture–algorithm mapping restricts the range of applicability of ASIC-based systems.

Reconfigurable computing systems represent a hybrid approach between the design paradigms of general-purpose systems and application-specific systems. They combine a software programmable processor and a reconfigurable hardware component which can be customized for different applications. This combination allows reconfigurable systems to achieve performance levels much higher than that obtained with general-purpose systems, with a wider flexibility than that offered by application-specific systems.

This paper introduces the MorphoSys parallel reconfigurable system. MorphoSys (*Morphoing System*) is primarily targeted to applications with inherent parallelism, high regularity, word-level granularity and computation-intensive nature. Some examples of such applications are video compression, image processing, multimedia and data security. However, MorphoSys is flexible enough to support bit-level and irregular applications.

P. Amestoy et al. (Eds.): Euro-Par'99, LNCS 1685, pp. 727–734, 1999.

The remainder of this paper is organized as follows. Section 2 presents the MorphoSys architecture and emphasizes its unique features. Section 3 discusses the status of the MorphoSys prototype currently under development. Section 4 shows performance figures for important applications mapped to MorphoSys. Finally, Section 5 presents the main conclusions.

2   The MorphoSys System

2.1   The Architecture

The basic architecture of a parallel reconfigurable system [1] comprises a software programmable core processor and a reconfigurable hardware component. The core processor executes sequential tasks of the application and controls data transfers between the programmable hardware and data memory. In general, the reconfigurable hardware is dedicated to exploitation of parallelism available in the application. This hardware typically consists of a collection of interconnected reconfigurable elements. Both the functionality of the elements and their interconnection are determined through a special configuration program.

Figure 1 shows the MorphoSys architecture. It comprises five components: the core processor, the Reconfigurable Cell Array (RC Array), the Context Memory, the Frame Buffer and a DMA Controller.

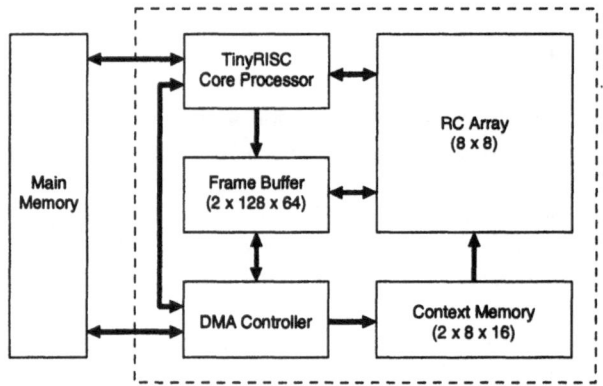

Fig. 1. Architecture of the MorphoSys reconfigurable system.

The core processor, also known as TinyRISC, is a MIPS-like processor with a 4-stage scalar pipeline. It has sixteen 32-bit registers and three functional units: a 32-bit ALU, a 32-bit shift unit and a memory unit. An on-chip data cache memory minimizes accesses to external main memory. In addition to typical RISC instructions, TinyRISC's ISA is augmented with specific instructions for controlling other MorphoSys components. DMA instructions initiate data transfers between main memory and the Frame Buffer, and configuration loading from main memory into the Context Memory. RC Array instructions specify one of the internally stored configuration programs and how it is broadcast to the RC Array. The TinyRISC processor is not intended to be used as a stand-alone, general-purpose processor. Although TinyRISC performs sequential tasks of the application, performance is mainly determined from data-parallel processing in the RC Array.

The RC Array consists of an 8×8 matrix of Reconfigurable Cells (RCs). An important feature of the RC Array is its three-layer interconnection network. The first layer connects the RCs in a two-dimensional mesh, allowing nearest neighbor data interchange. The second layer provides complete row and column connectivity within an array quadrant. It allows each RC to access data from any other RC in its row/column in the same quadrant. The third layer supports inter-quadrant connectivity. It consists of buses called *express lanes*, that run along the entire length of rows and columns, crossing the quadrant borders. An express lane carries data from any one of the four RCs in a quadrant's row (column) to the RCs in the same row (column) of the adjacent quadrant.

The Reconfigurable Cell (RC) is the basic programmable element in MorphoSys. As Figure 2 shows, each RC comprises: an ALU-Multiplier, a shift unit, input multiplexers, a register file with four 16-bit registers and the context register.

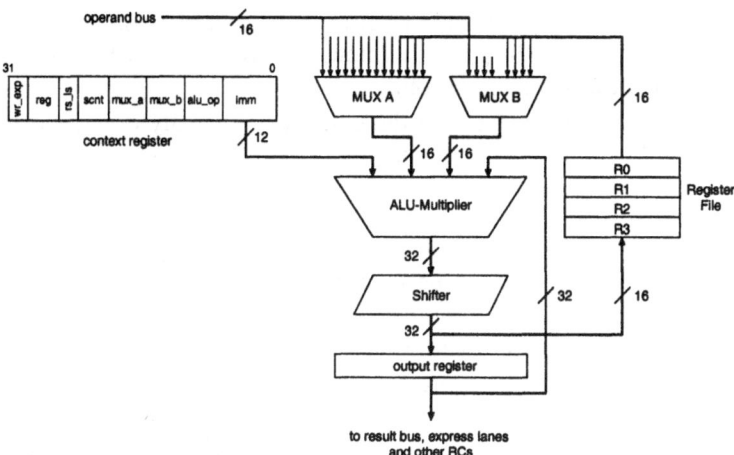

Fig. 2. Architecture of the Reconfigurable Cell (RC).

In addition to standard arithmetic and logical operations, the ALU-Multiplier can perform a multiply-accumulate operation in a single cycle. The input multiplexers select one of several inputs for the ALU-Multiplier. Multiplexer MUX A selects an input from: (1) one of the four nearest neighbors in the RC Array, or (2) other RCs in the same row/column within the same RC Array quadrant, or (3) the operand data bus, or (4) the internal register file. Multiplexer MUX

B selects one input from: (1) three of the nearest neighbors, or (2) the operand bus, or (3) the register file.

The context register provides control signals for the RC components through the *context word*. The bits of the context word directly control the input multiplexers, the ALU/Multiplier and the shift unit. The context word determines the destination of a result, which can be a register in the register file and/or the express lane buses. The context word also has a field for an immediate operand value.

The Context Memory stores the configuration program (the *context*) for the RC Array. It is logically organized into two partitions, called Context Block 0 and Context Block 1. By its turn, each Context Block is logically subdivided into eight partitions, called Context Set 0 to Context Set 7. Finally, each Context Set has a depth of 16 context words. A *context plane* comprises the context words at the same depth across the Context Block. As the Context Sets are 16 words deep, this means that up to 16 context planes can be simultaneously resident in each of the two Context Blocks. A context plane is selected for execution by the TinyRISC core processor, using the RC Array instructions.

Context words are broadcast to the RC Array on a row/column basis. Context words from Context Block 0 are broadcast along the rows, while context words from Context Block 1 are broadcast along the columns. Within Context Block 0 (1), Context Set n is associated with row (column) n, $0 \leq n \leq 7$, of the RC Array. Context words from a Context Set are sent to all RCs in the corresponding row (column). All RCs in a row (column) receive the same context word and therefore perform the same operation.

The Frame Buffer is an internal data memory logically organized into two sets, called Set 0 and Set 1. Each set is further subdivided into two banks, Bank A and Bank B. Each bank stores 64 rows of 8 bytes. Therefore, the entire Frame Buffer has 128×16 bytes. A 128-bit *operand bus* carries data operands from the Frame Buffer to the RC Array. This bus is connected to the RC Array columns, allowing eight 16-bit operands to be loaded into the eight cells of an RC Array column (i.e., one operand for each cell) in a single cycle. Therefore, the whole RC Array can be loaded in eight cycles.

In the current MorphoSys prototype, the operand bus has a single configuration mode, called *interleaved mode*. In this mode the operand bus carries data from the Frame Buffer banks in the order $A_0, B_0, A_1, B_1, ..., A_7, B_7$, where A_n and B_n denote the nth byte from Bank A and Bank B, respectively. Each cell in an RC Array column receives two bytes of data, one from Bank A and the other from Bank B. This operation mode is appropriate for image processing applications involving template matching, that compare two 8-bit operands.

The next MorphoSys implementation will allow a second configuration mode, called *contiguous mode*. In this mode, the operand bus carries data in the order $A_0, ..., A_7, B_0, ..., B_7$, where A_n and B_n denote the nth byte from Bank A and Bank B, respectively. Each cell in an RC Array column receives two consecutive bytes of data from either Frame Buffer Bank A or B. This additional mode satisfies application classes that require 16-bit data transfers.

Results from the RC Array are written back to the Frame Buffer through a separate 128-bit bus, called the *result bus*. Application mapping experience has indicated the need for flexibility in the result bus too. In the current MorphoSys prototype, the result bus operates in an *8-bit mode*. In this mode, each cell of an RC Array column provides an 8-bit result, forming a 64-bit word which is written into Frame Buffer Bank A or B. But, in some applications, it is necessary to write back 16-bit data from each RC. To satisfy this requirement, the result bus in the next MorphoSys implementation will have an additional *16-bit mode*. Here, 16-bit results from the first four cells of an RC Array column are written into Frame Buffer Bank A, while the 16-bit results from the remaining four cells are stored into Bank B.

The DMA controller performs data transfers between the Frame Buffer and the main memory. It is also responsible for loading contexts into the Context Memory. The TinyRISC core processor uses DMA instructions to specify the necessary data/context transfer parameters for the DMA controller.

2.2 Execution Flow Model

The execution model of MorphoSys is based on partitioning applications into sequential and data-parallel tasks. The former are handled by TinyRISC core processor whereas the latter are mapped to the RC Array. TinyRISC initiates all data and context transfers. RC Array execution is also enabled by TinyRISC, through the special context broadcast instructions. These instructions select one of the context planes and send the context words to the RC Array. While RC Array performs computations on data in one Frame Buffer set, fresh data may be loaded in the other set or Context Memory may receive new contexts.

2.3 Important Features of MorphoSys

MorphoSys is a coarse-grain, multiple-context reconfigurable system with considerable depth of programmability (32 context planes) and two different context broadcast modes. Bus configurability supports applications with different data sizes and data flow patterns. The hierarchical RC Array interconnection network contributes for algorithm mapping flexibility. Structures like the express lanes enhance global connectivity. Even irregular communication patterns, that otherwise require extensive interconnections, can be handled efficiently.

MorphoSys is a highly parallel system. First, MorphoSys is dynamically reconfigurable. While the RC Array is executing one of the sixteen contexts in row broadcast mode, the other sixteen contexts for column broadcast can be reloaded into the Context Memory (or vice-versa). Secondly, RC Array computations using data in one Frame Buffer set can proceed in parallel with data transfers from/to the other Frame Buffer set. The internal Frame Buffer and DMA controller, and the adoption of wide datapaths, allow high-bandwidth transfers for both data and configuration information.

3   Implementation of MorphoSys

MorphoSys is tightly-coupled reconfigurable system. The TinyRISC core processor, the RC Array and the remaining components are to be integrated into a single chip. The first implementation of MorphoSys is called the M1 chip. M1 is being designed for operation at 100 MHz clock frequency, using a 0.35 μm CMOS technology. The TinyRISC core processor and the DMA controller were synthesized from a VHDL model. The other components (RC Array, Context Memory and Frame Buffer) were completely custom designed. The final design will be obtained through the integration of both synthesized and custom parts.

There is a SUIF-based C compiler for the TinyRISC core processor and a simple assembler-like parser for context generation. A GUI tool called *mView* supports interactive programming and simulation. Using mView, the programmer can specify the functions and interconnections corresponding to each context for the application. mView then automatically generates the appropriate context file. As a simulation tool, mView reads a context file and displays the RC outputs and interconnection patterns at each cycle of the application execution.

4   Algorithm Mapping and Performance Analysis

4.1   Image Processing Application: DCT/IDCT

The Discrete Cosine Transform (DCT) and Inverse Discrete Cosine Transform (IDCT) are examples typifying the image processing area. Both transforms are part of the JPEG and MPEG standards.

A two-dimensional DCT (2-D DCT) can be performed on an 8 × 8 pixel matrix (the size in most image and video compressing standards) by applying one-dimensional DCT (1-D DCT) [2] to the rows of the pixel matrix, followed by 1-D DCT on the results along the columns. For high throughput, the eight row (column) 1-D DCTs can be computed in parallel.

When mapping 2-DCT to MorphoSys, the operand and result buses are configured for interleaved and 8-bit modes, respectively. For an 8 × 8 pixel matrix, each pixel is mapped to a RC. To perform eight 1-D DCTs along the rows (columns) of the pixel matrix, context is broadcast along the columns (rows) of the RC Array. As MorphoSys has the ability to broadcast context along both rows and columns of the RC Array, the need of transposing the pixel matrix is eliminated, thus saving a considerable amount of cycles.

The operand data bus allows the entire pixel matrix to be loaded in eight cycles. Once data is in the RC Array, two butterfly operations are performed to compute intermediate variables. Inter-quadrant connectivity provided by the express lanes enables one butterfly operation in three cycles. As the butterfly operations are also performed in parallel, only six cycles are necessary to accomplish the butterfly operations for the whole matrix. Row/column 1-D DCTs take 12 cycles. Two additional cycles are used for data re-arrangement. Finally, eight cycles are needed for result write back.

Table I. Performance comparison for DCT/IDCT application.

System	Performance
MorphoSys	36 cycles
sDCT	240 cycles
REMARC	54 cycles
V830	201 cycles
TMS	320 cycles

Table I shows performance figures for 2-D DCT on MorphoSys and other systems. Performance is given in number of cycles, in order to isolate influences from the different integration technologies employed to implement the systems. sDCT is a software implementation written in optimized Pentium assembly code using 64-bit special MMX instructions [3]. REMARC [4] is another reconfigurable system, targeting multimedia applications. V830R/AV [5] is a superscalar multimedia processor. TMS320C80 [6] is a commercial digital signal processor. Execution of DCT/IDCT on MorphoSys results in a speedup of 6X as compared to a Pentium MMX-based system. MorphoSys yields a throughput much better than that of the considered hardware designs.

4.2   Data Encryption/Decryption Application: IDEA

Today, data security is a key application domain. The International Data Encryption Algorithm (IDEA) [7] is a typical example of this application class. IDEA involves processing of plaintext data (i.e., data to be encrypted) in 64-bit blocks with a 128-bit encryption/decryption key. The algorithm performs eight iterations of a core function. After the eighth iteration, a final transformation step produces a 64-bit ciphertext (i.e., encrypted data) block. IDEA employs three operations: bitwise exclusive-or, addition modulo 2^{16} and multiplication modulo $2^{16} + 1$. Encryption/decryption keys are generated externally and then loaded once into the Frame Buffer.

When mapping IDEA to MorphoSys, the operand bus and the result bus are configured for the contiguous and 16-bit modes, respectively. Some operations of IDEA's core function can be performed in parallel, while others must be performed sequentially due to data dependencies. The maximum number of operations that can be performed in parallel is four. In order to exploit this parallelism, clusters of four cells in the RC Array columns are allocated to operate on each plaintext block. Thus, the whole RC Array can operate on sixteen plaintext blocks in parallel.

As two 64-bit plaintext blocks can be transferred simultaneously through the operand bus, it takes only eight clock cycles to load 16 plaintext blocks into the entire RC Array. Each of the eight iterations of the core function takes seven clock cycles to execute in a cell cluster. The final transformation step needs one additional cycle. Once the ciphertext blocks have been produced, eight cycles are necessary to write them back to the Frame Buffer before loading the next plaintext blocks. Therefore, it takes 73 cycles to produce 16 cyphertext blocks.

Table II compares the performance of MorphoSys for the IDEA algorithm. sIDEA is a software implementation on a Pentium II processor. Performance shown was measured by using the Intel VTune profiling tool. HiPCrypto [8] is a 7-stage pipelined ASIC chip that implements IDEA in hardware. A single HiPCrypto produces 7 cyphertext blocks every 56 cycles. Two HiPCrypto chips in a cascade produce 7 cyphertext blocks every 28 cycles. IDEA running on MorphoSys is much faster than running on an advanced superscalar processor. It is also faster that an IDEA ASIC processor in a single-chip configuration.

Table II. Performance comparison for IDEA application.

System	Performance
MorphoSys	16 blocks/73 cycles
sIDEA	1 block/357 cycles
HiPCrypto, single-chip	7 blocks/56 cycles
HiPCrypto, two-chips	7 blocks/28 cycles

5   Concluding Remarks

In this paper, we described the MorphoSys parallel reconfigurable system. We also presented results showing significant performance gains of MorphoSys for important applications in the imaging processing and data security domains. Overall, this paper demonstrates that the combination of general-purpose processors with reconfigurable hardware blocks represents a potential design paradigm to address the performance needs of future applications and for the microprocessors of the next decade.

References

[1] W. H. Mangione-Smith *et al.*, *Seeking Solutions in Configurable Computing*, IEEE Computer, Dec. 1997, pp. 38–43.
[2] W-H Chen, C. H. Smith, S. C. Fralick, *A Fast Computational Algorithm for the Discrete Cosine Transform*, IEEE Trans. on Communications, Vol. 25, No. 9, Sept. 1997, pp. 1004–1009.
[3] App. Notes for Pentium MMX, http://developer.intel.com/drg/mmx/appnotes.
[4] T. Miyamori, K. Olukotun, *A Quantitative Analysis of Reconfigurable Coprocessors for Multimedia Applications*, Proc. of the IEEE Symposium on Field Programmable Custom Computing Machines, 1998.
[5] T. Arai *et al.*, *V830R/AV: Embedded Multimedia Superscalar RISC Processor*, IEEE Micro, Mar./Apr. 1998, pp. 36–47.
[6] F. Bonimini *et al.*, *Implementing an MPEG2 Video Decoder Based on TMS320C80 MVP*, SPRA 332, Texas Instruments, Sept. 1996.
[7] B. Schneier, *Applied Cryptography*, John Wiley, New York, NY, 1996.
[8] S. Salomao, V. Alves, E. C. Filho, *HiPCrypto: A High Performance VLSI Cryptographic Chip*, Proc. of the 1998 IEEE ASIC Conference, pp. 7–13.

A Graph-Oriented Task Manager for Small Multiprocessor Systems

Xavier Verians[1], Jean-Didier Legat[1], Jean-Jacques Quisquater[1], and Benoit Macq[2]

[1] Microelectronics Laboratory, Université Catholique de Louvain,
[2] Telecommunications Laboratory, Université Catholique de Louvain,
place du Levant, 3; 1348 Louvain-la-Neuve; Belgium

Abstract. A task manager that dynamically decodes the data-dependent task graph is a key component of general multiprocessor systems. The emergence of small-scale parallel systems for multimedia and general-purpose applications requires the extraction of complex parallelism patterns. The small system size also allows the centralization of the task generation and synchronization. This paper proposes such a task manager. It uses a structured representation of the task dependence graph to issue and synchronize tasks. We describe several optimizations to extract more parallelism, discuss the software/hardware implementation issue and show it produces efficient parallelism exploitation in case of applications with complex parallelism patterns.
keywords: parallelism, dependence graph, synchronization, multiprocessors

1 Introduction

Processors embedded in personal computers intensively use the instruction-level parallelism with one-chip superscalar, VLIW or multithreaded architectures. However, new systems begin to exploit medium and coarse-grained parallelism. These systems are based on small-scale multiprocessors or one-chip "super-processors" able to execute several tasks simultaneously. It is especially true in small commercial computers where several processors begin to be used on the same board to boost performances. It is indeed simpler to design and duplicate a small processor than to develop a huge and complex uniprocessor with similar performances. These small-scale multiprocessors - i.e. with less than 32 *processing units* (*PUs*) - require an efficient parallelism extraction at the task level to achieve good performances. Currently, mainly software methods are used to exploit parallelism. These techniques, inspired from those used in larger systems, perform well in case of large threads or large data sets. However, general-purpose personal computers are generally used to execute applications on relatively small data sets. These applications present a reduced and complex parallelism that cannot easily or efficiently be described with a classical method.

Indeed, classical methods provide a powerful graph analysis to schedule tasks efficiently. But the complex algorithms require time-consuming tasks to hide

P. Amestoy et al. (Eds.): Euro-Par'99, LNCS 1685, pp. 735–744, 1999.

the graph management overhead. They are seldom optimized for small task granularity or complew parallelism structures. It is interesting to note that for example Hamidzadeh[1] does not study the data structure implementation to boost its algorithm execution. Johnson[2] bases its system on a task graph stored in a hash table. However, it does not explicitly conserve the graph structure to simplify the operations and dynamic branches are ignored.

This paper addresses this issue by proposing a task manager that can handle parallelism from complex applications at a small software or hardware cost. It is based on an original and structured task graph representation built from a separated parallelism description. The data structure is close to some used in static scheduling methods[3]. However, the proposed structure allows the manager to store the graph under a well-structured pattern, consequently simplifying the graph management operations and improving their efficiency. The underlying algorithm it uses is presented more theoretically in[4]. As the parallelism management is centralized, it is dedicated to small-scale systems. We also limit our study to shared-memory multiprocessors.

This paper is organized as follows. Section 2 recalls some definitions from the graph-theory field. An overview on the main task manager characteristics is provided in section 3. A more detailed description of the task manager is presented in section 4. It explains the data structure used to store the task graph and the way it is dynamically handled. Subsection 4.3 discusses the hardware/software implementation issue. Finally, the task manager is validated with simulations whose results are presented in section 5. They show that on complex applications, the proposed parallelism exploitation significantly improves performances with a reduced instruction overhead.

2 Definitions

Parallelism among program parts can be represented with a task dependence graph[2]. We define a task as an instruction flow that will be executed on a single processing unit. Tasks can comprise function calls and (un)conditional branches, provided all jumps are referencing an address included in the same task. They can begin in a function and end in another one.

A task dependence graph is a directed graph consisting of a set of nodes (tasks) and a set of arcs (dependences) on the nodes. The dependences encode the execution ordering constraints. A task A depends on task B if B has to be completed before A can be executed. The notation we use is the following: capitals to design tasks and lower-case letters for arcs. Arcs are identified by the same letter as the origin-task identifier. The arcs pointing on a task are called the *incoming arcs* of this task. The task graph defines a partial order among tasks. A task is said *eligible* when all tasks it depends on have been completed. At run-time, the task graph is unfolded to produce a directed acyclic graph (DAG). Due to its limited resources, the processor will only see a part of this graph, called the *task window*. It encloses all the tasks already decoded by the system, but waiting for their execution to be launched or completed.

a)

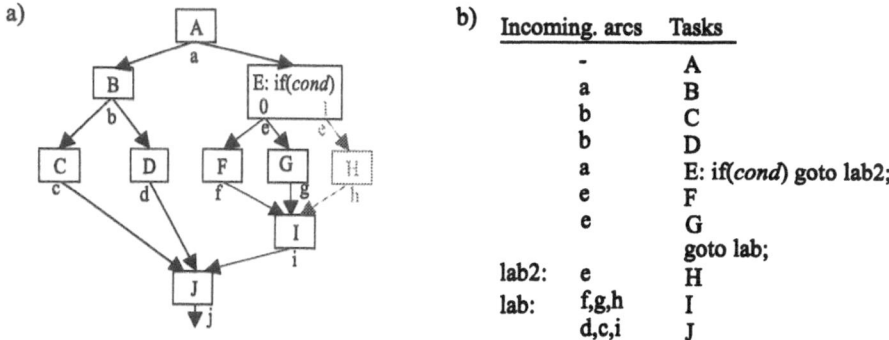

b)

Incoming. arcs	Tasks
-	A
a	B
b	C
b	D
a	E: if(cond) goto lab2;
e	F
e	G
	goto lab;
lab2: e	H
lab: f,g,h	I
d,c,i	J

Fig. 1. *left:* task dependence graph, *right:* task program. Syntax: each line (except the *goto instruction*) describes a task and its incoming arcs

3   Task Manager Characteristics

This section describes the main task manager characteristics that allow it to handle dynamic task dependence graph. It can be qualified as:

• *general:* General-purpose or new multimedia applications have a complex structure at the control-flow level. The main capability of a good parallelism management is its ability to handle various kinds of parallelism structures. As about all parallelism patterns can be represented by directed acyclic graphs, we use a parallelism description directly based on a topological description of the task dependence graph. Tasks are ordered in the topological description following the partial execution order [4]. This description, called the *task program*, avoids limiting the pattern diversity by the use of specific parallel directives.

• *dynamic:* The manager sequentially decodes the task program at run-time. It supports data-dependent control structures to dynamically modify the task graph. It is achieved by inserting structural forward or backward branches into the task program, as classical tasks. With branches, several instances of a task could exist at the same time in the system. To distinguish it, tasks and dependences are renamed at the decoding step. As the tasks in the task program are ordered, the arcs named *t* where *T* has not been renamed are ignored. This mechanism transforms the conservative and static dependence graph into a dynamic graph where valid dependences only are taken into account. An example of topological task graph description with a branch is provided in figure 1.

• *structured:* The task manager uses a queue bank to store the task graph in a structured fashion. This simplifies the graph management operations and increases their efficiency. Intelligent task scheduling policies based on a local analysis of the task graph can be supported.

• *separation of the parallelism from the program code:* The advantages of separating the parallelism description from the program code are discussed in [4]. It mainly uses the benefit of the centralized approach to simplify the parallelism extraction and enlarge the task manager view on the program structure.

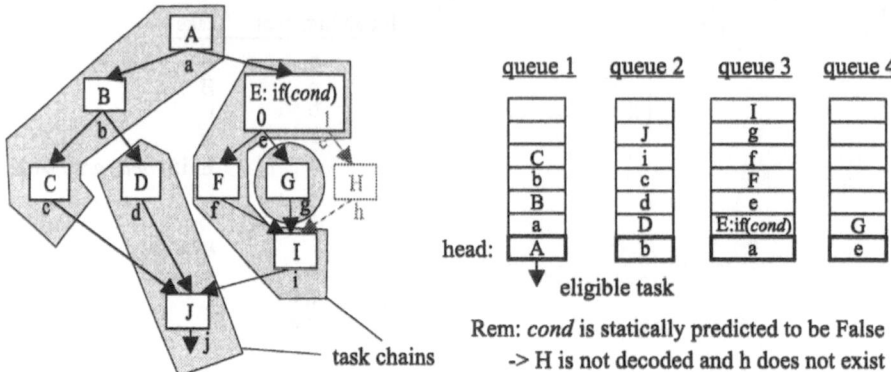

Fig. 2. *left:* task graph sliced in sequential task chained, *right:* filled queue bank

4   Task Manager Structure

This section describes how the task manager processes the task graph to extract the parallelism. The task manager has three main jobs. It first decodes the task program and internally builds an acyclic representation of the static dependence graph. Its second job is to transform the static graph into a dynamic one by the execution of data-dependent control commands. Finally, it has to select eligible tasks to send it to the PUs and to remove completed tasks. These operations are described in the first subsection. Subsection 4.2 presents some parallelism management optimizations, while the last subsection discusses the hardware/software implementation issue.

4.1   Graph Management

The whole task manager is based on a queue bank storing the task window. The parallelism available at a given time is equal to the number of parallel paths in the task window graph. To conserve the parallel structure of the task graph, parallel paths are stored in different queues. As a path is constituted by a chain of consecutive tasks, each queue holds tasks sequentially chained. If the task graph is drawn as a DAG oriented from top to bottom (fig. 2), the queue bank slices the task window into vertical task chains. The advantage of this structure is clear, as the maximum parallelism is equal to the number of non-empty queues.

To process the task graph at run-time, three basic operations need to be executed on the queue bank:

– *task addition:* To add a new decoded task T, the manager searches a queue Q whose tail contains a task on which T *depends* . If no queue satisfies the condition, Q points on an empty queue. Then all the incoming arcs are added to the queue followed by the task identifier T. If no more queue is available, the decoding waits for the execution to proceed further until a queue is

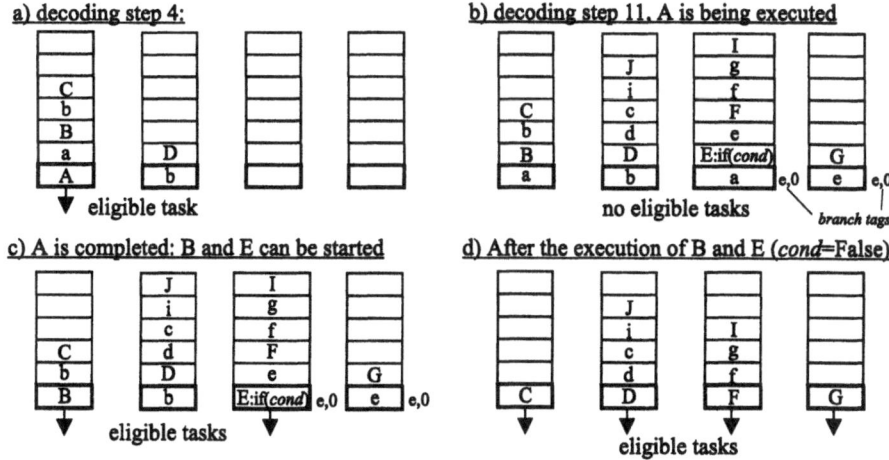

Fig. 3. Queue-bank management. The task program of fig. 1 is sequentially decoded. One step corresponds to the decoding of a line in the task program

released. The search operation is quite simple: it only needs to access the last element of the queues. The operation complexity is $O(\#queues)$ fetching and comparisons in a sequential system and $O(1)$ in a parallel implementation.

- *eligible task selection:* This operation is straightforward. If a task in a queue depends on other tasks of the task window, it is preceded by its incoming arcs. Eligible tasks only are not preceded by arcs or tasks and are stored in the queue head. Finding an eligible task simply requires testing if the queue head contains an arc or a task identifier. When selected, the task is removed. The operation requires $\#queues$ or one test(s) respectively in a sequential or parallel implementation.
- *dependence removal:* When a task T is completed, the arcs t are removed from the queues. It is the most expensive process. However, the critical operation is the removal of a dependence preventing a new task to become eligible. These dependences are found in the queue heads. The delay between a task completion and a dependent task launching is determined by the time needed to remove the arcs from the queue heads. This operation can be performed with a $0(\#queues)$ or $0(1)$ cost respectively with a sequential or parallel implementation. It simply consists for each queue in a comparison of the head element with the arc identifier. If they match, the head element is popped from the queue. Removing the arcs from the queue bodies takes a longer time. As these operations are not critical, the overhead can be hidden by the task execution.

Branches are processed in a similar way. When a branch is decoded, it is statically predicted to determine which side will be taken. The decoding will continue on the predicted side, while the address pointed by the other branch side is stored in a small table. The branch is written in an empty queue. A tag identifying the branch and the token side is attached to each queue storing tasks depending on this branch. When a branch becomes eligible, it is resolved. If the prediction was right, the depending queues are untagged. In the opposite case, the tagged queues are discarded and the decoding resumes on the other side. For forward branches with limited scope, both sides can be speculatively decoded.

An example of the task management is provided at figure 3. Actually, the task addition, selection and removal operations are executed concurrently during execution, tasks being added on one the queue tails, while others are removed from the queue heads.

4.2   Parallelism Management Optimizations

The described task manager can handle all the parallelism patterns that can be drawn as a task graph. However the task program is sequentially decoded and it can lead to inefficient executions. This subsection details how to solve this problem by parallelizing the task program decoding.

Applications are often based on small kernels easily parallelizable. These kernels can be described by a single task executed many times. We called it a low-level loop. The queue-based system would require writing a new task instance and a testing command for each iteration. To avoid this expensive solution, the task attached to the low-level loop is written once in the queue bank. When selected for execution, it is sent to a dedicated resource, the *task server*. The server generates several task instances to schedule it to the PUs and holds some loop state bits to detect the loop end. The queue bank considers low-level loops as a simple task, reducing sharply the management overheads. Eligible tasks can then be found in the queue head or in the task servers.

On the other side, the sequential decoding of the task program is less efficient in case of coarse-grained parallelism level. It often arises that several program parts can be executed in parallel. If each part is described by a large number of tasks, a sequential decoding cannot provide a view large enough on the global parallel structure. A simultaneous decoding of different task graph parts is necessary. High-level commands, such as classical *fork/join* or *doall* commands[5][6], are added to the task program. They generate several *task pointers* able to decode multiple task program parts simultaneously. The key difference with other systems is that these commands are only used to allow a simultaneous decoding and not necessary an unconditional simultaneous execution. Explicit commands can impose some dependences between task instances of different parts or consecutive iterations. As tasks are written in the queues by analyzing the dependence information, the task pointers could use the same queue bank. At least one queue will be statically allocated to each active task pointer to avoid deadlocks.

The *doall* command allows a speculative execution of tasks computing loop indexes. Indeed, a large number of loops have iterations only depending on each other by the loop index computations. A speculative execution of these computations allows the execution of several iterations in parallel. This loop management requires a small memory to hold some loop state bits. Such high-level loops can be overlapped to exploit parallelism among instances of several loop levels.

4.3   Task Manager Implementation

The task manager is aimed toward small-scale multiprocessor systems. As the communications between the task manager and the PUs are reduced, the management is centralized in a single resource. The task manager can either be a software program interlaced with task execution on a single PU as in[1] or a small additional resource. However, the task manager mainly requires additions and memory accesses. Considering the first solution, the interruption of a PU to execute the task manager wastes the complex PU resources and might incur some load unbalance with other PUs. Moreover, an additional mechanism has to ensure that the task manager program is executed quickly enough to generate eligible tasks when needed, and not too often to avoid the PU overloading. For these reasons, we took the second solution consisting of using a dedicated resource to handle the parallelism.

The task manager consists in a queue bank, several task servers, several task pointers, a command interpreter with its branch and loop tables and a control part. The queue bank has to be accessed for the task addition, selection and synchronization. The cost of these operations is proportional to the number of queue and thus the number of PUs. On the other hand, the task pointers, the command interpreter and the control part are complex operations. Their cost is proportional to the number of processed tasks. As they can be executed in parallel with task execution, their cost will be hidden, provided the mean time to process a task is lower than the time to execute it.

These considerations suggest to execute the task pointers, the command interpreter and the control parts in software, while the queue bank management and the task servers are processed in hardware. The queue bank is implemented as a set of circular buffers. It is coupled with a comparator to test the queue tail for the task addition or the queue head for the task selection and critical synchronization. With this architecture, a task can be added to the queue bank with about $3+2\times(\#\text{incoming arcs})$ cycles. An eligible task is selected and popped in 3 cycles, while the arc removal from a queue head requires about 4 cycles. The removal of an arc from the queue body will take about $4\times(\text{queue size})$ cycles. A software implementation would be far more expensive, as for each operation, the data should be fetched from memory, processed and then written again to memory. The tests required by the circular buffer management would also be quite expensive in computation time. The branch tags are stored in a 32-bits vector. Coupled with the branch table, it allows 8 task pointers to have each 7 simultaneous pending branches, which is enough in most cases for up to 64 PUs.

5   Simulations

Simulations have been performed on several classes of applications first to validate the proposed parallelism exploitation and secondly to verify the possibility to implement complex parallelism patterns with a low instruction overhead. The ease of programming is also taken into account. Simulations related to the cost of a software/hardware implementation are still in progress.

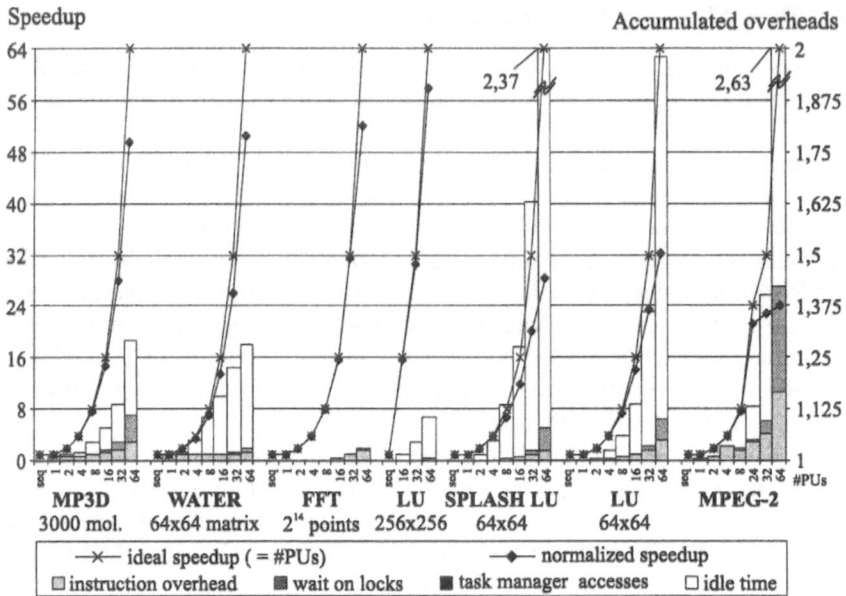

Fig. 4. Right axis: Normalized speedup, left axis: Accumulated overheads. Value "1" is equal to the sequential execution time. Results are obtained with our parallelism exploitation, except for SPLASH LU where the classical SPLASH parallelism exploitation has been used

The simulation system comprises an experimental compiler and a multiprocessor simulator. The compiler is based on the Stanford's SUIF system extended with the Harvard's MACHINE library. Several passes have been added to extract the parallelism, generate the task program and managing parallel stack frames. Currently, the tasks are defined by the programmer with parallel directives inserted in the program code. The multiprocessor simulator simulates a bus-based multiprocessor with simple RISC processors and a shared memory. The analysis is mainly based on the SPLASH benchmarks[7][8]. As these benchmarks have been evaluated on an ideal system, we simulate ideal RISC PUs, memories and task manager. It means that each PU executes one instruction per cycle. It allows us to compare our results with those from SPLASH and study the performance of the proposed parallelism extraction without the influence of other multiprocessor overheads. The speedup definition we use is the normalized speedup i.e. the ratio of the measured execution time to the execution time of an equivalent sequential program executed on a PU.

Simulation results on SPLASH benchmarks are provided in fig. 4. MP3D, WATER and FFT are simple algorithms with a parallelism based on several simple parallel loops. In this case, results are similar to the ideal SPLASH results, as classical multiprocessors are well suited to execute large parallel loops. Performances for the LU algorithm with 256x256 matrix are far better, despite a smaller data set used compared with SPLASH. The performance improvement is explained by a deeper exploitation of the available parallelism. Indeed, the comparison between the SPLASH parallelism exploitation (SPLASH LU) and ours (LU) on a small data set (64x64 matrix sliced in 256 data blocks) shows that the performance gain rises up to 12% for 64 PUs. While SPLASH only exploits the parallelism inside each iteration, our parallelism representation allows the exploitation of the partial parallelism available among several iterations. This improvement is clearly shown by the reduced idle time. The limited parallelism management overheads also allow smaller tasks to be processed. Finally, the instruction overhead is kept below 1% up to 16 PUs and 4% for 64 PU. The parallelism increase thus largely compensates the additional instructions and synchronization overheads due to the deeper parallelism exploitation. In all cases, the number of communications between the task manager and the PUs is less than 0,2% of the instruction number.

To test the system on actual applications, an MPEG-2 decoder has been simulated. The decoder program is sequential and directly derived from the MPEG-2 standard. It has been parallelized by introducing parallelization directives and rewriting minor code parts. It introduces an instruction overhead of less than 1%. Indeed, the use of an optimized parallel program will produce better results. On the last graph of fig. 4, the MPEG-2 decoder shows good performances up to 24 PUs. For larger systems, the curve flattening shows that the critical parts become dominant. It is interesting to note that the instruction and locking overheads mainly rises above 32 PUs. It means that as long as there is enough parallelism available, the PUs perform useful computations. When no more parallelism is available, the PUs begin to execute speculatively additional -but invalid- loop iterations. It increases the instruction and locking overheads.

The execution of complex programs shows that our parallelism management is efficient to describe and exploit complex structures. Moreover, this capability simplifies the programmer job, as he has not to thoroughly modify the sequential program to compel with some limiting parallel syntax. Indeed, we used for the SPLASH benchmarks the original version with slight modifications. Concerning the MPEG-2 decoder, we took a sequential program and inserted parallel commands. No modifications have been made to ensure that parallel tasks will be close together in the program. This is not important as the task manager rearranges the tasks in the queue bank.

6 Conclusion

A new centralized task manager based on a dynamic task-graph analysis has been presented. This manager extracts the complex parallelism patterns occurring in general applications. Graph-based manipulations are possible by the use of a queue bank storing the task graph in a structured pattern. This queue bank has been designed to simplify and optimize the task addition, selection and synchronization in a sequential or parallel system. The data-dependent parallelism is described by encoding in a task program the task dependences, the main control structures and some optimization commands. The decoding of this task program allows the task manager to fill the queues at a limited cost.

Current dynamic parallelism extraction methods are mainly designed by studying an efficient scheduling algorithm. It is then implemented in software on the processing units. Our approach is different. It has first studied how a task graph can be efficiently represented and processed. The queue-based structure is designed to optimize the graph management operations. Then, the parallelism description has been developed to be general, dynamic and compatible with the queue filling policy. The resulting architecture is a manager able to efficiently perform operations such as eligible task selection and task synchronization.

This paper has demonstrated the potentialities of a graph-based task manager. Currently, no particular load balancing or data locality techniques have been implemented. However, the queue bank stores a structured description of the dependence graph. Techniques based on a local graph analysis could be implemented in the manager to improve the performances. The data locality is a more critical problem than the load balancing. The load imbalance can be hidden is enough parallelism is extracted. The problem of data locality is that it decreases with the increase of the parallelism extraction. To overcome this problem, a data locality policy should be added to the task manager.

Future work will be oriented toward the actual implementation of the task manager. This paper has presented the global architecture and the hardware/ software options, but several implementation alternatives still need to be studied and validated to have an cheap but efficient task manager. Concerning the compiler, the programmer has to introduce commands into the program code to help it to partition the program into tasks. We will later develop passes to perform this task automatically on a sequential program.

References

[1] Hamidzadeh, B., Lilja, D.J.: Dynamic Scheduling Strategies for Shared-Memory Multiprocessors. Proc. of the 16th Int. Conf. on Distributed Computing Systems, Hong Kong, 1996, 208–215
[2] Johnson, T., Davis,T., Hadfield, S.: A Concurrent Dynamic Task Graph. Parallel Computing, Vol 22, No 2, Ferbruary 1996, 327–333
[3] Park, G.-L., Shirazi, B., Marquis J., Choo, H.: Decisive Path Scheduling: A New List Scheduling Method. Proc. of the Int. Conf. on Parallel Processing, 1997, 472–480
[4] Verians, X., Legat J.-D., Quisquater, J.-J., Macq, B.: A New Parallelism Management Scheme for Multiprocessor Systems Proc. of the 4th Int. ACPC Conf., Salzburg, Austria, Lecture Notes in Computer Science, Vol 1557, February 1996, 246–256
[5] Nikhil, R.S., Papadopoulos, G.M., Arvind: *T: A Multithreaded Massively Parallel Architecture. Int. Symp. on Computer Architecture, 1992, 156–167
[6] Kai Hwan, Zhiwei Xu: Scalable Parallel Computing. McGraw-Hill, 1998
[7] Singh, J.P., Weber W-D., Gupta, A.: SPLASH: Stanford Parallel Applications for Shared Memory. Computer Architecture News, 20(1), July 1994, 5–44
[8] Woo, S.C., Ohara, M., Torrie, E., Singh, J.P., Gupta, A.: The SPLASH-2 Programs: Characterization and Methodological Considerations. Proc. of the 22nd Ann. Symp. on Computer Architecture, June 1995, 24–36

Implementing Snoop-Coherence Protocol for Future SMP Architectures

Wissam Hlayhel[1], Jacques Collet[2], and Laurent Fesquet[2]

[1] IRIT-Universit de Paul Sabatier, 118 route de Narbonne
31062 Toulouse-France.
hlayhel@irit.fr
[2] LAAS-CNRS, 7 avenue du colonel Roche,31077 Toulouse-France.
{collet,fesquet}@laas.fr

Abstract. Maintaining the coherence is becoming one of the most serious problems faced when designing today's machines. Initially, this problem was relatively simple when the interconnection network of Symmetric MultiProcessors (SMP) was an atomic bus, which simplified the implementation of invalidation coherence protocols. However, due to the increasing bandwidth demand, atomic busses have been progressively replaced by split busses that uncoupled the request and response phases of a transaction. Split busses enable initiating new requests before receiving the response to those already in progress but make more complicated the preservation of the coherence. Indeed, a new request induces a conflict when it concerns a block address involved by another current request and when one of the requests is a WRITE miss. Several solutions exist to solve this problem. That one used in the SGI machine is based on a shared data bus which traces the completion of transactions. Unfortunately, it becomes impracticable in the recent machines which replace data busses by more efficient networks (again for bandwidth constrains), ultimately by a crossbar. This work describes and quantitatively evaluates two possible solutions to the coherence problem for the new architectures where all the data responses cannot be traced by each processor.

1 Introduction

The presence of caches in multiprocessors sets the problem of the data coherency. Each processor may load and subsequently modify a data block in its cache that makes incoherent the copies in the different caches. The coherence is preserved either by broadcasting each modification to immediately update all the other caches (write update protocol) or by invalidating the copies in the other caches (write invalidate protocol). This second solution is mostly used [1]. The way copies are invalidated depends on the network architecture. In a shared-bus multiprocessor, each memory transaction is snooped by all the processors. This simplifies the implementation of coherence protocols by global invalidation. The MESI (Illinois) protocol [2,3] is often used in this kind of architecture. In the first implementations, the bus and the memory were completely locked during

P. Amestoy et al. (Eds.): Euro-Par'99, LNCS 1685, pp. 745–752, 1999.

a transaction from the address transmission phase to the reception of the data. The bus was atomic. This operation mode simplified the preservation of the data coherence. Unfortunately, the atomic bus progressively became unable to satisfy the bandwidth demand because of the slow evolution of the bus-memory subsystem with respect to increase of the processor power. It was replaced by a 64-bit multiplexed and split bus in the XDBus from Sun [4], by two separate busses (one for the addresses, one for the data which was enlarged up to 256 bits) in the SGI Challenge [5] or in the AlphaServer 8x00 [6]. In the recent designs, data are transmitted through a crossbar network and addresses through a shared bus for preserving the cache coherence (ESCALA-Bull [7], STARFIRE-Sun [8]).

Uncoupling the address and data phases of transactions makes more complex the implementation of the MESI invalidation protocol as conflicting situations may occur when several progressing requests affect the same memory block. Consider for instance the insertion of a WRITE miss on the bus before completion of a preceding WRITE miss on the same block. A conflict occurs because the latest request must not invalidate the first one before completion of its WRITE operation and in the same time it cannot ignore the first request which is anterior in the order of insertion on the bus. A coherence conflict occurs when two (or more) current requests affect the same block and when at least one request is a WRITE miss.

Existing architectures surely solve this problem, but to our knowledge, the protocol has been only disclosed for the SGI machine. Based on the implementation of a shared data bus (see the next section), it does not work in the new data-crossbar SMPs. In the following we study the coherence conflicts and introduce two solutions free from restrictive data-network hypothesis. The work is organized as follows: We review in section 2 the SGI machine solution . Our two new proposals are described in section 3.1 and 3.2 . They are evaluated in section 4. We especially study the influence of the memory latency on the conflict number. The results are summarized in the conclusion.

2   The Solution of the SGI Challenge Machine

The SGI machine [5] is a Symmetric Multiprocessor (SMP) using a split transaction bus. Coherence conflicts are avoided by delaying the emission of the requests which can generate a conflict. For this purpose, each cache controller includes an associative table which lists and traces the current requests of all the caches. Sharing the address bus enables to detect the initiation of any new request that is added to the table. Sharing the data bus enables snooping the response and removing the request from the table. Thus, the cache controller delays the initiation of a new request when it is conflicting with one already in its table.

We stress that this solution is appropriate when the size of tables is small (8 entries are sufficient in the SGI machine where a maximum of 8 requests coming from all caches may be in progress) because all entries of tables must be searched at snooping cycle. The snooping cycle ought to be extended in nowadays non-blocking caches[9] (which accept up to 8 simultaneous misses) to allow searching in larger tables (as 256=8x32 entries are required for 32 processors).

3   New Solutions

The solution of the SGI machine is simple, cheap but inappropriate when the data network is not a shared-bus. In this case, a processor cannot snoop the data responses (i.e., the request completion) which are intented to the other processors. Therefore, it cannot foresee a conflict when emitting a new request. However, as the processors permanently snoop the address bus, they can detect that any new request is conflicting with a local one already in progress. Two actions are possible: The processor which detects the conflict may ask the other processor to postpone the request (so-called retry solution) or it may register the request and forward the block as soon as possible (notification solution). Nowadays SMPs probably use one of these two logical solutions to preserve the coherence. To our knowledge, no quantitative results have been so far published. We detail the operation of the two techniques now.

3.1   Retry Technique

In this technique, a controller detecting a new conflicting request on the address bus asks the emitter to postpone the request. The emitter will retry later with the hope that the conflict has disappeared. Sometimes, several retries are necessary. When probating this technique with our simulator (see section 4) we discovered that conflicting requests sometimes face a service equity issue, which may induce an execution deadlock. This situation occurs in the following conditions: With the "Round Robbin" arbitration mechanism implemented in the classical busses, a controller cannot resubmit a request before the scan of the bus by all the other controllers. Suppose initially that the request 'i' is conflicting with the progressing request 'a'. If no new conflicting request on the same bloc is meanwhile generated, an attempt to resubmit 'i' will be successfully done after the completion of 'a'. But contrarily, if a new request 'j' on the same bloc attempts the bus before 'i', it will be submitted successfully. In consequence, next attempt for request 'i' will fail, but because of request 'j' this time. If the race between the request 'i' and a new request 'j' iterate permanently, request 'i' will be never submitted successfully. We observed this phenomena during the execution of a LOCK by several processors. The processor which executes the UNLOCK never succeeds to release the LOCK as the other processors permanently attempt to get the LOCK.

This problem can be avoided by imposing a constraint on the arbitration mechanism : a controller which must resubmit a request keeps the bus ownership regarding the emission of miss request until the conflict disappears. During that period, only invalidation or block flush requests can be sent by the other controllers.

The retry technique simply solves the conflicts and requires a minimal hardware extension which reduces to adding one supplementary line to the bus for dispatching the retry order. However, it wastes a fraction of the bus bandwidth as a failed request gets the ownership of bus and thus prevents submitting other requests, even if they are independents.

3.2 Notification Technique

A second solution enables initiating new conflicting requests (not to waste the bus bandwidth) provided that a coherent response is warranted to each one. A response is said coherent if the block READ returns the latest copy of this block (latest in the appearance order of the WRITE requests on the bus). Thus, it suffices to register the apparition order of the requests on the bus and to propagate the responses in the same order. For instance, in the case of several WRITE misses on the same block, the first controller registers the request of the second controller, the second controller that from the third one... and so on. When the first controller gets the block, it modifies and forwards it to the second controller. As this block is only kept during a WRITE period (as it will be written by a second controller), this operation can be executed on the fly without having to write the block in the cache (this operation is carried out in a buffer). Then, the second controller similarly modifies and propagates the block to the third controller... The requests notification mechanism is executed in the cache controller tables. Each local request is stored in the table with two additional fields to specify whether the request must be forwarded and to which processor. These tables only store the local requests contrarily to the solution of the SGI architecture which keeps track of all the processor requests.

This technique not only enables optimizing the bus bandwidth, it also reduces the service latency of a conflicting request. When calculating the request latency of the second processor in our previous example, an evident reduction of the latency is observed (see the definitions below).

Access latency in the retry technique = latency to get the response for request 1 + load and write latency of the block in cache 1 + resubmission latency for request 2 + latency to read and transfer the block form cache1 to cache 2.

Access latency in the notification technique = latency to get the response for request 1 + block update time in the buffer + latency to transfer the block form cache1 to cache 2.

The notification mechanism also enables to propagate the responses of non-conflicting requests. In the case of consecutive READ miss on the same bloc, the block is fetched from the memory for the first READ, then forwarded from one controller to another one for the other requests. This optimization is not studied in the following.

4   Experiment

The two techniques were validated and evaluated with an instruction-driven simulator that we developed to study the network-memory subsystem in the next generation of processors. The development cycle was complicated and required numerous efforts to construct a simulator both accurate, fast and able to execute the SPLASH2-suite programs [10].

4.1 Validation

The occurrence of unexpected states represents the major difficulty for preserving the coherence. These states are hardly predictable considering the multiple possible executions of a parallel program and the complexity to analyze the execution. Thus, we resorted to the logical definition of an execution which preserves the coherence of a shared-bus architecture: Each block READ must return the latest copy of this block (latest in the appearance order of the WRITE requests on the bus). As a result, we systematically stored the latest copy of each block and the processor cycle number of the corresponding WRITE operation. For any READ, we tested whether the returned and the stored blocks were identical. If they were not, the number of the WRITE cycle was accessible to start a step by step code debugging. A lot of time was saved by this method when testing and implementing the two coherence solutions.

4.2 Evaluation

The simulated architecture is a symmetric multiprocessor (SMP). The MESI algorithm, extended by the Retry or Notification technique implementation, preserves the coherence of caches. The architecture is composed of a set of processor-cache units (PCU) connected by a network to a shared memory. The number and the latency of PCUs, the data network model and the memory are fully configurable. Contrarily, the address transactions are always transmitted through a shared bus. We consider 8-way-issue superscalar processors with a speculative and out-of-order execution. The functional units and the issue/retire buffers were configured following Ref. [11]. Each processor is connected to two levels of lock-up-free caches. The 32-Kbyte first-level cache (L1) is direct-mapped with write-through policy and a one-cycle access time. The second-level cache (L2) is 4-way associative with write-back policy. Its size is 512Kbyte. The size of the cache block is 64 bytes for the L1 and L2. We describe in the following a standard and an aggressive models for the network and the memory.

Standard Model: We consider a classical split bus, i.e., two separate busses for transmitting address and data. The latency yields 3 bus cycles both for address and data transactions. This corresponds to 9 processor cycles (busses and processors operating at 100 MHz and 300 MHz respectively). The 8-bank interleaved memory enables 8 simultaneous accesses with an access time of 20 processor cycles ($\sim$70ns). The request service latency (counted from the cycle when the address request leaves the cache through the cycle when data come back) can be much larger than the memory plus the network access time due to contentions.

Aggressive model: As for the *standard* model, addresses are transmitted through a shared bus to maintain the coherence of caches. The bus latency as previously yields 9 processor cycles. The "novelty" in the aggressive model is that we consider a fully-connected data architecture. We assume that the rest of the memory-network is perfect, that is to say, that there is no contention effects. The data service latency is constant. It yields 100 processor cycles when data come from the memory, and 30 processor cycles when they come from another cache (direct cache to cache transfer). We had to consider longer access times in this model as more hardware layers must be bypassed in strongly connected architectures (crossbar, complex memory controller...).

The simulator executed the following benchmarks: BARRIER: This simple program was written to estimat the topmost possible gain. Each processor executes a loop of synchronization barrier. LU, OCEAN: These applications from the SPLASH2 suite [10] have been parameterized with a reduced number of input data to increase the number of conflicts. Executing LU requires 6.7 millions of instructions including 17 synchronization BARRIERS. Executing OCEAN necessitates 27 millions of instructions, 491 BARRIERS and 114 LOCKS.

Benchmark	Execution time		Bus waste		Average latency	
BARRIER	25 %	27 %	28.8 %	33.4 %	23 %	23 %
LU	0.4 %	1.5 %	0.9 %	3.1 %	22 %	52 %
OCEAN	5 %	13 %	8 %	20.6 %	13 %	25 %
	16 Proc.	32 Proc.	16 Proc.	32 Proc.	16 Proc.	32 Proc.

Table 1 : Simulation results for the *Standard* Model

Benchmark	Execution time		Bus waste		Average latency	
BARRIER	31 %	42 %	43.4 %	51.3 %	27 %	32 %
LU	0.4 %	3.3 %	2 %	5.4 %	26 %	29 %
OCEAN	13 %	30 %	19.4 %	42.5 %	21 %	36 %
	16 Proc.	32 Proc.	16 Proc.	32 Proc.	16 Proc.	32 Proc.

Table 2 : Simulation results for the *Aggressive* Model

Tables 1 and 2 above display the simulation results for the two models considering 16 or 32 processors. Column 2 (Execution time) shows the execution time gain, namely $(N_R - N_N)/N_R$ where N_R and N_N are the numbers of cycles necessary for executing the application with the Retry or the Notification technique respectively. Column 3 displays the waste of bus bandwidth in the retry technique defined by (Number of conflicts x 9 cycles)/ N_R, where 9 cycles is the address transaction latency. Column 4 shows the relative reduction of the request service latency, namely $(T_R - T_N)/T_R$ where T_R and T_N are the average request latencies for the Retry and the Notification technique respectively.

It turns out that the performance gain varies from 0.4% to 27% in the standard model and from 0.4% to 42% in the aggressive model. The gain (column 2) depends on the conflict frequency during the execution which can be represented by the bus waste (column 3). The BARRIER application shows the topmost enhancement with the second technique. The OCEAN and LU applications exhibit a smaller (though not negligible) enhancement. When comparing

LU and OCEAN, it turns out that the gain for LU is clearly much smaller than that for OCEAN (for instance: 1.5% for LU and 13% for OCEAN, lines 3 and 4 in table 1 for 32 processors). This is because with the same network configuration and number of processors, the frequency of conflicts depends on the communication behavior of the benchmark (OCEAN exhibits more synchronization and communication phases than LU).

Now if we assume different network configurations for the same Benchmark, the frequency of conflicts mostly depends on three parameters. It linearly increases versus the number of processors (see for instance the Bus waste columns for 16 and 32 processors in the tables), the Memory service latency and the address bus bandwidth. The Bus waste is smaller in the standard than in the aggressive model (see tables 1 and 2) as the request service latency is longer in the aggressive case.

It must be stressed that the results displayed in the tables have been generated by "extreme" benchmarks which exhibit numerous conflicts. The performance enhancement is not so high for usual applications as the conflicts only occur in the "hot" phases of the execution characterized by the synchronization and the communications of data. In this case, improvement of the notification technique is applicable to 20-50% of the applications execution time.

4.3   Influence of the Latency on the Number of Conflicts

Evaluating the influence of the memory latency versus the conflict frequency is especially interesting as the memory latency has continuously increased in the last years versus the processor cycle. This paragraph describes this evaluation in the framework of the aggressive network model to consider service latencies independently of the contention effect. Figures (a) and (b) show the increase of the conflict frequency versus the latency for 16 and 32 processors. Dots in the two figures were calculated using the previously defined "Bus waste" formula. The latency notation (x-y) for the X-axis means that the block transfer and the memory READ latency last x and y processor cycles respectively. We used the retry technique to count the number of conflicts. A request which needs n retries is counted as n conflicts.

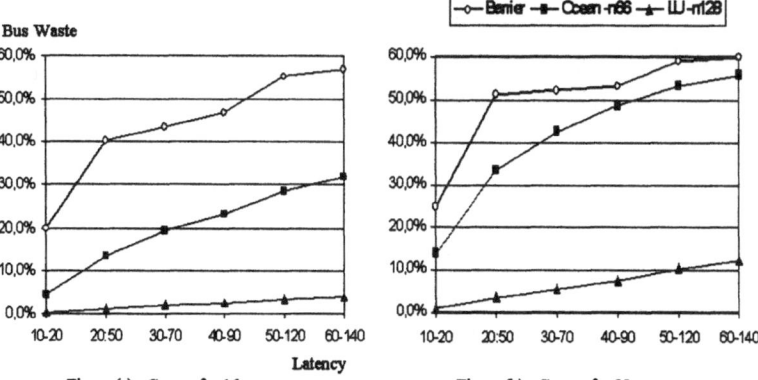

Figure (a) : Curves for 16 processors   Figure (b) : Curves for 32 processors

As shown in these curves, the number of conflicts will increase significantly versus the latency, thus future architecture must pay more attention to this problem.

5 Conclusion

The technological advances permanently widens the performance gap existing in the parallel machines between the processors, the interconnection network and the memories. New solutions are explored to reduce this gap, but every time, the implementation of caches sets the data coherence problem.

The introduction of split transactions enables using new non-shared data networks, but to the cost of an increased complexity for managing and preserving the coherence. We described and evaluated two new solutions to this problem : Retry and Notification techniques. Retry technique is less expensive than the second technique since no additional buffers are required to memorize notification data. We think that it is practical for small classical SMPs where coherence conflicts are not so important (limited number of processors and reduced memory latency). However, notification technique performs better on large scale SMPs, and provides a better implementation flexibility for future address medium because it only snoops the medium and does not require a direct intervention on it (to abord a current transaction).

References

[1]   D. Culler, J.P. Singh, and A. Gupta : Parallel Computer Architecture, A Hardware/Software Approach. Morgan Kaufmann Publishers, section 5.5.4

[2]   M. Papamarcos and J. Patel : A Low Overhead Coherence Solution for Multiprocessors with Private Cache Memories. The 11th Annual International Symposium on Computer Architecture, p. 348-354, June 1984

[3]   James Archibald and Jean-Loup Baer : Cache Coherence Protocols : Evaluation Using a Multiprocessor Simulation Model. ACM Transactions on Computer Systems, 4(4) p. 273-298, November 1986

[4]   Pradeep Sindhu, et al. XDBus : A High-Performance, Consistent, Packet Switched VLSI Bus. Proceedings of COMPCON, p. 338-344, Spring 1993

[5]   Mike Galles and Eric Williams : Performance Optimisations, Implementation, and verification of the SGI Challenge Multiprocessor. The 27th Annual Hawaii International Conference on Systems Sciences, January 1993

[6]   Technical information on the WEB : www.digital.com/alphaserver

[7]   Technical information on the WEB : http://www.bull.fr

[8]   Alan Charlesworth   STARFIRE : Extending the SMP Envelope. IEEE MICRO, p.39-49, JAN/FEB 1998

[9]   D. Kroft : Lockup-free Instruction Fetch/Prefetch Cache Organization. The 8th International Symposium on Computer Architecture, p. 81-87, May 81

[10]   S.C. Woo et al. : The SPLASH-2 Programs : Characterization and Methodological Consideration. The 25th International Symposium on Computer Architecture, p. 24-36, june 1995

[11]   S. Jourdan, P. Sainrat et D. Litaize : Exploring Configuration of Functional Units in an Out-Of-Order Superscalar Processor. The 25th International Symposium on Computer Architecture, p. 117-125, june 1995

An Adaptive Limited Pointers Directory Scheme for Cache Coherence of Scalable Multiprocessors

Cheol Ho Park[1], Jong Hyuk Choi[2], Kyu Ho Park[1], and Daeyeon Park[1]

[1] Department of Electrical Engineering,
Korea Advanced Institute of Science and Technology (KAIST), Taejon, Korea
[2] Digital Media Research Laboratory, LG Electronics, Seoul, Korea
{chpark,jhchoi,kpark,daeyeon}@coregate.kaist.ac.kr

Abstract. This paper presents an adaptive limited pointers directory scheme which works better than the broadcast and the non-broadcast limited pointers schemes. The former is good at workloads comprising mostly-read data and the latter is good at workloads comprising frequently read / write data. The proposed adaptive scheme dynamically switches between the two according to the application sharing patterns. It works well for both types of workloads, and outperforms the previous schemes for the workloads having both the mostly-read and the frequently read / write data at the same time. The adaptive limited pointers scheme reduces the traffic volume of real workloads by up to 33%. Further, it can be implemented with minor addition to the directory memory overhead.

1 Introduction

Most scalable shared memory multiprocessors (SSMP) of current generation use directory schemes to ensure cache coherence of shared data. Directory schemes avoid broadcast as in the snoopy cache coherence schemes. Among the directory schemes, the full map scheme [1] employs one N(# of processors)-bit full map vector per memory block. Because the directory has full knowledge about caching processors, the full map directory has the best performance a directory scheme can have. As the system size grows, however, the memory for a full map directory increases accordingly. This makes the full map directory scheme unacceptable for a very large scale multiprocessors.

The limited pointers directory schemes reduce the memory overhead by using a small number of pointers per directory. The pointer is an identifier pointing to one caching processor. The limited pointers schemes are based on the observation that most data objects are cached by only a small number of processors at any given time [2]. When pointers per block are running short, special actions have to be taken. We refer to this situation as the *pointer overflow*. The pointer overflow is the main source of the performance degradation in the limited pointers directory schemes.

Different limited pointers schemes take different actions upon directory overflows. The limited pointers without broadcast scheme [3] ($Dir_i NB$ [1]) makes a pointer available by invalidating one of the cached copies pointed to by the pointers. In the limited pointers with broadcast scheme [3] ($Dir_i B$), the broadcast bit is set as soon as the pointer

---

[1] i is the number of pointers per memory block.

P. Amestoy et al. (Eds.): Euro-Par'99, LNCS 1685, pp. 753–756, 1999.

overflow occurs, after which the directory does not keep the information on the sharing locations. Upon a subsequent write, it sends invalidation messages to all processors in the system because it can not identify the sharing processors.

In $Dir_i NB$, the performance degradation comes from the increased cache misses due to the extra invalidations upon pointer overflows. The increased misses, then, increases the number of pointer overflows again. Even the read-only or mostly-read data, to which little invalidating writes arrive, will experience a large number of invalidations because of the directory overflows. In $Dir_i B$, the performance degradation comes from the large invalidation volume upon invalidating writes. Even in the case that the number of sharing processors is slightly larger than the number of pointers, the invalidating writes cause the costly broadcast. Neither scheme does well for a wide range of applications having various sharing patterns. In general, $Dir_i NB$ exhibits a very unstable performance for many applications. On the other hand, $Dir_i B$ exhibits bounded performance but the performance for some applications is several times of the full map performance because of the costly broadcast.

In this paper, we propose an adaptive limited pointers scheme ($Dir_i A$) which is a hybrid of the two limited pointer schemes. $Dir_i A$ switches between the two schemes dynamically according to the behavior of the running application programs. At first, $Dir_i A$ behaves as $Dir_i NB$. After repeated directory overflows of a specified threshold, it switches to $Dir_i B$. Upon an invalidating write, the invalidations are to be broadcast to the entire system, as are in $Dir_i B$. If the number of reads arrived after the switch until the invalidation is smaller than another threshold, $Dir_i A$ will still act as $Dir_i B$ after the invalidation. Otherwise, it will start as $Dir_i NB$ again.

$Dir_i A$ does well for a wide range of applications. First, it does not generate a long sequence of pointer overflows and cache misses for the read-only and mostly-read data. The length of those sequences will be bounded within the first threshold. Second, it does not initiate the costly broadcast, when the number of cached copies is slightly larger than the number of pointers per memory block.

The rest of the paper is organized as follows. In Section 2, we present the $Dir_i A$ in detail. Section 3 gives the performance evaluation and Section 4 concludes the paper.

2   Adaptive Limited Pointers Scheme ($Dir_i A$)

First, we define the *read run* as a continuous sequence of processor reads that does not contain intervening writes in them. $Dir_i B$ performs well for applications whose read runs are mostly long, whereas $Dir_i NB$ does well for applications whose read runs are mostly short.

In $Dir_i A$, coherence protocol per memory block operates in one of the two modes: B and NB. During the execution of an application program, the protocol switches between the two modes dynamically by examining the length of the read run to the memory block after the first overflow.

Initially, all memory blocks operate under the NB mode. After i subsequent read requests, one more read generates the directory overflow, which will be serviced as in $Dir_i NB$. Once as many read requests as a certain threshold ($ThrNB$) arrive to the memory block, the protocol switches to the B mode, after which the locations of the

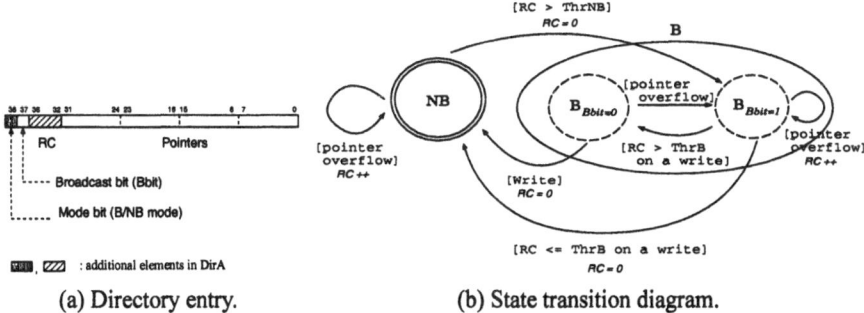

(a) Directory entry.   (b) State transition diagram.

Fig. 1. $Dir_i A$.

cached copies are not kept any more. Upon an invalidating write, invalidations are to be sent to all over the system. After the write, the block will go back to the *NB* mode if the read run length under the *B* mode is smaller than another threshold (*ThrB*). Otherwise, the protocol will remain in the *B* mode.

To implement the $Dir_i A$ protocol, a mode bit and a read counter (*RC*) is required for each directory block (Figure 1). The mode bit indicates whether the protocol for the corresponding memory block is in the *B* mode or in the *NB* mode. *RC* counts the length of the read runs after the first pointer overflow. The *RC* value is compared with *ThrB* and *ThrNB* upon the mode switch decision.

3   Performance Evaluation

We evaluate the performance of $Dir_i A$ from the execution-driven simulation of the target machine with a synthesized benchmark and two application programs from the SPLASH-2 benchmark [4]. We use the Tango Lite [5] execution-driven simulation environment to model a 64 node CC-NUMA (Cache Coherent - Non Uniform Memory Access) multiprocessor whose individual nodes are connected via the mesh wormhole routing network. The traffic is measured in number of bytes; the execution time is measured in number of cycles. They are normalized to those of the full map directory scheme.

The synthesized benchmark generates random-length read runs and the distribution is controlled by some parameters. For the configuration with mostly long read runs, $Dir_i NB$ suffered seriously. The configuration with mostly short read runs degraded $Dir_i B$ performance by much. In the both configuration, $Dir_i A$ showed robust performance. For the configuration with short and long read runs at the same time, $Dir_i A$ outperformed both $Dir_i NB$ and $Dir_i B$.

Barnes is a 3D hierarchical N body simulation program. Both short and long read runs coexist in Barnes. The long read runs degrade the performance of $Dir_i NB$. On the other hand, the short read runs degrade the performance of $Dir_i B$. $Dir_i A$ outperforms both of them, because it neither generates repeated invalidate / cache miss sequences nor initiates invalidation broadcasts, which are the main bottlenecks of the system performance with limited pointers directory schemes. $Dir_i A$ eliminates 33% of the traffic volume of $Dir_i B$, thereby accelerating the program execution by about 5%.

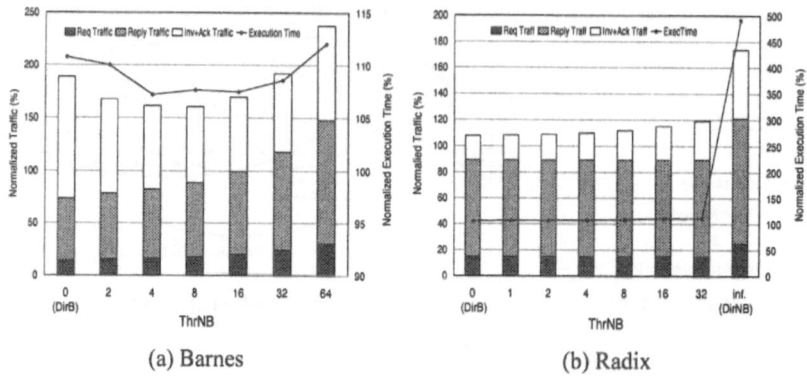

(a) Barnes   (b) Radix

Fig. 2. $Dir_i A$ results of Barnes and Radix

Radix is the integer radix sorting kernel. Each processor uses a global histogram residing on shared memory in each iteration. Most of the shared data structures in Radix are mostly-read. Therefore, the $Dir_i B$ performance is the best close in Figure 2(b) to the full-map one. However, the various $Dir_i A$ configurations perform as well as does $Dir_i B$ because the additional invalidations caused by pointer replacements are not serious because the sequence is bounded in a small number.

4   Conclusions

In this paper, we proposed an adaptive limited pointers directory scheme, denoted as $Dir_i A$, which is a hybrid of $Dir_i B$ and $Dir_i NB$. The simulation results prove that $Dir_i A$ is robust for different kinds of workloads. It performs as well as $Dir_i B$ and much better than $Dir_i NB$ for most workloads. For application programs having many short / long mixed read runs, such as Barnes, $Dir_i A$ performs better than $Dir_i B$ by much.

References

[1] L. M. Censier and P. Feautrier, "A new solution to coherence problems in multicache systems," *IEEE Trans. Computers*, vol. C-27, pp. 1112–1118, Dec. 1978.
[2] W.-D. Weber and A. Gupta, "Analysis of cache invalidation patterns in shared-memory multiprocessors," in *3rd International Conference on Architectural Support for Programming Languages and Operating Systems*, pp. 243–256, 1989.
[3] A. Agarwal, R. Simoni, J. Hennessy, and M. Horowitz, "An evaluation of directory schemes for cache coherence," in *15th Annual International Symposium on Computer Architecture*, pp. 280–289, 1988.
[4] S. C. Woo, M. Ohara, E. Torrie, J. P. Singh, and A. Gupta, "The SPLASH-2 programs: Characterization and methodological considerations," in *22nd Annual International Symposium on Computer Architecture*, pp. 24–36, 1995.
[5] S. Goldschmidt, *Simulation of Multiprocessors: Accuracy and Performance*. PhD thesis, Stanford University, 1993.

Two Schemes to Improve the Performance of a *Sort-Last* 3D Parallel Rendering Machine with Texture Caches

Alexis Vartanian, Jean-Luc Béchennec, and Nathalie Drach-Temam

LRI, Université Paris-Sud,
91405 Orsay France,
(alex,jlb,drach)@lri.fr
tel: 33 1 69 15 42 22
fax: 33 1 69 15 65 86

Abstract. A *sort-last* 3D parallel rendering machine distributes the triangles to draw to different processors. When building such a machine with each processor having a texture cache, the texture locality is worse and the performance is reduced. This article investigates two schemes to preserve this locality while keeping a good load balancing: triangle slicing and locality aware triangle distribution. With both schemes, the speedups are improved between 2 and 6 times.
Keywords: Cache memories, multiprocessing, application specific architecture, parallel rendering, texture mapping.

1 Introduction

Multimedia applications are now expected to be the main consumer of computing power in the coming years. Among these applications, real-time 3D image rendering has raised a lot of attention this year as most microprocessor manufacturers are releasing new instruction set extensions (Intel SSE) mainly aimed at accelerating geometry transformations of PC microprocessors. Such a trend reduces the performance gap between a high end PC with a 3D accelerator and a workstation with expensive ASICs. As 3D rendering contains a fair amount of parallelism, a parallel machine with PC 3D accelerators could offer a good performance/price ratio.

There are many ways to use parallelism in 3D rendering. Molnar have given a classification of all kinds of machines. In this paper, according to this classification, we focus on a *sort-last* machine based on PC 3D accelerators. In such a machine, the triangles to be drawn are distributed among independent pipelines that do both geometry transformations and texture mapping on the same data. The PixelFlow machine is the best known sort-last 3D computer. This architecture has two main advantages: it is not necessary to redistribute the triangles between the geometry stage and the texture mapping stage and dynamic load balancing can be used. However these architectures have a main drawback: as each pipeline can render anywhere on the screen, all processors need to transfer

P. Amestoy et al. (Eds.): Euro-Par'99, LNCS 1685, pp. 757–760, 1999.

their frame buffer to one processor. This part of the machine requires a high bandwidth.

This is the reason why most research on sort-last machines has focused on this part of the pipeline. However, when such a machine is based on PC 3D accelerators, they will have an internal texture cache. Hakura and Gupta have shown that a cache works very well in handling texture locality. But when texture caches are used in a 3D parallel machine, a problem arises. If two triangles are rendered in different pipelines and if some *texels* [1] were reused between the two triangles, there could be less locality on a parallel machine compared to a sequential machine. In our previous work, we have observed such a phenomenon. We have shown that a synchronous *sort-middle* architecture performance was strongly reduced if such a phenomenon was taken into account. We have also evaluated sort-last machines and shown that some benchmarks could suffer from this phenomenon.

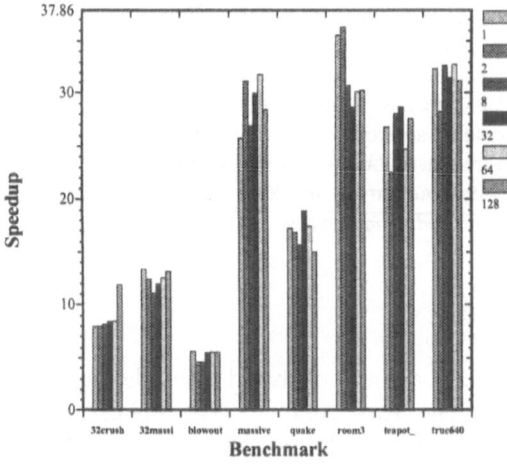

Fig. 1. *Speedup compared to a single processor machine with triangle set size, with a 64 processor machine having texture caches of 16KB.*

In this study, we evaluate two schemes to deal with this problem. These evaluations are based on event-driven simulations of the architecture on different 3D benchmarks. The first section presents the performance of the machine with the second scheme: we try to find out a distribution of the triangle that does not break the locality and we show that it generates load unbalance. The second section shows how we solve that problem with triangle slicing. The third section presents the overall performance when both schemes are used [2].

---

[1]   A texel is a pixel in the texture.

[2]   An more detailed version of this paper is available at http://www.lri.fr/~alex/.

2   The Impact of Load Unbalance

The main idea of our whole study is to find out a triangle distribution algorithm that improves the locality. The algorithm we use in this study is the following: each time a geometry engine has finished its work, it asks the distributor for a set of N consecutive (in the submission order) triangles. With this algorithm, we expect to increase the cache locality. However if we try to use this algorithm alone, we don't success in increasing the performance as seen in figure 1. We see that speedup is not increasing with N. Moreover, the best value for N is not the same for all the benchmarks. It means that when this algorithm is used, it is not clear whether it increases or decreases the performance. This is due to the fact that the load balancing heavily depends on the big triangles. And our algorithm might send a set of very big triangles to the same processor. We are going to present a solution to this issue.

3   Triangle Slicing

The suggested scheme is the following. Each time a triangle has more pixels than S, the distributor cuts it into two fair slices. It does that while any triangle is bigger than S. We have measured the impact of such a scheme on load balancing with a perfect cache. If we cut any triangle having more than 2500 pixels, the negative impact of the first scheme disappears and the speedup with 64 processors is the same with any set size.

However, such a scheme has an impact on the locality. When a big triangle is sliced, two effects are expected. The locality might be improved: as triangles are smaller, the lines could be shorter. If a texel is shared between two pixels on the same row, the duration between the two uses depends on the line size. If the line is too long, the texels might have been evicted from the cache because of a conflict or a capacity problem. On the other hand, when a triangle is sliced, the pixels of the side could have less locality. As far as these pixels are concerned, the duration between the texel uses is far longer with slicing. Before slicing, these texels were immediately reused. With slicing, they have to wait for the complete triangle to be drawn.

We observed that the impact of the two phenomena were highly dependent on the slicing method used. When a triangle is cut into two triangles, we have to choose on which side we add a vertex. We evaluated different algorithms and showed that it is better to choose the highest side than the biggest side. It generates less high triangles than the other. If the triangles are less high with the same size, they must have a greater width. As we said before, this is an important issue for the locality.

4   Evaluation of the Distribution Scheme

We now measure the performance of the architecture with triangle slicing. We observe on figure 2 that, for all the benchmarks, performance increases with N.

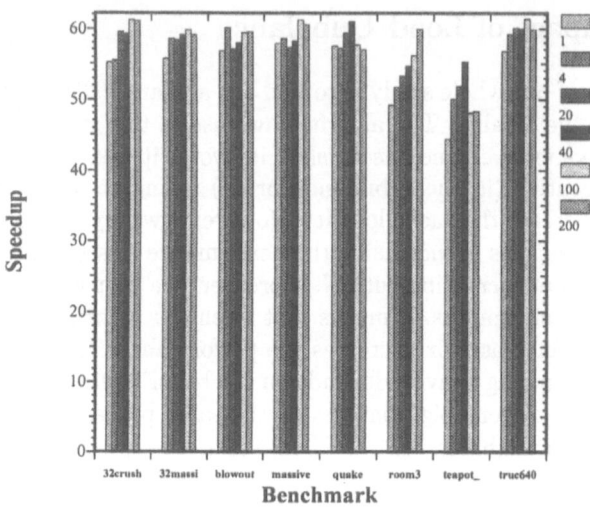

Fig. 2. *Speedup compared to a single processor machine with triangle set size, with a 64 processor machine having texture caches of 16KB. Triangle are sliced if bigger than 1000 pixels with the algorithm 2.*

This is what we expected: the locality is improved by our algorithm. However for *quake* and *teapot.full* performance starts to decrease if N is greater than 40. To understand that phenomenon we measured the average cache miss rate of all the processors. We saw that for all the benchmarks, the miss rate decreases between 1 and 40 and increases with N after 40. We have the following analysis of this behavior: if too many triangles are sent, the number of triangles to be redistributed at the end of the frame increases. The impact of this on performance depends on the bus load at this period. It shows that if we want to have the optimal performance with any benchmark, a set size of 40 should be used.

5   Conclusion

In this study, we have seen that the performance of a sort-last parallel machine aimed at real-time 3D rendering and based on PC 3D accelerators greatly depends on texture cache behavior and on load balancing. We have seen that any attempt to manage the locality with a good distribution was limited by the big triangles. We have presented a solution to this problem: triangle slicing. However, if this is done, locality can be greatly reduced. To avoid this behavior, we have suggested a slicing algorithm that keeps a good locality. With that slicing algorithm, we have shown that if the distributor sends 40 triangles to a pipeline each time it is free, the locality is increased and the performance is better. With these two schemes, we are able to increase the speedup between 2 and 6 times.

This paper is a summary of our work. A more detailed version with complete references is available at **http://www.lri.fr/~alex/**.